ANNALS OF THE NEW YORK ACADEMY OF SCIENCES

Volume 967

EDITORIAL STAFF

Executive Editor
BARBARA M. GOLDMAN

Managing Editor
JUSTINE CULLINAN

Associate Editor
STEFAN MALMOLI
STEVE BOHALL

The New York Academy of Sciences
2 East 63rd Street
New York, New York 10021

THE NEW YORK ACADEMY OF SCIENCES
(Founded in 1817)

BOARD OF GOVERNORS, September 2001 – September 2002

TORSTEN N. WIESEL, *Chairman of the Board*
JOHN F. NIBLACK, *Vice Chairman of the Board*
BILL GREEN, *Past Chairman*

Honorary Life Governors
WILLIAM T. GOLDEN   JOSHUA LEDERBERG
JOHN T. MORGAN, *Treasurer*

Governors

ELEANOR BAUM	D. ALLAN BROMLEY	KAREN E. BURKE
	LAWRENCE B. BUTTENWIESER PRAVEEN CHAUDHARI	
JOHN H. GIBBONS	MICHAEL GOLDEN	RONALD L. GRAHAM
JACQUELINE LEO	SANDRA PANEM	RICHARD A. RIFKIND
JOHN J. ROCHE		SARA LEE SCHUPF
JAMES H. SIMONS		LEE VANCE

HELENE L. KAPLAN, *Counsel* [ex officio]

LIPIDS AND INSULIN RESISTANCE
THE ROLE OF FATTY ACID METABOLISM AND FUEL PARTITIONING

ANNALS OF THE NEW YORK ACADEMY OF SCIENCES
Volume 967

LIPIDS AND INSULIN RESISTANCE

THE ROLE OF FATTY ACID METABOLISM AND FUEL PARTITIONING

Edited by Iwar Klimeš, Elena Šeböková, Barbara V. Howard, and Eric Ravussin

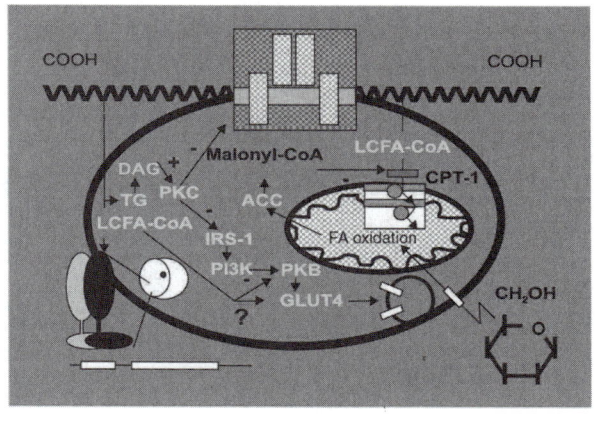

The New York Academy of Sciences
New York, New York
2002

Copyright © 2002 by the New York Academy of Sciences. All rights reserved. Under the provisions of the United States Copyright Act of 1976, individual readers of the Annals *are permitted to make fair use of the material in them for teaching or research. Permission is granted to quote from the* Annals *provided that the customary acknowledgment is made of the source. Material in the* Annals *may be republished only by permission of the Academy. Address inquiries to the Permissions Department (editorial@nyas.org) at the New York Academy of Sciences.*

Copying fees: *For each copy of an article made beyond the free copying permitted under Section 107 or 108 of the 1976 Copyright Act, a fee should be paid through the Copyright Clearance Center, Inc., 222 Rosewood Drive, Danvers, MA 01923 (www.copyright.com).*

⊚ *The paper used in this publication meets the minimum requirements of the American National Standard for Information Sciences—Permanence of Paper for Printed Library Materials, ANSI Z39.48-1984.*

Library of Congress Cataloging-in-Publication Data

International Smolenice Insulin Symposium on "Lipids and Insulin Resistance: the Role of Fatty Acid Metabolism and Fuel Partitioning" (2001).
  Lipids and insulin resistance: the role of fatty acid metabolism and fuel partitioning / Edited by Iwar Klimes ... [*et al.*].
    p. cm. — (Annals of the New York Academy of Sciences; v. 967)
  Includes bibliographical references and index.
  ISBN 1-57331-368-8 (cloth: alk. paper) — ISBN 1-57331-369-6 (paper: alk. paper)
  1. Insulin resistance—Pathophysiology—Congresses. 2. Fatty acids—Metabolism—Congresses.
    3. Energy metabolism—Congresses.
    [DNLM: 1. Insulin Resistance—physiology—Congresses. 2. Energy Metabolism—Congresses.
    3. Fatty Acids—metabolism—Congresses. 4. Lipids—metabolism—Congresses. WK 820 I6038L
    2002] I. Klimes, Iwar. II. Title. III. Series.
Q11.N5 vol. 967 RC662.4
500 s—dc21
[616.4/62   2002004276

GYAT/BM
Printed in the United States of America
ISBN 1-57331-368-8 (cloth)
ISBN 1-57331-369-6 (paper)
ISSN 0077-8923

ANNALS OF THE NEW YORK ACADEMY OF SCIENCES
Volume 967
June 2002

LIPIDS AND INSULIN RESISTANCE

THE ROLE OF FATTY ACID METABOLISM AND FUEL PARTITIONING

Editors
IWAR KLIMEŠ, ELENA ŠEBÖKOVÁ, BARBARA V. HOWARD,
AND ERIC RAVUSSIN

Conference Organizers
IWAR KLIMEŠ AND ELENA ŠEBÖKOVÁ

This volume is the result of the Fourth International Smolenice Insulin Symposium on "Lipids and Insulin Resistance: The Role of Fatty Acid Metabolism and Fuel Partitioning" held on August 29 to September 2, 2001, in Smolenice Castle, Slovak Republic.

CONTENTS

Preface. *By* IWAR KLIMEŠ AND ELENA ŠEBÖKOVÁ xiii

State-of-the-Art Lecture

Identification of Susceptibility Genes for Complex Metabolic Diseases. *By* CLIFTON BOGARDUS, LESLIE BAIER, PASKA PERMANA, MICHAL PROCHAZKA, JOHANNA WOLFORD, AND ROBERT HANSON 1

Part I. Nuclear Transcription Factors, Lipids, and Energy Balance

PPARs: Transcription Factors Controlling Lipid and Lipoprotein Metabolism. *By* VIRGINIE BOCHER, INÉS PINEDA-TORRA, JEAN-CHARLES FRUCHART, AND BART STAELS 7

SREBP-1: Gene Regulatory Key to Syndrome X? *By* DIRK MÜLLER-WIELAND AND JÖRG KOTZKA 19

PPARγ, an X-ceptor for Xs. *By* HANA KOUTNIKOVA AND JOHAN AUWERX ... 28

Sterol Regulatory Element-Binding Proteins (SREBPs) as Regulators of Lipid Metabolism: Polyunsaturated Fatty Acids Oppose Cholesterol-Mediated Induction of SREBP-1 Maturation. *By* HYOUN-JU KIM, MAKOTO MIYAZAKI, WENG CHI MAN, AND JAMES M. NTAMBI 34

Part II. Fatty Acid Oxidation, Fuel Sensing, and Insulin Resistance

Hyperglycemia and Insulin Resistance: Possible Mechanisms. *By* EVA TOMÁS, YEN-SHOU LIN, ZEINA DAGHER, ASISH SAHA, ZHIJUN LUO, YASUO IDO, AND NEIL B. RUDERMAN 43

Insulin Stimulation of Hepatic Triacylglycerol Secretion in the Insulin-Replete State: Implications for the Etiology of Peripheral Insulin Resistance. *By* VICTOR A. ZAMMIT 52

Free Fatty Acid Metabolism in Human Skeletal Muscle Is Regulated by PPARγ and RXR Agonists. *By* THEODORE P. CIARALDI, BONG-SOO CHA, KYONG-SOO PARK, LESLIE CARTER, SUNDER R. MUDALIAR, AND ROBERT R. HENRY 66

An Increase in Peroxisomal Fatty Acid Oxidation Is Not Sufficient to Prevent Tissue Lipid Accumulation in hHTg Rats. *By* J. UKROPEC, I. KLIMEŠ, D. GAŠPERÍKOVÁ, E. DEMCÁKOVÁ, C. A. DREVON, J. E. RESELAND, AND E. ŠEBÖKOVÁ 71

Part III. Knockout and Transgenic Animal Models of Obesity and Insulin Resistance

Adiposity and the Development of Diabetes in Mouse Genetic Models. *By* LESLIE P. KOZAK AND MARTIN ROSSMEISL 80

Modulation of Lipid Metabolism by Energy Status of Adipocytes: Implications for Insulin Sensitivity. *By* JAN KOPECKÝ, PAVEL FLACHS, KRISTINA BARDOVÁ, PETR BRAUNER, TOMÁŠ PRAŽÁK, AND JANA ŠPONAROVÁ 88

Transgenic Mice Overexpressing the Rate-Limiting Enzyme for Hexosamine Synthesis in Skeletal Muscle or Adipose Tissue Exhibit Total Body Insulin Resistance. *By* ROBERT C. COOKSEY AND DONALD A. MCCLAIN 102

Enhanced Diabetogenic Effect of Streptozotocin in Mice Overexpressing UCP-3 in Skeletal Muscle. *By* STEVEN WANG, MICHAEL A. CAWTHORNE, AND JOHN C. CLAPHAM 112

Part IV. Muscle Lipid Metabolism in Obesity and Type 2 Diabetes

From Receptor to Effector: Insulin Signal Transduction in Skeletal Muscle from Type II Diabetic Patients. *By* JULEEN R. ZIERATH AND HARRIET WALLBERG-HENRIKSSON 120

Skeletal Muscle Triglycerides: An Aspect of Regional Adiposity and Insulin Resistance. *By* DAVID E. KELLEY 135

Protein Kinase C and Lipid-Induced Insulin Resistance in Skeletal Muscle. *By* CARSTEN SCHMITZ-PEIFFER 146

Evaluation of Free Fatty Acid Metabolism *in Vivo*. *By* NICHOLAS D. OAKES AND STUART M. FURLER 158

Insulin Action in Skeletal Muscle: Isozyme-Specific Effects of Protein Kinase C. *By* ISKANDAR IDRIS, SAMUEL GRAY, AND RICHARD DONNELLY 176

Part V. Skeletal Muscle Lipid Composition and Insulin Signaling

Desaturation and Elongation of Fatty Acids and Insulin Action.
By B. VESSBY, I-B. GUSTAFSSON, S. TENGBLAD, M. BOBERG,
AND A. ANDERSSON 183

Muscle Long-Chain Acyl CoA Esters and Insulin Resistance. *By*
G. J. COONEY, A. L. THOMPSON, S. M. FURLER, J. YE, AND
E. W. KRAEGEN 196

Eicosanoids and the Regulation of Cardiac Glucose Transport. *By* OLAF
DRANSFELD, IRINI RAKATZI, SHLOMO SASSON, AND JÜRGEN ECKEL .. 208

Part VI. Skeletal Muscle Lipid Metabolism at Rest and during Exercise

Regulation of Fat Metabolism in Skeletal Muscle. *By* ASKER E. JEUKENDRUP. 217

The Sphingomyelin-Signaling Pathway in Skeletal Muscles and Its Role in
Regulation of Glucose Uptake. *By* JAN GÓRSKI, AGNIESZKA DOBRZYN,
AND MALGORZATA ZENDZIAN-PIOTROWSKA 236

Part VII. Susceptibility Genes for Complex Metabolic Disorders

Genetic and Molecular Analyses of Complex Metabolic Disorders: Genetic
Linkage. *By* S. MENZEL 249

Positional Cloning of an Obesity/Diabetes Susceptibility Gene(s) on
Chromosome 11 in Pima Indians. *By* LESLIE BAIER, PETER KOVACS,
CHRISTOPHER WIEDRICH, KIMBERLY CRAY, AMY SCHEMIDT,
GONG-QING SHEN, JEFFREY SUTHERLAND, PAMELA THUILLEZ,
YUNHUA LI MULLER, MICHAEL TRAURIG, AND CLIFTON BOGARDUS .. 258

Is a Pro12Ala Polymorphism of the PPARγ2 Gene Related to Obesity and
Type 2 Diabetes Mellitus in the Czech Population? *By* DANIELA
ŠRÁMKOVÁ, MARIE KUNEŠOVÁ, VOJTECH HAINER, MARTIN HILL,
JOSEF VCELÁK, AND BELA BENDLOVÁ 265

A Custom-Built Insulin Resistance Gene Chip. *By* KEN WALDER, DAVID
SEGAL, SAM CHEHAB, GUY AUGERT, DAVID CAMERON-SMITH, MARK
HARGREAVES, AND GREG R. COLLIER 274

Part VIII. Mono- and Polyunsaturated Fatty Acids in Prevention of the Insulin Resistance Syndrome

Fatty Acid Regulation of Gene Expression: A Genomic Explanation for the
Benefits of the Mediterranean Diet. *By* STEVEN D. CLARKE, DANIELA
GASPERIKOVA, CAROLANNE NELSON, ALEXANDRE LAPILLONNE, AND
WILLIAM C. HEIRD 283

Perinatal Supply and Metabolism of Long-Chain Polyunsaturated Fatty Acids:
Importance for the Early Development of the Nervous System. *By*
ELVIRA LARQUE, HANS DEMMELMAIR, AND BERTHOLD KOLETZKO 299

The Responses of Serum and Adipose Fatty Acids to a One-Year Weight Reduction Regimen in Female Obese Monozygotic Twins. *By* M. KUNEŠOVÁ, S. PHINNEY, V. HAINER, E. TVRZICKÁ, V. ŠTICH, J. PARÍZKOVÁ, A. ŽÁK, AND A. STUNKARD 311

Part IX. Dietary Fats, Diabetes, and Cardiovascular Disease

Dietary Fat as a Risk Factor for Type 2 Diabetes. *By* BARBARA V. HOWARD.. 324

Type of Dietary Fat and Insulin Resistance. *By* ANGELA A. RIVELLESE, CLAUDIA DE NATALE, AND STEFANIA LILLI 329

Treatment of Hypertriglyceridemia with Fenofibrate, Fatty Acid Composition of Plasma and LDL, and Their Relations to Parameters of Lipoperoxidation of LDL. *By* M. ZEMAN, A. ŽÁK, M. VECKA, E. TVRZICKÁ, S. ROMANIV, AND M. KONÁRKOVÁ 336

Effect of Iron Depletion on Cardiovascular Risk Factors: Studies in Carbohydrate-Intolerant Patients. *By* FRANCESCO S. FACCHINI AND KAMI L. SAYLOR 342

Erythrocyte Membrane Ion Transport in Offspring of Hypertensive Parents: Effect of Acute Hyperinsulinemia and Relation to Insulin Action. *By* GABRIELA SUCHÁNKOVÁ, ZUZANA VLASÁKOVÁ, JOSEF ZICHA, MARTINA VOKURKOVÁ, ZDENA DOBEŠOVÁ, AND TEREZIE PELIKÁNOVÁ 352

Part X. Lipids, Energy Balance, and Insulin Resistance

Increased Fat Intake, Impaired Fat Oxidation, and Failure of Fat Cell Proliferation Result in Ectopic Fat Storage, Insulin Resistance, and Type 2 Diabetes Mellitus. *By* ERIC RAVUSSIN AND STEVEN R. SMITH... 363

Leptin Signaling, Adiposity, and Energy Balance. *By* ERIC JÉQUIER 379

Neuroimaging and Obesity: Mapping the Brain Responses to Hunger and Satiation in Humans Using Positron Emission Tomography. *By* ANGELO DEL PARIGI, JEAN-FRANCOIS GAUTIER, KEWEI CHEN, ARLINE D. SALBE, ERIC RAVUSSIN, ERIC REIMAN, AND P. ANTONIO TATARANNI.. 389

Uncoupling Proteins, Leptin, and Obesity: An Updated Review. *By* JEAN-PAUL GIACOBINO 398

New Approaches to Gene Discovery with Animal Models of Obesity and Diabetes. *By* GREG COLLIER, KEN WALDER, ANDREA DE SILVA, JANETTE TENNE-BROWN, ANDREW SANIGORSKI, DAVID SEGAL, LAKSHMI KANTHAM, AND GUY AUGERT 403

Poster Papers

High-Fat High-Energy Feeding Impairs Fasting Glucose and Increases Fasting Insulin Levels in the Göttingen Minipig: Results from a Pilot Study. *By* MARIANNE OLHOLM LARSEN, BIDDA ROLIN, MICHAEL WILKEN, RICHARD DAVID CARR, AND OVE SVENDSEN 414

Comparison of the Extrapancreatic Action of BRX-220 and Pioglitazone in the High-Fat Diet–Induced Insulin Resistance. *By* ELENA ŠEBÖKOVÁ, MARIA KÜRTHY, T. MOGYOROSI, KAROLY NAGY, EDITA DEMCÁKOVÁ, JOZEF UKROPEC, LASZLO KORANYI, AND IWAR KLIMEŠ 424

Opposing Effects of Fatty Acids and Acyl-CoA Esters on Conformation and Cofactor Recruitment of Peroxisome Proliferator-Activated Receptors. *By* CLAUS JØRGENSEN, ANNE-M. KROGSDAM, IRINA KRATCHMAROVA, TIMOTHY M. WILLSON, JENS KNUDSEN, SUSANNE MANDRUP, AND KARSTEN KRISTIANSEN 431

Relationship between Insulin Resistance and Muscle Triglyceride Content in Nonobese and Obese Experimental Models of Insulin Resistance Syndrome. *By* JANA DIVIŠOVÁ, LUDMILA KAZDOVÁ, MIRIAM HUBOVÁ, AND ELEN MESCHIŠVILI 440

Insulin Resistance in the Hereditary Hypertriglyceridemic Rat Is Associated with an Impairment of Δ-6 Desaturase Expression in Liver. *By* D. GAŠPERÍKOVÁ, E. DEMCÁKOVÁ, J. UKROPEC, I. KLIMEŠ, AND E. ŠEBÖKOVÁ 446

Heart Remodeling in the Hereditary Hypertriglyceridemic Rat: Effect of Captopril and Nitric Oxide Deficiency. *By* F. SIMKO, I. LUPTAK, J. MATUSKOVA, P. BABAL, O. PECHANOVA, I. BERNATOVA, AND I. HULIN 454

Energy Transfer in Acute Diabetic Rat Hearts: Adaptation to Increased Energy Demands due to Augmented Calcium Transients. *By* A. ZIEGELHÖFFER, T. RAVINGEROVÁ, I. WACZULÍKOVÁ, J. CÁRSKY, J. NECKÁR, B. ZIEGELHÖFFER-MIHALOVICOVÁ, AND J. STYK 463

Impaired Endothelial Function of Thoracic Aorta in Hereditary Hypertriglyceridemic Rats. *By* J. TÖRÖK, P. BABÁL, J. MATUŠKOVÁ, I. LUPTÁK, I. KLIMEŠ, AND F. ŠIMKO 469

A Novel Mammalian Homologue of a Bacterial Citrate-Metabolizing Enzyme. *By* CHARLOTTE SÖDERBERG AND PETER LIND 476

Effect of BRX-220 against Peripheral Neuropathy and Insulin Resistance in Diabetic Rat Models. *By* MARIA KÜRTHY, TAMÁS MOGYORÓSI, KÁROLY NAGY, TIBOR KUKORELLI, ANDREA JEDNÁKOVITS, LÁSZLÓ TÁLOSI, AND KATALIN BÍRÓ 482

Terguride Treatment Attenuated Prolactin Release and Enhanced Insulin Receptor Affinity and GLUT 4 Content in Obese Spontaneously Hypertensive Female, but Not Male Rats. *By* S. ZORAD, V. GOLDA, M. FICKOVA, L. MACHO, L. PINTEROVA, AND J. JURCOVICOVA 490

Endocrine Regulation of Subcutaneous Fat Metabolism during Cold Exposure in Humans. *By* JURAJ KOSKA, LUCIA KSINANTOVA, ELENA ŠEBÖKOVÁ, RICHARD KVETNANSKY, IWAR KLIMEŠ, GEORGE CHROUSOS, AND KAREL PACAK 500

Electrical Stimulation Improves Insulin Responses in a Human Skeletal Muscle Cell Model of Hyperglycemia. *By* VIGDIS AAS, SIRI TORBLÅ, MERETHE H. ANDERSEN, JØRGEN JENSEN, AND ARILD CHR. RUSTAN .. 506

Differences in Oral Temperature and Body Shape in Two Populations with Different Propensities for Obesity. *By* B. VOZAROVA, C. WEYER, C. BOGARDUS, E. RAVUSSIN, AND P. A. TATARANNI 516

Effects of Selected Anthropometric Parameters on Plasma Lipoproteins, Fatty
  Acid Composition, and Lipoperoxidation. *By* ALEŠ ŽÁK, EVA TVRZICKÁ,
  MAREK VECKA, SEVERYN ROMANIV, MIROSLAV ZEMAN, AND
  MARTA KONÁRKOVÁ... 522

Intima-Media Thickness and Atherosclerotic Plaques in Familial Defective
  Apolipoprotein B-100 and Familial Hypercholesterolemia. *By*
  M. KAISER, T. TEMELKOVA-KURKTSCHIEV, AND M. HANEFELD....... 528

Acute Elevation of NEFA Causes Hyperinsulinemia without Effect on Insulin
  Secretion Rate in Healthy Human Subjects. *By* BEATE BALENT,
  GAYOTRI GOSWAMI, GEORGE GOODLOE, EDUARD ROGATSKY,
  OLIMPIA RAUTA, ROBERT NEZAMI, LISA MINTS, RUTH HOGUE
  ANGELETTI, AND DANIEL T. STEIN................................ 535

Trans Fatty Acids in Subcutaneous Fat of Pregnant Women and in Human
  Milk in the Czech Republic. *By* PAVEL DLOUHÝ, EVA TVRZICKÁ,
  BARBORA STANKOVÁ, MARTA BUCHTÍKOVÁ, RAJMUND POKORNÝ,
  OLGA WIEREROVÁ, DIANA BÍLKOVÁ, JOLANA RAMBOUSKOVÁ, AND
  MICHAL ANDEL... 544

Detection of a Promoter Polymorphism in the Gene of Intestinal Fatty Acid
  Binding Protein (I-FABP). *By* K. GESCHONKE, M. KLEMPT, N. LYNCH,
  S. SCHREIBER, S. FENSELAU, AND J. SCHREZENMEIR................ 548

Glucose Transporter 4 Gene: Association Studies Pertaining to Alleles of Two
  Polymorphisms in Extremely Obese Children and Adolescents and in
  Normal and Underweight Controls. *By* SUSANN FRIEDEL, BENJAMIN
  ANTWERPEN, ANNE HOCH, CONSTANZE VOGEL, WOLFGANG GRASSL,
  FRANK GELLER, JOHANNES HEBEBRAND, AND ANKE HINNEY......... 554

Association of Insulin Gene VNTR Polymorphism with Polycystic Ovary
  Syndrome. *By* MARKÉTA VANKOVÁ, JANA VRBÍKOVÁ, MARTIN HILL,
  ONDREJ CINEK, AND BELA BENDLOVÁ............................. 558

Association between a Variant at the $GABA_A\alpha6$ Receptor Subunit Gene,
  Abdominal Obesity, and Cortisol Secretion. *By* ROLAND ROSMOND,
  CLAUDE BOUCHARD, AND PER BJÖRNTORP........................ 566

Increased Abdominal Obesity in Subjects with a Mutation in the 5-HT_{2A}
  Receptor Gene Promoter. *By* ROLAND ROSMOND, CLAUDE BOUCHARD,
  AND PER BJÖRNTORP... 571

Lipids and Insulin Resistance: What We've Learned at the Fourth
  International Smolenice Symposium. *By* E. RAVUSSIN, I. KLIMEŠ,
  E. ŠEBÖKOVÁ, AND B. V. HOWARD................................ 576

Index of Contributors.. 581

Financial assistance was received from:

Principal Sponsors
- ASTRA ZENECA, SWEDEN
- THE EUROPEAN COMMISSION, BELGIUM

Sponsors
- BERLIN CHEMIE, SLOVAKIA
- COCA-COLA BEVERAGES, SLOVAKIA
- ELI LILLY, SLOVAKIA
- FOURNIER LABORATORIES, SLOVAKIA
- GLAXOSMITHKLINE, SLOVAKIA
- LIPHA MERCK, SLOVAKIA
- NEW YORK ACADEMY OF SCIENCES
- NOVO NORDISK, DENMARK
- PFIZER, SLOVAKIA
- ROCHE DIAGNOSTICS SYSTEMS, SLOVAKIA
- SERVIER, CZECH REPUBLIC
- SLOVAK DIABETES SOCIETY

The New York Academy of Sciences believes it has a responsibility to provide an open forum for discussion of scientific questions. The positions taken by the participants in the reported conferences are their own and not necessarily those of the Academy. The Academy has no intent to influence legislation by providing such forums.

Preface

IWAR KLIMEŠ AND ELENA ŠEBÖKOVÁ

Diabetes and Nutrition Research Laboratory, Institute of Experimental Endocrinology, Slovak Academy of Sciences, SK-83306 Bratislava, Slovak Republic

The Fourth International Meeting on "Lipids and Insulin Resistance" took place from August 29 to September 2, 2001, in the congress center of the Slovak Academy of Sciences located in the Smolenice Castle at the southeastern foot of the Small Carpathian Range close to the capital of the Slovak Republic. It was a continuation of the series of symposia on the relation between insulin action and insulin resistance that have been organized by the Diabetes and Nutrition Research Laboratory of the Institute of Experimental Endocrinology, Slovak Academy of Sciences, in Bratislava, Slovak Republic, in 4- to 5-year intervals since 1988. Special scientific emphasis was placed this time on the "role of fatty acid metabolism and fuel partitioning".

At the first meeting in 1988, most of the invited speakers addressed various aspects of the classical signaling pathways of intracellular insulin action and the possible allotment of lipids. The two meetings held in the 1990s were devoted to the role of lipids in the pathogenesis of insulin resistance and/or of the syndrome of insulin resistance. In particular, during the 1992 Smolenice Symposium, the focus was restricted to dietary lipids in relation to insulin action per se. In the interlude, firm evidence accumulated linking lipids in their multiple metabolic roles (major energy resources, cell structural elements, intracellular second messengers, gene regulators, etc.) to many aspects of the diseases clustering within the clinically known and recognized insulin resistance syndrome. Hence, these relationships were thoroughly discussed at the Third Smolenice Symposium held in 1996.

As the end of the millennium approached, new information (including our own data) have been emerging pointing to the important role of fatty acids and/or of their metabolism (oxidation) for regulation of insulin action in health and disease states. In that context, we initially planned a meeting aimed at providing a focused forum for analyzing the role of fatty acid metabolism in fuel partitioning, where the latter is tightly linked to insulin action.

If the 4-year interim had been maintained, the Fourth Symposium would have been organized in 2000. The last millennium year, however, was full of major international diabetes events held around the globe at places such as Jerusalem, Sydney, and Mexico City. Obviously, many potential sponsors had other priorities that year. Thus, in 1999, we decided to plan the fourth meeting for the early fall of 2001 and

---

Address for correspondence: Iwar Klimeš, Diabetes and Nutrition Research Laboratory, Institute of Experimental Endocrinology, Slovak Academy of Sciences, Vlárska 3, SK-83306 Bratislava, Slovak Republic. Voice/fax: 421-7-5466-2687.
ueeniwar@savba.savba.sk

applied to the European Commission in Brussels, Belgium, for a symposium sponsorship.

The Fourth International Smolenice Symposium on Lipids and Insulin Resistance was very successful, with up to 130 participants from 15 countries on 4 continents attending. There were 35 invited lectures, 19 free oral communications, and 29 poster presentations. Scientific sessions were held on nuclear transcription factors; lipids and energy balance; fatty acid oxidation, fuel sensing, and insulin resistance; knockout and transgenic animal models of obesity and insulin resistance; muscle lipid metabolism in obesity and type 2 diabetes; skeletal muscle lipid composition and insulin signaling; skeletal muscle lipid metabolism at rest and during exercise; susceptibility genes for complex metabolic disorders; the role of mono- and polyunsaturated fatty acids in prevention of the insulin resistance syndrome; dietary fats, diabetes, and cardiovascular disease; and lipids, energy balance, and insulin resistance. This volume documents the excellence of the individual contributions. Regrettably, the tragic events that hit the world just 9 days after the close of our conference have sparked a crisis and influenced priorities. This has led to a delay in collection of the manuscripts, followed by a delay in the reviewing process. Moreover, some of the manuscripts had not reached us before the last submission deadline.

Nonetheless, we are delighted to acknowledge the support that we received from our Cochair, Eric Ravussin (Pennington Biomedical Research Center, Baton Rouge, LA), and from the Honorary Chair, Barbara V. Howard (MedStar Research Institute, Washington, D.C.), in the organization of the symposium by sharing their professional experience and serving as our ambassadors on the American subcontinent. Very valuable and substantial assistance also came from Leonard H. Storlien (Astra Zeneca, Mölndal, Sweden), one of the spiritual fathers of the "lipid-induced insulin resistance" concept and one of the cofounders of the Smolenice symposia. Personal events, however, prevented him from attending the meeting this time.

Thanks are also due to all the people who made this symposium possible: the members of the Local Organizing Committee and, particularly, the invited speakers who traveled long distances to participate. The unexpected cessation of international air traffic, which occurred shortly after our meeting, underlined the potential risk of foreign travel at that time. In fact, several invited speakers had to remain in Europe for up to two weeks.

We also wish to acknowledge the generous support of the Institute of Experimental Endocrinology of the Slovak Academy of Sciences in Bratislava as led by Richard Kvetnansky. The institutional support gained a new significance as our Institute became, in 2000, a "Center of Excellence" partially supported by the European Commission in Brussels, Belgium. The Fourth Smolenice Insulin Resistance Symposium work-package was a part of the general application for this prestigious quality trademark. Resources obtained via this channel have significantly contributed to the coverage of overall expenses of Europe-based invited speakers and young presenters. In addition, most of the other activities within the symposium would not have been imaginable without the significant support of the institutions listed at the end of the Table of Contents. Their support as well as the patronage received for this meeting from the Slovak Diabetes Association and its chair, Juraj Vozár, are greatly appreciated.

Further, we would like to extend our sincere thanks to the New York Academy of Sciences, which graciously agreed—for the third time over the last 10 years—to publish the manuscripts from the symposium. Particular thanks go to Barbara M. Goldman, Executive Editor of the *Annals* of the Academy, for her effective cooperation and to Stefan Malmoli, Associate Editor, for his expert editorial skills.

Finally, we stress that we are indebted to our spouses, Zuzana Jezerská and Peter Žák, respectively, for their enthusiasm and understanding to accompany us, for the fourth time, through the odyssey of organizing this symposium.

The Fifth International Smolenice Insulin Symposium should be held, if we remain at the 4-year intervals, in the summer/early fall of 2005. At the farewell party in 2001, some voices urged us to shorten the interim to 3 years. The future will determine whether this idea finds larger support within the European research area. Whatever the date of the Fifth Smolenice Symposium, we sincerely believe that as many outstanding colleagues will again attend it from all over the world as this time.

Identification of Susceptibility Genes for Complex Metabolic Diseases

CLIFTON BOGARDUS, LESLIE BAIER, PASKA PERMANA, MICHAL PROCHAZKA, JOHANNA WOLFORD, AND ROBERT HANSON

Phoenix Epidemiology and Clinical Research Branch, National Institute of Diabetes and Digestive and Kidney Diseases, National Institutes of Health, Phoenix, Arizona, USA

ABSTRACT: There are few successful attempts to identify genes for common, non-Mendelian diseases such as diabetes, hyperlipidemia, hypertension, etc. Such common disorders are typically both metabolically and genetically complex and the genetic technologies to identify their underlying susceptibility genes are still in their infancy. Nonetheless, genetic strategies have emerged that, when the technologies are fully developed, should allow similar success rates as for Mendelian diseases.

KEYWORDS: Mendelian/non-Mendelian disease; susceptibility genes; genetic; linkage; positional cloning; allele; subphenotype

INTRODUCTION

Specific disease-causing mutations have been identified for over 100 genetic disorders. In most cases, these are rare disorders with Mendelian modes of inheritance in which a single gene is necessary and sufficient to result in disease. These successes in positional cloning, previously called "reverse genetics", were possible because of recent technological advances and availability of genetic tools such as genome-wide maps of highly polymorphic microsatellite markers. Many more successful positional cloning efforts for uncommon Mendelian diseases can be anticipated in the years to come.

In contrast, there are few successful attempts to identify genes for common, non-Mendelian diseases such as diabetes, hyperlipidemia, hypertension, etc. Such common disorders are typically both metabolically and genetically complex and the genetic technologies to identify their underlying susceptibility genes are still in their infancy. Nonetheless, genetic strategies have emerged that, when the technologies are fully developed, should allow similar success rates as for Mendelian diseases.

Address for correspondence: Dr. Clifton Bogardus, Phoenix Epidemiology and Clinical Research Branch, National Institute of Diabetes and Digestive and Kidney Diseases, National Institutes of Health, 4212 North 16th Street, Room 541, Phoenix, AZ 85016. Voice: 1-602-200-5300; fax: 1-602-200-5335.
cbogardu@mail.nih.gov

POSITIONAL CLONING STRATEGY FOR MENDELIAN DISEASE

Successful positional cloning efforts for rare Mendelian diseases have generally employed a common genetic strategy. Large kindreds with affected members are collected and a segregation analysis is performed to determine the mode of inheritance (dominant, codominant, recessive, sex-linked) and to estimate the disease allele frequency. These parameters are used in a linkage analysis to localize the disease allele to a chromosomal region. The linkage analysis is done on genotypes determined on as many members of the kindred as possible, using highly polymorphic genetic markers, typically microsatellites, evenly spaced at ~10-cM intervals across all the autosomes and sex chromosomes. There are sets of commercially available genetic markers that are highly genetically informative (i.e., ~70% of individuals are heterozygous) and, therefore, specific allelic inheritance can be traced reliably through large pedigrees.

A successful linkage analysis indicates a most likely genomic region linked to the disease allele. A statistically significant result is generally accepted as a LOD (logarithm of the odds) score of greater than 3.3, corresponding to a genome-wide significance level of 4.5×10^{-5}.[1] The linked genomic region may be as broad as ~10 cM.

To further narrow the linked region, more highly polymorphic markers within the region are genotyped on members of the kindreds. By analyzing recombination events in the kindreds, the region harboring the putative disease allele can then be narrowed to a region containing only a few genes. Each of these genes will then be sequenced in affected and unaffected individuals to identify the specific genetic mutation, or mutations, associated with the disease.

Because of the availability of well-characterized, highly polymorphic genetic markers and the availability of automated means of genotyping and sequencing large numbers of samples, this approach has been both successful and relatively cost-effective. A very time-consuming and costly part of the strategy is in collecting DNA samples from several large kindreds of affected individuals with a rare disorder.

PROBLEMS IN IDENTIFYING GENES FOR COMMON, NON-MENDELIAN DISEASES

There are several limitations to applying this Mendelian-type genetic strategy to find genes for common, complex diseases such as diabetes, hyperlipidemia, etc. These disorders are not characterized by the simple genetic architecture in which a single gene is necessary and sufficient to cause the disease. The genetic determinants of these common, metabolically complex diseases are susceptibility alleles rather than disease alleles. This means that there are individuals who carry the susceptibility allele, but are unaffected by the disease, either because they lack another allele (or alleles) that is necessary for disease expression (i.e., a gene-gene interaction) or because they lack an environmental exposure necessary for disease expression (i.e., gene-environment interaction). These interactions, and the relatively high frequency of a given susceptibility allele, make it difficult to trace the allele through kindreds. Hence, the mode of inheritance is unclear, or non-Mendelian, and an estimate of the frequency of the susceptibility allele cannot be reliably ascertained by segregation analysis. In addition, the phenotypic effect of an individual susceptibility allele may be quite small.

These characteristics of complex diseases necessitate a different positional cloning strategy than that used for rare Mendelian disorders.

GENETIC STRATEGY TO FIND SUSCEPTIBILITY GENES FOR COMPLEX DISEASES

The strategy to find complex disease susceptibility alleles begins with linkage analysis. Compared to that of a Mendelian disease, this linkage analysis is considerably less powerful since it is performed without a known mode of inheritance and an estimated allele frequency (results of a segregation analysis), and many unaffected individuals may carry the susceptibility allele. To circumvent these limitations, the linkage analysis is often performed on many affected sibling pairs rather than a few large kindreds. In many complex metabolic diseases, this is also necessary because of the late age of onset of disease, which makes collection of DNA samples from parents of affected individuals difficult.

The linkage analysis is used to estimate the extent of allele sharing of sibling pairs at each polymorphic marker or, in multipoint analyses, of several adjacent markers. Affected sibling pairs are expected to share more than half their alleles identical by descent at, or near, the susceptibility locus. The likelihood of excess (>½) allele sharing being greater than chance is estimated, and a LOD score greater than 3.6 is considered statistically significant evidence for a linked locus.[1]

Due to the limited power of this linkage strategy, a genomic region linked to a complex disease is generally very broad (often 20–40 cM) and of borderline statistical significance (i.e., $1 < LOD < 3.6$). Putative linkage results in this range of significance should be confirmed in a separate group of affected sibling pairs before being considered indicative of localization of a disease susceptibility locus.

If putative linkage is replicated, fine mapping to narrow the genomic region harboring the putative susceptibility allele is undertaken. In this non-Mendelian case, analysis of recombination events is not useful due to unaffected carriers of the susceptibility allele; therefore, fine mapping is done using linkage disequilibrium, or association testing, between genetic markers and disease. The linkage disequilibrium mapping is done using single nucleotide polymorphisms (SNPs) since these are more common than microsatellites, occurring about every 1000 base pairs, and are less mutable than microsatellites.[2] SNPs also are potentially more amenable to automated means of genotyping. Automated genotyping would be a major advantage since localization of a susceptibility allele that has only a small phenotypic effect will require genotyping a large number of SNPs in a large number of individuals.

The aim of "SNPing" is to narrow the region of a putative susceptibility locus to a region of DNA that is technically feasible to sequence in a reasonable time. In other words, DNA from several affected and unaffected individuals must ultimately be sequenced and compared to identify a specific disease susceptibility allele. The SNPing strategy should narrow a broad region of linkage (~10–40 cM) to a physically small region of association (~1,000,000 base pairs) for intensive and thorough analysis.

This general approach to identify a susceptibility gene for a complex disease has been used to find a putative type 2 diabetes susceptibility allele[2,3] and, more recently, to identify polymorphisms in the NOD2 gene causing Crohn's disease.[4] However, these successes are the exceptions and it is clear that SNPing has its problems.

SNPs can now be obtained for most genomic regions from public databases,[5] but many of these SNPs (possibly as much as 50%) have not been validated as being polymorphic (i.e., >10% frequency for the less common allele). Every SNP obtained from databases has to be validated in a test sample of a particular population. Also, gaps remain in the human genome draft sequence, leaving some genomic regions with no known SNPs. To fill in these gaps with novel SNPs for a particular positional cloning project is very time-consuming and costly. In addition, some genomic regions contain high densities of repeat sequences that interfere with reliable genotyping of SNPs in that area.

There is also the very difficult and incompletely resolved issue as to how many SNPs need to be genotyped to definitively identify a disease-associated locus.[6] In some genes, there are multiple SNPs that are in varying degrees of linkage disequilibrium with one another, yet in other areas a high degree of linkage disequilibrium extends over large genomic regions (over 1 megabase). Thus, it remains unclear exactly what SNP density is required to find a disease-associated SNP (or SNPs). One answer may be that, because of variability across the genome, the appropriate SNP density can only be determined empirically for any specific region, within some broad guidelines of extreme possibilities. These broad guidelines will differ from one population to another depending on its genetic history.

For the present, until the human genome sequence is completed, an iterative approach may be the best way to proceed. A SNP density of 1 SNP/50 kb is approximated as closely as possible and an analysis of disease association and apparent extent of linkage disequilibrium in the specific region is estimated. A decision can then be made whether a greater SNP density (1 SNP/25 kb) is potentially useful and/or necessary, and the process repeated. As more SNPs become available in various databases and the human genome is closer to completion, each iteration should be more efficient and informative.

COMPLEMENTARY STRATEGY TO POSITIONALLY CLONE A COMPLEX DISEASE GENE

Until the human genome sequence is complete and there are large numbers of fully validated SNPs available in public databases, the use of linkage disequilibrium analyses of SNPs to positionally clone complex disease genes will remain a promising, but costly and risky strategy. In the meantime, it is worth considering complementing it with parallel studies.

One parallel approach is to sequence known candidate genes near the peak of linkage. A gene is selected as a good candidate if it has a known or suspected role in biologic pathways physiologically relevant to the disease under study. Depending on how much genomic sequence of these candidates is available in databases, this can be a relatively easy or a more difficult and time-consuming undertaking. It is also worthwhile to remember that the rate of success of researchers to identify genes for diseases based on current understanding of their pathophysiology is very low. No one would have predicted, for example, that hepatocyte nuclear factor-1 alpha is a diabetes susceptibility gene.[7] Also, the general intent of positional cloning has been to overcome this shortcoming by finding a gene linked to a disease and working backwards toward understanding the pathophysiology—hence, its earlier name of "reverse genetics". Thus, it is somewhat incongruous to use a candidate gene approach as part of a positional cloning strategy. Nonetheless, the linkage results will have narrowed the list of potentially biologically relevant genes to a relatively small number, and it would be unwise to ignore a gene near the peak of linkage that is thought to have biological effects directly related to the pathophysiology of the disease under study.

Another complementary approach to a positional cloning effort to find complex disease genes is the use of microarrays to identify large numbers of expressed sequences [either full-length cDNA or expressed sequence tags (ESTs)] for differential levels of expression in tissues obtained from affected and unaffected individuals. The assumption of this approach is that the susceptibility allele influences the disease phenotype by increasing or decreasing the level of the gene transcript in a particular tissue.

There are two types of microarrays: (1) full-length cDNAs and/or ESTs printed on either glass slides or nylon membranes and (2) oligonucleotide arrays, on which oligomers representing the gene transcript are synthesized *in situ* on glass wafers using photolithography.[8] These microarrays are probed with RNA extracted from particular tissues obtained from affected and unaffected individuals. In the resulting list of differentially expressed genes, one hopes to find a cDNA or EST that is encoded by a gene that maps near the peak of linkage.

This method has the advantage that large numbers of expressed sequences can be probed in a short period of time. However, there are many limitations of the method. There are, as yet, no commercially available microarrays that contain all the known cDNAs from the human genome. It would be possible to prepare one's own slides from the known ESTs on the region of linkage, but this list of ESTs will be incomplete until the human genome sequence is completed and fully analyzed. Another limitation of the microarray approach is that the arrays are probed using RNA extracted from biopsied tissues of living donors. This greatly limits the tissues that can be studied. Although it is theoretically possible to collect autopsy material, it would be logistically very difficult to obtain well-preserved RNA. Hence, only tissues amenable to biopsy in living donors can be analyzed for differential gene expression, and these may not be the tissue(s) in which the disease-associated differential gene expression potentially exists. Thus, in the short term, this approach has major limitations and probably should be considered as a "long shot". In the long term, when cDNA microarrays for a particular chromosome are commercially available, this method should be a very valuable complement to a positional cloning effort.

USE OF SUBPHENOTYPES

It is often debated whether it is preferable to attempt to positionally clone a complex disease susceptibility gene using the disease itself as the studied phenotype or to use a subphenotype. Analyses of subphenotypes (such as the insulin resistance subphenotype of type 2 diabetes) are potentially advantageous because they have fewer genetic determinants than the full disease phenotype. There is also a closer one-to-one correlation between a genotype and subphenotype than between a genotype and the full disease, making positional cloning a simpler task. Also, subphenotypes are generally quantitative traits that may be more informative than the dichotomy of affected versus unaffected by disease.

There are potential disadvantages to the subphenotype approach, though. Subphenotypes are often more difficult to measure *in vivo*—for example, determining the level of insulin resistance as a subphenotype of type 2 diabetes or syndrome X. This often greatly limits the numbers of relative pairs that can be obtained for a genomic scan and linkage analysis. Given the limited power of linkage analysis to localize genes for non-Mendelian traits, a reduction in the number of relative pairs for analyses is a major drawback.

It is also debatable whether a positional cloning effort should be undertaken to find susceptibility alleles for a subphenotype trait rather than for a disease. Given the enormity of the effort to positionally clone a non-Mendelian trait, it is questionable, at present, whether such a large task should be undertaken to find genes for any other trait than a disease that is closely associated with major morbidity and/or mortality.

Perhaps in the long term, when the human genome sequence is completed, the SNP databases are completed, and SNP genotyping has been automated, a search for susceptibility alleles for a variety of traits can be justifiably undertaken and successfully completed in reasonable periods of time.

REFERENCES

1. LANDER, E. & L. KRUGLYAK. 1995. Genetic dissection of complex traits: guidelines for interpreting and reporting linkage results. Nat. Genet. **11:** 241–247.
2. WANG, D.G., J-B. FAN, C-J. SIAO *et al.* 1998. Large-scale identification, mapping, and genotyping of single-nucleotide polymorphisms in the human genome. Science **280:** 1077–1082.
3. HORIKAWA, Y., N. ODA, N.J. COX *et al.* 2000. Genetic variation in the gene encoding calpain-10 is associated with type 2 diabetes mellitus. Nat. Genet. **26:** 163–175.
4. HUGOT, J-P., M. CHAMAILLARD, H. ZOUALL *et al.* 2001. Association of NOD2 leucine-rich repeat variants with susceptibility to Crohn's disease. Nature **411:** 599–603.
5. THE INTERNATIONAL SNP MAP WORKING GROUP. 2001. A map of human genome sequence variation containing 1.42 million single nucleotide polymorphisms. Nature **409:** 928–933.
6. KRUGLYAK, L. 1999. Prospects for whole-genome linkage disequilibrium mapping of common disease genes. Nat. Genet. **22:** 139–144.
7. YAMAGATA, K., N. ODA, P.J. KAISAKI *et al.* 1996. Mutations in the hepatocyte nuclear factor-1 alpha gene in maturity-onset diabetes of the young (MODY3). Nature **384:** 455–458.
8. PEASE, A.C., D. SOLAS, E.J. SULLIVAN *et al.* 1994. Light-generated oligonucleotide arrays for rapid DNA sequence analysis. Proc. Natl. Acad. Sci. U.S.A. **91:** 5022–5026.

PPARs: Transcription Factors Controlling Lipid and Lipoprotein Metabolism

VIRGINIE BOCHER, INÉS PINEDA-TORRA, JEAN-CHARLES FRUCHART, AND BART STAELS

U.545 INSERM, Département d'Athérosclérose, Institut Pasteur de Lille, Lille, France, and Faculté de Pharmacie, Université de Lille II, Lille, France

> ABSTRACT: Nuclear receptors are transcription factors that are activated by ligands and subsequently bind to regulatory regions in target genes, thereby modulating their expression. Nuclear receptors thus allow the organism to integrate signals coming from the environment and to adapt by modifying the expression levels of relevant genes. The peroxisome proliferator-activated receptors (PPARs) α, β/δ, and γ constitute a subfamily of nuclear receptors. PPARα has been shown to bind and to be activated by leukotriene B4 and the hypolipidemic drugs of the fibrate class; PPARβ/δ ligands are polyunsaturated fatty acids and prostaglandins; while prostaglandin J2 derivatives and the antidiabetic glitazones are, respectively, natural and synthetic ligands for PPARγ. Upon binding and activation by their ligands, they regulate the transcription of numerous genes involved in intracellular lipid metabolism, lipoprotein metabolism, and reverse cholesterol transport in a subtype- and tissue-specific manner. PPARs therefore constitute interesting targets for the development of therapeutic compounds useful in the treatment of disorders of lipid and lipoprotein metabolism.
>
> KEYWORDS: transcription factors; gene expression; fatty acids; lipoprotein metabolism

INTRODUCTION

The control of lipid and lipoprotein metabolism and glucose homeostasis is a complex process implying numerous genes, the expression of which is regulated in a coordinated manner. By modulating the expression of these genes, the organism can adapt its metabolism to changes in energy requirements. The expression of these genes can be regulated through ligand-activated transcription factors, called nuclear receptors, that are capable of responding to small lipophilic signaling molecules. Among these nuclear receptors, the peroxisome proliferator-activated receptors (PPARs) play a major role since they are activated by fatty acids and derivatives. These nuclear receptors were first identified in 1990 when the isoform α (PPARα) was shown to be the receptor of xenobiotics capable of inducing peroxisome prolif-

---

Address for correspondence: Bart Staels, U.545 INSERM, Département d'Athérosclérose, Institut Pasteur de Lille, 1, rue Calmette BP245, 59019 Lille, France. Voice: 33-3-20-87-73-88; fax: 33-3-20-87-73-60.
  bart.staels@pasteur-lille.fr

eration in rodent liver. Subsequently, two other PPAR genes, PPARγ and PPARβ/δ, have been identified. Studies on the physiological functions of these receptors have expanded to such diverse fields as lipid and glucose homeostasis, cellular proliferation and differentiation, and control of inflammation.

TISSUE DISTRIBUTION OF THE PPAR GENES

PPARα is mainly expressed in tissues where fatty acid catabolism is important, such as liver, kidney, heart, and muscles, whereas PPARγ is preferentially expressed in brown and white adipose tissue and intestine, where it controls cellular differentiation and lipid storage and modulates the action of insulin.[1] Both isoforms are also expressed in the different cell types of the vascular wall: endothelial cells,[2] smooth muscle cells,[3] and monocytes/macrophages.[4–6] PPARα and PPARγ are present in the lipid core of the atherosclerotic lesion of humans, where they colocalize with markers specific for endothelial cells, macrophages, foam cells, and smooth muscle cells.[7,8] In contrast, the expression of PPARβ/δ is more ubiquitous since it is present in a number of tissues including heart, adipose tissue, brain, intestine, muscle, spleen, lung, and adrenal glands.[9] Although PPARβ/δ is still the less well studied PPAR, it was suggested to be involved in the control of brain lipid metabolism,[9] HDL metabolism,[10] epidermal cell proliferation,[11] fatty acid (FA)–induced adipogenesis, and preadipocyte proliferation.[9,12–14]

MOLECULAR MECHANISM OF ACTION

Transactivation

After binding of their ligands and heterodimerization with the retinoid X receptor (RXR), PPARs are capable of recognizing specific DNA sequences in the promoter region of target genes. These short sequences called peroxisome proliferator response elements (PPREs) usually consist of a direct repeat of the hexanucleotide AGGTCA sequence separated by one or two spacers (DR1 or DR2).[1]

Transrepression

PPARs are also able to repress the expression of certain genes by interfering negatively with the NF-κB, STAT, AP-1, and C/EBP signaling pathways in a DNA binding–independent manner. This mechanism of transrepression involves protein-protein interactions and cofactor competition, as well as the induction of IκBα, which is the major inhibitor of the NF-κB pathway.[2–5,15–17] This mechanism is currently thought to explain the anti-inflammatory properties of the PPARs.

Regulation of PPAR Activity

Since many cell types express more than one isoform of PPARs, different mechanisms have been proposed to explain the specificity of a subtype for a particular target gene such as differential cofactor recruitment, receptor posttranslational modifications, or endogenous ligand availability. Coactivators, including members of p160/CBP/p300 and DRIP/TRAP families and corepressors, such as N-CoR or SMRT, mediate contact between the PPAR-RXR heterodimer and the transcriptional machinery, thus promoting activation or repression of gene expression.[18] Although there are no known receptor-specific cofactors, some PPAR ligands can selectively recruit cofactors, thereby resulting in ligand-specific target gene regulation.[18] Additionally, it was shown that both PPARα and PPARγ activity can be modulated either positively or negatively by phosphorylation and that PPARγ ligands induce ubiquitination and subsequent degradation of the receptor.[18] Finally, receptor activity can be modified by increasing or decreasing ligand availability. Thus, albumin may be an important regulator of PPARγ function by sequestering PG-J2 and effectively reducing the concentration available to PPARγ.[19] In contrast, the liver FA-binding protein (L-FABP), which could serve as a cytosolic gateway for transport of FAs and other PPARα agonists to the nucleus, interacts directly with PPARα.[20]

Natural and Synthetic Ligands

A wide range of compounds have been identified as PPAR ligands. FAs and FA-derived compounds are natural ligands for PPARα, PPARγ, and PPARβ/δ.[21,22]

PPARα is activated by natural eicosanoids derived from arachidonic acid via the lipoxygenase pathway, such as 8-S-hydroxytetraenoic acid (8-S-HETE)[21] and leukotriene B4 (LTB4),[23] as well as by oxidized phospholipids from oxidized lipoproteins (OxLDL).[24] Fibrates, a class of hypolipidemic drugs, are synthetic agonists of PPARα.[21] PPARγ is activated by various metabolites derived from arachidonic acid through the lipoxygenase and cyclooxygenase pathways, such as 15-deoxy prostaglandin J2 (PG-J2) and 15-HETE.[21,25] FA-derived compounds from oxidized LDL (9-HODE, 13-HODE)[25] are also natural PPARγ ligands. More recently, two other prostaglandins, PG-H(1) and PG-H(2), were reported to be weak ligands for PPARγ, albeit as potent as PG-J2, suggesting new pathways of PPARγ-linked gene activation.[26] In addition, the antidiabetic glitazones are high-affinity ligands for PPARγ.[21] Other pharmacological compounds—in particular, nonsteroidal anti-inflammatory drugs—have been reported as agonists for PPARα and PPARγ.[27] Finally, PPARβ/δ ligands are polyunsaturated FAs and prostaglandins, such as PG-J2 or the synthetic ligand carbaprostacyclin.[22] Recently, a potent, subtype-selective, synthetic PPARβ/δ agonist has also been reported.[10]

PPARs AND THE REGULATION OF INTRACELLULAR LIPID METABOLISM

PPARα plays a key role in FA catabolism. In rodents, but not in humans, PPARα activation causes a peroxisomal proliferation, leading to hepatomegaly and hepatocarcinogenesis. The species differences in response to peroxisome proliferators could be due to distinct regulations of the enzymes involved in peroxisomal β-oxidation, such as acyl-CoA oxidase (ACO). Indeed, ACO is induced by fibrates in rodents, whereas no major effect is observed in human primary hepatocyte cultures.[28] Other enzymes of the peroxisomal β-oxidation pathway have also been shown to be regulated by PPARα, such as multifunctional enzyme and 3-ketoacyl-CoA thiolase.[29] Recent studies using PPARα-overexpressing human hepatoma HepG2 cells confirmed that peroxisome proliferation-associated genes are not significantly induced by PPARα.[30] In contrast, in both human and rodents, PPARα regulates the expression of genes involved in FA uptake, activation to acyl-CoA esters, mitochondrial β-oxidation, and ketone body synthesis[29,31] (FIG. 1). Intracellular FA concentrations are partly controlled by the activity of FA transport proteins such as FAT/CD36 and FATP1, which control the entry of FA through the cell membrane, and by acyl-CoA synthetase (ACS), which traps FAs inside the cells by their conversion to ester derivatives. PPARα induces FATP expression in liver and intestine, and ACS in liver and kidney.[32] Finally, by inducing muscle- and liver-type α-carnitine palmitoyl-transferase I,[33–36] PPARα activators favor mitochondrial FA uptake and metabolism. Hence, through their effect on the expression of FA transporters and FA oxidation genes, PPARα agonists increase the β-oxidation of FA and therefore decrease the pool of FA available for incorporation into triglyceride-rich lipoproteins (FIG. 1). Furthermore, fibrates and long-chain FAs (LCFAs) upregulate the expression of the L-FABP, a protein highly expressed in liver and small intestine where it is thought to play a key role in cellular FA flux. In contrast to what is known in liver, PPARα appears not to significantly regulate L-FABP gene expression in the small intestine.[37] However, GW 2433, an agonist for both PPARα and PPARβ/δ, is able to increase L-FABP expression in the intestine of PPARα-null mice. Thus, PPARβ/δ may contribute to adaptation of the intestine to changes in the dietary lipid composition.[37]

Whereas PPARα regulates hepatic lipid metabolism, PPARγ modulates lipid homeostasis through its role in adipocyte differentiation and lipid storage in adipose tissue (FIG. 1). The induction of LPL[38] by PPARγ promotes FA delivery to adipocytes, while induction of FATP[32] and ACS[39,40] results in enhanced FA uptake by the adipocyte. These actions contribute to enhanced TG synthesis and accumulation in adipose tissue (FIG. 1).

PPARs AND THE REGULATION OF LIPOPROTEIN METABOLISM

High plasma TG and low HDL cholesterol levels are important risk factors for coronary artery disease (CAD), especially in patients with the metabolic syndrome. Fibrates were shown to lower plasma levels of TG and cholesterol, while increasing HDL cholesterol levels via PPARα.[41] PPARγ synthetic agonists, in addition to their antidiabetic action, also modulate plasma lipid levels.

PPARs and TG

PPARα in liver, muscle, and heart and PPARγ in adipose tissue increase the expression and activity of LPL[38] (FIG. 1). This enzyme is involved in lipolysis and therefore contributes to increasing the clearance of TG-rich lipoproteins, thus decreasing the circulating TG levels. Furthermore, fibrates inhibit the expression of ApoCIII, which is a natural inhibitor of LPL activity (FIG. 1). However, saturated FAs could also induce macrophage LPL expression via PPARα, a potential atherogenic activity that may counteract the beneficial lipid-lowering effect.[42]

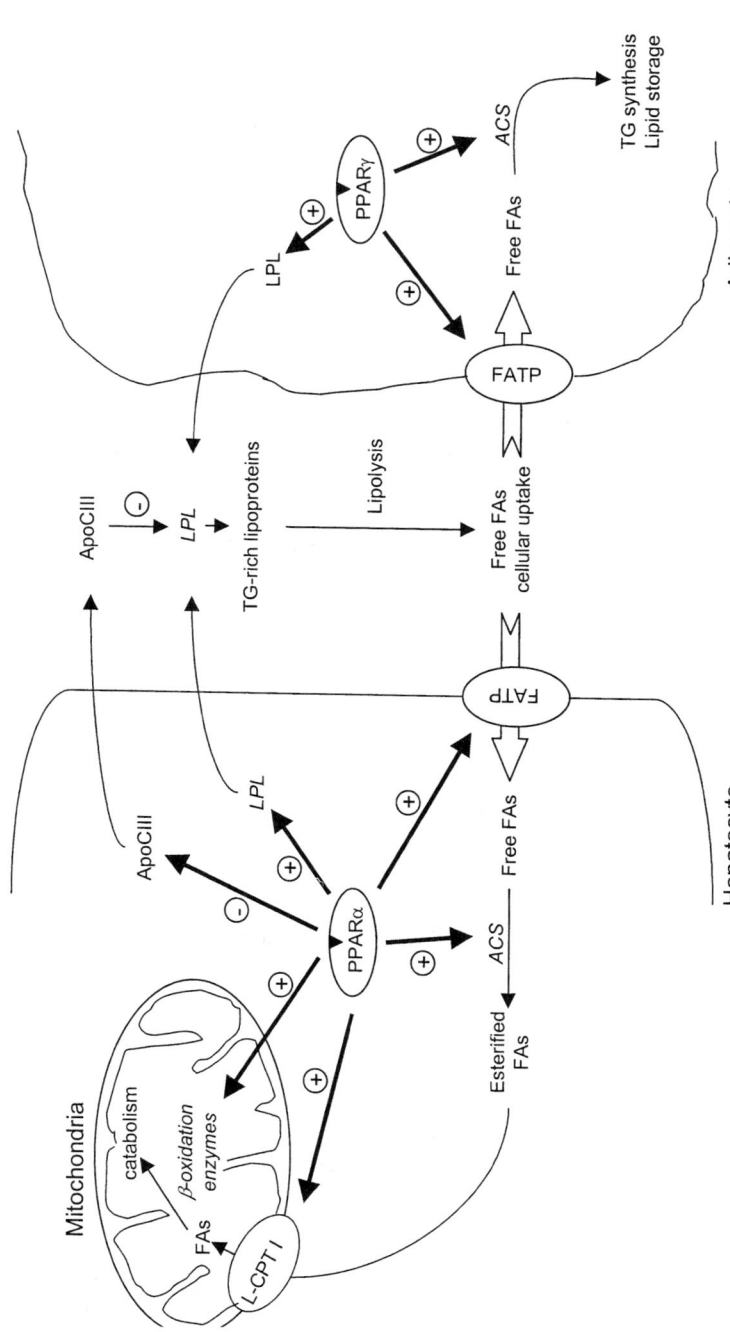

FIGURE 1. Role of PPARs in the control of triglyceride (TG) and fatty acid (FA) metabolism. In the liver, PPARα regulates the expression of genes involved in FA uptake (FATP), activation to acyl-CoA esters (ACS), mitochondrial uptake (L-CPT I), and β-oxidation. In the adipose tissue, PPARγ promotes FA uptake (via FATP) and TG synthesis and storage. By inducing the expression of lipoprotein lipase (LPL), both PPARα and PPARγ favor lipolysis and clearance of TG-rich lipoproteins.

PPARs and Foam Cell Formation

Initially, PPARs were thought to promote foam cell formation. In particular, PPARγ was suggested to play a role in monocyte differentiation to macrophages and to induce the expression of FAT/CD36, an FA transporter and receptor for oxidized LDL, in macrophages.[43] This would lead to intracellular accumulation of cholesterol, and the provision of natural PPARγ ligands derived from the phospholipids of oxidized LDL would create a positive feedback loop, thus propelling the foam cell formation process.[25] However, recent studies using PPARγ-deficient cells have demonstrated that PPARγ does not promote monocyte to macrophage differentiation.[44] Although PPARγ agonists induce CD36 levels in macrophages,[44,45] they have opposed effects on the expression of another scavenger receptor, SR-A, involved in acetylated LDL uptake.[5] Indeed, studies in embryonic stem cell–derived macrophages from PPARγ-deficient mice demonstrated that troglitazone inhibits SR-A protein expression in a PPARγ-dependent manner.[44] In contrast, troglitazone and rosiglitazone do not seem to have an effect on SR-A mRNA levels, suggesting a mechanism of posttranscriptional regulation. Similarly, *in vivo*, treatment of LDL receptor–deficient mice with specific PPARγ agonists (rosiglitazone and GW 7845) does not modify SR-A mRNA expression in the atherosclerotic lesion.[46] Thus, through its opposed effects on CD36 and SR-A, PPARγ activation does not promote an accumulation of cholesterol in macrophages. This was further confirmed using different animal models of atherosclerosis. Indeed, PPARγ agonists prevent the progression of atherosclerotic lesions and decrease macrophage accumulation in plaques of LDL receptor–deficient mice.[47] In addition, in ApoE-knockout mice, the antidiabetic PPARγ agonists, thiazolidinediones, inhibit fatty streak formation, while enhancing HDL cholesterol levels and increasing FAT/CD36 mRNA levels in the aorta.[48]

PPARs AND REVERSE CHOLESTEROL TRANSPORT

PPARs and Cholesterol Efflux

Not only do PPARs not favor foam cell formation, they on the contrary have beneficial effects on macrophage cholesterol homeostasis. Cholesterol efflux is the first step of the reverse cholesterol transport pathway that transports cholesterol back to the liver. Cholesterol efflux from the peripheral cells can occur passively via diffusion or actively through the scavenger receptor class B type I (SR-B1/CLA-1) or via the ATP-binding cassette A1 (ABCA1) pathways. Recent data have shown that ABCA1, in an ApoA1-dependent manner, allows the removal of intracellular cholesterol from peripheral cells towards pre-β-HDL or HDL[49] (FIG. 2). Both PPARα and PPARγ activators were shown to enhance the expression of SR-BI/CLA-1[7] and ABCA1[50] in human macrophages. The induction of ABCA1 occurs indirectly via the liver X receptor α (LXRα), another nuclear receptor for which ABCA1 is a known target gene[50,51] (FIG. 2). Moreover, a selective PPARβ/δ agonist also induces ABCA1 expression and ApoAI-dependent cholesterol efflux in macrophages, fibroblasts, and intestinal cells, leading to increased HDL levels in insulin-resistant monkeys.[10]

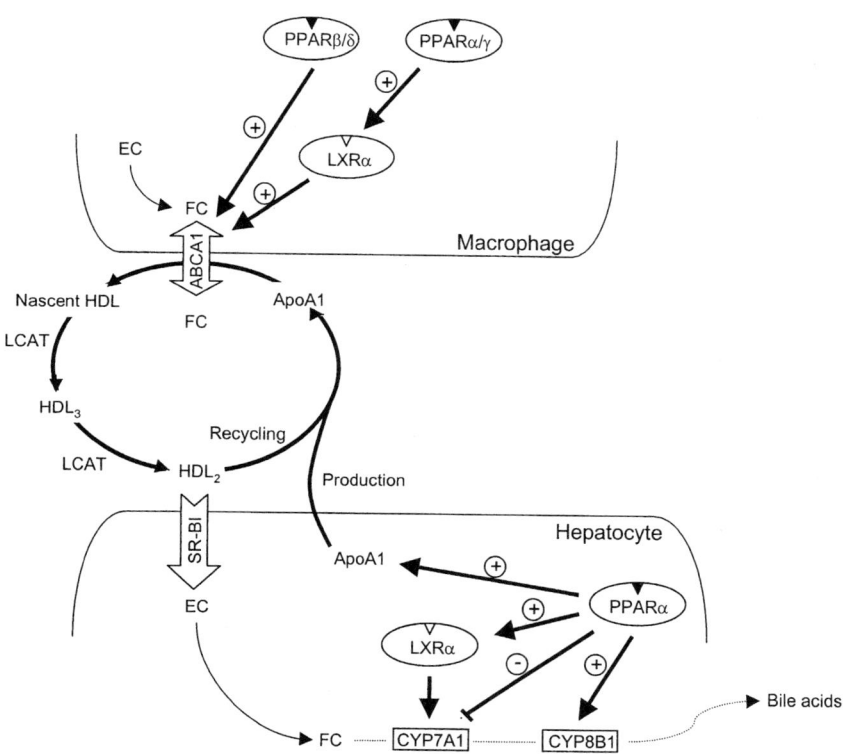

FIGURE 2. Role of PPARs in the regulation of the reverse cholesterol transport pathway. ATP-binding cassette A1 (ABCA1) mediates apolipoprotein (Apo) A1–dependent cholesterol efflux from macrophages. Nascent HDL particles accept excess cholesterol, which is then esterified by lecithin:cholesterol acyl transferase (LCAT). The hepatic scavenger receptor (SR)–BI ensures the uptake of cholesteryl esters in the liver, where cholesterol is excreted into bile either directly or after enzymatic conversion to bile acids. Positive or negative regulation of these different steps of the reverse cholesterol transport pathway by PPARs is shown as "+" or "–", respectively. EC, esterified cholesterol; FC, free cholesterol; LXR, liver X receptor.

PPARα also enhances the transcription of the major HDL apolipoproteins, ApoAI and ApoAII, in humans in a PPRE-dependent manner, thereby increasing HDL synthesis[52] (FIG. 2). On the contrary, PPARα activation in rodents leads to a decrease in circulating HDL as well as ApoAI and ApoAII levels. It was demonstrated that three differing nucleotides in the rat ApoAI promoter render it unresponsive to PPARα.[53] Furthermore, Rev-erbα, an orphan nuclear receptor, inhibits ApoAI transcription through a negative response element situated next to the TATA box in the rat, but not in the human promoter.[53] PPARα, which induces Rev-erbα expression, could thus be responsible for the repressive effect of fibrates on rat ApoAI expression. In addition, fibrate treatment of patients with CAD increases their ApoAII plasma levels.[54] This effect, due to an increase in hepatic ApoAII production, was shown *in vitro* in human primary hepatocytes and HepG2 cells and to be mediated via a PPRE in ApoAII promoter site J.[54] Finally, PPARβ/δ agonists were shown to increase HDL cholesterol plasma levels in insulin-resistant monkeys[10] and mice.[55] The mechanism through which this effect is achieved needs to be further determined since an increase in HDL cholesterol could result either from an increased production or from a decreased catabolism and elimination of HDL particles by the liver. In contrast, it has also been suggested that PPARβ/δ is a powerful promoter of lipid accumulation in cultured macrophages and might thus play a detrimental role in the process of atherogenesis.[56] Therefore, the role of PPARβ/δ in the reverse cholesterol transport pathway remains to be clarified.

PPARs and Bile Acid Metabolism

The cholesterol removed from peripheral cells is taken up in the liver and excreted in bile, either directly or after conversion to bile acids. This pathway constitutes the major pathway of cholesterol elimination from the body. Interestingly, PPARα is also involved in the control of this last step of the reverse cholesterol transport pathway (FIG. 2). Indeed, PPARα was shown to induce expression of LXRα in liver.[57] This oxysterol nuclear receptor is known to upregulate the expression of cholesterol 7α-hydroxylase (CYP7A), the rate-limiting enzyme in bile acid synthesis, in rodents, but not in humans.[58,59] However, fibrates unexpectedly reduce CYP7A activity in both rodents and humans.[60–62] Therefore, the role of PPARα in CYP7A regulation is controversial: although PPARα was shown to inhibit CYP7A expression in an indirect manner,[63,64] it was also reported to positively regulate murine and, to a much lesser extent, human CYP7A promoter activity.[65] A recent study shows that fibrates decrease bile acid synthesis and CYP7A and sterol 27-hydroxylase (CYP27) activities and mRNA expressions in cultured rat hepatocytes as well as in rats and mice, but not in PPARα-null mice.[66] Such a decrease in bile acid production could result in increased gallstone formation in patients under fibrate treatment. In addition, PPARα induces the expression of cholesterol 12α-hydroxylase, a branch-point enzyme in the bile acid biosynthetic pathway, and increases the ratio between cholic and chenodeoxycholic acid[67] (FIG. 2). As cholic acid is more hydrophobic, this might allow enhanced sterol and fatty acid resorption in the intestine under fasting conditions that induce PPARα activation, thus allowing the body to adapt to the diet. However, in the case of a hypolipidemic treatment, it might also adversely contribute to the increased risk of gallstone formation.

CONCLUSIONS

PPARs are nuclear receptors that, upon activation by FAs and FA-derived compounds, are capable of regulating not only FA metabolism, but also whole body lipid and glucose homeostasis. Numerous studies provide a molecular mechanism explaining the lipid-lowering action of PPARα agonists observed in clinical trials, thus providing a rationale for their use in diseases linked to atherosclerosis such as stroke, myocardial infarction, and unstable angina. Moreover, recent data suggest that PPARγ and PPARβ/δ could also constitute interesting therapeutical targets in the treatment of cardiovascular pathologies.

REFERENCES

1. DESVERGNE, B. & W. WAHLI. 1999. Peroxisome proliferator-activated receptors: nuclear control of metabolism. Endocr. Rev. **20:** 649–688.
2. DELERIVE, P., F. MARTIN-NIZARD, G. CHINETTI et al. 1999. Peroxisome proliferator-activated receptor activators inhibit thrombin-induced endothelin-1 production in human vascular endothelial cells by inhibiting the activator protein-1 signaling pathway. Circ. Res. **85:** 394–402.
3. STAELS, B., W. KOENIG, A. HABIB et al. 1998. Activation of human aortic smooth-muscle cells is inhibited by PPARα, but not by PPARγ activators. Nature (Lond.) **393:** 790–793.
4. CHINETTI, G., S. GRIGLIO, M. ANTONUCCI et al. 1998. Activation of peroxisome proliferator-activated receptors α and γ induces apoptosis of human monocyte-derived macrophages. J. Biol. Chem. **273:** 25573–25580.
5. RICOTE, M., A.C. LI, T.M. WILLSON et al. 1998. The peroxisome proliferator-activated receptor-gamma is a negative regulator of macrophage activation. Nature (Lond.) **391:** 79–82.
6. JIANG, C., A.T. TING & B. SEED. 1998. PPAR-γ agonists inhibit production of monocyte inflammatory cytokines. Nature (Lond.) **391:** 82–86.
7. CHINETTI, G., G.F. GBAGUIDI, S. GRIGLIO et al. 2000. CLA-1/SR-BI is expressed in atherosclerotic lesion macrophages and regulated by activators of peroxisome proliferator-activated receptors. Circulation **101:** 2411–2417.
8. MARX, N., G. SUKHOVA, C. MURPHY et al. 1998. Macrophages in human atheroma contain PPARgamma: differentiation-dependent peroxisomal proliferator-activated receptor gamma (PPARgamma) expression and reduction of MMP-9 activity through PPAR-gamma activation in mononuclear phagocytes *in vitro*. Am. J. Pathol. **153:** 17–23.
9. BASU-MODAK, S., O. BRAISSANT, P. ESCHER et al. 1999. Peroxisome proliferator-activated receptor beta regulates acyl-CoA synthetase 2 in reaggregated rat brain cell cultures. J. Biol. Chem. **274:** 35881–35888.
10. OLIVER, W.R., JR., J.L. SHENK, M.R. SNAITH et al. 2001. A selective peroxisome proliferator-activated receptor delta agonist promotes reverse cholesterol transport. Proc. Natl. Acad. Sci. U.S.A. **98:** 5306–5311.
11. MICHALIK, L., B. DESVERGNE, N.S. TAN et al. 2001. Impaired skin wound healing in peroxisome proliferator-activated receptor (PPAR) alpha and PPAR beta mutant mice. J. Cell Biol. **154:** 799–814.
12. JEHL-PIETRI, C., C. BASTIE, I. GILLOT et al. 2000. Peroxisome-proliferator-activated receptor delta mediates the effects of long-chain fatty acids on post-confluent cell proliferation. Biochem. J. **350**(part 1)**:** 93–98.
13. HANSEN, J.B., H. ZHANG, T.H. RASMUSSEN et al. 2001. Peroxisome proliferator-activated receptor delta (PPARdelta)-mediated regulation of preadipocyte proliferation and gene expression is dependent on cAMP signaling. J. Biol. Chem. **276:** 3175–3182.
14. PETERS, J.M., S.S. LEE, W. LI et al. 2000. Growth, adipose, brain, and skin alterations resulting from targeted disruption of the mouse peroxisome proliferator-activated receptor beta(delta). Mol. Cell. Biol. **20:** 5119–5128.
15. DELERIVE, P., P. GERVOIS, J-C. FRUCHART et al. 2000. Induction of IkappaBalpha expression as a mechanism contributing to the anti-inflammatory activities of peroxisome proliferator-activated receptor-alpha activators. J. Biol. Chem. **275:** 36703–36707.
16. GERVOIS, P., N. VU-DAC, R. KLEEMANN et al. 2001. Negative regulation of human fibrinogen gene expression by peroxisome proliferator-activated receptor alpha agonists via inhibition of CCAAT box/enhancer-binding protein beta. J. Biol. Chem. **276:** 33471–33477.

17. ZHOU, Y.C. & D.J. WAXMAN. 1999. Cross-talk between Janus kinase–signal transducer activator of transcription (JAK-STAT) and peroxisome proliferator-activated receptor α (PPARα) signaling pathways. J. Biol. Chem. **274:** 2672–2681.
18. PINEDA-TORRA, I., G. CHINETTI, C. DUVAL *et al.* 2001. Peroxisome proliferator-activated receptors: from transcriptional control to clinical practice. Curr. Opin. Lipidol. **12:** 245–254.
19. PERSON, E.C., L.L. WAITE, R.N. TAYLOR *et al.* 2001. Albumin regulates induction of peroxisome proliferator-activated receptor-gamma (PPARgamma) by 15-deoxy-delta(12–14)-prostaglandin J(2) *in vitro* and may be an important regulator of PPAR-gamma function *in vivo*. Endocrinology **142:** 551–556.
20. WOLFRUM, C., C.M. BORRMANN, T. BORCHERS *et al.* 2001. Fatty acids and hypolipidemic drugs regulate peroxisome proliferator-activated receptors alpha– and gamma–mediated gene expression via liver fatty acid binding protein: a signaling path to the nucleus. Proc. Natl. Acad. Sci. U.S.A. **98:** 2323–2328.
21. WILLSON, T.M. & W. WAHLI. 1997. Peroxisome proliferator-activated receptor agonists. Curr. Opin. Chem. Biol. **1:** 235–241.
22. DESVERGNE, B. & W. WAHLI. 1994. PPAR: a key nuclear factor in nutrient/gene interactions. *In* Inducible Gene Expression. Vol. 1, pp. 142–176. Birkhäuser. Boston.
23. DEVCHAND, P.R., H. KELLER, J.M. PETERS *et al.* 1996. The PPARα–leukotriene B4 pathway to inflammation control. Nature (Lond.) **384:** 39–43.
24. DELERIVE, P., C. FURMAN, E. TEISSIER *et al.* 2000. Oxidized phospholipids activate PPARα in a phospholipase A2–dependent manner. FEBS Lett. **471:** 34–38.
25. NAGY, L., P. TONTONOZ, J.G.A. ALVAREZ *et al.* 1998. Oxidized LDL regulates macrophage gene expression through ligand activation of PPARγ. Cell **93:** 229–240.
26. FERRY, G., V. BRUNEAU, P. BEAUVERGER *et al.* 2001. Binding of prostaglandins to human PPARgamma: tool assessment and new natural ligands. Eur. J. Pharmacol. **417:** 77–89.
27. LEHMANN, J.M., J.M. LENHARD, B.B. OLIVER *et al.* 1997. Peroxisome proliferator-activated receptors α and γ are activated by indomethacin and other non-steroidal anti-inflammatory drugs. J. Biol. Chem. **272:** 3406–3410.
28. STAELS, B., N. VU-DAC, V.A. KOSYKH *et al.* 1995. Fibrates downregulate apolipoprotein C-III expression independent of induction of peroxisomal acyl coenzyme A oxidase. J. Clin. Invest. **95:** 705–712.
29. SCHOONJANS, K., B. STAELS & J. AUWERX. 1996. The peroxisome proliferator activated receptors (PPARs) and their effects on lipid metabolism and adipocyte differentiation. Biochim. Biophys. Acta **1302:** 93–109.
30. LAWRENCE, J.W., Y. LI, S. CHEN *et al.* 2001. Differential gene regulation in human versus rodent hepatocytes by peroxisome proliferator-activated receptor (PPAR) alpha: PPAR alpha fails to induce peroxisome proliferation-associated genes in human cells independently of the level of receptor expression. J. Biol. Chem. **276:** 31521–31527.
31. AOYAMA, T., J.M. PETERS, N. IRITANI *et al.* 1998. Altered constitutive expression of fatty acid–metabolizing enzymes in mice lacking the peroxisome proliferator-activated receptor alpha (PPARalpha). J. Biol. Chem. **273:** 5678–5684.
32. MARTIN, G., K. SCHOONJANS, A. LEFEBVRE *et al.* 1997. Coordinate regulation of the expression of the fatty acid transport protein and acyl-CoA synthetase genes by PPARalpha and PPARgamma activators. J. Biol. Chem. **272:** 28210–28217.
33. MASCARO, C., E. ACOSTA, J. ORTIZ *et al.* 1998. Control of human muscle-type carnitine palmitoyltransferase I gene transcription by peroxisome proliferator-activated receptor. J. Biol. Chem. **273:** 8560–8563.
34. BRANDT, J., F. DJOUADI & D. KELLY. 1998. Fatty acids activate transcription of the muscle carnitine palmitoyltransferase I gene in cardiac myocytes via the peroxisome proliferator-activated receptor alpha. J. Biol. Chem. **273:** 23786–23792.
35. YU, G.S., Y.C. LU & T. GULICK. 1998. Co-regulation of tissue-specific alternative human carnitine palmitoyltransferase Ibeta gene promoters by fatty acid enzyme substrate. J. Biol. Chem. **273:** 32901–32909.
36. KERSTEN, S., J. SEYDOUX, J.M. PETERS *et al.* 1999. Peroxisome proliferator-activated receptor alpha mediates the adaptive response to fasting. J. Clin. Invest. **103:** 1489–1498.

37. POIRIER, H., I. NIOT, M.C. MONNOT et al. 2001. Differential involvement of peroxisome-proliferator-activated receptors alpha and delta in fibrate and fatty-acid-mediated inductions of the gene encoding liver fatty-acid-binding protein in the liver and the small intestine. Biochem. J. **355:** 481–488.
38. SCHOONJANS, K., J. PEINADO-ONSURBE, A-M. LEFEBVRE et al. 1996. PPARα and PPARγ activators direct a distinct tissue-specific transcriptional response via a PPRE in the lipoprotein lipase gene. EMBO J. **15:** 5336–5348.
39. SCHOONJANS, K., B. STAELS, P. GRIMALDI et al. 1993. Acyl-CoA synthetase mRNA expression is controlled by fibric-acid derivatives, feeding, and liver proliferation. Eur. J. Biochem. **216:** 615–622.
40. SCHOONJANS, K., M. WATANABE, H. SUZUKI et al. 1995. Induction of the acyl-coenzyme A synthetase gene by fibrates and fatty acids is mediated by a peroxisome proliferator response element in the C promoter. J. Biol. Chem. **270:** 19269–19276.
41. STAELS, B., J. DALLONGEVILLE, J. AUWERX et al. 1998. Mechanism of action of fibrates on lipid and lipoprotein metabolism. Circulation **98:** 2088–2093.
42. MICHAUD, S.E. & G. RENIER. 2001. Direct regulatory effect of fatty acids on macrophage lipoprotein lipase: potential role of PPARs. Diabetes **50:** 660–666.
43. TONTONOZ, P., L. NAGY, J. ALVAREZ et al. 1998. PPARγ promotes monocyte/macrophage differentiation and uptake of oxidized LDL. Cell **93:** 241–252.
44. MOORE, K.J., E.D. ROSEN, M.L. FITZGERALD et al. 2001. The role of PPARγ in macrophage differentiation and cholesterol uptake. Nat. Med. **7:** 41–47.
45. CHAWLA, A., Y. BARAK, L. NAGY et al. 2001. PPARγ dependent and independent effects on macrophage-gene expression in lipid metabolism and inflammation. Nat. Med. **7:** 48–52.
46. LI, A.C., K.K. BROWN, M.J. SILVESTRE et al. 2000. Peroxisome proliferator-activated receptor γ ligands inhibit development of atherosclerosis in LDL receptor–deficient mice. J. Clin. Invest. **106:** 523–531.
47. COLLINS, A.R., W.P. MEEHAN, U. KINTSCHER et al. 2001. Troglitazone inhibits formation of early atherosclerotic lesions in diabetic and nondiabetic low density lipoprotein receptor–deficient mice. Arterioscler. Thromb. Vasc. Biol. **21:** 365–371.
48. CHEN, Z., S. ISHIBASHI, S. PERREY et al. 2001. Troglitazone inhibits atherosclerosis in apolipoprotein E–knockout mice: pleiotropic effects on CD36 expression and HDL. Arterioscler. Thromb. Vasc. Biol. **21:** 372–377.
49. LAWN, R.M., D.P. WADE, M.R. GARVIN et al. 1999. The Tangier disease gene product ABC1 controls the cellular apolipoprotein-mediated lipid removal pathway. J. Clin. Invest. **104:** R25–R31.
50. CHINETTI, G., S. LESTAVEL, V. BOCHER et al. 2001. PPARα and PPARγ activators induce cholesterol removal from human macrophage foam cells through stimulation of the ABCA1 pathway. Nat. Med. **7:** 53–58.
51. COSTET, P., Y. LUO, N. WANG et al. 2000. Sterol-dependent transactivation of the human ABC1 promoter by LXR/RXR. J. Biol. Chem. **275:** 28240–28245.
52. VU-DAC, N., K. SCHOONJANS, B. LAINE et al. 1994. Negative regulation of the human apolipoprotein A-I promoter by fibrates can be attenuated by the interaction of the peroxisome proliferator-activated receptor with its response element. J. Biol. Chem. **269:** 31012–31018.
53. VU-DAC, N., S. CHOPIN-DELANNOY, P. GERVOIS et al. 1998. The nuclear receptor peroxisome proliferator-activated receptors α and Rev-erbα mediate the species-specific regulation of apolipoprotein A-I expression by fibrates. J. Biol. Chem. **273:** 25713–25720.
54. VU-DAC, N., K. SCHOONJANS, V. KOSYKH et al. 1995. Fibrates increase human apolipoprotein A-II expression through activation of the peroxisome proliferator-activated receptor. J. Clin. Invest. **96:** 741–750.
55. LEIBOWITZ, M.D., C. FIEVET, N. HENNUYER et al. 2000. Activation of PPARdelta alters lipid metabolism in db/db mice. FEBS Lett. **473:** 333–336.
56. VOSPER, H., L. PATEL, T.L. GRAHAM et al. 2001. The peroxisome proliferator-activated receptor delta promotes lipid accumulation in human macrophages. J. Biol. Chem. **276:** 44258–44265.
57. TOBIN, K.A., H.H. STEINEGER, S. ALBERTI et al. 2000. Cross-talk between fatty acid and cholesterol metabolism mediated by liver X receptor-alpha. Mol. Endocrinol. **14:** 741–752.

58. LEHMANN, J.M., S.A. KLIEWER, L.B. MOORE et al. 1997. Activation of the nuclear receptor LXR by oxysterols defines a new hormone response pathway. J. Biol. Chem. **272:** 3137–3140.
59. CHIANG, J.Y., R. KIMMEL & D. STROUP. 2001. Regulation of cholesterol 7alpha-hydroxylase gene (CYP7A1) transcription by the liver orphan receptor (LXRalpha). Gene **262:** 257–265.
60. STAHLBERG, D., E. REIHNER, M. RUDLING et al. 1995. Influence of bezafibrate on hepatic cholesterol metabolism in gallstone patients: reduced activity of cholesterol 7alpha-hydroxylase. Hepatology **21:** 1025–1030.
61. STAHLBERG, D., B. ANGELIN & K. EINARSSON. 1989. Effects of treatment with clofibrate, bezafibrate, and ciprofibrate on the metabolism of cholesterol in rat liver microsomes. J. Lipid Res. **30:** 953–958.
62. BERTOLOTTI, M., M. CONCARI, P. LORIA et al. 1995. Effects of different phenotypes of hyperlipoproteinemia and of treatment with fibric acid derivatives on the rates of cholesterol 7alpha-hydroxylation in humans. Arterioscler. Thromb. Vasc. Biol. **15:** 1064–1069.
63. MARRAPODI, M. & J.Y. CHIANG. 2000. Peroxisome proliferator-activated receptor alpha (PPARalpha) and agonist inhibit cholesterol 7alpha-hydroxylase gene (CYP7A1) transcription. J. Lipid Res. **41:** 514–520.
64. PATEL, D.D., B.L. KNIGHT, A.K. SOUTAR et al. 2000. The effect of peroxisome-proliferator-activated receptor-alpha on the activity of the cholesterol 7alpha-hydroxylase gene. Biochem. J. **351**(part 3): 747–753.
65. CHEEMA, S.K. & L.B. AGELLON. 2000. The murine and human cholesterol 7alpha-hydroxylase gene promoters are differentially responsive to regulation by fatty acids mediated via peroxisome proliferator-activated receptor alpha. J. Biol. Chem. **275:** 12530–12536.
66. POST, S.M., H. DUEZ, P.P. GERVOIS et al. 2001. Fibrates suppress bile acid synthesis via peroxisome proliferator-activated receptor-alpha–mediated downregulation of cholesterol 7alpha-hydroxylase and sterol 27-hydroxylase expression. Arterioscler. Thromb. Vasc. Biol. **21:** 1840–1845.
67. HUNT, M.C., Y.Z. YANG, G. EGGERTSEN et al. 2000. The peroxisome proliferator-activated receptor alpha (PPARalpha) regulates bile acid biosynthesis. J. Biol. Chem. **275:** 28947–28953.

SREBP-1: Gene Regulatory Key to Syndrome X?

DIRK MÜLLER-WIELAND AND JÖRG KOTZKA

Klinische Biochemie und Pathobiochemie, Deutsches Diabetes-Forschungsinstitut, Heinrich-Heine Universität Düsseldorf, 40225 Düsseldorf, Germany

> ABSTRACT: Combined appearance of different cardiovascular risk factors seems to be more prevalent in individuals with decreased insulin sensitivity and increased visceral obesity, thereby being components of the so-called metabolic syndrome or syndrome X. Alterations in the abundance and activity of transcription factors lead to complex dysregulation of gene expression, which might be a key to understand insulin resistance–associated clinical clustering of coronary risk factors at the cellular or gene regulatory level. Recent examples are members of the nuclear hormone receptor superfamily—for example, peroxisome proliferator-activated receptors (PPARs) and sterol regulatory element-binding proteins (SREBPs). Besides their regulation by metabolites and nutrients, these transcription factors are also targets of hormones (like insulin and leptin), growth factors, inflammatory signals, and drugs. Major signaling pathways coupling transcription factors to extracellular stimuli are the MAP kinase cascades. We have recently shown that SREBPs appear to be substrates of MAP kinases and propose that SREBP-1 might play a role in the development of cellular features belonging to lipid toxicity and possibly syndrome X. Thus, the metabolic syndrome appears to be not only a disease or state of altered glucose tolerance, plasma lipid levels, blood pressure, and body fat distribution, but rather a complex clinical phenomenon of dysregulated gene expression.
>
> KEYWORDS: SREBP-1; gene expression; regulation

SREBPs: A GENE REGULATORY LINK FOR DIFFERENT CARDIOVASCULAR RISK FACTORS

Cardiovascular risk factors, like arterial hypertension, obesity, diabetes mellitus type 2, or glucose intolerance and disorders of lipid metabolism, are multifactorial polygenic diseases, which are frequently found in a single individual. This clinical clustering of cardiovascular risk factors appears to be associated with decreased insulin sensitivity or insulin resistance. Therefore, one current hypothesis is that mechanisms might exist linking decreased insulin sensitivity with the development of these coronary risk factors at the molecular or gene regulatory level. Transcription

---

Address for correspondence: Prof. Dr. Dirk Müller-Wieland, Klinische Biochemie und Pathobiochemie, Deutsches Diabetes Forschungsinstitut, Heinrich-Heine Universität Düsseldorf, Auf'm Hennekamp 65, 40225 Düsseldorf, Germany. Voice: +49-211-3382-240; fax: +49-211-3382-430.

mueller-wieland@ddfi.uni-duesseldorf.de

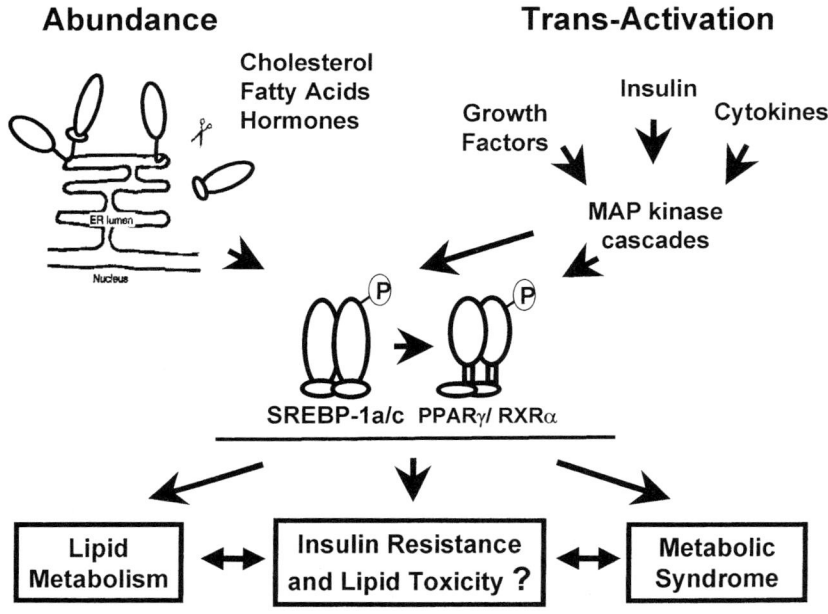

FIGURE 1. Role of SREBP-1 as a link between lipid metabolism, insulin action, and clinical features of the metabolic syndrome or syndrome X. Abundance of SREBP-1a/c is regulated by intracellular cholesterol levels, nutrients (like fatty acids), and hormones. The latter also appear to stimulate transactivity of these transcription factors by phosphorylation via MAP kinase cascades. SREBPs, in concert with other transcription factors, like PPARs, affect the expression of many genes controlling lipid metabolism and insulin sensitivity, and possibly genes involved in the development of visceral obesity, blood pressure control, inflammation, and other features of the metabolic syndrome. For further details, see text.

factors are such candidates because they regulate the expression of many genes and are targets of different intracellular signaling cascades, coupling stimuli from the cell surface to gene regulatory events. Well-known examples are peroxisome proliferator-activated receptors (PPARs) and sterol regulatory element-binding proteins (SREBPs). Apparently, SREBPs are key players in the control of intracellular lipid accumulation. One interesting aspect arising is that increased intracellular lipid accumulation might impair the function of the corresponding cell—for example, insulin secretion in the case of pancreatic β-cells, or insulin-stimulated glucose uptake or insulin sensitivity in the case of adipose tissue, skeletal muscle, and liver.[1-5] In this respect, intracellular lipid accumulation, called lipid toxicity, might be a link between insulin resistance, visceral obesity, and increased lipid deposition in non-adipose tissue, perhaps even including cells of the arterial vessel wall, which is a feature of atherosclerosis. Therefore, it is interesting to note that SREBPs not only regulate lipid metabolism, but also appear to be a target of insulin action and thus possibly a key link for different features of the metabolic syndrome. (See FIGURE 1.)

SREBPs: STRUCTURE AND FUNCTION

There are three SREBP isoforms, the so-called SREBP-1a, SREBP-1c, and SREBP-2, of which two are products of a single SREBP-1 gene.[6–8] SREBP-1a and SREBP-1c are generated by the recruitment of two distinct promoters and different first exons. Exon-1 of SREBP-1a codes for 29 amino acids (aa), whereas SREBP-1c contains only 5 amino acids. These alternative exons are connected to a common second exon in the same reading frame by mRNA splicing.

The SREBPs (SREBP-1a: 1147 aa; SREBP-1c: 1134 aa; SREBP-2: 1141 aa) have a sequence homology of 47% at the protein level.[9] The protein structure contains three essential domains: the N-terminal domain (SREBP-1a: ca. 460 aa; SREBP-1c: ca. 435 aa; SREBP-2: ca. 460 aa), two transmembrane domains containing a short loop of approximately 80 aa, and a C-terminal domain (SREBP-1a: ca. 600 aa; SREBP-1c: ca. 585 aa; SREBP-2: ca. 600 aa). The N-terminal domain contains several regions, beginning with a so-called acidic region (SREBP-1a: 51 aa; SREBP-1c: 27 aa; SREBP-2: 61 aa) exhibiting a sequence homology of 33%. This region is typical for transcription factors and appears to have a transactivating role.[10] According to this hypothesis, SREBP-1a, which contains a 24-aa-longer sequence in this region than SREBP-1c, is a stronger activator of transcription in liver.

There is a growing number of genes that appear to be regulated by SREBPs. These include (i) enzymes involved in the synthesis of cholesterol and fatty acids, (ii) proteins playing a role in lipid catabolism and intracellular transport, and (iii) secretory proteins, like leptin.[11–13] Furthermore, investigations in transgenic mice indicate that SREBP-1a and SREBP-1c regulate predominantly fatty acid metabolism and *de novo* synthesis, whereas SREBP-2 appears to be the master regulator of cholesterol homeostasis. Transgenic mice overexpressing the N-terminal domain of SREBP-1a under the control of the phosphoenolpyruvate carboxykinase (PEPCK) promoter exhibit a 4-fold increase of liver size after a protein-rich diet. This is a consequence of an increased accumulation of cholesterol and triglycerides.[14] Cholesterol synthesis increased 5-fold and fatty acid synthesis by 25-fold in the liver of these animals, whereas the concentration of plasma lipids was unaltered. Only crossing of these mice with LDL receptor knockout mice led to a dramatic increase of plasma lipids.[15] From these observations, it was concluded that the lack of increase in plasma lipids in animals overexpressing SREBP-1a was due to a dramatic upregulation of the hepatic LDL receptor. Overexpression of SREBP-1c led only to an increase of liver size of 33%.[16] The triglyceride content was slightly increased, but the cholesterol content was unaltered. Transgenic mice overexpressing the transactive domain of SREBP-2 under the PEPCK promoter showed a 28-fold increase of cholesterol synthesis, but only a 4-fold stimulation of fatty acid synthesis.[17] Further experiments of these three transgenic mice models indicate a dominant role of SREBP-1 in the regulation of enzymes involved in lipid metabolism, whereas SREBP-2 appears to play a dominant role in the regulation of cholesterol metabolism.[18] Feeding experiments showed that there appears to be a relation between nutrition, SREBP regulation, and fatty acid and cholesterol metabolism.[19]

Spiegelman's group showed that ADD-1/SREBP-1c plays an essential role in adipocyte differentiation.[20,21] Ectopic expression of ADD-1/SREBP-1c in 3T3-L1 preadipocytes leads to a dramatic increase of differentiation markers, like expression of adipocyte protein (aP) 2, lipoprotein lipase (LPL), PPARγ2, and CAAT/enhancer binding protein (C/EBP).[21–23] Although SREBP-1a appears to be the predominant isoform in different cell culture experiments,[24] SREBP-1c appears to be the predominant isoform in white adipose tissue of mice and humans.[22,24] Ectopic expression of a negative form of ADD-1/SREBP-1c prevents the differentiation of 3T3-L1 preadipocytes into adipocytes.[21] In contrast, the expression of the wild type activates genes, which are involved in fat cell differentiation and lipid accumulation. In this process, ADD-1/SREBP-1c appears to cooperate with the nuclear receptor PPARγ2, which is the essential transcription factor of adipocyte differentiation.[21,25,26] Under addition of a PPARγ2 activator to preadipocytes, about 2–3% of the cells differentiate into adipocytes. Transfection of the cells with ADD-1/SREBP-1c enhances the effect of PPARγ2 activators dramatically, that is, 15–20% of preadipocytes differentiate into adipocytes. This cooperative act appears to be related to the ADD-1/SREBP-1c-dependent generation of endogenous PPARγ2 ligands.[27] Additional experiments have led to the hypothesis that ADD-1/SREBP-1c plays a role in not only adipocyte differentiation, but also in the nutritional and hormonal control of genes involved in lipid metabolism and leptin signaling.[22] Transgenic mice overexpressing SREBP-1a in the liver have not only massive lipid accumulation in this organ, but also a dramatic reduction of white adipose tissue.[14] In transgenic mice overexpressing ADD-1/SREBP-1c in the liver, this reduction in white adipose tissue is not observed.[16] However, animals overexpressing ADD-1/SREBP-1c in adipocytes by using the aP2 promoter lack fat tissue.[18] Interestingly, these animals develop severe insulin resistance with hyperglycemia and hypertriglyceridemia, resembling the clinical picture of generalized lipodystrophy. Further evidence that SREBPs are required for cellular lipid synthesis and are targets of insulin action has been obtained in a mice model where the SREBP cleavage-activating protein (SCAP) has been conditionally deleted in the liver.[28] These liver-specific SCAP-deficient mice exhibit an 80% reduction in the basal synthesis rates of cholesterol and fatty acids, as well as blunted gene regulatory responses to cholesterol deprivation, fasting, and refeeding.

REGULATION OF SREBPs BY MAP KINASES

The release of SREBPs from the endoplasmic reticulum or nuclear envelope is a complex cholesterol-regulated proteolytic cascade and a key step in gene regulation by metabolites.[29] However, besides mechanisms controlling intracellular abundance of the mature form of SREBPs, there is growing evidence that an additional major mechanism of control is regulating transactivity of SREBPs directly, for example, via posttranslational modification.[22,30–41]

A general mechanism of fast and specific signal transduction by receptors at the cell surface to the nucleus is the recruitment of protein kinase cascades; defined endpoints of these pathways are transcription factors.[42–45] The mitogen-activated protein (MAP) kinases, which are members of such gene regulatory protein phosphorylation cascades, are one of the best-investigated groups of protein kinases. At this point, it is necessary to focus on these signaling cascades in some more detail. SREBPs are substrates of MAP kinases and thereby appear to be a gene regulatory point of convergence between metabolites, hormones, and cytokines.

The so-called MAP kinase signaling pathways represent fairly well characterized sets of parallel signaling cascades, which are present in all eukaryotic organisms as evolutionally highly conserved units.[46] The general structure of these cascades is typically a signaling module consisting of three kinases, which are activated by sequential phosphorylation.[47,48] The first kinase of these modules, the MAP kinase kinase kinase (MKKK), is activated via phosphorylation by an additional kinase (MKKKK) or by interaction with a small GTP-binding protein belonging to the Ras or Rho family. As serine/threonine kinases, MKKKs activate consecutive kinases, MAP kinase kinases (MKKs), by phosphorylation. These latter dual-specific kinases activate the final kinases of this signaling module, the MAP kinases (MAPKs), by recognizing and phosphorylating the TXY motif in the activation loop of the MAPK. MAPKs are proline-directed kinases—that is, they phosphorylate their substrates at serine and threonine residues in the consensus motif PXS/TP or PXXS/TP.[49–51]

Presently, many different MKKKs, MKKs, and MAPKs have been identified.[47] Phylogenetic investigations indicate that these kinases can be classified in various subfamilies.[46] However, the extent to which these kinases generate functional units in mammalian cells has been carefully shown only for three different MAPK signaling pathways—the extracellular regulated kinase (ERK) cascade, the cJun N-terminal kinase (JNK) cascade, and the p38HOG MAPK cascade.[47] It has been shown that the specificity of these signaling pathways is achieved by so-called scaffolding proteins. These anchoring proteins function as backbones of MAPK signal modules and prevent possible cross talk between these pathways.[52–54]

The ERK signaling pathway mediates mainly the information of proliferation as well as differentiation stimuli, whereas the JNK and p38 signaling pathways mediate stress and cytokine stimuli to the nucleus.[42,55] In each of these pathways, different MAPKs have been identified. Therefore, the ERK family consists of two members (ERK-1, ERK-2), the JNK family of three members (JNK-1, JNK-2, JNK-3), and the p38 family of four members (p38-α, p38-β, p38-γ, p38-δ).[36,56–62] In addition, many splice variants have been identified for the three JNK isoforms.[56] The specific role of these different or analogous MAPKs is still not entirely clear, especially with regard to the extent to which the spectrum of different MAPKs provides redundancy versus highly specific response by differential phosphorylation and regulation of nuclear substrates. Combinations of *in vitro* as well as *in vivo* approaches have identified various nuclear targets of MAPK signaling pathways. Recently, we have shown that SREBPs are substrates of the ERK, JNK, and p38, which regulate lipid metabolism.[27,37,38,63]

Experiments using SREBP-deficient cells show that SREBPs are essential for the regulation of the LDL receptor gene by insulin and growth factors.[35,38] Furthermore, these regulatory effects of insulin and PDGF on the LDL receptor gene were completely abolished by pharmacological inhibition of the MAP kinase cascade. Accordingly, ectopic expression of constitutively active members of the MAP kinase cascades, like MEKK1 or MEK1, leads to a manyfold stimulation of the LDL receptor promoter activity, which is stimulated synergistically by coexpression of the N-terminal domains of the SREBPs. Evidence that activation of MAP kinase cascades affects transactivity of the SREBPs directly and does not affect proteolytic cleavage was obtained recently.

The most likely mechanism whereby MAP kinases regulate transactivity of SREBPs is phosphorylation. In accordance with this, it has been shown that SREBP-1a, SREBP-1c, and SREBP-2 are substrates of the ERK family of MAP kinases *in vitro*.[37,38,63] Further studies have shown that the N-terminal domain of SREBP-1c is phosphorylated on serine and threonine residues, whereas the homologous region of SREBP-2 is phosphorylated exclusively on serine residues. The best characterization of phosphorylation sites has been obtained for SREBP-1a. The N-terminal domain was investigated after phosphorylation by ERK-2 *in vitro*. A protein chemistry approach using mass spectrometry and peptide sequencing identified serine 117 as the major phosphorylation site, which was verified by the corresponding mutation of this site to alanine. Functional characterization of this site *in vivo* by using the Gal4 system revealed that this site is a key regulator of transactivity of SREBP-1a. In accordance with this, analyses of the LDL receptor promoter reporter gene have shown that mutation of serine 117 to alanine abolishes the stimulatory effect of insulin and PDGF.[63]

CONCLUSIONS

Discovery and elucidation of complex gene regulatory networks will ultimately lead to the identification of master regulators like transcription factors, which will give new insights into the pathophysiology of complex clinical phenotypes like obesity and will provide medicine with new potential drug targets. SREBPs are a novel protein family appearing to play an essential role in molecular signaling mechanisms of cholesterol lowering (along with growth factors and cytokines), transcription of various genes, and probably still many to come. SREBPs are new intracellular mediators linking metabolic alterations to gene regulatory events, some of which might play an important part in the concert of plaque biology. Perhaps, SREBPs built the bridge between dyslipidemia and other cardiovascular risk factors like obesity and the syndrome of insulin resistance.

REFERENCES

1. DOBBINS, R.L., L.S. SZCZEPANIAK, B. BENTLEY *et al.* 2001. Prolonged inhibition of muscle carnitine palmitoyltransferase-1 promotes intramyocellular lipid accumulation and insulin resistance in rats. Diabetes **50:** 123–130.
2. JACOB, S., J. MACHANN, K. RETT *et al.* 1999. Association of increased intramyocellular lipid content with insulin resistance in lean nondiabetic offspring of type 2 diabetic subjects. Diabetes **48:** 1113–1119.
3. MARCHESINI, G., *et al.* 2001. Nonalcoholic fatty liver disease: a feature of the metabolic syndrome. Diabetes **50:** 1844–1850.
4. UNGER, R.H. & Y.T. ZHOU. 2001. Lipotoxicity of β-cells in obesity and in other causes of fatty acid spillover. Diabetes **50**(suppl. 1): S118–S121.
5. UNGER, R.H. & L. ORCI. 2001. Diseases of liporegulation: new perspective on obesity and related disorders. FASEB J. **15:** 312–321.
6. YOKOYAMA, C., X. WANG, M.R. BRIGGS *et al.* 1993. SREBP-1, a basic-helix-loop-helix-leucine zipper protein that controls transcription of the low density lipoprotein receptor gene. Cell **75:** 187–197.
7. HUA, X., J. WU, J.L. GOLDSTEIN *et al.* 1995. Structure of the human gene encoding sterol regulatory element binding protein-1 (SREBF1) and localization of SREBF1 and SREBF2 to chromosomes 7p11.2 and 22q13. Genomics **25:** 667–673.
8. SHIMANO, H., J.D. HORTON, I. SHIMOMURA *et al.* 1997. Isoform 1c of sterol regulatory element binding protein is less active than isoform 1a in livers of transgenic mice and in cultured cells. J. Clin. Invest. **99:** 846–854.

9. HUA, X., C. YOKOYAMA, J. WU et al. 1993. SREBP-2, a second basic-helix-loop-helix-leucine zipper protein that stimulates transcription by binding to a sterol regulatory element. Proc. Natl. Acad. Sci. U.S.A. **90:** 11603–11607.
10. MA, J. & M. PTASHNE. 1987. A new class of yeast transcriptional activators. Cell **51:** 113–119.
11. KOTZKA, J., W. KRONE & D. MÜLLER-WIELAND. 2001. Sterol-regulatory element binding proteins (SREBPs): gene regulatory target of statin action. In Milestones in Drug Therapy: HMG CoA Reductase Inhibitors. Birkhäuser. Basel. In press.
12. EDWARDS, P.A., et al. 2000. Regulation of gene expression by SREBP and SCAP. Biochim. Biophys. Acta **1529:** 103–113.
13. BROWN, M.S. & J.L. GOLDSTEIN. 1997. The SREBP pathway: regulation of cholesterol metabolism by proteolysis of a membrane-bound transcription factor. Cell **89:** 331–340.
14. SHIMANO, H., J.D. HORTON, R.E. HAMMER et al. 1996. Overproduction of cholesterol and fatty acids causes massive liver enlargement in transgenic mice expressing truncated SREBP-1a. J. Clin. Invest. **98:** 1575–1584.
15. HORTON, J.D., H. SHIMANO, R.L. HAMILTON et al. 1999. Disruption of LDL receptor gene in transgenic SREBP-1a mice unmasks hyperlipidemia resulting from production of lipid-rich VLDL. J. Clin. Invest. **103:** 1067–1076.
16. SHIMANO, H., I. SHIMOMURA, R.E. HAMMER et al. 1997. Elevated levels of SREBP-2 and cholesterol synthesis in livers of mice homozygous for a targeted disruption of the SREBP-1 gene. J. Clin. Invest. **100:** 2115–2124.
17. HORTON, J.D., I. SHIMOMURA, M.S. BROWN et al. 1998. Activation of cholesterol synthesis in preference to fatty acid synthesis in liver and adipose tissue of transgenic mice overproducing sterol regulatory element-binding protein-2. J. Clin. Invest. **101:** 2331–2339.
18. SHIMOMURA, I., R.E. HAMMER, J.A. RICHARDSON et al. 1998. Insulin resistance and diabetes mellitus in transgenic mice expressing nuclear SREBP-1c in adipose tissue: model for congenital generalized lipodystrophy. Genes Dev. **12:** 3182–3194.
19. HORTON, J.D., Y. BASHMAKOV, I. SHIMOMURA & H. SHIMANO. 1998. Regulation of sterol regulatory element binding proteins in livers of fasted and refed mice. Proc. Natl. Acad. Sci. U.S.A. **95:** 5987–5992.
20. TONTONOZ, P., J.B. KIM, R.A. GRAVES & B.M. SPIEGELMAN. 1993. ADD1: a novel helix-loop-helix transcription factor associated with adipocyte determination and differentiation. Mol. Cell. Biol. **13:** 4753–4759.
21. KIM, J.B. & B.M. SPIEGELMAN. 1996. ADD1/SREBP1 promotes adipocyte differentiation and gene expression linked to fatty acid metabolism. Genes Dev. **10:** 1096–1107.
22. KIM, J.B., H.M. WRIGHT, M. WRIGHT & B.M. SPIEGELMAN. 1998. ADD1/SREBP1 activates PPARγ through the production of endogenous ligand. Proc. Natl. Acad. Sci. U.S.A. **95:** 4333–4337.
23. ROSEN, E.D., C.J. WALKEY, P. PUIGSERVER & B.M. SPIEGELMAN. 2000. Transcriptional regulation of adipogenesis. Genes Dev. **14:** 1293–1307.
24. SHIMOMURA, I., H. SHIMANO, J.D. HORTON et al. 1997. Differential expression of exons 1a and 1c in mRNAs for sterol regulatory element binding protein-1 in human and mouse organs and cultured cells. J. Clin. Invest. **99:** 838–845.
25. TONTONOZ, P., E. HU & B.M. SPIEGELMAN. 1994. Stimulation of adipogenesis in fibroblasts by PPARγ2, a lipid-activated transcription factor. Cell **79:** 1147–1156.
26. BRUN, R.P., J.B. KIM, E. HU & B.M. SPIEGELMAN. 1997. Peroxisome proliferator-activated receptor γ and the control of adipogenesis. Curr. Opin. Lipidol. **8:** 212–218.
27. KIM, J.B., P. SARRAF, M. WRIGHT et al. 1998. Nutritional and insulin regulation of fatty acid synthetase and leptin gene expression through ADD1/SREBP1. J. Clin. Invest. **101:** 1–9.
28. MATSUDA, M., B.S. KORN, R.E. HAMMER et al. 2001. SREBP cleavage-activating protein (SCAP) is required for increased lipid synthesis in liver induced by cholesterol deprivation and insulin elevation. Genes Dev. **15:** 1206–1216.
29. BROWN, M.S. & J.L. GOLDSTEIN. 1999. A proteolytic pathway that controls the cholesterol content of membranes, cells, and blood. Proc. Natl. Acad. Sci. U.S.A. **96:** 11041–11048.

30. SALTER, A.M., S.C. FISHER & D.N. BRINDLEY. 1987. Binding of low-density lipoprotein to monolayer cultures of rat hepatocytes is increased by insulin and decreased by dexamethasone. FEBS Lett. **220:** 159–162.
31. BRINDLEY, D.N., N.F. BROWN, A.M. SALTER et al. 1989. Role of insulin and counterregulatory hormones in the control of hepatic glycerolipid synthesis and low-density-lipoprotein catabolism in diabetes. Biochem. Soc. Trans. **17:** 43–46.
32. WADE, D.P., B.L. KNIGHT & A.K. SOUTAR. 1988. Hormonal regulation of the low-density lipoprotein (LDL) receptor activity in human hepatoma HepG2 cells: insulin increases LDL receptor activity and diminishes its suppression by exogenous LDL. Eur. J. Biochem. **174:** 213–218.
33. WADE, D.P., B.L. KNIGHT & A.K. SOUTAR. 1989. Regulation of low-density-lipoprotein-receptor mRNA by insulin in humans. Eur. J. Biochem. **181:** 727–731.
34. LLOYD, D.B. & J.F. THOMPSON. 1995. Transcriptional modulators affect *in vivo* protein binding to the low density lipoprotein receptor and 3-hydroxy-3-methylglutaryl coenzyme A reductase promoter. J. Biol. Chem. **270:** 25812–25818.
35. STREICHER, R., J. KOTZKA, D. MÜLLER-WIELAND et al. 1996. SREBP-1 mediates activation of the low density lipoprotein receptor promoter by insulin and insulin-like growth factor-I. J. Biol. Chem. **271:** 7128–7133.
36. WANG, D. & H.S. SUL. 1997. Upstream stimulatory factor binding to the E-box at −65 is required for insulin regulation of the fatty acid synthase promoter. J. Biol. Chem. **272:** 26367–26374.
37. KOTZKA, J., D. MÜLLER-WIELAND, A. KOPONEN et al. 1998. ADD1/SREBP-1c mediates insulin-induced gene expression linked to the MAP kinase pathway. Biochem. Biophys. Res. Commun. **249:** 75–79.
38. KOTZKA, J., D. MÜLLER-WIELAND, G. ROTH et al. 2000. Sterol regulatory element binding proteins SREBP-1a and SREBP-2 are linked to the MAP kinase cascade. J. Lipid Res. **41:** 99–108.
39. KUMAR, A., A. MIDDLETON, T.C. CHAMBERS & K.D. MEHTA. 1998. Differential roles of extracellular signal-regulated kinase-1/2 and p38^{MAPK} in interleukin-1β– and tumor necrosis factor-α–induced low density lipoprotein receptor expression in HepG2 cells. J. Biol. Chem. **273:** 15742–15748.
40. SINGH, R.P., P. DHAWAN, C. GOLDEN et al. 1999. One-way cross-talk between p38(MAPK) and p42/44(MAPK): inhibition of p38(MAPK) induces low density lipoprotein receptor expression through activation of the p42/44(MAPK) cascade. J. Biol. Chem. **274:** 19593–19600.
41. GIERENS, H., M. NAUCK, M. ROTH et al. 2000. Interleukin-6 stimulates LDL receptor gene expression via activation of sterol-responsive and Sp1 binding elements. Arterioscler. Thromb. Vasc. Biol. **20:** 1777–1783.
42. TREISMAN, R. 1996. Regulation of transcription by MAP kinase cascades. Curr. Opin. Cell Biol. **8:** 205–215.
43. SU, B. & M. KARIN. 1996. Mitogen-activated protein kinase cascades and regulation of gene expression. Curr. Opin. Immunol. **8:** 402–411.
44. WASYLYK, B., J. HAGMAN & A. GUTIERREZ-HARTMANN. 1998. Ets transcription factors: nuclear effectors of the Ras-MAP-kinase signaling pathway. TIBS **23:** 213–216.
45. WHITMARSH, A.J. & R.J. DAVIS. 2000. Regulation of transcription factor function by phosphorylation. Cell. Mol. Life Sci. **57:** 1172–1183.
46. CAFFREY, D.R., L.A. O'NEILL & D.C. SHIELDS. 1999. The evolution of the MAP kinase pathways: coduplication of interacting proteins leads to new signaling cascades. J. Mol. Evol. **49:** 567–582.
47. WIDMANN, C., S. GIBSON, M.B. JARPE & G.L. JOHNSON. 1999. Mitogen-activated protein kinase: conservation of a three-kinase module from yeast to human. Physiol. Rev. **79:** 143–180.
48. GARRINGTON, T.P. & G.L. JOHNSON. 1999. Organization and regulation of mitogen-activated protein kinase signaling pathways. Curr. Opin. Cell Biol. **11:** 211–218.
49. CLARK-LEWIS, I., J.S. SANGHERA & S.L. PELECH. 1991. Definition of a consensus sequence for peptide substrate recognition by p44^{mpk}, the meiosis-activated myelin basic protein kinase. J. Biol. Chem. **266:** 15180–15184.

50. ALVAREZ, E., I.C. NORTHWOOD, F.A. GONZALEZ et al. 1991. Pro-Leu-Ser/Thr-Pro is a consensus primary sequence for substrate protein phosphorylation. J. Biol. Chem. **266:** 15277–15285.
51. GONZALEZ, F.A., D.L. RADEN & R.J. DAVIS. 1991. Identification of substrate recognition determinants for human Erk1 and Erk2 protein kinases. J. Biol. Chem. **266:** 22159–22163.
52. PAWSON, T. & J.D. SCOTT. 1997. Signaling through scaffold, anchoring, and adaptor proteins. Science **278:** 2075–2080.
53. WHITMARSH, A.J., J. CAVANAGH, C. TOURNIER et al. 1998. A mammalian scaffold complex that selectively mediates MAP kinases activity. Science **281:** 1671–1674.
54. SCHAEFFER, H.J., A.D. CATLING, S.T. EBLEN et al. 1998. MP1: a MEK binding partner that enhances enzymatic activation of the MAP kinase cascade. Science **281:** 1668–1671.
55. WHITMARSH, A.J. & R.J. DAVIS. 1996. Transcription factor AP-1 regulation by mitogen-activated protein kinase signal transduction pathways. J. Mol. Med. **74:** 589–607.
56. GUPTA, S., T. BARRETT, A.J. WHITMARSH et al. 1996. Selective interaction of JNK protein kinase isoforms with transcription factors. EMBO J. **15:** 2760–2770.
57. JIANG, Y., C. CHEN, Z. LI et al. 1996. Characterization of the structure and function of a new mitogen-activated protein kinase (p38beta). J. Biol. Chem. **271:** 17920–17926.
58. LECHNER, C., M.A. ZAHALKA, J.F. GIOT et al. 1996. ERK6, a mitogen-activated protein kinase involved in C2C12 myoblast differentiation. Proc. Natl. Acad. Sci. U.S.A. **93:** 4355–4359.
59. STEIN, B., M.X. YANG, D.B. YOUNG et al. 1997. p38-2, a novel mitogen-activated protein kinase with distinct properties. J. Biol. Chem. **272:** 19509–19517.
60. CUENDA, A., P. COHEN, V. BUEE-SCHERRER & M. GOEDERT. 1997. Activation of stress-activated protein kinase-3 (SAPK3) by cytokines and cellular stresses is mediated via SAPKK3 (MKK6); comparison of the specificities of SAPK3 and SAPK2 (RK/p38). EMBO J. **16:** 295–305.
61. GOEDERT, M., A. CUENDA, M. CRAXTON et al. 1997. Activation of the novel stress-activated protein kinase SAPK4 by cytokines and cellular stresses is mediated by SKK3 (MKK6); comparison of its substrate specificity with that of other SAP kinases. EMBO J. **16:** 3563–3571.
62. ENSLEN, H., J. RAINGEAUD & R.J. DAVIS. 1998. Selective activation of p38 mitogen-activated protein (MAP) kinase isoforms by the MAP kinase kinases MKK3 and MKK6. J. Biol. Chem. **273:** 1741–1748.
63. ROTH, G., J. KOTZKA, L. KREMER et al. 2000. MAP kinases Erk1/2 phosphorylate sterol regulatory element-binding protein (SREBP)–1a at serine 117 *in vitro*. J. Biol. Chem. **275:** 33302–33307.

PPARγ, an X-ceptor for Xs

HANA KOUTNIKOVA AND JOHAN AUWERX

Institut de Génétique et de Biologie Moléculaire et Cellulaire (IGBMC), CNRS/INSERM/Université Louis Pasteur, F-67404 Illkirch, France

ABSTRACT: Evidence from both human genetic studies and characterization of peroxisome proliferator-activated receptor gamma (PPARγ) knockout mice suggested that the prime function of PPARγ is fat formation and that its role in insulin sensitization might be secondary to this function. The thrifty function of PPARγ was most likely evolutionary beneficial, but might in "times of plenty" contribute to the pathogenesis of disorders, such as obesity, insulin resistance, type 2 diabetes, and hyperlipidemia, often commonly referred to as "syndrome X". This role of PPARγ in these diseases also questions the eventual therapeutic benefits of pure PPARγ activation, which is associated with an increase in adipose tissue mass. We characterized a new chemical class of PPARγ agonists, that is, FMOC-L-leucine (FLL). FLL induces a different conformation of PPARγ relative to classical PPARγ ligands. Mass spectrometry indicates that two molecules of FLL bind to a single PPARγ molecule, making its mode of receptor interaction distinctive. FLL recruits a different set of co-activators and activates PPARγ with a lower potency, but a similar maximal efficacy, relative to known PPARγ ligands. In contrast, FLL is a more effective insulin sensitizer than current PPARγ agonists, an effect potentially linked to its weak adipogenic activity. These data make a strong point for potential therapeutic benefits of PPARγ modulation rather than activation.

KEYWORDS: adipocyte; differentiation; transcription; regulation; agonist; modulator

INTRODUCTION

PPARγ belongs to the nuclear hormone receptor family of transcription factors that act together with its heterodimerization partner, the retinoid X receptor, RXR.[1] Upon binding to small lipophilic molecules, PPARγ undergoes a conformational change that stabilizes its interaction with both RXR and cofactors, resulting in facilitation of the transcription of target genes. Fatty acid–derived molecules such as 15-deoxy-delta prostaglandin J2 (PG-J2)[2,3] and 13-hydroxyoctadecadienoic acid (13-HODE)[4] are natural ligands for PPARγ. PPARγ is also activated by a series of synthetic compounds, including the antidiabetic thiazolidinediones,[5] L-tyrosine-based compounds,[6] and some nonsteroidal anti-inflammatory agents.[7] Ectopic expression of PPARγ in NIH-3T3 fibroblasts in the presence of its natural or synthetic

Address for correspondence: Johan Auwerx, Institut de Génétique et de Biologie Moléculaire et Cellulaire (IGBMC), CNRS/INSERM/Université Louis Pasteur, B.P. 163, F-67404 Illkirch, France. Voice: (33) 388 65 34 25; fax: (33) 388 65 32 01.
auwerx@igbmc.u-strasbg.fr

ligands leads to adipogenesis[2,3,8] and adipose tissue is considered the major site of action for PPARγ.[9] In fact, PPARγ coordinates the terminal step of adipocyte differentiation and regulates expression of genes involved in lipid metabolism and glucose transport (reviewed in reference 10).

In humans, several mutations in the PPARγ gene have been described. The most prevalent P12A mutation in the adipose tissue–specific PPARγ2 isoform is associated with a decreased body mass index and improved insulin sensitivity.[11,12] This phenotype is also affected by nutrition since an intake of food rich in polyunsaturated fatty acids results in higher insulin sensitivity.[13] Two mutations in the ligand-binding domain of PPARγ, the P467L and V290M, lead in contrast to severe insulin resistance, type 2 diabetes, and hypertension.[14] In mice, PPARγ deficiency is lethal due to developmental defects of placenta and heart.[15,16] When the PPARγ –/– embryos are rescued, animals suffer from severe lipodystrophy, confirming the important role of PPARγ in adipogenesis. Additional evidence for a role of PPARγ in adipogenesis was obtained by the lack of contribution of PPARγ –/– embryonic stem cells to white adipose tissue formation in chimeric animals.[16] Furthermore, the heterozygote PPARγ +/– animals are protected from adipocyte hypertrophy and insulin resistance induced by high-fat diet.[17] In combination, all these data hence suggest that PPARγ serves as a thrifty gene by enabling energy storage in "times of plenty" and by safeguarding energy stores at times of deprivation.[10] This thrifty function of PPARγ was most likely evolutionary beneficial, but may now contribute to the pathogenesis of "syndrome X".

A CASE FOR PPARγ MODULATION INSTEAD OF ACTIVATION

The thiazolidinediones (TZDs), a class of synthetic PPARγ ligands, are used in the treatment of type 2 diabetes.[5] This class of drugs induces fat formation, more particularly the formation of small adipocytes.[10] TZDs also promote a redistribution of fat from intra- to extramuscular stores, such as adipose tissue. However, upon longer treatment, the TZD-induced increase in adipose mass will lead ultimately to insulin resistance. This is, in fact, one of the main potential side effects of commercially available TZD-based drugs. In addition, these drugs have other potential side effects, such as an increase in LDL-cholesterol levels, fluid retention, hepatotoxicity, and cardiac hypertrophy.[18]

Recently, we have identified a derivative of L-tyrosine-based PPARγ ligands[6] that has a significantly improved pharmacological profile.[19] This compound, a derivative of the N-(9-fluorenylmethyloxycarbonyl)-amino acid, that is, FMOC-L-leucine (FLL), has a potent insulin-sensitizing action with unique PPARγ-activating and PPARγ-binding properties (FIG. 1). FLL transactivates PPARγ in transfection assay, with a maximal transactivation achieved at a concentration of 10^{-5} M. This effect is specific to PPARγ since FLL is unable to activate other nuclear receptors, such as the PPARα or the liver X receptor α. In addition, the L-isomer (FLL) activates the receptor significantly better than the D-isomer (FDL), demonstrating some stereoselectivity of these compounds. The optimal concentration for PPARγ activation by FLL is similar to that of PG-J2 and about 100-fold higher than the concentration of TZDs, such as rosiglitazone or pioglitazone. In addition, RXR agonists have an additive effect with FLL on the activation of the PPARγ/RXR heterodimer.

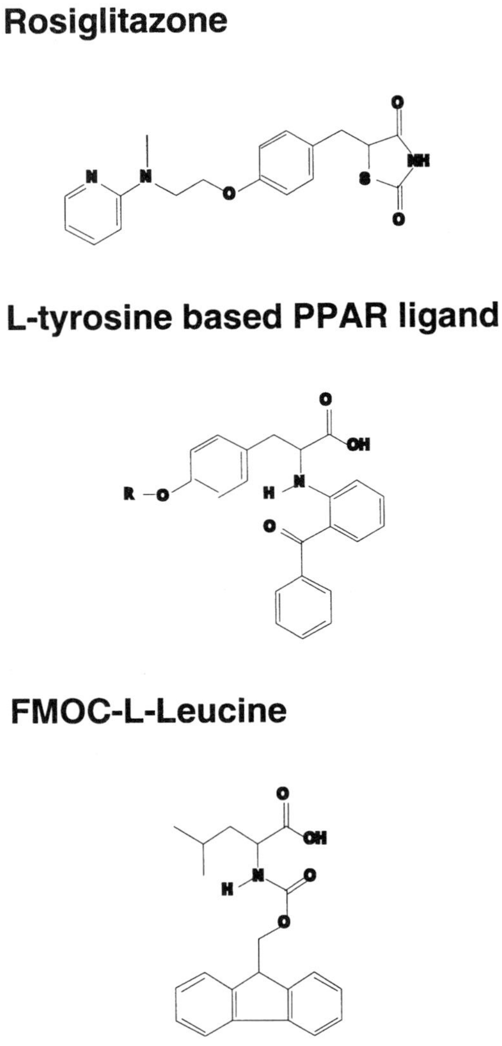

Rosiglitazone

L-tyrosine based PPAR ligand

FMOC-L-Leucine

FIGURE 1. Schematic representation of the structure of various synthetic PPARγ ligands. FMOC: *N*-(9-fluorenylmethyloxycarbonyl).

TZDs can induce a conformational change in PPARγ as assessed by generation of protease-resistant bands following partial trypsin digestion of recombinant receptor[20,21] and, interestingly, FLL produces a protease protection pattern similar to that observed with rosiglitazone. FLL binds PPARγ *in vitro* as demonstrated by scintillation proximity assays and also electrospray ionization mass spectrometry. Interestingly, this last experiment suggested the existence of two high-affinity binding sites for FLL within the PPARγ molecule. Upon binding to the PPARγ DE domain,

FLL induces a conformational change permitting the association of the cofactor p300 and SRC-1, but not that of another cofactor of the p160 family, that is, TIF2. This condition is in contrast to the conformational change induced by the TZD, rosiglitazone, which preferentially results in TIF2 recruitment to the PPARγ DE domain. These results suggest that PPARγ activation by different ligands induces the specific docking of cofactors.

FLL has the ability to stimulate adipocyte differentiation of murine preadipocyte 3T3-L1 cells and this effect is less potent in comparison to classical PPARγ ligands as assessed by gene expression analysis or morphological changes associated with the differentiation process. Most importantly, FLL can improve insulin sensitivity as evaluated by intraperitoneal glucose tolerance or by meal tolerance testing in normal C57BL/6J mice, while food intake is not affected. Interestingly, the structure-activity relationship to activate a PPAR response element in transfection assays, with an activity order of FLL > FDL > FMOC-D-tyrosine, was also reproduced *in vivo*. FLL also improves glucose tolerance at least to the same extent as rosiglitazone in diabetic *db/db* mice, an animal model of type 2 diabetes, and in diet-induced insulin-resistant mice. Yet, unlike the TZDs, FLL does not induce weight gain in these diabetic models. However, since the pharmacokinetic and organ distributions of FLL are at present unknown, a head-to-head comparison of *in vivo* efficacy with that of the TZDs is at present impossible.

CONCLUSIONS

All these physical and biological properties, which overlap with those of classical PPARγ ligands, suggest that modulation of PPARγ activity could explain part of the mechanisms of action of FLL. Since FLL is clearly structurally different from TZDs, L-tyrosine-based PPARγ ligands,[6,22,23] or other partial PPARγ agonists and antagonists,[21,24–26] it defines a chemically new class of PPARγ ligands. FLL alters glucose homeostasis through modulation of PPARγ activity, while it has a lower tendency to increase body weight, a phenomenon clearly beneficial in the treatment of type 2 diabetes. These results hence indicate that FLL selectively modulates PPARγ activity, triggering only certain PPARγ-controlled pathways, such as insulin sensitization, yet being much less efficient in activating other pathways, such as the stimulation of adipocyte differentiation and fat accretion. The different PPARγ conformation, which is induced by FLL binding, relative to classical PPARγ ligands, most likely underpins the distinct biological activity of this PPARγ modulator. In light of these results on FLL and the observations made in PPARγ +/− mice, we argue that partial PPARγ agonists and perhaps PPARγ antagonists might hold promise as antidiabetic agents.

ACKNOWLEDGMENTS

This work was supported by grants of CNRS, INSERM, CHU de Strasbourg, ARC (No. 9943); the Juvenile Diabetes Foundation (No. 1-1999-819); the European Community RTD Program (No. QLG1-CT-1999-00674); the National Institutes of Health (No. 1 P01 DK59820-01); the Ligue Nationale Contre le Cancer; and the Human Frontier Science Program (No. RG0041/1999-M).

REFERENCES

1. KERSTEN, S., B. DESVERGNE & W. WAHLI. 2000. Roles of PPARs in health and disease. Nature **405:** 421–424.
2. FORMAN, B.M., et al. 1995. 15-Deoxy-Δ12,14 prostaglandin J2 is a ligand for the adipocyte determination factor PPARγ. Cell **83:** 803–812.
3. KLIEWER, S.A., et al. 1995. A prostaglandin J2 metabolite binds peroxisome proliferator-activated receptor γ and promotes adipocyte differentiation. Cell **83:** 813–819.
4. NAGY, L., et al. 1998. Oxidized LDL regulates macrophage gene expression through ligand activation of PPARγ. Cell **93:** 229–240.
5. LEHMANN, J.M., et al. 1995. An antidiabetic thiazolidinedione is a high affinity ligand for peroxisome proliferator-activated receptor γ (PPARγ). J. Biol. Chem. **270:** 12953–12956.
6. HENKE, B.R., et al. 1998. N-(2-Benzoylphenyl)-L-tyrosine PPARgamma agonists: 1. Discovery of a novel series of potent antihyperglycemic and antihyperlipidemic agents. J. Med. Chem. **41:** 5020–5036.
7. LEHMANN, J.M., et al. 1997. Peroxisome proliferator-activated receptors α and γ are activated by indomethacin and other non-steroidal anti-inflammatory drugs. J. Biol. Chem. **272:** 3406–3410.
8. TONTONOZ, P., E. HU & B.M. SPIEGELMAN. 1994. Stimulation of adipogenesis in fibroblasts by PPARγ2, a lipid-activated transcription factor. Cell **79:** 1147–1156.
9. TONTONOZ, P., et al. 1994. mPPARγ2: tissue-specific regulator of an adipocyte enhancer. Genes Dev. **8:** 1224–1234.
10. AUWERX, J. 1999. PPARγ, the ultimate thrifty gene. Diabetologia **42:** 1033–1049.
11. DEEB, S., et al. 1998. A Pro 12 Ala substitution in the human peroxisome proliferator-activated receptor gamma2 is associated with decreased receptor activity, improved insulin sensitivity, and lowered body mass index. Nat. Genet. **20:** 284–287.
12. ALTSHULER, D., et al. 2000. The common PPARgamma Pro12Ala polymorphism is associated with decreased risk of type 2 diabetes. Nat. Genet. **26:** 76–80.
13. LUAN, J., et al. 2001. Evidence for gene-nutrient interaction at the PPARgamma locus. Diabetes **50:** 686–689.
14. BARROSO, I., et al. 1999. Dominant negative mutations in human PPARgamma associated with severe insulin resistance, diabetes mellitus, and hypertension. Nature **402:** 880–883.
15. BARAK, Y., et al. 1999. PPARgamma is required for placental, cardiac, and adipose tissue development. Mol. Cell **4:** 585–595.
16. ROSEN, E.D., et al. 1999. PPARgamma is required for the differentiation of adipose tissue in vivo and in vitro. Mol. Cell **4:** 611–617.
17. KUBOTA, N., et al. 1999. PPARγ mediates high-fat diet–induced adipocyte hypertrophy and insulin resistance. Mol. Cell **4:** 597–609.
18. SCHOONJANS, K. & J. AUWERX. 2000. Thiazolidinediones: an update. Lancet **355:** 1008–1010.
19. ROCCHI, S., et al. 2001. A unique PPARγ ligand with potent insulin-sensitizing, yet weak adipogenic activity. Mol. Cell. **8:** 737–747.
20. BERGER, J., et al. 1999. Novel peroxisome proliferator activated receptor (PPAR) γ and PPARδ ligands produce distinct biological effects. J. Biol. Chem. **274:** 6718–6725.
21. ELBRECHT, A., et al. 1999. L-764406 is a partial agonist of human peroxisome proliferator-activated receptor gamma: the role of Cys313 in ligand binding. J. Biol. Chem. **274:** 7913–7922.
22. COBB, J.E., et al. 1998. N-(2-Benzoylphenyl)-L-tyrosine PPARgamma agonists: 3. Structure-activity relationship and optimization of the N-aryl substituent. J. Med. Chem. **41:** 5055–5069.
23. COLLINS, J.L., et al. 1998. N-(2-Benzoylphenyl)-L-tyrosine PPARgamma agonists: 2. Structure-activity relationship and optimization of the phenyl alkyl ether moiety. J. Med. Chem. **41:** 5037–5054.
24. OBERFIELD, J.L., et al. 1999. A peroxisome proliferator-activated receptor gamma ligand inhibits adipocyte differentiation. Proc. Natl. Acad. Sci. U.S.A. **96:** 6102–6106.

25. WRIGHT, H.M., *et al.* 2000. A synthetic antagonist for the peroxisome proliferator-activated receptor gamma inhibits adipocyte differentiation. J. Biol. Chem. **275:** 1873–1877.
26. MUKHERJEE, R., *et al.* 2000. A selective peroxisome proliferator-activated receptor-gamma (PPARgamma) modulator blocks adipocyte differentiation, but stimulates glucose uptake in 3T3-L1 adipocytes. Mol. Endocrinol. **14:** 1425–1433.

Sterol Regulatory Element-Binding Proteins (SREBPs) as Regulators of Lipid Metabolism

Polyunsaturated Fatty Acids Oppose Cholesterol-Mediated Induction of SREBP-1 Maturation

HYOUN-JU KIM, MAKOTO MIYAZAKI, WENG CHI MAN, AND JAMES M. NTAMBI

Departments of [a]Biochemistry and [b]Nutritional Sciences, University of Wisconsin, Madison, Wisconsin 53706, USA

ABSTRACT: Cellular cholesterol and fatty acid metabolism in mammals is controlled by a family of transcription factors called sterol regulatory element-binding protein isoforms, three of which (SREBP-1a, 1c, and 2) are well characterized. These proteins, which are synthesized as precursors, are inserted into the endoplasmic reticulum (ER) membrane with both the amino and carboxylic acid domains facing the cytosolic face of the membrane. In sterol-deficient cells, proteolytic cleavage of SREBPs occurs, thereby releasing their N-terminal mature and active forms and enabling them to enter the nucleus, where they bind to the sterol regulatory response element (SRE) and/or E-box sequences and activate genes involved in cholesterol, triglyceride, and fatty acid biosynthesis. Of the three SREBP isoforms, SREBP-1c gene expression is induced by cholesterol and repressed by polyunsaturated fatty acids (PUFA). We have examined the changes in SREBP-1c mRNA and protein levels as well as the mRNA levels of several SREBP-1c target genes when a high-cholesterol diet is combined with diets rich in PUFA of the n-6 series. Our studies show that PUFA oppose the cholesterol-mediated SREBP-1 maturation without affecting the cholesterol-mediated increase of SREBP-1c mRNA and precursor protein. The decrease in SREBP-1 mature protein paralleled the decrease in mRNAs for genes of fatty acid and cholesterol biosynthesis, such as HMG-CoA synthase and fatty acid synthase, but interestingly gene expression of stearoyl-CoA desaturase 1 (SCD1) was instead induced. These studies suggest that the main point of control of PUFA-mediated suppression of lipogenic gene expression is the inhibition of SREBP-1 maturation. The studies also reveal that the induction of SCD1 gene expression by cholesterol occurs through a mechanism independent of SREBP-1 maturation.

KEYWORDS: SREBP; PUFA; cholesterol; protein; gene; metabolism

Address for correspondence: James M. Ntambi, Department of Biochemistry, University of Wisconsin, 433 Babcock Drive, Madison, WI 53706. Voice: 608-265-3700; fax: 608-265-3272.
ntambi@biochem.wisc.edu

INTRODUCTION

Sterol regulatory element-binding proteins (SREBPs) are important transcription factors that regulate fatty acid and cholesterol metabolism.[1–6] Three major SREBP isoforms, SREBP-1a, SREBP-1c, and SREBP-2, are encoded by two different genes and are well characterized.[5–8] These proteins are synthesized as precursors and inserted into the endoplasmic reticulum (ER) membrane, where they are anchored through a two-pass membrane-spanning domain with both the amino and carboxylic acid domains facing the cytosolic face of the membrane.[5,6] In sterol-deficient cells, proteolytic cleavage of the SREBP-specific proteases (S1P and S2P)[6,9–11] occurs in the Golgi body, thereby releasing their N-terminal mature and active forms from the membrane and enabling them to enter the nucleus, where they bind to the sterol regulatory response elements (SREs) and/or E-box sequences and activate genes involved in cholesterol, triglyceride, and fatty acid biosynthesis.[2–6,12,13] In contrast, when cellular sterol levels are high, the proteolytic processing of SREBPs is inhibited, the nuclear levels of mature proteins are decreased, and transcription of target genes is reduced. Thus, both SREBP gene expression and the processing of the SREBPs regulate target gene expression. The target genes of SREBPs involved in cholesterol metabolism include LDL receptor, 3-hydroxy-3-methylglutaryl-CoA (HMG-CoA) reductase, HMG-CoA synthase (HMG-S), and the SREBPs themselves.[12–14] The genes involved in fatty acid and triglyceride synthesis that are regulated by SREBPs include acetyl-CoA carboxylase (ACC), fatty acid synthase (FAS), glycerol-3-phosphate acyltransferase (GPAT), stearoyl-CoA desaturase (SCD1), and the Δ-6 and Δ-5 desaturases.[2–4,14]

A significant role for SREBPs in both lipogenesis and cholesterol metabolism has been established by numerous studies through both genetic manipulation of SREBP levels and feeding experiments in animals.[14–17] Several expression studies of the individual isoforms demonstrate that the SREBP-1a isoform, which is the dominant form in cell lines, is the regulator of genes encoding proteins involved in both lipogenesis and cholesterol biosynthesis.[5,6] SREBP-1c isoform, which constitutes more than 90% of the *in vivo* SREBP-1, is a key regulator of early events in the liver's response to insulin and is a major determinant of lipogenic gene transcription.[18–23] SREBP-2 selectively regulates genes of cholesterol metabolism.[11]

Polyunsaturated fatty acid (PUFA)–rich diets repress the transcription of lipogenic genes by suppressing SREBP-1c gene transcription or by reducing the maturation of SREBP-1 protein,[24] while cholesterol feeding on the other hand results in liver X receptor (LXR)–mediated upregulation of SREBP-1c gene expression.[25,26] In this study, when PUFA and cholesterol are combined and fed to mice, we demonstrate that the inductive effect of dietary cholesterol on SREBP-1c maturation is opposed by PUFA. The cholesterol-mediated induction of SREBP-1c mRNA and precursor protein is not affected. The decrease in the SREBP-1 mature form paralleled the decrease in mRNAs for SREBP-1c target genes, such as HMG-CoA synthase and fatty acid synthase, but mRNAs for genes of stearoyl-CoA desaturase were increased. These studies suggest that the main point of control of PUFA-mediated suppression of lipogenic gene expression is the inhibition of SREBP-1 maturation. The studies also reveal that the induction of SCD1 gene expression by cholesterol occurs through a mechanism independent of SREBP-1 maturation.

EXPERIMENTAL PROCEDURES

Materials

Radioactive [α-^{32}P]dCTP (3000 Ci/mmol) was obtained from Dupont (Wilmington, DE). Immobilon-P transfer membranes were from Millipore (Danvers, MA). ECL Western blot detection kit was from Amersham-Pharmacia Biotech (Piscataway, NJ). Diets and oil were from Harlan Teklad (Madison, WI). All other chemicals were purchased from Sigma Chemical (St. Louis, MO).

Animals and Diets

C57BL6/129J female mice were bred and maintained in our animal room facilities of the Department of Biochemistry at the University of Wisconsin-Madison. The breeding of these animals is in accordance with the protocols approved by the animal care research committee (ACRC) of the University of Wisconsin-Madison. Mice were maintained on a 12-h dark/light cycle and, at 6 weeks, were fed a standard mouse/rat diet (no. 7001, Harlan Teklad) containing 21% linoleic acid or semipurified diets (Harlan Teklad) containing 4% fat in the form of soybean oil with or without 2% cholesterol. The semipurified diets contained (by weight) 21% casein, 14% maltodextrin, 55% sucrose, 5% cellulose, 3% mineral mix (AIN-93G-MX), and 1% vitamin mix (AIN-93G-MX). Soybean oil contained 18% oleic acid (18:1 n-9) and 54% linoleic acid (18:2 n-6) as total fatty acids.

Isolation and Analysis of RNA

Total RNA was isolated from livers using the acid guanidinium-phenol-chloroform extraction method.[27] Twenty μg of total RNA was separated by 1.0% agarose/2.2 M formaldehyde gel electrophoresis and transferred onto nylon membrane. The DNA fragments for mouse SREBP-1 and HMG-CoA synthetase were obtained by reverse transcriptase–polymerase chain reaction from first-strand cDNA derived from mouse liver total RNA. The amplified products were subcloned into pGEM-T Easy vector (Promega, WI). John Chiang at Northeastern Ohio University kindly provided the cDNA for rat cholesterol 7α-hydroxylase (7αHase) and Hei Sook Sul at the University of California, Berkeley, provided FAS cDNA. The mouse SCD1 cDNA is from our laboratory.[28] These cDNAs were used as probes for Northern analysis.

Immunoblotting

Nuclear extracts of mice liver were prepared according to the methods described by Sheng and coworkers.[29] To prevent proteolysis of mature SREBPs, all buffers contained 50 μg/mL N-acetyl-leucyl-norleucinal, 1 mM PMSF, 10 μg/mL leupeptin, and 2 μg/mL aprotinin. Fresh livers were homogenized in 30 mL of buffer A [10 mM HEPES (pH 7.6), 25 mM KCl, 1 mM EDTA-Na, 2 M sucrose, 10% glycerol, 0.15 mM spermine, 2 mM spermidine, and protease inhibitors]. The homogenate was layered over 10 mL of buffer A and centrifuged at 75,000g for 1 h at 4°C. The pellets were resuspended with 1 mL of buffer B [10 mM HEPES (pH 7.6), 100 mM KCl, 2 mM spermidine, and protease inhibitors]. After the addition of 4 M ammonium sulfate (pH 7.9), the suspension was agitated gently for 1 h at 4°C and centrifuged at 257,000g for 45 min at 4°C. The supernatant was collected as nuclear extract. Aliquots of nuclear extracts (30 µg) from mice liver were mixed with SDS loading buffer, subjected to SDS/PAGE on an 8% gel, transferred, and immobilized on Immobilon-P transfer membranes. After blocking with 3% BSA in TBS buffer (pH 8.0) plus Tween 20 at 4°C overnight, the membrane was washed and incubated with monoclonal anti-SREBP-1 (IgG-2A4) as the primary antibody and antimouse IgG-HRP conjugate as the secondary antibody. Visualization of the SREBP-1 protein was performed by using the ECL Western blotting detection system kit (Amersham-Pharmacia Biotech). The bands were quantified by Canon scanning (Canon, Tokyo, Japan). Osamu Ezaki at the National Institute of Health and Nutrition in Japan kindly provided the monoclonal antibodies for SREBP-1.

A. SREBP-1 mRNA level

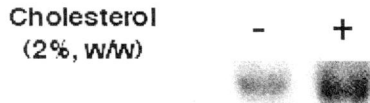

B. SREBP-1 precursor form

C. SREBP-1 mature form

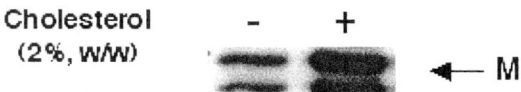

FIGURE 1. Northern blot and Western blot analysis for the expression of SREBP-1 mRNA and protein in liver of mice fed standard chow diet or chow diet supplemented with 2% cholesterol. (**A**) Total RNA (10 µg) was subjected to Northern analysis followed by hybridization with labeled cDNA probes specific for SREBP-1. (**B**) Thirty-µg aliquots of membrane fractions and nuclear extracts were subjected to 8% SDS/PAGE and transferred to Hybond-P membranes, followed by detection with 5 µg/mL of mouse monoclonal antibody IgG-2A4 against amino acids 301–407 of human SREBP-1. (**C**) SREBP-1 mature form. P, precursor protein; M, mature form.

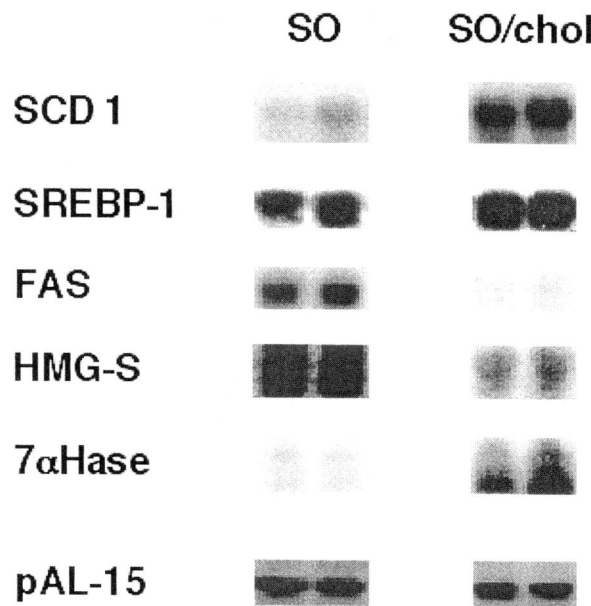

FIGURE 2. Northern blot analysis for the expression of SCD1, SREBP-1, FAS, HMG-S, and 7α-hydroxylase mRNA from liver of mice fed 5% soybean oil diet (SO) with or without 2% cholesterol (SO/chol). Total RNA (10 μg) was subjected to Northern analysis followed by hybridization with labeled cDNA probes specific for SCD1, SREBP-1, FAS, HMG-S, and 7α-hydroxylase. pAL-15 cDNA was used as a loading control.[27]

RESULTS

FIGURE 1 shows a Northern blot and Western blot analysis of the expression of SREBP-1 mRNA and protein, respectively, in liver of mice fed the 7001 standard chow diet or the chow diet supplemented with 2% cholesterol. Compared with chow-fed mice, chow/cholesterol-fed mice showed increases in SREBP-1c mRNA (A), the SREPB-1 precursor protein (B), and the mature protein (C).

FIGURE 2 shows a Northern blot analysis of the expression of lipogenic genes in liver of mice fed soybean oil (SO) or soybean oil supplemented with 2% cholesterol (SO/chol). Compared with SO-fed mice, SO/chol-fed mice showed increases in the levels of SREBP-1 and SCD1 mRNAs, whereas FAS and HMG-CoA synthetase mRNA levels were decreased. The 7α-hydroxylase converts cholesterol to 7α-hydroxycholesterol in liver[30,31] and, in agreement with several studies,[25,31] mRNA levels of 7α-hydroxylase were increased by a high-cholesterol diet.

FIGURE 3 shows immunoblot analysis of the precursor and mature SREBP-1 in nuclear extracts from livers of mice fed SO and SO/chol. As shown in FIGURE 3A, the levels of the SREBP-1 precursor protein were not decreased by the SO/chol diet; however, as shown in FIGURE 3B, there was a greater than 40% reduction in the SREBP-1 mature form of SREBP-1 protein by the SO/chol diet. The fish oil diet

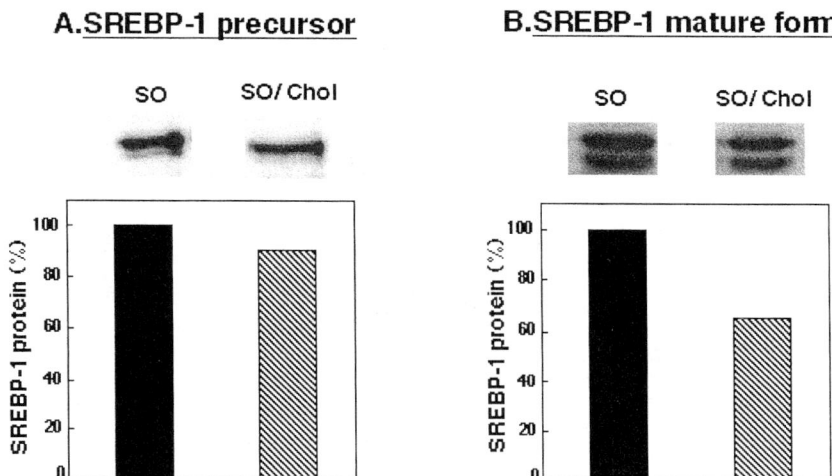

FIGURE 3. Immunoblot analysis of SREBP-1 in membrane fractions and nuclear extracts for the expression of SREBP-1 protein in liver of mice fed soybean oil (SO) or soybean with 2% cholesterol (SO/chol). Thirty-µg aliquots of membrane fractions and nuclear extracts were subjected to 8% SDS/PAGE and transferred to Hybond-P membranes, followed by detection with human SREBP-1. **(A)** SREBP-1 precursor protein. **(B)** SREBP-1 mature form.

supplemented with 2% cholesterol showed an increase in the SREBP-1 precursor protein, but a dramatic reduction in the level of the mature form of the SREBP-1 protein (data not shown).

DISCUSSION

Cholesterol suppresses the expression of genes of cholesterol synthesis and of fatty acid synthesis in several cultured cell lines through the inhibition of proteolytic processing of SREBP.[32] *In vivo* studies also showed that high-cholesterol feeding decreased the levels of the mature forms of SREBPs in liver of hamsters, and a corresponding suppression in the expression of the genes of cholesterol metabolism such as LDL receptor, HMG-CoA reductase, and HMG-CoA synthetase was observed.[33] There are, however, *in vivo* studies that show that, unlike *in vitro*, high levels of cholesterol can induce SREBP-1c gene expression as well as SCD1 gene expression in mouse liver.[25,26] Recently, the detailed mechanism of cholesterol-mediated induction of SREBP-1c expression was reported to be through the oxysterol nuclear receptor, LXRα.[26] In these studies, it was suggested that the induction of SCD1 gene expression by cholesterol was indirect and was beneficial to the cell because it would increase oleoyl-CoA, the preferred substrate of ACAT for the esterification of toxic cholesterol and its subsequent storage. However, the results presented in FIGURE 2 indicate that SCD1 gene expression is induced by the SO/chol diet despite a reduction in the mature levels of SREBP-1 protein, suggesting that cholesterol overrides the well-known PUFA-mediated suppression of SCD1 gene expression.[34]

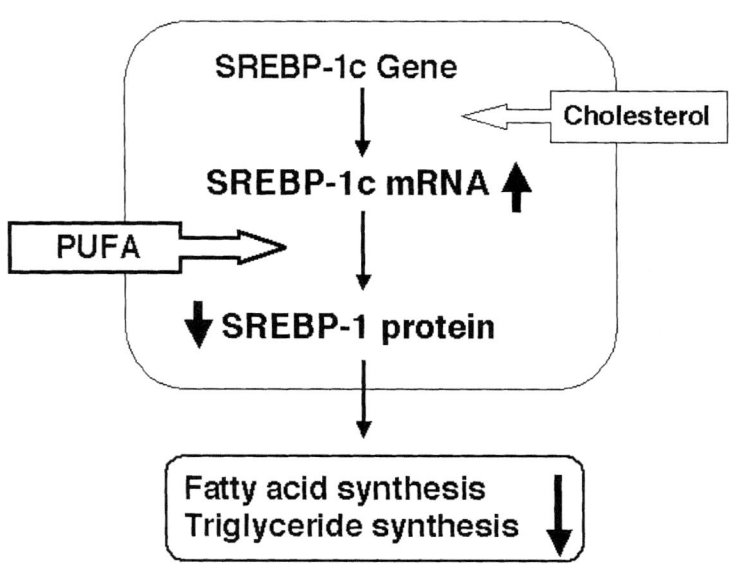

FIGURE 4. Model depicting cholesterol-mediated induction of SREBP-1c gene expression and PUFA-mediated inhibition of SREBP-1 maturation. It is proposed that cholesterol induces SREBP-1c gene expression by increasing SREBP-1 mRNA and precursor protein, but the processing of the precursor protein into the mature form is inhibited by PUFA. Reduction in expression of lipogenic genes results in a decrease in fatty acid and triglyceride synthesis.

Because SREBPs are involved in fatty acid metabolism, studies have been conducted to determine whether they are directly regulated by fatty acids. PUFA reduce the nuclear content of SREBP-1 via a two-phase mechanism.[24] The first phase is believed to be rapid and involves the inhibition of the proteolytic release step.[24] The second phase involves a reduction in SREBP-1c mRNA that is subsequently followed by a reduction in the amount of precursor SREBP-1c protein.[15,24] The mechanism by which PUFA inhibit SREBP maturation is not known. Nuclear run-on studies suggested that PUFA reduce the hepatic levels of SREBP-1c mRNA through post-transcriptional mechanisms.[24] The suppression of SREBP-1c gene expression by PUFA and a corresponding reduction in the levels of the mature SREBP-1 protein result in a decrease in the expression of genes encoding enzymes and proteins involved in fatty acid biosynthesis.

The regulation of SREBP-1 gene expression by dietary factors therefore occurs at several levels and through different mechanisms. When fed individually, PUFA suppress SREBP-1 gene expression and inhibit SREBP-1 protein maturation. High-cholesterol diets, on the other hand, induce SREBP-1c gene expression through LXRα-mediated transcription, resulting in increased levels of SREBP-1 mature protein. We propose that, as depicted in FIGURE 4, when PUFA are combined with cholesterol, the cholesterol opposes the PUFA-mediated reduction in SREBP-1c mRNA and precursor protein, whereas the PUFA override the cholesterol-mediated induction of SREBP-1 maturation. The strong inhibitory effect that PUFA exert on

SREBP-1 maturation results in suppression of SREBP-1 target genes such as FAS and GPAT, resulting in reduced fatty acid and triglyceride synthesis.

ACKNOWLEDGMENTS

We thank Alan Attie and M. P. Gary-Keller for useful discussions. This work was supported in part from a grant from the American Heart Association and in part by funds from Xenon Biogenetics to J. M. Ntambi.

REFERENCES

1. BROWN, M.S. & J.L. GOLDSTEIN. 1997. The SREBP pathway: regulation of cholesterol metabolism by proteolysis of a membrane bound transcription factor. Cell **89:** 331–340.
2. YAHAGI, N., H. SHIMANO, A.H. HASTY et al. 1999. A crucial role of sterol regulatory element-binding protein-1 in the regulation of lipogenic gene expression by polyunsaturated fatty acids. J. Biol. Chem. **274:** 35840–35844.
3. SHIMANO, H., J.D. HORTON, R.E. HAMMER et al. 1996. Overproduction of cholesterol and fatty acids causes massive liver enlargement in transgenic mice expressing truncated SREBP-1a. J. Clin. Invest. **98:** 1575–1584.
4. HORTON, J.D., I. SHIMOMURA, M.S. BROWN et al. 1998. Activation of cholesterol synthesis in preference to fatty acid synthesis in liver and adipose tissue of transgenic mice overproducing sterol regulatory element-binding protein-2. J. Clin. Invest. **101:** 2331–2339.
5. OSBORNE, T.F. 2001. CREating a SCAP-less liver keeps SREBPs pinned in the ER membrane and prevents increased lipid synthesis in response to low cholesterol and high insulin. Genes Dev. **15:** 1873–1878.
6. OSBORNE, T.F. 2000. Sterol regulatory element-binding proteins (SREBPs): key regulators of nutritional homeostasis and insulin action. J. Biol. Chem. **275:** 32379–32382.
7. HUA, X., J. WU, J.L. GOLDSTEIN et al. 1995. Structure of human gene encoding sterol regulatory element binding protein-1 (SREBP-1) and localization of SREBP-1 and SREBP-2 to chromosomes 17p11.2 and 22q13. Genomics **25:** 667–673.
8. MISEREZ, A.R., G. CAO, L.C. PROBST & H.H. HOBBS. 1997. Structure of the human gene encoding sterol regulatory element binding protein 2 (SREBP-2). Genomics **40:** 31–40.
9. HUA, X., A. NOHTURFFT, J.L. GOLDSTEIN & M.S. BROWN. 1996. Sterol resistance in CHO cells traced to point mutation in SREBP cleavage activating protein (SCAP). Cell **87:** 415–426.
10. WANG, X., R. SATO, M.S. BROWN et al. 1994. SREBP-1, a membrane-bound transcription factor released by sterol-regulated proteolysis. Cell **77:** 53–62.
11. SAKAI, J., E.A. DUNCAN, R.B. RAWSON et al. 1996. Sterol regulated release of SREBP-2 from cell membrane requires two sequential cleavages, one within a transmembrane segment. Cell **85:** 1037–1046.
12. BRIGGS, M.R., C. YOKOYAMA, X. WANG et al. 1993. Nuclear protein that binds sterol regulatory element of low density lipoprotein receptor promoter: I. Identification of the protein and delineation of its target nucleotide sequence. J. Biol. Chem. **268:** 14490–14496.
13. WANG, X., M.R. BRIGGS, X. HUA et al. 1993. Nuclear protein that binds sterol regulatory element of low density lipoprotein receptor promoter: II. Purification and characterization. J. Biol. Chem. **268:** 14497–14504.
14. KIM, H-J., M. TAKAHASHI & O. EZAKI. 1999. Fish oil feeding decreases mature sterol regulatory element-binding protein-1 by down-regulation of SREBP-1c mRNA in mouse liver: a possible mechanism for down-regulation of lipogenic enzyme mRNAs. J. Biol. Chem. **274:** 25892–25898.

15. XU, J., M.T. NAKAMURA, H.P. CHO & S.D. CLARKE. 1999. Sterol regulatory element binding protein-1 expression is suppressed by dietary polyunsaturated fatty acid. J. Biol. Chem. **274:** 23577–23583.
16. EDWARDS, P.A. & J. ERICSSON. 1999. Sterols and isoprenoids: signaling molecules derived from the cholesterol biosynthetic pathway. Annu. Rev. Biochem. **68:** 157–185.
17. EDWARDS, P.A., D. TABOR, H.R. KAST & A. VENKATESWARAN. 2000. Regulation of gene expression by SREBP and SCAP. Biochim. Biophys. Acta **1529:** 103–113.
18. SHIMOMURA, I., H. SHIMANO, J.D. HORTON et al. 1997. Differential expression of exons 1a and 1c in mRNAs for sterol regulatory element binding protein-1 in human and mouse organs and cultured cells. J. Clin. Invest. **99:** 838–845.
19. SHIMOMURA, I., Y. BASHMAKOV, S. IKEMOTO et al. 1999. Insulin selectively increases SREBP-1c mRNA in the livers of rats with streptozotocin-induced diabetes. Proc. Natl. Acad. Sci. U.S.A. **96:** 13656–13661.
20. SHIMOMURA, I., R.E. HAMMER, S. IKEMOTO et al. 1999. Leptin reverses insulin resistance and diabetes mellitus in mice with congenital lipodystrophy. Nature **401:** 73–76.
21. HORTON, J.D., Y. BASHMAKOV, I. SHIMOMURA & H. SHIMANO. 1998. Regulation of sterol regulatory element binding proteins in livers of fasted and refed mice. Proc. Natl. Acad. Sci. U.S.A. **95:** 5987–5992.
22. FORETZ, M., C. PACOT, I. DUGAIL et al. 1999. ADD1/SREBP-1c is required in the activation of hepatic lipogenic gene expression by glucose. Mol. Cell. Biol. **19:** 3760–3768.
23. FORETZ, M., C. GUICHARD, P. FERRÉ & F. FOUFELLE. 1999. Sterol regulatory element binding protein-1c is a major mediator of insulin action on the hepatic expression of glucokinase and lipogenesis-related genes. Proc. Natl. Acad. Sci. U.S.A. **96:** 12737–12742.
24. CLARKE, S.D. 2001. Polyunsaturated fatty acid regulation of gene transcription: a molecular mechanism to improve the metabolic syndrome. J. Nutr. **131:** 1129–1132.
25. PEET, D.J., S.D. TURLEY, W. MA et al. 1998. Cholesterol and bile acid metabolism are impaired in mice lacking the nuclear oxysterol receptor LXR alpha. Cell **93:** 693–704.
26. REPA, J.J., G. LIANG, J. OU et al. 2000. Regulation of mouse sterol regulatory element-binding protein-1c gene (SREBP-1c) by oxysterol receptors, LXRα and LXRβ. Genes Dev. **14:** 2819–2830.
27. BERNLOHR, D.A., M.A. BOLANOWSKI, T.J. KELLY, JR. & M.D. LANE. 1985. Evidence for an increase in transcription of specific mRNAs during differentiation of 3T3-L1 preadipocytes. J. Biol. Chem. **260:** 5563–5567.
28. MIYAZAKI, M., Y.C. KIM, M.P. GARY-KELLER et al. 2000. The biosynthesis of hepatic cholesterol esters and triglycerides is impaired in mice with a disruption of the gene for stearoyl-CoA desaturase 1. J. Biol. Chem. **275:** 30132–30138.
29. SHENG, Z., H. OTANI, M.S. BROWN & J.L. GOLDSTEIN. 1995. Independent regulation of sterol regulatory element-binding proteins 1 and 2 in hamster liver. Proc. Natl. Acad. Sci. U.S.A. **92:** 935–938.
30. SCHWARZ, M., E.G. LUND, R. LATHE et al. 1997. Identification and characterization of a mouse oxysterol 7α-hydroxylase cDNA. J. Biol. Chem. **272:** 23995–24001.
31. RUSSELL, D.W. & K.D.R. SETCHELL. 1992. Bile acid biosynthesis. Biochemistry **31:** 4737–4749.
32. SHIMOMURA, I., Y. BASHMAKOV, H. SHIMANO et al. 1997. Cholesterol feeding reduces nuclear forms of sterol regulatory element binding proteins in hamster liver. Proc. Natl. Acad. Sci. U.S.A. **94:** 12354–12359.
33. HUA, X., J. SAKAI, M.S. BROWN & J.L. GOLDSTEIN. 1996. Regulated cleavage of sterol regulatory element binding proteins (SREBPs) requires sequences on both sides of the endoplasmic reticulum membrane. J. Biol. Chem. **271:** 10379–10384.
34. NTAMBI, J.M. 1999. Regulation of stearoyl-CoA desaturase by polyunsaturated fatty acids and cholesterol. J. Lipid Res. **40:** 1549–1558.

Hyperglycemia and Insulin Resistance: Possible Mechanisms

EVA TOMÁS, YEN-SHOU LIN, ZEINA DAGHER, ASISH SAHA, ZHIJUN LUO, YASUO IDO, AND NEIL B. RUDERMAN

Diabetes and Metabolism Research Unit, Departments of Medicine and Physiology, and Section of Endocrinology, Boston University School of Medicine, Boston, Massachusetts 02118, USA

ABSTRACT: Sustained hyperglycemia impairs insulin-stimulated glucose utilization and glycogen synthesis in human and rat skeletal muscles, a phenomenon referred to clinically as glucose toxicity. In rat extensor digitorum longus (EDL) muscle preparations preincubated for 2–4 h in a hyperglycemic medium (25 mM vs. 0 mM glucose), we have shown that the ability of insulin to stimulate glucose incorporation into glycogen is impaired. Interestingly, this was associated with a decreased activation of Akt/PKB, but not its upstream regulator, PI3-kinase. A similar pattern of signaling abnormalities has been observed in adipocytes, L6 muscle cells, C2C12 cells, and (as reported here) EDL incubated with C_2-ceramide. On the other hand, no increase was observed in ceramide mass in EDL incubated with 25 mM glucose. Hyperglycemia-induced insulin resistance also has been described in adipocytes, where it has been linked to activation of novel and conventional protein kinase C isoforms that phosphorylate the insulin receptor and IRS. In addition, we have recently shown that hyperglycemia causes insulin resistance in cultured human umbilical vein endothelial cells (HUVEC). Here, it was associated with an increased propensity to apoptosis and, as in muscle, with an impaired ability of insulin to activate Akt. Interestingly, these effects of hyperglycemia and an increase in diacylglycerol synthesis, which is also caused, were prevented by adding AICAR, an activator of AMP-activated protein kinase (AMPK), to the incubation medium. These results suggest that hyperglycemia causes insulin resistance in cells other than those in classic insulin target tissues. Whether AMPK activation can reverse or prevent insulin resistance in all of these cells remains to be determined.

KEYWORDS: AMP-protein kinase; apoptosis; ceramide; endothelium; muscle

INSULIN RESISTANCE AND TYPE 2 DIABETES

A reduced capacity for insulin to elicit increases in glucose uptake and glycogen synthesis in target tissues such as skeletal muscle and adipose tissue is a common feature in patients with obesity, type 2 diabetes, hypertension, and cardiovascular disease.[1] This diminished ability of a given concentration of insulin to exert its usual

Address for correspondence: Dr. Neil B. Ruderman, Diabetes Unit, Boston Medical Center, Room 825, 650 Albany Street, Boston, MA 02118.
nruderman@medicine.bu.edu

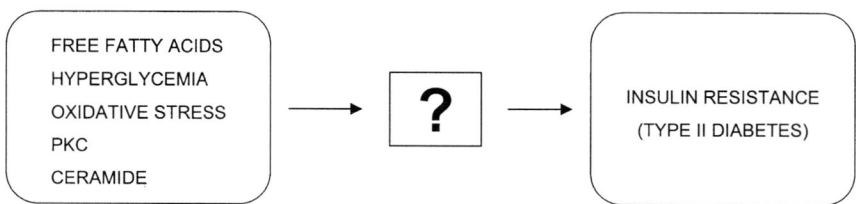

FIGURE 1. Insulin resistance. Some extracellular and intracellular factors that have been implicated in causing insulin resistance in skeletal muscle.

biological effects is referred to as insulin resistance. A multitude of often interrelated factors have been linked to the pathogenesis of insulin resistance in humans and experimental animals. They include primary abnormalities in the insulin signal transduction pathway, extracellular factors such as TNFα and free fatty acids, and intracellular molecules including PKC, ceramide, and oxidative stress (FIG. 1). Another factor that has gained recognition as a cause of insulin resistance is sustained hyperglycemia, also referred as glucose toxicity.[2] A number of studies have shown that insulin resistance can be produced by hyperglycemia in muscle *in vitro*.[3] However, precisely how it does so is not known. Several possible mechanisms will be discussed in this manuscript. In addition, we will review studies from our laboratory suggesting that the inhibition of Akt/PKB, a key molecule in the insulin signaling cascade, is a common event in hyperglycemia-induced insulin resistance. Finally, we will briefly describe preliminary data suggesting that hyperglycemia can alter Akt activation and cause insulin resistance in vascular endothelium.

INSULIN SIGNALING STIMULATION

Insulin stimulates glucose transport into adipose tissue, cardiac, and skeletal muscle by causing the redistribution of the insulin-responsive glucose transporter, GLUT4, from its intracellular location to cell surface membranes. It also activates numerous metabolic pathways, including these that promote glycogen and glycerolipid synthesis and prevent apoptosis (FIG. 2). These actions of insulin occur after it binds to its receptor, an event that activates a tyrosine kinase intrinsic to the receptor and secondarily sets in motion a cascade of intracellular signaling events.[4] One of these events is the phosphorylation of adaptor proteins of the insulin receptor substrate (IRS) family, which recruit and activate downstream effector molecules such as phosphatidylinositol 3-kinase (PI3-kinase). Thus, treatment with PI3-kinase inhibitors, such as wortmanin or LY294002, blocks essentially all of the effects of insulin on glucose transport and metabolism,[5] whereas expression of constitutively active PI3-kinase stimulates them.[6]

Another protein that plays a key role in insulin signaling is the serine/threonine kinase, Akt/PKB. Akt/PKB is activated by insulin and other growth factors in a number of cell types, often in a manner dependent on PI3-kinase.[7] It has been shown to play a crucial role in the control of such diverse cellular functions as apoptosis and glycogen metabolism. Other evidence has implicated it in the regulation by insulin

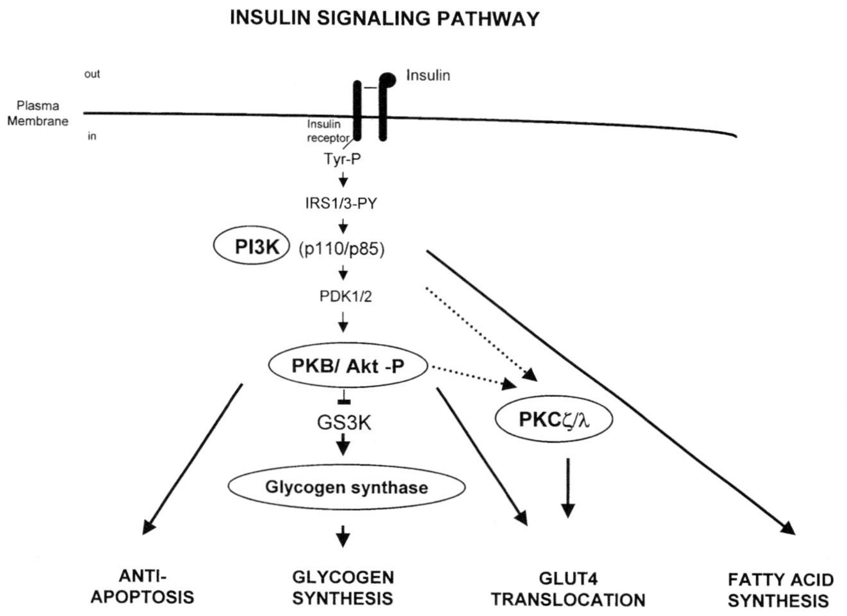

INSULIN SIGNALING PATHWAY

FIGURE 2. The PI3-kinase wing of the insulin-signaling pathway and some of the processes it regulates.

TABLE 1. Effects of preincubation for 4 h with 0 mM or 25 mM glucose on insulin-stimulated glucose uptake and disposition in rat EDL muscles

	Preincubation conditions	
	0 mM glucose	25 mM glucose
DG uptake	169 ± 13 (10)	167 ± 9 (10)
Glucose to glycogen	42 ± 3 (10)	22 ± 2* (10)
Glucose to total lipids	2 ± 0.1 (10)	2 ± 0.1 (10)
Glucose to CO_2	4 ± 0.3 (10)	4 ± 0.3 (9)

NOTE: The uptake of 2-deoxyglucose (DG) and glucose disposition were assessed over 30 min in muscles incubated in media containing 6 mM glucose and 10 mU/mL of insulin. Data are means ± SE (number of observations) and are given as nmol · g^{-1} · ms · min^{-1}. *Significantly different from muscles preincubated in the absence of added glucose ($p < 0.0001$). Adapted from reference 3.

of GLUT4 translocation;[8] indeed, many states of cellular insulin resistance have been associated with impaired Akt/PKB activation and a decrease in glucose transport. Evidence to the contrary has also been reported, however, and the precise role of Akt/PKB in regulating glucose transport and metabolism remains controversial.[9] The activation of Akt/PKB requires its phosphorylation at both Ser473 and Thr308 by a

3-phosphoinositide-dependent kinase identified as PDK1.[10] One of the first physiological targets of Akt/PKB is glycogen synthase kinase-3 (GSK3), an enzyme that has been implicated in the control of many cellular processes, including glycogen and protein synthesis, and the modulation of transcription. When insulin stimulates glycogen synthesis, it concurrently causes the phosphorylation and inhibition of GSK3.

Recent studies in rat adipocytes have suggested that atypical PKC ζ/λ isoforms also play a role in insulin-stimulated glucose transport by promoting the redistribution of GLUT4 from its intracellular location to cell surface membrane.[11]

STUDIES OF HYPERGLYCEMIA-INDUCED INSULIN RESISTANCE IN THE INCUBATED RAT EXTENSOR DIGITORUM LONGUS MUSCLE (EDL)

Rat EDL were incubated for 4 h in media containing 25 mM or 0 mM glucose, after which they were washed in a glucose-free medium and reincubated in a medium containing 6 mM glucose and 10 mU/mL of insulin for 20 min. As shown in TABLE 1, the ability of insulin to stimulate glucose incorporation into glycogen was diminished by over 50% in muscles preincubated with 25 mM glucose; however, glucose oxidation to CO_2, its conversion to lipid, and glucose transport (as reflected by 2-deoxyglucose uptake) were unaffected. The diminished ability of insulin to stimulate glycogen synthesis was paralleled by a decrease in its ability to phosphorylate and activate Akt/PKB (TABLE 2). Interestingly, tyrosine phosphorylation of the insulin receptor and PI3-kinase activation, two events thought to lead to Akt/PKB activation by insulin, were not depressed. Likewise, MAP kinase activation in muscles preincubated with the two glucose concentrations did not differ (TABLE 2).

Possible mechanisms by which hyperglycemia could lead to insulin resistance are depicted in FIGURE 3. We initially focused our attention on ceramide since increases

TABLE 2. Effect of preincubation with 0 mM or 25 mM glucose on insulin signaling events

	Preincubation conditions	
	0 mM glucose	25 mM glucose
PI3-kinase (a)	302 ± 41 (10)	341 ± 83 (12)
PI3-kinase (b)	559 ± 127 (5)	527 ± 94 (5)
Akt/PKB 3	352 ± 42 (9)	214 ± 43* (10)
ERK2	269 ± 91 (4)	315 ± 136 (5)

NOTE: Data are means ± SE (number of observations) and are expressed as the percent of activity determined in contralateral muscle incubated in the absence of insulin. Experimental conditions were as described in TABLE 1, except that incubations were for only 5 min. The symbol (a) indicates that PI3-kinase was immunoprecipitated with antiphosphotyrosine antibody, and (b) with anti-IRS-1 antibody. Akt/PKB was immunoprecipitated with an antibody that recognizes Akt/PKB2 as well as Akt/PKB1. Akt/PKB and PI3-kinase activities were determined by using standard kinase assays, and ERK2 by immunoblotting with a phospho-antibody. *Significantly different from muscles preincubated in the absence of added glucose ($p < 0.05$). Adapted from reference 3.

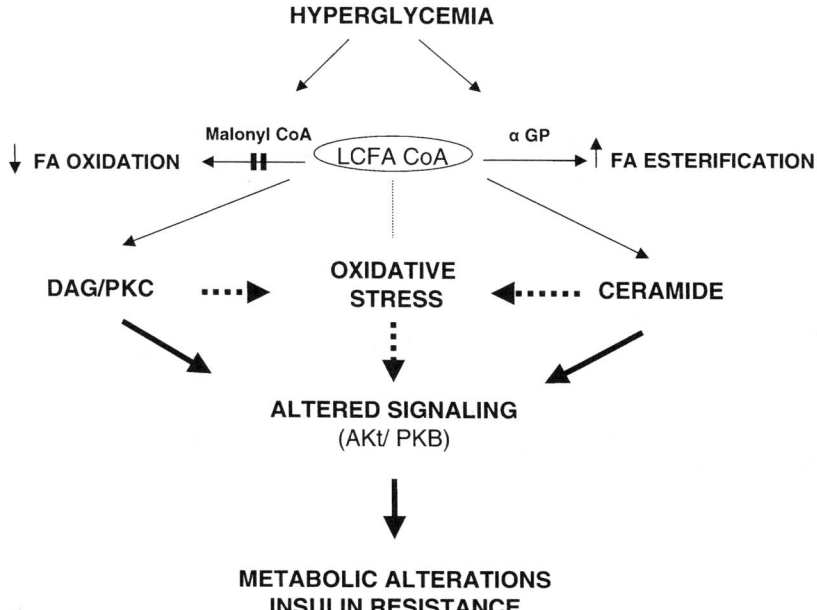

FIGURE 3. Hypothetical mechanisms by which hyperglycemia could cause insulin resistance.

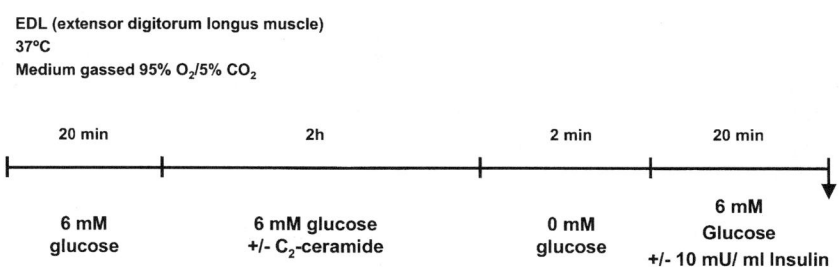

FIGURE 4. Experimental protocol used to examine the effect of C_2-ceramide on the insulin signaling cascade in incubated rat extensor digitorum longus.

in its *de novo* synthesis have been associated with an impaired ability of insulin to stimulate glycogen synthesis and activate Akt/PKB, but not earlier events in the insulin signaling pathway, in C2C12 cells incubated with palmitate by Schmitz-Peiffer and coworkers.[12] In addition, earlier work by Summers *et al.*[13] demonstrated that incubation with C_2-ceramide produces similar signaling changes and inhibits insulin-stimulated glucose transport in 3T3-L1 adipocytes. Identical findings have been reported in L6 skeletal muscle cells.[14] To examine the effect of ceramide on mammalian skeletal muscle, EDL were preincubated with or without C_2-ceramide, and insulin-stimulated glucose uptake and disposition were assayed during a subse-

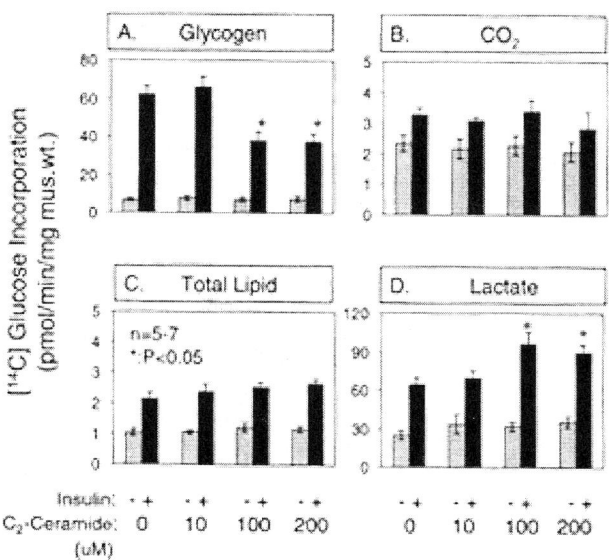

FIGURE 5. Effect of preincubation with media containing various C_2-ceramide concentrations and 6 mM glucose on glucose metabolism in rat EDL muscle. Results are the means ± SEM. *Significantly different ($p < 0.05$) versus insulin-stimulated muscle preincubation in the absence of added C_2-ceramide.

quent incubation (FIG. 4). As shown in FIGURES 5 and 6, the pattern of events was very similar to that observed when muscle was preincubated with 25 mM vs. 0 mM glucose. Thus, the stimulation of glucose incorporation into glycogen by insulin was diminished by nearly 50%, as was Akt/PKB phosphorylation. Furthermore, the uptake of 2-deoxyglucose and the conversion of glucose to products other than glycogen were unchanged or even increased. We had previously shown that incubation of muscles with 25 mM vs. 0 mM glucose caused a marked increase in the concentration of malonyl-CoA, an inhibitor of carnitine palmitoyl transferase, which regulates long-chain fatty acid transfer into mitochondria.[15] Thus, by inhibiting fatty acid oxidation and causing the accumulation of palmitoyl-CoA, hyperglycemia could lead to an increase in *de novo* ceramide synthesis. To date, we have only measured ceramide mass in EDL incubated with media containing 25 mM or 0 mM glucose and have found no difference. Measurements of *de novo* ceramide synthesis and studies with inhibitors of this pathway are necessary before its role in causing insulin resistance can be rejected, however.

STUDIES OF HYPERGLYCEMIA-INDUCED INSULIN RESISTANCE IN VASCULAR ENDOTHELIUM

A number of reports have shown that incubation with a hyperglycemia medium causes multiple abnormalities in cultured endothelium, including increases in oxida-

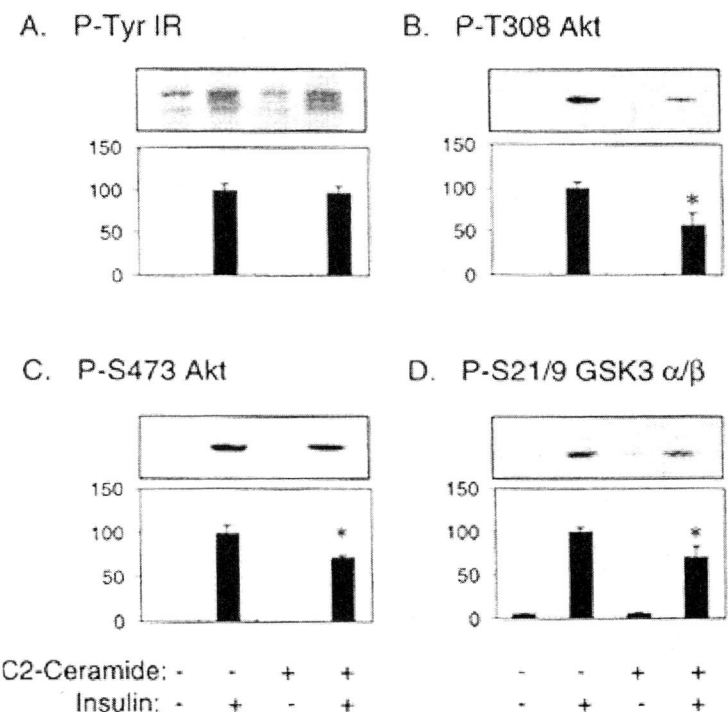

FIGURE 6. Effect of 2-h preincubation with media containing 0 μM versus 200 μM C_2-ceramide on insulin signaling events in rat EDL muscle: **(A)** tyrosine phosphorylation of insulin receptor; **(B)** serine phosphorylation of Akt/PKB; **(C)** threonine phosphorylation of Akt/PKB; **(D)** GSK3 phosphorylation by insulin.

tive stress,[16,17] apoptosis,[18] and increases in DAG mass and PKC activity.[19] For this reason, human umbilical vein endothelial cells (HUVEC) were incubated in media containing 5 or 30 mM glucose and various parameters were studied. As reported by Ido and coworkers,[20] cells incubated for 72 h with 30 mM vs. 5 mM glucose showed an increase in apoptosis, which was preceded at 24 h by an increase in DAG synthesis and an impaired ability of insulin, at a physiological concentration (150 μU/mL), to activate Akt/PKB (TABLE 3). Furthermore, all of these changes were prevented by incubating the cells with the AMPK activator, AICAR. No increase in ceramide mass was observed (TABLE 3).

DISCUSSION

The mechanism by which hyperglycemia causes insulin resistance is still uncertain, but in at least two tissues, skeletal muscle and cultured vascular endothelium, it is associated with impaired activation of Akt. In muscle, the insulin resistance was

TABLE 3. Effects of incubation for 24–72 h with media containing 30 mM glucose or 30 mM glucose + AICAR (1 mM) on diacylglycerol (DAG) synthesis, Akt activation by insulin, and apoptosis

	DAG synthesis	Akt/PKB activation by insulin	Apoptosis
30 mM glucose	Increased	Decreased	Increased
30 mM glucose + AICAR	Decreased	Increased	Decreased

NOTE: Results are expressed relative to findings in cells incubated in media containing 5 mM glucose and no AICAR. Adapted from reference 20.

manifested as impaired glycogen synthesis and in the vascular endothelium as enhanced apoptosis. The precise mechanisms by which insulin action is impaired in the two cells remain to be determined. In light of these findings and previous reports in other cell types that linked Akt inhibition with ceramide accumulation, the latter initially seemed a likely mediator. The failure to find ceramide mass increased in the two models studied by us goes against this theory; however, the possibility that increases in a small pool of ceramide, produced by *de novo* synthesis, played a causative role has not been ruled out. The possibility that insulin resistance in these cells is related to alterations in DAG-PKC signaling, oxidative stress, and/or hexosamine formation also needs to be examined.

An intriguing finding was that incubation with AICAR, an AMPK activator, prevented both the apoptosis and the impaired activation of Akt by a physiological concentration of insulin in the HUVEC. In a preliminary report,[21] we have shown that AICAR treatment, *in vivo*, appears to reverse the insulin resistance in both liver and muscle of fat-fed rats. How AMPK works in these settings to diminish insulin resistance should provide novel insights into both the mechanisms by which the insulin resistance occurs and how it might be treated.

ACKNOWLEDGMENTS

This work was supported by research grants from the Juvenile Diabetes Foundation (1-2000-292) (N.R.), and a grant from the Kilo Foundation and a Center Grant from the JDRF (996004) (Y.I.).

REFERENCES

1. SEELY, L.B. & J.M. OLEFSKY. 1993. Potential cellular and genetic mechanisms for insulin resistance in the common disorders of diabetes and obesity. *In* Insulin Resistance, pp. 187–252. Wiley. New York.
2. ROSSETTI, L., D. SMITH, G.I. SHULMAN *et al.* 1987. Correction of hyperglycemia with phlorizin normalizes tissue sensitivity to insulin in diabetes rats. J. Clin. Invest. **79:** 1510–1515.
3. KUROWSKI, T.G., Y. LIN, Z. LUO *et al.* 1999. Hyperglycemia inhibits insulin activation of Akt/protein kinase B, but not phosphatidylinositol 3-kinase in rat skeletal muscle. Diabetes **48:** 658–663.
4. SUMMERS, S.A., V.P. YIN, E.L. WHITEMAN *et al.* 1999. Signaling pathways mediating insulin-stimulated glucose transport. Ann. N.Y. Acad. Sci. **892:** 169–186.

5. CLARKE, J.F., P.W. YOUNG, K. YONEZAWA et al. 1994. Inhibition of the translocation of GLUT1 and GLUT4 in 3T3-L1 cells by the phosphatidylinositol 3-kinase inhibitor, wortmanin. Biochem. J. **300:** 631–635.
6. FREVERT, E.U. & B.B. KAHN. 1997. Differential effects of constitutively active phosphatidylinositol 3-kinase on glucose transport, glycogen synthase activity, and DNA synthesis in 3T3L1-adipocytes. Mol. Cell. Biol. **17:** 190–198.
7. KOHN, A.D., A. BOGE, A. BARTHEL et al. 1998. Construction and characterization of a conditionally active version of the ser/thr kinase Akt. J. Biol. Chem. **273:** 11937–11943.
8. HAJDUCH, E., G.J. LITHERLAND & H.S. HUNDAL. 2001. Protein kinase B (PKB/Akt)—a key regulator of glucose transport? FEBS Lett. **492:** 199–203.
9. KIM, Y.B., S.E. NIKOULINA, T.P. CIARALDI et al. 1999. Normal insulin-dependent activation of Akt/protein kinase B, with diminished activation of phosphoinositide 3-kinase, in muscle in type 2 diabetes. J. Clin. Invest. **104:** 733–741.
10. ALESSI, D.R., S.R. JAMES, C.P. DOWNES et al. 1997. Characterization of a 3-phosphoinositide-dependent protein kinase which phosphorylates and activates protein kinase B-alpha. Curr. Biol. **7:** 261–269.
11. BANDYOPADHYAY, G., M.L. STANDAERT, M.P. SAJAN et al. 1999. Dependence of insulin-stimulated glucose transporter 4 translocation on 3-phosphoinositide-dependent protein kinase-1 and its target threonine-410 in the activation loop of protein kinase C-zeta. Mol. Endocrinol. **13:** 1766–1772.
12. SCHMITZ-PEIFFER, C., D.L. CRAIG & T.J. BIDEN, 1999. Ceramide generation is sufficient to account for the inhibition of the insulin-stimulated PKB pathway in C2C12 skeletal muscle cells pretreated with palmitate. J. Biol. Chem. **274:** 24202–24210.
13. SUMMERS, S.A., L.A. GARZA, H. ZHOU & M.J. BIRNBAUM. 1998. Regulation of insulin-stimulated glucose transporter GLUT4 translocation and Akt kinase activity by ceramide. Mol. Cell. Biol. **18:** 5457–5464.
14. HAJDUCH, E., A. BALENDRAN, I.H. BATTY et al. 2001. Ceramide impairs the insulin-dependent membranes recruitment of protein kinase B leading to loss in downstream signaling in L6 skeletal muscle cells. Diabetologia **44:** 173–183.
15. SAHA, A.K., D.R. LAYBUTT, D. DEAN et al. 1999. Cytosolic citrate and malonyl-CoA regulation in rat muscle *in vivo*. Am. J. Physiol. **276:** E1030–E1037.
16. HO, F.M., S.H. LIU, C.S. LIAU et al. 2000. High glucose-induced apoptosis in human endothelial cell is mediated by sequential activations of c-Jun NH_2-terminal kinase and caspase-3. Circulation **101:** 2618–2624.
17. INOGUCHI, T., P. LI, F. UMEDA et al. 2000. High glucose level and free fatty acid stimulate reactive oxygen species production through protein kinase C–dependent activation of NAD(P)H oxidase in cultured vascular cells. Diabetes **49:** 1939–1945.
18. BAUMGARTNER-PARZER, S.M., L. WAGNER, M. PETTERMAN et al. 1995. High-glucose-triggered apoptosis in cultured endothelial cells. Diabetes **44:** 1323–1327.
19. INOGUCHI, T., P. XIA, M. KUNISAKI et al. 1994. Insulin's effect on protein kinase C and diacylglycerol induced by diabetes and glucose in vascular tissues. Am. J. Physiol. **267:** E369–E379.
20. IDO, Y., D. CARLING & N.B. RUDERMAN. 2001. Hyperglycemia induced apoptosis in human umbilical vein endothelial cells (HUVEC): inhibition by the AMP-activated protein kinase, activator AICAR. Diabetes **51:** 159–167.
21. KRAEGEN, E.W., M.A. IGLESIAS, G. FRANGIOUDAKIS et al. 2001. AICAR administration causes a prolonged enhancement of muscle insulin action in insulin resistance high fat fed rats. Diabetes **50**(suppl. 2): A80.

Insulin Stimulation of Hepatic Triacylglycerol Secretion in the Insulin-Replete State

Implications for the Etiology of Peripheral Insulin Resistance

VICTOR A. ZAMMIT

Cell Biochemistry, Hannah Research Institute, Ayr KA6 5HL, Scotland, United Kingdom

ABSTRACT: Observations on humans, on rats *in vivo*, and on isolated perfused rat livers indicate that insulin stimulates hepatic very-low-density lipoprotein (VLDL)–TAG secretion when the liver is chronically exposed to the hormone. They suggest that frequent stimulation of insulin secretion throughout the diurnal cycle may result in a chronic stimulation of VLDL secretion and increased delivery of acyl moieties to the periphery, particularly to muscle, the most important site of insulin-sensitive glucose disposal. If acyl groups are provided in excess of the oxidative needs of the tissue, this may lead to induction of insulin resistance, irrespective of whether obesity is established concomitantly. Dietary factors that stimulate hepatic VLDL secretion may have the same effect and contribute to the induction of a vicious spiral leading to the development of the full-blown Metabolic Syndrome and its pathological consequences, including type-2 diabetes, stroke, and cardiovascular disease.

KEYWORDS: liver; very-low-density lipoproteins (VLDL); triglycerides; insulin resistance; secretion; muscle; adipose tissue

INTRODUCTION

The central question posed in this review is as follows: does hepatic very-low-density lipoproteins (VLDL) secretion have the potential to initiate insulin resistance in muscle, the major site of insulin-sensitive glucose disposal, after which hypertrophic obesity merely locks the individual into the insulin-resistant state through the increased release of nonesterified fatty acids (NEFA) and factors such as adiponectin?[1] In focusing on the causative link between obesity and insulin resistance, do we ignore the etiological role of the liver with respect to insulin resistance even under conditions of overall energy balance? The same questions arise about type-2 diabetes and cardiovascular disease. Would our view of the etiology of insulin resistance, and its pathological consequences, be different were clinicians able to measure the rate

---

Address for correspondence: Victor A. Zammit, Cell Biochemistry, Hannah Research Institute, Ayr KA6 5HL, Scotland, United Kingdom. Voice: (44) 1292 674058; fax: (44) 1292 674059.

zammitv@hri.sari.ac.uk

Ann. N.Y. Acad. Sci. 967: 52–65 (2002). © 2002 New York Academy of Sciences.

of hepatic VLDL-triacylglycerol (TAG) secretion as easily as they can calculate BMIs or measure waistlines?

In the preobese state, skeletal muscle accounts for the greatest proportion (>80%) of insulin-dependent glucose uptake and metabolism. Therefore, whole-body insulin resistance can be established well ahead of that in adipose tissue and, in particular, of the latter's resistance to the antilipolytic effect of the hormone. This applies to other peripheral cell types too, such that the pathological manifestations of insulin resistance would not necessarily depend on the prior emergence of hypertrophic obesity.[2] Each cell type gives rise to distinct functional defects resulting in the clinical syndrome associated with insulin resistance.[3] Extensive studies have established the occurrence of changes in protein expression that accompany insulin resistance in different tissues, ranging from those directly involved in insulin signaling[4] to the expression of transport proteins such as Glut 4.[4,5]

FATTY ACIDS AND INSULIN RESISTANCE IN MUSCLE

In muscle cells, a crucial causative metabolic factor involved in the induction of insulin resistance has emerged (but one that does not necessarily have to be restricted to this cell type), namely, the oversupply of long-chain acyl moieties in excess of the ability of the tissue to oxidize them. Thus, whole-body insulin resistance (due mostly to muscle) can be induced simply by administering chronically a low dose of an inhibitor of the overt carnitine palmitoyltransferase of mitochondria (CPT I) even in the absence of any change in the type or quantity of dietary intake.[6] Many studies have shown that oversupply of saturated fatty acids results in the induction of muscle insulin resistance and vice versa.[7,8] In turn, this is associated with increased intramyocellular TAG,[9–13] presumably as a consequence of the diversion of long-chain acyl-CoA metabolism towards glyceride synthesis rather than oxidation. However, it is not thought that TAG per se interferes with insulin signaling, but rather that acyl-CoA esters and/or other products of acyl-CoA metabolism [diacylglycerol (DAG), ceramides, hexosamines] are responsible.[6,14,15] Fatty acids and their CoA esters are potent effectors of gene transcription and vesicular membrane trafficking.[16–18] DAG species are activators of members of the PKC family that, through the serine/threonine phosphorylation of proteins involved in the protein kinase cascades of insulin signaling, have the ability to interfere with the tyrosine phosphorylation–dependent process.[14] Indeed, saturated fatty acids, which are metabolized less readily to TAG, giving rise to a higher level of accumulation of DAG and ceramide, are the most potent inducers of insulin resistance in muscle fibers *in vitro*.[15]

Two important corollaries emerge. First, the delivery of fatty acyl moieties as hydrolysis products of lipoprotein lipase (LPL) action on triacylglycerol-rich lipoproteins (TRLs)—chylomicrons and VLDL—may also induce insulin resistance in muscle and other tissues that express LPL. Second, it should be possible to prevent or alleviate muscle insulin resistance by promoting fatty acid oxidation. This is borne out by observations that moderate hyperleptinemia depletes intramyocellular TAG, promotes fatty acid β-oxidation,[19] and concomitantly reverses diet-induced insulin resistance.[20] It also provides one plausible mechanism through which exercise improves insulin sensitivity, both through direct utilization of acyl-CoA esters as substrates for mitochondrial β-oxidation[21] and through the increase in the overall oxidative capacity of the tissue.[22] Although exercise increases the expression of LPL in muscle, the concomitant increase in oxidative capacity of the tissue still ensures an improvement in insulin sensitivity. Moreover, increased LPL expression after exercise contributes towards a more rapid TRL clearance in the postprandial period.[23,24] Conversely, in the spontaneously hypertensive rat, muscle insulin resistance is associated with a profound decrease in muscle fatty acid oxidation, to a much greater extent than that of esterification to glycerides,[25] again emphasizing the importance of the balance between the two pathways.

This role of TRLs in delivering acyl moieties to peripheral tissues may be very important in rationalizing the impact of diet, both directly through intestinal chylomicron-TAG secretion and indirectly through its effects on the partitioning of hepatic fatty acid metabolism, on the etiology of insulin resistance in muscle. It may explain why whole-body insulin resistance can be induced before the onset of obesity, for example, after fructose feeding[26] and, as in the preobese *fa/fa* rat,[27] even at a time of enhanced sensitivity of adipose tissue to the hormone.[28] The induction of muscle insulin resistance before that of antilipolysis in adipose tissue may also be related by the fact that the latter is normally much more sensitive to lower concentrations of the hormone than, for example, glycogen synthesis in muscle.[29] Interestingly, lipodystrophy results in extreme insulin resistance,[30,31] presumably because increased delivery of TRL to muscle, owing to the absence of an effective adipose "sink", is sufficient to induce an oversupply of acyl moieties to muscle. In the preobese state, insulin resistance in muscle is associated with hypertriglyceridemia before resistance develops in adipose tissue or in the liver,[32,33] a sequence mimicked in fructose-fed rats[34] in which euglycemic clamp glucose disposal correlates negatively with hypertriglyceridemia.[35] In humans, insulin resistance can be detected in nonobese subjects before any elevation of plasma NEFA concentrations, but at a time when VLDL-TAG secretion rate is already increased.[36] The particular importance of the role of hepatic VLDL production is demonstrated by the fact that the development of insulin resistance does not depend necessarily on the ingestion of a high-fat diet; indeed, the degree of insulin resistance is exacerbated by high-carbohydrate diets containing refined sugars,[37,38] which result in increased rates of hepatic lipogenesis and VLDL-TAG secretion.[39–41] Moreover, the degree of postprandial hypertriglyceridemia in response to a standard meal is markedly increased after a short period of adaptation to a refined-carbohydrate diet.[42] Conversely, feeding diets rich in *n-3* fatty acids, which suppress hepatic VLDL secretion, prevents the induction of insulin resistance.[9] These observations all support the above suggestion that overemphasis on the concept that insulin resistance is necessarily the consequence of obesity may lead to the neglect of the possibility that dietary composition and diurnal feeding patterns may have a deleterious effect on insulin sensitivity through their effects on liver metabolism.

ASYMMETRY OF OUTCOME BETWEEN GLUCOSE UPTAKE BY MUSCLE AND BY ADIPOSE TISSUE

Depending on muscle fiber type, a greater or smaller proportion of glucose is used either as a substrate to generate ATP oxidatively or for the synthesis of muscle glycogen that is stored until required. It is glucose utilization for this nonoxidative component that is largely affected by insulin resistance, through the blunted ability of insulin to stimulate glycogen synthase, one of the earliest markers of insulin resistance.[43] Initially, euglycemia is maintained owing to compensatory hyperinsulinemia.[7,44] However, diversion of glucose from muscle to adipose tissue may contribute towards adipocyte hypertrophy, initially in the absence of net body weight gain,[45] a situation mimicked experimentally by the muscle-specific knockout of the insulin receptor gene in mice.[46,47] The preferential channeling of lipogenic substrates towards adipose tissue and the altered expression of lipogenic enzymes in muscle and adipose tissue during NEFA-induced muscle insulin resistance have been demonstrated experimentally.[5] Interestingly, the adipocyte hyperplasia that accompanies the adipose-specific overexpression of Glut 4 results in oversensitization of the muscle to insulin in spite of the overall increase in adiposity,[48] confirming that the provision of additional, newly differentiated adipocytes as a sink for lipogenic substrates is sufficient to improve insulin sensitivity in muscle. Conversely, failure of adipocyte differentiation results in type-2 diabetes in humans.[49] It is tempting to speculate, therefore, that the existence of a subset of moderately obese individuals who are not insulin-resistant[50] may be related to a higher capacity in these individuals for adipocyte differentiation, acquired either genetically or as a result of nutritional patterns earlier in their lives.

Importantly, there is an asymmetry between the glucose taken up and converted by muscle into glycogen and its fate upon diversion to adipose tissue. Whereas the carbon skeleton of glycogen stored by muscle is eventually released from the tissue mostly as lactate, when glucose is diverted to adipose tissue its carbons reenter the circulation mostly as NEFA, only to cause further insulin resistance, lipotoxicity of pancreatic β-cells, and type-2 diabetes.[51] This asymmetry explains the observations made on knockout mice with a muscle-specific disruption of the Glut 4 gene, which become insulin-resistant from an early age.[52]

THE ROLE OF HEPATIC FATTY ACID PARTITIONING IN THE ETIOLOGY OF INSULIN RESISTANCE

The TAGs in chylomicrons reflect, both in quantity and in acyl chain composition, those of dietary fats. Therefore, high-fat diets, especially those rich in saturated fatty acids, would be expected to result directly in the induction of muscle insulin resistance in sedentary individuals. By contrast, rates of hepatic VLDL secretion depend largely on the partitioning of preformed fatty acids between (i) oxidation and esterification and (ii) the retention or secretion, by the liver, of the synthesized TAG.[53,54] Synthesis *de novo* of fatty acids within the liver makes a minor quantitative contribution towards the overall TAG secretion, but there is a strong positive relationship between the rate of VLDL-TAG secretion and the *de novo* lipogenic rate.[39,55,56] When *de novo* hepatic lipogenesis is increased through the feeding of low-fat, high-carbohydrate diets containing sucrose (fructose), there is an enrichment of saturated fatty acid in VLDL-TAG, reflecting the predominant nature of the products of hepatic *de novo* lipogenesis.[41] However, it is noteworthy that the disruption of the stearoyl-CoA desaturase 1 (SCD 1) gene in mice results in the inhibition of VLDL-TAG secretion, accompanied by impaired synthesis of TAG and cholesteryl esters.[57,58]

Dietary constituents that favor the hepatic esterification of fatty acids, and the secretion of the resulting TAGs, increase the rate of appearance of VLDL.[53] These include saturated fatty acids, alcohol,[59] and fructose (sucrose).[60,61] By contrast, diets rich in *n-3* fatty acids divert fatty acid partitioning away from VLDL-TAG secretion.[62] These opposing actions are mirrored in the effects that these respective diets have on the development and maintenance of insulin resistance in humans and animals, although direct effects through altered phospholipid composition of tissue membranes may also be involved.[9,15]

The intake of sucrose may be particularly important. The fructose moiety is the molecular half that accounts for the insulin resistance–inducing effect of the disaccharide;[35] dietary fructose, but not glucose or starch, induces hypertriglyceridemia in humans, especially in males.[63–65] High-carbohydrate diets containing sucrose are even more potent inducers of insulin resistance than high-fat diets;[38] an increased rate of hepatic VLDL secretion[66] and the deterioration of lipoprotein profiles occur in parallel,[38,67] before any induction of obesity or rise in plasma NEFA concentrations occurs. Also, test-meal-induced postprandial hyperlipidemia is exacerbated in subjects previously fed high-refined-carbohydrate diets,[38,42] although a lower rate of clearance of TRLs in such individuals may also contribute to the overall lipidemia.[68] The central importance of an increased rate of hepatic VLDL secretion rate in the etiology of insulin resistance is demonstrated by the fact that feeding high-starch diets to rats does not induce insulin resistance, in spite of the increased adiposity.[69] The potentially health-related role of diet-induced increases in the rate of hepatic VLDL secretion under these conditions is indicated by the higher rates of cardiovascular disease and type-2 diabetes that occur in individuals deriving a greater percentage of energy from refined grains and sucrose rather than from complex carbohydrates.[70]

In terms of cardiovascular disease, VLDL remnants or LDLs are potentially of greater pathological importance, particularly in the postprandial period. Chylomicrons outcompete VLDL particles for LPL action, resulting in a longer residence time for LDL in the circulation and in the increased formation of small dense LDL and of oxidized LDL, both of which are highly atherogenic. For this reason, the stimulation of hepatic VLDL secretion concomitantly with absorption of fat from the gut may explain why postprandial lipemia is so strongly associated with atherosclerosis.[71]

REGULATION OF THE RATE OF HEPATIC VLDL SECRETION

Dietary factors that modulate the partitioning of hepatic fatty acids and of synthesized TAG do so through effects at a multiplicity of metabolic steps within the hepatocyte. For example, a diet rich in *n-3* fatty acids decreases the rate of hepatic VLDL secretion by (i) increasing the rate of fatty acid oxidation through an increase in the expression of key fatty acid oxidation enzymes, (ii) altering the composition of the mitochondrial outer membrane such that the sensitivity of CPT I to malonyl-CoA is highly diminished,[72] (iii) lowering lipogenic enzyme expression and the concentration of malonyl-CoA, (iv) increasing the diversion of DAG into phospholipids, and (v) increasing the proportion of the TAG synthesized that is diverted into the secretory pathway, as previously reviewed.[62] By contrast, fructose feeding results in higher lipogenic enzyme expression, a greater diversion of acyl chains into glyceride synthesis, and facilitation of TAG secretion through increased expression of microsomal transfer protein (MTP).[60,61] Insulin is the hormone that would be expected to be most closely associated with the regulation of VLDL secretion. Interest in its effects has been long-standing, but has yielded conflicting data, which are only now being resolved. It has long been known that there is a strong correlation between hyperinsulinemia and hypertriglyceridemia in humans and in animal models. Reaven[73] demonstrated that the higher the degree of insulinemia in rats *in vivo*, the higher is the maximal rate of VLDL secretion by the isolated livers; prior *in vivo* hyperinsulinemia also increases the affinity for fatty acid substrate,[73] through an as yet undetermined mechanism. Insulin stimulation of VLDL-TAG secretion by the isolated perfused liver was demonstrated by several groups in the 1970s and 1980s.[72,74–79] Such stimulation agrees with the above-mentioned correlation between hyperinsulinemia and hypertriglyceridemia in humans,[80] and the observation that chronic hyperinsulinemia in rats results in an increased rate of hepatic VLDL secretion.[81] However, the advent of the use of cultured hepatocyte preparations to study hepatic metabolism resulted in a seismic change in the accepted view (as reviewed previously[54]). Many laboratories found that insulin, when added to cultured rat hepatocytes, inhibited TAG secretion, in spite of activating TAG synthesis. Therefore, the consensus grew, and was extrapolated to the situation *in vivo*, that insulin always inhibits hepatic TAG secretion through a direct intrahepatic effect.[82] Importantly, however, prolongation of the *in vitro* culture of rat hepatocytes beyond 48 h reverses the direction of this effect,[83] and in a highly lipogenic hepatocyte culture system (avian hepatocytes) insulin always stimulates both TAG synthesis and TAG secretion, even during short-term culture.[84]

Teologically, it is important that insulin should inhibit VLDL-TAG secretion during the absorptive and early postprandial phases. First, it would minimize the competition for LPL action between the two types of TRL and hence lower the degree of postprandial lipemia.[85] Second, it would occur in the same direction as that of the concomitant inhibition of adipose tissue lipolysis.[86] Although this latter effect may be considered to be sufficient to ensure low rates of hepatic TAG secretion by decreasing the rate of delivery of NEFA to the liver, it is to be emphasized that significant release of NEFA into the circulation occurs during LPL action[86] and that a degree of portal delivery of acyl moieties may occur during intestinal fat absorption.[87] The importance of the rate of hepatic VLDL secretion to the overall postprandial lipemia in humans is demonstrated by observations[59] that ethanol consumed with a mixed meal acutely elevates postprandial VLDL primarily through an increase in the secretion of large VLDL particles. Direct intrahepatic inhibition of TAG secretion occurring after acute insulin administration in humans has been demonstrated in overnight-fasted subjects; insulin treatment results in a greater inhibition of hepatic VLDL secretion than that calculated to be due to the concomitant decrease in plasma NEFA concentrations,[88–90] although more recent data have failed to find any effect of a glucose load on net splanchnic output of VLDL.[91]

Experiments conducted in the author's laboratory, in which the partitioning of hepatic fatty acid metabolism was studied *in vivo* in awake rats during food intake, after a 24-h fast,[92] showed that an acute inhibition of the partitioning of fatty acid flux towards TAG secretion occurs during the first 4 h after the commencement of food intake, with maximum inhibition occurring at 2 h. Moreover, during this period, the pathway of fatty acid oxidation remains largely active, thus contributing to the lowering of the overall flux of fatty acids towards TAG secretion during the prandial/absorptive period.[93,94] It is important to note that the design of these experiments was such that the quantification of the partitioning of hepatic fatty acid metabolism was independent of any simultaneous fluctuations in the rates of delivery of NEFA to the liver, thus avoiding the interference that could have been introduced by changes in the rate of adipose tissue lipolysis.[94] These observations show that, during the meal-induced surge in insulin secretion, there is an acute and concerted diversion of the flux of fatty acids away from pathways leading to the secretion of TAG, thus explaining the direct inhibition of VLDL secretion by acute insulin administration in overnight *fasted* humans.[90,95]

However, a permanently inhibitory intrahepatic effect of insulin on TAG secretion would conflict with (i) the correlations between hypertriglyceridemia and hyperinsulinemia (above) and (ii) the stimulation of VLDL-TAG secretion by chronic insulin treatment of rats[79,96] or by mild hyperinsulinemia in humans.[39] In order to counter these difficulties, it has been suggested that the increased rate of VLDL secretion associated with hypertriglyceridemic states is due to a double-negative phenomenon, that is, that hepatic insulin resistance leads to a lowered degree of insulin-mediated inhibition.[82] Indeed, the potency of inhibition of VLDL secretion in fasted humans is blunted in patients in whom the liver is already insulin-resistant, for example, type-2 diabetics.[97,98] However, as pointed out by Steiner,[79] an attenuation of insulin-mediated inhibition cannot result in a higher absolute rate of secretion of VLDL-TAG than basal. Observations in humans also do not support the view that hypertriglyceridemia is secondary to a failure of insulin to inhibit TAG secretion.[99]

BIDIRECTIONAL EFFECTS OF INSULIN ON HEPATIC TAG SECRETION

These apparently conflicting results and interpretations appear to have been resolved recently by the observations, from this laboratory, that the direction of the effect of insulin on hepatic TAG secretion is dependent on the prior metabolic (possibly insulinemic) state.[100] In experiments conducted *in vivo* using the same techniques described above,[92] the partitioning of hepatic fatty acid metabolism was quantified in fed normoinsulinemic, insulin-deficient, or 24-h-fasted rats before or after acute insulinization of the liver.[101] It was found that insulin increased the proportion of metabolized fatty acids secreted as TAG in the fed normoinsulinemic animals, but markedly lowered it in the fasted and insulin-deficient animals.[101] This switch in direction of insulin action was also observed in perfused livers isolated from animals in the same three insulinemic states. Insulin stimulated the absolute rate of TAG secretion by livers from fed normoinsulinemic rats, but inhibited it in livers from fasted or insulin-deficient animals.[102] The biochemical mechanism responsible for this switch in direction of insulin action is still to be determined; however, it is possible that insulin (i) permanently antagonizes VLDL secretion through stimulation of apoB degradation,[82] but (ii) acutely activates and (iii) induces the expression of a protein that has a very short half-life and that is required for the synthesis of a component(s) VLDL. A candidate target protein is HMG-CoA reductase.

The possibility that it is the prior insulinemic state that determines the direction of the subsequent response of TAG secretion to insulin emerges strongly from the data of Heimberg and coworkers.[78] These authors found that making animals insulin-deficient, by the infusion of anti-insulin serum 8 h prior to perfusion of the livers, induces insulin inhibition of TAG secretion. The additional observation that chronic experimental hyperinsulinemia *in vivo*, both in humans[39] and in rats,[75,81] has an effect opposite to that of hypoinsulinemia points towards the direct involvement of prior insulin levels in determining the direction of the insulin effect on hepatic TAG secretion. This would explain the intrahepatic inhibition of VLDL-TAG secretion by acute insulin treatment of human subjects that had been fasted overnight,[89,90] a protocol anticipated to predispose the liver towards inhibition of hepatic VLDL-TAG secretion. Interestingly, in the livers of animals fed a fish oil–rich diet, insulin does not stimulate VLDL secretion, even when the donor animals are in the fed state.[103] This may provide an additional mechanism through which *n-3* fatty acids prevent insulin resistance.[9]

A ROLE FOR DIURNAL INSULIN SECRETION PATTERNS IN INCREASED HEPATIC VLDL SECRETION

A role for prior insulinemia in determining the hepatic response to the hormone suggests that repeated diurnal surges of insulin secretion may be the cause of hypertriglyceridemia and peripheral insulin resistance. This is pertinent to the dietary pattern adopted in modern affluent societies in which a near-constant postprandial state is achieved for most of the day due to frequent intake of high-glycemic-index snacks and sucrose-rich beverages.[104] Such a pattern may result in the liver being set in a positive response mode, to the effect of insulin on TAG secretion, for most of the diurnal cycle. It is suggested, therefore, that diurnal patterns of insulin release may be an important determinant of the role of the liver in the etiology of insulin resistance, even when individuals are in overall energy balance but (i) consuming a diet that stimulates hepatic VLDL-TAG secretion and/or (ii) maintaining high interprandial insulin levels. A positive energy balance would exacerbate the situation by inducing hypertrophic obesity with concomitant insulin resistance in adipocytes and raised plasma NEFA, thus completing the vicious cycle of the Metabolic Syndrome.[54] These interactions may provide an explanation for the observed positive correlations between interprandial plasma insulin concentration, postprandial lipemia, and the development of CVD.[105]

RELATIVE ROLES OF NEFA AND TRLs IN THE ETIOLOGY AND MAINTENANCE OF INSULIN RESISTANCE

The observations that pravastatin treatment of hypertriglyceridemic patients results in a marked decrease in the incidence of type-2 diabetes over a 10-year follow-up period,[106] combined with the knowledge that statins inhibit apoB secretion by the liver,[107] suggest a role for an elevated rate of VLDL-TAG secretion in the etiology of the Metabolic Syndrome and the ensuing loss of insulin secretory capacity. In accordance with this, it was shown[108] that gemfibrozil treatment of severely insulin-resistant patients alleviates, in parallel, both their pronounced hypertriglyceridemia and their insulin resistance. Although others[109] have not observed such a correlation in mildly hypertriglyceridemic patients, it is possible that, in this cohort of subjects, insulin resistance may have been maintained by NEFA delivery to muscle as the plasma NEFA levels were not decreased by the treatment.[109] These apparently contradictory observations indicate that the relative importance of NEFA and TRL in delivering acyl moieties may change during the progression of the syndrome. Although there may be variable improvement of insulin sensitivity after treatment with drugs that inhibit VLDL secretion, in obese individuals with insulin-resistant adipose tissue, the complete reversal of their insulin resistance will depend on the normalization of adipocyte size and insulin sensitivity. Physiologically, this can be achieved by reduction of caloric intake. Pharmacologically, it can be achieved through the use of PPARγ agonists. One of the multiplicity of effects of thiazolidinediones (TZDs) on different tissues is to promote the differentiation of small, insulin-sensitive (highly lipogenic) adipocytes into which fatty acids can be turned over from larger adipocytes, together with excess dietary substrates.[110] This increases whole-body insulin sensitivity despite overall weight gain and increased adiposity.[111] The fact that PPARγ agonist–stimulated differentiation of adipocytes occurs preferentially in subcutaneous rather than abdominal depots[110,112]—which drain into the hepatic portal vein—is also anticipated to diminish the concentration of NEFA (important in determining TAG synthetic and secretion rates[113]) to which the liver is exposed, for any given rate of subsequent adipose TAG lipolysis.

CONCLUSIONS

The rate of secretion of VLDL-TAG by the liver may be an important determinant of the etiology of insulin resistance even in nonobese individuals. When the partitioning of fatty acid metabolism towards the secretion of TAG is enhanced by such dietary components as sucrose, alcohol, and saturated fats, the secretion of VLDL may be raised sufficiently to induce insulin resistance. Moreover, stimulation of hepatic VLDL secretion by insulin under conditions in which the liver is exposed to high-basal insulin levels interprandially (e.g., by frequent snacking on high-glycemic-index foods and beverages) may lead to increased overall diurnal VLDL secretion and especially to exacerbated postprandial hypertriglyceridemia, and of extended residence times for LDL within the circulation. Recommendations that include the moderation of the intake of dietary constituents that stimulate hepatic VLDL secretion and that encourage the adequate spacing of meals to ensure an inhibitory action of insulin on hepatic VLDL secretion during the prandial/early postprandial period may be appropriate.

ACKNOWLEDGMENTS

Work in the author's laboratory was supported by Diabetes UK, the British Heart Foundation, and the Scottish Executive.

REFERENCES

1. YAMAUCHI, T., et al. 2001. The fat-derived hormone adiponectin reverses insulin resistance. Nat. Med. **7:** 941–946.
2. DESPRES, J-P. & A. MARETTE. 1999. Obesity and insulin resistance. In Epidemiological, Metabolic, and Molecular Aspects, pp. 51–82. Humana Press. Totowa, New Jersey.
3. VERMA, S., et al. 1997. Vascular insulin resistance in fructose-hypertensive rats. Eur. J. Pharmacol. **322:** R1–R2.
4. BEZERRA, R.M., et al. 2000. A high fructose diet affects the early steps of insulin action in muscle and liver of rats. J. Nutr. **130:** 1531–1535.
5. FABRIS, R., et al. 2001. Preferential channeling of energy fuels toward fat rather than muscle during high free fatty acid availability in rats. Diabetes **50:** 601–608.
6. DOBBINS, R.L., et al. 2001. Prolonged inhibition of muscle carnitine palmitoyl-transferase-1 promotes intramyocellular lipid accumulation and insulin resistance in rats. Diabetes **50:** 123–130.
7. BODEN, G., et al. 1995. Effects of a 48-h fat infusion on insulin secretion and glucose utilization. Diabetes **44:** 1239–1242.
8. SANTOMAURO, A.T., et al. 1999. Overnight lowering of free fatty acids with acipimox improves insulin resistance and glucose tolerance in obese diabetic and nondiabetic subjects. Diabetes **48:** 1836–1841.
9. STORLIEN, L.H., et al. 1991. Influence of dietary fat composition on development of insulin resistance in rats: relationship to muscle triglyceride and omega-3 fatty acids in muscle phospholipid. Diabetes **40:** 280–289.
10. PAN, D.A., et al. 1997. Skeletal muscle triglyceride levels are inversely related to insulin action. Diabetes **46:** 983–988.
11. JACOB, S., et al. 1999. Association of increased intramyocellular lipid content with insulin resistance in lean nondiabetic offspring of type 2 diabetic subjects. Diabetes **48:** 1113–1119.
12. PERSEGHIN, G., et al. 1999. Intramyocellular triglyceride content is a determinant of in vivo insulin resistance in humans: a ^{1}H-^{13}C nuclear magnetic resonance spectroscopy assessment in offspring of type 2 diabetic parents. Diabetes **48:** 1600–1606.
13. SHULMAN, G.I. 2000. Cellular mechanisms of insulin resistance. J. Clin. Invest. **106:** 171–176.
14. SCHMITZ-PEIFFER, C. 2000. Signalling aspects of insulin resistance in skeletal muscle: mechanisms induced by lipid oversupply. Cell. Signalling **12:** 583–594.
15. MONTELL, E., et al. 2001. DAG accumulation from saturated fatty acids desensitizes insulin stimulation of glucose uptake in muscle cells. Am. J. Physiol. Endocrinol. Metab. **280:** E229–E237.
16. SLEEMAN, M.W., et al. 1998. Association of acyl-CoA synthetase-1 with GLUT4-containing vesicles. J. Biol. Chem. **273:** 3132–3135.
17. HERTZ, R., et al. 1998. Fatty acyl–CoA thioesters are ligands of hepatic nuclear factor-4α. Nature **392:** 512–516.
18. JUMP, D.B. & S.D. CLARKE. 1999. Regulation of gene expression by dietary fat. Annu. Rev. Nutr. **19:** 63–90.
19. WANG, M.Y., et al. 1999. Novel form of lipolysis induced by leptin. J. Biol. Chem. **274:** 17541–17544.
20. BUETTNER, R., et al. 2000. Correction of diet-induced hyperglycemia, hyperinsulinemia, and skeletal muscle insulin resistance by moderate hyperleptinemia. Am. J. Physiol. Endocrinol. Metab. **278:** E563–E569.
21. OAKES, N.D., et al. 1997. Diet-induced muscle insulin resistance in rats is ameliorated by acute dietary lipid withdrawal or a single bout of exercise: parallel relationship between insulin stimulation of glucose uptake and suppression of long-chain fatty acyl-CoA. Diabetes **46:** 2022–2028.
22. KELLEY, D.E., et al. 1999. Skeletal muscle fatty acid metabolism in association with insulin resistance, obesity, and weight loss. Am. J. Physiol. **277:** E1130–E1141.
23. TSETSONIS, N.V. & A.E. HARDMAN. 1996. Reduction in postprandial lipemia after walking: influence of exercise intensity. Med. Sci. Sports Exercise **28:** 1235–1242.

24. GILL, J.M. & A.E. HARDMAN. 2000. Postprandial lipemia: effects of exercise and restriction of energy intake compared. Am. J. Clin. Nutr. **71:** 465–471.
25. HAJRI, T., *et al.* 2001. Defective fatty acid uptake in the spontaneously hypertensive rat is a primary determinant of altered glucose metabolism, hyperinsulinemia, and myocardial hypertrophy. J. Biol. Chem. **276:** 23661–23666.
26. PAGLIASSOTTI, M.J., *et al.* 1996. Changes in insulin action, triglycerides, and lipid composition during sucrose feeding in rats. Am. J. Physiol. **271:** R1319–R1326.
27. PENICAUD, L., *et al.* 1987. Development of obesity in Zucker rats: early insulin resistance in muscles, but normal sensitivity in white adipose tissue. Diabetes **36:** 626–631.
28. CZECH, M.P., *et al.* 1978. Insulin response in skeletal muscle and fat cells of the genetically obese Zucker rat. Metabolism **27:** 1967–1981.
29. REAVEN, G.M. 1993. Role of insulin resistance in human disease (syndrome X): an expanded definition. Annu. Rev. Med. **44:** 121–131.
30. SPECKMAN, R.A., *et al.* 2000. Mutational and haplotype analyses of families with familial partial lipodystrophy (Dunnigan variety) reveal recurrent missense mutations in the globular C-terminal domain of lamin A/C. Am. J. Hum. Genet. **66:** 1192–1198.
31. SHACKLETON, S., *et al.* 2000. LMNA, encoding lamin A/C, is mutated in partial lipodystrophy. Nat. Genet. **24:** 153–156.
32. STEINER, G., *et al.* 1980. Resistance to insulin, but not to glucagon in lean human hypertriglyceridemics. Diabetes **29:** 899–905.
33. MCKANE, W.R., *et al.* 1990. The assessment of hepatic and peripheral insulin sensitivity in hypertriglyceridemia. Metabolism **39:** 1240–1245.
34. HOLNESS, M.J. 1994. Hypertriglyceridaemia precedes impaired muscle glucose utilization during fructose feeding. Biochem. Soc. Trans. **22:** 105S.
35. THORBURN, A.W., *et al.* 1989. Fructose-induced *in vivo* insulin resistance and elevated plasma triglyceride levels in rats. Am. J. Clin. Nutr. **49:** 1155–1163.
36. GUERCI, B., *et al.* 2000. Relationship between altered postprandial lipemia and insulin resistance in normolipidemic and normoglucose tolerant obese patients. Int. J. Obes. Related Metab. Disord. **24:** 468–478.
37. REAVEN, G.M. 1997. Do high carbohydrate diets prevent the development or attenuate the manifestations (or both) of syndrome X? A viewpoint strongly against. Curr. Opin. Lipidol. **8:** 23–27.
38. CHEN, Y.D., *et al.* 1995. Why do low-fat high-carbohydrate diets accentuate postprandial lipemia in patients with NIDDM? Diabetes Care **18:** 10–16.
39. AARSLAND, A., *et al.* 1996. Contributions of *de novo* synthesis of fatty acids to total VLDL-triglyceride secretion during prolonged hyperglycemia/hyperinsulinemia in normal man. J. Clin. Invest. **98:** 2008–2017.
40. AARSLAND, A., *et al.* 1997. Hepatic and whole-body fat synthesis in humans during carbohydrate overfeeding. Am. J. Clin. Nutr. **65:** 1774–1782.
41. HUDGINS, L.C. 2000. Effect of high-carbohydrate feeding on triglyceride and saturated fatty acid synthesis. Proc. Soc. Exp. Biol. Med. **225:** 178–183.
42. KOUTSARI, C., *et al.* 2000. Postprandial lipemia after short-term variation in dietary fat and carbohydrate. Metabolism **49:** 1150–1155.
43. STERN, M. & B. MITCHELL. 1999. Genetics of insulin resistance. *In* Epidemiological, Metabolic, and Molecular Aspects, pp. 3–18. Humana Press. Totowa, New Jersey.
44. REAVEN, G.M., *et al.* 1993. Insulin resistance and insulin secretion are determinants of oral glucose tolerance in normal individuals. Diabetes **42:** 1324–1332.
45. ODELEYE, O.E., *et al.* 1997. Fasting hyperinsulinemia is a predictor of increased body weight gain and obesity in Pima Indian children. Diabetes **46:** 1341–1345.
46. BRUNING, J.C., *et al.* 1998. A muscle-specific insulin receptor knockout exhibits features of the metabolic syndrome of NIDDM without altering glucose tolerance. Mol. Cell **2:** 559–569.
47. KIM, J.K., *et al.* 2000. Redistribution of substrates to adipose tissue promotes obesity in mice with selective insulin resistance in muscle. J. Clin. Invest. **105:** 1791–1797.
48. SHEPHERD, P.R., *et al.* 1993. Adipose cell hyperplasia and enhanced glucose disposal in transgenic mice overexpressing GLUT4 selectively in adipose tissue. J. Biol. Chem. **268:** 22243–22246.

49. DANFORTH, E., JR. 2000. Failure of adipocyte differentiation causes type II diabetes mellitus? Nat. Genet. **26:** 13.
50. GROOP, L.C. & T. TUOMI. 1997. Non-insulin-dependent diabetes mellitus—a collision between thrifty genes and an affluent society. Ann. Med. **29:** 37–53.
51. LEE, Y., et al. 1994. Beta-cell lipotoxicity in the pathogenesis of non-insulin-dependent diabetes mellitus of obese rats: impairment in adipocyte-beta-cell relationships. Proc. Natl. Acad. Sci. U.S.A. **91:** 10878–10882.
52. ZISMAN, A., et al. 2000. Targeted disruption of the glucose transporter 4 selectively in muscle causes insulin resistance and glucose intolerance. Nat. Med. **6:** 924–928.
53. ZAMMIT, V.A. 1996. Role of insulin in hepatic fatty acid partitioning: emerging concepts. Biochem. J. **314:** 1–14.
54. ZAMMIT, V., et al. 2001. Insulin stimulation of hepatic triacylglycerol secretion and the aetiology of insulin resistance. J. Nutr. **131:** 2074–2077.
55. SCHWARZ, J.M., et al. 1995. Short-term alterations in carbohydrate energy intake in humans: striking effects on hepatic glucose production, de novo lipogenesis, lipolysis, and whole-body fuel selection. J. Clin. Invest. **96:** 2735–2743.
56. PARK, J., et al. 1997. Chronic exogenous insulin and chronic carbohydrate supplementation increase de novo VLDL triglyceride fatty acid production in rats. J. Lipid Res. **38:** 2529–2536.
57. MIYAZAKI, M., et al. 2000. The biosynthesis of hepatic cholesterol esters and triglycerides is impaired in mice with a disruption of the gene for stearoyl-CoA desaturase 1. J. Biol. Chem. **275:** 30123–30128.
58. MIYAZAKI, M., et al. 2001. A lipogenic diet in mice with a disruption of the stearoyl-CoA desaturase 1 gene reveals a stringent requirement of endogenous monounsaturated fatty acids for triglyceride synthesis. J. Lipid Res. **42:** 1018–1024.
59. FIELDING, B.A., et al. 2000. Ethanol with a mixed meal increases postprandial triacylglycerol, but decreases postprandial non-esterified fatty acid concentrations. Br. J. Nutr. **83:** 597–604.
60. MAYES, P.A. 1993. Intermediary metabolism of fructose. Am. J. Clin. Nutr. **58:** 754S–765S.
61. TAGHIBIGLOU, C., et al. 2000. Mechanisms of hepatic very low density lipoprotein overproduction in insulin resistance: evidence for enhanced lipoprotein assembly, reduced intracellular ApoB degradation, and increased microsomal triglyceride transfer protein in a fructose-fed hamster model. J. Biol. Chem. **275:** 8416–8425.
62. MOIR, A.M.B., et al. 1995. Quantification in vivo of the effects of different types of dietary fat on the loci of control involved in hepatic triacylglycerol secretion. Biochem. J. **208:** 537–542.
63. GEORGOPOULOS, A., et al. 2000. A high carbohydrate versus a high monounsaturated fatty acid diet lowers the atherogenic potential of big VLDL particles in patients with type 1 diabetes. J. Nutr. **130:** 2503–2507.
64. HALLFRISCH, J. 1990. Metabolic effects of dietary fructose. FASEB J. **4:** 2652–2660.
65. HOLLENBECK, C.B. 1993. Dietary fructose effects on lipoprotein metabolism and risk for coronary artery disease. Am. J. Clin. Nutr. **58:** 800S–809S.
66. BLADES, B. & A. GARG. 1995. Mechanisms of increase in plasma triacylglycerol concentrations as a result of high carbohydrate intakes in patients with non-insulin-dependent diabetes mellitus. Am. J. Clin. Nutr. **62:** 996–1002.
67. DREON, D.M., et al. 1999. A very low-fat diet is not associated with improved lipoprotein profiles in men with a predominance of large, low-density lipoproteins. Am. J. Clin. Nutr. **69:** 411–418.
68. BAUM, C.L. & M. BROWN. 2000. Low-fat, high-carbohydrate diets and atherogenic risk. Nutr. Rev. **58:** 148–151.
69. PAWLAK, D.B., et al. 2001. High glycemic index starch promotes hypersecretion of insulin and higher body fat in rats without affecting insulin sensitivity. J. Nutr. **131:** 99–104.
70. MORRIS, K.L. & M.B. ZEMEL. 1999. Glycemic index, cardiovascular disease, and obesity. Nutr. Rev. **57:** 273–276.
71. ZILVERSMIT, D.B. 1995. Atherogenic nature of triglycerides, postprandial lipidemia, and triglyceride-rich remnant lipoproteins. Clin. Chem. **41:** 153–158.

72. WONG, S.H., *et al.* 1984. The adaptive effects of dietary fish and safflower oil on lipid and lipoprotein metabolism in perfused rat liver. Biochim. Biophys. Acta **792:** 103–109.
73. REAVEN, G.M. & C.E. MONDON. 1984. Effect of *in vivo* plasma insulin levels on the relationship between perfusate free fatty acid concentration and triglyceride secretion by perfused rat livers. Horm. Metab. Res. **16:** 230–232.
74. LAKER, M.E. & P.A. MAYES. 1984. Investigations into the direct effects of insulin on hepatic ketogenesis, lipoprotein secretion, and pyruvate dehydrogenase activity. Biochim. Biophys. Acta **795:** 427–430.
75. STEINER, G. & G.F. LEWIS. 1996. Hyperinsulinemia and triglyceride-rich lipoproteins. Diabetes **45**(suppl. 3): S24–S26.
76. TOPPING, D.L. & P.A. MAYES. 1970. Direct stimulation by insulin and fructose of very-low-density lipoprotein secretion by the perfused liver. Biochem. J. **119:** 48P.
77. TOPPING, D.L. & P.A. MAYES. 1972. The immediate effects of insulin and fructose on the metabolism of the perfused liver: changes in lipoprotein secretion, fatty acid oxidation and esterification, lipogenesis, and carbohydrate metabolism. Biochem. J. **126:** 295–311.
78. WOODSIDE, W.F. & M. HEIMBERG. 1976. Effects of anti-insulin serum, insulin, and glucose on output of triglycerides and on ketogenesis by the perfused rat liver. J. Biol. Chem. **251:** 13–23.
79. STEINER, G. 1991. Insulin regulation of triglyceride metabolism. Atheroscler. Rev. **22:** 27–32.
80. OLEFSKY, J.M., *et al.* 1974. Reappraisal of the role of insulin in hypertriglyceridemia. Am. J. Med. **57:** 551–560.
81. STEINER, G., *et al.* 1984. Hyperinsulinemia and *in vivo* very-low-density lipoprotein-triglyceride kinetics. Am. J. Physiol. **246:** E187–E192.
82. SPARKS, J.D. & C.E. SPARKS. 1994. Insulin regulation of triacylglycerol-rich lipoprotein synthesis and secretion. Biochim. Biophys. Acta **1215:** 9-32.
83. BARTLETT, S.M. & G.F. GIBBONS. 1988. Short- and longer-term regulation of very-low-density lipoprotein secretion by insulin, dexamethasone, and lipogenic substrates in cultured hepatocytes: a biphasic effect of insulin. Biochem. J. **249:** 37–43.
84. LEGRAND, P., *et al.* 1996. Effect of insulin on triacylglycerol synthesis and secretion by chicken hepatocytes in primary culture. Int. J. Biochem. Cell. Biol. **28:** 431–440.
85. POTTS, J.L., *et al.* 1991. Peripheral triacylglycerol extraction in the fasting and postprandial states. Clin. Sci. **81:** 621–626.
86. COPPACK, S.W., *et al.* 1992. Adipose tissue metabolism in obesity: lipase action *in vivo* before and after a mixed meal. Metabolism **41:** 264–272.
87. MANSBACH, C.M., II, *et al.* 1991. Portal transport of absorbed lipids in rats. Am. J. Physiol. **261:** G530–G538.
88. LEWIS, G.F., *et al.* 1993. Effects of acute hyperinsulinemia on VLDL triglyceride and VLDL apoB production in normal weight and obese individuals. Diabetes **42:** 833–842.
89. LEWIS, G.F., *et al.* 1995. Interaction between free fatty acids and insulin in the acute control of very low density lipoprotein production in humans. J. Clin. Invest. **95:** 158–166.
90. MALMSTROM, R., *et al.* 1998. Effects of insulin and acipimox on VLDL1 and VLDL2 apolipoprotein B production in normal subjects. Diabetes **47:** 779–787.
91. BULOW, J., *et al.* 1999. Co-ordination of hepatic and adipose tissue lipid metabolism after oral glucose. J. Lipid Res. **40:** 2034–2043.
92. MOIR, A.M.B. & V.A. ZAMMIT. 1992. Selective labelling of hepatic fatty acids *in vivo*: studies on the synthesis and secretion of glycerolipids in the rat. Biochem. J. **283:** 145–149.
93. MOIR, A.M.B. & V.A. ZAMMIT. 1993. Monitoring of changes in hepatic fatty acid and glycerolipid metabolism during the starved-to-fed transition *in vivo*: studies on awake, unrestrained rats. Biochem. J. **289:** 49–55.
94. ZAMMIT, V.A. & A.M.B. MOIR. 1994. Monitoring the partitioning of hepatic fatty acids *in vivo*: keeping track of control. Trends Biochem. Sci. **19:** 313–317.

95. LEWIS, G.F. & G. STEINER. 1996. Acute effects of insulin in the control of VLDL production in humans: implications for the insulin-resistant state. Diabetes Care **19:** 390–393.
96. SPARKS, J.D., *et al.* 2000. Insulin-treated Zucker diabetic fatty rats retain the hypertriglyceridemia associated with obesity. Metabolism **49:** 1424–1430.
97. MALMSTROM, R., *et al.* 1997. Defective regulation of triglyceride metabolism by insulin in the liver in NIDDM. Diabetologia **40:** 454–462.
98. MALMSTROM, R., *et al.* 1997. Metabolic basis of hypotriglyceridemic effects of insulin in normal men. Arterioscler. Thromb.Vasc. Biol. **17:** 1454–1464.
99. MCLAUGHLIN, T., *et al.* 2000. Carbohydrate-induced hypertriglyceridemia: an insight into the link between plasma insulin and triglyceride concentrations. J. Clin. Endocrinol. Metab. **85:** 3085–3088.
100. ZAMMIT, V.A. 2000. Use of *in vivo* and *in vitro* techniques for the study of the effects of insulin on hepatic triacylglycerol secretion in different insulinaemic states. Biochem. Soc. Trans. **28:** 103–109.
101. RENNIE, S.M., *et al.* 2000. A switch in the direction of the effect of insulin on the partitioning of hepatic fatty acids for the formation of secreted triacylglycerol occurs *in vivo*, as predicted from studies with perfused livers. Eur. J. Biochem. **267:** 935–941.
102. ZAMMIT, V.A., *et al.* 1999. Insulin stimulates triacylglycerol secretion by perfused livers from fed rats, but inhibits it in livers from fasted or insulin-deficient rats: implications for the relationship between hyperinsulinaemia and hypertriglyceridaemia. Eur. J. Biochem. **263:** 859–864.
103. TOPPING, D.L., *et al.* 1987. Failure of insulin to stimulate lipogenesis and triacylglycerol secretion in perfused livers from rats adapted to dietary fish oil. Biochim. Biophys. Acta **927:** 423–428.
104. FLODIN, N.W. 1986. Atherosclerosis: an insulin-dependent disease? J. Am. Coll. Nutr. **5:** 417–427.
105. HAFFNER, S.M., *et al.* 1990. Cardiovascular risk factors in confirmed prediabetic individuals: does the clock for coronary heart disease start ticking before the onset of clinical diabetes? JAMA **263:** 2893–2898.
106. FREEMAN, D.J., *et al.* 2001. Pravastatin and the development of diabetes mellitus: evidence for a protective treatment effect in the West of Scotland Coronary Prevention Study. Circulation **103:** 357–362.
107. BATTULA, S.B., *et al.* 2000. Postprandial apolipoprotein B48- and B100-containing lipoproteins in type 2 diabetes: do statins have a specific effect on triglyceride metabolism? Metabolism **49:** 1049–1054.
108. STEINER, G. 1991. Altering triglyceride concentrations changes insulin-glucose relationships in hypertriglyceridemic patients: double-blind study with gemfibrozil with implications for atherosclerosis. Diabetes Care **14:** 1077–1081.
109. SANE, T., *et al.* 1995. Decreasing triglyceride by gemfibrozil therapy does not affect the glucoregulatory or antilipolytic effect of insulin in nondiabetic subjects with mild hypertriglyceridemia. Metabolism **44:** 589–596.
110. OKUNO, A., *et al.* 1998. Troglitazone increases the number of small adipocytes without the change of white adipose tissue mass in obese Zucker rats. J. Clin. Invest. **101:** 1354–1361.
111. SCHOONJANS, K. & J. AUWERX. 2000. Thiazolidinediones: an update. Lancet **355:** 1008–1010.
112. KELLY, I.E., *et al.* 1999. Effects of a thiazolidinedione compound on body fat and fat distribution of patients with type 2 diabetes. Diabetes Care **22:** 288–293.
113. ZAMMIT, V.A. & D.L. LANKESTER. 2001. Oleate acutely stimulates the secretion of triacylglycerol by cultured rat hepatocytes by accelerating the emptying of the secretory compartment. Lipids **36:** 607–612.

Free Fatty Acid Metabolism in Human Skeletal Muscle Is Regulated by PPARγ and RXR Agonists

THEODORE P. CIARALDI, BONG-SOO CHA, KYONG-SOO PARK, LESLIE CARTER, SUNDER R. MUDALIAR, AND ROBERT R. HENRY

Veterans Affairs San Diego Healthcare System, San Diego, California 92161, and Department of Medicine, University of California, San Diego, La Jolla, California 92093, USA

ABSTRACT: Free fatty acid (FFA) oxidation in human skeletal muscle cells can be stimulated, both independently and in a synergistic manner, by agonists for PPARγ and RXR. Increased FFA disposal in muscle through augmented oxidation could reduce intramyocellular lipid accumulation. The abilities of such agents to improve glucose tolerance and insulin action may thus involve effects on both glucose and FFA metabolism.

KEYWORDS: free fatty acids (FFA); metabolism; oxidation; muscle; PPARγ; RXR; lipid; insulin

The defining characteristics of patients with type 2 diabetes are impaired glucose metabolism and insulin resistance. However, such individuals are often dyslipidemic as well and display insulin resistance for control of lipid metabolism. Free fatty acids (FFA) represent the major fuel utilized in skeletal muscle in the fasting and exercising states, so derangements in FFA utilization could have a major impact on energy homeostasis. It has been reported that obese individuals utilize lower than normal amounts of fatty acid in muscle in the fasting state, when fatty acid utilization should be greatest.[1,2] The same investigators have reported that lipid oxidation is indeed reduced in type 2 diabetic subjects.[3]

Recently, attention has focused on lipid accumulation in skeletal muscle and its impact on insulin action. A strong inverse correlation between muscle lipid content and insulin action has been established across a wide weight range of nondiabetic subjects.[4,5] The processes responsible for triglyceride accumulation in muscle could be numerous. One possibility is the observation that obese individuals favor fat esterification over oxidation in skeletal muscle.[2]

Address for correspondence: Robert R. Henry, M.D., VA San Diego Healthcare System, 3350 La Jolla Village Drive, San Diego, CA 92161. Voice: 858-552-8585, ext. 3648; fax: 858-642-6242.

rrhenry@vapop.ucsd.edu

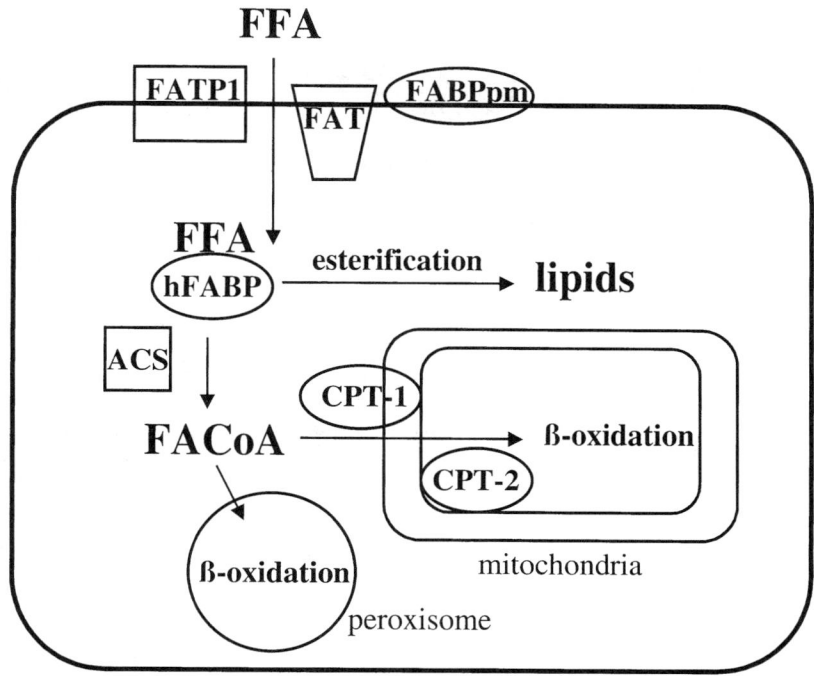

FIGURE 1. Schematic summary of major pathways of FFA utilization in muscle cells. Key proteins are identified in the text.

Lipid metabolism—synthesis and disposal—is a highly complex and tightly regulated process, key events of which are summarized in FIGURE 1. FFA may diffuse passively across the cell membrane depending on the rate of cellular metabolism.[6] Alternatively, it is suggested that a portion of transport is protein-mediated. Possible mediators of FFA uptake include fatty acid translocase (FAT/CD36), fatty acid transport protein (FATP), and plasma membrane fatty acid binding protein (FABPpm). Once inside the muscle cell, there are two major fates for FFA: incorporation into lipids by esterification for storage and structural purposes, and β-oxidation in mitochondria and peroxisomes. After production of fatty acyl CoA derivatives (FACoA) by acetyl CoA synthase (ACS), entry of FACoA into mitochondria is mediated by carnitine palmitoyl transferase-1 (CPT-1), a crucial regulatory point in lipid and glucose metabolism.[7] Control of the expression of key genes involved in fatty acid metabolism has been shown to occur through members of the peroxisome proliferator-activated (PPAR) subfamily of nuclear receptors, acting as heterodimeric partners with the retinoid X receptor (RXR).[8]

To investigate FFA metabolism in muscle, we have utilized human skeletal muscle cells in culture. When differentiated, cultured human skeletal muscle cells display the morphologic, biochemical, and metabolic characteristics of mature skeletal muscle.[9] Most importantly, cells from diabetic subjects display defects in glucose

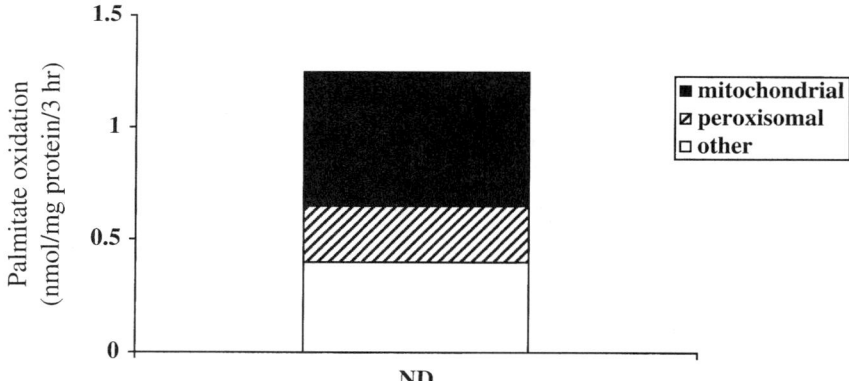

FIGURE 2. Palmitate oxidation in human muscle cells. Myotubes were incubated with ^{3}H-palmitate (5 µM final concentration) for 3 h at 37°C. Products of oxidation were monitored as ^{3}H$_2$O released to the media. Components of oxidation are defined by sensitivity to etomoxir and are described in the text. Results are expressed as average ± SEM, $n = 10$.

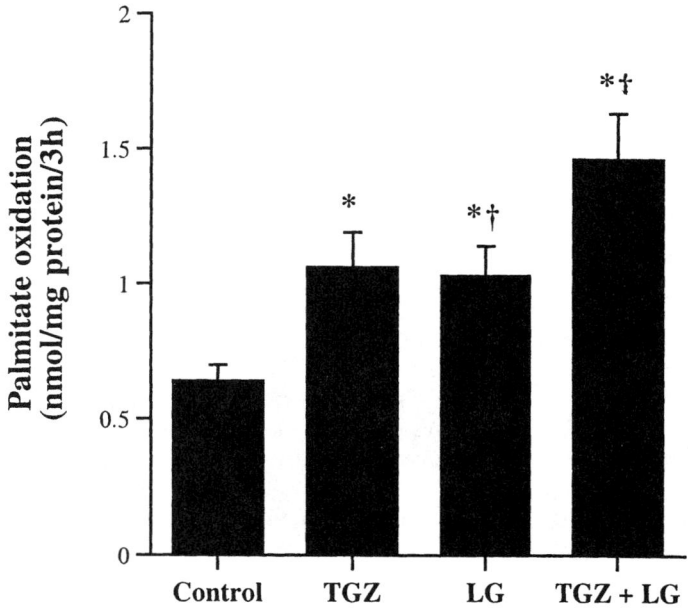

FIGURE 3. Regulation of palmitate oxidation in skeletal muscle cells. Cells were treated for 4 days, during fusion/differentiation with troglitazone (TGZ, 4.6 µM), LG100268 (LG, 4 µM), or a combination of the two, before assay of palmitate oxidation. Results are expressed as average ± SEM, $n = 22$. *$p < 0.05$ vs. control for same subject group; $^†p < 0.05$ vs. either treatment alone. Reprinted from reference 12.

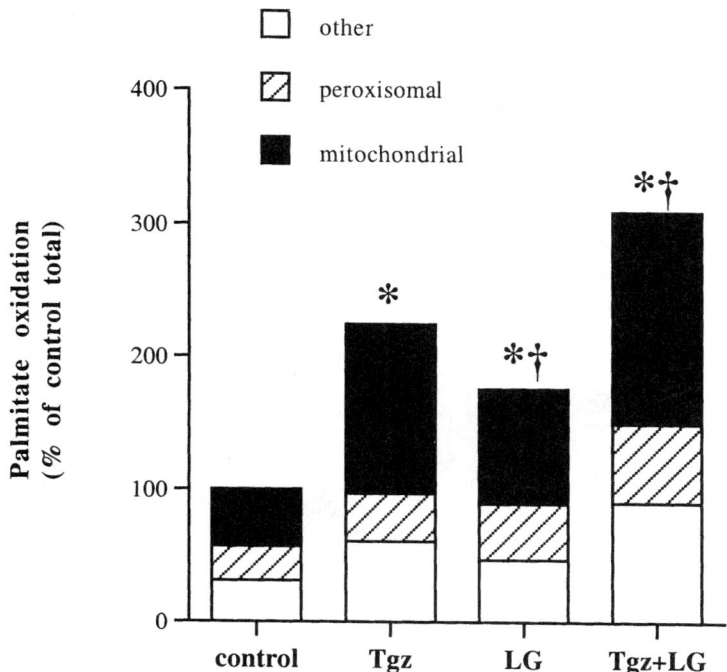

FIGURE 4. Regulation of palmitate oxidation in muscle cells. Cells treated as described in FIGURE 3. Components of oxidation are as defined in FIGURE 2 and the text. Results are expressed as average, $n = 8$–13. *$p < 0.05$ vs. control for same group; †$p < 0.05$ vs. either treatment alone. Reprinted from reference 12.

metabolism that are reflective of the subject's *in vivo* metabolic status.[10] Cells were treated with selective agonists for either the gamma isoform of PPAR (PPARγ) (troglitazone) or RXR (LG100268) during the differentiation period and then β-oxidation of palmitate was monitored.

To discriminate between the mitochondrial and peroxisomal pathways of FFA β-oxidation, we exploited the differential sensitivity of FACoA uptake across the organelle membrane for inhibition by etomoxir.[11] The oxidation of ^{3}H-palmitate measured in untreated cells was considered as total fatty acid oxidation. The portion of palmitate oxidation occurring in the mitochondria is irreversibly inhibited by a low concentration of etomoxir (1 μM). The peroxisomal component of palmitate oxidation is reversibly inhibited by a high dose (50 μM) of etomoxir. Residual palmitate oxidation is considered to occur by other undefined processes. Under baseline conditions, ~50% of palmitate oxidation occurs in mitochondria (FIG. 2), with the rest nearly equally divided between peroxisome and "other" undefined compartments.

Chronic treatment (2–4 days) of muscle cells with either troglitazone or LG100268 resulted in stimulation of palmitate oxidation (FIG. 3). On a molar basis, LG100268 was the more potent compound.[12] Combined treatment with submaximal concentrations of both drugs resulted in a synergistic stimulation of FFA oxidation

(FIG. 3). The increase in palmitate oxidation in response to troglitazone was due entirely to a change in the mitochondrial component (FIG. 4). Neither peroxisomal proliferation nor augmentation of peroxisomal β-oxidation of FFAs was observed after treatment with the PPARγ agonist. The RXR agonist, LG100268, stimulated both mitochondrial and peroxisomal FFA oxidation, although the major effect was on the mitochondrial component.

In summary, FFA oxidation in human skeletal muscle cells can be stimulated, both independently and in a synergistic manner, by agonists for PPARγ and RXR. Increased FFA disposal in muscle through augmented oxidation could reduce intramyocellular lipid accumulation. The abilities of such agents to improve glucose tolerance and insulin action may therefore involve effects on both glucose and FFA metabolism.

REFERENCES

1. KELLEY, D.E., et al. 1999. Skeletal muscle fatty acid metabolism in association with insulin resistance, obesity, and weight loss. Am. J. Physiol. **277:** E1130–E1141.
2. SIMONEAU, J-A., et al. 1999. Markers of capacity to utilize fatty acids in human skeletal muscle: relation to insulin resistance and obesity and effects of weight loss. FASEB J. **13:** 2051–2060.
3. KELLEY, D.E. & J-A. SIMONEAU. 1994. Impaired free fatty acid utilization by skeletal muscle in non-insulin-dependent diabetes mellitus. J. Clin. Invest. **94:** 2349–2356.
4. PHILLIPS, D.I.W., et al. 1996. Intramuscular triglyceride and muscle insulin sensitivity: evidence for a relationship in nondiabetic subjects. Metabolism **45:** 947–950.
5. PERSEGHIN, G., et al. 1999. Intramyocellular triglyceride content is a determinant of *in vivo* insulin resistance in humans. Diabetes **48:** 1600–1606.
6. HAMILTON, J.A. & F. KAMP. 1999. How are free fatty acids transported in membranes? Diabetes **48:** 2255–2269.
7. MCGARRY, J.D. & N.F. BROWN. 1997. The mitochondrial carnitine palmitoyltransferase system. Eur. J. Biochem. **244:** 1–14.
8. AUWERX, J. 1999. PPARgamma, the ultimate thrifty gene. Diabetologia **42:** 1033–1049.
9. HENRY, R.R., et al. 1995. Insulin action and glucose metabolism in non-diabetic control and NIDDM subjects: comparison using human skeletal muscle cell cultures. Diabetes **44:** 936–946.
10. CIARALDI, T.P., et al. 1995. Glucose transport in cultured human skeletal muscle cells: regulation by insulin and glucose in nondiabetic and non-insulin-dependent diabetes mellitus subjects. J. Clin. Invest. **96:** 2820–2827.
11. BHUIYAN, A.K.M., M.S.R. MURTHY & S.V. PANDE. 1994. Some properties of the malonyl-CoA sensitive carnitine long/medium chain acyltransferase activities of peroxisomes and microsomes of rat liver. Biochem. Mol. Biol. Int. **34:** 493–503.
12. CHA, B-S., et al. 2001. Peroxisome proliferator-activated receptor (PPAR) γ and retinoid X receptor (RXR) agonists have complementary effects on glucose and lipid metabolism in human skeletal muscle. Diabetologia **44:** 444–452.

An Increase in Peroxisomal Fatty Acid Oxidation Is Not Sufficient to Prevent Tissue Lipid Accumulation in hHTg Rats

J. UKROPEC,[a] I. KLIMEŠ,[a] D. GAŠPERÍKOVÁ,[a] E. DEMCÁKOVÁ,[a]
C. A. DREVON,[b] J. E. RESELAND,[b] AND E. ŠEBÖKOVÁ[a]

[a]*Diabetes and Nutrition Research Laboratory, Institute of Experimental Endocrinology, Slovak Academy of Sciences, Bratislava, Slovak Republic*

[b]*Institute for Nutrition Research, University of Oslo, Oslo, Norway*

ABSTRACT: We observed earlier that increased skeletal muscle lipid content in the hereditary hypertriglyceridemic (hHTg) rat is accompanied by a decline in plasma leptin. Leptin has recently been shown to enhance peripheral insulin sensitivity by decreasing the tissue triglyceride accumulation, possibly through regulation of fatty acid oxidation and lipogenesis. Thus, to test the hypothesis that insulin resistance and increased skeletal muscle lipid accumulation in hHTg rats are due to a defect in lipid catabolism, we measured mitochondrial and peroxisomal fatty acid oxidation and malonyl-CoA and acetyl-CoA carboxylase-2 content in skeletal muscles of these animals. In addition, we investigated possible molecular mechanisms responsible for the lower leptin levels in hHTg rats by measuring leptin and leptin-receptor (Ob-Ra) mRNA levels. We found the following: (1) in spite of a higher skeletal muscle malonyl-CoA content and an increased sensitivity of carnitine palmitoyltransferase-1 to malonyl-CoA, carnitine palmitoyltransferase-1 activity in muscle of hHTg rats was normal; (2) increased peroxisomal fatty acid oxidation did not seem to be sufficient to prevent the tissue lipid accumulation in these animals; (3) both lower leptin production by white adipose tissue and increased leptin uptake seem to be responsible for lower circulating leptin levels and therefore lower fatty acid catabolism.

KEYWORDS: fatty acid; lipid; oxidation; hHTg rats; skeletal muscle; leptin

INTRODUCTION

The hereditary hypertriglyceridemic (hHTg) rat is a nonobese animal model of hypertriglyceridemia and insulin resistance.[1–3] These animals are characterized by elevated lipogenesis and hepatic VLDL secretion,[4,5] and increased skeletal muscle triglyceride (TG) accumulation.[6] The aforementioned TG accumulation is proposed to play a causative role in the pathogenesis of insulin resistance.[7–11]

Address for correspondence: Elena Šebökövá, Ph.D., D.Sc., Diabetes and Nutrition Research Laboratory, Institute of Experimental Endocrinology, Slovak Academy of Sciences, Vlárska 3, 833 06 Bratislava, Slovak Republic. Voice/fax: (+4212) 54 77 26 87.
ueensebo@savba.savba.sk

It has been previously established that a defect in fatty acid oxidation leading to tissue TG accumulation may contribute to insulin resistance.[9–12] Fatty acids are essentially oxidized in mitochondria and in peroxisomes. Mitochondrial oxidation of the long-chain fatty acid is determined by the rate of its transport across the inner mitochondrial membrane, which is guarded by the malonyl-CoA inhibitory action on the carnitine palmitoyltransferase-1 (CPT-1).[9–16] Malonyl-CoA is generated by acetyl-CoA carboxylases (ACC1 and ACC2). It is likely that the ACC2—the skeletal muscle predominating isoform of this enzyme[17]—could play an important role for muscle malonyl-CoA production.[18,19] Peroxisomal fatty acid oxidation, however, is not inhibited by malonyl-CoA, and the oxidation rate in the peroxisomes is determined by enzyme activity of the acyl-CoA oxidase (AOX).[20,21] It is likely that the lipid-lowering effect of "energy-dissipating" peroxisomal fatty acid oxidation may be exerted only in cooperation with mitochondria.[22,23] Efficiency of mitochondrial oxidation of acyl-carnitines released after oxidation in peroxisomes is proportional to CPT-2 activity.[22–24] In an effort to test the hypothesis that insulin resistance and increased skeletal muscle lipid stores in hHTg rats are associated with a defect in lipid catabolism, we investigated the key regulatory points of mitochondrial and peroxisomal fatty acid oxidation.

Moreover, leptin has recently been shown to enhance peripheral insulin sensitivity by decreasing tissue TG accumulation, possibly through the regulation of fatty acid synthesis and oxidation.[7,25–27] We earlier observed that increased skeletal muscle lipid content in the hHTg rat is accompanied by a decline in plasma leptin concentration. Therefore, we investigated the potential molecular mechanism responsible for the low circulating leptin level in the hHTg rat by measuring the expression of the leptin and leptin-receptor (Ob-Ra) genes in adipose tissue, skeletal muscle, and hypothalamus.

The results presented here showed that increased peroxisomal fatty acid oxidation in the skeletal muscle does not seem to be sufficient to prevent the muscle lipid accumulation in hHTg animals. This is most likely due to a defect in CPT-2-dependent mitochondrial fatty acid oxidation. Higher muscle malonyl-CoA content and increased sensitivity of CPT-1 to malonyl-CoA were not sufficient to depress CPT-1 activity in hHTg rats. Lower leptin production in white adipose tissue and increased leptin uptake by tissues may be responsible for lower leptin levels in hHTg rats.

ANIMALS AND METHODS

Animals

All the experiments reported were approved by the Institute of Experimental Endocrinology Animal House Ethics Committee. Eight male rats of hHTg strain[1,2] and eight male Wistar Charles River rats (AnLab, Prague, Czech Republic) were housed in wire-mesh cages in a temperature ($22 \pm 2°C$) and light–controlled room (12-h light-dark cycle; lights off at 1800 h). The animals were fed a standard laboratory chow (Velaz, Prague, Czech Republic). Half of the control and hHTg rat groups were decapitated in the fed state, and the other half were subjected to a euglycemic hyperinsulinemic clamp (EHC). Selected tissues were immediately dissected and either used for fatty acid oxidation rate measurement or freezen in liquid nitrogen and stored at −70°C.

Euglycemic Clamp Studies

Animals were anesthetized and fitted with chronic artery and jugular cannulae.[28] The euglycemic hyperinsulinemic clamp studies were conducted 72 h after catheter implantation as previously described.[29,30]

Fatty Acid Oxidation Rate

Fatty acid oxidation rate was measured in a postnuclear fraction of gastrocnemius muscle as a conversion of [1-^{14}C]-palmitoyl-CoA (NEN, Boston) into acetyl-CoA. Quantity of the radioactive "acid-soluble product" was measured radiometrically.[31]

CPT-1 and CPT-2 Enzyme Activity

CPT-1 and CPT-2 enzyme activities were measured as described by Bremer.[32] Radioactive [1-^{14}C]-L-carnitine (NEN, Boston) was used as a reaction substrate, and production of labeled fatty acyl-carnitine was measured after its extraction to butanol.

AOX Activity

Enzyme activity of peroxisomal AOX was measured in tissue extract by an assay coupled with the β-oxidation multienzyme complex composed of 2-enoyl-acyl-CoA hydratase, 3-hydroxyacyl-CoA dehydrogenase, and 3-oxoacyl-CoA thiolase, (HDT, Asahi Chemical Industry, Ohito, Japan) and with [1-^{14}C]-palmitoyl-CoA as a reaction substrate.[21]

AOX Gene Expression

The relative abundance of AOX mRNA was determined by Northern blot. Briefly, RNA was size-fractionated by agarose gel electrophoresis and by capillary blotting transferred to membrane (HYBOND C-extra, Amersham, Buckinghamshire, United Kingdom). Abundance of the transcript of interest was determined by hybridization with the specific cDNA probe. Probes were radiolabeled by using PRIME-IT RmT (Stratagene, La Jolla, CA) and [α-^{32}P]-dCTP (Amersham, Buckinghamshire, United Kingdom). Results were corrected to G3PDH mRNA level.

Leptin and Leptin-Receptor Gene Expression

Rt-PCR reactions were carried out using a Gene Amp Rt-PCR kit (Perkin Elmer, Norwalk, CT), where 3 μg of total RNA was used to synthesize first-strand cDNA. The Rt reaction was carried out by MuLV-RT (42°C, 15 min). First-strand cDNA was then amplified by the AmpliTaq® DNA Polymerase (Perkin Elmer, Foster City, CA). Leptin mRNA in rat tissue was amplified using primers to murine leptin, estimated product size, 250 bp: 5′-AGC-AGT-GCC-TAT-CCA-GAA-AGT-3′, 5′-ATT-CTC-CAG-GTC-ATT-GGC-TAT-3′; Ob-Ra cDNA, estimated product size, 236 bp: 5′-ACA-CTG-TTA-ATT-TCA-CAC-CAG-AG-3′, 5′-AGT-CAT-TCA-AAC-CAT-AGT-TTA-GG-3′. G3PDH was used as a housekeeping gene, estimated product size,

452 bp: 5'-ACC-ACA-GTC-CAT-GCC-ATC-AC-3', 5'-TCC-ACC-ACC-CTG-TTG-CTG-TA-3' (Clontech Laboratories, Palo Alto, CA). cDNAs were amplified for 33 (leptin and Ob-Ra) and 28 (G3PDH) cycles in all the tissues studied (liver, skeletal muscle, white adipose tissue, and hypothalamus) using the following parameters: 94°C for 30 s, 60°C for 30 s, 72°C for 45 s, with a final extension at 72°C for 7 min. Specific incorporation of ^{32}P-dCTP was measured radiometrically and corrected to G3PDH mRNA level.

ACC Immunoprotein Level

The proteins from cytosolic fraction were resolved by SDS-PAGE and transferred to nitrocellulose membrane (HYBOND C-extra, Amersham, Buckinghamshire, United Kingdom). Proteins fixed to the membrane were hybridized with streptavidin-labeled horseradish peroxidase (Molecular Probes Europe, Leiden, Holland) and specific protein content was determined using a detection system based on enhanced chemiluminescence (ECL, Amersham, Buckinghamshire, United Kingdom).

TG

TG levels in blood and in muscle lipid extracts were measured by a spectrophotometric method using a specific, commercially available enzymatic set (Triglyceridy DST-P, DOT-Diagnostics, Prague, Czech Republic).

Statistical Analyses

Results were expressed as mean ± SEM. Differences between control and hHTg animals were evaluated by the unpaired, one-tailed t test at an overall significance threshold of $p < 0.05$.

RESULTS

In this study, we confirm our earlier observation that hypertriglyceridemia and increased skeletal muscle TG accumulation is accompanied by impaired *in vivo* insulin action in the hHTg rat (TABLE 1). The disturbance in skeletal muscle lipid oxidative metabolism was indicated by elevated muscle malonyl-CoA content (90%, $p < 0.001$), by increased CPT-1 sensitivity to malonyl-CoA (93%, $p < 0.01$), and by increased ACC-2 protein content (77%, $p < 0.05$) (TABLE 2). However, neither palmitoyl-CoA oxidation rate nor CPT-1 activity was lower in hHTg rats than in

TABLE 1. TG and *in vivo* insulin sensitivity in hHTg rats

	Control	hHTg
Serum TG (mmol/L)	2.4 ± 0.1^a	4.3 ± 0.2^b
Skeletal muscle TG (µmol/g)	1.7 ± 0.4^a	3.5 ± 0.3^b
Glucose infusion rate (mg/kg/min)	28.4 ± 5.8^a	20.2 ± 0.4^b

NOTE: Values without a common superscript are significantly different ($p < 0.05$). hHTg, hereditary hypertriglyceridemic rat; TG, triglyceride.

TABLE 2. Skeletal muscle malonyl-CoA content and fatty acid oxidation

	Control	hHTg
Malonyl-CoA content (nmol/g)	1.1 ± 0.2^a	2.1 ± 0.1^b
ACC-2 immunoprotein level (AU)	41 ± 8^a	72 ± 13^b
CPT-1 inhibition with malonyl-CoA (%)	31 ± 4^a	60 ± 2^b
CPT-1 activity (pmol/mg/min)	253 ± 19^a	297 ± 42^a
Palmitoyl-CoA oxidation rate (pmol/mg/min)	50 ± 16^a	52 ± 13^a
Ratio of CPT-2 over CPT-1 activity	2.5 ± 0.2^a	1.3 ± 0.1^b

NOTE: Values without a common superscript are significantly different ($p < 0.05$). ACC-2, acetyl-CoA carboxylase-2; AU, arbitrary units; CPT-1 & -2, carnitine palmitoyltransferase-1 & -2; hHTg, hereditary hypertriglyceridemic rat.

TABLE 3. Leptin and leptin-receptor gene expression

	Control	hHTg
Circulating leptin (ng/mL)	3.1 ± 0.4^a	1.4 ± 0.1^b
Leptin mRNA in adipose tissue (AU)	1.0 ± 0.14^a	0.3 ± 0.05^b
Ob-Ra mRNA in skeletal muscle (AU)	1.0 ± 0.11^a	2.5 ± 0.4^b
Ob-Ra mRNA in adipose tissue (AU)	1.0 ± 0.04^a	2.0 ± 0.2^b
Ob-Ra mRNA in hypothalamus (AU)	1.0 ± 0.08^a	1.5 ± 0.06^b

NOTE: Values without a common superscript are significantly different ($p < 0.05$). AU, arbitrary units; hHTg, hereditary hypertriglyceridemic rat; Ob-Ra, short form of leptin receptor.

controls (TABLE 2). Surprisingly, we observed an increase of peroxisomal fatty acid oxidation (AOX enzyme activity, 155%, $p < 0.05$; AOX mRNA, 65%, $p < 0.005$) in hHTg rats (FIG. 1). The ratio between CPT-2 (an enzyme essential for mitochondrial acyl-carnitine oxidation) and CPT-1 enzyme activities was about 45% ($p < 0.005$) lower in hHTg rats when compared to controls (TABLE 2).

We also found that the decline of circulating leptin (54%, $p < 0.005$) levels in the hHTg rat was likely due to a decreased gene expression for leptin in the epididymal white adipose tissue (WAT) (70%, $p < 0.05$). The increased uptake of circulating leptin by its receptors in various tissues, as evidenced by increased gene expression for the short form of the leptin receptor (TABLE 3), seemed to contribute to lower levels of circulating leptin as well (TABLE 3).

DISCUSSION

Elevated skeletal muscle malonyl-CoA content and production (as suggested by higher ACC-2 immunoprotein level) and high CPT-1 sensitivity to malonyl-CoA both suggest a possible downregulation of the mitochondrial β-oxidation in hHTg rats. Direct measurement of the fatty acid oxidation rate and CPT-1 activity, though, did not confirm the above assumption. However, CPT-1 activity seemed to be lower

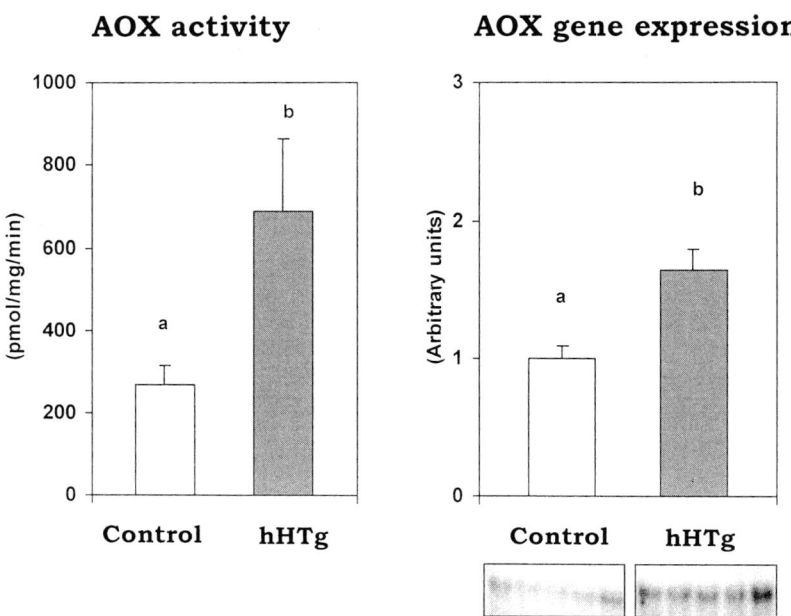

FIGURE 1. Activity and gene expression for peroxisomal acyl-CoA oxidase (AOX). Values without a common superscript are significantly different ($p < 0.05$). hHTg: hereditary hypertriglyceridemic rat.

than expected from excessive lipid availability in hHTg animals. Based on these results, it seems likely that the tissue TG accumulation and hypertriglyceridemia in hHTg rats develop due to a higher fatty acid synthesis[5] and relatively low mitochondrial fatty acid oxidation. Recent data from our laboratory showed that the hypertriglyceridemia might also be related to a decrease in hepatic mitochondrial CPT-1 activity in hHTg rats (unpublished data). A surprising finding was the increase in peroxisomal fatty acid oxidation in the gastrocnemius muscle of hHTg animals. An increase of the "energy-dissipating" peroxisomal oxidation could be related to the lower body weight and the relative resistance to diet-induced obesity in these animals.[5] In addition, peroxisomal fatty acid oxidation may be effectively coupled to mitochondrial β-oxidation.[22,23] An effective coupling would facilitate the mitochondrial oxidation of short-chain acyl-carnitines released from the peroxisomes. In particular, short-chain acyl-carnitines can enter the mitochondrial matrix, bypassing the CPT-1 regulatory step. Once inside the mitochondria, acyl-carnitines are then converted to acyl-CoAs (a reaction catalyzed by CPT-2) and completely oxidized.[22-24] However, the lower ratio of CPT-2/CPT-1 activities indicated that hHTg rats may have limited capacity for mitochondrial oxidation in skeletal muscle by the aforementioned mechanism. The latter may substantially blunt a possible lipid-lowering effect of peroxisomal fatty acid oxidation in hHTg rats.

It has been previously shown that chronic administration of leptin severely depletes nonadipose tissue TG content.[7,25-28] We found that the decline in circulating leptin in the hHTg rat is most likely due to a decreased leptin gene expression in WAT, that is, lower "leptin production". This decrease in circulating leptin was further supported by an increase of gene expression for the leptin receptor, that is, "leptin uptake" in WAT, skeletal muscle, and hypothalamus. Leptin has also been recently hypothesized to protect nonadipocytes from the nonoxidative metabolic products of long-chain fatty acids by increasing β-oxidation and reducing lipogenesis.[7] In addition, it was proposed that TG content is a useful index of overall nonoxidative metabolism. Thus, it is feasible to speculate that the lack of the effect of leptin may contribute to tissue TG accumulation and insulin resistance in hHTg rats.

Thus, the following can be concluded:

(1) In spite of a higher skeletal muscle malonyl-CoA content and an increased sensitivity of CPT-1 to malonyi-CoA, no change in the CPT-1 activity in muscle of hHTg rats was found. A decreased ratio of CPT-2 over CPT-1 in skeletal muscle of hHTg rats might be indicative of a lower mitochondrial fatty acid oxidation via the CPT-2 pathway.
(2) Moreover, raised peroxisomal fatty acid oxidation does not seem to be sufficient to prevent the tissue lipid accumulation in these animals.
(3) Both the lower leptin production in WAT and the increased leptin uptake in several tissues seem to be responsible for lower leptin levels in circulation, a potential cause of lower fatty acid catabolism in hHTg rats.

ACKNOWLEDGMENTS

This work was supported by a grant from the Slovak Diabetes Society, a European Commission grant in a frame of COST Action B17, and a VEGA grant (no. 2/7210/21). We are grateful to Shigeyuki Imamura, Asahi Chemical Industry, Japan, for generously providing the HDT multienzyme complex. The skillful technical assistance of Alica Mitkova and Anne Randi Alvestad is greatly appreciated.

REFERENCES

1. VRÁNÁ, A. & L. KAZDOVÁ. 1990. The hereditary hypertriglyceridemic nonobese rat: an experimental model of nonobese hypertriglyceridemia. Transplant. Proc. **22:** 2579.
2. KLIMEŠ, I., et al. 1995. The hereditary hypertriglyceridemic rat, a new model of the insulin resistance syndrome. In Lessons from Animal Diabetes, pp. 271–283. Smith-Gordon. United Kingdom.
3. GAŠPERÍKOVÁ, D., E. ŠEBÖKOVÁ & I. KLIMEŠ. 1996. Intracellular insulin signaling, glucose transport, and in vivo insulin action in selected rat models of insulin resistance. Endocr. Regul. **30:** 207–228.
4. VRÁNA, A., et al. 1990. Nutrients and glucose homeostasis: effects of the type of dietary carbohydrates and the n-3 fatty acids on insulin secretion and action. In Insulin and the Cell Membrane, pp. 397–411. Harwood Academic Pub. New York.
5. ŠEBÖKOVÁ, E., et al. 1996. Regulation of gene expression for lipogenic enzymes in the liver and adipose tissue of hereditary hypertriglyceridemic, insulin-resistant rats: effect of dietary sucrose and marine fish oil. Biochim. Biophys. Acta **1303:** 56–62.
6. ŠEBÖKOVÁ, E. & I. KLIMEŠ. 1997. Molecular and cellular determinants of triglyceride availability. Ann. N.Y. Acad. Sci. **827:** 200–214.

7. UNGER, R.H. & Y-T. ZHOU. 2001. Lipotoxicity of β-cells in obesity and in other causes of fatty acid spillover. Diabetes **50:** S118–S121.
8. KLIMEŠ, I., *et al.* 1991. Dietary fish oil in hyperlipidemia and impaired insulin action— the experimental approach: a minireview. *In* Lipoprotein and Atherosclerosis, pp. 56–62. Fisher Verlag. Jena.
9. RUDERMAN, N.B., *et al.* 1997. Lipid abnormalities in muscle of insulin-resistant rodents: the malonyl-CoA hypothesis. Ann. N.Y. Acad. Sci. **827:** 221–230.
10. RUDERMAN, N.B., *et al.* 1999. Malonyl-CoA, fuel sensing, and insulin resistance. Am. J. Physiol. **276:** E1–E18.
11. KELLEY, D.E. & B.H. GOODPASTER. 2001. Skeletal muscle triglyceride: an aspect of regional adiposity and insulin resistance. Diabetes Care **24:** 933–941.
12. SAHA, A.K., *et al.* 1999. Cytosolic citrate and malonyl-CoA regulation in rat muscle *in vivo*. Am. J. Physiol. **276:** E1030–E1037.
13. DOBBINS, R.L., *et al.* 2001. Prolonged inhibition of muscle carnitine palmitoyl-transferase-1 promotes intramyocellular lipid accumulation and insulin resistance in rats. Diabetes **50:** 123–130.
14. MCGARRY, J.D., *et al.* 1983. Observations on the affinity for carnitine and malonyl-CoA sensitivity of carnitine palmitoyl-transferase-I in animal and human tissues. Biochem. J. **214:** 21–28.
15. SIDOSSIS, L.S., *et al.* 1996. Glucose plus insulin regulate fat oxidation by controlling the rate of fatty acid entry into the mitochondria. J. Clin. Invest. **98:** 2244–2250.
16. MCGARRY, J.D. & N.F. BROWN. 2000. Reconstitution of purified, active, and malonyl-CoA-sensitive rat liver carnitine palmitoyltransferase-I: relationship between membrane environment and malonyl-CoA sensitivity. Biochem. J. **349:** 179–187.
17. HA, J., *et al.* 1996. Cloning of human acetyl-CoA carboxylase-beta and its unique features. Proc. Natl. Acad. Sci. U.S.A. **93:** 11466–11470.
18. ABU-ELHEIGA, L., *et al.* 2001. Continuous fatty acid oxidation and reduced fat storage in mice lacking acetyl-CoA carboxylase 2. Science **291:** 2613–2616.
19. ABU-ELHEIGA, L., *et al.* 2000. The subcellular localization of acetyl-CoA carboxylase 2. Proc. Natl. Acad. Sci. U.S.A. **97:** 1444–1449.
20. IMAMURA, S., *et al.* 1990. Purification of the multienzyme complex for fatty acid oxidation from *Pseudomonas fragi* and reconstitution of the fatty acid oxidation system. J. Biochem. **107:** 184–189.
21. SOURI, M., T. AOYAMA & T. HASHIMOTO. 1994. A sensitive assay of acyl-CoA oxidase by coupling with beta-oxidation multienzyme complex. Anal. Biochem. **221:** 362–367.
22. MADSEN, L., *et al.* 1999. Mitochondrial 3-hydroxy-3methylglutaryl CoA synthase and carnitine palmitoyltransferase-2 as potential control sites for ketogenesis during mitochondrion and peroxisome proliferation. Biochem. Pharmacol. **57:** 1011–1019.
23. WANDERS, R.J.A., *et al.* 2001. Peroxisomal fatty acid α- and β-oxidation in humans: enzymology, peroxisomal metabolite transporters, and peroxisomal diseases. Biochem. Soc. Trans. **29:** 250–267.
24. AOYAMA, T., *et al.* 1998. Altered constitutive expression of fatty acid–metabolizing enzymes in mice lacking the peroxisome proliferator-activated receptor α. J. Biol. Chem. **273:** 5678–5684.
25. MUOIO, D.M., *et al.* 1997. Leptin directly alters lipid partitioning in skeletal muscle. Diabetes **46:** 1360–1363.
26. SHIMABUKURO, M., *et al.* 1997. Direct antidiabetic effect of leptin through triglyceride depletion of tissues. Proc. Natl. Acad. Sci. U.S.A. **94:** 4637–4641.
27. ZHOU, Y.T. 1998. Enhanced *de novo* lipogenesis in the leptin-unresponsive pancreatic islets of prediabetic Zucker diabetic fatty rats: role in the pathogenesis of lipotoxic diabetes. Diabetes **47:** 1904–1908.
28. KOOPMANS, S.J., *et al.* 1992. *In vivo* insulin responsiveness for glucose uptake at eu- and hyperglycemic levels in diabetic rats. Biochim. Biophys. Acta **1115:** 2130–2138.
29. KRAEGEN, E.W., *et al.* 1983. *In vivo* insulin sensitivity determined by euglycemic clamp. Am. J. Physiol. **245:** E1–E7.
30. KLIMEŠ, I., *et al.* 1998. The effect of the new oral hypoglycemic agent A-4166 on glucose turnover in the high fat diet–induced and/or in the hereditary insulin resistance of rats. Arch. Physiol. Biochem. **106:** 325–332.

31. WILLUMSEN, N., *et al.* 1993. Docosahexaenoic acid shows no triglyceride-lowering effects, but increases the peroxisomal fatty acid oxidation in liver of rats. J. Lipid Res. **34:** 13–22.
32. BREMER, J. 1981. The effect of fasting on the activity of liver carnitine palmitoyltransferase and its inhibition by malonyl-CoA. Biochim. Biophys. Acta **665:** 628–631.

Adiposity and the Development of Diabetes in Mouse Genetic Models

LESLIE P. KOZAK AND MARTIN ROSSMEISL

Pennington Biomedical Research Center, Baton Rouge, Louisiana 70808, USA

ABSTRACT: While it is certain from the analysis of transgenic models of lipodystrophy that a critical mass of adipose tissue is necessary to prevent the development of diabetes, the reasons why diabetes develops in one obese individual and not in another need to be further investigated. This is also one of the major questions associated with the current obesity epidemic and the development of the metabolic syndrome. The severe lipodystrophic models do not really address this big problem. In this presentation, we point out that models exist among inbred strains of mice that can contribute towards finding answers to the diabetes found in the metabolic syndrome. The differences responsible for the phenotypic variations are undoubtedly multifactorial and involve many genes, but there are powerful genetic resources to investigate these problems.

KEYWORDS: adiposity; diabetes; transgenic; mouse; lipodystrophy

INTRODUCTION

There is a dynamic relationship between adiposity and energy metabolism that is extremely complex and varied. While in some individuals the primary perturbation that leads to diabetes and obesity may be impaired glucose tolerance, in others the development of adiposity sets in motion changes in energy and lipid metabolism that lead to insulin resistance and diabetes. Although the accumulation of lipid into muscle, liver, and pancreas appears to be a condition for the development of insulin resistance and diabetes, the mechanisms that lead to this pattern of calorie partitioning are not understood. Recently, Danforth proposed that type II diabetes is the result of the inability of the adipose organ to expand to accommodate excess energy and that type II diabetes in the centrally obese, in spite of their unlikely phenotype, is a form of lipodystrophy.[1] In other words, if the adipocyte cell number could expand to store excess calories rather than the adipocyte expanding with fat to the point that it became insulin-resistant, then the deposition of excess calories to other sites, such as the liver and muscle, would be avoided.

One line of evidence in support of the adipocyte morphology mechanism is found in a study by Weyer *et al.* in which adipocyte size in obese Pima Indians was correlated with the incidence of impaired glucose tolerance and overt diabetes.[2] Statistical

Address for correspondence: Leslie P. Kozak, Pennington Biomedical Research Center, 6400 Perkins Road, Baton Rouge, LA 70808. Voice: 225-763-2771; fax: 225-763-3030.
kozaklp@pbrc.edu

analysis of the data suggested that individuals with smaller adipocytes had less impaired glucose tolerance. While in principle one should be able to test this hypothesis simply by characterizing the morphology of adipocytes in fat depots of diabetic and nondiabetic individuals, in fact it is not so simple to test the hypothesis, particularly when the progression to diabetes varies between two apparently similar obese individuals. A major limitation stems from the difficulty in being able experimentally to establish rigorously the minimal difference in adipocyte size and number that initiates the diabetic condition. Accordingly, the best evidence showing that reduced adiposity causes the development of diabetes comes from the phenotypes of transgenic mice with severe lipodystrophy.

DIABETES IN TRANSGENIC MODELS OF LIPODYSTROPHY

TABLE 1 lists four transgenic models and one knockout model that have reductions in adipocyte mass. The most severe lipodystrophies are observed in aP2-A-ZIP-F1 and aP2-SREBP-1c transgenic mice.[3,4] Since the phenotypes of these two transgenic strains are comparable, we describe only the former. The aP2-A-ZIP-F1 transgenic mice express a dominant negative protein, selectively in fat, that forms heterodimers with B-ZIP transcription factors. Strong interactions with C/EBP and JUN, factors that regulate early steps in adipocyte differentiation, result in mice with severe lipodystrophy. It is known that mice in which the C/EBPα gene has been inactivated by gene targeting lack adipose tissue, but they die shortly after birth.[5] These neonatal mice have severe hypoglycemia, presumably because of the effects of the C/EBPα deficiency on gluconeogenesis in the liver. The selective expression of A-ZIP in fat tissue avoids premature death, thereby providing animals that are severely deficient in adipose tissue. In effect, Moitra *et al.* have designed a novel genetic approach to create a mouse model in which transcription factor deficiency is established in specific tissues. The effect of lipodystrophy in the transgenic mice is a severe form of diabetes that is characterized by hyperglycemia, hyperinsulinemia, hypertriglyceridemia, elevated free fatty acid levels, and insulin resistance. Proof that this phenotype is due to the absence of adequate fat depots came from an experiment in which fat depots from nontransgenic mice of the same strain were transplanted subcutaneously into the aP2-A-ZIP-F1 mice.[6] Essentially, a complete reversal of the diabetic phenotype was achieved within 2 weeks of receiving the transplants and lasted for at least 13 weeks. These experiments clearly demonstrated that diabetes develops in animals with a severe deficiency of adipose tissue.

The aP2-DTA transgenic mice with a low level of the transgene expression are viable and show an age-dependent loss of fat tissue that begins at about 5 months of age and results in over 90% loss of adipose tissue by 8 months of age.[7,8] By 8 months of age, the mice also show frank diabetes with characteristics similar to that of aP2-A-ZIP-F1 and aP2-SREBP-1c mice.[8] A second, more mild form of lipodystrophy occurs in transgenic mice expressing the cytoplasmic glycerol-3-phosphate dehydrogenase gene (*Gdc-1*) from its own promoter. In this model, the pattern of expression of the transgene characteristic of the endogenous gene is maintained, but expression is very high, particularly in brown fat.[9] Brown fat is hypertrophied with reduced Ucp1 expression, but fully functional when mice are exposed to the cold. By 10 months of age, the amount of white fat is as low as in the aP2-DTA mouse, that

TABLE 1. Models of lipodystrophy

Transgene	Molecular mechanism	Adiposity phenotype	Reference
aP2-A-ZIP-F1	Dominant negative protein prevents the binding of B-ZIP transcription factors (e.g., C/EBP and JUN families)	aP2-A-ZIP-F1 mice have no fat; C/EBPα KO mice die shortly after birth and have no discernible fat	3, 5
aP2-SREBP-1c	Sterol regulatory element binding protein-1c activates promoters of genes involved in cholesterol and fatty acid biosynthesis	Overexpression of SREBP-1c in 3T3-L1 cells promotes adipocyte differentiation; in adipose tissue, it paradoxically inhibits WAT differentiation, but increases BAT differentiation	4
aP2-DTA	Expression of diphtheria toxin A chain causes cell death in target cells	Prevention of chemically induced obesity	7, 8
Gdc-1	Overexpression of NAD-glycerol-P dehydrogenase causes increased thermogenesis by substrate cycling	Repression of WAT expression; BAT hypertrophy; tg mice on *db/db* background are resistant to cold	9; Kozak, unpublished results
Hmgic	High mobility group proteins are essential components of the enhancesome	Reduced adipocyte cellularity that prevents obesity caused by defects in leptin signaling or a high-fat diet	10

is, adipose tissues are reduced by 95%,[9] but there is no evidence of diabetes when the animal is on a common inbred background (unpublished observations). We interpret the phenotype as indicating that overexpression of Gdc-1 sets in motion thermogenic substrate cycling that is centered around the glycerol phosphate cycle that consumes fat, but without destroying the adipocyte population. This thermogenic cycle is capable of reducing reliance on nonshivering thermogenesis to protect the animal from the cold at ambient temperature.

The last model of reduced adiposity is found in mice in which the Hmgic gene has been inactivated.[10] This mouse has a severe depletion of adipocytes; however, a comparably severe reduction in body size indicates that the effects of the mutant gene are not restricted to adipose tissue. It is not known how the reduction of adiposity in this mouse affects glucose homeostasis.

The phenotypes of transgenic mice with severe lipodystrophies confirm the vital role of adipose tissue in the etiology of the accompanying diabetes. Furthermore, the ability of the transplanted fat depots to reverse the diabetes also suggests a therapeutic strategy for treatment of lipodystrophic humans. It is questionable, though, whether one can extrapolate from a mechanism of diabetes that occurs in lipodystrophic humans and transgenic animals to proposing that the hyperglycemia and insulin resistance of type II diabetes in general are due to variation in the ratios of small to large adipocytes. Nonetheless, this is an important concept and should be tested with additional genetically controlled models. However, the question we are asking needs to be modified to address the more important problem: that is, why do some obese individuals develop diabetes, but others do not? Is the accumulation of triglycerides in liver, muscle, and pancreas due to differences in adipocyte numbers or is there some other mechanism underlying this phenotype? To address these questions experimentally, it is important to find genetic models in which animals with similar high levels of obesity vary in the development of diabetes. Genetic and environmental variation among humans is simply too great to permit any real test of the relation between the number and size of the adipocyte pool to the development of diabetes.

DIABETES IN INBRED STRAINS OF MICE

C57BL/6J (B6) versus C57BL/Ks

Models exist in the mouse that can test the question of adipocyte size and number in the etiology of diabetes, as well as other mechanisms leading to diabetes. The enormous influence of different genetic backgrounds of common inbred strains in the development of diabetes was first described by Coleman and Hummel when they showed that the difference in the diabetic phenotype of *lep* and *lepr* homozygotes was not due to the phenotype of the mutant genes, but rather to the genetic background of B6 and C57BL/Ks strains upon which the *lep* and *lepr* mutants resided.[11] The genetic basis of the strain differences has not been defined, although the C57BL/Ks has increased pancreatic beta cell destruction as a consequence of the stress caused by the uncontrolled food intake.[12]

B6 versus BTBR

More recently, an interesting model involving B6 and the BTBR strain has been described.[13] Both strains have normal insulin responsiveness on a basal diet (6.5% fat); however, F1 progeny produced severe insulin resistance, particularly on a 15% fat diet. The genetic basis of this background difference is not known, but is likely to be multifactorial.

NON versus NZO

One of the most important problems in the development of type II diabetes has to do with the fact that obesity does not always lead to diabetes. Why is it that some individuals develop diabetes and others do not? A mouse model that could provide insight into this problem is the NZO strain. NZO mice develop both obesity and diabetes spontaneously when fed a standard low-fat laboratory chow. Variation among individual mice occurs for body weight and blood glucose and immunoreactive insulin levels; however, the range of variation is higher for diabetes than obesity (TABLE 2). All animals have excessive body weight; however, some have normal blood glucose and insulin levels and others are characteristically diabetic. Accordingly, the NZO mouse by itself provides a model in which variation in both obesity and diabetes must be dependent on environmental factors. However, in order to

TABLE 2. Diabetogenic subphenotypes in obese NZO and (NZO × NON) F1 mice

Subphenotypes	36-week BW (g)	36-week BG (mg/dL)	36-week IRI (µU/mL)
NON			
BG normal, IRI normal (20)	47.5 ± 0.8	158 ± 9	29 ± 4
NZO			
BG normal, IRI high (13)	57.8 ± 1.5	189 ± 10	92 ± 16
BG high, IRI high (16)	66.4 ± 1.3	267 ± 15	110 ± 16
BG high, IRI normal (7)	67.3 ± 2.7	366 ± 12	60 ± 22
BG very high, IRI low (3)	40.7 ± 6.1	615 ± 64	15 ± 8
(NZO × NON) F1			
BG normal, IRI high (5)	59.6 ± 2.5	239 ± 27	73 ± 22
BG high, IRI high (23)	64.7 ± 0.5	337 ± 12	163 ± 12
BG high, IRI normal (13)	66.6 ± 0.9	439 ± 13	70 ± 9
BG very high, IRI low (11)	58.2 ± 1.3	509 ± 27	22 ± 4
(NON × NZO) F1			
BG normal, IRI high (7)	64.6 ± 2.3	208 ± 30	109 ± 16
BG high, IRI very high (9)	69.7 ± 1.2	314 ± 24	297 ± 35
BG high, IRI high (45)	68.8 ± 0.6	356 ± 12	162 ± 13
BG very high, IRI normal (8)	62.9 ± 2.0	538 ± 38	32 ± 4

NOTE: Data are means ± SD. Numbers of animals analyzed are given in parentheses. Terms: BG, blood glucose; IRI, immunoreactive insulin. Adapted from: Leiter, E.H., *Diabetes*, 1998.

identify metabolic pathways affecting the development of obesity and diabetes, an intercross was made between the NZO strain and NON, a strain that has a high body weight, although lower than those of the NZO strain, and normal blood glucose and immunoreactive insulin levels. Animals from both the F1 and F2 generations showed a range of phenotypes described in TABLE 2, but with more variation in the F2 than the F1 generation. The (NZO × NON) F1 mice were particularly interesting since body weights were essentially indistinguishable, while a large range of values existed for both blood glucose and immunoreactive insulin (TABLE 2). The increased variance in the phenotypes of the F2 progeny suggested that variant genes were segregating in this generation to affect the observed phenotypes. Accordingly, a genome-wide scan of the F2 progeny with polymorphic microsatellite chromosome markers identified three quantitative trait loci (QTL) with highly significant LOD scores: *Nidd1* on Chr 4, *Nidd2* on Chr 18, and *Nidd3* on Chr 11.

The adiposity phenotype in the NZO/NON system has been characterized only with respect to total body weight. However, it is clear that the NZO strain itself and F1 and F2 generations could provide systems to determine the relationship of adipocyte number and size to the development of hyperglycemia and insulin resistance. In addition, the analysis of lipid deposition in liver, muscle, and pancreas would establish whether variation in this parameter correlated with insulin resistance. An additional advantage is that it is also possible to evaluate the influence of environmental factors that affect the diabetic phenotype by analysis of NZO and (NZO × NON) F1 mice. Similarly, analysis of the F2 generation would provide an opportunity to identify genes that are associated with a causal relationship between adipocyte morphology and type II diabetes.

B6 versus A/J

A second diabesity model, based on the B6 and A/J strains, emerged from efforts to use genetic variation among inbred strains to analyze the development of obesity and diabetes that was dependent on a high-fat diet.[14] While A/J mice are resistant to the development of both obesity and diabetes when fed a high-fat diet, B6 mice become obese and become mildly insulin-resistant and diabetic. Efforts have been made to link the development of the obesity and diabetes to differences in leptin expression,[15] regulation of Ucp1 and Ucp2,[16] and induction of brown adipocytes in white fat depots.[17] Differences in the induction of Ucp2 between B6 and A/J mice were not reproduced by others[18] and the capacity for inducing brown adipocytes in white fat depots by adrenergic stimulation does not influence the development of obesity by high-fat diets in the absence of an adrenergic stimulation.[19]

SWR/J versus AKR/J

Independently, West and colleagues initiated a QTL analysis for susceptibility to the development of obesity by the high-fat diet.[20] This work was based on previous experiments with a dietary obesity in nine inbred strains of mice, including the B6 and AKR/J, both of which were susceptible to obesity.[21] The analysis of intercrosses between SWR/J and AKR/J strains showed that the sensitivity to dietary obesity was determined by genetic variation at several loci.[20] However, their work did not evaluate whether there were differences in the development of diabetes among these strains. When we became interested in the question of why two equally obese individuals varied in the subsequent development of diabetes, it was obvious that the inbred strains of mice, particularly the B6 and AKR/J combination, could provide a model system to address this issue.

Work in progress using obesity-prone B6 and AKR/J mouse strains has established the following features:

(1) Both B6 and AKR/J develop a comparable level of obesity on a high-fat diet (58 kcal%).
(2) Hyperglycemia is present only in the B6 strain after 8 weeks on the high-fat diet. The B6 mice have impaired glucose tolerance. AKR/J mice also have a mildly impaired glucose intolerance that is less than that observed in the B6 strain.
(3) No differences in the morphological features of the adiposity exist between B6 and AKR/J mice; in other words, the cellularity of adipose tissue in B6 and AKR/J mice is similar.
(4) B6 mice showed increased deposition of lipid in muscle and liver.
(5) Glut4 is reduced in the plasma membrane of B6 mice.

Our current working hypothesis is that the increased hyperglycemia appears to be due to increased insulin resistance as evidenced by Glut4 levels in the muscle. The role of an increase in lipid deposition in the development of the insulin resistance needs to be evaluated.

CONCLUDING REMARKS

While it is certain from the analysis of transgenic models of lipodystrophy that a critical mass of adipose tissue is necessary to prevent the development of diabetes, the reasons why diabetes develops in one obese individual and not in another need to be further investigated. This is also one of the major questions associated with the current obesity epidemic and the development of the metabolic syndrome. The severe lipodystrophic models do not really address this big problem. In this presentation, we have pointed out that models exist among inbred strains of mice that can contribute towards finding answers to the diabetes found in the metabolic syndrome. The differences responsible for the phenotypic variations are undoubtedly multifactorial and involve many genes, but there are powerful genetic resources to investigate these problems. Recombinant inbred lines, recombinant congenics, and congenic and consomic lines are readily available to resolve the genetics not only for diabetes, but also for other diseases of the metabolic syndrome. (For further info, see http://www.informatics.jax.org/.)

REFERENCES

1. DANFORTH, E., JR. 2000. Failure of adipocyte differentiation causes type II diabetes mellitus? Nat. Genet. **26:** 13.
2. WEYER, C., J.E. FOLEY, C. BOGARDUS *et al.* 2000. Enlarged subcutaneous abdominal adipocyte size, but not obesity itself, predicts type II diabetes independent of insulin resistance. Diabetologia **43:** 1498–1506.
3. MOITRA, J., M.M. MASON, M. OLIVE *et al.* 1998. Life without white fat: a transgenic mouse. Genes Dev. **12:** 3168–3181.
4. SHIMOMURA, I., R.E. HAMMER, J.A. RICHARDSON *et al.* 1998. Insulin resistance and diabetes mellitus in transgenic mice expressing nuclear SREBP-1c in adipose tissue: model for congenital generalized lipodystrophy. Genes Dev. **12:** 3182–3194.
5. WANG, N.D., M.J. FINEGOLD, A. BRADLEY *et al.* 1995. Impaired energy homeostasis in C/EBP alpha knockout mice. Science **269:** 1108–1112.
6. GAVRILOVA, O., B. MARCUS-SAMUELS, D. GRAHAM *et al.* 2000. Surgical implantation of adipose tissue reverses diabetes in lipoatrophic mice. J. Clin. Invest. **105:** 271–278.
7. ROSS, S.R., R.A. GRAVES & B.M. SPIEGELMAN. 1993. Targeted expression of a toxin gene to adipose tissue: transgenic mice resistant to obesity. Genes Dev. **7:** 1318–1324.
8. BURANT, C.F., S. SREENAN, K-I. HIRANO *et al.* 1997. Troglitazone action is independent of adipose tissue. J. Clin. Invest. **100:** 2900–2908.
9. KOZAK, L.P., U.C. KOZAK & G.T. CLARKE. 1991. Abnormal brown and white fat development in transgenic mice overexpressing glycerol 3-phosphate dehydrogenase. Genes Dev. **5:** 2256–2264.
10. ANAND, A. & K. CHADA. 2000. *In vivo* modulation of Hmgic reduces obesity. Nat. Genet. **24:** 377–380.
11. COLEMAN, D.L. & K.P. HUMMEL. 1973. The influence of genetic background on the expression of the obese (ob) gene in the mouse. Diabetologia **9:** 287–293.
12. LEITER, E.H. 1989. The genetics of diabetes susceptibility in mice. FASEB J. **3:** 2231–2241.

13. RANHEIM, T., C. DUMKE, K.L. SCHUELER *et al.* 1997. Interaction between BTBR and C57BL/6J genomes produces an insulin resistance syndrome in (BTBR × C57BL/6J) F1 mice. Arterioscler. Thromb. Vasc. Biol. **17:** 3286–3293.
14. SURWIT, R.S., C.M. KUHN, C. COCHRANE *et al.* 1988. Diet-induced type II diabetes in C57BL/6J mice. Diabetes **37:** 1163–1167.
15. SURWIT, R.S., A.E. PETRO, P. PAREKH *et al.* 1997. Low plasma leptin in response to dietary fat in diabetes- and obesity-prone mice. Diabetes **46:** 1516–1520. [Erratum. 1997 (November). Diabetes **46**(11)**:** 1920.]
16. SURWIT, R.S., S. WANG, A.E. PETRO *et al.* 1998. Diet-induced changes in uncoupling proteins in obesity-prone and obesity-resistant strains of mice. Proc. Natl. Acad. Sci. U.S.A. **95:** 4061–4065.
17. COLLINS, S., K.W. DANIEL, A.E. PETRO *et al.* 1997. Strain-specific response to beta 3-adrenergic receptor agonist treatment of diet-induced obesity in mice. Endocrinology **138:** 405–413.
18. GONG, D.W., Y. HE & M.L. REITMAN. 1999. Genomic organization and regulation by dietary fat of the uncoupling protein 3 and 2 genes. Biochem. Biophys. Res. Commun. **256:** 27–32.
19. GUERRA, C., R.A. KOZA, H. YAMASHITA *et al.* 1998. Emergence of brown adipocytes in white fat in mice is under genetic control: effects on body weight and adiposity. J. Clin. Invest. **102:** 412–420.
20. WEST, D.B., J. GOUDEY-LEFEVRE, B. YORK *et al.* 1994. Dietary obesity linked to loci on chromosomes 9 and 15 in a polygenic mouse model. J. Clin. Invest. **94:** 1410–1416.
21. WEST, D.B., C.N. BOOZER, D.L. MOODY *et al.* 1992. Dietary obesity in nine inbred strains of mice. Am. J. Physiol. **262:** R1025–R1032.

Modulation of Lipid Metabolism by Energy Status of Adipocytes

Implications for Insulin Sensitivity

JAN KOPECKÝ, PAVEL FLACHS, KRISTINA BARDOVÁ, PETR BRAUNER, TOMÁŠ PRAŽÁK, AND JANA ŠPONAROVÁ

Department of Adipose Tissue Biology, Institute of Physiology, Academy of Sciences of the Czech Republic, Prague, Czech Republic

ABSTRACT: It is becoming evident that insulin resistance of white adipose tissue is a major factor underlying the cardiovascular risk of obesity. Impaired fat storage rather than altered glucose metabolism in adipocytes probably contributes to development of insulin resistance in muscle and other tissues, in particular via increased delivery of nonesterified fatty acids into circulation. Lipid metabolism of adipose tissue is affected by the energy status of fat cells. *In vitro* experiments indicated the dependence of both lipogenesis and lipolysis on ATP levels in adipocytes. Thus, respiratory uncoupling in adipocytes that results in stimulation of energy dissipation and depression of ATP synthesis may contribute to the control of lipid metabolism, adiposity, and insulin sensitivity. This notion is supported by the expression of UCPs in adipocytes, for example, UCP2, UCP5, as well as some protonophoric anion transporters, and by induction of UCP1 and UCP3 in white fat by pharmacological treatments that reduce adiposity. A negative correlation between expression of UCPs in adipocytes and accumulation of white fat was also found. Expression of UCP1 from the adipose-specific promoter in the aP2-*Ucp1* transgenic mice mitigated obesity induced by genetic or dietary factors. The obesity resistance, accompanied by respiratory uncoupling in adipocytes and increased energy expenditure, resulted from ectopic expression of UCP1 in white, but not brown fat. Probably due to depression of the ATP/ADP ratio, both fatty acid synthesis and lipolytic action of norepinephrine in adipocytes of transgenic mice were relatively low. Expression of regulatory G-proteins, which are essential for both catecholamine and insulin signaling in adipocytes, was also altered by ectopic UCP1. These results support the role of protonophoric proteins in adipocytes in the control of adiposity and insulin sensitivity. Antidiabetic effects of thiazolidinediones, fibrates, β_3-adrenoreceptor agonists, dietary n-3 PUFAs, and leptin may be explained at least partially by their effects on the energy and hence also the lipid metabolism of fat cells.

KEYWORDS: uncoupling protein; adipose tissue; lipogenesis; lipolysis; obesity

Address for correspondence: Jan Kopecký, Institute of Physiology, Academy of Sciences of the Czech Republic, Vídeňská 1083, 142 20 Prague, Czech Republic. Voice: (+420-2) 4106 2554; fax: (+420-2) 4106 2599.
kopecky@biomed.cas.cz

In lean subjects, adipocytes are sensitive to insulin, which stimulates uptake of nonesterified fatty acids (FA) via lipoprotein lipase (LPL) and inhibits lipolysis. By these two mechanisms, insulin increases clearance of triacylglycerols by adipose tissue and mitigates release of FA from adipose tissue into circulation. On the other hand, hypertrophic adipocytes are becoming resistant to insulin, resulting in both lower clearance of plasma triacylglycerols by adipose tissue and higher FA release from adipose tissue into blood.[1] It is becoming apparent that insulin resistance of white adipose tissue is a major factor underlying the cardiovascular risk of obesity. Compared with the rest of the body, glucose flux into the adipose tissue is relatively low.[1] Accordingly, impaired fat storage rather than altered glucose metabolism in adipocytes probably contributes to the development of insulin resistance in muscle and other tissues, in particular via increased delivery of FA into circulation.[1,2] Recent studies show that insulin resistance also develops as a consequence of the lack of white adipose tissue in mice, documenting the important role of adipocytes in the development of diabetes.[3,4] Due to both lipodystrophy and obesity, high levels of circulating FA result in abnormally high deposition of lipids in various organs. The excessive lipid infiltration of organs leads to an impaired insulin sensitivity, especially in liver, skeletal muscle, and pancreas, with all the consequences typical of a metabolic syndrome.[1]

Many pieces of evidence also suggest that metabolism of adipose tissue contributes to the control of body fat content. First, most of the candidate genes for obesity[5] have important functions in adipocytes.[6] Second, mice that are prone[7–11] or resistant[12–16] to obesity were created by transgenic modification of adipose tissue. An important conclusion emerging from a comparison of these transgenic models (TABLE 1) is that metabolic changes of white (but not brown) adipose tissue are mostly responsible for the altered accretion of body fat in these animals, highlighting the importance of lipid metabolism in adipocytes of white fat.

Under all circumstances, any change in body fat content always results from a change in the balance between the rates of accumulation and breakdown of triacylglycerols occurring in adipocytes. The accumulation depends on the extraction of lipids from circulation mediated by LPL. *In situ* FA synthesis from glucose is also important. In humans, adipose tissue may account for up to 40% of whole-body lipogenesis.[17,18] High lipogenic rate in adipose tissue may contribute to the development of obesity[19] and may explain the high rate of relapse in patients treated by caloric restriction.[20]

The breakdown of triacylglycerols is initiated by hormone-sensitive lipase (HSL). Released FA may be reesterified or oxidized in adipocytes, or they are exported to other tissues, especially to muscles. FA oxidation is intense in brown fat, where FA serve as fuel for thermogenesis, depending on the protonophoric activity of the uncoupling protein 1 (UCP1), which controls proton leak in mitochondria.[21] Regulatable proton leak probably exists also in white fat cells, but its function is not clear (see below). Lipid metabolism in adipocytes is under complex neurohormonal control, with insulin and catecholamines representing the two most important regulatory factors.[22]

Evidently, treatment strategies for metabolic syndrome should include specific modifications of the metabolism of white adipose tissue. Insulin resistance may be counteracted by treatments that will decrease adiposity and/or increase clearance of triacylglycerols by white adipose tissue. This paper aims to characterize possible

TABLE 1. Consequences of transgenic modification of metabolism of adipose tissue in mice

Gene modification	Location	Adiposity	Reference
Induction			
GLUT4	WF, BF	↑	7
Agouti	WF, BF	↑	8
huα_2-AR/β_3-AR-KO	WF, BF	↑	9
Angiotensinogen	WF, BF	↑	97
UCP1	WF, (BF)	↓	12, 13
β_1-AR	WF, BF	↓	14
LPL	WF, BF	no change	98
Disruption			
β_3-AR	WF, BF	no change or ↑[b]	11, 99
$G_s\alpha$	WF	↑[c]	10
HSL	WF, BF	no change[d]	100
PKA	BF[a]	↓	15
PPARγ	WF, BF	↓	16
Dgat	B, H	↓	101
GLUT4	WF	no change[e]	102
UCP1	BF	no change[f]	103

TERMS: β_1-AR, β_1-adrenergic receptor; β_3-AR, β_3-adrenergic receptor; huα_2-AR/β_3-AR-KO, expression of human α_2-AR in mice with disrupted β_3-AR gene; Dgat, acyl CoA:diacylglycerol transferase; GLUT4, isoform 4 of glucose transporter; $G_s\alpha$, stimulatory G-protein α-subunits; PKA, subunit RIIβ of protein kinase A; WF, white fat; BF, brown fat.
[a]Obesity resistance results from overexpression of UCP1 in brown fat.
[b]Depending on genetic background.
[c]Obesity develops only in the case of the transfer of the maternal allele, while the paternal allele is silenced by genomic imprinting.
[d]Hypertrophy of adipocytes.
[e]Insulin resistance.
[f]Defect of thermogenesis.

involvement of energy metabolism in the control of lipid metabolism in adipocytes, with the main focus on the role of mitochondrial proton leak in such process. Various treatments that are known to decrease adiposity and increase insulin sensitivity will be reexamined with respect to their influence on adipose tissue metabolism.

IN VITRO STUDIES INDICATE MODULATION OF LIPID METABOLISM, MITOCHONDRIAL BIOGENESIS, AND HORMONAL SIGNALING BY ENERGY METABOLISM IN ADIPOCYTES

Enzymic equipment of both white and brown fat mitochondria promotes FA synthesis,[23,24] which requires cooperation between mitochondrial and cytoplasmic enzymes. Respiratory uncoupling in adipocytes could affect *in situ* FA synthesis as was suggested three decades ago by Rognstad and Katz,[25,26] who studied the effect of 2,4-dinitrophenol (DNP), an uncoupler of oxidative phosphorylation, on glucose metabolism in epididymal fat from rats. Addition of DNP resulted in depressed FA synthesis and increased production of lactate.[25] Inhibition of FA synthesis probably resulted from limited availability of intramitochondrial ATP for the carboxylation of pyruvate.[25,27] The level of both ATP and NADPH in the cytoplasm remained sufficiently high.[25,28] Activity of pyruvate carboxylase is 3-fold greater in white fat than in liver mitochondria, and the ATP/ADP ratio directly affects the enzyme activity.[27,29] Also, in our experiments on 3T3-L1 adipocytes differentiated in cell cultures, DNP depressed activity of FA synthesis by 4-fold, while FA oxidation was increased very little.[30]

Importantly, it has been shown *in vitro*[31] that the uncoupling induced by ectopic UCP1 in HeLa cells could induce mitochondrial biogenesis by upregulating its coordinating factor, the nuclear respiratory factor-1 (NRF-1). In order to assess whether also respiratory uncoupling in adipocytes could induce mitochondrial biogenesis, gene expression was characterized in 3T3-L1 adipocytes incubated with DNP. It was found that the chemical uncoupler upregulated genes for both NRF-1 and cytochrome c oxidase subunit IV (P. Flachs, unpublished results).

Also, lipolysis depends on the energy status of adipocytes.[32] A decrease in intracellular ATP, elicited in white adipocytes *in vitro* by uncouplers, inhibitors of the mitochondrial respiratory chain, or ionophores, counteracted the stimulation of lipolysis by catecholamines.[32,33] In turn, incubation of isolated adipocytes with lipolytic hormones resulted in up to 50% decrease of their intracellular ATP level[34] and in inhibition of lipolysis itself.[34] Metabolism of glucose, lipogenesis, and protein synthesis in adipocytes, as well as oxygen consumption, were also affected.[35] Very recently, ATP was shown to be required for translocation of HSL from cytoplasm to the surface of lipid storage droplets and for phosphorylation of HSL and other proteins that are involved in lipolysis in adipocytes.[36] These new data provide the explanation for the dependence of the adrenergically stimulated lipolysis on the energy status of fat cells. The intracellular ATP is also essential for insulin signaling, including the antilipolytic effect of this hormone.[37,38] The signaling is altered at the level of the interaction of insulin with its receptor,[38] as well as downstream from the receptor.[37]

Experiments performed *in vitro* suggested strong links between energy and lipid metabolism in adipocytes, as well as the effect of the energy status on hormonal signaling in these cells. However, without complementary evidence from studies *in vivo*, the significance of the above findings would be very limited. Therefore, the following paragraphs will focus on the relevant studies in humans and in experimental animals.

RESPIRATORY UNCOUPLING IN WHITE FAT *IN VIVO* MAY PREVENT DEVELOPMENT OF OBESITY

Experiments of Brand and colleagues demonstrated a mitochondrial proton leak in muscle and liver cells,[39] suggesting a general occurrence of proton leak in mitochondria *in vivo*. Several candidate genes for increasing the leak are expressed in adipocytes, namely, the genes for UCPs. In brown fat, UCP1, UCP2, UCP3, and UCP5 are expressed. In white fat, only UCP2 and UCP5 genes are normally active[21,40] and UCP2 antigen could be detected.[41] Similarly to UCP1, also UCP2 and UCP3 probably enhance the proton leak and decrease ATP synthesis.[21,42–44] Moreover, some mitochondrial anion carriers such as the adenine nucleotide[45,46] and 2-oxoglutarate carriers[47] could also mediate the proton leak. It should be stressed, however, that not all UCPs may function as simple protonophores in the inner mitochondrial membrane and their involvement in FA transport cannot be ruled out at present.[48,49]

Substantial amount of evidence has accumulated, indicating a link between mitochondrial UCPs in white fat and control of adiposity. Expression of the UCP1 gene could be induced in white fat depots of experimental animals by pharmacological compounds that reduce adiposity, for example, by β_3-adrenoreceptor agonists,[50–52] nicotine,[53] or leptin.[20] The β_3-agonists also upregulate UCP3 gene expression,[54] and induction of both UCP1 and UCP3 in white fat is associated with recruitment of multilocular adipocytes that are rich in mitochondria. Importantly, most multilocular cells in white adipose tissue of rats treated with β_3-adrenergic agonists originated from unilocular adipocytes and contained UCP3, while only a small fraction of novel multilocular adipocytes contained UCP1, suggesting that the multilocular cells arouse by transdifferentiation of white adipocytes.[51] Even in adult humans, relatively low levels of the UCP1 transcript could be detected in various fat depots. In abdominal fat, UCP1 mRNA levels are negatively correlated with obesity,[55] as well as those of UCP2.[56] A common polymorphism in the promoter of the UCP2 gene is associated with a decreased risk of obesity in middle-aged humans.[57] In humans, a negative correlation between heat production in adipocytes and body fat has also been found.[58]

In adult mice of the obesity-prone B6/J strain, UCP2 mRNA levels in white adipose tissue were lower than in A/J mice, which are resistant to dietary obesity. Also, induction of UCP2 by high-fat diet in white fat was higher in A/J than in B6/J mice.[59,60] Therefore, it has been suggested that induction of UCP2 in white fat protected the animals against obesity.[59,60] However, our experiments in postweaning mice also revealed an induction of UCP1 in subcutaneous white fat that was higher in A/J than in B6/J mice and could explain the obesity resistance of the former mice; in contrast to the adult mice, no differential effect of the high-fat diet on UCP2 expression in white fat could be detected in the postweaning mice of the two genotypes (unpublished results).

All these data strongly support the idea that respiratory uncoupling mediated by UCPs in white fat can mitigate obesity and that developmental factors are engaged in such control. However, the detailed mechanism of the obesity resistance remains unclear. The major paradox is related to the fact that specific metabolic rate and oxidative capacity of white fat are relatively low. Contribution of white fat to resting metabolic rate of a lean human subject is close to 5%,[58] while thermogenesis in brown fat in rodents may account for more than 50% of total metabolic rate.[61] The low specific metabolic rate of white fat reflects ultrastructural features of unilocular adipocytes that are filled with triacylglycerols, while all organelles including mitochondria are present in a thin periplasmic rim. In contrast, cytoplasmic space in multilocular brown fat cells is relatively large and contains numerous mitochondria.[62] White fat mitochondria are well equipped for oxidative phosphorylation, with pyruvate serving as the main source of energy for ATP synthesis. Due to low activity of carnitine palmitoyl transferase-1 in the inner mitochondrial membrane, oxidation of FA is relatively slow and FA are directed towards esterification,[23,24] unless the transferase is activated by leptin.[63] This is in contrast with mitochondria in brown fat, where FA serve as the main substrate for oxidation during activation of thermogenesis and activity of the transferase is not rate-limiting.[23] In tissues with high oxidative capacity (such as brown fat, liver, and muscle), stimulation of respiration by the uncoupling may augment significantly energy expenditure and shift energy balance of the whole organism. However, the mechanism by which respiratory uncoupling in white fat induces the obesity resistance is probably different.

MECHANISM OF OBESITY RESISTANCE INDUCED BY RESPIRATORY UNCOUPLING IN WHITE FAT *IN VIVO*: aP2-*Ucp1* MICE

The mechanism by which respiratory uncoupling may reduce accumulation of fat can be analyzed in transgenic mice in which the UCP1 gene is driven by the fat-specific aP2 promoter to achieve enhanced expression in both brown and white fat.[12] These mice with aP2-*Ucp1* transgene are partially resistant to obesity related to age, induced by genetic background,[12] or induced by feeding a high-fat-containing diet.[13,64] The resistance to obesity reflects lower accumulation of triacylglycerols in all fat depots, except for gonadal fat, which becomes relatively large.[12,13,64] Interestingly, reduction in total body weight becomes apparent when the animals are getting obese, but not under standard conditions,[12,13,64] similar to other models of obesity resistance induced by transgenic modification of adipose tissue[14-16] or muscle.[65,66] All these models indicate a strong systemic defense against weight loss in lean organisms that could not be compromised by metabolic changes in fat or muscle tissues. However, such metabolic changes may be sufficient for mitigation of obesity.

Transgenic UCP1 is present in both brown and white fat, while expression of the UCP1 endogene in brown fat is greatly reduced.[12] However, the obesity resistance results only from the transgenic modification of white fat because (i) transgenic mice exhibit atrophy of brown adipose tissue as indicated by reduction of its size and DNA content;[67] (ii) the levels of transcriptional coactivator PGC-1 mRNA, the activator of mitochondrial biogenesis, were about 5-fold lower in interscapular brown fat of transgenic mice, while no such effect of the genotype was observed in white fat (unpublished data); (iii) oxygen consumption by white fat fragments from transgenic mice was higher than that from control mice, while an opposite effect of the transgene was observed in brown fat;[64] and (iv) the dose of the transgene was inversely correlated to the thermogenic response to norepinephrine injections or to cold.[67] The mechanism of brown fat atrophy in the transgenic mice is not clear.[67]

Consequences of the expression of transgenic UCP1 in white fat, which induces the obesity resistance, have been studied in great detail. Transgenic UCP1 is contained in all the unilocular adipocytes, but it cannot induce conversion of these cells into multilocular adipocytes.[12] Expression of the transgene differs in various fat depots, with gonadal fat showing a relatively low expression.[30] This may explain in part the lack of the effect of the transgene on the accumulation of lipids in gonadal fat (see above). However, even in the gonadal fat, transgenic UCP1 could decrease mitochondrial membrane potential in adipocytes[68] and elevate oxygen consumption by 2-fold.[64] UCP1 also induces mitochondrial biogenesis in unilocular adipocytes (unpublished results), although the induction is much smaller than in adipocytes of mice treated with a β_3-adrenoceptor agonist.[51] In adult mice, the total content of transgenic UCP1 in white fat does not exceed 2% of the UCP1 in interscapular brown fat.[12] Apparently, only minute amounts of ectopic UCP1 in unilocular adipocytes of white fat can uncouple oxidative phosphorylation[68] and reduce accumulation of fat.

In agreement with the low oxidative capacity of white fat, the 2-fold increase of oxygen consumption brought about by transgenic UCP1 in this tissue (see above) results in only a marginal stimulation of the resting metabolic rate of mice.[67] The strong mitigation of obesity in aP2-*Ucp1* mice and the fact that UCP1 acts locally to reduce adiposity (see above and reference 50) indicate that the reduction results not only from increased energy expenditure, but also from a differential modification of lipid metabolism in various fat depots. Indeed, a strong diminution of FA synthesis was found in subcutaneous, but not gonadal fat of transgenic mice. The diminution was up to 4-fold, reflecting the magnitude of UCP1 expression and the decrease of adiposity in different fat depots,[30] as well as the drop in ATP/ADP ratio, that was observed only in the subcutaneous and not gonadal fat of transgenic mice (unpublished results). The decrease of FA synthesis was accompanied by downregulation of acetyl-CoA carboxylase and FA synthase in white fat, reflecting the decrease of metabolic flux through the lipogenic pathway.[30] The depression of FA synthesis by transgenic UCP1 thus confirmed the results of the *in vitro* experiments with DNP in 3T3-L1 adipocytes and in white fat fragments (see above), indicating a strong reduction of FA synthesis by respiratory uncoupling in adipocytes.

It was also tested whether transgenic UCP1 could affect norepinephrine-induced lipolysis in adipocytes as suggested by the *in vitro* experiments (see above). It was found that the maximum lipolytic effect of the catecholamine, measured as glycerol release from isolated adipocytes, was suppressed by 50% in the subcutaneous, but not gonadal fat of the transgenics. In parallel, UCP1 downregulated the expression of HSL, lowered its activity, and altered the expression of G-proteins in adipocytes, while upregulating G_s α-subunits and downregulating G_i α-subunits (P. Flachs, unpublished results).

The effect of respiratory uncoupling on triacylglycerol clearance mediated by LPL in white adipose tissue was also characterized. The activity of LPL was found to be higher in transgenic mice than in control mice, while the depression of the LPL activity by the transgene was significantly higher in animals fed high fat as compared with standard diet (M. Rossmeisl, unpublished results). Thus, the uncoupling stimulates clearance of triacylglycerols by adipose tissue and this effect is potentiated under obesity-promoting conditions. Accordingly, it was also found that plasma triacylglycerol levels were lower in transgenic than in control mice[13] and that the homozygous transgenic animals fed high-fat diet had the lowest triacylglycerol levels. In contrast, the levels of nonesterified FA in plasma, which were also depressed by the transgene, were not affected by the type of diet (unpublished results).

The experiments with aP2-*Ucp1* mice suggest that the main function of mitochondrial protonophores in white fat is the modulation of lipogenesis, oxidation of substrates, and hormonal control of lipid metabolism through the effect on ATP/ADP ratio in adipocytes. In accordance with the low oxidative capacity of white adipocytes, augmentation of energy expenditure may be of relatively small importance for the control of total body fat content. A reciprocal link between FA oxidation and synthesis may exist. It is well established that in several cell types, including adipocytes,[69] FA oxidation is depressed by malonyl-CoA (the first committed intermediate in FA synthesis), which inhibits the transfer of FA into mitochondria. On the other hand, according to a phenomenological theory considering the output of oxidative phosphorylation,[70] oxidation of all substrates in mitochondria requires (and therefore induces) a certain degree of respiratory uncoupling. Thus, FA oxidation per se may limit FA synthesis by inducing uncoupling of oxidative phosphorylation.

RESPIRATORY UNCOUPLING AND ENERGY METABOLISM IN ADIPOCYTES: PROMISING TARGETS FOR IMPROVING INSULIN SENSITIVITY AND MITIGATING OBESITY

All experimental evidence summarized above strongly suggests that specific modifications of the metabolism of white adipose tissue may be extremely beneficial for both improving insulin sensitivity and reducing obesity. This notion is supported by the results of the experiments both *in vitro*, on adipocytes incubated with DNP, and *in vivo*, with respiratory uncoupling induced by ectopic UCP1 in the aP2-*Ucp1* mice (see above). The latter experimental model especially shows that the respiratory uncoupling in adipocytes reduces accumulation of body fat by increasing FA oxidation and reducing lipogenesis in adipocytes. The uncoupling also decreases the levels of circulating triacylglycerols and FA by increasing the clearance of triacylglycerols via the stimulation of LPL in adipocytes and counteracting stimulation of lipolysis by adrenergic agonists in adipocytes. The inhibition of lipolysis can reduce circulating FA levels, but it does not override the beneficial effect of the uncoupling on fat accumulation since the transgenic mice are resistant to obesity.

Concerning the insulin sensitivity, it has been indeed demonstrated that aP2-*Ucp1* mice exhibited a better oral glucose tolerance than control mice. Importantly, the tolerance was always better in transgenic than in control mice, regardless of the type of diet or body weight, indicating that respiratory uncoupling in white fat improved insulin sensitivity independently of the weight-reducing effect of the transgenic modification.[13] Important questions are whether the efficiency of mitochondrial energy conversion and energy metabolism of adipocytes may be affected by other means besides genetic manipulation and whether such changes occur in adipocytes during treatments that affect adiposity and insulin resistance. Several typical situations known to improve insulin sensitivity can be compared with that in aP2-*Ucp1* mice. Especially relevant in this respect is the peripheral administration of leptin,[71] dietary n-3 polyunsaturated fatty acids (PUFAs),[72] or treatments by fibrates,[73] thiazolidinediones (glitazones),[73] and β_3-adrenoreceptor agonists.[74] In all these situations, insulin sensitivity is improved and fat accumulation is reduced, except for the thiazolidinediones, which increase lipogenesis,[75,76] FA uptake in adipocytes,[77–79] and adiposity.[73]

The antidiabetic and weight-reducing effects of leptin are well pronounced in animals and humans with defects in leptin signaling. These effects are mediated both centrally and at the periphery.[71] Leptin affects directly the metabolism of white fat by modulating expression of several clusters of genes, and its effects on the gene expression only partially overlap with those induced by fasting.[80] The fibrates, n-3 PUFAs, and thiazolidinediones are known to influence gene expression through their interactions with peroxisomal proliferator-activated receptor (PPAR).[73] All the PPAR subtypes, PPARα, PPARβ, and PPARγ, are present in white fat, with PPARγ predominating.[73] The PPARα, which is typical for liver,[73] can be induced in adipocytes by leptin.[20,80] The PPARγ induces adipocyte differentiation, while both PPARα and PPARβ probably upregulate genes of FA oxidation and inhibit FA synthesis.[73,79,81] Therefore, thiazolidinediones, which are specific ligands of the PPARγ, stimulate lipogenesis and differentiation of adipocytes (see above). Both fibrates and n-3 PUFAs, which inhibit lipogenesis and increase FA oxidation through PPARα in liver,[73] may do the same also in white fat. However, the effects on lipid metabolism in adipocytes are probably mediated by the PPARβ, but not the PPARα;[79,81] further studies are required to characterize the direct effects of fibrates and n-3 PUFAs in adipocytes.

In spite of the different mechanisms of action in adipocytes, in all situations leading to improvement of insulin resistance (due to n-3 PUFAs, fibrates, thiazolidinediones, and β$_3$-adrenoreceptor agonists; see above), several metabolic features of adipocytes are strikingly similar. In all the situations, clearance of triacylglycerols by the adipocytes is probably increased due to elevated LPL activity.[20,77–79,81,82] The increase in clearance may reflect the stimulation of FA oxidation in adipocytes as elicited by leptin,[20,83] fibrates,[81] and β$_3$-adrenoreceptor agonists.[51] In all the situations, upregulation of various UCPs occurs in adipocytes, specifically of UCP1,[20,77,84–87] UCP2,[20,77,81,85,88,89] and UCP3.[54,81,84,87,88,90] Also, *in situ* lipogenesis is mostly suppressed.[20,72,80,91–93] All the changes are similar to those induced by respiratory uncoupling in white fat of aP2-*Ucp1* mice (see above), except for lipolysis, which is inhibited in the aP2-*Ucp1* mice, while it is stimulated under all the other circumstances.[63,77,81,94–96] Thus, it is tempting to suggest that the almost uniform response of lipid metabolism in adipocytes in all situations leading to improvement of insulin sensitivity may involve mitochondrial UCPs and their effects on energy metabolism. The effects of leptin and various pharmaceuticals on gene expression will diversify the overall metabolic response of fat cells.

In spite of all the indirect evidence summarized above, the role of respiratory uncoupling in white fat cells in control of lipid metabolism, adiposity, and insulin resistance remains to be firmly established. New studies focused on the energy metabolism in adipocytes both *in vivo* and *in vitro* are needed. Nevertheless, it is apparent that energy status of fat cells has an important impact on various pathways of lipid metabolism that affect adiposity, clearance of circulating triacylglycerols, and release of FA from adipocytes. New strategies for treatment of the metabolic syndrome will benefit from understanding the differential effects of diets, leptin, antidiabetics, and hypolipidemics on both energy metabolism of adipocytes and gene expression in these cells. Understanding of these effects will provide a background for defining optimal combinations of therapeutical approaches in the treatment of the metabolic syndrome.

ACKNOWLEDGMENTS

This work was supported by the Ministry of Education of the Czech Republic (COST-918 and B17), the Grant Agency of the Academy of Sciences of the Czech Republic (A5011710), and the Internal Grant Agency of the Charles University of the Czech Republic (235/1999/B-CH/PrF). We thank A. Kotyk (Institute of Physiology, Prague, Czech Republic) for critical reading of the manuscript.

REFERENCES

1. Frayn, K.N. & L.K.M. Summers. 1998. Substrate fluxes in skeletal muscle and white adipose tissue and their importance in the development of obesity. *In* Clinical Obesity, pp. 129–157. Blackwell Science. Oxford.
2. Danforth, E., Jr. 2000. Failure of adipocyte differentiation causes type II diabetes mellitus? Nat. Genet. **26:** 13.
3. Moitra, J., M.M. Mason, M. Live et al. 1998. Life without white fat: a transgenic mouse. Genes Dev. **12:** 3168–3181.
4. Gavrilova, O., B. Marcus-Samuels, D. Graham et al. 2000. Surgical implantation of adipose tissue reverses diabetes in lipoatrophic mice. J. Clin. Invest. **105:** 271–278.
5. Rankinen, T., L. Perusse, S.J. Weisnagel et al. 2002. The human obesity gene map: the 2001 update. Obes. Res. **10:** 196–243.
6. Arner, P. 2000. Obesity—a genetic disease of adipose tissue? Br. J. Nutr. **83**(suppl. 1): S9–S16.
7. Shepherd, P.R., L. Gnudi, E. Tozzo et al. 1993. Adipose cell hyperplasia and enhanced glucose disposal in transgenic mice overexpressing GLUT4 selectively in adipose tissue. J. Biol. Chem. **268:** 22243–22246.
8. Mynatt, R.L., R.J. Miltenberger, M.L. Klebig et al. 1997. Combined effects of insulin treatment and adipose tissue-specific agouti expression on the development of obesity. Proc. Natl. Acad. Sci. U.S.A. **94:** 919–922.
9. Valet, P., D. Grujic, J. Wade et al. 2000. Expression of human α_2-adrenergic receptors in adipose tissue of β_3-adrenergic receptor deficient mice promotes diet-induced obesity. J. Biol. Chem. **275:** 34797–34802.
10. Yu, S., O. Gavrilova, H. Chen et al. 2000. Paternal versus maternal transmission of a stimulatory G-protein alpha subunit knockout produces opposite effects on energy metabolism. J. Clin. Invest. **105:** 615–623.
11. Revelli, J.P., F. Preitner, S. Samec et al. 1997. Targeted gene disruption reveals a leptin-independent role for the mouse β_3-adrenoceptor in the regulation of body composition. J. Clin. Invest. **100:** 1098–1106.
12. Kopecký, J., G. Clarke, S. Enerback et al. 1995. Expression of the mitochondrial uncoupling protein gene from the aP2 gene promoter prevents genetic obesity. J. Clin. Invest. **96:** 2914–2923.
13. Kopecký, J., Z. Hodný, M. Rossmeisl et al. 1996. Reduction of dietary obesity in the aP2-*Ucp* transgenic mice: physiology and adipose tissue distribution. Am. J. Physiol. **270:** E768–E775.
14. Soloveva, V., R.A. Graves, M.M. Rasenick et al. 1997. Transgenic mice overexpressing the β_1-adrenergic receptor in adipose tissue are resistant to obesity. Mol. Endocrinol. **11:** 27–38.
15. Cummings, D.E., E.P. Brandon, J.V. Planas et al. 1996. Genetically lean mice result from targeted disruption of the RIIβ subunit of protein kinase A. Nature **382:** 622–626.
16. Kubota, N., Y. Terauchi, H. Miki et al. 1999. PPARγ mediates high-fat diet–induced adipocyte hypertrophy and insulin resistance. Mol. Cell **4:** 597–609.
17. Chascione, C., D.H. Elwyn, M. Davila et al. 1987. Effect of carbohydrate intake on de novo lipogenesis in human adipose tissue. Am. J. Physiol. **253:** E664–E669.
18. Swierczynski, J., E. Goyke, L. Wach et al. 2000. Comparative study of the lipogenic potential of human and rat adipose tissue. Metabolism **49:** 594–599.
19. Belfiore, F., V. Borzi, E. Napoli & A.M. Rabuazzo. 1976. Enzymes related to lipogenesis in the adipose tissue of obese subject. Metabolism **25:** 483–493.
20. Zhou, Y-T., Z-W. Wang, M. Higa et al. 1999. Reversing adipocyte differentiation: implications for treatment of obesity. Proc. Natl. Acad. Sci. U.S.A. **96:** 2391–2395.
21. Ricquier, D. & F. Bouillaud. 2000. The uncoupling protein homologues: UCP1, UCP2, UCP3, StUCP, and AtUCP. Biochem. J. **345:** 161–179.

22. KOPELMAN, P.G. 1998. Effects of obesity on fat topography: metabolic and endocrine determinants. *In* Clinical Obesity, pp. 158–175. Blackwell Science. Oxford.
23. NEDERGAARD, J. & B. CANNON. 1979. Overview—preparation and properties of mitochondria from different sources. *In* Methods in Enzymology. Vol. 55, pp. 1–17. Academic Press. New York.
24. MARTIN, B.R. & R.M. DENTON. 1970. The intracellular localization of enzymes in white-adipose-tissue fat-cells and permeability properties of fat-cell mitochondria. Biochem. J. **117:** 861–877.
25. ROGNSTAD, R. & J. KATZ. 1969. The effect of 2,4-dinitrophenol on adipose-tissue metabolism. Biochem. J. **111:** 431–444.
26. KATZ, J., P.A. WALS & R. ROGNSTAD. 1974. ATP balance and the effect of 2,4-dinitrophenol on fatty acid synthesis. Biochim. Biophys. Acta **337:** 313–317.
27. MARTIN, B.R. & R.M. DENTON. 1971. Metabolism of pyruvate and malate by isolated fat-cell mitochondria. Biochem. J. **125:** 105–113.
28. BASHAN, N., E. BURDETT, A. GUMA *et al.* 1993. Mechanisms of adaptation of glucose transporters to changes in the oxidative chain of muscle and fat cells. Am. J. Physiol. **264:** C430–C440.
29. PATEL, M.S. & R.W. HANSON. 1970. Carboxylation of pyruvate by isolated rat adipose tissue mitochondria. J. Biol. Chem. **245:** 1302–1310.
30. ROSSMEISL, M., I. SYROVÝ, F. BAUMRUK *et al.* 2000. Decreased fatty acid synthesis due to mitochondrial uncoupling in adipose tissue. FASEB J. **14:** 1793–1800.
31. LI, B., J.O. HOLLOSZY & C.F. SEMENKOWICH. 1999. Respiratory uncoupling induces delta-aminolevulinate synthase expression through a nuclear respiratory factor-1–dependent mechanism in HeLa cells. J. Biol. Chem. **274:** 17534–17540.
32. FASSINA, G., P. DORIGO & R.M. GAION. 1974. Equilibrium between metabolic pathways producing energy: a key factor in regulating lipolysis. Pharmacol. Res. Commun. **6:** 1–21.
33. HUBER, C.T., W.C. DUCKWORTH & S.S. SOLOMON. 1981. The reversible inhibition by carbonyl cyanide *m*-chlorophenyl hydrazone of epinephrine-stimulated lipolysis in perfused isolated fat cells. Biochim. Biophys. Acta **666:** 462–467.
34. ANGEL, A., K. DESAI & M.L. HALPERIN. 1971. Reduction in adipocyte ATP by lipolytic agents: relation to intracellular free fatty acid accumulation. J. Lipid Res. **12:** 203–213.
35. ANGEL, A., K. DESAI & M.L. HALPERIN. 1971. Free fatty acid and ATP levels in adipocytes during lipolysis. Metabolism **20:** 87–99.
36. BRASAEMLE, D.L., D.M. LEVIN, D.C. ADLER-WAILES & C. LONDOS. 2000. The lipolytic stimulation of 3T3-L1 adipocytes promotes the translocation of hormone-sensitive lipase to the surfaces of lipid storage droplets. Biochim. Biophys. Acta **1483:** 251–262.
37. HARING, H.U., F. RINNINGER & W. KEMMLER. 1981. Decreased insulin sensitivity due to a postreceptor defect as a consequence of ATP-deficiency in fat cells. FEBS Lett. **132:** 235–238.
38. STEINFELDER, H.J. & H.G. JOOST. 1983. Reversible reduction of insulin receptor affinity by ATP depletion in rat adipocytes. Biochem. J. **214:** 203–207.
39. BRAND, M.D., K.M. BRINDLE, J.A. BUCKINGHAM *et al.* 1999. The significance and mechanism of mitochondrial proton conductance. Int. J. Obes. Metab. Disord. **23**(suppl. 6): S4–S11.
40. YU, X.X., W. MAO, A. ZHONG *et al.* 2000. Characterization of novel UCP5/BMCP1 isoforms and differential regulation of UCP4 and UCP5 expression through dietary or temperature manipulation. FASEB J. **14:** 1611–1618.
41. PECQUEUR, C., M.C. ALVES-GUERRA, C. GELLY *et al.* 2001. Uncoupling protein 2: *in vivo* distribution, induction upon oxidative stress, and evidence for translational regulation. J. Biol. Chem. **276:** 8705–8712.
42. VIDAL-PUIG, A.J., D. GRUJIC, C-Y. ZHANG *et al.* 2000. Energy metabolism in uncoupling protein 3 gene knockout mice. J. Biol. Chem. **275:** 16258–16266.
43. GONG, D.W., S. MONEMDJOU, O. GAVRILOVA *et al.* 2000. Lack of obesity and normal response to fasting and thyroid hormone in mice lacking uncoupling protein-3. J. Biol. Chem. **275:** 16251–16257.
44. ECHTAY, K.S., E. WINKLER, K. FRISCHMUTH & M. KLINGENBERG. 2001. Uncoupling proteins 2 and 3 are highly active H^+ transporters and highly nucleotide sensitive

when activated by coenzyme Q (ubiquinone). Proc. Natl. Acad. Sci. U.S.A. **98:** 1416–1421.
45. SKULACHEV, V.P. 1999. Anion carriers in fatty acid–mediated physiological uncoupling. J. Bioenerg. Biomembr. **31:** 431–445.
46. CADENAS, S., J.A. BUCKINGHAM, J. ST. PIERRE et al. 2000. AMP decreases the efficiency of skeletal-muscle mitochondria. Biochem. J. **351:** 307–311.
47. YU, X.X., D.A. LEWIN et al. 2001. Overexpression of the human 2-oxoglutarate carrier lowers mitochondrial membrane potential in HEK-293 cells: contrast with the unique cold-induced mitochondrial carrier CGI-69. Biochem. J. **353:** 369–375.
48. GARCIA-MARTINEZ, C., B. SIBILLE, G. SOLANES et al. 2001. Overexpression of UCP3 in cultured human muscle lowers mitochondrial membrane potential, raises ATP/ADP ratio, and favors fatty acid versus glucose oxidation. FASEB J. **15:** 2033–2035.
49. HIMMS-HAGEN, J. & M.E. HARPER. 2001. Physiological role of UCP3 may be export of fatty acids from mitochondria when fatty acid oxidation predominates: an hypothesis. Exp. Biol. Med. **226:** 78–84.
50. GUERRA, C., R.A. KOZA, H. YAMASHITA et al. 1998. Emergence of brown adipocytes in white fat in mice is under genetic control: effects on body weight and adiposity. J. Clin. Invest. **102:** 412–420.
51. HIMMS-HAGEN, J., A. MELNYK, M.C. ZINGARETTI et al. 2000. Multilocular fat cells in WAT of CL-316243-treated rats derive directly from white adipocytes. Am. J. Physiol. Cell. Physiol. **279:** C670–C681.
52. CHAMPIGNY, O., D. RICQUIER, O. BLONDEL et al. 1991. Beta3-adrenergic receptor stimulation restores message and expression of brown-fat mitochondrial uncoupling protein in adult dogs. Proc. Natl. Acad. Sci. U.S.A. **88:** 10774–10777.
53. YOSHIDA, T., N. SAKANE, T. UMEKAWA et al. 1999. Nicotine induced uncoupling protein 1 in white adipose tissue of obese mice. Int. J. Obes. **23:** 570–575.
54. GONG, D.W., Y. HE, M. KARAS & M. REITMAN. 1997. Uncoupling protein-3 is a mediator of thermogenesis regulated by thyroid hormone, β_3-adrenergic agonists, and leptin. J. Biol. Chem. **272:** 24129–24132.
55. OBERKOFLER, H., G. DALLINGER, Y.M. LIU et al. 1997. Uncoupling protein gene: quantification of expression levels in adipose tissues of obese and non-obese humans. J. Lipid Res. **38:** 2125–2133.
56. OBERKOFLER, H., Y.M. LIU, H. ESTERBAUER et al. 1998. Uncoupling protein-2 gene: reduced mRNA expression in intraperitoneal adipose tissue of obese humans. Diabetologia **41:** 940–946.
57. ESTERBAUER, H., C. SCHNEITLER, H. OBERKOFLER et al. 2001. A common polymorphism in the promoter of UCP2 is associated with decreased risk of obesity in middle-aged humans. Nat. Genet. **28:** 178–183.
58. BOTTCHER, H. & P. FURST. 1997. Decreased white fat cell thermogenesis in obese individuals. Int. J. Obes. **21:** 439–444.
59. FLEURY, C., M. NEVEROVA, S. COLLINS et al. 1997. Uncoupling protein-2: a novel gene linked to obesity and hyperinsulinemia. Nat. Genet. **15:** 269–272.
60. SURWIT, R.S., S. WANG, A.E. PETRO et al. 1998. Diet-induced changes in uncoupling proteins in obesity-prone and obesity-resistant strains of mice. Proc. Natl. Acad. Sci. U.S.A. **95:** 4061–4065.
61. JÁNSKÝ, I. 1995. Humoral thermogenesis and its role in maintaining energy balance. Physiol. Rev. **75:** 237–259.
62. CINTI, S. 1999. The Adipose Organ. Editrice Kurtis. Milan.
63. WANG, M.Y., Y. LEE & R.H. UNGER. 1999. Novel form of lipolysis induced by leptin. J. Biol. Chem. **274:** 17541–17544.
64. KOPECKÝ, J., M. ROSSMEISL, Z. HODNÝ et al. 1996. Reduction of dietary obesity in the aP2-*Ucp* transgenic mice: mechanism and adipose tissue morphology. Am. J. Physiol. **270:** E776–E786.
65. JENSEN, D.R., I.R. SCHLAEPFER, C.L. MORIN et al. 1997. Prevention of diet-induced obesity in transgenic mice overexpressing skeletal muscle lipoprotein lipase. Am. J. Physiol. **273:** R683–R689.
66. CLAPHAM, J.C., J.R. ARCH et al. 2000. Mice overexpressing human uncoupling protein-3 in skeletal muscle are hyperphagic and lean. Nature **406:** 415–418.

67. ŠTEFL, B., A. JANOVSKÁ, Z. HODNÝ et al. 1998. Brown fat is essential for cold-induced thermogenesis, but not for obesity resistance in aP2-Ucp mice. Am. J. Physiol. **274:** E527–E533.
68. BAUMRUK, F., P. FLACHS, M. HORÁKOVÁ et al. 1999. Transgenic UCP1 in white adipocytes modulates mitochondrial membrane potential. FEBS Lett. **444:** 206–210.
69. SAGGERSON, E.D. & C.A. CARPENTER. 1983. The effect of malonyl-CoA on overt and latent carnitine acyltransferase activities in rat liver and adipocyte mitochondria. Biochem. J. **210:** 591–597.
70. STUCKI, J.W. 1980. The optimal efficiency and the economic degrees of coupling of oxidative phosphorylation. Eur. J. Biochem. **109:** 269–283.
71. UNGER, R.H. 2000. Leptin physiology: a second look. Regul. Pept. **92:** 87–95.
72. PRICE, P.T., C.M. NELSON & S.D. LARKE. 2000. Omega-3 polyunsaturated fatty acid regulation of gene expression. Curr. Opin. Lipidol. **11:** 3–7.
73. CHINETTI, G., J.C. FRUCHART & B. STAELS. 2000. Peroxisome proliferator-activated receptors (PPARs): nuclear receptors at the crossroads between lipid metabolism and inflammation. Inflamm. Res. **49:** 497–505.
74. CAWTHORNE, M.A. 1992. Thermogenic drugs. In Obesity, pp. 762–777. Lippincott. Philadelphia.
75. SANDOUK, T., D. REDA & C. HOFMANN. 1993. Antidiabetic agent pioglitazone enhances adipocyte differentiation of 3T3-F442A cells. Am. J. Physiol. **264:** C1600–C1608.
76. TONTONOZ, P., E. HU & B.M. SPIEGELMAN. 1994. Stimulation of adipogenesis in fibroblasts by PPARgamma2, a lipid-activated transcription factor. Cell **79:** 1147–1156.
77. STROBEL, A., K. SIQUIER, V. ZILBERFARB et al. 1999. Effect of thiazolidinediones on expression of UCP2 and adipocyte markers in human PAZ6 adipocytes. Diabetologia **42:** 527–533.
78. MARTIN, G., K. SCHONJANS, A.M. LEFEBVRE et al. 1997. Coordinate regulation of the expression of the fatty acid transport protein and acyl-CoA synthetase genes by PPARalpha and PPARgamma activators. J. Biol. Chem. **272:** 28210–28217.
79. LEFEBVRE, A.M., J. PEINADO-ONSURBE, I. LEITERSDORF et al. 1997. Regulation of lipoprotein metabolism by thiazolidinediones occurs through a distinct but complementary mechanism relative to fibrates. Arterioscler. Thromb. Vasc. Biol. **17:** 1756–1764.
80. SOUKAS, A., P. COHEN, N.D. SOCCI & J.M. FRIEDMAN. 2000. Leptin-specific patterns of gene expression in white adipose tissue. Genes Dev. **14:** 963–980.
81. CABRERO, A., M. ALEGRET, R.M. SANCHEZ et al. 2001. Bezafibrate reduces mRNA levels of adipocyte markers and increases fatty acid oxidation in primary culture of adipocytes. Diabetes **50:** 1883–1890.
82. BENHIZIA, F., I. HAINAULT et al. 1994. Effects of fish oil–lard diet on rat plasma lipoproteins, liver FAS, and lipolytic enzymes. Am. J. Physiol. **267:** E975–E982.
83. ZHOU, Y.T., M. SHIMABUKURO, K. KOYAMA et al. 1997. Induction by leptin of uncoupling protein-2 and enzymes of fatty acid oxidation. Proc. Natl. Acad. Sci. U.S.A. **94:** 6386–6390.
84. CABRERO, A., G. LLAVERIAS, N. ROGLANS et al. 1999. Uncoupling protein-3 mRNA levels are increased in white adipose tissue and skeletal muscle of bezafibrate-treated rats. Biochem. Biophys. Res. Commun. **260:** 547–556.
85. AUBERT, J., O. CHAMPIGNY, P. SAINT-MARC et al. 1997. Up-regulation of UCP-2 gene expression by PPAR agonists in preadipose and adipose cells. Biochem. Biophys. Res. Commun. **238:** 606–611.
86. DIGBY, J.E., C.T. MONTAGUE, C.P. SEWTER et al. 1998. Thiazolidinedione exposure increases the expression of uncoupling protein 1 in cultured human preadipocytes. Diabetes **47:** 138–141.
87. YOSHITOMI, H., K. YAMAZAKI, S. ABE & I. TANAKA. 1998. Differential regulation of mouse uncoupling proteins among brown adipose tissue, white adipose tissue, and skeletal muscle in chronic beta 3 adrenergic receptor agonist treatment. Biochem. Biophys. Res. Commun. **253:** 85–91.
88. HUN, C.S., K. HASEGAWA, T. KAWABATA et al. 1999. Increased uncoupling protein 2 mRNA in white adipose tissue, and decrease in leptin, visceral fat, blood glucose, and cholesterol in KK-Ay mice fed with eicosapentaenoic and docosahexaenoic acid in addition to linolenic acid. Biochem. Biophys. Res. Commun. **259:** 85–90.

89. EKHTERAE, D., H.J. TAE, S. DANIEL et al. 1996. Regulation of acetyl coenzyme-A carboxylase gene in a transgenic animal model. Biochem. Biophys. Res. Commun. **227:** 547–552.
90. MATSUDA, J., K. HOSODA, H. ITOH et al. 1998. Increased adipose expression of the uncoupling protein-3 gene by thiazolidinediones in Wistar fatty rats and in cultured adipocytes. Diabetes **47:** 1809–1814.
91. CEDDIA, R.B., W.N. WILLIAM, F.B. LIMA et al. 2000. Leptin stimulates uncoupling protein-2 mRNA expression and Krebs cycle activity and inhibits lipid synthesis in isolated rat white adipocytes. Eur. J. Biochem. **267:** 5952–5958.
92. NAKAMURA, M.T., H.P. CHO & S.D. CLARKE. 2000. Regulation of hepatic delta-6 desaturase expression and its role in the polyunsaturated fatty acid inhibition of fatty acid synthase gene expression in mice. J. Nutr. **130:** 1561–1565.
93. RODBELL, M. 1964. Metabolism of isolated fat cells. I. Effects of hormones on glucose metabolism and lipolysis. J. Biol. Chem. **239:** 375–380.
94. FRUHBECK, G., M. AGUADO, J. GOMEZ-AMBRISI & J.A. MARTINEZ. 1998. Lipolytic effect of *in vivo* leptin administration on adipocytes of lean and *ob/ob* mice, but not *db/db* mice. Biochem. Biophys. Res. Commun. **250:** 99–102.
95. RACLOT, T. & H. OUDART. 1999. Selectivity of fatty acids on lipid metabolism and gene expression. Proc. Nutr. Soc. **58:** 633–646.
96. GALITZKY, J., D. LANGIN, P. VERWAERDE et al. 1997. Lipolytic effects of conventional β_3-adrenoceptor agonists and of CGP 12,177 in rat and human fat cells: preliminary pharmacological evidence for a putative β_4-adrenoceptor. Br. J. Pharmacol. **122:** 1244–1250.
97. AILHAUD, G., A. FUKAMIZU, F. MASSIERA et al. 2000. Angiotensinogen, angiotensin II, and adipose tissue development. Int. J. Obes. Relat. Metab. Disord. **24**(suppl. 4): S33–S35.
98. SHIMADA, M., S. ISHIBASHI, K. YAMAMOTO et al. 1995. Overexpression of human lipoprotein lipase increases hormone-sensitive lipase activity in adipose tissue of mice. Biochem. Biophys. Res. Commun. **211:** 761–766.
99. SUSULIC, V.S., R.C. FREDERICH, J. LAWITTS et al. 1995. Targeted disruption of the β_3-adrenergic receptor gene. J. Biol. Chem. **270:** 29483–29492.
100. OSUGA, J., S. ISHIBASHI, T. OKA et al. 2000. Targeted disruption of hormone-sensitive lipase results in male sterility and adipocyte hypertrophy, but not in obesity. Proc. Natl. Acad. Sci. U.S.A. **97:** 787–792.
101. SMITH, S.J., S. CASES, D.R. JENSEN et al. 2000. Obesity resistance and multiple mechanisms of triglyceride synthesis in mice lacking Dgat. Nat. Genet. **25:** 87–90.
102. ABEL, E.D., O. PERONI, J.K. KIM et al. 2001. Adipose-selective targeting of the GLUT4 gene impairs insulin action in muscle and liver. Nature **409:** 729–733.
103. ENERBACK, S., A. JACOBSSON, E.M. SIMPSON et al. 1997. Mice lacking mitochondrial uncoupling protein are cold-sensitive, but not obese. Nature **387:** 90–94.

Transgenic Mice Overexpressing the Rate-Limiting Enzyme for Hexosamine Synthesis in Skeletal Muscle or Adipose Tissue Exhibit Total Body Insulin Resistance

ROBERT C. COOKSEY AND DONALD A. McCLAIN

VA Medical Center and Department of Medicine, University of Utah, Salt Lake City, Utah 84132, USA

ABSTRACT: High concentrations of glucose induce insulin resistance and impair insulin secretion in a manner that mirrors type 2 diabetes, a phenomenon known as glucose toxicity. High concentrations of hexosamines mimic these effects, leading to the hypothesis that cells use hexosamine flux as a glucose- and satiety-sensing pathway. Overexpression of the rate-limiting enzyme for hexosamine synthesis (glutamine:fructose-6-phosphate amidotransferase, GFA) in muscle and fat results in insulin resistance and hyperleptinemia. GFA overexpression targeted to liver results in hyperlipidemia and to the beta cell in increased insulin secretion. Thus, excess hexosamine flux leads to a coordinated response whereby fuel is shunted toward long-term storage, mirroring the "thrifty phenotype". The results suggest a mechanism by which chronic overnutrition leads to the phenotype of type 2 diabetes.

KEYWORDS: hexosamine(s); insulin resistance; adipose tissue; transgenic

Although there is a major genetic contribution to type 2 diabetes, the largest predisposing factor remains caloric excess and/or obesity. The importance of excess nutrients is underlined by the fact that the pathophysiologic hallmarks of the disease—insulin resistance and beta cell failure—not only contribute to, but also can *result from* excess glucose and lipids (so-called "glucose toxicity" and "lipotoxicity"). Diabetes is also characterized by abnormalities in several metabolic pathways and in several organs, and this multitude of abnormalities may be more easily explained in terms of a common mechanism involving normal physiologic responses to excess nutrients. A fuel-sensing pathway has been discovered that may explain some of the complexity of type 2 diabetes in terms of normal signaling pathways that control the partitioning of ingested calories for long-term storage. Namely, chronically excessive

Address for correspondence: Donald A. McClain, Department of Medicine 4C116, 50 North Medical Drive, University of Utah School of Medicine, Salt Lake City, UT 84132. Voice: 801-585-0954; fax: 801-585-0356.
donald.mcclain@hsc.utah.edu

flux through one glucose-sensing pathway, the hexosamine biosynthetic pathway, leads to many of the phenotypic characteristics of diabetes.

Hexosamine synthesis begins with the amination of fructose-6-phosphate to glucosamine-6-phosphate. This step is rate-limiting for hexosamine synthesis and is catalyzed by glutamine:fructose-6-phosphate amidotransferase (GFA).[1,2] The chief product of the pathway is uridine diphospho-*N*-acetyl glucosamine (UDP-GlcNAc). Marshall originally demonstrated the pathway's importance in causing insulin resistance in cultured adipocytes exposed to high glucose.[3] This was followed by the demonstration that overexpression of GFA in fibroblasts resulted in insulin resistance for stimulation of glycogen synthase[4] and that glucosamine caused insulin resistance in muscle cells[5] and in animals infused with glucosamine.[6–10] We have constructed transgenic mice with tissue-specific overexpression of GFA to further study these pathways. These models avoid potential untoward effects of treating cells or animals with high concentrations of glucosamine by increasing hexosamine flux physiologically and without major perturbation of other metabolic pathways.

METABOLIC CONSEQUENCES OF INCREASED HEXOSAMINE FLUX IN TRANSGENIC ANIMAL MODELS

Several results support the hypothesis that hexosamines play a role in nutrient signaling:

- Overexpression of GFA in liver leads to increased glycogen (FIG. 1) and increased triglyceride levels (FIG. 2) even in the presence of subnormal plasma glucose concentrations.[11]
- Overexpression of GFA in muscle plus fat results in decreased insulin-stimulated glucose uptake (FIG. 3).[12,13]
- Overexpression of GFA in the beta cell results in hyperinsulinemia (FIG. 4).[14]

Importantly, the effects of hexosamines are probably physiologic. They are seen with ~2-fold changes in hexosamine flux rates (within the physiologic range) in the absence of significant changes in ATP or other intracellular metabolites. Furthermore, results of experimental manipulation of animals correlate well with observations in nonmanipulated humans: namely, GFA activity is correlated with insulin sensitivity and adiposity.[15,16] Of note, the changes in GFA activity are manifest as changes in the end product of the pathway, UDP-GlcNAc;[11–14] levels of glucosamine-6-phosphate do not rise appreciably because that intermediate is rapidly acetylated. Thus, in these models, the changes seen cannot be attributed to inhibition of glucokinase by glucosamine.

In the short run, all of the effects of hexosamines enumerated above can be seen as adaptations that allow for cells to sense excess nutrients so that those nutrients can be saved as fat for the future. However, chronically increased hexosamine flux results in phenotypes that mirror type 2 diabetes, including insulin resistance when GFA is expressed in the liver, muscle, and/or fat;[4–6,8,9,12] impaired suppression of hepatic glucose output by hyperglycemia with increased hexosamine flux in the liver;[17] obesity with GFA overexpression in liver or beta cells;[11] hyperlipidemia with GFA overexpression in the liver;[11] and impaired insulin secretion with GFA

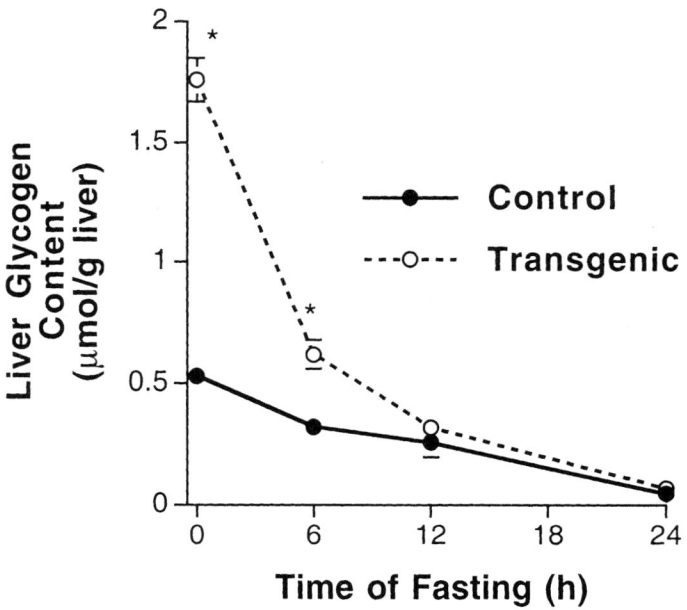

FIGURE 1. Hepatic glycogen content in control and PEPCK-GFA-1 mice. Glycogen content was measured in random-fed animals and after the indicated times of fasting in random-fed PEPCK-GFA-1 animals (●) compared to nontransgenic littermates (○). Glycogen content was 3.2-fold elevated in random-fed PEPCK-GFA-1 animals ($n = 9$ animals per data point; *$p < 0.01$ compared to controls).

overexpression or increased hexosamine flux in beta cells.[9,18,19] When GFA is targeted to beta cells, animals are initially hyperinsulinemic. This results in insulin resistance and eventual beta cell failure.[14] These results are consistent with the hypothesis that hexosamines play a role in how nutrients control fuel partitioning and in the detrimental consequences of chronic overnutrition.

MECHANISMS OF HEXOSAMINE-DRIVEN INSULIN RESISTANCE

The ultimate mediator of insulin resistance in models of hexosamine excess is similar to that seen in type 2 diabetes, namely, decreased recruitment of GLUT4 by insulin.[8,13] What causes that decreased recruitment is less clear, however. Glucosamine treatment leads to defects in early steps of insulin signal transduction pathways, particularly in the stimulation of PI-3 kinase activity.[20,21] However, whether these results accurately model the defects observed in hyperglycemic conditions is controversial (e.g., see references 22 and 23). Such controversy arises, at least partly, from variations among the different models and potential differences between acute treatments with either glucose or glucosamine compared to chronic models of more physiologic increases in hexosamine flux.

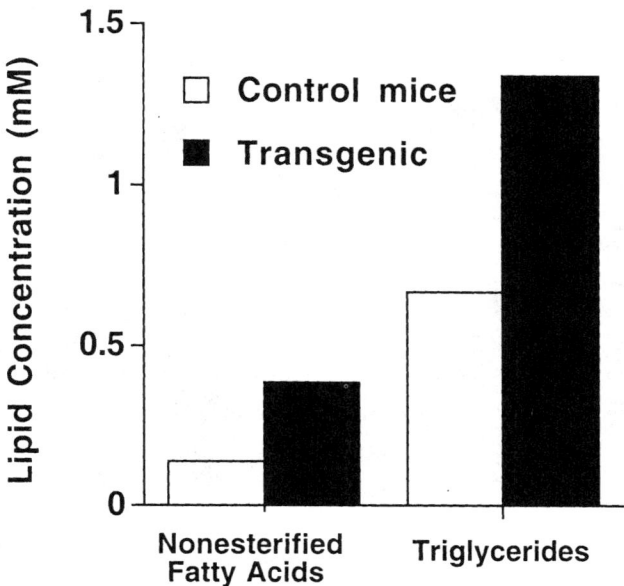

FIGURE 2. FFA and triglyceride levels in blood samples of random-fed control (open bars) and transgenic (PEPCK-GFA-1, solid bars) mice overexpressing GFA in liver. FFA and triglyceride levels were 4.9- and 3.5-fold elevated in transgenic compared to control mice, respectively. Mice were aged 2–4 months. Results are the means (± SEM) of 9 determinations. $p < 0.01$ for differences between control and transgenic animals for each lipid.

An interesting finding in several transgenic models has been that metabolic alterations targeted to fat alone are sufficient for the organism to develop insulin resistance as well. For example, GFA targeted to fat alone (FIG. 5) or targeted disruption of the GLUT4 gene in fat only[24] causes insulin resistance in muscle as well as in fat. This dominant role of fat may explain the effect of PPARγ agonists in the treatment of insulin resistance. Although the effects of pharmacologic PPARγ agonists to sensitize insulin-stimulated glucose uptake are quantitatively seen mainly in skeletal muscle,[25] PPARγ is predominantly expressed in fat. In addition, PPARγ agonists "cured" GFA-induced models of insulin resistance.[13] Thus, the effects of PPARγ agonists to improve insulin resistance may be explained in terms of changes in adipocyte metabolism and/or PPARγ-mediated production of proteins from fat (leptin, TNFα, or the more recently described mediators, resistin[26] and adiponectin[27]) that could affect insulin resistance in muscle through an endocrine mechanism. If the production of these proteins could be shown to be stimulated by increased hexosamine levels, then a pathway linking dietary excess to insulin resistance could be defined.

In further support of either the endocrine or nutrient-induced mechanism of adipocyte-induced insulin resistance in muscle, we had earlier noted that insulin resistance in animals overexpressing GFA in fat plus muscle was reversed upon explanting those muscles and stimulating with insulin *in vitro* (unpublished data).

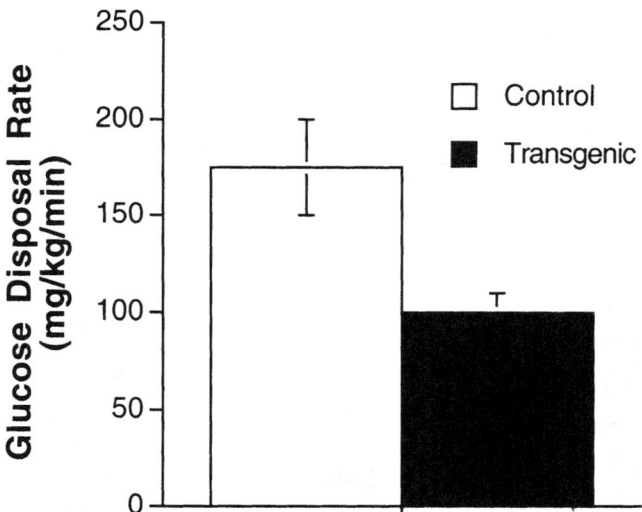

FIGURE 3. Insulin resistance in mice overexpressing GFA in muscle and fat under control of the GLUT4 promoter. Hyperinsulinemic-euglycemic clamp studies were performed on 2- to 4-month-old mice. Values represent the glucose infusion rates used to maintain a constant blood glucose value of 165 mg/dL during the procedure ($n = 4$/group; *$p < 0.05$).

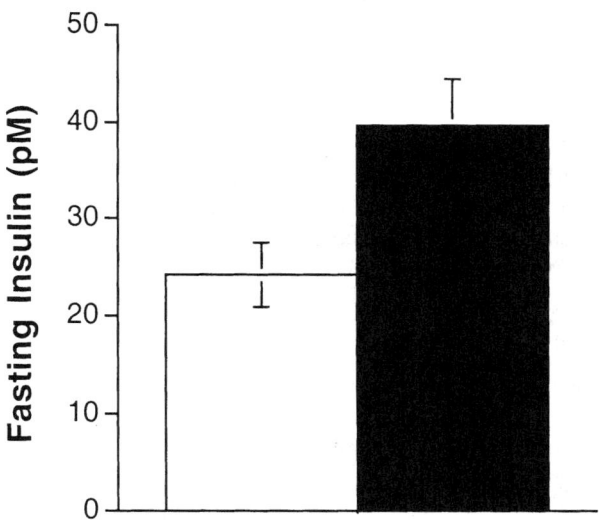

FIGURE 4. Plasma insulin levels in transgenic mice overexpressing GFA in beta cells under control of the rat insulin promoter. Insulin levels were determined by radioimmunoassay in sera of 12 mice [2- to 4-month-old control (open bar) and transgenic (closed bar) mice] per group ($p < 0.05$ transgenic vs. control).

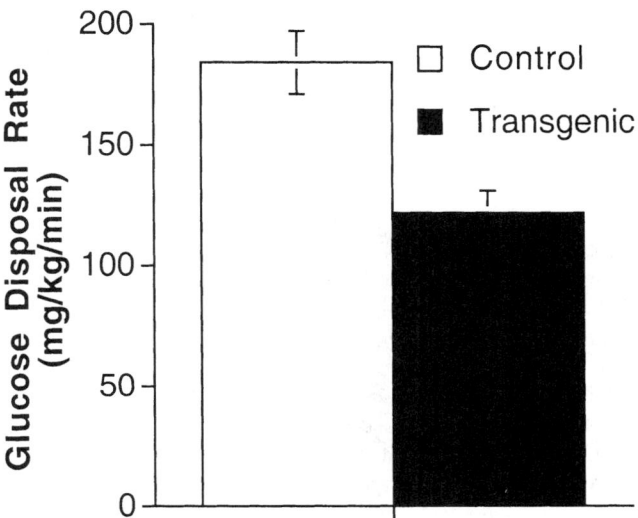

FIGURE 5. Transgenic mice overexpressing GFA under control of the aP2 promoter are insulin-resistant for total body and skeletal muscle glucose uptake. Glucose disposal rates were determined by the hyperinsulinemic-euglycemic clamp technique. Results are the means (± SE) of 13 control and 11 transgenic females. Steady-state glucose levels during the clamp were comparable in the two groups (control, 7.0 ± 1.5 mM; transgenic, 7.7 ± 1.5 mM). Glucose uptake into skeletal muscle, determined in animals at the end of the clamp by infusing a bolus of 2-deoxy-D-[^{3}H]glucose, paralleled the whole body glucose uptake data.

Similar results were noted in mice wherein adipose-specific disruption of the GLUT4 gene led to *in vivo* insulin resistance in muscle and liver, whereas insulin resistance was not seen in explanted muscle.[24] The results support the hypothesis that nutrient-induced insulin resistance in muscle is a physiologic and adaptive response whereby muscle can normalize nutrient flux in the face of delivery of excess calories to the organism.[28–30] Consistent with this idea, work from Yki-Jarvinen has shown that glucose uptake into muscle is relatively normal in poorly controlled diabetes.[30] That is, hyperglycemia or hyperinsulinemia is balanced by insulin resistance so that net flux into muscle is normal.

CYTOSOLIC *O*-LINKED GLYCOSYLATION

How might changes in protein function or transcription be signaled by hexosamines? Production of the substrates for protein glycosylation by the hexosamine pathway suggests a possible mechanism. In 1986, a pathway was described in which cytosolic proteins are modified by the *O*-linked addition of single GlcNAc residues.[31] Catalyzed by a specific *O*-glycosyltransferase (OGT[32,33]), this process is dynamic and highly regulated.[34,35] Several transcription factors, signal transduction proteins, and cytoskeletal proteins are *O*-glycosylated.[34] These proteins are generally multimeric, which suggests that this modification may be important in regulating the assembly of large protein complexes. The importance of the pathway is manifest by the embryonic lethality of deletion of OGT in mice.[32] However, the precise functional consequences of O-glycosylation are unclear and are currently being studied. Identification of the O-glycosylation sites on several proteins, such as c-myc,[36] suggests a possible reciprocal functional relationship of O-glycosylation with phosphorylation.

Among the factors modified by the addition of O-linked GlcNAc is Sp1, which we have implicated in the regulation of the TGFα gene by glucose and glucosamine.[37] It has been reported that O-linked GlcNAc modification of Sp1 may affect its interaction with DNA and slow proteasomal degradation.[38,39] Recently, hexosamine-driven O-glycosylation of Sp1 has also been linked with the stimulation of TGFβ and PAI-1 in hyperglycemia.[40]

The hexosamine and O-glycosylation pathway is an attractive candidate for glucose sensing. It is not intrinsically highly regulated, so physiologic changes in the levels of glucose are reflected in changes in UDP-GlcNAc and O-linked glycosylation.[41] In contrast, levels of N-linked or lipid glycosylation are affected little by modulation of hexosamine availability.[42] It has recently been reported that homologues of the mammalian O-GlcNAc transferase exist in plants. Deletion of this gene (*SPINDLY*) results in plants that are the precise opposite of "thrifty": they are tall, skinny, and do not store starch in their roots.[43] Thus, O-glycosylation is a candidate mechanism by which cells might modify transcriptional activity—and consequently hormone action and metabolism—in response to changes in fuel availability.

SUMMARY AND CONCLUSIONS

In summary, results from a number of laboratories demonstrate that hexosamines serve as glucose sensors in a wide variety of cells and tissues. Hexosamine flux is used therefore as an indicator of cellular satiety: in conditions of free availability of food, tissues divert the excess incoming fuel to storage pathways. Additionally, fat cells use hexosamine flux to translate cellular satiety signals into a satiety signal for the rest of the organism, that is, leptin. Activation of these physiologic storage pathways is adaptive in the short run. In fact, being "quick on the draw" in their activation would be analogous to "thrift", that is, preferentially storing calories for future lean times even at the expense of instantaneously available calories. However, with chronic caloric excess and saturation of storage depots, the same pathways lead to the "metabolic syndrome" of obesity, hyperlipidemia, insulin resistance, and (eventually) type 2 diabetes. Tissues such as vascular endothelium that are not involved in fuel economy are also responding, in parallel, to these pathways with activation of growth factors and other proteins. Further investigation of how these pathways function on a molecular level and how they impact growth and metabolism should be of value in understanding obesity, diabetes, and other nutrient-dependent processes.

ACKNOWLEDGMENTS

This work was supported by the Research Service of the Veterans Administration, the NIH (No. DK 43526), and the Ben and Iris Margolis Foundation.

REFERENCES

1. McKnight, G.L., S.L. Mudri, S.L. Mathewes et al. 1992. Molecular cloning, cDNA sequence, and bacterial expression of human glutamine:fructose-6-phosphate amidotransferase. J. Biol. Chem. **267:** 25208–25212.
2. Zhou, J., J.L. Neidigh, R. Espinosa III et al. 1995. Human glutamine:fructose-6-phosphate amidotransferase: characterization of mRNA and chromosomal assignment to 2p13. Hum. Genet. **96:** 99–101.
3. Marshall, S., V. Bacote & R.R. Traxinger. 1991. Discovery of a metabolic pathway mediating glucose-induced desensitization of the glucose transport system: role of hexosamine biosynthesis in the induction of insulin resistance. J. Biol. Chem. **266:** 4706–4712.
4. Crook, E.D., M.C. Daniels, T.M. Smith & D.A. McClain. 1993. Regulation of insulin-stimulated glycogen synthase activity by overexpression of glutamine:fructose-6-phosphate amidotransferase in rat-1 fibroblasts. Diabetes **42:** 1289–1296.
5. Robinson, K.A., D.A. Sens & M.G. Buse. 1993. Pre-exposure to glucosamine induces insulin resistance of glucose transport and glycogen synthesis in isolated rat skeletal muscles: study of mechanisms in muscle and in rat-1 fibroblasts overexpressing the human insulin receptor. Diabetes **42:** 1333–1346. [1993. Erratum. Diabetes **42**(10): 1547.]
6. Rossetti, L., M. Hawkins, W. Chen et al. 1995. In vivo glucosamine infusion induces insulin resistance in normoglycemic, but not in hyperglycemic conscious rats. J. Clin. Invest. **96:** 132–140.
7. Wang, J., R. Liu, M. Hawkins et al. 1998. A nutrient-sensing pathway regulates leptin gene expression in muscle and fat. Nature **393:** 684–688.
8. Baron, A.D., J.S. Zhu, J.H. Zhu et al. 1995. Glucosamine induces insulin resistance in vivo by affecting GLUT 4 translocation in skeletal muscle: implications for glucose toxicity. J. Clin. Invest. **96:** 2792–2801.
9. Giaccari, A., L. Morviducci, D. Zorretta et al. 1995. In vivo effects of glucosamine on insulin secretion and insulin sensitivity in the rat: possible relevance to the maladaptive responses to chronic hyperglycaemia. Diabetologia **38:** 518–524.
10. Virkamaki, A., M.C. Daniels, S. Hamalainen et al. 1997. Activation of the hexosamine pathway by glucosamine in vivo induces insulin resistance in multiple insulin sensitive tissues. Endocrinology **138:** 2501–2507.
11. Veerababu, G., J. Tang, R. Hoffman et al. 2000. Overexpression of glutamine:fructose-6-phosphate amidotransferase in the liver of transgenic mice results in enhanced glycogen storage, hyperlipidemia, obesity, and impaired glucose tolerance. Diabetes **49:** 2070–2078.
12. Hebert, L.F., Jr., M.C. Daniels, J. Zhou et al. 1996. Overexpression of glutamine:fructose-6-phosphate amidotransferase in transgenic mice leads to insulin resistance. J. Clin. Invest. **98:** 930–936.
13. Cooksey, R.C., L.F. Hebert, Jr., J.H. Zhu et al. 1999. Mechanism of hexosamine-induced insulin resistance in transgenic mice overexpressing glutamine:fructose-6-phosphate amidotransferase: decreased glucose transporter GLUT4 translocation and reversal by treatment with thiazolidinedione. Endocrinology **140:** 1151–1157.
14. Tang, J., J.L. Neidigh, R.C. Cooksey & D.A. McClain. 2000. Transgenic mice with increased hexosamine flux specifically targeted to beta-cells exhibit hyperinsulinemia and peripheral insulin resistance. Diabetes **49:** 1492–1499.
15. Daniels, M.C., T.P. Ciaraldi, S. Nikoulina et al. 1996. Glutamine:fructose-6-phosphate amidotransferase activity in cultured human skeletal muscle cells: relationship to glucose disposal rate in control and non-insulin-dependent diabetes mellitus subjects and regulation by glucose and insulin. J. Clin. Invest. **97:** 1235–1241.
16. Considine, R.V., R.C. Cooksey, L.B. Williams et al. 2000. Hexosamines regulate leptin production in human subcutaneous adipocytes. J. Clin. Endocrinol. Metab. **85:** 3551–3556.
17. Barzilai, N., M. Hawkins, I. Angelov et al. 1996. Glucosamine-induced inhibition of liver glucokinase impairs the ability of hyperglycemia to suppress endogenous glucose production. Diabetes **45:** 1329–1335.

18. SHANKAR, R.R., J.S. ZHU & A.D. BARON. 1998. Glucosamine infusion in rats mimics the beta-cell dysfunction of non-insulin-dependent diabetes mellitus. Metabolism **47:** 573–577.
19. MONAUNI, T., M.G. ZENTI, A. CRETTI et al. 2000. Effects of glucosamine infusion on insulin secretion and insulin action in humans. Diabetes **49:** 926–935.
20. KIM, Y.B., J.S. ZHU, J.R. ZIERATH et al. 1999. Glucosamine infusion in rats rapidly impairs insulin stimulation of phosphoinositide 3-kinase, but does not alter activation of Akt/protein kinase B in skeletal muscle. Diabetes **48:** 310–320.
21. PATTI, M.E., A. VIRKAMAKI, E.J. LANDAKER et al. 1999. Activation of the hexosamine pathway by glucosamine *in vivo* induces insulin resistance of early postreceptor insulin signaling events in skeletal muscle. Diabetes **48:** 1562–1571.
22. KIM, Y.B., S.E. NIKOULINA, T.P. CIARALDI et al. 1999. Normal insulin-dependent activation of Akt/protein kinase B, with diminished activation of phosphoinositide 3-kinase, in muscle in type 2 diabetes [see comments]. J. Clin. Invest. **104:** 733–741.
23. KUROWSKI, T.G., Y. LIN, Z. LUO et al. 1999. Hyperglycemia inhibits insulin activation of Akt/protein kinase B, but not phosphatidylinositol 3-kinase in rat skeletal muscle. Diabetes **48:** 658–663.
24. ABEL, E.D., O. PERONI, J.K. KIM et al. 2001. Adipose-selective targeting of the GLUT4 gene impairs insulin action in muscle and liver. Nature. In press.
25. SALTIEL, A.R. & J.M. OLEFSKY. 1996. Thiazolidinediones in the treatment of insulin resistance and type II diabetes. Diabetes **45:** 1661–1669.
26. STEPPAN, C.M., S.T. BAILEY, S. BHAT et al. 2001. The hormone resistin links obesity to diabetes. Nature **409:** 307–311.
27. YAMAUCHI, T., J. KAMON, H. WAKI et al. 2001. The fat-derived hormone adiponectin reverses insulin resistance associated with both lipoatrophy and obesity. Nat. Med. **7:** 941–946.
28. MCCLAIN, D.A. & E.D. CROOK. 1996. Hexosamines and insulin resistance. Diabetes **45:** 1003–1009.
29. ROSSETTI, L. 2000. Perspective: hexosamines and nutrient sensing. Endocrinology **141:** 1922–1925.
30. YKI-JARVINEN, H. 1997. Acute and chronic effects of hyperglycaemia on glucose metabolism: implications for the development of new therapies. Diabet. Med. **14**(suppl. 3): S32–S37.
31. HOLT, G.D. & G.W. HART. 1986. The subcellular distribution of terminal *N*-acetyl-glucosamine moieties: localization of a novel protein-saccharide linkage, *O*-linked GlcNAc. J. Biol. Chem. **261:** 8049–8057.
32. SHAFI, R., S.P. IYER, L.G. ELLIES et al. 2000. The *O*-GlcNAc transferase gene resides on the X chromosome and is essential for embryonic stem cell viability and mouse ontogeny. Proc. Natl. Acad. Sci. U.S.A. **97:** 5735–5739.
33. LUBAS, W.A. & J.A. HANOVER. 2000. Functional expression of *O*-linked GlcNAc transferase: domain structure and substrate specificity. J. Biol. Chem. **275:** 10983–10988.
34. COMER, F.I. & G.W. HART. 1999. *O*-GlcNAc and the control of gene expression. Biochim. Biophys. Acta **1473:** 161–171.
35. COMER, F.I. & G.W. HART. 2000. *O*-Glycosylation of nuclear and cytosolic proteins: dynamic interplay between *O*-GlcNAc and *O*-phosphate. J. Biol. Chem. **275:** 29179–29182.
36. CHOU, T.Y., G.W. HART & C.V. DANG. 1995. c-Myc is glycosylated at threonine 58, a known phosphorylation site and a mutational hot spot in lymphomas. J. Biol. Chem. **270:** 18961–18965.
37. MCCLAIN, D.A., A.J. PATERSON, M.D. ROOS et al. 1992. Glucose and glucosamine regulate growth factor gene expression in vascular smooth muscle cells. Proc. Natl. Acad. Sci. U.S.A. **89:** 8150–8154.
38. ROOS, M.D., K. SU, J.R. BAKER & J.E. KUDLOW. 1997. *O* glycosylation of an Sp1-derived peptide blocks known Sp1 protein interactions. Mol. Cell. Biol. **17:** 6472–6480.
39. HAN, I. & J.E. KUDLOW. 1997. Reduced O glycosylation of Sp1 is associated with increased proteasome susceptibility. Mol. Cell. Biol. **17:** 2550–2558.
40. DU, X.L., D. EDELSTEIN, L. ROSSETTI et al. 2000. Hyperglycemia-induced mitochondrial superoxide overproduction activates the hexosamine pathway and induces

plasminogen activator inhibitor-1 expression by increasing Sp1 glycosylation. Proc. Natl. Acad. Sci. U.S.A. **97:** 12222–12226.
41. YKI-JARVINEN, H., C. VOGT, P. LOZZO *et al.* 1997. UDP-*N*-acetylglucosamine transferase and glutamine:fructose 6-phosphate amidotransferase activities in insulin-sensitive tissues. Diabetologia **40:** 76–81.
42. BOEHMELT, G., A. WAKEHAM, A. ELIA *et al.* 2000. Decreased UDP-GlcNAc levels abrogate proliferation control in EMeg32-deficient cells. EMBO J. **19:** 5092–5104.
43. JACOBSEN, S.E., K.A. BINKOWSKI & N.E. OLSZEWSKI. 1996. SPINDLY, a tetratricopeptide repeat protein involved in gibberellin signal transduction in *Arabidopsis*. Proc. Natl. Acad. Sci. U.S.A. **93:** 9292–9296.

Enhanced Diabetogenic Effect of Streptozotocin in Mice Overexpressing UCP-3 in Skeletal Muscle

STEVEN WANG,[a] MICHAEL A. CAWTHORNE,[a] AND JOHN C. CLAPHAM[b,c]

[a]*Clore Laboratory, University of Buckingham, Buckingham MK18 1EG, United Kingdom*

[b]*GlaxoSmithKline, Neurology-CEDD, New Frontiers Science Park, Pinnacles, Harlow CM19 5AW, United Kingdom*

ABSTRACT: Diabetic patients exhibit varying degrees of increased muscle UCP-3 expression in skeletal muscle and, in rodents, the pancreatoxin streptozotocin (STZ) upregulates UCP-3 mRNA in skeletal and cardiac muscles. We have investigated the development of STZ-induced diabetes in transgenic mice overexpressing UCP-3 in skeletal muscle in order to provide further insight on the functional role of muscle UCP-3. UCP-3 transgenic mice treated with STZ (UCP3-STZ) showed a significant increase in blood glucose concentration 3 days after the last dose of STZ with a progressive induction of diabetes, attaining blood glucose concentrations of 24.7 ± 1.5 mmol/L on day 17. Wild-type mice treated with STZ (WT-STZ) only started to show an increase in blood glucose concentration 6 days after the last dose of STZ and peaked on day 17 at a lower concentration than in the UCP-STZ mice. The pancreatic insulin content of UCP-3 control mice (UCP3-CON) was decreased relative to wild-type control mice (WT-CON), and STZ reduced the total pancreatic insulin content by 72% in WT-STZ mice and by 88% in UCP3-STZ mice. In an insulin tolerance test, blood glucose concentrations declined more in the UCP-3 transgenic mice than in the wild-type mice. Mice overexpressing UCP-3 in skeletal muscle have a lower pancreatic insulin content, but tend to be more insulin-sensitive. These twin actions result in an increased susceptibility to STZ-induced diabetes in UCP-3 transgenic mice.

KEYWORDS: uncoupling proteins; UCP-3; transgenic mice; streptozotocin (STZ); insulin

INTRODUCTION

Uncoupling protein-3 (UCP-3)[1,2] was given its designation as a result of its mRNA sequence homology with UCP-1. It is predominantly expressed in skeletal muscle and brown adipose tissue, and this expression pattern led to the hypothesis that UCP-3 might have an uncoupling role leading to thermogenesis. However, this

---

Address for correspondence: Steven Wang, Clore Laboratory, University of Buckingham, Buckingham MK18 1EG, United Kingdom.Voice: +44-1280-820242; fax: +44-1280-820261.
steven.wang@buckingham.ac.uk

[c]Present address: Cell Biology and Biochemistry, AstraZeneca R&D Mölndal, Pepparedsleden 1, SE431 83 Mölndal, Sweden.

Ann. N.Y. Acad. Sci. 967: 112–119 (2002). © 2002 New York Academy of Sciences.

role has not been supported so far by experimental data. Its functional role remains unidentified and a recent hypothesis suggests that it could be a fatty acid transporter to facilitate fatty acid oxidation.[3] Thermogenic agents such as the β3-adrenoceptor agonist CL316243[4] and rexinoids[5] have shown induction of UCP-3 mRNA.

UCP-3 gene expression in skeletal muscle is also increased by treatment with a high-fat diet, exercise, and fasting. In uncontrolled diabetes, there is also an increase in fatty acid oxidation, and streptozotocin (STZ)–induced diabetic rats showed an upregulation in UCP-3 mRNA expression in skeletal muscle and heart.[6,7] Although it is now well documented that fatty acids regulate UCP-3 gene expression,[8] it has also been reported that an increased glucose uptake by skeletal muscle is associated with increased UCP-3 gene expression.[9,10] Moreover, a positive correlation between UCP-3 mRNA expression in skeletal muscle and whole-body insulin-mediated glucose utilization has been found among NIDDM patients.[11] Thus, UCP-3 has been identified as a potential molecular target for the treatment of both obesity and diabetes.

Further insight on the functional role of UCP-3 might be obtained from overexpression studies. Thus, overexpression of UCP-3 in L6 myotubes or H_9C_2 cardiomyoblasts results in an increase in glucose uptake.[12] Transgenic mice overexpressing human UCP-3 in skeletal muscle[13] have increased whole-body energy expenditure, increased insulin sensitivity (as shown by increased glucose clearance rate), and reduced plasma total cholesterol. The UCP-3 transgenic mice were leaner and weighed less than wild-type controls. In the present study, we investigated the development of STZ-induced diabetes in transgenic mice overexpressing hUCP-3 in skeletal muscle in order to provide further insight on the role of muscle UCP-3 in fuel partitioning and the potential role of UCP-3 in the development of diabetes.

MATERIALS AND METHODS

Mice and Treatment

Transgenic mice overexpressing hUCP-3 in skeletal muscle were generated by Clapham *et al.*[13] Breeding pairs of the transgenic mice on the C57BL/6 × CBA background were provided and the progeny from the breeding program were used for the present study. Wild-type mice on a similar background strain were purchased from Charles River (Manston, United Kingdom). All experimental procedures were performed according to national and institutional guidelines and regulations for care and use of laboratory animals.

Eight-week-old female UCP-3 transgenic mice and wild-type mice were treated with multiple low doses of STZ to induce diabetes. STZ (Sigma, Gillingham, United Kingdom) at 50 mg/kg body weight in 0.1 M citrate buffer (pH 4.5) was given intraperitoneally daily for 5 consecutive days. Control mice received 0.1 M citrate buffer. The mice were monitored over a period of 25 days. Blood glucose concentrations of fed mice and insulin concentrations in 2-h fasted mice were measured regularly. Daily food and water intake were also recorded. At the end of the experiment, the mice were killed by cervical dislocation. Terminal whole blood was collected for measurements of plasma insulin. Various tissues were quickly dissected out of the animal, snap-frozen in liquid nitrogen, and stored at −80°C.

Blood glucose concentrations were determined using Trinder's reagent (Sigma). A rat insulin ELISA kit (Crystal Chem, IL) was used to measure plasma insulin concentrations according to the manufacturer's instructions.

Total Pancreatic Insulin Content

Pancreatic insulin was extracted in acid ethanol (165 mM HCl in 75% ethanol). Whole pancreas was homogenized in 10 mL/g acid ethanol and the homogenate was left overnight at 4°C. The homogenate was then centrifuged at 2000g for 5 min and the supernatant was removed for determination of insulin concentration. Insulin was assayed using a rat insulin [^{125}I]-radioimmunoassay kit (Amersham International, Little Chalfont, United Kingdom).

Insulin Tolerance Test

Four-month-old male UCP-3 transgenic mice and wild-type mice were tested for their tolerance to insulin. After a 5-h fast, the mice were anesthetized with pentobarbitone (Rhône Mérieux, Dagenham, United Kingdom) ip at 70 mg/kg body weight and then given a single ip bolus of bovine insulin (Sigma) at 0.75 U/kg body weight.[14] Blood (10 µL) from the tail vein was sampled at 2-min intervals for the measurement of blood glucose concentrations. All mice were killed by cervical dislocation without recovery from the anesthetic.

RESULTS

Blood Glucose and Plasma Insulin Concentration

At the start of the study, significantly lower blood glucose concentrations were found in UCP-3 transgenic mice compared with wild-type controls, being 8.0 ± 0.2 mmol/L and 8.9 ± 0.1 mmol/L, respectively ($p < 0.0005$, $n = 15$). Hyperphagia was observed in both groups of diabetic mice and was exacerbated in the UCP3-STZ group (FIG. 1). Fed blood glucose concentrations (FIG. 2) in UCP3-STZ mice were significantly increased compared with UCP3-CON mice 3 days after the last dose of STZ (12.9 ± 0.9 mmol/L vs. 7.7 ± 0.1 mmol/L, $p < 0.01$). Blood glucose concentrations in the former group subsequently reached 24.7 ± 1.5 mmol/L compared with 7.8 ± 0.4 mmol/L in the UCP3-CON group ($p < 0.01$) from day 17. The wild-type mice treated with STZ (WT-STZ) were only mildly diabetic 6 days after the last dose of STZ (10.5 ± 0.6 mmol/L vs. 8.4 ± 0.3 mmol/L, $p < 0.05$). The blood glucose concentrations in this group peaked on day 17 (12.1 ± 1.0 mmol/L vs. 8.2 ± 0.2 mmol/L, $p < 0.01$), but at a lower concentration than in the UCP3-STZ group. Plasma insulin concentrations (FIG. 3) in UCP3-STZ mice were elevated on day 12 relative to UCP3-CON mice (877 ± 120 pg/mL vs. 506 ± 58 pg/mL, $p < 0.05$). However, by day 19, they had declined to just below control levels, whereas wild-type mice treated with STZ had values significantly below controls. Interestingly, although the plasma insulin concentration in UCP3-STZ mice was significantly increased on day 12, there was no reduction in the blood glucose concentration.

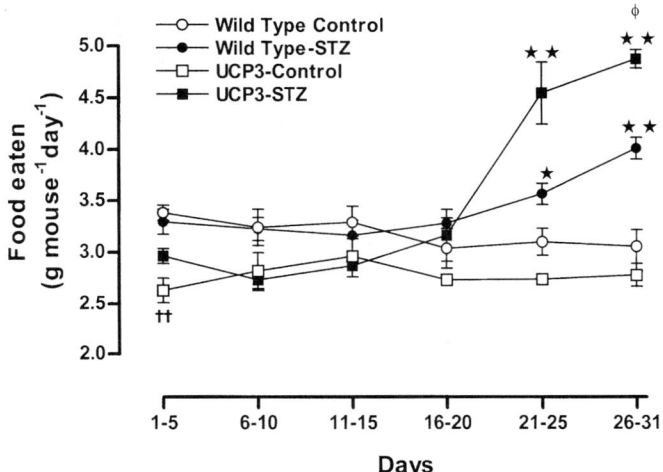

FIGURE 1. Food intake of wild-type and UCP-3 transgenic mice treated with STZ. Values are recorded as g/mouse/day and expressed as mean ± SEM. STZ (50 mg/kg) was given on days 1–5 inclusive. Significant difference between STZ-treated and control groups with same genotype: $^{\star}p < 0.05$; $^{\star\star}p < 0.01$. Significant difference between UCP-3 transgenic and wild-type groups: $^{\dagger\dagger}p < 0.01$. Significant difference between UCP-3 transgenic mice given STZ and wild-type groups given STZ: $^{\phi}p < 0.05$.

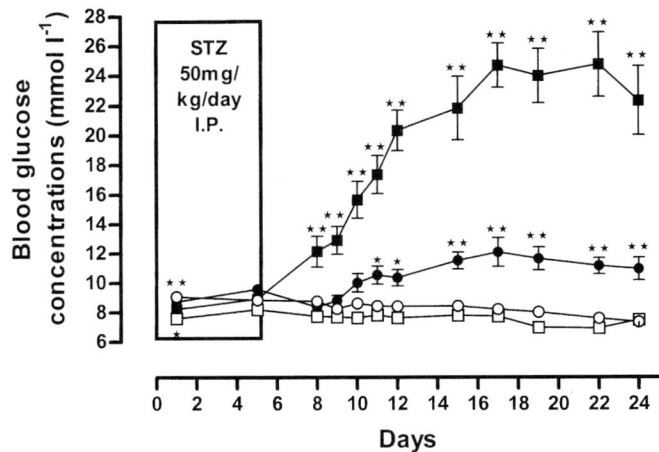

FIGURE 2. Blood glucose concentrations of wild-type and UCP-3 transgenic mice treated with STZ (mmol/L ± SEM). Wild-type mice are denoted by circles. UCP-3 transgenic mice are denoted by squares. Open symbols are control groups. Full symbols are STZ groups. Significant difference from control groups of same genotype not given STZ: $^{\star}p < 0.05$; $^{\star\star}p < 0.01$.

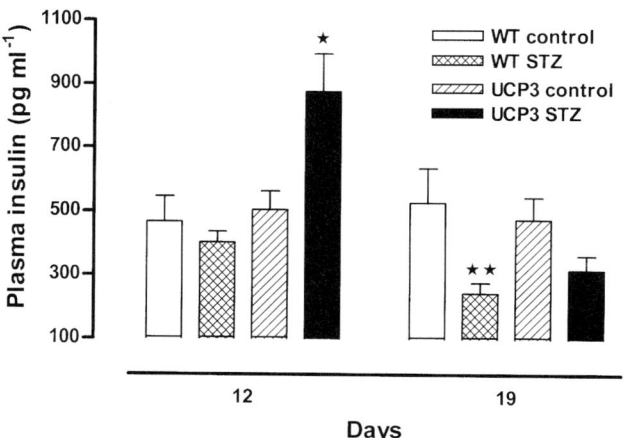

FIGURE 3. Plasma insulin (pg/mL) of wild-type and UCP-3 transgenic mice treated with STZ. STZ (50 mg/kg) was given on days 1–5 inclusive. Mice were fasted for 2 h prior to plasma sampling on days 12 and 19. Significant difference between STZ-treated and control mice of the same genotype: ★$p < 0.05$; ★★$p < 0.01$.

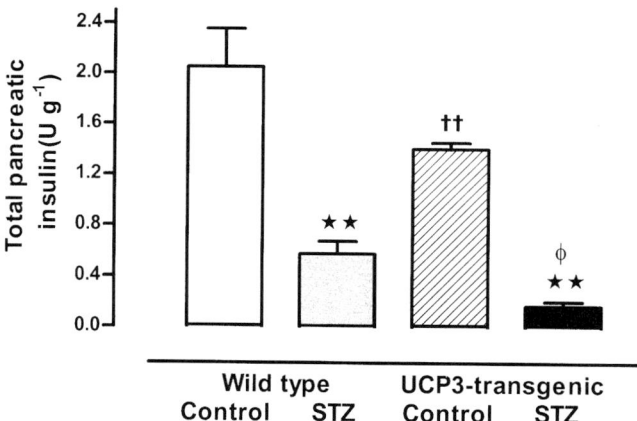

FIGURE 4. Total pancreatic insulin content (U/g pancreas) in wild-type and UCP-3 transgenic mice treated with STZ. Significant difference between STZ and control mice of the same genotype: ★★$p < 0.01$. Significant difference between UCP3-CON and WT-CON groups: ††$p < 0.01$. Significant difference between UCP3-STZ and WT-STZ groups: ϕ$p < 0.05$.

Total Pancreatic Insulin Content

UCP-3 control mice had lower pancreatic insulin content (FIG. 4) compared with wild-type control mice (1.40 ± 0.05 U/g vs. 2.05 ± 0.30 U/g, $p < 0.01$). STZ treatment significantly lowered the pancreatic insulin content in WT-STZ and UCP3-STZ groups by 72% and 88%, respectively.

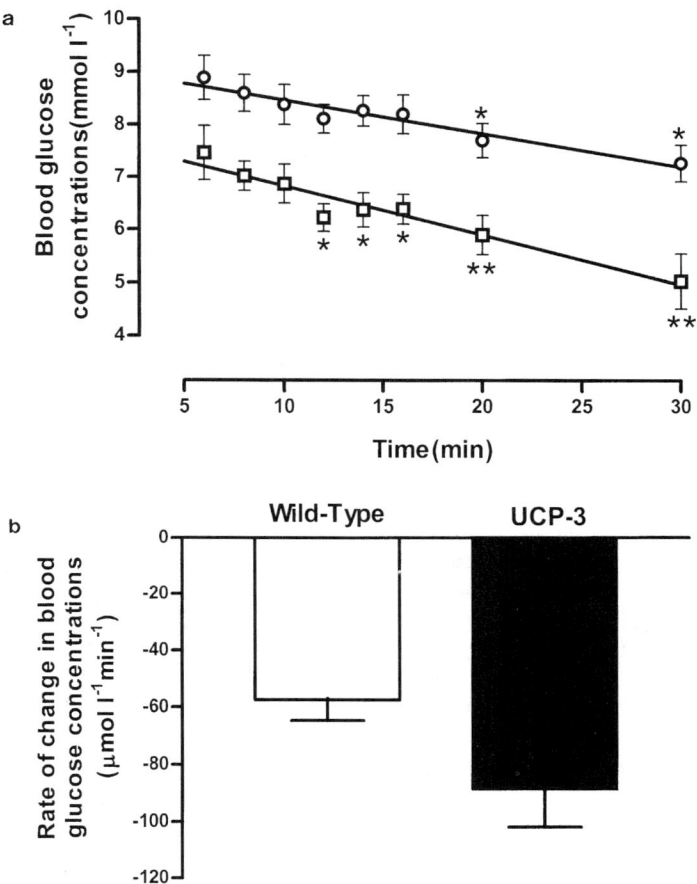

FIGURE 5. Insulin tolerance test in wild-type and UCP-3 transgenic mice. Male UCP-3 transgenic and wild-type mice were anesthetized and given an ip bolus of insulin. **(a)** Blood glucose concentrations were monitored over time. Open circles and open squares represent wild-type and UCP-3 mice, respectively. Significant difference: $*p < 0.05$; $**p < 0.01$. **(b)** Rate of change in blood glucose concentrations expressed as μmol/L/min ± SEM ($p = 0.07$). The fall in blood glucose was monitored over 30 min.

Insulin Tolerance Test

The blood glucose concentrations (FIG. 5a) in the UCP-3 transgenic mice fell significantly ($p < 0.05$) 12 min after the bolus injection of insulin and they continued to fall until 30 min ($p < 0.01$). The wild-type mice responded similarly, but the starting blood glucose concentration was significantly greater in the wild-type mice ($p = 0.0015$).

Analysis of the rate of change in the blood glucose concentration with time in the wild-type and transgenic mice indicated a trend towards increased insulin sensitivity ($p = 0.07$) in the mice overexpressing UCP-3 (FIG. 5b).

DISCUSSION

Previous studies[13] have shown that mice overexpressing human UCP-3 in skeletal muscle had lower fasting plasma glucose, greater glucose clearance following an oral glucose load, and lower fasting plasma insulin concentration, suggesting that the UCP-3 transgenic mice might be more insulin-sensitive. In the present study, we have found significantly lower blood glucose concentrations in UCP-3 transgenic mice compared with wild-type mice and a trend towards greater glucose clearance following an intraperitoneal insulin bolus. These findings, together with earlier results, confirm the tendency of UCP-3 transgenic mice to be more insulin-sensitive.

In vitro studies have also shown that adenovirus-mediated overexpression of UCP-3 in L6 myotubes and H_9C_2 cardiomyoblasts results in increased glucose uptake.[12] Thus, the overexpression of UCP-3 in skeletal muscle in mice might result in increased rates of glucose utilization and would ameliorate any diabetic condition. In contrast, we found that STZ was more diabetogenic in UCP-3 transgenic mice than in wild-type mice.

A second surprising finding from the present study was that the pancreatic insulin content of UCP-3 transgenic mice was significantly lower than that of wild-type mice. There are several lines of evidence to suggest that a low islet cell mass is a predisposing factor in the development of diabetes. Thus, C57BL/6 mice have twice the islet cell mass of C57BL/Ks mice and are more resistant to the diabetogenic action of STZ.[15] Similarly, perinatally malnourished rats have reduced islet cell mass and show less regenerative capacity than islets from control rats.[16,17] It seems possible that the reduced pancreatic insulin content in the UCP-3 transgenic mice relative to controls might be a factor in the increased sensitivity to STZ.

The mechanism by which increased expression of human UCP-3 in skeletal muscle alters pancreatic insulin content is not clear. It is also not known yet whether the reduced insulin content is directly associated with a reduced pancreatic islet cell mass, but this seems likely. Pancreatic islet cell mass is determined predominantly during fetal and early life. However, these UCP-3 transgenic mice were generated using an α-skeletal actin promoter, which is only expressed postpartum.[13] Thus, any effect of the transgene on islet cell mass must have occurred postpartum or it must be the result of the maternally expressed transgene on *in utero* development, perhaps through alterations in fuel supplies to the fetuses.

The effect of the transgene on the diabetogenic effects of STZ might also be affected by changes in fuel utilization. It is established that STZ-induced diabetic rats show hyperphagia, increased plasma fatty acid concentration, and an upregulation in UCP-3 mRNA.[6,7] Himms-Hagen and Harper[3] recently proposed that the possible role of UCP-3 was as a carrier for fatty acid anions during fat oxidation. Thus, in a situation where insulin action is reduced and fatty acid is the predominant fuel, the flux of fatty acid anions through UCP-3 might limit fatty acid oxidation. Consequently, UCP3-STZ mice might have a greater potential than WT-STZ mice for fatty acid oxidation, which will indirectly drive increased food intake. The increased hyperglycemia in UCP3-STZ mice may indirectly be the result of increased food intake to support the increased fatty acid oxidation. Further work is required to explore this hypothesis.

ACKNOWLEDGMENTS

We thank David Hislop and Anita Roberts for technical support. Experimental advice and guidance by Matthew V. Sennitt are appreciated. We also thank Ed Wargent for help with measurements of pancreatic insulin content.

REFERENCES

1. Boss, O., S. Samec, A. Paoloni-Giacobino et al. 1997. Uncoupling protein-3: a new member of the mitochondrial carrier family with tissue-specific expression. FEBS Lett. **408:** 39–42.
2. Vidal-Puig, A., G. Solanes, D. Grujic et al. 1997. UCP-3: an uncoupling protein homologue expressed preferentially and abundantly in skeletal muscle and brown adipose tissue. Biochem. Biophys. Res. Commun. **235:** 79–82.
3. Himms-Hagen, J. & M.E. Harper. 2001. Physiological role of UCP3 may be export of fatty acids from mitochondria when fatty acid oxidation predominates: an hypothesis. Exp. Biol. Med. (Maywood) **226:** 78–84.
4. Nakamura, Y., I. Nagase, A. Asano et al. 2001. Beta3-adrenergic agonist up-regulates uncoupling proteins 2 and 3 in skeletal muscle of the mouse. J. Vet. Med. Sci. **63:** 309–314.
5. Emilsson, V., J. O'Dowd, S. Wang et al. 2000. The effects of rexinoids and rosiglitazone on body weight and uncoupling protein isoform expression in the Zucker fa/fa rat. Metabolism **49:** 1610–1615.
6. Kageyama, H., A. Suga, M. Kashiba et al. 1998. Increased uncoupling protein-2 and -3 gene expressions in skeletal muscle of STZ-induced diabetic rats. FEBS Lett. **440:** 450–453.
7. Hidaka, S., T. Kakuma, H. Yoshimatsu et al. 1999. Streptozotocin treatment upregulates uncoupling protein 3 expression in the rat heart. Diabetes **48:** 430–435.
8. Weigle, D.S., L.E. Selfridge, M.W. Schwartz et al. 1998. Elevated free fatty acids induce uncoupling protein 3 expression in muscle: a potential explanation for the effect of fasting. Diabetes **47:** 298–302.
9. Tsuboyama-Kasaoka, N., N. Tsunoda, K. Maruyama et al. 1998. Up-regulation of uncoupling protein 3 (UCP3) mRNA by exercise training and down-regulation of UCP3 by denervation in skeletal muscles. Biochem. Biophys. Res. Commun. **247:** 498–503.
10. Tsuboyama-Kasaoka, N., N. Tsunoda, K. Maruyama et al. 1999. Overexpression of GLUT4 in mice causes up-regulation of UCP3 mRNA in skeletal muscle. Biochem. Biophys. Res. Commun. **258:** 187–193.
11. Krook, A., J. Digby, S. O'Rahilly et al. 1998. Uncoupling protein 3 is reduced in skeletal muscle of NIDDM patients. Diabetes **47:** 1528–1531.
12. Huppertz, C., B.M. Fischer, Y.B. Kim et al. 2001. Uncoupling protein 3 (UCP3) stimulates glucose uptake in muscle cells through a phosphoinositide 3-kinase–dependent mechanism. J. Biol. Chem. **276:** 12520–12529.
13. Clapham, J.C., J.R. Arch, H. Chapman et al. 2000. Mice overexpressing human uncoupling protein-3 in skeletal muscle are hyperphagic and lean. Nature **406:** 415–418.
14. Klaman, L.D., O. Boss, O.D. Peroni et al. 2000. Increased energy expenditure, decreased adiposity, and tissue-specific insulin sensitivity in protein-tyrosine phosphatase 1B-deficient mice. Mol. Cell. Biol. **20:** 5479–5489.
15. Swenne, I. & A. Andersson. 1984. Effect of genetic background on the capacity for islet cell replication in mice. Diabetologia **27:** 464–467.
16. Garofano, A., P. Czernichow & B. Breant. 1997. In utero undernutrition impairs rat beta-cell development. Diabetologia **40:** 1231–1234.
17. Garofano, A., P. Czernichow & B. Breant. 2000. Impaired beta-cell regeneration in perinatally malnourished rats: a study with STZ. FASEB J. **14:** 2611–2617.

From Receptor to Effector: Insulin Signal Transduction in Skeletal Muscle from Type II Diabetic Patients

JULEEN R. ZIERATH[a] AND HARRIET WALLBERG-HENRIKSSON[a,b]

[a]*Department of Clinical Physiology, Karolinska Hospital, Karolinska Institutet, Stockholm, Sweden*

[b]*Department of Physiology and Pharmacology, Karolinska Institutet, Stockholm, Sweden*

ABSTRACT: Insulin resistance is a characteristic feature of type II diabetes mellitus and obesity. Although defects in glucose homeostasis have been recognized for decades, the molecular mechanisms accounting for impaired whole body glucose uptake are still not fully understood. Skeletal muscle constitutes the largest insulin-sensitive organ in humans; thus, insulin resistance in this tissue will have a major impact on whole body glucose homeostasis. Intense efforts are under way to define the molecular mechanisms that regulate glucose metabolism and gene expression in insulin-sensitive tissues. Knowledge of the human genome sequence, used in concert with gene and/or protein array technology, will provide a powerful means to facilitate efforts in revealing molecular targets that regulate glucose homeostasis in type II diabetes mellitus. This will offer quicker ways forward to identifying gene expression profiles in insulin-sensitive and insulin-resistant human tissue. This review will present our current understanding of potential defects in insulin signal transduction pathways, with an emphasis on mechanisms regulating glucose transport in skeletal muscle from people with type II diabetes mellitus. Elucidation of the pathways involved in the regulation of glucose homeostasis will offer insight into the causation of insulin resistance and type II diabetes mellitus. Furthermore, this will identify biochemical entry points for drug intervention to improve glucose homeostasis.

KEYWORDS: insulin receptor; insulin receptor substrates; phosphatidylinositol 3-kinase; mitogen-activated protein kinase; glucose transporter; non-insulin-dependent diabetes mellitus; obesity; polycystic ovary syndrome; gestational diabetes mellitus; hyperglycemia free fatty acids

INTRODUCTION

The incidence of type II (non-insulin-dependent) diabetes mellitus is growing at an astronomical rate as millions of people are diagnosed with this profound metabolic disorder every year. Type II diabetes mellitus is a progressive metabolic disorder that

Address for correspondence: Juleen R. Zierath, Ph.D., Professor of Physiology, Department of Clinical Physiology and Integrative Physiology, Karolinska Institutet, von Eulers väg 4, II, SE-171 77 Stockholm, Sweden. Voice: +46-8-728-7580; fax: +46-8-33-54-36.
Juleen.Zierath@fyfa.ki.se

develops from both environmental and undefined genetic factors.[1] Although the pattern of inheritance is complex, a genetic component of type II diabetes mellitus has been demonstrated. Importantly, environmental factors have a major impact on the development of the disease. Type II diabetic patients are characterized by fasting hyperglycemia and by elevated, normal, or low levels of insulin.[2] Either of these factors may be a cause or consequence of defects in insulin secretion from the β-cell and/or peripheral insulin resistance in skeletal muscle, adipose tissue, or liver.[2] Although the primary defect in the development of type II diabetes mellitus is unclear, insulin resistance in skeletal muscle is a hallmark feature of the disease.[3] In normal glucose-tolerant relatives of type II diabetic patients, insulin resistance in skeletal muscle has been observed several years before the development of overt diabetes.[4] Thus, impaired insulin action in skeletal muscle constitutes an early defect in the pathogenesis of type II diabetes mellitus.

During recent years, major advances have been made in our understanding of the molecular mechanisms of insulin action, bringing closer the goal of identifying the defect(s) leading to insulin resistance. Understanding the regulation of pathways that mediate insulin action may lead to the identification of molecular targets for therapy. This review will present the basic understanding of intercellular signaling mechanisms by which insulin regulates metabolic responses. Special emphasis will be placed on studies of insulin action in skeletal muscle from people with type II diabetes mellitus.

SIGNALING PATHWAYS REGULATING GLUCOSE METABOLISM

Insulin Receptor

Defects in the insulin signaling pathway are linked with the pathogenesis of type II (non-insulin-dependent) diabetes mellitus. Glucose transport and metabolism, protein synthesis, and gene expression are all regulated by activation of signal transduction pathways mediated via the insulin receptor (FIG. 1). The insulin receptor is a heterotetrameric glycoprotein membrane composed of two α and two β subunits, linked together by disulfide bonds.[5,6] Insulin binds to the extracellular α subunits and this leads to activation of the transmembrane β subunits and autophosphorylation of the receptor.[7] Multiple tyrosine phosphorylation sites present on the β subunit of the insulin receptor play important functional roles in promoting receptor kinase activity, mediating differential responses along mitogenic and metabolic pathways, and facilitating the interaction between the receptor and its intracellular substrates.[8] Rare mutations in the insulin receptor have been associated with severe insulin resistance, such as leprechaunism or type A syndrome of insulin resistance and acanthosis nigricans.[9] However, these mutations account for less than 1% of type II diabetic patients.[1] While insulin receptor mutations are not likely to be the cause of type II diabetes mellitus in the majority of patients, impaired signal transduction via the insulin receptor pathway is likely to have a profound effect on important metabolic responses such as the regulation of glucose homeostasis.

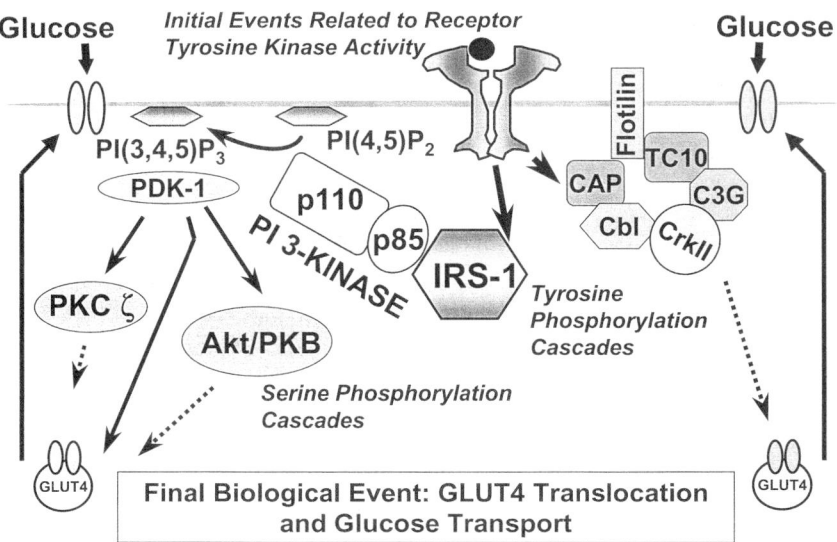

FIGURE 1. Simplified outline of insulin signaling pathways based on literature reviewed in the text.

Insulin Receptor Substrates

Insulin-receptor substrate isoforms (IRS-1 to -4),[10–13] Gab-1,[14] and Cbl[15] link the initial event of insulin receptor signaling cascade to downstream events. Signal transduction via the different IRS molecules partly accounts for the diverse effects of insulin along metabolic and mitogenic pathways. IRS molecules contain multiple tyrosine phosphorylation sites that become phosphorylated after insulin stimulation.[16] These phosphorylation sites form binding sites for downstream signaling molecules containing *src* homology 2 (SH2) domains.[17] IRS-1 and IRS-2 play selective roles in the regulation of metabolic and mitogenic responses. Genetic ablation of IRS-1 is associated with severe growth retardation and mild insulin resistance,[18,19] whereas ablation of IRS-2 leads to insulin resistance in peripheral tissues and impaired growth of the pancreatic β-cells.[20] The difference in the phenotype between these two animal models may be explained in part by the ability of IRS-2 to compensate for the lack of IRS-1[21] or from variations in the tissue distribution of the IRS isoforms.[22] Clearly, both molecules are important for insulin action; however, IRS-1 appears to be the predominant isoform mediating signal transduction in skeletal muscle,[18,19] IRS-2 appears to be important in β-cell development,[20] and both isoforms are important for regulation of metabolism in liver.[22]

PI 3-Kinase–Dependent Pathways

Selection and differentiation of the insulin signal toward further metabolic or mitogenic events can be achieved through interaction of proteins with SH2 domains to tyrosine-phosphorylated motifs on IRSs. One of the most characterized intermediate effector molecules that associate with IRS-1 is phosphatidylinositol (PI) 3-kinase. PI 3-kinase associates with tyrosine-phosphorylated IRS-1 after insulin stimulation and catalyzes the formation of phosphatidylinositol-3,4,5-trisphosphate (PI-3,4,5-P$_3$), which serves as an allosteric regulator of phosphoinositide-dependent kinase (PDK).[23] PDK activates protein kinase B (PKB/Akt) and the atypical protein kinase C isoforms PKCζ and PKCλ.[23–25] PI 3-kinase plays an important role in insulin-stimulated glucose transport and GLUT4 translocation in rat[26,27] and human[28] skeletal muscle. PI 3-kinase regulates glucose transport presumably in part via PKB/Akt and/or PKCζ.[25,29–32] However, the precise role of PKB/Akt and PKCζ in glucose transport has not been fully clarified. PI 3-kinase also mediates insulin signaling to p70 S6 kinase, DNA synthesis, and glycogen synthesis.[33] Thus, PI 3-kinase can be considered to be a central player in insulin signal transduction to downstream targets.

PI 3-Kinase–Independent Pathways

The network of insulin signal transduction is complex as multiple effector proteins are believed to orchestrate diverse cellular responses. Adding to this complexity, the signaling pathways are not necessarily linear, and there is a high degree of cross-talk between the signal transducers. For example, PI 3-kinase activation is necessary, but not sufficient for metabolic actions of insulin.[34–36] Thus, PI 3-kinase–independent pathways have been implicated in the regulation of glucose transport. Recent attention has been given to the CAP/TC10 pathway as an alternative means to induce glucose transport.[37,38] In response to insulin stimulation, the intracellular substrate Cbl, a proto-oncogene product, is recruited to the insulin receptor by interaction with the SH3 domain of the Cbl-associated (CAP) adapter protein.[37] Upon phosphorylation of Cbl, the Cbl/CAP complex dissociates from the insulin receptor and recruits the CrkII-C3G complex to lipid rafts, where the guanine nucleotide exchange factor, C3G, activates the small GTP-binding protein TC10.[38] TC10 is expressed in skeletal muscle and adipocytes, and activation of this G protein appears to be important for insulin-stimulated glucose uptake and GLUT4 translocation in cultured cell systems.[38] The TC10 pathway functions in parallel with PI 3-kinase to fully stimulate GLUT4 translocation in response to insulin. To date, the physiological role of this novel pathway in mediating insulin signal transduction in skeletal muscle and adipose tissue from animal models or humans with type II diabetes has not been elucidated.

Glucose Transport

GLUT4 is the predominant glucose transporter isoform expressed in skeletal muscle.[39–41] In response to insulin,[42] muscle contractions,[43] or hypoxia,[44] the transport of glucose into skeletal muscle is mediated by a translocation of GLUT4 from an intracellular site to the plasma membrane. Early studies provide evidence that insulin-stimulated glucose transport is impaired in isolated skeletal muscle from type II diabetic patients.[45] Given that glucose transport is the first rate-limiting step in glucose metabolism,[46,47] a defect in any of the steps involved in glucose transport may play a role in the development of peripheral insulin resistance. Thus, in addition to insulin-signaling defects, insulin resistance may also be due to a failure of GLUT4 vesicles to translocate, dock, or fully fuse with the plasma membrane.

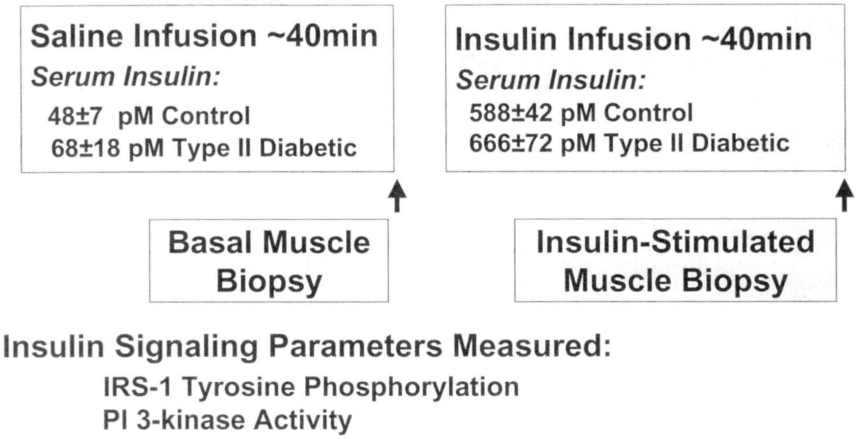

Saline Infusion ~40min	Insulin Infusion ~40min
Serum Insulin:	**Serum Insulin:**
48±7 pM Control	588±42 pM Control
68±18 pM Type II Diabetic	666±72 pM Type II Diabetic

Basal Muscle Biopsy ↑	Insulin-Stimulated Muscle Biopsy ↑

Insulin Signaling Parameters Measured:
  IRS-1 Tyrosine Phosphorylation
  PI 3-kinase Activity

FIGURE 2. Schematic representation of the study design to obtain skeletal muscle following insulin infusion. Serum insulin concentration under basal and insulin-stimulated conditions is indicated. Arrows represent the time at which biopsy sampling was performed.

INSULIN SIGNALING DEFECTS IN TYPE II DIABETES MELLITUS

The earliest studies performed to test the hypothesis that impaired insulin signal transduction contributes to altered glucose homeostasis were performed in animal models of type II diabetes. In skeletal muscle from obese (*ob/ob* or dietary-induced) mice, dexamethasone-treated rats, and streptozotocin-induced diabetic rats, insulin receptor kinase activity, IRS-1 tyrosine phosphorylation, PI 3-kinase activity, and glucose transport were shown to be markedly decreased in response to *in vivo* or *in vitro* insulin stimulation.[48–54] These findings in animal models of diabetes prompted us to consider whether insulin signal transduction is altered in skeletal muscle from people with type II diabetes. Skeletal muscle is quantitatively the most important tissue involved in maintaining glucose homeostasis and accounts for ~80% of glucose disposal following a glucose infusion or ingestion.[2] Thus, alterations in signal transduction to glucose transport should have a profound effect on whole body glucose homeostasis.

In Vivo *Studies of Insulin Action in Skeletal Muscle from Type II Diabetic Patients*

In our initial study,[55] a modification of the hyperinsulinemic clamp procedure was used in conjunction with an open-muscle biopsy to obtain vastus lateralis muscle from people with type II diabetes and healthy subjects (FIG. 2). Muscle biopsies were obtained before (basal) and after (insulin-stimulated) a 40 min *in vivo* insulin infusion, sufficient to increase fasting insulin levels approximately 10-fold. Insulin induced a 6-fold increase in IRS-1 tyrosine phosphorylation in skeletal muscle from control subjects, whereas no significant effect was noted in type II diabetic subjects. The lack of insulin-stimulated IRS-1 phosphorylation in skeletal muscle from type II diabetic patients is not related to changes in IRS-1 protein content,[55,56] suggesting a functional defect from either impaired insulin receptor phosphorylation or serine phosphorylation of IRS-1. Similar results have also been observed in skeletal muscle from women with polycystic ovary syndrome (PCOS), a condition associated with severe insulin resistance secondary to postbinding defects in insulin signaling.[57] These results in insulin-resistant skeletal muscle are in contrast to observations in adipocytes from lean and obese type II diabetic subjects.[58] In adipocytes, impaired insulin action on IRS-1 tyrosine phosphorylation is associated with a corresponding decrease in IRS-1 protein content, highlighting tissue-specific differences between skeletal muscle and adipose tissue in type II diabetic patients.

We next assessed PI 3-kinase activity in either antiphosphotyrosine or anti-IRS-1 immunoprecipitates of basal or insulin-stimulated muscle from control and type II diabetic subjects. Physiological hyperinsulinemia increased PI 3-kinase activity by 2-fold (measured as either tyrosine-associated or IRS-1-associated) in control subjects. In contrast, insulin failed to increase either tyrosine- or IRS-1-associated PI 3-kinase activity in skeletal muscle from type II diabetic subjects. This finding was consistent with the impaired IRS-1 tyrosine phosphorylation in skeletal muscle from type II diabetic subjects. Subsequent studies performed in obese insulin-resistant and obese type II diabetic subjects[56,59] support our initial observation of impaired insulin action on PI 3-kinase in skeletal muscle from moderately obese type II diabetic subjects.[55]

Since PI 3-kinase is linked to glucose transport, collectively these studies identify this step as a potential molecular candidate that may account for impaired glucose uptake characteristic of skeletal muscle from obese insulin-resistant and type II diabetic subjects.[45,60] Before removal of the non-insulin-stimulated (basal) biopsy, an open-muscle biopsy was excised, and isolated muscle strips were prepared for *in vitro* incubation and assessment of basal and insulin-stimulated (600 pmol/L) 3-*O*-methylglucose transport. Consistent with our previous studies,[45,61,62] insulin-stimulated 3-*O*-methylglucose transport was 40% lower in isolated muscle from type II diabetic patients. These findings couple reduced insulin-stimulated IRS-1 tyrosine phosphorylation and PI 3-kinase activity to impaired insulin-stimulated glucose transport in skeletal muscle from lean to moderately obese type II diabetic subjects. Furthermore, they reveal that physiological hyperinsulinemia is sufficient to increase IRS-1 tyrosine phosphorylation and PI 3-kinase activity *in vivo* in skeletal muscle from healthy subjects.

Factors Contributing to Impaired Signal Transduction

Under *in vivo* conditions, hyperglycemia may induce insulin resistance via increased serine phosphorylation on the insulin receptor or IRS-1, although this hypothesis has yet to be tested in human skeletal muscle. We have reported that insulin action on glucose transport is normalized in isolated muscle strips from type II diabetic patients after a 2-h *in vitro* incubation in the presence of 5 mM glucose.[63] Furthermore, in diabetic GK rats, an animal model of lean type II diabetes mellitus, normalization of glycemia by phlorizin treatment improves glucose tolerance, insulin signaling, and glucose transport in skeletal muscle.[64,65] Similarly, a recent report provides evidence that proximal insulin signaling parameters elicit a normal response in cultured myotubes prepared from muscle biopsies from insulin-resistant nondiabetic subjects.[66] These finding support the hypothesis that, in some type II diabetic patients, insulin resistance in skeletal muscle is secondary to an altered metabolic milieu.

Several factors associated with type II diabetes, including hyperglycemia, hyperinsulinemia, elevated free fatty acids, and possibly cytokines, may impair insulin signal transduction, causing serine phosphorylation of IRS-1.[67–70] Elevated glucose concentrations inhibit insulin receptor kinase activity in Rat1 fibroblasts by inducing serine phosphorylation of the insulin receptor through a PKC-mediated mechanism.[67] Exposure of fibroblasts to high glucose led to a greater impairment at the level of IRS-1 tyrosine phosphorylation, which resulted in attenuated PI 3-kinase activity. Interestingly, the glucose-induced defects were not observed in cells incubated in the presence of PKC inhibitors. Furthermore, infusion of FFA leads to serine phosphorylation of IRS-1 in skeletal muscle.[69] Considerable evidence suggests that increased membrane-associated PKC activity impairs insulin action in liver, adipocytes, and skeletal muscle from rodents and humans.[71] Thus, defects in insulin signal transduction may occur in response to high levels of glucose or FFA through a PKC-mediated induction of serine phosphorylation of IRS-1.

IN VITRO STUDIES OF INSULIN ACTION IN SKELETAL MUSCLE

In our initial study,[55] we were intrigued by the fact that we did not detect a significant insulin-stimulated increase in either IRS-1 tyrosine phosphorylation or PI 3-kinase activity in skeletal muscle from type II diabetic subjects. Thus, we undertook a more extensive study to fully characterize the metabolic and mitogenic signal transduction networks in skeletal muscle from type II diabetic patients and well-matched control subjects.[72] Biopsies were obtained under local anesthesia from the vastus lateralis portion of the quadriceps femoris muscle, and approximately 15–20 smaller muscle samples (20 mg) were dissected free and placed on Plexiglas clamps (FIG. 3). Skeletal muscle strips from type II diabetic and control subjects were incubated in the presence of 5 mM glucose, thus normalizing levels between the groups in an effort to minimize or exclude any potential negative influence of hyperglycemia on signal transduction.

Time Course Studies

First, we established the time course for the insulin-stimulated (60 nM) induction of key components of the insulin-signaling cascade[72] in skeletal muscle from control subjects (FIG. 4). Peak tyrosine phosphorylation of the insulin receptor and IRS-1 was noted at 8 min and 8–15 min, respectively, with phosphorylation sustained at all time points studied (up to 40 min). A time course experiment revealed that both anti-phosphotyrosine- and IRS-1-associated PI 3-kinase activity occurred in parallel, suggesting that IRS-1 is the predominant tyrosine-phosphorylated molecule transmitting the insulin signal to PI 3-kinase in human skeletal muscle. Peak PI 3-kinase activity occurred between 15 and 20 min, and activity was sustained. Insulin-stimulated IRS-2-associated PI 3-kinase activity was highest after 20 min and, in contrast to antiphosphotyrosine- and IRS-1-associated PI 3-kinase activity, decreased

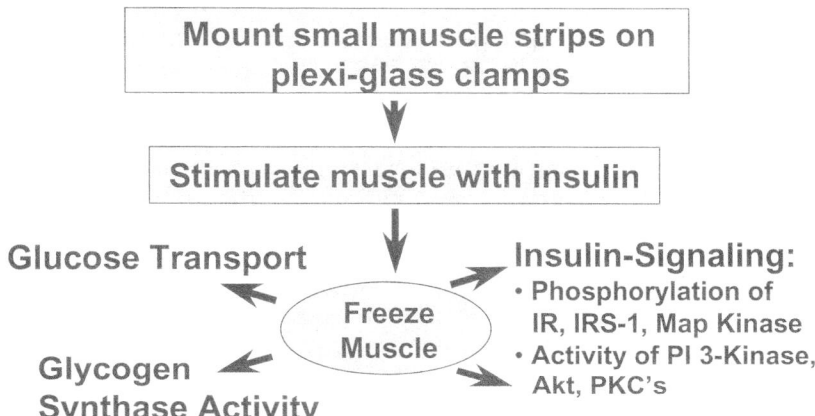

FIGURE 3. Schematic representation of the isolated human skeletal muscle procedure to assess *in vitro* effects of insulin on signal transduction and glucose metabolism.

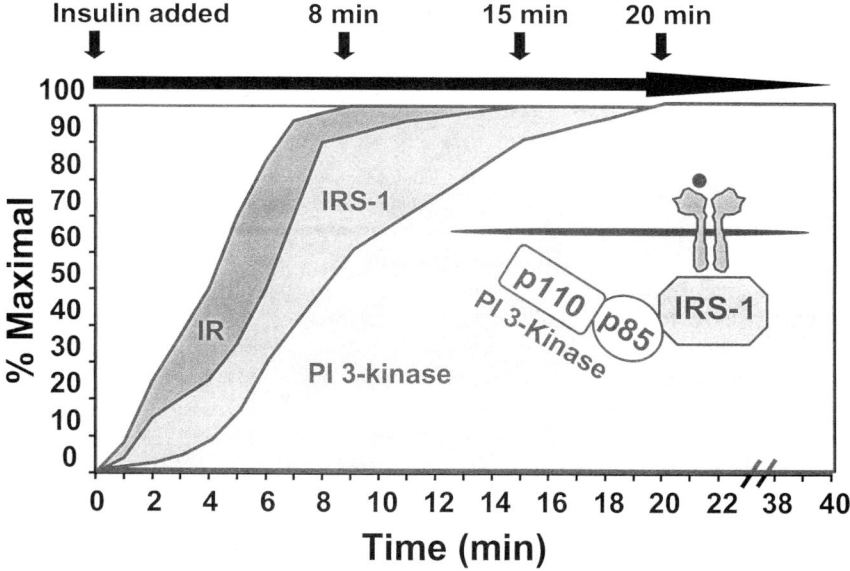

FIGURE 4. Time course for the insulin-stimulated (60 nM) induction of key components of the insulin signaling cascade in skeletal muscle from healthy subjects. Time (min) represents the point at which insulin-stimulated phosphorylation/activity is maximal.

at 40 min. Thus, we provide evidence for sequential induction of the insulin receptor, IRS-1, and PI 3-kinase in skeletal muscle from healthy humans. In rodent skeletal muscle, induction of these signaling intermediates occurs at earlier time points (between 5 and 10 min for IRS-1 and PI 3-kinase), with a rapid and transient activation curve.[73] The difference in the time course of insulin action on IR, IRS-1, and PI

3-kinase between human and rat skeletal muscle is not apparent, but may be related to muscle fiber–type composition, relative expression of the signaling proteins, tissue preparation, or other differences between rat and human skeletal muscle studied.

Studies in Type II Diabetic Subjects

We next established insulin dose-response curves for the activation of the insulin receptor, IRS-1/2, PI 3-kinase, glucose transport, glycogen synthase, and MAP kinase in skeletal muscle from control and type II diabetic subjects. Insulin receptor tyrosine kinase defects have been observed in skeletal muscle from type II diabetic and obese insulin-resistant subjects.[56,74–76] However, this is not a consistent finding as some studies provide evidence that insulin receptor function is normal in skeletal muscle from type II diabetic patients.[77,78] In moderately obese type II diabetic and control subjects (matched for BMI and VO_{2max}), a similar dose-dependent increase in insulin receptor tyrosine phosphorylation in skeletal muscle was noted,[72] suggesting that postreceptor signal transduction defects are likely to account for impaired insulin action. However, in skeletal muscle from morbidly obese subjects[79] or in women who develop gestational diabetes mellitus (GDM),[80] insulin-receptor phosphorylation is impaired. Protein expression of the insulin receptor is not reduced in skeletal muscle from type II diabetic, GDM, or PCOS subjects,[57,72,80] suggesting functional defects at the level of the receptor.

Insulin induced a concentration-dependent increase in IRS-1 tyrosine phosphorylation and phosphotyrosine-associated PI 3-kinase activity in skeletal muscle from type II diabetic and control subjects. However, in contrast to our findings at the level of the insulin receptor, insulin action on IRS-1 and PI 3-kinase, in response to pharmacological concentrations of ligand (2.4 and 60 nM insulin), was impaired in skeletal muscle from type II diabetic subjects. Thus, there is an intrinsic defect in insulin signal transduction in skeletal muscle from type II diabetic subjects. These defects are not associated with reduced protein expression of the insulin receptor, IRS-1, or the p85 subunit of PI 3-kinase.[72] Similar results have been observed in skeletal muscle from women with PCOS.[57] However, in skeletal muscle from morbidly obese insulin-resistant subjects[79] or from women who develop GDM,[80] reduced protein expression of the IRS-1 has been observed. Interestingly, in muscle from women with PCOS[57] or GDM,[80] IRS-2 expression is increased, suggesting this isoform plays a compensatory role in signal transduction. However, skeletal muscles from nonobese (BMI < 28 kg/m^2)[72] and obese (BMI > 30 kg/m^2)[56,59] type II diabetic subjects show parallel decreases in IRS-1- and IRS-2-associated PI 3-kinase activity, challenging this hypothesis. Thus, several studies provide evidence that insulin signaling defects in skeletal muscle are associated with insulin resistance and type II diabetes mellitus (TABLE 1).

Defects in insulin signal transduction in skeletal muscle from moderately obese subjects are likely to be exacerbated by a deleterious *in vivo* environment that may exacerbate insulin resistance in muscle, thereby revealing further signaling defects at more physiological insulin levels. For example, hyperglycemia,[67,68] FFA,[69] and TNFα[68,70] exposure increase serine phosphorylation of IRS-1 and inhibit further downstream insulin signaling. Importantly, IRS-2 does not compensate for reduced antiphosphotyrosine- or IRS-1-associated PI 3-kinase activity in skeletal muscle from type II diabetic subjects. This is in contrast to findings in adipocytes from an

TABLE 1. Regulation of insulin signaling in skeletal muscle from insulin resistance and type II diabetic subjects

	Insulin receptor		IRS-1		PI 3-kinase	
	Binding or protein level	Phosphorylation or activity	Protein level	Phosphorylation	p85 protein level	Activity
Nonobese type II diabetic	⇔72,74 ⇓76	⇔72,78 ⇓74,76	⇔55,72	⇓55,72	⇔72	⇓55,72
Obese type II diabetic	⇔74	⇓56,74,75	⇔56	⇓56	⇔56,59	⇓56,59
Obese insulin-resistant	⇔56,75 ⇓79a	⇔75 ⇓56,79a	⇔56	⇔56 ⇓79a	⇔56,59 ⇓79a	⇔59 ⇓59,79a
Obese gestational diabetic	⇔80	⇓80	⇓80b	⇓80	⇑80	
Obese polycystic ovary syndrome			⇔57b		⇔57	⇓57

[a]Changes observed in morbidly obese individuals (BMI > 50 kg/m^2).
[b]Increase in IRS-2 protein expression was observed.

insulin-resistant type II diabetic subject,[58] where increased signaling through IRS-2 to PI 3-kinase compensates for impairments at the level of IRS-1.

Insulin signaling defects appear to be selective for metabolic pathways.[56,72] Normal signal transduction along the MAP kinase pathway has been reported, despite profound insulin resistance along IRS/PI 3-kinase pathways in skeletal muscle from obese insulin-resistant and moderately obese type II diabetic subjects.[56,72] This may not be surprising since insulin receptor tyrosine phosphorylation was normal[72] or perhaps defects at the level of the receptor were not severe enough to prevent normal Shc phosphorylation for full activation of MAP kinase.[56]

SUMMARY

Collectively, our studies of skeletal muscle from lean to moderately obese subjects provide evidence that initial signaling events related to insulin receptor tyrosine kinase activity are normal in type II diabetic patients. In contrast, early and intermediate postreceptor signal events are impaired. However, insulin-mediated glucose transport and cell-surface GLUT4 content are profoundly reduced in skeletal muscle from type II diabetic patients. Thus, the signal transduction defects observed in skeletal muscle from people with type II diabetes may not fully account for the severe reduction in glucose transport activity. Nevertheless, IRS-1/2 and PI 3-kinase are targeted to intracellular locations, such as the actin cytoskeleton, and thus signals for glucose transport may be compartmentalized within the cell. Hence, the possibility remains that signaling transduction may be impaired to a greater extent in critical compartments that may be important for activation of glucose transport. Detailed studies of intracellular location of the insulin signal will be a challenge to perform in human skeletal muscle. Future studies are warranted to determine whether defects in the insulin signaling pathway precede the development of insulin resistance. Since type II diabetes mellitus is a progressive metabolic disorder, studies in newly diagnosed type II diabetic patients, glucose-intolerant individuals, and even first-degree relatives to type II diabetic patients may reveal whether these defects are causally or directly related to the development of insulin resistance in humans. Furthermore, the clinical effectiveness of current antidiabetic therapies to improve insulin signal transduction should be assessed.

ACKNOWLEDGMENTS

This work was supported by grants from the Swedish Medical Research Council, the Swedish Diabetes Association, and the Novo-Nordisk Foundation.

REFERENCES

1. KAHN, C.R. 1994. Insulin action, diabetogenes, and the cause of type II diabetes. Diabetes **43**: 1066–1084.
2. DEFRONZO, R.A. 1988. The triumvirate: β-cell, muscle, or liver—a collusion responsible for NIDDM. Diabetes **37**: 667–687.
3. ZIERATH, J.R., et al. 2000. Insulin action and insulin resistance in human skeletal muscle. Diabetologia **43**: 821–835.

4. VAAG, A., et al. 1992. Decreased insulin activation of glycogen synthase in skeletal muscle in young nonobese Caucasian first-degree relatives of patients with non-insulin-dependent diabetes mellitus. J. Clin. Invest. **89:** 782–788.
5. MASSAGUE, J., et al. 1981. Electrophoretic resolution of three major insulin receptor structures with unique subunit stoichiometries. Proc. Natl. Acad. Sci. U.S.A. **77:** 7137–7141.
6. KASUGA, M., et al. 1982. Structure of insulin receptor and its subunits. J. Biol. Chem. **257:** 10392–10399.
7. KASUGA, M., et al. 1982. Insulin stimulates the phosphorylation of the 95,999-dalton subunit of its own receptor. Science **215:** 185–187.
8. TAVARE, J.M. & K. SIDDLE. 1993. Mutational analysis of insulin receptor function: consensus and controversy. Biochim. Biophys. Acta **1178:** 21–39.
9. KROOK, A. & S. O'RAHILLY. 1996. Mutant alleles of the insulin receptor in syndromes of insulin resistance. Bailliere's Clin. Endocrinol. Metab. **10:** 97–122.
10. SUN, X.J., et al. 1991. Structure of the insulin receptor substrate IRS-1 defines a unique signal transduction protein. Nature **352:** 73–77.
11. SUN, X.J., et al. 1995. Role of IRS-2 in insulin and cytokine signalling. Nature **377:** 173–177.
12. LAVAN, B.E., et al. 1997. The 60-kDa phosphotyrosine protein in insulin-treated adipocytes is a new member of the insulin receptor substrate family. J. Biol. Chem. **272:** 11439–11443.
13. LAVAN, B.E., et al. 1997. A novel 160-kDa phosphotyrosine protein in insulin-treated embryonic kidney cells is a new member of the insulin receptor substrate family. J. Biol. Chem. **272:** 21403–21407.
14. HOLGADO-MADRUGA, M., et al. 1996. A Grb-2-associated docking protein in EGF- and insulin-receptor signalling. Nature **379:** 560–564.
15. RIBON, V. & A.R. SALTIEL. 1997. Insulin stimulates tyrosine phosphorylation of the proto-oncogene product of c-Cbl in 3T3-L1 adipocytes. Biochem. J. **324:** 839–845.
16. WHITE, M.F. & C.R. KAHN. 1994. The insulin signaling system. J. Biol. Chem. **269:** 1–4.
17. WHITE, M.F. 1998. The insulin signalling system: a network of docking proteins that mediate insulin action. Mol. Cell. Biochem. **182:** 3–11.
18. ARAKI, E., et al. 1994. Alternative pathway of insulin signalling in mice with targeted disruption of the IRS-1 gene. Nature **372:** 186–190.
19. TAMEMOTO, H., et al. 1994. Insulin resistance and growth retardation in mice lacking insulin receptor substrate-1. Nature **372:** 128–129.
20. WHITHERS, D.J., et al. 1998. Disruption of IRS-2 causes type 2 diabetes in mice. Nature **391:** 900–904.
21. PATTI, M.E., et al. 1995. 4PS/insulin receptor substrate (IRS)–2 is the alternative substrate of the insulin receptor in IRS-1-deficient mice. J. Biol. Chem. **270:** 24670–24673.
22. KEROUZ, N.J., et al. 1997. Differential regulation of insulin receptor substrates-1 and -2 (IRS-1 and IRS-2) and phosphatidylinositol 3-kinase isoforms in liver and muscle of the obese diabetic (ob/ob) mouse. J. Clin. Invest. **100:** 3164–3172.
23. ALESSI, D.R., et al. 1997. Characterization of a 3-phosphoinositide-dependent protein kinase which phosphorylates and activates protein kinase Bα. Curr. Biol. **7:** 261–269.
24. NAKANISHI, H., et al. 1993. Activation of the ζ isoenzyme of protein kinase C by phosphatidylinositol 3,4,5-trisphosphate. J. Biol. Chem. **268:** 13–16.
25. BANDYOPADHYAY, G., et al. 1997. Activation of protein kinase C (α, β, and ζ) by insulin in 3T3/L1 cells: transfection studies suggest a role for PKC-ζ in glucose transport. J. Biol. Chem. **272:** 2551–2558.
26. LUND, S., et al. 1995. Contraction stimulates translocation of glucose transporter GLUT4 in skeletal muscle through a mechanism distinct from that of insulin. Proc. Natl. Acad. Sci. U.S.A. **92:** 5817–5821.
27. YEH, J-I., et al. 1995. The effects of wortmannin on rat skeletal muscle. J. Biol. Chem. **270:** 2107–2111.
28. SHEPHERD, P.R., et al. 1997. Involvement of phosphoinositide 3-kinase in insulin stimulation of MAP-kinase and phosphorylation of protein kinase-B in human skeletal muscle: implications for glucose metabolism. Diabetologia **40:** 1172–1177.

29. HAJDUCH, E., et al. 1998. Constitutive activation of protein kinase Bα by membrane targeting promotes glucose and system A amino acid transport, protein synthesis, and inactivation of glycogen synthase kinase 3 in L6 muscle cells. Diabetes **47:** 1006–1013.
30. KOHN, A.D., et al. 1996. Expression of a constitutively active AKT ser/thr kinase in 3T3-L1 adipocytes stimulates glucose uptake and glucose transporter 4 translocation. J. Biol. Chem. **271:** 31372–31378.
31. TANTI, J-F., et al. 1997. Potential role of protein kinase B in glucose transporter 4 translocation in adipocytes. Endocrinology **138:** 2005–2010.
32. KOTANI, K., et al. 1998. Requirement of atypical protein kinase C λ for insulin stimulation of glucose uptake, but not for AKT activation in 3T3-L1 adipocytes. Mol. Cell. Biol. **18:** 6971–6982.
33. VIRKAMÄKI, A., et al. 1999. Protein-protein interactions in insulin signaling and the molecular mechanisms of insulin resistance. J. Clin. Invest. **103:** 931–943.
34. KROOK, A., et al. 1997. Two naturally occurring insulin receptor tyrosine kinase domain mutants provide evidence that phosphatidylinositol 3-kinase activation alone is not sufficient for the meditation of insulin's metabolic and mitogenic effects. J. Biol. Chem. **272:** 30208–30214.
35. JIANG, T., et al. 1998. Membrane-permeant esters of phosphatidylinositol 3,4,5-trisphosphate. J. Biol. Chem. **273:** 11017–11024.
36. MARTIN, S.S., et al. 1996. Activated phosphatidylinositol 3-kinase is sufficient to mediate actin rearrangement and GLUT4 translocation in 3T3-L1 adipocytes. J. Biol. Chem. **271:** 17605–17608.
37. BAUMANN, C.A., et al. 2000. CAP defines a second signalling pathway required for insulin-stimulated glucose transport. Nature **407:** 147–148.
38. CHIANG, S.H., et al. 2001. Insulin-stimulated GLUT4 translocation requires the CAP-dependent activation of TC10. Nature **410:** 944–948.
39. JAMES, D.E., et al. 1989. Molecular cloning and characterization of an insulin-regulatable glucose transporter. Nature **338:** 83–87.
40. BIRNBAUM, M.J. 1989. Identification of a novel gene encoding an insulin-responsive glucose transporter protein. Cell **57:** 305–315.
41. CHARRON, M.J., et al. 1989. A glucose transport protein expressed predominately in insulin-sensitive tissues. Proc. Natl. Acad. Sci. U.S.A. **86:** 2535–2539.
42. HIRSHMAN, M.F., et al. 1990. Identification of an intracellular pool of glucose transporters from basal and insulin-stimulated rat skeletal muscle. J. Biol. Chem. **265:** 987–991.
43. DOUEN, A.G., et al. 1990. Exercise induces recruitment of the "insulin responsive" glucose transporter: evidence for distinct intracellular insulin- and exercise-recruitable transporter pools in skeletal muscle. J. Biol. Chem. **265:** 13427–13430.
44. CARTEE, G.D., et al. 1991. Stimulation of glucose transport in skeletal muscle by hypoxia. J. Appl. Physiol. **70:** 1593–1600.
45. ANDRÉASSON, K., et al. 1991. Decreased insulin-stimulated 3-O-methylglucose transport in in vitro incubated muscle strips from type II diabetic subjects. Acta Physiol. Scand. **142:** 255–260.
46. REN, J-M., et al. 1993. Evidence from transgenic mice that glucose transport is rate limiting for glycogen deposition and glycolysis in skeletal muscle. J. Biol. Chem. **268:** 16113–26115.
47. MARSHALL, B.A. & M.M. MUECKLER. 1994. Differential effects of GLUT1 or GLUT4 overexpression on insulin responsiveness in transgenic mice. Am. J. Physiol. **267:** E738–E744.
48. SAAD, M.J.A., et al. 1992. Regulation of insulin receptor substrate-1 in liver and muscle of animal models of insulin resistance. J. Clin. Invest. **90:** 1839–1849.
49. SAAD, M.J.A., et al. 1993. Modulation of insulin receptor, insulin receptor substrate-1, and phosphatidylinositol 3-kinase in liver and muscle of dexamethasone-treated rats. J. Clin. Invest. **92:** 2065–2072.
50. FOLLI, F., et al. 1993. Regulation of phosphatidylinositol 3-kinase activity in liver and muscle of animal models of insulin-resistant and insulin-deficient diabetes mellitus. J. Clin. Invest. **92:** 1787–1794.

51. HEYDRICK, S.J., et al. 1993. Defect in skeletal muscle phosphatidylinositol-3-kinase in obese insulin-resistant mice. J. Clin. Invest. **91:** 1358–1366.
52. HEYDRICK, S.J., et al. 1995. Early alteration of insulin stimulation of PI 3-kinase in muscle and adipocyte from gold thioglucose obese mice. Am. J. Physiol. **268:** E604–E612.
53. GIORGINO, F., et al. 1992. Changes in tyrosine phosphorylation of insulin receptor and a 170,000 molecular weight non-receptor protein *in vivo* in skeletal muscle of streptozotocin-induced diabetic rats: effects of insulin and glucose. Endocrinology **130:** 1433–1444.
54. ZIERATH, J.R., et al. 1997. High fat feeding impairs insulin-stimulated GLUT4 recruitment in muscle via an early insulin signaling defect. Diabetes **46:** 215–223.
55. BJÖRNHOLM, M., et al. 1997. Insulin receptor substrate-1 phosphorylation and phosphatidylinositol 3-kinase activity are decreased in skeletal muscle from NIDDM subjects following *in vivo* insulin stimulation. Diabetes **46:** 524–527.
56. CUSI, K., et al. 2000. Insulin resistance differentially affects the PI 3-kinase- and MAP kinase-mediated signaling in human muscle. J. Clin. Invest. **105:** 311–320.
57. DUNAIF, A., et al. 2001. Defects in insulin receptor signaling *in vivo* in the polycystic ovary syndrome (PCOS). Am. J. Physiol. **281:** E392–E399.
58. RONDINONE, C.M., et al. 1997. Insulin receptor substrate (IRS) 1 is reduced and IRS-2 is the main docking protein for phosphatidylinositol 3-kinase in adipocytes from subjects with non-insulin-dependent diabetes mellitus. Proc. Natl. Acad. Sci. U.S.A. **94:** 4171–4175.
59. KIM, Y.B., et al. 1999. Normal insulin-dependent activation of AKT/protein kinase B, with diminished activation of phosphoinositide 3-kinase, in muscle in type 2 diabetes. J. Clin. Invest. **104:** 733–741.
60. DOHM, G.L., et al. 1988. An *in vitro* human skeletal muscle preparation suitable for metabolic studies: decreased insulin stimulation of glucose transport in muscle from morbidly obese and diabetic subjects. J. Clin. Invest. **82:** 486–494.
61. GALUSKA, D., et al. 1994. Effect of metformin on glucose transport in isolated skeletal muscle obtained from type II diabetic patients and healthy individuals. Diabetologia **37:** 872–879.
62. ZIERATH, J.R., et al. 1996. Insulin action on glucose transport and plasma membrane GLUT4 content in skeletal muscle from patients with NIDDM. Diabetologia **39:** 1180–1189.
63. ZIERATH, J.R., et al. 1994. Effects of glycemia on glucose transport in isolated skeletal muscle from patients with NIDDM: *in vitro* reversal of muscular insulin resistance. Diabetologia **37:** 270–277.
64. KROOK, A., et al. 1997. Improved glucose tolerance restores insulin-stimulated AKT kinase activity and glucose transport in skeletal muscle from diabetic Goto-Kakizaki (GK) rats. Diabetes **46:** 2110–2114.
65. SONG, X.M., et al. 1999. Muscle fiber–type specific defects in insulin signal transduction to glucose transport in diabetic Goto-Kakizaki rats. Diabetes **48:** 664–670.
66. KRUTZFELDT, J., et al. 2000. Insulin signaling and action in cultured skeletal muscle cells from lean healthy humans with high and low insulin sensitivity. Diabetes **49:** 992–998.
67. PILLAY, T.S., et al. 1996. Glucose-induced phosphorylation on the insulin receptor. J. Clin. Invest. **97:** 613–620.
68. KRODER, G., et al. 1995. Tumor necrosis factor-α- and hyperglycemia-induced insulin resistance: evidence for different mechanisms and different effects on insulin signaling. J. Clin. Invest. **97:** 1471–1477.
69. DRESNER, A., et al. 1999. Effects of free fatty acids on glucose transport and IRS-1-associated phosphatidylinositol 3-kinase activity. J. Clin. Invest. **103:** 253–259.
70. HOTAMISLIGIL, G.S., et al. 1996. IRS-1 mediated inhibition of insulin receptor tyrosine kinase activity in TNF-α- and obesity-induced insulin resistance. Science **271:** 665–668.
71. CONSIDINE, R.V., et al. 1995. Protein kinase C is increased in the liver of humans and rats with non-insulin-dependent diabetes mellitus: an alteration not due to hyperglycemia. J. Clin. Invest. **95:** 2938–2944.

72. KROOK, A., et al. 2000. Characterization of signal transduction and glucose transport in skeletal muscle from type 2 diabetic patients. Diabetes **49:** 284–292.
73. SONG, X.M., et al. 1999. Muscle fiber type specificity in insulin signal transduction. Am. J. Physiol. **277:** R1690–R1696.
74. ARNER, P., et al. 1987. Defective insulin receptor tyrosine kinase in human skeletal muscle in obesity and type 2 (non-insulin-dependent) diabetes mellitus. Diabetologia **30:** 437–440.
75. NOLAN, J.J., et al. 1994. Role of human skeletal muscle insulin receptor kinase in the *in vivo* insulin resistance of noninsulin-dependent diabetes mellitus and obesity. J. Clin. Endocrinol. Metab. **78:** 471–477.
76. MAEGAWA, H., et al. 1991. Impaired autophosphorylation of insulin receptors from abdominal skeletal muscles in nonobese subjects with NIDDM. Diabetes **40:** 815–819.
77. CARO, J.F., et al. 1987. Insulin receptor kinase in human skeletal muscle from obese subjects with and without non-insulin dependent diabetes. J. Clin. Invest. **79:** 1330–1337.
78. KLEIN, H.H., et al. 1995. Elevation of serum insulin concentration during euglycemic hyperinsulinemic clamp studies leads to similar activation of insulin receptor kinase in skeletal muscle of subjects with and without NIDDM. Diabetes **44:** 1310–1317.
79. GOODYEAR, L.J., et al. 1995. Insulin receptor phosphorylation, insulin receptor substrate-1 phosphorylation, and phosphatidylinositol 3-kinase activity are decreased in intact skeletal muscle strips from obese subjects. J. Clin. Invest. **95:** 2195–2204.
80. FRIEDMAN, J.E., et al. 1999. Impaired glucose transport and insulin receptor tyrosine phosphorylation in skeletal muscle from obese women with gestational diabetes. Diabetes **48:** 1807–1814.

Skeletal Muscle Triglycerides

An Aspect of Regional Adiposity and Insulin Resistance

DAVID E. KELLEY

Department of Medicine, Montefiore University Hospital, University of Pittsburgh, Pittsburgh, Pennsylvania 15213, USA

ABSTRACT: The composition and biochemistry of skeletal muscle are altered in obesity and type 2 diabetes mellitus (DM) as compared to nonobese individuals. In health, skeletal muscle has a clear capacity to utilize both carbohydrate and lipid fuels and to transition between these in response to hormonal, chiefly insulin, and substrate signals. This metabolic flexibility is key for the major role that skeletal muscle can have in overall fuel balance. In obesity and type 2 DM, there is a loss of this plasticity and, instead, there is metabolic inflexibility. Rates of lipid oxidation do not suppress effectively in response to insulin, but neither do rates of lipid oxidation effectively increase during the transition to fasting conditions. An important morphological characteristic of skeletal muscle in obesity and type 2 DM is an increased content of triglyceride. The accretion of fat within muscle tissues appears to strongly correlate with insulin resistance and may not be simply a passive process, paralleling fat storage in other tissues. Instead, and of particular metabolic interest, a concept is emerging that biochemical characteristics of skeletal muscle in obese individuals dispose to fat accumulation in muscle. An effort to modify skeletal muscle in individuals with obesity and type 2 DM so that the capacity for fat oxidation and metabolic flexibility is improved should be among the goals of treatment for these disorders.

KEYWORDS: skeletal muscle; adiposity; triglyceride; lipid; obesity; insulin resistance

INTRODUCTION

In obesity, one manifestation of a change in the metabolic profile of skeletal muscle is an increased lipid content within and around muscle fibers. How this occurs is an important question. Altered composition of skeletal muscle may arise only as a consequence of having become obese, reflecting the general increase in adiposity in multiple organs. Yet, there are strong data that suggest that changes in the physiology and biochemistry of skeletal muscle in obesity dispose to an accumulation of lipid within muscle. Indeed, these changes in muscle in fuel partitioning of lipid

Address for correspondence: David E. Kelley, M.D., Professor of Medicine, 810N Montefiore University Hospital, University of Pittsburgh, 3459 Fifth Avenue, Pittsburgh, PA 15213. Voice: 412-692-2158; fax: 412-692-2165.
kelley@msx.dept-med.pitt.edu

between oxidation and storage of fat calories may contribute to the pathogenesis of obesity and precede its development. This hypothesis could be of central importance to our understanding of this chronic disease. A related theme is that skeletal muscle insulin resistance, a well-recognized metabolic complication of obesity, entails perturbations not only of glucose, but also of fatty acid metabolism. In metabolic health, skeletal muscle physiology is characterized by the capacity to utilize either lipid or carbohydrate fuels and to effectively transition between these fuels. We will review recent findings that indicate that, in obesity, skeletal muscle manifests a loss of the capacity for transition between lipid and carbohydrate fuels. This inflexibility in fuel selection by skeletal muscle, as well as differences in fuel partitioning, will be key pathophysiological characteristics that contribute to an altered composition of muscle in obesity and to the insulin resistance of muscle.

SKELETAL MUSCLE QUALITY: IMAGING OF FAT AND LEAN COMPARTMENTS

While muscle mass may be higher in obesity, a considerable body of literature has accumulated that shows that, qualitatively, this muscle mass is quite different in obesity and type 2 diabetes mellitus (DM), with one of the features being different levels and localization of fats. Computed tomography (CT) is a powerful *in vivo* imaging technique of growing sophistication (see reference 1 for a recent review of the technique). CT is particularly effective at distinguishing between water and lipid and this is because the attenuation values measured by CT reflect the chemical composition of tissue. Using water as a reference value (set to 0 Hounsfield units), adipose tissue imaged by CT has a strongly negative attenuation value, generally ranging from −200 to −1 Hounsfield units. Thus, for example, a finding of lower attenuation in skeletal muscle in obesity is indicative of increased fat deposition within muscle, and recent chemical phantom studies by Goodpaster *et al.*[1] have confirmed the relationship between lipid content and CT attenuation values.

One of the earlier studies in this area was from our laboratory using CT of the thigh for regional analysis of skeletal muscle composition.[2] In that study and as later reaffirmed by Simoneau[3] and then Goodpaster *et al.*,[4] in obesity the CT attenuation value (expressed as Hounsfield units) of skeletal muscle is lower than in lean individuals. It is of interest that, following weight loss, the density of skeletal muscle increases and approaches the range of values found in lean individuals.[5] This change in muscle reflects the fact that not only the quantity of muscle mass, but also its composition are altered during periods of weight gain and weight loss. Reduced attenuation values for skeletal muscle are associated with aging and with gender, with lower values in older individuals and among women.[6] Muscle lipid content is a strong determinant of the CT attenuation values and likely accounts for the association between CT attenuation values, insulin sensitivity, and aspects of physical performance. Recent chemical phantom studies by Goodpaster *et al.*[1] confirm the relationship between lipid content and CT attenuation values. There is an important physiological significance to the altered composition of skeletal muscle as revealed by CT imaging. Across several studies, it has been found that the altered composition of skeletal muscle in obesity, as reflected in the lower CT attenuation values, is a correlate of the severity of insulin resistance.[3,4,7] This association between muscle density and insulin sensitivity is at least partially independent of the adverse effect of visceral adiposity to aggravate insulin resistance,[3,4,7] a point to which we will return.

REGIONAL ADIPOSE TISSUE DISTRIBUTION ADJACENT TO SKELETAL MUSCLE

There is another aspect of muscle composition, from the whole organ perspective, that has been learned from the application of regional CT imaging and this concerns the distribution of adipose tissue outside skeletal muscle. There can be substantial subcutaneous adipose tissue located near muscle and this is especially true with respect to the lower extremities. In general and as opposed to depots such as visceral adipose tissue, adipose tissue located in the extremities has been regarded as a relatively benign depot with respect to insulin resistance. However, recent studies suggest that this perspective needs to be modified to account for adipose tissue distribution or sublocations. In the lower extremities, the majority of adipose tissue is in a subcutaneous location, above the muscle fascia. In CT imaging of the midthigh, the fascial plane formed by the fascia lata can be discerned and it has been found that adipose tissue located beneath the fascia lata is significantly correlated with insulin sensitivity (negatively). In contrast, the much greater depot located above this fascia is not significantly related to insulin sensitivity in either men or women. The mechanism(s) that accounts for the association of subfascial AT, but not subcutaneous AT, with IR is not well understood.

MICROSCOPIC STUDIES OF SKELETAL MUSCLE FIBERS IN OBESITY

Another approach to the study of lipid content within skeletal muscle is to directly examine tissue obtained by muscle biopsy. A number of studies have used muscle obtained by percutaneous biopsy of the vastus lateralis. Pan *et al.* extracted lipid from the biopsy samples. Triglyceride was greater in skeletal muscle in obesity[9] and was related to the severity of insulin resistance. However, even with careful removal of visibly identifiable adipose tissue, a potential limitation of performing lipid extraction from muscle biopsy samples is that the respective contributions of intra- and extramyocyte triglyceride cannot be determined. Direct visualization of muscle fibers using microscopy can circumvent this issue.[10]

In situ staining of neutral lipid with Oil Red O has been used not only to ascertain that lipid droplets are contained within muscle fibers, but (by the use of contemporary computer-assisted image analyses) this method can be used to obtain quantitative assessments. Several ultrastructural investigations of human skeletal muscle have shown that lipid droplets account for ~1% of cell volume within muscle cells in lean, healthy individuals. Goodpaster *et al.*[11] observed that the volume of lipid droplets in skeletal muscle is increased in obesity and type 2 DM. In that study, the approximate volume of myocytes occupied by lipid droplets was 1.5% in lean volunteers, 3% to 4% in obesity, and slightly greater in type 2 DM.[11] Also, following weight loss, the microscopic analysis of muscle tissue revealed less triglyceride as would be consistent with the data from CT imaging of skeletal muscle. Another study using light microscopy and the Oil Red O method[12] noted that not only the volume, but also the cellular distribution of lipid droplets may differ in muscle from obese compared to lean individuals. In muscle from obese volunteers, the size of the lipid droplets did not appear to differ from those observed in muscle from nonobese individuals: simply, the muscle from obese individuals had more lipid droplets. However, in muscle from obese individuals, a higher proportion of lipid droplets appeared to be located more centrally within the muscle fiber.

One of the striking features that can be noted when examining the Oil Red O staining patterns in skeletal muscle, especially human skeletal muscle, is that there is considerable heterogeneity between muscle fibers in the amount of lipid staining. This heterogeneity is related to muscle fiber type. It is well known, based on prior work with rat skeletal muscle, that muscle fiber types differ in their respective content of lipid.[13–15] In general, type 1 or slow-twitch oxidative (endurance) fibers contain greater lipid than type 2 fibers; and within the type 2 fibers, fast-twitch oxidative (sprint) fibers contain more lipid than do fast-twitch glycolytic (intermediate) fibers. Therefore, the question arises as to whether the increase in muscle triglyceride content in human obesity reflects an interdependence upon difference in muscle fiber type that might occur with obesity. To address this issue, He et al.[16] performed single fiber analyses in which serial sections of a muscle biopsy sample were stained for muscle lipid content (Oil Red O staining), muscle fiber type, muscle oxidative enzyme activity, and muscle glycolytic enzyme activity. By serially measuring these characteristics for each muscle fiber, an overall profile for each fiber type could be ascertained. In the study by He et al.,[16] vastus lateralis muscles from lean, obese, and obese type 2 diabetic men and women were examined. Lipid content was noted to be highest in type 1 fibers and lowest in type 2b fibers, with an intermediate value in type 2a fibers. This pattern was observed in all three groups of subjects, but skeletal muscle from obese and type 2 diabetic individuals was found to have increased lipid content regardless of fiber type. In each fiber type, muscle from obese individuals had greater lipid content than lean individuals, and this was also found for individuals with type 2 DM.

An additional and equally important finding from the study by He et al.[16] was that skeletal muscle from obese individuals and from those with type 2 DM had a reduced oxidative enzyme activity as determined by standard histochemical methods. As would be expected, type 1 fibers had the highest oxidative enzyme activity, followed in order by types 2a and 2b. Within each fiber type, oxidative enzyme activity was lower in obesity and type 2 DM. The ratio of oxidative enzyme activity to lipid content was also examined. In lean individuals, this ratio was relatively consistent across fiber types, despite substantial differences between fiber types in content of lipid and oxidative enzyme activities. In muscle from obese or type 2 diabetic individuals, the ratio of lipid content to oxidative capacity was also relatively consistent across fiber types, but this ratio differed markedly from that found in lean individuals. The physiological meaning of this proportionality is uncertain, although it might be speculated that a homeostatic balance between fuel stores and capacity for oxidative metabolism is present in lean individuals. If this is so, then the pattern in the muscle of individuals with obesity and type 2 DM suggests that lipid storage is increased out of proportion to the capacity of these myocytes for substrate oxidation.

In studies with Simoneau and others, we found that the activities of several marker enzymes of oxidative and glycolytic capacity were altered in obesity.[3,17,18] Citrate synthase (CS), an enzyme of the TCA cycle activity and a strong marker of oxidative capacity, was shown to be negatively correlated with visceral obesity ($r = -0.51$, $p < 0.05$).[3] This was confirmed in an entirely different population (Pima Indians) both for central adiposity and for % total body fat ($r = -43$, $p < 0.01$).[19] Conversely, CS activity was positively correlated with rates of lipid oxidation across the leg during fasting conditions and positively correlated with both whole-body insulin action and rates of glucose uptake during insulin-stimulated conditions.[17,20] These data indicate that oxidative capacity, as exemplified by CS activity, influences both postabsorptive utilization of FFA and insulin sensitivity. Conversely, glycolytic potential of muscle, as reflected by activity of phosphofructokinase (PFK), a regulatory enzyme in the glycolytic pathway, is increased in individuals with visceral obesity. In particular, the ratio of PFK/CS activity was a strong marker of insulin resistance. Taken together, this pattern of enzyme activity indicates that insulin-resistant muscle is disposed toward anaerobic and glycolytic generation of energy. Our interpretation of the enzyme and metabolic data is that obesity adversely affects substrate metabolism by skeletal muscle during both basal and insulin-stimulated conditions.

Several clinical investigations suggest that the capacity for lipid oxidation is reduced in human skeletal muscle. Ferraro *et al.* found that skeletal muscle lipoprotein lipase activity was decreased in obesity and that this was related to a decreased reliance on fat oxidation, as measured in a whole-body calorimetry chamber.[21] Zurlo *et al.*, using a similar approach, found that marker enzymes of the beta-oxidation pathway are reduced in obesity.[22] Simoneau *et al.* found reduced activity of CPT in skeletal muscle in obesity.[18] The reduction in skeletal muscle CPT activity was of approximately the same proportion as the decrease in activity of other marker enzymes of mitochondria, such as CS for the TCA cycle and cytochrome C oxidase of the electron transport chain.[18] This suggests a decrease in mitochondria number or function, or both. Interestingly, following weight loss, the subjects in the study by Simoneau did not have improvement in the capacity for fat oxidation as measured by activities of CPT, CS, and cytochrome C oxidase. In that same study, an increased content of cytosolic fatty acid binding protein was found in obesity, with an additional gender-related effect of higher concentrations in women.[18] However, other groups have found diminished content of cytosolic fatty acid binding protein in skeletal muscle in obese individuals with type 2 DM.[23]

CLINICAL INVESTIGATIONS OF FATTY ACID METABOLISM BY SKELETAL MUSCLE IN TYPE 2 DM AND OBESITY

Skeletal muscle can oxidize either lipid or carbohydrate to yield energy. During postabsorptive conditions, as occur after an overnight fast, skeletal muscle predominately relies upon lipid oxidation. This is reflected in a respiratory quotient (RQ) across the forearm in lean individuals of approximately 0.71 to 0.82.[24–26] There is also a high rate of extraction of plasma FFA by skeletal muscle during fasting conditions, of approximately 40%.[25] Oxidation of plasma FFA taken up by muscle, if these were to be completely oxidized, would account for nearly 80% of resting oxygen consumption by muscle. Thus, it is clear that skeletal muscle can have an important role in systemic patterns of fatty acid utilization, especially during postabsorptive metabolism.

In obesity and type 2 DM, skeletal muscle has an increased content of triglyceride, as was emphasized in the preceding section. Accordingly, it is important to inquire as to the mechanisms that could account for increased skeletal muscle lipid deposition in obesity. Rates of *de novo* lipogenesis are low within skeletal muscle.[27] Accordingly, skeletal muscle accretion of triglyceride in obesity and type 2 DM would seem to arise as a consequence of an imbalance between the "importation" of plasma fatty acids and rates of fatty acid oxidation. Such a putative imbalance might result from increased fatty acid uptake, perhaps driven by increased plasma concentrations of fatty acids. Recent work from Boden and coworkers has provided strong support for this possibility. In healthy, young volunteers, they increased FFA levels by a combination of intralipid and heparin during a prolonged hyperinsulinemic, euglycemic clamp. They then measured both insulin action and accumulation of intramyocellular triglyceride by NMR spectroscopy. Even within 3–4 hours, there was significant accumulation of intramyocellular triglyceride that related significantly both to the magnitude of elevation of the plasma FFA levels and, importantly, to the FFA-induced insulin resistance.[28] Alternatively (or additionally), the excess accumulation of skeletal muscle triglyceride in obesity might arise from diminished rates of fat oxidation. Given the normal high reliance of skeletal muscle on lipid oxidation during postabsorptive conditions, it would seem logical to inquire whether any defect in lipid oxidation in obesity is evident in this physiological context.

During the past decade and more, a number of studies have begun to address whether patterns of lipid utilization by skeletal muscle differ in obese compared to nonobese individuals. Ravussin and colleagues found that obesity is associated with an impaired capacity for oxidation of fat calories.[21,29] These studies included data that suggest that higher values for RQ predict weight gain over several subsequent years.[29] Two of the principal areas of investigation of our laboratory, especially in those studies carried out in collaboration with the late Jean-Aime Simoneau, have been to examine whether skeletal muscle capacity for lipid oxidation is reduced in obesity and type 2 DM and whether there is a link between fasting patterns of muscle FFA utilization and insulin resistance. In patients with type 2 DM, during fasting hyperglycemia, our laboratory has observed reduced lipid oxidation, and inefficiency in fractional extraction of plasma fatty acids by skeletal muscle has been found.[30,31] More recently, the impaired uptake of fatty acid by muscle during fasting conditions in type 2 DM has been found by other laboratories.[23] A relatively small study among nondiabetic women, ranging from lean to obese, found that, in obesity and in particular visceral obesity, there were inefficiencies in the uptake and oxidation of plasma fatty acids by skeletal muscle during fasting conditions.[20] Yet, in a seeming paradox, artificial elevations of plasma fatty acids induce a pattern of skeletal muscle IR in healthy volunteers that is similar to that of individuals with type 2 DM.[32] Recent studies from Boden *et al.*[28] suggest that this response is related to accumulation of muscle triglyceride or related lipid metabolites. Moreover, from our laboratory, using positron emission tomography to study IR in type 2 DM, we have observed that a strong determinant is elevated plasma fatty acids.[33] Thus, there is at least the appearance of a paradox of the findings of both reduced lipid oxidation in skeletal muscle in obesity and type 2 DM and increased lipid oxidation in skeletal muscle in these disorders leading to IR.

METABOLIC INFLEXIBILITY OF SUBSTRATE UTILIZATION BY SKELETAL MUSCLE IN OBESITY

Along with Simoneau, we undertook a clinical investigation designed to address this paradox, and the study involved both nondiabetic men and women.[7] The first objective was to examine fasting patterns of lipid metabolism in order to test the hypothesis that reliance on lipid oxidation is reduced in obesity and whether fasting patterns of lipid metabolism were associated with the phenotype of insulin resistance in obesity. The second objective was to determine how fasting patterns of lipid metabolism were related to lipid metabolism during insulin-stimulated conditions.

Volunteers for this study were approximately 60 healthy, young adults. One-third of the group was lean. The rest were overweight or obese with BMIs greater than 25 kg/m^2 and ranging to an upper limit of 40. Leg balance measurements (product of arteriovenous differences and blood flow) for glucose and FFA uptake (based on the fractional extraction of 9,10-^{3}H-oleate) were carried out during fasting and insulin-stimulated conditions. In addition, indirect calorimetry across the leg was performed to estimate substrate oxidation during fasting and insulin-stimulated conditions. The constant infusion of labeled oleate permitted measurement of uptake of plasma FFA across the leg, despite the negative net balance of plasma FFA that occurs during postabsorptive conditions.

During fasting conditions, there was robust fractional extraction (FEX) of labeled FFA across the leg, of approximately 40%, indicative of the uptake of plasma FFA by leg tissues. FEX was similar in lean and obese subjects and was approximately 10- to 15-fold higher during fasting conditions than corresponding FEX for glucose. Rates of FFA uptake across the leg were similar in lean and obese subjects. However, despite similar rates of FFA uptake across the leg, rates of fat oxidation across the leg during fasting conditions were less in obese compared to lean subjects ($p < 0.01$). During fasting conditions, obese subjects had an elevated leg RQ (0.83 ± 0.02 vs. 0.90 ± 0.01; $p < 0.01$). The values for the RQ across the leg in obesity denoted a reduced reliance upon lipid oxidation such that only a third of energy production was accounted for by fat oxidation, while nearly twice this proportion was found in muscle of lean volunteers. In both lean and obese volunteers, the fasting rates of fatty acid uptake across the leg were greater than the fasting rates of lipid oxidation, indicating a modest net surplus of fatty acid uptake, as has been well described in animal studies.[14,15] These rates of "net storage" of fatty acids were greater in obesity. Thus, a paradigm suggested by these findings is that, in obesity, skeletal muscle accrues triglyceride due to a reduced rate of lipid oxidation in the face of rates of fatty acid uptake that are equivalent to those of lean individuals.

A further key objective of our aforementioned study[7] was to address the potential relation between insulin-resistant glucose metabolism and patterns of fatty acid uptake and oxidation during both fasting and insulin-stimulated conditions. Considering all subjects, lean and obese, a decreased reliance on lipid oxidation during fasting conditions was associated with resistance to insulin stimulation of glucose metabolism. Fasting values for leg RQ were negatively correlated with insulin sensitivity ($r = -0.57$, $p < 0.001$). Thus, in addition to group differences in fasting leg RQ, there was significant correlation between the individual variation in the severity of obesity-related insulin resistance and fasting patterns of glucose and lipid oxidation in leg tissues. This observation extends the phenotype of skeletal muscle insulin resistance in obesity to entail a broader concept of an organ system poorly performing its homeostatic function of substrate utilization during both fasting and insulin-stimulated conditions.

The hypothesis that insulin resistance is associated with decreased fasting fatty acid oxidation is a novel reformulation of the concept that perturbed skeletal muscle fatty acid metabolism may contribute to skeletal muscle insulin resistance. The more classic concept of substrate competition in relation to insulin resistance is that "excessive" lipid oxidation reduces glucose utilization by skeletal muscle. It is appropriate to ask whether the finding that reduced fat oxidation during fasting conditions is related to insulin resistance of obesity is a contradiction to the "classic" Randle hypothesis of substrate competition and insulin resistance. Several recent studies indicate that glucose inhibits fat oxidation,[34,35] a so-called "reverse Randle cycle"; this could be pertinent to the observation that insulin-resistant skeletal muscle in animal models of obesity has increased malonyl CoA[36] and that inhibition of ACC2, which decreases malonyl CoA, results in both lower body fatness and improved glucose tolerance in mice.[37]

In our study,[7] insulin-stimulated conditions were examined and, therefore, the role of fat oxidation during this physiological context was assessed in obesity and in relation to insulin-stimulated glucose metabolism. Under the stimulation of insulin, utilization of fatty acids by skeletal muscle is normally suppressed,[30] although this can be disturbed by increased availability of plasma fatty acids.[32,38] In lean subjects in these studies,[7] infusion of insulin stimulated a significant increase in leg RQ ($p < 0.001$); in contrast, in obese subjects, the insulin-stimulated values for leg RQ did not differ from fasting values of leg RQ. In lean subjects, infusion of insulin also stimulated a significant increase in the rates of energy expenditure across the leg ($p < 0.01$); in contrast, in obese subjects, the rates of energy expenditure across the leg were unchanged compared to fasting conditions. Insulin-stimulated values for leg RQ were significantly greater in lean compared to obese subjects (0.99 ± 0.03 vs. 0.91 ± 0.02; $p < 0.01$). Thus, during insulin-stimulated conditions, obese subjects manifested a failure to suppress lipid oxidation, and rates of lipid oxidation were unchanged from fasting conditions. During insulin infusions, rates of leg lipid oxidation were negatively correlated to insulin sensitivity ($r = -0.45$, $p < 0.001$)—that is, greater lipid oxidation during insulin-stimulated conditions predicted insulin-resistant glucose metabolism, whereas lower rates of lipid oxidation during postabsorptive conditions predicted insulin-resistant glucose metabolism.

These findings from fasting and insulin-stimulated conditions are not disparate, but instead are interconnected pieces of the puzzle of how insulin resistance is manifest within skeletal muscle in obesity. The concept that links these two findings is one of metabolic flexibility as a component of insulin sensitivity in lean individuals and metabolic inflexibility as a component of insulin resistance in obesity. Obese subjects had less change in leg RQ in response to insulin infusion than did lean subjects. Across the entire cohort, the amplitude of insulin-stimulated change in leg RQ (Δ leg RQ: insulin-stimulated leg RQ − fasting leg RQ) correlated significantly with insulin-stimulated increases in glucose metabolism ($r = 0.66$, $p < 0.001$). This indicates that responsiveness to insulin in modulation of leg RQ is related to capacity to respond to insulin stimulation of glucose uptake. In obesity, the effect of insulin to suppress lipid oxidation was blunted, as has been previously reported,[39,40] and this clearly fits with the classic concept of fatty acid–induced insulin resistance.[41] Not only was incomplete suppression of lipid oxidation during insulin stimulation observed among the obese volunteers in this study, but rates of muscle lipid oxidation during insulin infusion were correlated with the severity of insulin-resistant glucose metabolism. Yet, these observations do not indicate that fatty acid oxidation within insulin-resistant muscle is persistently "increased". While insulin infusion did not suppress muscle lipid oxidation in obesity (compared to strong suppression in lean individuals), these rates of fat oxidation were unchanged from fasting conditions. During fasting conditions, rates of fat oxidation in skeletal muscle were lower in obese compared to lean individuals. Thus, in regard to the nature of substrate competition, muscle in obesity manifested a severe inflexibility in the modulation of fatty acid oxidation, with neither suppression by insulin infusion nor an appropriate enhancement in response to an overnight fast.

SUMMARY AND CONCLUSIONS

In summary, the composition and biochemistry of skeletal muscle are altered in obesity and type 2 diabetes mellitus (DM) as compared to nonobese individuals. In health, skeletal muscle has a clear capacity to utilize both carbohydrate and lipid fuels and to transition between these in response to hormonal, chiefly insulin, and substrate signals. This metabolic flexibility is key for the major role that skeletal muscle can have in overall fuel balance. In obesity and type 2 DM, there is a loss of this plasticity and, instead, there is metabolic inflexibility. Rates of lipid oxidation do not suppress effectively in response to insulin, but neither do rates of lipid oxidation effectively increase during the transition to fasting conditions. An important morphological characteristic of skeletal muscle in obesity and type 2 DM is an increased content of triglyceride. The accretion of fat within muscle tissues appears to strongly correlate with insulin resistance and may not be simply a passive process, paralleling fat storage in other tissues. Instead, and of particular metabolic interest, a concept is emerging that biochemical characteristics of skeletal muscle in obese individuals dispose to fat accumulation in muscle. An effort to modify skeletal muscle in individuals with obesity and type 2 DM so that the capacity for fat oxidation and metabolic flexibility is improved should be among the goals of treatment for these disorders.

REFERENCES

1. GOODPASTER, B.H., D.E. KELLEY, F.L. THAETE *et al.* 2000. Skeletal muscle attenuation determined by computed tomography is associated with skeletal muscle lipid content. J. Appl. Physiol. **89:** 104–110.
2. KELLEY, D.E., S. SLASKY & J. JANOSKY. 1991. Effects of obesity and non-insulin dependent diabetes mellitus. Am. J. Clin. Nutr. **54:** 509–515.
3. SIMONEAU, J-A., S.R. COLBERG, F.L. THAETE & D.E. KELLEY. 1995. Skeletal muscle glycolytic and oxidative enzyme capacities are determinants of insulin sensitivity and muscle composition in obese women. FASEB J. **9:** 273–278.
4. GOODPASTER, B.H., F.L. THAETE, J-A. SIMONEAU & D.E. KELLEY. 1997. Subcutaneous abdominal fat and thigh muscle composition predict insulin sensitivity independently of visceral fat. Diabetes **46:** 1579–1585.

5. GOODPASTER, B.H., D.E. KELLEY, R.R. WING et al. 1999. Effects of weight loss on insulin sensitivity in obesity: influence of regional adiposity. Diabetes **48:** 839–847.
6. GOODPASTER, B.H., C.L. CARLSON, M. VISSER et al. 2001. Attenuation of skeletal muscle and strength in the elderly: the Health ABC Study. J. Appl. Physiol. **90:** 2157–2165.
7. KELLEY, D.E., B.H. GOODPASTER, R.R. WING & J-A. SIMONEAU. 1999. Skeletal muscle fatty acid metabolism in association with insulin resistance, obesity, and weight loss. Am. J. Physiol. Endocrinol. Metab. **277**(40): E1130–E1141.
8. GOODPASTER, B.H., F.L. THAETE & D.E. KELLEY. 2000. Thigh adipose tissue distribution is associated with insulin resistance in obesity and in type 2 diabetes mellitus. Am. J. Clin. Nutr. **71:** 885–892.
9. PAN, D.A., S. LILLIOJA, M.R. MILNER et al. 1995. Skeletal muscle membrane lipid composition is related to adiposity and insulin action. J. Clin. Invest. **96:** 2802–2808.
10. PHILLIPS, D.W., S. CADDY, V. LLIC et al. 1996. Intramuscular triglycerides and muscle insulin sensitivity: evidence for a relationship in nondiabetic subjects. Metabolism **45:** 947–950.
11. GOODPASTER, B.H., R. THERIAULT, S.C. WATKINS & D.E. KELLEY. 1999. Intramuscular lipid content is increased in obesity and decreased by weight loss. Metabolism **49:** 467–472.
12. MALENFANT, P., R. THERIAULT, B.H. GOODPASTER et al. 2001. Fat content in individual muscle fibers of lean and obese subjects. Int. J. Obes. **25:** 1316–1321.
13. BONEN, A., J.J. LUIKEN, S. LIU et al. 1998. Palmitate transport and fatty acid transporters in red and white muscles. Am. J. Physiol. **275:** E471–E478.
14. BUDOHOSKI, L., J. GORSKI, K. NAZAR et al. 1996. Triacylglycerol synthesis in the different skeletal muscle fiber sections of the rat. Am. J. Physiol. Endocrinol. Metab. **271**(34): E574–E581.
15. DYCK, D.J., S.J. PETERS, J. GLATZ et al. 1997. Functional differences in lipid metabolism in resting skeletal muscle of various fiber types. Am. J. Physiol. Endocrinol. Metab. **272**(35): E340–E351.
16. HE, J., S. WATKINS & D.E. KELLEY. 2001. Skeletal muscle lipid content and oxidative enzyme activity in relation to muscle fiber type in type 2 diabetes and obesity. Diabetes **50:** 817–823.
17. SIMONEAU, J-A. & D.E. KELLEY. 1997. Altered glycolytic and oxidative capacities of skeletal muscle contribute to insulin resistance in NIDDM. J. Appl. Physiol. **83:** 166–171.
18. SIMONEAU, J-A., J.H. VEERKAMP, L.P. TURCOTTE & D.E. KELLEY. 1999. Markers of capacity to utilize fatty acids in human skeletal muscle: relation to insulin resistance and obesity and effects of weight loss. FASEB J. **13:** 2051–2060.
19. KRIKETOS, A.D., D.A. PAN, S. LILLIOJA et al. 1996. Interrelationships between muscle morphology, insulin action, and adiposity. Am. J. Physiol. **270:** R1332–R1339.
20. COLBERG, S.R., J-A. SIMONEAU, F.L. THAETE & D.E. KELLEY. 1995. Skeletal muscle utilization of free fatty acids in women with visceral obesity. J. Clin. Invest. **95:** 1846–1853.
21. FERRARO, R., R. ECKEL, E. LARSON et al. 1993. Relationship between skeletal muscle lipoprotein lipase activity and 24-hour macronutrient oxidation. J. Clin. Invest. **92:** 441–445.
22. ZURLO, F., P.M. NEMETH, R.M. CHOSKI et al. 1994. Whole body energy metabolism and skeletal muscle biochemical characteristics. Metabolism **43:** 481–486.
23. BLAAK, E.E., A.J.M. WAGENMAKERS, J.F.C. GLATZ et al. 2000. Plasma FFA utilization and fatty acid–binding protein content are diminished in type 2 diabetic muscle. Am. J. Physiol. **279:** 146–154.
24. ANDRES, R., G. CADER & K. ZIERLER. 1956. The quantitatively minor role of carbohydrate in oxidative metabolism by skeletal muscle in intact man in the basal state: measurement of oxygen and glucose uptake and carbon dioxide and lactate production in the forearm. J. Clin. Invest. **35:** 671–682.
25. DAGENAIS, G., R. TANCREDI & K. ZIERLER. 1976. Free fatty acid oxidation by forearm muscle at rest, and evidence for an intramuscular lipid pool in the human forearm. J. Clin. Invest. **58:** 421–431.

26. BALTZAN, M., R. ANDRES, G. CADER & K. ZIERLER. 1962. Heterogeneity of forearm metabolism with special reference to free fatty acids. J. Clin. Invest. **41:** 116–125.
27. SCHWARZ, J., S. TURNER, D. DARE & M. HELLERSTEIN. 1995. Short-term alterations in carbohydrate energy intake in humans: striking effects on hepatic glucose production, *de novo* lipogenesis, lipolysis, and whole-body fuel selection. J. Clin. Invest. **96:** 2735–2743.
28. BODEN, G., B. LEBED, M. SCHATZ *et al.* 2001. Effects of acute changes of plasma free fatty acids on intramyocellular fat content and insulin resistance in healthy subjects. Diabetes **50:** 1612–1617.
29. ZURLO, F., S. LILLIOJA, A. ESPOSITO–DEL PUENTE *et al.* 1990. Low ratio of fat to carbohydrate oxidation as a predictor of weight gain: a study of 24-h RQ. Am. J. Physiol. Endocrinol. Metab. **259**(22): E650–E657.
30. KELLEY, D.E., J. REILLY, T. VENEMAN & L.J. MANDARINO. 1990. The influence of physiologic hyperinsulinemia on skeletal muscle glucose storage, oxidation, and glycolysis in man. Am. J. Physiol. Endocrinol. Metab. **258**(21): E923–E929.
31. KELLEY, D.E. & J-A. SIMONEAU. 1994. Impaired free fatty acid utilization by skeletal muscle in non-insulin-dependent diabetes mellitus. J. Clin. Invest. **94:** 2349–2356.
32. KELLEY, D.E., M. MOKAN, J-A. SIMONEAU & L.J. MANDARINO. 1993. Interaction between glucose and free fatty acid metabolism in human skeletal muscle. J. Clin. Invest. **92:** 93–98.
33. KELLEY, D., K. WILLIAMS, J. PRICE *et al.* 2001. Plasma fatty acids, adiposity, and variance of skeletal muscle insulin resistance in type 2 diabetes mellitus. J. Clin. Endocrinol. Metab. In press.
34. MANDARINO, L.J., A. CONSOLI, A. JAIN & D.E. KELLEY. 1996. Interaction of carbohydrate and fat fuels in human skeletal muscle: impact of obesity and NIDDM. Am. J. Physiol. Endocrinol. Metab. **270**(33): E463–E470.
35. SIDOSSIS, L.S., C.A. STUART, G.I. SHULMAN *et al.* 1996. Glucose plus insulin regulate fat oxidation by controlling the rate of fatty acid entry into the mitochondria. J. Clin. Invest. **98:** 2244–2250.
36. RUDERMAN, N.B., A.K. SAHA, D. VAVVAS & L.A. WITTERS. 1999. Malonyl-CoA, fuel sensing, and insulin resistance. Am. J. Physiol. **276:** E1–E18.
37. ABU-ELHEIGA, L., M.M. MATZUK, K.A.H. ABO-HASHEMA & S.J. WAKIL. 2001. Continuous fatty acid oxidation and reduced fat storage in mice lacking acetyl-CoA carboxylase 2. Science **291:** 2613–2616.
38. BODEN, G., F. JADALI, J. WHITE *et al.* 1991. Effect of fat on insulin-stimulated carbohydrate metabolism in normal men. J. Clin. Invest. **88:** 960–966.
39. FELBER, J.P., E. FERRANNINI, A. GOLAY *et al.* 1987. Role of lipid oxidation in the pathogenesis of insulin resistance of obesity and type II diabetes. Diabetes **36:** 1341–1350.
40. LILLIOJA, S., C. BOGARDUS, D. MOTT *et al.* 1985. Relationship between insulin-mediated glucose disposal and lipid metabolism in man. J. Clin. Invest. **75:** 1106–1115.
41. RANDLE, P.J., P.B. GARLAND *et al.* 1963. The glucose fatty acid cycle: its role in insulin sensitivity and the metabolic disturbances of diabetes mellitus. Lancet, pp. 785–789.

Protein Kinase C and Lipid-Induced Insulin Resistance in Skeletal Muscle

CARSTEN SCHMITZ-PEIFFER

Cell Signalling Group, Garvan Institute of Medical Research, Darlinghurst, NSW 2010, Australia

ABSTRACT: Insulin resistance of skeletal muscle in humans, animals, and cells is often strongly correlated with increased lipid availability. The elevation of certain intracellular lipid species can lead to the activation of signal transduction pathways that inhibit normal insulin action. Thus, increased diacylglycerol levels in muscle are associated with the activation of one or more isoforms of the protein kinase C family, which is known to attenuate insulin signaling, especially at the level of IRS-1. In addition, *de novo* synthesis of ceramide can inhibit more distal sites by the activation of protein phosphatase 2A and hence promote the dephosphorylation and inactivation of protein kinase B. Such mechanisms may account at least in part for the reduced insulin sensitivity occurring in obesity and type 2 diabetes where lipid oversupply is a major factor.

KEYWORDS: insulin resistance; diacylglycerol; ceramide; protein kinase C

INTRODUCTION

Insulin resistance is a major feature of type 2 diabetes and is also often present in other components of "syndrome X", such as obesity, hypertension, and cardiovascular disease. Peripheral tissues, especially fat and skeletal muscle, fail to respond adequately to insulin, exhibiting diminished glucose disposal. Since muscle is the major site of insulin-stimulated glucose uptake, the insulin resistance seen in this tissue is of particular importance. Many studies have correlated lipid oversupply with muscle insulin resistance, implying that there is a mechanistic link between the two. In addition, a number of lesions in insulin signal transduction have recently been described in human, animal, and cell-based studies and, since several lipids are signaling molecules, it has been proposed that increased lipid availability, by activation of inhibitory signaling pathways, leads to attenuation of insulin action.

For the utilization of lipids as fuel, free fatty acids (FFA) are converted intracellularly to long-chain acyl-CoAs (LCACoA), imported into mitochondria by carnitine palmitoyltransferases and subjected to β-oxidation. In addition, LCACoA are esterified during synthesis of triglycerides and phospholipids, and contribute to

---

Address for correspondence: Carsten Schmitz-Peiffer, Cell Signalling Group, Garvan Institute of Medical Research, 384 Victoria Street, Darlinghurst, NSW 2010, Australia. Voice: +61-2-9295-8212; fax: +61-2-9295-8201.

c.schmitz-peiffer@garvan.org.au

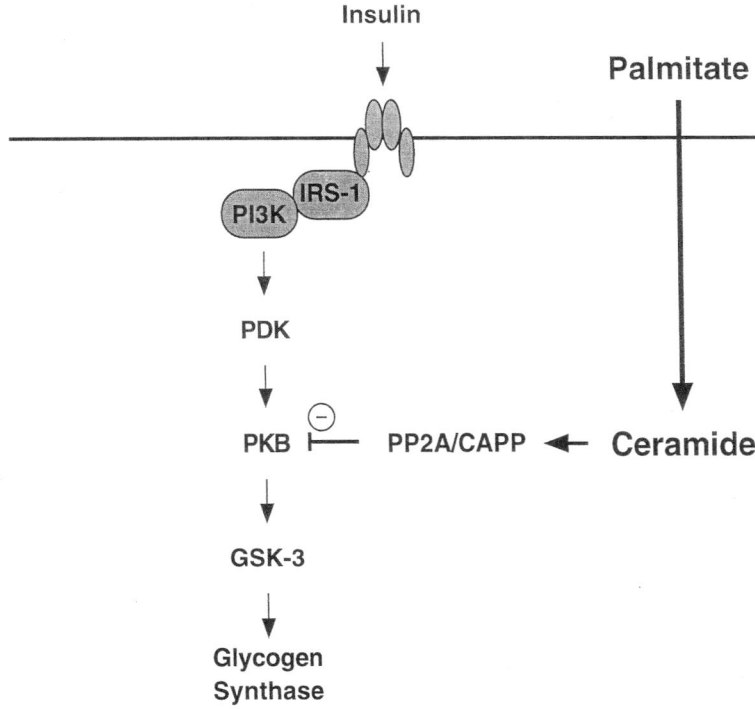

FIGURE 1. Schematic representation of the mechanism of inhibition of glycogen synthesis by chronic treatment with the FFA palmitate. See text for details.

the source of second messengers such as diacylglycerol (DAG), either by *de novo* synthesis or by acute generation upon activation of phospholipase C and phospholipid hydrolysis (see FIG. 1).

PROTEIN KINASE C AND INSULIN RESISTANCE

The family of protein kinase C (PKC) isoforms are activated by lipids, especially DAG, making them candidates for playing a causative role in lipid-induced insulin resistance. PKC isoforms are classified as conventional (cPKC α, β, and γ), novel (nPKC δ, ε, θ, η, and μ), or atypical (aPKC ζ and ι). The cPKCs are calcium- and DAG-dependent; the nPKCs are calcium-independent, but DAG-dependent; and the aPKCs are both calcium- and DAG-independent (reviewed in reference 1). FFA and LCACoA can also modulate PKC activity, either directly[2-5] or through conversion to DAG.[6] While all PKCs appear to require phosphorylation by PDK1 for full activity, this may occur as a constitutive posttranslational modification for the cPKC and nPKC isoforms,[7] while aPKCs are acutely activated in this way.[8] PKCs also differ in their interaction with protein binding partners that may determine their cellular location and hence substrate specificity.[9] Upon activation, the kinases redistribute within the cell such that membrane-associated PKC is considered to be the active form. After chronic activation, PKC can also be downregulated by proteolysis.

Several studies have shown associations between increased PKC activity, especially nPKCs, and insulin resistance in insulin target tissues upon lipid oversupply. PKCδ, PKCε, and PKCθ were chronically activated in skeletal muscle from the fat-fed rat, as assessed by translocation, and changes in the cellular distribution of these isoforms correlated with muscle triglyceride and DAG content.[10] Total levels of, but not membrane-associated, PKCδ and PKCθ were also reduced, while, in agreement with the relative resistance of PKCε to degradation, this isoform was not downregulated. Treatment of fat-fed rats with rosiglitazone, a thiazolidinedione that lowers lipid availability to muscle in this model and improves insulin action,[11] was able to reverse these alterations in PKC,[12] further supporting a role for these kinases in the generation of insulin resistance. PKCε translocation was also observed in muscles of the glucose-infused rat, which again exhibits increased lipid levels and insulin resistance.[13] Increased PKCε and PKCθ translocation has been reported in obese Zucker rats, together with elevated PKC activity and DAG content.[14] Increased expression of PKCε also plays a role in the onset of muscle insulin resistance caused by a high-energy diet in the diabetes-prone sand rat, *Psammomys obesus*.[15]

In humans, skeletal muscle PKCθ translocation has been linked to insulin resistance caused by acute FFA infusion.[16] Alterations in several PKC isoforms were observed in skeletal muscle from obese insulin-resistant humans,[17] while PKCθ protein content and activity were increased in the particulate fraction in muscle from diabetic patients.[18] Regarding the effect of different FFA on PKC activation, reports are conflicting. In C2C12 myotubes[19] and smooth muscle cells,[20] palmitate treatment had no effect on PKC, while oleate promoted PKC translocation. In human skeletal muscle cells, however, palmitate treatment caused a 2-fold increase in PKC activity measured in the membrane fraction of these cells.[21]

Thus, a number of studies have demonstrated a correlation between lipid-induced insulin resistance and nPKC activation, especially PKCε and PKCθ. While persistent activation of PKCε is not accompanied by degradation of this isoform,[10,13,14,22] a decrease in cytosolic, but not membrane-associated PKCθ has been observed under the same conditions.[10,22] PKCε is known to be relatively resistant to proteolysis upon long-term activation[23–25] and these observations therefore support a role for this kinase in mediating the inhibitory effects of lipids on insulin resistance. Although PKCθ exhibited signs of downregulation, this was limited to the inactive cytosolic pool of the enzyme, and it is tempting to speculate that this indicates a constant movement to the membrane in the presence of activating lipid and hence increased substrate phosphorylation. Subsequent proteolysis may thus be manifest as a reduction in cytosolic levels only, such that this kinase can still have an inhibitory effect on insulin action. Alternatively, PKCθ may be responsible for acute insulin resistance, such as upon lipid infusion, whereas PKCε may be more important in chronic lipid oversupply, such as in high-fat feeding.

PKC activation has also been associated with insulin resistance in models not concerned with lipids.[22,26–29] Hyperinsulinemia and hyperglycemia are other factors that lead to reduced insulin sensitivity in muscle and can also induce PKC activation, probably through elevation of intracellular DAG levels.[30–35] Hence, PKC is likely to be involved both in primary muscle insulin resistance, such as that caused by lipids, and in secondary insulin resistance induced by high insulin or glucose levels.

DIACYLGLYCEROL

The lipid species responsible for the activation of PKC upon lipid oversupply have not been clearly defined and, although FFA and LCACoA have been reported to modulate PKC activity,[2,3] DAG is a prime candidate. DAG levels have been found to be elevated in a number of models of insulin resistance,[10,14,22,30,36,37] in many cases with elevated PKC activity.

Quantification in lipid-pretreated cells has indicated that specific FFA lead to distinct effects on total DAG levels. In lipid-treated myotubes, DAG was only elevated by the saturated FFA palmitate; unsaturated FFA such as oleate had no effect.[19] This specific effect of palmitate on DAG has also been observed in smooth muscle cells[20,38,39] and cultured human muscle.[21] It has been suggested that this is due to the different affinities for saturated and unsaturated DAG of acyl-CoA:DAG transferase, the enzyme catalyzing the conversion of DAG to triglyceride.[21] Some of these studies,[21,38,39] but not all,[19,20] also reported increased membrane-associated PKC activity together with the palmitate-induced increase in DAG. In direct contrast, unsaturated FFA, but not palmitate, have been shown to stimulate isoform translocation, especially PKCε, determined by immunoblotting for specific PKC isoforms.[19] This is in agreement with the currently held view that polyunsaturated DAG species are the true activators of PKC.[40] It is therefore possible that treatment of cells with unsaturated FFA increases a subset of DAG with polyunsaturated acyl groups. Indeed, evidence for distinct pools of DAG has been reported.[31] Differences in the observed abilities of saturated and unsaturated FFA to promote PKC activity in such studies may be related to the specific lipid treatments and PKC measurements used since activity measurements in crude membrane fractions will not yield information concerning the lipid effects on different PKC isoforms.

LESIONS IN INSULIN SIGNALING: POTENTIAL EFFECTS OF PKC ACTIVATED BY LIPID OVERSUPPLY

Normal Insulin Signaling

Activation of the insulin receptor tyrosine kinase leads to recruitment and tyrosine phosphorylation of docking proteins, such as insulin receptor substrate-1 (IRS-1), which in turn binds the p85 adaptor subunit of phosphoinositide 3-kinase (PI3K), activating the lipid kinase. The subsequent increase in phosphatidylinositol 3,4-bisphosphate and phosphatidylinositol 3,4,5-trisphosphate (PIP_3) levels enables the activation of protein kinase B (PKB) and atypical protein kinase C (aPKC) by phosphoinositide-dependent kinase-1 (PDK1). PKB and aPKC may each play a role in insulin-stimulated glucose uptake through the translocation of GLUT4 glucose transporters to the plasma membrane, although the mechanisms involved are not fully understood. PKB also phosphorylates and inactivates glycogen synthase kinase-3 (GSK-3), thus inhibiting the phosphorylation of glycogen synthase and so stimulating glycogen synthesis. Dephosphorylation of the metabolic enzyme may also be promoted by the activation of protein phosphatase-1 (PP1), although the pathway upstream of PP1 remains to be determined.

Human and Animal Studies

Alterations in insulin signaling have been observed in type 2 diabetic and/or obese subjects and also in human and animal models of lipid-induced insulin resistance. Although inhibition of insulin receptor kinase activity does not appear to be lipid-mediated,[41,42] the interaction between the insulin receptor and IRS-1 was often found to be perturbed. Thus, lipid oversupply and insulin resistance in skeletal muscle are frequently associated with diminished IRS-1 tyrosine phosphorylation and recruitment of PI3K.[16,43-51]

Reduced IRS-1 tyrosine phosphorylation has been linked to its increased serine/threonine phosphorylation.[52,53] Insulin itself appears to induce such phosphorylation as a form of negative feedback and this could be inappropriately activated in diabetes.[54] Recent work indicates that aPKCζ, which is activated downstream of PI3K, acts in this way[55,56] and so can play both positive and negative roles in insulin action. cPKC and/or nPKC activation also leads to IRS-1 phosphorylation and hence diminished insulin receptor association[54] and decreased PI3K stimulation.[57] PKC isoforms may not directly phosphorylate IRS-1, but several observations suggest that they could act through the stimulation of MAP kinase.[58-60]

Fewer studies have addressed more distal insulin signaling components. The activation of PKB by insulin in muscle from type 2 diabetic patients is controversial.[43,61] PKB was found to be reduced in muscle from fat-fed rats despite normal PDK1 phosphorylation.[62] By contrast, basal aPKC activity was elevated in the same muscle, but refractory to further activation by insulin.[62] The only other reports of aPKC concern a nonobese model of diabetes, the Goto-Kakizaki rat, in which the kinase was found to be diminished together with IRS-1-dependent PI3K, while PKB activation was unaffected.[63,64] Further work is required to determine the significance of changes in these effectors of PI3K. MAP kinase, which lies downstream of Shc tyrosine phosphorylation on a separate insulin-activated pathway to IRS-1 and PI3K, has been reported to show normal activation in insulin-resistant muscle,[44,48,65] suggesting that IRS-1 signaling is selectively affected by increased lipid availability.

Cell Models

Lipid treatment of cultured cells to study insulin resistance can avoid problems of confounding hyperglycemia and hyperinsulinemia associated with human and animal studies and can also determine the distinct effects of different lipids. Thus, palmitate pretreatment has been shown to inhibit glucose disposal and PKB activation in myotubes.[66,67] In C2C12 cells, this occurs despite normal IRS-1 tyrosine phosphorylation and PI3K activation.[67] In direct contrast, pretreatment of these cells with the unsaturated FFA oleate and linoleate does inhibit PI3K activation as well as glycogen synthesis, although PKB activation was unaffected.[67] Therefore, as also suggested by human studies,[43,61] it appears that the activation of PI3K and PKB may not be tightly coupled. In other work employing cultured human muscle, basal or insulin-stimulated 2-deoxyglucose uptake was unaltered in cells preincubated with oleate, whereas basal glucose uptake was increased and insulin response was impaired in palmitate- and stearate-loaded cells.[68] Taken together, these studies demonstrate that different lipids affect insulin action by different mechanisms.

IRS-1-Independent Effects of PKC

In addition to the serine/threonine phosphorylation of IRS-1, PKC may also affect insulin signaling at more distal sites, providing further possible mechanisms for the above observations. aPKCζ has been shown to act as a negative regulator of PKB.[69,70] PKC can phosphorylate glycogen synthase[71] and also a novel inhibitor of PP1, PHI-1,[72] *in vitro*, which in each case would be expected to inhibit glycogen synthesis in intact cells, although this has not been shown. Overexpression of the "phosphoprotein enriched in diabetes" (PED/PEA15) in L6 skeletal muscle cells inhibits insulin-stimulated glucose transport through the activation of PKCα and to a lesser extent PKCβ.[73,74] Together with the serine/threonine phosphorylation of IRS-1, these forms of negative interaction between PKC and insulin signaling are potential mechanisms for the generation of insulin resistance, but their importance is yet to be established.

PKC AND INHIBITION OF GLYCOGEN SYNTHESIS BY UNSATURATED FFA IN C2C12 CELLS

In C2C12 myotubes, the basal rate of glucose uptake is 10-fold higher than that of glycogen synthesis and is not greatly affected by either insulin or FFA pretreatment.[67] Thus, lipid effects on glycogen synthesis can be determined in the absence of complications caused by reduced glucose influx. The use of PKC inhibitors or overexpression of dominant negative PKC isoforms indicates that PKC does not regulate glycogen synthesis in this model, despite the fact that unsaturated FFA also activate PKC isoforms.[19] However, this does not exclude the possibility that one or more PKC isoforms could inhibit insulin-stimulated glucose transport in intact muscle. Indeed, other studies have reported PKC-mediated inhibition of glucose transport in insulin-sensitive models.[21,75] While some studies have suggested the involvement of PKC in glycogen synthesis independently of glucose uptake, this may be due in part to the shorter time course of PKC activation studied[76] or to the use of nonspecific PKC inhibitors that have been shown to stimulate glycogen synthesis independently of PKC.[77]

CERAMIDE AND THE INHIBITION OF GLYCOGEN SYNTHESIS BY THE SATURATED FFA PALMITATE

Treatment of cells with palmitate has been shown to increase ceramide levels.[67,78] This signaling molecule can be generated in muscle by *de novo* synthesis from serine and palmitoyl-CoA[79] or by acute hydrolysis of sphingomyelin, and has previously been linked to reduced insulin sensitivity in response to TNFα treatment.[80]

Investigation of insulin signaling components in several systems has demonstrated that ceramide can specifically prevent activation of PKB despite normal PI3K activity and IRS-1 tyrosine phosphorylation.[67,81,82] Ceramide could exert its effects through several agents, including a ceramide-activated protein kinase (CAPK) that activates the ras-MAPK cascade to inhibit PKB,[83] and one or more PKC isoforms, including aPKCζ,[84–86] also a known inhibitor of PKB.[69] Although activation of both MAPK and aPKCζ was observed in palmitate-treated myotubes, inhibition of these kinases did not protect against the inhibitory effects of the FFA on glycogen synthesis.[67,87]

A further effector of ceramide is the ceramide-activated protein phosphatase (CAPP), which is a member of the protein phosphatase 2A (PP2A) family. Inhibition of PP2A by insulin, but activation by ceramide, has been reported[88] and PP2A can mediate the dephosphorylation of PKB.[89,90] Particulate fractions from palmitate-treated myotubes exhibit increased PP2A activity that is insensitive to inhibition by insulin, and the decreased insulin-stimulated PKB phosphorylation caused by the lipid can be prevented by the presence of the phosphatase inhibitor, okadaic acid.[87] Furthermore, palmitate is unable to reduce PKB activity and glycogen synthesis in cells overexpressing a constitutively activated PKB, (T308D,S473D)-PKB, which is independent of phosphatase. Palmitate was still able to exert its effects, however, in cells overexpressing a second constitutively activated PKB mutant, myristoylated PKB, which is still phosphatase-sensitive.[87] These data indicate that palmitate inhibits insulin action in C2C12 muscle cells by activating PP2A, which dephosphorylates PKB and hence reduces glycogen synthesis, as illustrated in FIGURE 1.

SUMMARY

A role for PKC in the generation of insulin resistance by lipid oversupply is supported by two lines of evidence. First, activation of certain PKC isoforms, especially PKCε and PKCθ, is associated with increased DAG levels and reduced insulin sensitivity and glucose disposal. Second, mechanistic approaches indicate that PKC activity can interfere with insulin signaling at the level of IRS-1, but also at other sites. The roles of specific isoforms are currently under investigation and the use of molecular techniques, such as overexpression of dominant negative mutants, is expected to overcome the problem of specifically inhibiting the individual enzymes. In addition, elevation of other lipid second messengers, such as ceramide, can also promote insulin resistance independently of PKC.

REFERENCES

1. MELLOR, H. & P.J. PARKER. 1998. The extended protein kinase C superfamily. Biochem. J. **332:** 281–292.
2. KASAHARA, K. & U. KIKKAWA. 1995. Distinct effects of saturated fatty acids on protein kinase C subspecies. J. Biochem. (Tokyo) **117:** 648–653.
3. ORELLANA, A., P.C. HIDALGO, M.N. MORALES *et al.* 1990. Palmitoyl-CoA and the acyl-CoA thioester of the carcinogenic peroxisome-proliferator ciprofibrate potentiate diacylglycerol-activated protein kinase C by decreasing the phosphatidylserine requirement of the enzyme. Eur. J. Biochem. **190:** 57–61.
4. MAJUMDAR, S., M.W. ROSSI, T. FUJIKI *et al.* 1991. Protein kinase C isotypes and signaling in neutrophils. J. Biol. Chem. **266:** 9285–9294.
5. NESHER, M. & A. BONEH. 1994. Effect of fatty acids and their acyl-CoA esters on protein kinase c activity in fibroblasts—possible implications in fatty acid oxidation defects. Biochim. Biophys. Acta **1221:** 66–72.

6. NISHIZUKA, Y. 1995. Protein kinases: 5. Protein kinase C and lipid signaling for sustained cellular responses. FASEB J. **9:** 484–496.
7. DUTIL, E.M. & A.C. NEWTON. 2000. Dual role of pseudosubstrate in the coordinated regulation of protein kinase C by phosphorylation and diacylglycerol. J. Biol. Chem. **275:** 10697–10701.
8. STANDAERT, M.L., L. GALLOWAY, P. KARNAM et al. 1997. Protein kinase C-zeta as a downstream effector of phosphatidylinositol 3-kinase during insulin stimulation in rat adipocytes: potential role in glucose transport. J. Biol. Chem. **272:** 30075–30082.
9. JAKEN, S. & P.J. PARKER. 2000. Protein kinase C binding partners. BioEssays **22:** 245–254.
10. SCHMITZ-PEIFFER, C., C.L. BROWNE, N.D. OAKES et al. 1997. Alterations in the expression and cellular localization of protein kinase C isozymes epsilon and theta are associated with insulin resistance in skeletal muscle of the high-fat-fed rat. Diabetes **46:** 169–178.
11. OAKES, N.D., C.J. KENNEDY, A.B. JENKINS et al. 1994. A new antidiabetic agent, BRL 49653, reduces lipid availability and improves insulin action and glucoregulation in the rat. Diabetes **43:** 1203–1210.
12. SCHMITZ-PEIFFER, C., N.D. OAKES, C.L. BROWNE et al. 1997. Reversal of chronic alterations of skeletal muscle protein kinase c from fat-fed rats by BRL-49653. Am. J. Physiol. **273:** E915–E921.
13. LAYBUTT, D.R., C. SCHMITZ-PEIFFER, A.K. SAHA et al. 1999. Muscle lipid accumulation and protein kinase C activation in the insulin-resistant chronically glucose-infused rat. Am. J. Physiol. **277:** E1070–E1076.
14. QU, X., J.P. SEALE & R. DONNELLY. 1999. Tissue and isoform-selective activation of protein kinase C in insulin-resistant obese Zucker rats—effects of feeding. J. Endocrinol. **162:** 207–214.
15. IKEDA, Y., G.S. OLSEN, E. ZIV et al. 2001. Cellular mechanism of nutritionally induced insulin resistance in *Psammomys obesus*: overexpression of protein kinase C epsilon in skeletal muscle precedes the onset of hyperinsulinemia and hyperglycemia. Diabetes **50:** 584–592.
16. GRIFFIN, M.E., M.J. MARCUCCI, G.W. CLINE et al. 1999. Free fatty acid–induced insulin resistance is associated with activation of protein kinase C theta and alterations in the insulin signaling cascade. Diabetes **48:** 1270–1274.
17. ITANI, S.I., Q. ZHOU, W.J. PORIES et al. 2000. Involvement of protein kinase C in human skeletal muscle insulin resistance and obesity. Diabetes **49:** 1353–1358.
18. ITANI, S.I., W.J. PORIES, K.G. MACDONALD et al. 2001. Increased protein kinase C theta in skeletal muscle of diabetic patients. Metabolism **50:** 553–557.
19. CAZZOLLI, R., D.L. CRAIG, T.J. BIDEN et al. 2002. Inhibition of glycogen synthesis in C2C12 skeletal muscle cells by unsaturated free fatty acid is independent of PKC alpha, PKC epsilon, PKC theta, and beta-oxidation. Am. J. Physiol. In press.
20. LU, X., X.Y. YANG, R.L. HOWARD et al. 2000. Fatty acids modulate protein kinase C activation in porcine vascular smooth muscle cells independently of their effect on *de novo* diacylglycerol synthesis. Diabetologia **43:** 1136–1144.
21. MONTELL, E., M. TURINI, M. MAROTTA et al. 2001. DAG accumulation from saturated fatty acids desensitizes insulin stimulation of glucose uptake in muscle cells. Am. J. Physiol. **280:** E229–E237.
22. AVIGNON, A., K. YAMADA, X.P. ZHOU et al. 1996. Chronic activation of protein kinase C in soleus muscles and other tissues of insulin-resistant type II diabetic Goto-Kakizaki (GK), obese/aged, and obese/Zucker rats—a mechanism for inhibiting glycogen synthesis. Diabetes **45:** 1396–1404.
23. HUWILER, A., D. FABBRO & J. PFEILSCHIFTER. 1991. Differential recovery of protein kinase C-alpha and -epsilon isozymes after long-term phorbol ester treatment in rat renal mesangial cells. Biochem. Biophys. Res. Commun. **180:** 1422–1428.
24. BORNER, C., S.N. GUADAGNO, W.W.L. HSIAO et al. 1992. Expression of 4 protein kinase-C isoforms in rat fibroblasts: differential alterations in ras-transformed, src-transformed, and fos-transformed cells. J. Biol. Chem. **267:** 12900–12910.
25. ZHAO, L., M.L. STANDAERT, D.R. COOPER et al. 1994. Effects of insulin on protein kinase-C (PKC) in HIRC-B cells: specific activation of PKC epsilon and its resistance to phorbol ester–induced down-regulation. Endocrinology **135:** 2504–2510.

26. TANG, E.Y., P.J. PARKER, J. BEATTIE *et al.* 1993. Diabetes induces selective alterations in the expression of protein kinase-C isoforms in hepatocytes. FEBS Lett. **326:** 117–123.
27. DONNELLY, R., M.J. REED, S. AZHAR *et al.* 1994. Expression of the major isoenzyme of protein kinase-C in skeletal muscle, nPKC theta, varies with muscle type and in response to fructose-induced insulin resistance. Endocrinology **135:** 2369–2374.
28. CONSIDINE, R.V., M.R. NYCE, L.E. ALLEN *et al.* 1995. Protein kinase C is increased in the liver of humans and rats with noninsulin-dependent diabetes mellitus: an alteration not due to hyperglycemia. J. Clin. Invest. **95:** 2938–2944.
29. QU, X.Q., J.P. SEALE & R. DONNELLY. 1999. Tissue- and isoform-specific effects of aging in rats on protein kinase C in insulin-sensitive tissues. Clin. Sci. **97:** 355–361.
30. HEYDRICK, S.J., N.B. RUDERMAN, T.G. KUROWSKI *et al.* 1991. Enhanced stimulation of diacylglycerol and lipid synthesis by insulin in denervated muscle: altered protein kinase-C activity and possible link to insulin resistance. Diabetes **40:** 1707–1711.
31. CHEN, K.S., S.J. HEYDRICK, M.L. BROWN *et al.* 1994. Insulin increases a biochemically distinct pool of diacylglycerol in the rat soleus muscle. Am. J. Physiol. **266:** E479–E485.
32. STANDAERT, M.L., G. BANDYOPADHYAY, X.P. ZHOU *et al.* 1996. Insulin stimulates phospholipase D–dependent phosphatidylcholine hydrolysis, Rho translocation, *de novo* phospholipid synthesis, and diacylglycerol/protein kinase C signaling in L6 myotubes. Endocrinology **137:** 3014–3020.
33. YAMADA, K., A. AVIGNON, M.L. STANDAERT *et al.* 1995. Effects of insulin on the translocation of protein kinase C-theta and other protein kinase C isoforms in rat skeletal muscles. Biochem. J. **308:** 177–180.
34. AVIGNON, A., M.L. STANDAERT, K. YAMADA *et al.* 1995. Insulin increases mRNA levels of protein kinase C-alpha and -beta in rat adipocytes and protein kinase C-alpha, -beta, and -theta in rat skeletal muscle. Biochem. J. **308:** 181–187.
35. KOYA, D. & G.L. KING. 1998. Protein kinase C activation and the development of diabetic complications. Diabetes **47:** 859–866.
36. COOPER, D.R., J.E. WATSON & M.L. DAO. 1993. Decreased expression of protein kinase-C alpha, beta, and epsilon in soleus muscle of Zucker obese (fa/fa) rats. Endocrinology **133:** 2241–2247.
37. SAHA, A.K., T.G. KUROWSKI, J.R. COLCA *et al.* 1994. Lipid abnormalities in tissues of the KKA(y) mouse: effects of pioglitazone on malonyl-CoA and diacylglycerol. Am. J. Physiol. **267:** E95–E101.
38. INOGUCHI, T., P. LI, F. UMEDA *et al.* 2000. High glucose level and free fatty acid stimulate reactive oxygen species production through protein kinase C–dependent activation of NAD(P)H oxidase in cultured vascular cells. Diabetes **49:** 1939–1945.
39. YU, H.Y., T. INOGUCHI, M. KAKIMOTO *et al.* 2001. Saturated non-esterified fatty acids stimulate *de novo* diacylglycerol synthesis and protein kinase C activity in cultured aortic smooth muscle cells. Diabetologia **44:** 614–620.
40. WAKELAM, M. 1998. Diacylglycerol—when is it an intracellular messenger? Biochim. Biophys. Acta **1436:** 117–126.
41. NOLAN, J.J., G. FREIDENBERG, R. HENRY *et al.* 1994. Role of human skeletal muscle insulin receptor kinase in the *in vivo* insulin resistance of noninsulin-dependent diabetes mellitus and obesity. J. Clin. Endocrinol. Metab. **78:** 471–477.
42. KRUSZYNSKA, Y.T. & J.M. OLEFSKY. 1996. Cellular and molecular mechanisms of non-insulin dependent diabetes mellitus. J. Invest. Med. **44:** 413–428.
43. KIM, Y.B., S.E. NIKOULINA, T.P. CIARALDI *et al.* 1999. Normal insulin-dependent activation of Akt/protein kinase B, with diminished activation of phosphoinositide 3-kinase, in muscle in type 2 diabetes. J. Clin. Invest. **104:** 733–741.
44. KROOK, A., M. BJORNHOLM, D. GALUSKA *et al.* 2000. Characterization of signal transduction and glucose transport in skeletal muscle from type 2 diabetic patients. Diabetes **49:** 284–292.
45. GOODYEAR, L.J., F. GIORGINO, L.A. SHERMAN *et al.* 1995. Insulin receptor phosphorylation, insulin receptor substrate-1 phosphorylation, and phosphatidylinositol 3-kinase activity are decreased in intact skeletal muscle strips from obese subjects. J. Clin. Invest. **95:** 2195–2204.

46. BJORNHOLM, M., Y. KAWANO, M. LEHTIHET et al. 1997. Insulin receptor substrate-1 phosphorylation and phosphatidylinositol 3-kinase activity in skeletal muscle from NIDDM subjects after in vivo insulin stimulation. Diabetes **46:** 524–527.
47. DRESNER, A., D. LAURENT, M. MARCUCCI et al. 1999. Effects of free fatty acids on glucose transport and IRS-1-associated phosphatidylinositol 3-kinase activity. J. Clin. Invest. **103:** 253–259.
48. CUSI, K., K. MAEZONO, A. OSMAN et al. 2000. Insulin resistance differentially affects the PI3-kinase- and MAP-kinase-mediated signaling in human muscle. J. Clin. Invest. **105:** 311–320.
49. LE MARCHAND–BRUSTEL, Y. 1999. Molecular mechanisms of insulin action in normal and insulin-resistant states. Exp. Clin. Endocrinol. Diabetes **107:** 126–132.
50. ANAI, M., M. FUNAKI, T. OGIHARA et al. 1999. Enhanced insulin-stimulated activation of phosphatidylinositol 3-kinase in the liver of high-fat-fed rats. Diabetes **48:** 158–169.
51. ZIERATH, J.B., K.L. HOUSEKNECHT, L. GNUDI et al. 1997. High-fat feeding impairs insulin-stimulated GLUT4 recruitment via an early insulin-signaling defect. Diabetes **46:** 215–223.
52. TANTI, J.F., T. GREMEAUX, E. VAN OBBERGHEN et al. 1994. Serine/threonine phosphorylation of insulin receptor substrate 1 modulates insulin receptor signaling. J. Biol. Chem. **269:** 6051–6057.
53. KANETY, H., R. FEINSTEIN, M.Z. PAPA et al. 1995. Tumor necrosis factor alpha–induced phosphorylation of insulin receptor substrate-1 (IRS-1): possible mechanism for suppression of insulin-stimulated tyrosine phosphorylation of IRS-1. J. Biol. Chem. **270:** 23780–23784.
54. PAZ, K., R. HEMI, D. LEROITH et al. 1997. A molecular basis for insulin resistance: elevated serine/threonine phosphorylation of IRS-1 and IRS-2 inhibits their binding to the juxtamembrane region of the insulin receptor and impairs their ability to undergo insulin-induced tyrosine phosphorylation. J. Biol. Chem. **272:** 29911–29918.
55. LIU, Y.F., K. PAZ, A. HERSCHKOVITZ et al. 2001. Insulin stimulates PKC zeta–mediated phosphorylation of insulin receptor substrate-1 (IRS-1): a self-attenuated mechanism to negatively regulate the function of IRS proteins. J. Biol. Chem. **276:** 14459–14465.
56. RAVICHANDRAN, L.V., D.L. ESPOSITO, J. CHEN et al. 2001. Protein kinase C-zeta phosphorylates insulin receptor substrate-1 and impairs its ability to activate phosphatidylinositol 3-kinase in response to insulin. J. Biol. Chem. **276:** 3543–3549.
57. CHIN, J.E., F. LIU & R.A. ROTH. 1994. Activation of protein kinase C alpha inhibits insulin-stimulated tyrosine phosphorylation of insulin receptor substrate-1. Mol. Endocrinol. **8:** 51–58.
58. MOTHE, I. & E. VANOBBERGHEN. 1996. Phosphorylation of insulin receptor substrate-1 on multiple serine residues, 612, 632, 662, and 731, modulates insulin action. J. Biol. Chem. **271:** 11222–11227.
59. DE FEA, K. & R.A. ROTH. 1997. Protein kinase C modulation of insulin receptor substrate-1 tyrosine phosphorylation requires serine 612. Biochemistry **36:** 12939–12947.
60. DE FEA, K. & R.A. ROTH. 1997. Modulation of insulin receptor substrate-1 tyrosine phosphorylation and function by mitogen-activated protein kinase. J. Biol. Chem. **272:** 31400–31406.
61. KROOK, A., R.A. ROTH, X.J. JIANG et al. 1998. Insulin-stimulated Akt kinase activity is reduced in skeletal muscle from NIDDM subjects. Diabetes **47:** 1281–1286.
62. TREMBLAY, F., C. LAVIGNE, H. JACQUES et al. 2001. Defective insulin-induced GLUT4 translocation in skeletal muscle of high fat–fed rats is associated with alterations in both Akt/protein kinase B and atypical protein kinase C (zeta/lambda) activities. Diabetes **50:** 1901–1910.
63. KANOH, Y., G. BANDYOPADHYAY, M.P. SAJAN et al. 2000. Thiazolidinedione treatment enhances insulin effects on protein kinase C-zeta/lambda activation and glucose transport in adipocytes of nondiabetic and Goto-Kakizaki type II diabetic rats. J. Biol. Chem. **275:** 16690–16696.
64. KANOH, Y., G. BANDYOPADHYAY, M.P. SAJAN et al. 2001. Rosiglitazone, insulin treatment, and fasting correct defective activation of protein kinase C-zeta/lambda by

insulin in vastus lateralis muscles and adipocytes of diabetic rats. Endocrinology **142:** 1595–1605.
65. JIANG, Z.Y., Q.L. ZHOU, A. CHATTERJEE *et al.* 1999. Endothelin-1 modulates insulin signaling through phosphatidylinositol 3-kinase pathway in vascular smooth muscle cells. Diabetes **48:** 1120–1130.
66. STORZ, P., H. DOPPLER, A. WERNIG *et al.* 1999. Cross-talk mechanisms in the development of insulin resistance of skeletal muscle cells—palmitate rather than tumour necrosis factor inhibits insulin-dependent protein kinase B (PKB)/Akt stimulation and glucose uptake. Eur. J. Biochem. **266:** 17–25.
67. SCHMITZ-PEIFFER, C., D.L. CRAIG & T.J. BIDEN. 1999. Ceramide generation is sufficient to account for the inhibition of the insulin-stimulated PKB pathway in C2C12 skeletal muscle cells pretreated with palmitate. J. Biol. Chem. **274:** 24202–24210.
68. MONTELL, E., M. TURINI, M. MAROTTA *et al.* 2001. DAG accumulation from saturated fatty acids desensitizes insulin stimulation of glucose uptake in muscle cells. Am. J. Physiol. **280:** E229–E237.
69. DOORNBOS, R.P., M. THEELEN, P.C. VAN DER HOEVEN *et al.* 1999. Protein kinase C zeta is a negative regulator of protein kinase B activity. J. Biol. Chem. **274:** 8589–8596.
70. MAO, M.L., X.J. FANG, Y.L. LU *et al.* 2000. Inhibition of growth-factor-induced phosphorylation and activation of protein kinase B/Akt by atypical protein kinase C in breast cancer cells. Biochem. J. **352:** 475–482.
71. AHMAD, Z., F.T. LEE, R.A. DEPAOLI *et al.* 1984. Phosphorylation of glycogen synthase by the Ca^{2+}- and phospholipid-activated protein kinase (protein kinase C). J. Biol. Chem. **259:** 8743–8747.
72. ETO, M., A. KARGINOV & D.L. BRAUTIGAN. 1999. A novel phosphoprotein inhibitor of protein type-1 phosphatase holoenzymes. Biochemistry **38:** 16952–16957.
73. CONDORELLI, G., G. VIGLIOTTA, C. IAVARONE *et al.* 1998. PED/PEA-15 gene controls glucose transport and is overexpressed in type 2 diabetes mellitus. EMBO J. **17:** 3858–3866.
74. CONDORELLI, G., G. VIGLIOTTA, A. TRENCIA *et al.* 2001. Protein kinase C (PKC)–alpha activation inhibits PKC-zeta and mediates the action of PED/PEA-15 on glucose transport in the L6 skeletal muscle cells. Diabetes **50:** 1244–1252.
75. CORTRIGHT, R.N., J.L. AZEVEDO, Q. ZHOU *et al.* 2000. Protein kinase C modulates insulin action in human skeletal muscle. Am. J. Physiol. **278:** E553–E562.
76. LIN, Y.S., S.I. ITANI, T.G. KUROWSKI *et al.* 2001. Inhibition of insulin signaling and glycogen synthesis by phorbol dibutyrate in rat skeletal muscle. Am. J. Physiol. **281:** E8–E15.
77. STANDAERT, M.L., G. BANDYOPADHYAY, E.K. ANTWI *et al.* 1999. RO 31-8220 activates c-Jun N-terminal kinase and glycogen synthase in rat adipocytes and L6 myotubes: comparison to actions of insulin. Endocrinology **140:** 2145–2151.
78. SHIMABUKURO, M., Y.T. ZHOU, M. LEVI *et al.* 1998. Fatty acid–induced beta cell apoptosis: a link between obesity and diabetes. Proc. Natl. Acad. Sci. U.S.A. **95:** 2498–2502.
79. MERRILL, A., JR. & D.D. JONES. 1990. An update of the enzymology and regulation of sphingomyelin metabolism. Biochim. Biophys. Acta **1044:** 1–12.
80. MURASE, K., H. ODAKA, M. SUZUKI *et al.* 1998. Pioglitazone time-dependently reduces tumour necrosis factor-alpha level in muscle and improves metabolic abnormalities in Wistar fatty rats. Diabetologia **41:** 257–264.
81. SUMMERS, S.A., L.A. GARZA, H.L. ZHOU *et al.* 1998. Regulation of insulin-stimulated glucose transporter GLUT4 translocation and akt kinase activity by ceramide. Mol. Cell. Biol. **18:** 5457–5464.
82. ZHOU, H.L., S.K. SUMMERS, M.J. BIRNBAUM *et al.* 1998. Inhibition of akt kinase by cell-permeable ceramide and its implications for ceramide-induced apoptosis. J. Biol. Chem. **273:** 16568–16575.
83. BASU, S., S. BAYOUMY, Y. ZHANG *et al.* 1998. Bad enables ceramide to signal apoptosis via Ras and Raf-1. J. Biol. Chem. **273:** 30419–30426.
84. MULLER, G., M. AYOUB, P. STORZ *et al.* 1995. PKC zeta is a molecular switch in signal transduction of TNF-alpha, bifunctionally regulated by ceramide and arachidonic acid. EMBO J. **14:** 1961–1969.

85. HUWILER, A., D. FABBRO & J. PFEILSCHIFTER. 1998. Selective ceramide binding to protein kinase c-alpha and -delta isoenzymes in renal mesangial cells. Biochemistry **37:** 14556–14562.
86. VAN BLITTERSWIJK, W.J. 1998. Hypothesis: ceramide conditionally activates atypical protein kinases C, Raf-1, and KSR through binding to their cysteine-rich domains. Biochem. J. **331:** 679–680.
87. CAZZOLLI, R., L. CARPENTER, T.J. BIDEN et al. 2001. A role for protein phosphatase 2a–like activity, but not atypical protein kinase C zeta, in the inhibition of protein kinase B/akt and glycogen synthesis by palmitate. Diabetes **50:** 2210–2218.
88. BEGUM, N., L. RAGOLIA & M. SRINIVASAN. 1996. Effect of tumor necrosis factor-alpha on insulin-stimulated mitogen-activated protein kinase cascade in cultured rat skeletal muscle cells. Eur. J. Biochem. **238:** 214–220.
89. ANDJELKOVIC, M., T. JAKUBOWICZ, P. CRON et al. 1996. Activation and phosphorylation of a pleckstrin homology domain containing protein kinase (RAC-PK/PKB) promoted by serum and protein phosphatase inhibitors. Proc. Natl. Acad. Sci. U.S.A. **93:** 5699–5704.
90. SALINAS, M., R. LOPEZ-VALDALISO, D. MARTIN et al. 2000. Inhibition of PKB/Akt1 by C2-ceramide involves activation of ceramide-activated protein phosphatase in PC12 cells. Mol. Cell. Neurosci. **15:** 156–169.

Evaluation of Free Fatty Acid Metabolism in Vivo

NICHOLAS D. OAKES[a] AND STUART M. FURLER[b]

[a]*AstraZeneca R&D Mölndal, Mölndal, Sweden*

[b]*Garvan Institute of Medical Research, Sydney, Australia*

> ABSTRACT: In order to enable detailed studies of free fatty acid (FFA) metabolism, we recently introduced a method for the evaluation of tissue-specific FFA metabolism *in vivo*. The method is based on the simultaneous use of ^{14}C-palmitate (^{14}C-P) and the non-β-oxidizable FFA analogue, [9,10-^{3}H]-(R)-2-bromopalmitate (^{3}H-R-BrP). Indices of total FFA utilization and incorporation into storage products are obtained from tissue concentrations of ^{3}H and ^{14}C, respectively, following intravenous administration of ^{3}H-R-BrP and ^{14}C-P and their disappearance from plasma into tissues. This review covers the basis for, and developments in, the methodology, as well as some of the applications to date. In the rat, the method has been used to characterize tissue-specific alterations in FFA metabolism in various situations, including skeletal muscle contraction, fasting, hyperinsulinemia, and various pharmacological manipulations. The results of all these studies clearly demonstrate tissue-level control of FFA utilization and metabolic fate, refuting the traditional view that FFA utilization is simply supply-driven. Recent developments enable the simultaneous evaluation of both tissue-specific FFA and glucose metabolism by integrating the use of 2-deoxyglucose and stable isotope-labeled glucose tracers. In conclusion, the ^{3}H-R-BrP methodology, especially in combination with other tracers, represents a powerful tool for elucidation of tissue-specific fatty acid metabolism *in vivo*.
>
> KEYWORDS: tracer; kinetics; turnover; liver; heart; muscle; adipose tissue

INTRODUCTION

Disturbances in fatty acid metabolism appear to play important roles in the pathogenesis of several key features of the insulin resistance syndrome, including impaired glucose regulation, dyslipidemia, and obesity. In terms of glucose regulation, ample evidence has accumulated that systemic fatty acid oversupply can decrease insulin-stimulated glucose uptake in skeletal muscle,[1] reduce the ability of insulin to suppress hepatic glucose production,[2–4] and alter glucose-stimulated insulin secretion.[5] In terms of involvement in dyslipidemia, an oversupply of fatty acids to liver may cause hypertriglyceridemia as well as the atherogenic lipoprotein profile.[6] The problem may not be limited solely to control of systemic fatty acid availability. There is also evidence for defective fatty acid oxidation in conditions of

---

Address for correspondence: Nick Oakes, Integrative Pharmacology, AstraZeneca R&D Mölndal, S-431 83 Mölndal, Sweden.

insulin resistance and the ability to appropriately switch between glucose and lipid fuels, in fed and fasting states, respectively.[7] These defects may have an important role in the development of obesity.

Accumulation of lipids in nonadipose tissues has been implicated in virtually every major organ-specific feature associated with insulin-resistant states. In skeletal muscle, several studies in animals and humans[8–11] have documented a close correlation between tissue levels of lipids (usually triglycerides) and glucose metabolic insulin resistance. In the liver, increased triglyceride (TG) stores[12] may cause hypertriglyceridemia by driving increased VLDL production rates since virtually all VLDL TG is derived from the lipolysis and subsequent reesterification of cytosolic TG.[13] In studies of insulin-resistant animal models, lipotoxicity has been implicated as a cause of pancreatic failure, leading to diabetes[14] and cardiomyopathy.[15] Some advances have been made towards elucidating mechanistic links between lipid accumulation and these manifestations of the insulin resistance syndrome. The glucose–fatty acid cycle[16] provides one mechanism by which elevated fatty acid availability is able to decrease glucose utilization in skeletal muscle. More recently, evidence has emerged for the existence of lipid signaling pathways that may link elevated fatty acid availability to inhibition of very early steps in the insulin-signaling cascade (as recently reviewed in reference 17). Despite these advances, however, the basic question remains as to whether the lipid accumulation is a cause or consequence of the insulin resistance. In addition, the roles of supply and local utilization in the overaccumulation of lipid are unclear. These are important issues, especially in the context of the development of rational therapeutic strategies.

A major limitation in the elucidation of lipid metabolism has been the lack of available methodology for detailed assessment of *in vivo* lipid transfer and metabolism. While established techniques exist for assessing whole body metabolism using tracer dilution[18] or calorimetry,[19] studies at the individual tissue level have been conducted largely *in vitro*. In this respect, methodological advances in the study of fatty acid metabolism have lagged behind those for the study of glucose metabolism. The classic approach to assessing *in vivo* substrate metabolism at the tissue level is based on the mass balance or inflow-outflow principle. In theory, this approach requires only a few simple assumptions and has the additional advantage that it is based on the direct measurement of the native substance under study. In practice, flux estimates derived from these methods can be unreliable if arteriovenous substrate concentration differences are relatively small (e.g., see reference 20). Moreover, these methods are limited by the accessibility and degree of dedication of the venous drainage. This latter problem is exemplified in efforts to characterize disturbances in skeletal muscle fatty acid metabolism based on arteriovenous exchange across the forearm or leg in humans (e.g., see references 21 and 22). Interpretation of such studies is complicated by the regional nature of these assessments. Thus, there is an unknown contribution of tissues other than muscle and, perhaps more importantly, the probable large heterogeneity among different skeletal muscle fiber types, suggested by *in vitro* studies,[23] is completely obscured.

An alternative means of assessing tissue substrate utilization *in vivo* is based on the principle of metabolic trapping. This indirect approach is based on the use of an analogue tracer that is transported and undergoes initial metabolic steps of the native substrate, but then, because of a unique chemical structure, an analogue product is formed that cannot undergo further metabolism, leaving the associated radiolabel trapped within the tissue. Depending on the radioisotope used, tissue entrapment is assessed by external detection (γ-emitters), imaging (γ- or e$^+$-emitters), or (in experimental animals) tissue dissection followed by either autoradiography or extraction and liquid scintillography (e$^-$-emitters). Tissue accessibility and precise definition are major advantages of techniques based on the metabolic trapping principle. For evaluation of *in vivo* glucose metabolism, 2-deoxyglucose tracers have been applied extensively since 1977,[24] both in the clinical setting, largely for the assessment of regional glucose utilization in the brain,[25] as well as in experimental animals, for the assessment of glucose utilization in a number of tissues, including heart, skeletal muscle, and fat.[26] For evaluation of fatty acid metabolism, much of the effort has focused on the production of analogues suitable for myocardial imaging in humans using external detection. A critical condition for prolonged cardiac retention is that the analogue cannot undergo β-oxidization. This property has been achieved by introducing methyl branching on either the α- or β-positions of the fatty acid[27] or by insertion of a sulfur into the fatty acid chain.[28,29] Although analogues have been developed with excellent properties for clinical assessment of cardiac metabolism, several factors make these unsuitable for more general application, particularly in studies of small experimental animals. The analogues are not readily available, their synthesis requires nontrivial chemistry, requiring radioisotopes for external imaging, and they have not been extensively validated for use in tissues other than myocardium. Against this background, we developed a method for assessing free fatty acid (FFA) utilization *in vivo* using the (R)-2-bromopalmitate tracer.[30]

THE 2-BROMOPALMITATE TRACER METHOD FOR ASSESSING TISSUE-SPECIFIC FFA METABOLISM *IN VIVO*

Here, we review the initial validation work of the (R)-2-bromopalmitate methodology, some applications, as well as some more recent developments in the method that enable the simultaneous assessment of tissue-specific glucose and fatty acid metabolism.

Initial consideration is given to the general theoretical principle upon which the 2-bromopalmitate tracer method is based, that is, metabolic trapping of a partially metabolizable substrate analogue. The metabolic properties of an ideal FFA analogue are then compared to the actual properties of the 2-bromopalmitate tracer.

General Principles for Assessment of In Vivo *Substrate Metabolism Using a Partially Metabolizable Substrate Analogue*

FIGURE 1 represents the cellular uptake of a native substrate (S) and a structurally related substrate analogue (S^*). It is assumed that both the substrate and analogue compete for the same transport processes involved in delivery to the cell, traversal of the plasma membrane, and cytoplasmic transfer. The substances then compete for one or more serially arranged enzymatic transformation processes, at least one of which must be effectively irreversible, resulting in the formation of the native product (P_i) and analogue product (P_i^*). The fates of the native and analogue substances then diverge. In the case of the native substance, further metabolism can give rise to products that are lost from the cell (e.g., P_m). In the case of the analogue,

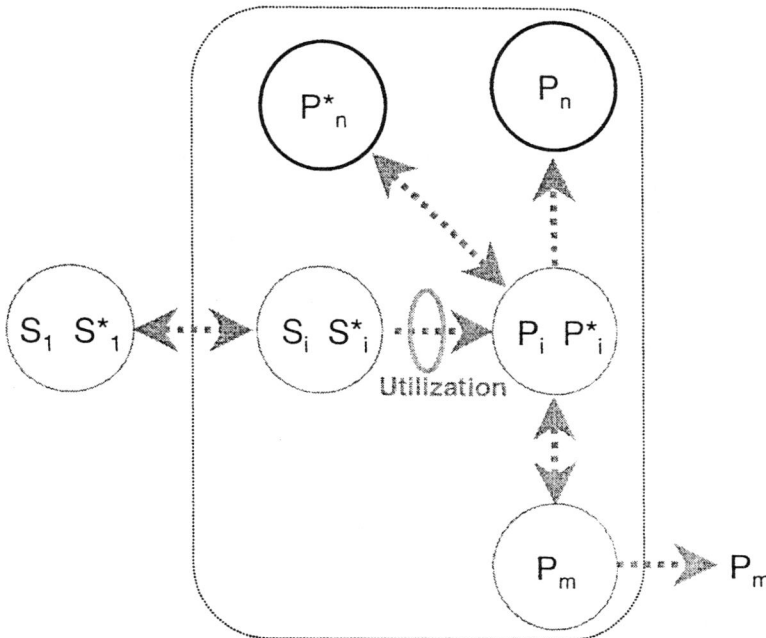

FIGURE 1. Cellular uptake and metabolism of a substrate (S) and a substrate analogue (S^*). P and P^* represent metabolic products of the substrate and analogue, respectively. *Subscripts* refer to distinct kinetic pools or distinct products.

structural modification(s) prevents complete metabolism and results in product trapping within the cell. Under specific experimental conditions, the sum of the analogue products ($\Sigma^n_{j=i} P_i^*$) can be used to estimate the flux of native molecules through the irreversible transformation step, here defined as the substrate utilization rate. Finally, it is also assumed that the metabolism of the native substance is not perturbed by the presence of the analogue substance. In practice, this means limiting the analogue product concentrations to very low or tracer levels.

Properties of the Ideal FFA Analogue

Long-chain FFAs are transported in blood bound to albumin. The precise mechanisms of transport of FFAs across plasma membranes are still uncertain, although recent evidence suggests that it may involve both passive diffusion across the lipid bilayer as well as protein-mediated transport.[31] Once inside the cell, FFAs are bound to fatty acid binding proteins (FABP). Metabolic sequestration of FFAs is effected by long-chain fatty acyl CoA synthetase (ACS), which activates the FFA to its CoA derivative. This activation is a necessary step for the major pathways of fatty acid metabolism, including β-oxidation, as well as phospholipid and acylglyceride synthesis. Flux through this ACS-mediated step, therefore, represents FFA utilization. It is necessary that an FFA analogue competes with native FFA for all transport and metabolic steps, up to and including activation by ACS. Ideally, the analogue product(s) should be known and measurable and must not be rapidly exported from the cell. In particular, the analogue must not be a substrate for complete β-oxidation.

Properties of 2-Bromopalmitate

Binding to FFA Transport Proteins

Albumin apparently possesses distinct ligand binding sites for various hydrophobic molecules, including FFAs and bilirubin.[32] Our binding studies[30] not only showed similar binding affinity compared to palmitate, but also strongly implied that the analogue occupies the same high-affinity binding sites as the native FFA. Likewise, studies with intestinal FABP indicated similar binding affinity to the same binding site for both the analogue and the native fatty acid. These results confirmed that 2-bromopalmitate and palmitate compete for the same ligand binding sites on extracellular and intracellular FFA transport proteins.

Activation by the Fatty Acid Sequestering Enzyme ACS

Bromination of the alpha carbon (C2) of palmitate introduces an active center of asymmetry to the molecule. The resulting 2-bromopalmitate is a racemic mixture of optical enantiomers. Using liver microsomal preparations, it was determined that only one of the enantiomers, (R)-2-bromopalmitate, was a substrate for ACS.[30] This finding is consistent with previous reports of stereoselectivity of ACS for other molecules.[33,34]

Metabolic Trapping

The issue of how well the products of ^{3}H-R-BrP are trapped can be examined by looking at the *in vivo* retention of all ^{3}H products. This has been assessed by examining the change in tissue ^{3}H levels over extended periods following administration and plasma disappearance of ^{3}H-R-BrP.[30] In fact, all tissues examined, with the exception of brain, showed a reduction of ^{3}H activity between 16 and 60 min following ^{3}H-R-BrP administration. First-order rate constants for this process were estimated from these data and used to calculate retention of ^{3}H activity under conditions of the standard 16-min experiment (see below). TABLE 1 summarizes this analysis for a range of different tissues. Retention appears acceptable (>84%) in the majority of tissues, including brain, brown adipose tissue, skeletal muscles, kidney, and white adipose tissue. It is also very reassuring that, in skeletal muscle, retention is apparently not influenced by the contraction status and, therefore, presumably not influenced by the metabolic (oxidative vs. nonoxidative) status. In contrast to the majority of tissues, higher rates of loss are observed in the liver and heart. Some caution may be required in the interpretation of results for these particular tissues.

Further Metabolism

The CoA derivative of 2-bromopalmitate is not a substrate for β-oxidation, but alternative oxidative pathways, particularly ω-oxidation, would theoretically not be prevented by the analogue structure. Indeed, ^{3}H$_2$O is produced *in vivo* from ^{3}H-R-BrP. However, the small fractional conversion (~5% of ^{3}H dose, 16 min following a systemic ^{3}H-R-BrP bolus in normal rats[30]) demonstrates that these are relatively

TABLE 1. Tissue retention of ^{3}H products of ^{3}H-R-BrP

	λ (min^{-1})	16-min retention (%)
Leg muscles		
Control	0.0087	90
Contracting	0.0072	92
Heart	0.0267	73
WAT	0.0143	84
BAT	0.0057	93
Kidney	0.0126	86
Liver	0.0224	77
Brain	−0.0040	105

NOTE: Tissue ^{3}H concentrations were determined in two separate groups of Wistar rats ($n = 6$ per group), with tissues collected either 16 or 60 min following intravenous administration of ^{3}H-R-BrP. First-order rate constants for loss of activity (λ) were calculated from the ratio of activities for the two time points. Retention of ^{3}H products under conditions of the standard 16-min experiment was calculated from the λ value and assuming direct and instantaneous delivery of the plasma ^{3}H-R-BrP profile for metabolism. Unilateral sciatic nerve stimulation was used to elicit sustainable twitch contractions in the lower leg muscles. *Contracting* muscles were compared with *control* muscles collected from the contralateral nonstimulated leg. WAT, white adipose tissue; BAT, brown adipose tissue.

minor pathways. At this time, the major intracellular products of (R)-2-bromopalmitate have not been characterized. The available evidence suggests that the rate of incorporation into TG is small. Virtually no neutral ^{3}H-labeled lipid appears in plasma following a systemic ^{3}H-R-BrP bolus. This contrasts with the response to administration of labeled palmitate tracer, which results in substantial plasma appearance of labeled neutral lipid: presumably, palmitate tracer esterified in the liver and exported in VLDL. One potential metabolic trap is the formation of an effectively irreversible ternary complex with carnitine and CPT I.[35,36] An important ongoing task is the identification of the major metabolites of ^{3}H-R-BrP. This could enable both refinements in methodology as well as a general insight into the robustness of the metabolic trapping assumption.

Tracer Assumption

The specific activity of [9,10-^{3}H]-(R)-2-bromopalmitate used in the standard ^{3}H-R-BrP experiment described below is ~2×10^{12} Bq/mmol. In practice, this means that only ~250 pmol of the analogue is administered to a rat in connection with the standard experiment. These trace amounts of analogue are not expected to have any significant impact on native FFA metabolism. In contrast, however, pharmacological quantities of 2-bromopalmitate (µmol–mmol/rat) inhibit fatty acid metabolism (e.g., see reference 37). Fatty acid oxidation is inhibited, at least partly, by an effectively irreversible interaction between 2-bromopalmitoyl-CoA, carnitine, and carnitine palmitoyl transferase I, which prevents entry of native long-chain FFAs into the mitochondria.[35,36] 2-Bromopalmitate also specifically inhibits ACS.[38,39] This latter effect is illustrated in previously unpublished data given in FIGURE 2, showing the activation of ^{3}H-R-BrP and ^{14}C-P by skeletal muscle microsomal preparations.

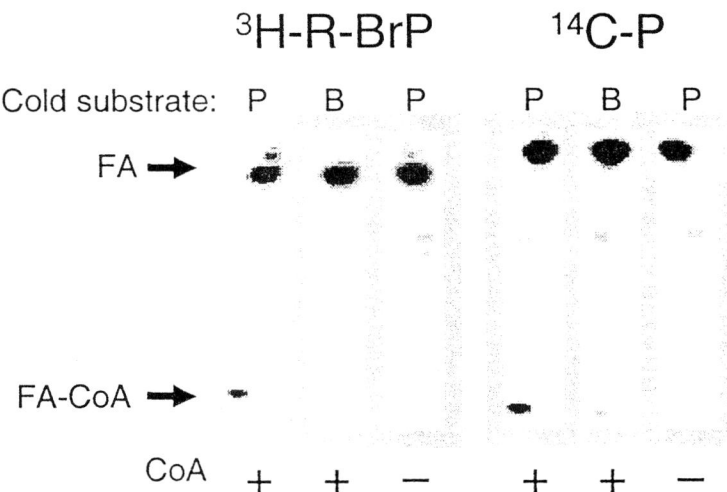

FIGURE 2. Conversion of ^{3}H-R-BrP and ^{14}C-P to their respective CoA derivatives by a skeletal muscle microsomal preparation of acyl-CoA synthetase. *Arrows* to the left indicate the approximate positions of the spots containing unreacted fatty acid (FA) and CoA derivatives on the thin-layer chromatography plates. Incubations performed in the absence of added CoA (−) confirm the identity of the CoA derivatives. In addition to the tracers, incubations were performed in the presence of cold substrate: 750 μM unlabeled palmitate (P) or (R,S)-2-bromopalmitate (B). Microsomes were obtained from red gastrocnemius, red quadriceps, and soleus muscles from Wistar rats. The methods used were taken from reference 50.

While both tracers are readily convertible to their respective CoA derivatives in the presence of a physiological concentration of palmitate (750 μM), their conversion is virtually arrested in the presence of an equal concentration of unlabeled 2-bromopalmitate. This example illustrates the fundamentally divergent situations of pharmacologic (~mM) and tracer concentrations (~nM) of the analogue and, related to this, the importance of using R-2-BrP tracer of sufficiently high specific activity.

THE ^{3}H-R-BrP METHOD FOR ASSESSING FFA METABOLISM IN THE RAT

Method Description

A detailed description of the basic methodology for the rat is given in reference 30 and therefore only a brief summary is given here.

Preparation of Tracers for In Vivo *Use*

The tracers ^{3}H-R-BrP and ^{14}C-P are administered simultaneously as an albumin-palmitate-tracer complex.

Experimental Protocol

It is assumed that the experiment is conducted under metabolically stable (steady state) conditions. Immediately before commencing tracer administration, an arterial plasma sample is collected for determination of FFA concentration (C_P). The tracer solution is then administered intravenously as a constant infusion for 4 min. To obtain plasma time courses of ^{3}H-R-BrP [$c_B^*(t)$] and ^{14}C-P [$c_P^*(t)$], frequent arterial blood samples are collected until 16 min after the start of tracer infusion. Immediately following collection of the 16-min sample, at $t = T$, the rat is sacrificed and tissues collected.

Plasma Processing

In order to discriminate ^{3}H-R-BrP and ^{14}C-P from labeled metabolites that appear in the plasma, lipid extractions followed by a polarity separation step are performed.

Tissue Processing

Total tissue concentrations of ^{3}H [$m_B^*(T)$] and ^{14}C [$m_P^*(T)$] are determined. In studies published to date, this has been achieved by complete oxidation of tissue samples, but extraction in organic solvents is also possible.

Calculations

Using the terms defined above, clearance rates of ^{3}H-R-BrP by individual tissues (K_f^*) are calculated as

$$K_f^* = m_B^*(T) / \int_0^T c_B^*(t)\, dt. \tag{1}$$

A tissue-specific index of plasma FFA utilization (R_f^*) is calculated as

$$R_f^* = C_P \cdot K_f^*. \tag{2}$$

The relationship of this index to genuine FFA utilization (R_f) is theoretically

$$R_f^* \approx LC^* \cdot R_f \tag{3}$$

(see reference 40 for a general theoretical treatment), where, in terminology borrowed from 2-deoxyglucose methodology,[24] the "lumped constant" for ^{3}H-R-Br, LC^*, represents the ratio of the probability that a molecule of arterial ^{3}H-R-BrP will undergo activation to ^{3}H-R-BrP-CoA to the probability that a molecule of palmitate will undergo activation to palmitoyl-CoA in the tissue of interest. This parameter is a function of several individual mass-transfer processes.

Tissue-specific clearance (K_{fs}) and flux (R_{fs}) parameters are also calculated from the ^{14}C data using analogous expressions to the above equations. Here, it is assumed that all ^{14}C label derived from locally metabolized ^{14}C-P that is directed into oxidative metabolism would be lost from the tissue as $^{14}CO_2$ by the time of tissue sampling. The tissue ^{14}C content would therefore represent storage products (predominantly acylglyceride and phospholipids). Consequently, R_{fs} is an index of the rate of nonoxidative metabolism.

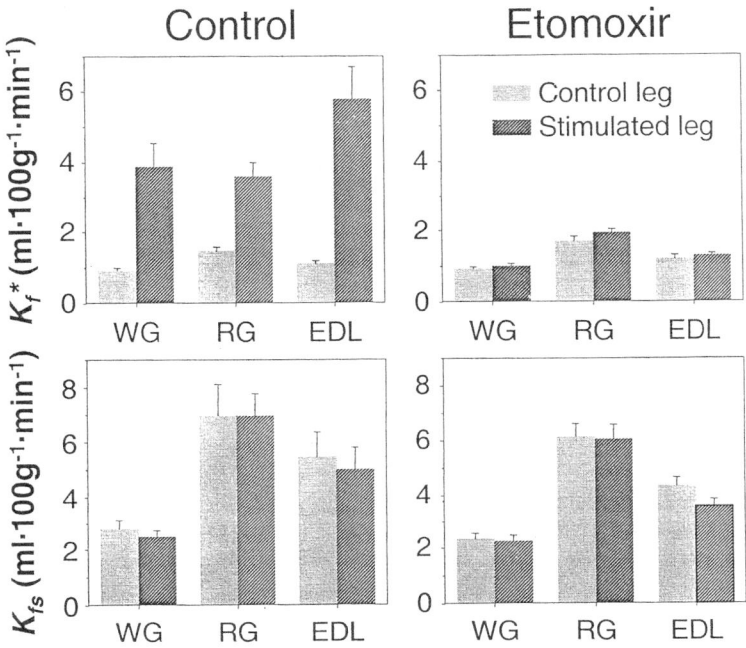

FIGURE 3. Effect of contraction and fatty acid oxidation blockade on ^{3}H-R-BrP and ^{14}C-P clearance into storage in hindlimb muscles of anesthetized Wistar rats. Untreated control group data are shown in the *left panels*. Data from animals treated with the β-oxidation inhibitor, etomoxir (20 μmol/kg), are shown in the *right panels*. Unilateral sciatic nerve stimulation elicited muscle contractions in the lower leg muscles of the stimulated side: WG, white gastrocnemius; RG, red gastrocnemius; EDL, extensor digitorum longus. *Uniformly filled bars* represent the control leg; *hatched bars* represent the stimulated leg.

Interpretation of Key Clearance Parameters

Interpretation of the tissue-specific clearance parameters (K_f^* and K_{fs}) can be illustrated using the results shown in FIGURE 3. These studies were performed in anesthetized Wistar rats in which sustained twitch contractions were induced in several lower leg muscles by efferent electrical stimulation of the transected sciatic nerve of one leg. Comparison with the contralateral, nonstimulated leg revealed the contraction effect. Two groups were studied: one control group and a second group in which β-oxidation was blocked by acute administration of etomoxir. In the control group (*left panels*, FIG. 3), contraction induced a large increase in K_f^*—our index of the ability to use FFAs for both oxidative and nonoxidative metabolism. This occurred without any alteration in K_{fs}, which describes the ability to use FFA for nonoxidative metabolism. It thus can be concluded that contraction induced a substantial increase in muscle FFA utilization that was directed selectively into fatty acid oxidation without perturbing nonoxidative FFA metabolism. This singular observation is consistent with a recently emerging view that differing pathways of fatty

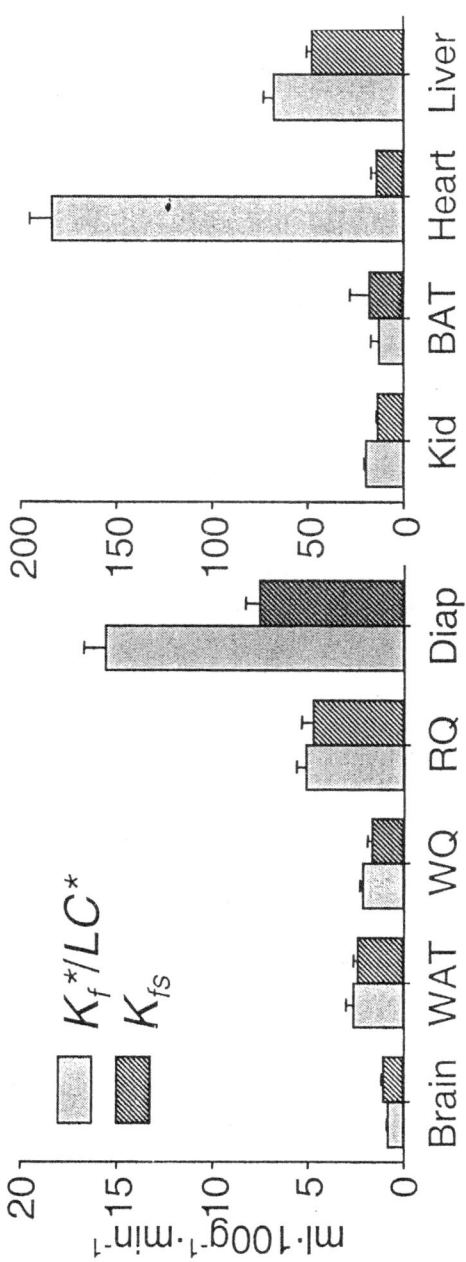

FIGURE 4. Absolute and storage-directed ^{14}C-P clearance rates in individual tissues of anesthetized rats. K_f^*/LC^* values represent estimates of absolute rates of ^{14}C-P clearance. The difference between the K_f^*/LC^* and K_{fs} provides an estimate of clearance into oxidative metabolism.

acid metabolism may be independently controlled.[41,42] Results obtained during β-oxidation blockade with etomoxir (*right panels*, FIG. 3) corroborate this interpretation. Etomoxir had little impact on K_f^* values in noncontracting muscles, consistent with the quiescent status of the muscles in anesthetized rats. In contracting muscles, though, etomoxir completely abolished the contraction-induced increase in K_f^*.

Lumped Constant

In the preceding analysis, conclusions were drawn concerning the roles of nonoxidative and oxidative FFA metabolism by comparing the response patterns of the K_f^* and K_{fs} parameters. This was done without referring to the absolute magnitudes of the different parameters (i.e., K_f^* vs. K_{fs}). This exemplifies the most conservative use of the data. However, it is possible (with additional assumptions) to extend the analysis to estimate rates of authentic FFA clearance and utilization. This is done by applying lumped constant corrections according to equation 3 above. Fatty acid oxidation blockade with etomoxir provides one situation where estimates of individual tissue lumped constants can be made. In this circumstance, a direct comparison of the behavior of ^{14}C-P and ^{3}H-R-BrP is valid since there is minimal loss of ^{14}C label as $^{14}CO_2$. Therefore, under these conditions, the ratio of K_f^* to K_{fs} provides an estimate of the lumped constant (LC^*). LC^* estimates obtained in this way are, for most tissues, significantly less than unity,[30,43] reflecting slower uptake/metabolism of ^{3}H-R-BrP than palmitate tracer. While the LC^* values are relatively constant for given tissue types (e.g., skeletal muscles), they vary from one tissue class to another. In order to compare across tissue classes, LC^* corrections need to be made to the data. FIGURE 4 shows the result of such calculations for a broad range of tissues of anesthetized Wistar rats. Tissue K_f^*/LC^* values, which are indices of genuine FFA clearance rates, are plotted alongside K_{fs} values. This analysis demonstrates the very large range, of at least two orders of magnitude, in the ability of different tissues of the body to use FFA. In addition, the level of FFA oxidation occurring in each tissue can be seen by comparing the magnitudes of K_f^* and LC^* for individual tissues. The level of FFA oxidation is low in the majority of tissues, with the exception of the working muscles, diaphragm, and heart.

AN APPLICATION STUDY

The ^{3}H-R-BrP technique has also been applied in conscious Wistar rats.[43] In these studies, nonobese animals were fitted with chronically indwelling catheters 7–10 days before study with ^{3}H-R-BrP. Apart from the fact that rats were awake, the experimental protocol was identical to that previously described.

In this study also, etomoxir was used to study the effects of acute β-oxidation blockade. In addition, in some rats, a euglycemic-hyperinsulinemic clamp[44] was used to elevate circulating insulin levels to high physiological values (~100 mU/L).

The effect of etomoxir administration was as expected (FIG. 5), with decreased ^{3}H-R-BrP clearance in oxidative muscles (heart and diaphragm), but no change in nonoxidative white gastrocnemius muscle. These results indicate a close association between FFA uptake and oxidation and suggest that the early steps of fatty acid uptake are closely coupled to the transfer of fatty acids to the mitochondria. This

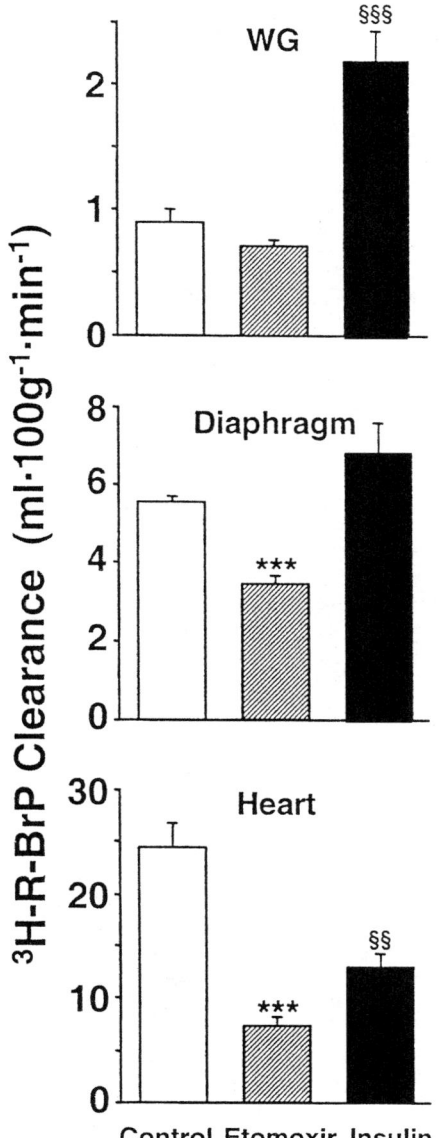

FIGURE 5. Tissue clearance (K_f^*) of ^{3}H-R-BrP tracer. *Unfilled columns* represent data from untreated rats. *Hatching* indicates etomoxir treatment. *Solid shading* indicates data from hyperinsulinemic animals. * and § indicate a significant effect of etomoxir and insulin administration, respectively: §§$p < 0.01$; ***,§§§$p < 0.001$.

contrasts with glucose metabolism where inhibition of glucose oxidation by infusion of fatty acids[45] does not immediately decrease glucose uptake, but diverts glucose metabolism to lactate or glycogen formation.

It was expected that the two treatment regimens would have similar effects since increased glucose oxidation induced by hyperinsulinemia is thought to indirectly inhibit lipid oxidation by elevation of cytosolic malonyl-CoA.[46] (Malonyl-CoA, like

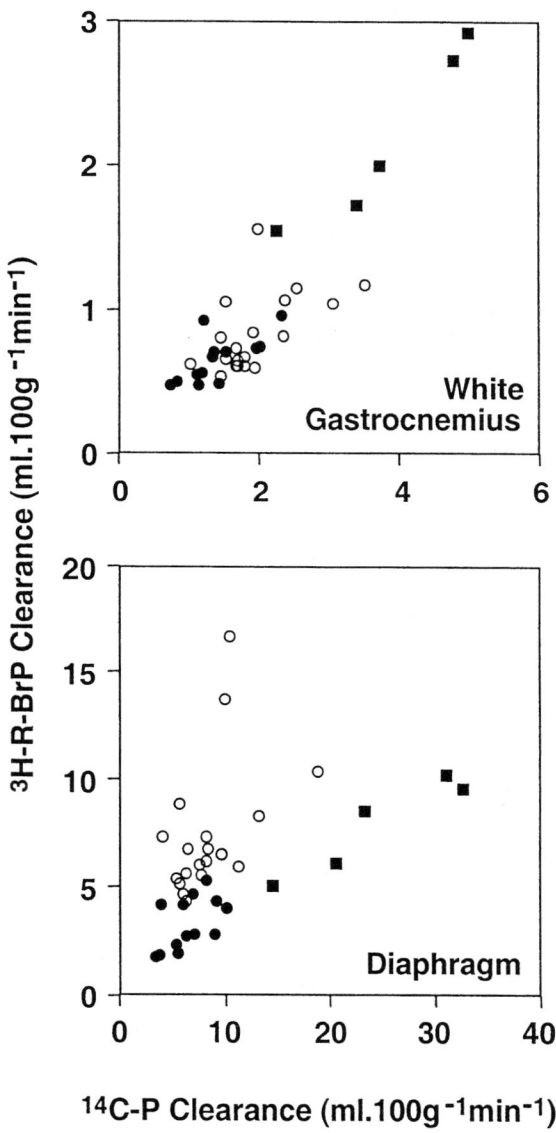

FIGURE 6. Comparison of clearance estimates derived from ^{3}H-R-BrP and ^{14}C-P accumulation. *Filled circles* indicate etomoxir administration. *Solid squares* indicate data from hyperinsulinemic animals. *Open circles* represent no treatment other than fasting.

etomoxir, inhibits the action of CPT I to transport acyl-CoA into the mitochondria.) However, this expected effect was only seen in highly oxidative heart muscle, where insulin administration reduced ^{3}H-R-BrP clearance by this tissue. In diaphragm, little effect of insulin was observed, while insulin increased ^{3}H-R-BrP clearance into white gastrocnemius muscle. These results suggest direct actions of insulin, which are only apparent in nonoxidative tissue (white gastrocnemius), to promote fatty acid uptake. Indeed, there is evidence that insulin directly promotes lipid storage (tending to increase FFA clearance) by stimulating acylglyceride formation.[47]

These results illustrate the full advantage of the tissue-specific ^{3}H-R-BrP technique. It is likely that there are two opposing actions of insulin on FFA clearance, which only become apparent because multiple tissues with different oxidative capacities could be studied simultaneously. The action of insulin to block FFA oxidation predominated in highly oxidative heart muscle; the direct action of insulin to promote FFA uptake was only evident in nonoxidative muscle where oxidation blockade effects were minimal; in the diaphragm, the effects canceled each other out and no net insulin effect was observed. Although individual treatment effects can be studied in other systems (e.g., *in vitro* preparations), the exact balance between two or more effects, attained in a specific tissue under specific physiological conditions, can only be determined by direct measurement *in vivo*.

More insight into the effect of oxidation blockade can be obtained by considering the relative behavior of ^{3}H-R-BrP and the authentic ^{14}C-P. Radiolabeled palmitate, unlike ^{3}H-R-BrP, is transported into the mitochondria where it is oxidized. The labeled product, ^{14}CO$_2$, is rapidly exported from the tissues, while the ^{14}C label that is trapped can be used to quantify storage processes. These processes are illustrated in FIGURE 6, which includes additional data from fasted rats (with and without acute etomoxir treatment).

In nonoxidative white muscle (*upper panel*), there is a good correlation ($R^2 = 0.80$) between clearance of ^{3}H-R-BrP (K_f^*) and ^{14}C-P (K_{fs}) across all treatment groups. The relationship was not substantially altered by oxidation blockade by either etomoxir or insulin administration, consistent with only a small degree of FFA oxidation in this tissue under basal conditions. However, this was not the case for the oxidative diaphragm (*lower panel*). Here, there was poor correlation ($R^2 = 0.23$) between clearance of ^{3}H-R-BrP and ^{14}C-P when all data were included, but a good association ($R^2 = 0.88$) was apparent when including only groups where β-oxidation was impeded (*solid symbols*). Under other conditions (*open symbols*), a substantial fraction of ^{14}C label taken up by the tissue was exported as ^{14}CO$_2$. In quantitative terms, the ratio ($LC^* \cdot K_{fs})/K_f^*$ describes the fraction of total FFA uptake that is stored rather than oxidized. In the postabsorptive state, this fraction was 95% in the white gastrocnemius, but was reduced to 58% in the diaphragm.

METHOD DEVELOPMENTS

Conditions of insulin resistance clearly involve dysregulation of both glucose and fatty acid metabolism. To optimally document the metabolism of the individual substrates, as well as explore the metabolic relationships between them, a natural extension to the basic ^{3}H-R-BrP methodology was to develop methods that enabled additional, simultaneous assessment of glucose metabolism. Techniques for studying *in vivo* tissue glucose metabolism are well established. Following the introduction of the 2-deoxyglucose tracer approach for assessing regional glucose utilization in the brain,[24] methodological developments were made enabling the assessment of glucose utilization and metabolic fate in a range of peripheral tissues, including skeletal muscles and adipose tissue.[26,48] Simultaneous estimates of tissue-specific FFA and glucose utilization can be made in experiments combining the use of ^{3}H-R-BrP and ^{14}C-2-deoxyglucose (^{14}C-2DG). Implementation of this design is straightforward, achieved by simply merging the 45-min 2DG protocol (described in reference 26) and the 16-min ^{3}H-R-BrP protocol (above). To do this, the individual tracers must be administered separately—a ^{14}C-2DG bolus given 45 min prior to tissue collection and a ^{3}H-R-BrP infusion commenced 16 min before tissue collection. A recently developed method, using tracers labeled with stable isotopes, for simultaneous determination of glucose and FFA utilization, as well as their oxidative versus nonoxidative fates, affords an unprecedented level of *in vivo* metabolic information at the cost of a substantially increased analytic burden.[49]

CONCLUSIONS

Methodology based on the simultaneous use of [9,10-^{3}H]-(*R*)-2-bromopalmitate and ^{14}C-palmitate tracers provides a unique opportunity for the *in vivo* study of tissue-specific plasma FFA utilization and metabolic fate, especially in skeletal muscle and adipose tissue. Particularly with the recent widespread application of molecular techniques (including differential display and proteomics), tissue-specific information about mRNA- and protein-level control of enzymes and transport molecules involved in FFA metabolism has been amassing. The [9,10-^{3}H]-(*R*)-2-bromopalmitate tracer method confers the capability to explore the functional consequences of such fundamental alterations.

ACKNOWLEDGMENTS

We are indebted to the following people at AstraZeneca Mölndal: Shalini Andersson and Gun-Britt Forsberg, for their crucial development of the method for chiral separation of the 2-bromopalmitate enantiomers; Lars Löfgren and Margit Wettesten, for developing the analytical procedures for the simultaneous application of 2-bromopalmitate and 2-deoxyglucose methods; Tony Clementz, for the *in vitro* ACS work; Ann Kjellstedt, for performing the *in vivo* studies; and Bengt Ljung, for his leadership and scientific input. Studies at the Garvan Institute were performed by Bronwyn Hegarty, Megan Lim-Fraser, and Gregory Cooney. We are indebted to Arthur Jenkins for his pioneering investigations into the feasibility of using 2-bromopalmitate as an FFA tracer. We also thank Edward Kraegen for his involvement in the project from its inception to the present.

REFERENCES

1. RANDLE, P.J. 1998. Regulatory interactions between lipids and carbohydrates: the glucose fatty acid cycle after 35 years. Diabetes Metab. Rev. **14:** 263–283.
2. CHEN, X., N. IQBAL & G. BODEN. 1999. The effects of free fatty acids on gluconeogenesis and glycogenolysis in normal subjects. J. Clin. Invest. **103:** 365–372.

3. SINDELAR, D.K., C.A. CHU, M. ROHLIE et al. 1997. The role of fatty acids in mediating the effects of peripheral insulin on hepatic glucose production in the conscious dog. Diabetes **46:** 187–196.
4. REBRIN, K., G.M. STEIL, L. GETTY & R.N. BERGMAN. 1995. Free fatty acid as a link in the regulation of hepatic glucose output by peripheral insulin. Diabetes **44:** 1038–1045.
5. MCGARRY, J.D. & R.L. DOBBINS. 1999. Fatty acids, lipotoxicity, and insulin secretion. Diabetologia **42:** 128–138.
6. LEWIS, G.F. 1997. Fatty acid regulation of very low density lipoprotein production. Curr. Opin. Lipidol. **8:** 146–153.
7. KELLEY, D.E. & L.J. MANDARINO. 2000. Fuel selection in human skeletal muscle in insulin resistance: a reexamination. Diabetes **49:** 677–683.
8. STORLIEN, L.H., A.B. JENKINS, D.J. CHISHOLM et al. 1991. Influence of dietary fat composition on development of insulin resistance in rats: relationship to muscle triglyceride and omega-3 fatty acids in muscle phospholipid. Diabetes **40:** 280–289.
9. PAN, D.A., S. LILLIOJA, A.D. KRIKETOS et al. 1997. Skeletal muscle triglyceride levels are inversely related to insulin action. Diabetes **46:** 983–988.
10. KRSSAK, M., K. FALK PETERSEN, A. DRESNER et al. 1999. Intramyocellular lipid concentrations are correlated with insulin sensitivity in humans: a 1H NMR spectroscopy study. Diabetologia **42:** 113–116.
11. ELLIS, B.A., A. POYNTEN, A.J. LOWY et al. 2000. Long-chain acyl-CoA esters as indicators of lipid metabolism and insulin sensitivity in rat and human muscle. Am. J. Physiol. Endocrinol. Metab. **279:** E554–E560.
12. MARCHESINI, G., M. BRIZI, A.M. MORSELLI-LABATE et al. 1999. Association of nonalcoholic fatty liver disease with insulin resistance. Am. J. Med. **107:** 450–455.
13. GIBBONS, G.F., K. ISLAM & R.J. PEASE. 2000. Mobilisation of triacylglycerol stores. Biochim. Biophys. Acta **1483:** 37–57.
14. UNGER, R.H. & Y.T. ZHOU. 2001. Lipotoxicity of beta-cells in obesity and in other causes of fatty acid spillover. Diabetes **50:** S118–S121.
15. ZHOU, Y.T., P. GRAYBURN, A. KARIM et al. 2000. Lipotoxic heart disease in obese rats: implications for human obesity. Proc. Natl. Acad. Sci. U.S.A. **97:** 1784–1789.
16. RANDLE, P.J., P.B. GARLAND, C.N. HALES & E.A. NEWSHOLME. 1963. The glucose–fatty acid cycle: its role in insulin sensitivity and metabolic disturbances of diabetes mellitus. Lancet **1:** 785–794.
17. SCHMITZ-PEIFFER, C. 2000. Signalling aspects of insulin resistance in skeletal muscle: mechanisms induced by lipid oversupply. Cell. Signalling **12:** 583–594.
18. MILES, J.M., M.G. ELLMAN, K.L. MCCLEAN & M.D. JENSEN. 1987. Validation of a new method for determination of free fatty acid turnover. Am. J. Physiol. **252:** E431–E438.
19. FERRANNINI, E. 1988. The theoretical bases of indirect calorimetry: a review. Metab. Clin. Exp. **37:** 287–301.
20. ELHAMRI, M., M. MARTIN, B. FERRIER & G. BAVEREL. 1993. Substrate uptake and utilization by the kidney of fed and starved rats *in vivo*. Renal Physiol. Biochem. **16:** 311–324.
21. BLAAK, E.E., D.P. VAN AGGEL-LEIJSSEN, A.J. WAGENMAKERS et al. 2000. Impaired oxidation of plasma-derived fatty acids in type 2 diabetic subjects during moderate-intensity exercise. Diabetes **49:** 2102–2107.
22. KELLEY, D.E., B. GOODPASTER, R.R. WING & J.A. SIMONEAU. 1999. Skeletal muscle fatty acid metabolism in association with insulin resistance, obesity, and weight loss. Am. J. Physiol. **277:** E1130–E1141.
23. BONEN, A., J.J. LUIKEN, S. LIU et al. 1998. Palmitate transport and fatty acid transporters in red and white muscles. Am. J. Physiol. **275:** E471–E478.
24. SOKOLOFF, L., M. REIVICH, C. KENNEDY et al. 1977. The [^{14}C]deoxyglucose method for the measurement of local cerebral glucose utilization: theory, procedure, and normal values in the conscious and anesthetized albino rat. J. Neurochem. **28:** 897–916.
25. GRAFTON, S.T. 2000. PET: activation of cerebral blood flow and glucose metabolism. Adv. Neurol. **83:** 87–103.
26. KRAEGEN, E.W., D.E. JAMES, A.B. JENKINS & D.J. CHISHOLM. 1985. Dose-response curves for *in vivo* insulin sensitivity in individual tissues in rats. Am. J. Physiol. **248:** E353–E362.

27. KNAPP, F.F., JR. & J. KROPP. 1999. BMIPP-design and development. Int. J. Cardiol. Imaging **15:** 1–9.
28. TAYLOR, M., T.R. WALLHAUS, T.R. DEGRADO *et al.* 2001. An evaluation of myocardial fatty acid and glucose uptake using PET with [^{18}F]fluoro-6-thia-heptadecanoic acid and [^{18}F]FDG in patients with congestive heart failure. J. Nucl. Med. **42:** 55–62.
29. DEGRADO, T.R., S. WANG, J.E. HOLDEN *et al.* 2000. Synthesis and preliminary evaluation of (18)F-labeled 4-thia-palmitate as a PET tracer of myocardial fatty acid oxidation. Nucl. Med. Biol. **27:** 221–231.
30. OAKES, N.D., A. KJELLSTEDT, G.B. FORSBERG *et al.* 1999. Development and initial evaluation of a novel method for assessing tissue-specific plasma free fatty acid utilization *in vivo* using (*R*)-2-bromopalmitate tracer. J. Lipid Res. **40:** 1155–1169.
31. GLATZ, J.F. & J. STORCH. 2001. Unravelling the significance of cellular fatty acid–binding proteins. Curr. Opin. Lipidol. **12:** 267–274.
32. WILTON, D.C. 1990. The fatty acid analogue 11-(dansylamino)undecanoic acid is a fluorescent probe for the bilirubin-binding sites of albumin and not for the high-affinity fatty acid–binding sites. Biochem. J. **270:** 163–166.
33. WEANER, L.E. & D.C. HOERR. 1987. A high-performance liquid chromatographic method for the enantiomeric analysis of the enzymatically synthesized coenzyme A ester of 2-tetradecylglycidic acid. Anal. Biochem. **160:** 316–322.
34. CHEN, C.S., W.R. SHIEH, P.H. LU *et al.* 1991. Metabolic stereoisomeric inversion of ibuprofen in mammals. Biochim. Biophys. Acta **1078:** 411–417.
35. CHASE, J.F. & P.K. TUBBS. 1972. Specific inhibition of mitochondrial fatty acid oxidation by 2-bromopalmitate and its coenzyme A and carnitine esters. Biochem. J. **129:** 55–65.
36. MCGARRY, J.D., A. SEN, V. ESSER *et al.* 1991. New insights into the mitochondrial carnitine palmitoyltransferase enzyme system. Biochimie **73:** 77–84.
37. HAGVE, T.A., M. NARCE, M. GRONN *et al.* 1989. The effect of dietary alpha-bromopalmitate on blood lipids in the rat. Biochim. Biophys. Acta **1004:** 143–146.
38. MAYOREK, N. & J. BAR-TANA. 1985. Inhibition of diacylglycerol acyltransferase by 2-bromooctanoate in cultured rat hepatocytes. J. Biol. Chem. **260:** 6528–6532.
39. VANDEN HEUVEL, J.P., B.I. KUSLIKIS, E. SHRAGO & R.E. PETERSON. 1991. Inhibition of long-chain acyl-CoA synthetase by the peroxisome proliferator perfluorodecanoic acid in rat hepatocytes. Biochem. Pharmacol. **42:** 295–302.
40. PATLAK, C.S. 1981. Derivation of equations for the steady-state reaction velocity of a substance based on the use of a second substance. J. Cereb. Blood Flow Metab. **1:** 129–131.
41. MUOIO, D.M., T.M. LEWIN, P. WIEDMER & R.A. COLEMAN. 2000. Acyl-CoAs are functionally channeled in liver: potential role of acyl-CoA synthetase. Am. J. Physiol. Endocrinol. Metab. **279:** E1366–E1373.
42. LEWIN, T.M., J.H. KIM, D.A. GRANGER *et al.* 2001. Acyl-CoA synthetase isoforms 1, 4, and 5 are present in different subcellular membranes in rat liver and can be inhibited independently. J. Biol. Chem. **276:** 24674–24679.
43. FURLER, S.M., G.J. COONEY, B.D. HEGARTY *et al.* 2000. Local factors modulate tissue-specific NEFA utilization: assessment in rats using 3H-(*R*)-2-bromopalmitate. Diabetes **49:** 1427–1433.
44. KRAEGEN, E.W., D.E. JAMES, S.P. BENNETT & D.J. CHISHOLM. 1983. *In vivo* insulin sensitivity in the rat determined by euglycemic clamp. Am. J. Physiol. **245:** E1–E7.
45. JENKINS, A.B., L.H. STORLIEN, D.J. CHISHOLM & E.W. KRAEGEN. 1988. Effects of nonesterified fatty acid availability on tissue-specific glucose utilization in rats *in vivo*. J. Clin. Invest. **82:** 293–299.
46. SIDOSSIS, L.S., C.A. STUART, G.I. SHULMAN *et al.* 1996. Glucose plus insulin regulate fat oxidation by controlling the rate of fatty acid entry into the mitochondria. J. Clin. Invest. **98:** 2244–2250.
47. VILA, M.C., G. MILLIGAN, M.L. STANDAERT & R.V. FARESE. 1990. Insulin activates glycerol-3-phosphate acyltransferase (*de novo* phosphatidic acid synthesis) through a phospholipid-derived mediator: apparent involvement of Gi alpha and activation of a phospholipase C. Biochemistry **29:** 8735–8740.

48. JAMES, D.E., A.B. JENKINS & E.W. KRAEGEN. 1985. Heterogeneity of insulin action in individual muscles *in vivo*: euglycemic clamp studies in rats. Am. J. Physiol. **248:** E567–E574.
49. OAKES, N.D., A. KJELLSTEDT, P.G. THALEN *et al.* 2001. Tissue-specific effects of AZ 242, a novel PPARa/g agonist, on glucose and fatty acid metabolism in obese Zucker rats: an *in vivo* multi-tracer assessment. Diabetes **50:** A121.
50. TANAKA, T., K. HOSAKA & S. NUMA. 1981. Long-chain acyl-CoA synthetase from rat liver. Methods Enzymol. **71:** 334–341.

Insulin Action in Skeletal Muscle
Isozyme-Specific Effects of Protein Kinase C

ISKANDAR IDRIS, SAMUEL GRAY, AND RICHARD DONNELLY

Division of Vascular Medicine, School of Medical and Surgical Sciences, University of Nottingham, Derbyshire Royal Infirmary, Derby DE1 2QY, United Kingdom

ABSTRACT: Protein kinase C (PKC) is a family of multifunctional isozymes that plays an important role in the regulation of intracellular insulin signal transduction in various insulin-sensitive tissues. This article highlights current understanding on the mechanism of PKC-induced insulin resistance in skeletal muscle, a major target site for insulin-mediated glucose disposal. Initial, apparently contradictory findings on the role of PKC on insulin action can be explained on the basis that certain PKC isoforms (e.g., -ζ and -λ) have been identified as downstream targets of PI3-kinase activation, while DAG-sensitive PKCs (e.g., -θ and -ϵ) have negative regulatory effects on insulin signaling. Hence, pharmacological therapies targeting specific PKC isoforms could enhance insulin action and improve glycemic control in patients with impaired glucose tolerance and overt diabetes.

KEYWORDS: insulin; protein kinase C (PKC); isozyme; skeletal muscle

PROTEIN KINASE C: A FAMILY OF MULTIFUNCTIONAL ISOZYMES

Protein kinase C (PKC), first identified over 20 years ago as a single proteolytically activated kinase in rat brain,[1] is a family of serine/threonine kinases that plays an important role in regulating various biochemical processes. Twelve isozymes of PKC have so far been cloned and characterized[2] with each isoform showing a different pattern of tissue distribution, substrate specificity, and cofactor requirements. These isoforms are classified into three groups: (1) classical isoforms (cPKC-α, -β_I, -β_{II}, and -γ) activated by phosphatidylserine (PS), Ca^{2+}, and DAG (or phorbol ester); (2) novel isoforms (nPKC-δ, -ϵ, -η, and -θ), which are calcium-independent, but require PS and DAG for activation; and (3) atypical isoforms (aPKC-ι, -λ, and -ξ), which are both calcium- and phospholipid-independent.

The signaling cascade leading to the metabolic effects of insulin in stimulating glucose uptake involves insulin stimulation of the insulin receptor tyrosine kinase, which in turn stimulates a number of downstream signaling factors, among which are the insulin receptor substrates (IRS-1, -2, -3, and -4), phosphatidylinositol 3-kinase (PI3-kinase), mitogen-activated protein kinase, and other protein kinases.[3] This sub-

---

Address for correspondence: Dr. I. Idris, Division of Vascular Medicine, University of Nottingham, Derbyshire Royal Infirmary, Derby DE1 2QY, United Kingdom. Voice: [+44] (0) 1332-254966; fax: [+44] (0) 1332-254968.
is.idris@nottingham.ac.uk

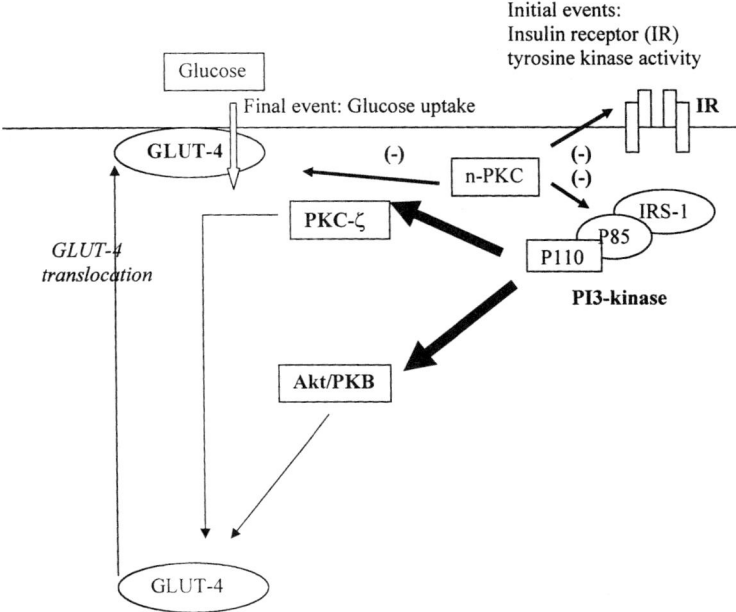

FIGURE 1. Simplified outline of insulin signaling in skeletal muscle: n-PKC, novel protein kinase C; PKB, protein kinase B; (–), inhibition.

sequently induces glucose uptake into several tissues via translocation of GLUT proteins from the cytoplasm to the plasma membrane[4] (FIG. 1).

PKC-MEDIATED PHOSPHORYLATION AND INSULIN RECEPTOR FUNCTION

The notion that PKC activation might, via phosphorylation, downregulate or transduce one or more steps involved in glucose transport and metabolism emerged in the mid-1990s,[5] supported by preliminary evidence that PKC phosphorylates the insulin receptor[6] and glycogen synthase.[7] In subsequent studies, changes in DAG-PKC signaling in liver, muscle, and adipose tissues have been associated with insulin resistance due to aging,[8] obesity,[9] high-fructose and high-fat feeding,[10,11] and muscle denervation.[12] More recent work has shown that PKC-β_I and -β_{II} exert strong inhibitory effects on the IR tyrosine kinase–autophosphorylation step[13] and that serine residues 994 and 1023/25 on the IR are particularly important for PKC-θ- and PKC-β_{II}-mediated inhibition of IR autophosphorylation.[14] In the immediate downstream signaling events following IR stimulation, PKC phosphorylates and modulates the IRS proteins[15,16] and inhibits insulin-stimulated Akt1 and Akt3 activity.[17]

Although some PKC isoforms seem to exert a negative regulatory effect on insulin signaling, others, including PKC-ζ and PKC-λ, are among several serine/threonine kinases (including Akt/PKB) that have been identified as downstream targets of PI3-kinase and intermediate steps in insulin signaling.[18,19] There is also evidence that PKC-related phosphorylation events, perhaps involving intermediate substrates, indirectly influence glucose uptake via modulation of GLUT4 protein activity. For example, pharmacological studies have shown increased glucose transport (secondary to increased GLUT1 and GLUT4 expression) in parallel with decreased PKC activity,[20] and GLUT4 trafficking pathways may be regulated by PKC.[21]

SKELETAL MUSCLE INSULIN RESISTANCE: ROLE OF PKC?

Skeletal muscle is the principal site of insulin-mediated glucose uptake, accounting for about 75% of glucose disposal[22] after glucose infusion *in vivo*. Hence, dysregulation of skeletal muscle insulin signaling is essential in the development of whole body insulin resistance. The observation that muscle triglyceride (TG) content is inversely related to *in vivo* insulin sensitivity[23] implicates local lipid availability as a putative mediator of muscle insulin sensitivity. DAG and long-chain acyl-CoAs, both of which are increased in the high-fat-fed rat, activate classical and novel PKCs[24] and have therefore been linked to reduced insulin action in skeletal muscle. Thus, accumulation of lipids and/or their metabolites could inadvertently modulate PKC activity.

Two novel PKCs, PKC-θ and PKC-ε, have been particularly implicated in this hypothesis. Both high-fructose and high-fat feeding are associated with increases in circulating and intramuscular TG content in rodents, and recent studies have shown isoform-selective increases in muscle PKC signaling in these dietary-induced models of insulin resistance.[10,11] Moreover, treatment of high-fat-fed rats with an antilipolytic drug, BRL49653, which reduced muscle DAG content and nonesterified fatty acid mass, improved insulin sensitivity and reversed the diet-induced changes in nPKC expression.[25] Similarly, in *ex vivo* experiments using rectus-abdominis muscle strips, Cortright *et al.* showed that PKC inhibition enhanced, whereas activation of PKC reduced, insulin-mediated glucose uptake.[26]

NOVEL PKC ISOFORMS AND SKELETAL MUSCLE INSULIN RESISTANCE

PKC-θ is one of the more recently identified PKC isoforms[27] with a unique pattern of tissue distribution.[28] PKC-θ is expressed predominantly (and almost exclusively) in skeletal muscle, hematopoietic tissues, testis, and platelets, and it is the most abundantly expressed isoform in skeletal muscle.[27] Donnelly and Reaven showed that expression of PKC-θ is 3-fold higher in tensor fascia latae (TFL, an insulin-resistant white muscle) compared with soleus muscle (an insulin-sensitive red muscle) in the hindlimb of normal rats[10] (FIG. 2). In addition, selective activation of muscle PKC-θ and PKC-ε has been associated with dietary-induced models of insulin resistance in normal rats[10,11] and DAG-PKC activation after food intake in the muscles of obese Zucker rats.[9] Thus, specific changes in PKC-θ and PKC-ε seem to be associated with feeding and diet-related changes in muscle lipid content and

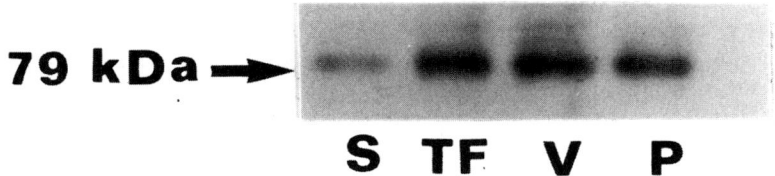

FIGURE 2. Western blots comparing protein expression of PKC-θ in four hindlimb muscles of fructose-fed rats. There was 2- to 3-fold more PKC-θ (per unit of total protein) in white tensor fascia latae (TF) muscle as compared with red soleus (S) muscle, whereas the two mixed muscles, vastus intermedius (V) and plantaris (P), showed intermediate values.[10]

insulin sensitivity. PKC-θ has also been particularly implicated in the inhibitory effects of fatty acids on insulin-induced PI3-kinase activation.[29]

Reports on the inhibitory effects of insulin action due to increased PKC-θ expression and activation have previously been observed in various rodent models of insulin resistance. Recently, we have shown that skeletal muscle expression of PKC-θ mRNA was significantly higher in patients with diabetes compared with nondiabetic controls after a glucose load.[30] Furthermore, skeletal muscle PKC-θ mRNA expression showed a significant inverse correlation with total body insulin sensitivity as measured by the euglycemic hyperinsulinemic clamp technique.[30] Similarly, Itani et al.[31] showed increased PKC-θ protein activation in skeletal muscle of patients with diabetes compared to controls, confirming previous animal data implicating the PKC-θ isoform and the development of skeletal muscle insulin resistance.

EVIDENCE FROM PHARMACOLOGICAL STUDIES

Pharmacological studies of phorbol ester–induced PKC activation and the effects of nonspecific PKC inhibitors have also provided important information. For example, several studies have shown that phorbol esters antagonize the effects of insulin[32] and inactivate the insulin receptor tyrosine kinase,[33] whereas PKC inhibitors seem to enhance glucose-induced activation of hepatic glycogen synthase[34] and promote glucose uptake in isolated adipocytes.[35] This stimulatory effect of PKC inhibition (via GF109203X) on skeletal muscle insulin action seemed to be associated with increased insulin receptor tyrosine kinase activity and PI3-kinase.[26] Similarly, okadaic acid (a protein phosphatase inhibitor) promotes an insulin-resistant state in adipocytes via sustained serine/threonine phosphorylation of IRS-1.[15]

PKC: MEDIATOR OR INHIBITOR OF INSULIN ACTION?

Previous review articles found some difficulty reconciling apparently contradictory research showing roles for PKC as both a mediator and inhibitor of insulin action,[36] but it now seems clear that distinct PKC isoforms serve widely different functions and in a tissue-dependent manner. Thus, whereas the Ca^{2+}-independent isozymes, PKC-θ and PKC-ε, exert negative regulatory effects on insulin action, especially in skeletal muscle, the atypical PKCs such as PKC-ζ, which are not activated by DAG, appear to be stimulated by phosphoinositides derived from PI3-kinase activation by insulin, for example, in adipose tissue.[18,19,37,38] Insulin may also affect mRNA expression of individual PKC isozymes.[39] It is therefore likely that some PKC isoforms (e.g., PKC-ζ) function as mediators of insulin action downstream of PI3-kinase, while other PKCs (e.g., PKC-θ and PKC-ε) have negative regulatory effects on insulin signaling.[18,19,37,38] The role of the PKC-β isoform on skeletal muscle insulin-mediated glucose uptake, however, remains controversial, with some studies showing this isoform to be an important component of insulin signaling,[40,41] while other studies reporting this isoform to be unnecessary for insulin-stimulated glucose transport.[18,38] Further understanding on the effect of this latter PKC isoform is important as PKC-β activation has widely been implicated in mediating diabetic vascular complications.[42]

REFERENCES

1. INOUE, M., A. KISHIMOTO, Y. TAKAI & Y. NISHIZUKA. 1977. Studies on a cyclic nucleotide–independent protein kinase and its proenzyme in mammalian tissues—II. J. Biol. Chem. **252:** 7610–7616.
2. NEWTON, A.C. 1995. Protein kinase C: structure, function, and regulation—minireview. J. Biol. Chem. **270:** 28495–28498.
3. WHITE, M.F. & R.C. KAHN. 1994. The insulin signalling system. J. Biol. Chem. **269:** 1–4.
4. KLIP, A., T. RAMLAL, P.J. BILAN et al. 1993. What signals are involved in the stimulation of glucose transport by insulin in muscle cells? Cell. Signalling **5:** 519–529.
5. SHMUELI, E., K.G.M. ALBERTI & C.O. RECORD. 1993. Diacylglycerol/protein kinase C signalling: a mechanism for insulin resistance? J. Intern. Med. **234:** 397–400.
6. BOLLAG, G.B., R.A. ROTH, J. BEAUDOIN et al. 1986. PKC directly phosphorylates the insulin receptor in vitro and reduces its protein–tyrosine kinase activity. Proc. Natl. Acad. Sci. U.S.A. **83:** 5822–5824.
7. BLACKMORE, P.F., W.G. STRICKLAND & J.H. EXTON. 1986. Phosphorylation of glycogen synthase by protein kinase C. Biochem. J. **237:** 235–242.
8. QU, X., J.P. SEALE & R. DONNELLY. 1999. Tissue- and isoform-specific effects of ageing in rats on protein kinase C in insulin-sensitive tissues. Clin. Sci. **97:** 355–361.
9. QU, X., J.P. SEALE & R. DONNELLY. 1999. Tissue and isoform-selective activation of protein kinase C in insulin-resistant obese Zucker rats—effects of feeding. J. Endocrinol. **162:** 207–214.
10. DONNELLY, R., M. REED, S. AZHAR & G.M. REAVEN. 1994. Expression of the major isoenzyme of PKC in skeletal muscle: PKC-θ varies with fibre-type composition and in response to fructose-induced insulin resistance. Endocrinology **135:** 2369–2374.
11. SCHMITZ-PEIFFER, C., C.L. BROWNE, N.D. OAKES et al. 1997. Alterations in the expression and cellular localisation of protein kinase C isozymes ε and θ are associated with insulin resistance in skeletal muscle of the high-fat fed rat. Diabetes **46:** 169–178.
12. HEYDRICK, S.J., N.B. RUDERMAN, T.G. KUROWSKI et al. 1991. Enhanced stimulation of diacylglycerol and lipid synthesis by insulin in denervated muscle. Diabetes **40:** 1707–1711.
13. BASSENMAIER, B., L. MOSTHAF, H. MISCHAK et al. 1997. Protein kinase C isoforms beta-1 and beta-2 inhibit the tyrosine kinase activity of the insulin receptor. Diabetologia **40:** 863–866.
14. STRACK, V., A.M. HENNIGE, J. KRUTZFELDT et al. 2000. Serine residues 994 and 1023/25 are important for insulin receptor kinase inhibition by protein kinase C isoforms β2 and θ. Diabetologia **43:** 443–449.

15. TANTI, J-F., T. GREMEAUX, E.V. OBBERGHEN & Y.L. MARCHAND-BRUSTEL. 1994. Serine/threonine phosphorylation of insulin receptor substrate-1 modulates insulin receptor signaling. J. Biol. Chem. **269:** 6051–6057.
16. BUSCH, A.K., I. CASTAN, E. DEGERMAN *et al.* 1998. Protein kinase C mediates hyperglycaemia-induced insulin resistance through serine phosphorylation of IRS-1. Diabetologia **41**(suppl. 1)**:** A10.
17. BARTHEL, A., K. NAKATANI, A.A. DANDEKAR & R.A. ROTH. 1998. Protein kinase C modulates the insulin-stimulated increase in Akt1 and Akt3 activity in 3T3-L1 adipocytes. Biochem. Biophys. Res. Commun. **243:** 509–513.
18. STANDAERT, M., G. BANDYOPADHYAY, L. GALLOWAY *et al.* 1999. Effects of knockout of the protein kinase C β gene on glucose transport and glucose homeostasis. Endocrinology **140:** 4470–4477.
19. STANDAERT, M.L., G. BANDYOPADHYAY, L. PEREZ *et al.* 1999. Insulin activates protein kinases C-ζ and -λ by an autophosphorylation-dependent mechanism and stimulates their translocation to GLUT4 vesicles and other membrane fractions in rat adipocytes. J. Biol. Chem. **274:** 25308–25316.
20. BAHR, M., M. SPENLIEKEN, M. BOCK *et al.* 1996. Acute and chronic effects of troglitazone on isolated rat ventricular cardiomyocytes. Diabetologia **39:** 766–774.
21. ZOROZANO, A., P. MUNOZ, M. CAMPS *et al.* 1996. Insulin-induced redistribution of GLUT4 glucose carriers in the muscle fibre: in search of GLUT4 trafficking pathways. Diabetes **45:** S70–S81.
22. DEFRONZO, R.A., E. JOCOT, E. JEQUIER *et al.* 1981. The effect of insulin on the disposal of intravenous glucose. Diabetes **30:** 1000–1007.
23. PHILLIPS, D.I.W., S. CADDY, V. IIIC *et al.* 1996. Intramuscular triglyceride and muscle insulin sensitivity: evidence for a relationship in nondiabetic subjects. Metabolism **45:** 947–950.
24. BRONFMAN, M., M.N. MORALES & A. ORELLANA. 1988. Diacylglycerol activation of protein kinase C is modulated by long chain acyl-CoA. Biochem. Biophys. Res. Commun. **83:** 5822–5824.
25. SCHMITZ-PEIFFER, C., N.D. OAKES, C.L. BROWNE *et al.* 1997. Reversal of chronic alterations of skeletal muscle protein kinase C from fat-fed rats by BRL49653. Am. J. Physiol. **273:** E915–E921.
26. CORTRIGHT, R.N., J.L. AZEVEDO, Q. ZHOU *et al.* 2000. Protein kinase C modulates insulin action in human skeletal muscle. Am. J. Physiol. **278:** E553–E562.
27. OSADA, S-I., K. MIZUNO, T.C. SAIDO *et al.* 1992. A new member of the protein kinase C family, nPKC-θ, predominantly expressed in skeletal muscle. Mol. Cell. Biol. **12:** 3930–3938.
28. CHANG, J.D., Y. XU, M.K. RAYCHOWDHURY & J.A. WARE. 1993. Molecular cloning and expression of a cDNA encoding a novel isoenzyme of protein kinase C (nPKCθ). J. Biol. Chem. **268:** 14208–14214.
29. GRIFFIN, M.E., M.J. MARCUCCI, G.W. CLINE *et al.* 1999. Free fatty acid–induced insulin resistance is associated with activation of protein kinase C-θ and alterations in the insulin signaling cascade. Diabetes **48:** 1270–1274.
30. GRAY, S., I. IDRIS, K. DAVIS & R. DONNELLY. 2001. Skeletal muscle expression of protein kinase C-θ is increased following an oral glucose load in insulin resistant patients with type 2 diabetes. Diabetes **50**(S2)**:** A314.
31. ITANI, S.I., W.J. PORIES, K.G. MCDONALD & G.L. DOHM. 2001. Increased protein kinase C-θ in skeletal muscle of diabetic patients. Metabolism **50:** 553–557.
32. VAN DE WERVE, G., J. PROIETTO, B. JEANRENAUD *et al.* 1985. Tumour-promoting phorbol esters increase basal and inhibit insulin-stimulated lipogenesis in rat adipocytes without decreasing insulin binding. Biochem. J. **225:** 523–527.
33. CARO, J.F., M. JENQUIN & S. LONG. 1992. Effects of phorbol esters on insulin receptor function and insulin action in hepatocytes. Mol. Cell. Biochem. **109:** 115–118.
34. GALI, R.R., S. PUGAZHENTHI & R.L. KHANDELWAL. 1993. Reciprocal effects of the protein kinase C inhibitors staurosporine and H-7 on the regulation of glycogen synthase and phosphorylase in the primary culture of hepatocytes. Metabolism **42:** 1475–1480.
35. SALTIS, J., A.D. HABBERFIELD, J.J. EGAN *et al.* 1991. Role of protein kinase C in the regulation of glucose transport in the rat adipose cell. J. Biol. Chem. **266:** 261–267.

36. CONSIDINE, R.V. & J.F. CARO. 1993. Protein kinase C: mediator or inhibitor of insulin action. J. Cell. Biochem. **52:** 8–13.
37. BANDYOPADHYAY, G., M.L. STANDAERT, L. ZHAO *et al.* 1997. Activation of protein kinase C (α, β, and ζ) by insulin in 3T3/L1 cells. J. Biol. Chem. **272:** 2551–2558.
38. BANDYOPADHYAY, G., M. STANDAERT, L. GALLOWAY *et al.* 1999. Evidence for involvement of protein kinase C (PKC)–ζ and noninvolvement of diacylglycerol-sensitive PKCs in insulin-stimulated glucose transport in L6 myotubes. Endocrinology **138:** 4721–4731.
39. AVIGNON, A., M.L. STANDAERT, K. YAMADA *et al.* 1995. Insulin increases mRNA levels of protein kinase C-α and -β in rat adipocytes and protein kinase C-α, -β, and -θ in rat skeletal muscle. Biochem. J. **308:** 181–187.
40. BRAIMAN, L., L. SHEFFI-FRIEDMAN, A. BAK *et al.* 1999. Tyrosine phosphorylation of specific protein kinase C isoenzymes participates in insulin stimulation of glucose transport in primary cultures of rat skeletal muscle. Diabetes **48:** 1922–1929.
41. SRINAVASAN, M. & N. BEGUM. 1994. Stimulation of protein phosphatase-1 activity by phorbol esters: evaluation of the role of protein kinase C in insulin action. J. Biol. Chem. **269:** 16662–16667.
42. ISHII, H., M.R. JIROUSEK, D. KOYA *et al.* 1996. Amelioration of vascular dysfunctions in diabetic rats by an oral PKC-β inhibitor. Science **272:** 728–731.

Desaturation and Elongation of Fatty Acids and Insulin Action

B. VESSBY, I-B. GUSTAFSSON, S. TENGBLAD, M. BOBERG, AND A. ANDERSSON

Unit for Clinical Nutrition Research, Department of Public Health and Caring Sciences, University of Uppsala, Uppsala, Sweden

> ABSTRACT: Insulin resistance is characterized by specific changes of the composition of fatty acids in the serum lipids and in the skeletal muscle membranes. Impaired insulin sensitivity is associated with high proportions of palmitic (16:0) acid and low levels of linoleic (18:2 n-6) acid in serum. In addition, there are apparent changes of the fatty acid desaturase activities, suggesting an increased activity of the $\Delta 9$ and $\Delta 6$ desaturases and a decreased activity of the $\Delta 5$ desaturase. The activity of the fatty acid desaturases is regulated by long-chain polyunsaturated fatty acids and insulin and is probably also dependent on the degree of physical activity. A high ratio between arachidonic (20:4 n-6) and dihomo-gamma linolenic (20:3 n-6) acid, as a measure of $\Delta 5$ desaturase activity, in the skeletal muscle phospholipids has been related to good insulin sensitivity. Available knowledge seems to indicate that the degree of saturation of the body lipids, and especially the proportion of palmitic acid in the lipid membranes, may be critical for insulin sensitivity. The strong relationships between the $\Delta 5$ desaturase activity, a high content of long-chain polyunsaturated fatty acids in the skeletal muscle, and insulin sensitivity may be due to parallel effects of diet and/or physical activity on the fatty acid composition and on insulin sensitivity.
>
> KEYWORDS: insulin; sensitivity; composition; fatty acids; lipids; desaturation; elongation; diabetes; skeletal muscle

Impaired insulin action, with a reduced insulin-mediated glucose uptake in skeletal muscles, is thought to be the major pathogenetic mechanism behind the insulin resistance syndrome, which is characterized by lipid abnormalities, impaired glucose tolerance and hypertension (often associated with abdominal obesity), and an increased risk to develop type 2 diabetes and atherosclerotic cardiovascular disease. The prevalence of obesity and insulin resistance, as well as type 2 diabetes, is rapidly increasing all over the world, underlining the importance of environmental factors, in addition to genetic disposition, for this development. The major lifestyle factors

---

Address for correspondence: Bengt Vessby, M.D., Ph.D., Unit for Clinical Nutrition Research, Department of Public Health and Caring Sciences, University of Uppsala, P. O. Box 609, SE 75125 Uppsala, Sweden.

bengt.vessby@pubcare.uu.se

involved are probably physical inactivity and diet, especially the quality of fatty acids and the type of carbohydrate-rich foods in the diet.[1–3]

FATTY ACIDS IN THE DIET

The composition of fatty acids in the tissues partly reflects the dietary fat composition,[4,5] but also the efficiency of the metabolism of fatty acids in the body, with intraindividual variations due to genetic disposition and probably early intrauterine and perinatal programming.[6] A change of the proportions of fatty acids in the diet is reflected in the serum triglycerides within the first hours, while the fatty acid composition of the serum cholesterol esters and phospholipids is related to the average dietary fatty acid composition during the last 3 to 6 weeks. The fatty acid profile of the erythrocyte membrane phospholipids reflects the dietary fat composition during the preceding months, and the fatty acids in the adipose tissue triglycerides reflect that during many months or years.[7]

The type of fat in the diet as reflected in the body tissues has profound effects on physiological and pathophysiological processes in the body. Thus, the fatty acid composition affects membrane properties, gene expression, metabolic signaling, eicosanoid production, and energy expenditure. The effects of the fatty acids are modulated by chain length, degree of unsaturation, isomeric form, and background diet.

METABOLISM OF FATTY ACIDS IN THE BODY

About one-third to one-half of the fatty acids in the Western diet are saturated, while the proportion of polyunsaturated fatty acids usually varies inversely with the saturated from one-third down to about one-tenth of the dietary fat. The major polyunsaturated fatty acid in the diet is linoleic acid (LA, 18:2 n-6), with lower amounts of the fatty acids from the n-3 series, alpha linolenic acid (ALA, 18:3 n-3) and the long-chain fatty acids eicosapentaenoic acid (EPA, 20:5 n-3) and docosahexaenoic acid (DHA, 22:6 n-3). The dominating monounsaturated fatty acid is oleic acid (18:1 n-9), with smaller amounts of monounsaturated *trans* fatty acids and palmitoleic acid (16:1 n-7).

The fatty acids are used as a source of energy in the body, but they are also metabolized by desaturation and elongation to longer and more unsaturated fatty acids with specific properties (FIG. 1). The desaturation steps are catalyzed by specific enzymes (desaturases), which are shared by all the fatty acid families, but with different affinities for the different fatty acid series. The preferred substrate for the $\Delta 6$ desaturase is 18:3 n-3 > 18:2 n-6 > 18:1 n-9.[8] Due to competition between the substrates and due to product inhibition, the efficiency of the desaturation of a certain fatty acid is dependent not only on the amount in the diet, but also on the content of other, competing fatty acids in the diet and in the tissues and on the activity of the enzymes. Thus, the desaturation and elongation of ALA to long-chain polyunsaturated n-3 fatty acids (EPA, DHA) may be severely inhibited by the presence of a high proportion of LA in the diet, while a diet with a low proportion of n-6 fatty acids and much saturated fatty acids allows for an efficient processing of ALA.

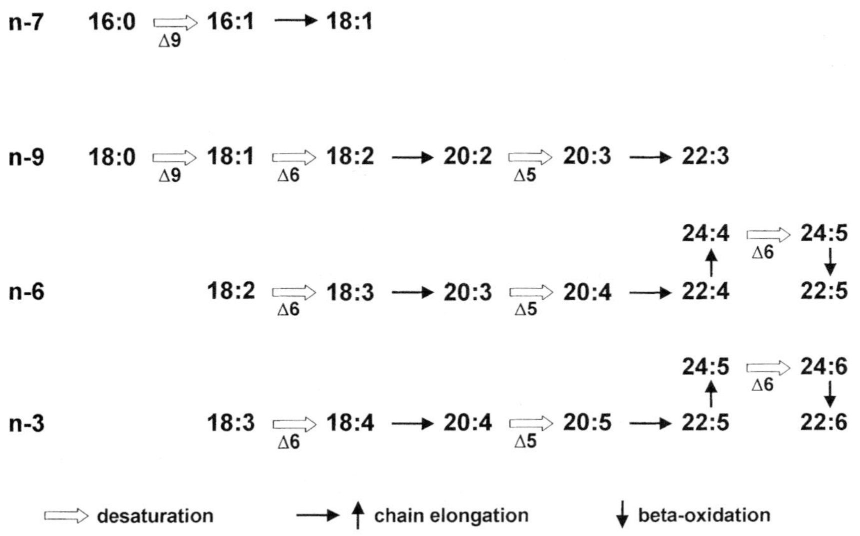

```
n-7    16:0 ⇒ 16:1 → 18:1
            Δ9

n-9    18:0 ⇒ 18:1 ⇒ 18:2 → 20:2 ⇒ 20:3 → 22:3
            Δ9      Δ6           Δ5

                                              24:4 ⇒ 24:5
                                               ↑    Δ6  ↓
n-6         18:2 ⇒ 18:3 → 20:3 ⇒ 20:4 → 22:4      22:5
                 Δ6           Δ5

                                              24:5 ⇒ 24:6
                                               ↑    Δ6  ↓
n-3         18:3 ⇒ 18:4 → 20:4 ⇒ 20:5 → 22:5      22:6
                 Δ6           Δ5

⇒ desaturation      →↑ chain elongation      ↓ beta-oxidation
```

FIGURE 1. Fatty acid metabolism.

REGULATION OF FATTY ACID DESATURASE ACTIVITIES

The activities of the fatty acid desaturases are influenced not only by the amount of and relationship between different substrate fatty acids, but also by the levels of the products in the tissues. Thus, the activities of Δ9 and Δ6 desaturase are down-regulated by polyunsaturated long-chain fatty acids (with more than two double bonds).[9,10] High proportions of arachidonic acid (AA, 20:4 n-6) or long-chain n-3 fatty acids in the tissues will inhibit the activity of the Δ6 desaturase and hence the conversion of the essential fatty acids to their longer and more unsaturated metabolites.

Experimental studies have indicated that insulin activates the Δ9 and Δ6 desaturases. In experimental diabetes and in spontaneously diabetic rats, there are reduced activities of Δ9, Δ6, and Δ5 liver microsomal desaturases, which are restored after insulin treatment.[11,12] Insulin-deficient patients with type 1 diabetes have high levels of LA and low levels of the metabolites including AA in their serum lipids, with an increase of AA and a normalization of the polyunsaturated fatty acids after insulin treatment.[13,14] Insulin-resistant states are accompanied by increased fasting levels of insulin, with an impaired early insulin secretion after glucose, but with a delayed and exaggerated late insulin response. It is unclear whether the activities of the desaturases are affected by variations of insulin concentrations also within a physiological range. Addition of insulin to human liver cells in culture did not change the Δ5 desaturase activity.[15] Supplementation with exogenous insulin to patients with type 2 diabetes did not significantly affect the desaturase activities (unpublished data). Neither do the proportions of fatty acids in serum lipids seem to change during hyperinsulinemic euglycemic clamp studies where high or supraphysiological insulin

TABLE 1. Relationships (correlation coefficients) between the serum cholesterol ester fatty acid composition and peripheral insulin sensitivity (M value at hyperinsulinemic euglycemic clamp) in 70-year-old men ($n = 92$)

Fatty acid	M value
Palmitic acid, 16:0	−0.29**
Palmitoleic acid, 16:1 n-7	−0.27*
Linoleic acid, 18:2 n-6	+0.33**
Dihomo-gamma linolenic acid, 20:3 n-6	+0.32**

NOTE: M = insulin sensitivity (glucose uptake during the last 60 min of the euglycemic hyperinsulinemic clamp test; mg/kg body wt/min).

levels are maintained during several hours. Improvement of insulin sensitivity by treatment with troglitazone of fructose-fed rats decreased the fatty acids associated with greater insulin sensitivity in skeletal muscle phospholipids.[16]

INSULIN RESISTANCE AND FATTY ACID COMPOSITION IN THE BODY

Insulin resistance and related disorders are characterized by a specific fatty acid pattern in the serum lipid esters, as well as in storage lipids and in cell membranes of the skeletal muscle. A comparison between the serum lipid fatty acid composition in patients with disorders related to insulin resistance (diabetes, obesity, coronary heart disease) and matched healthy control subjects[17–22] has revealed an aberrant fatty acid pattern common for these disorders characterized by an increased proportion of palmitic (16:0) and palmitoleic acid, low levels of LA, and a high proportion of dihomo-gamma linolenic acid (DHLA, 20:3 n-6). There is usually no difference regarding AA, EPA, and DHA between patients and controls.

When insulin sensitivity was measured in a group of apparently healthy subjects, significant relationships were seen between the fatty acid composition in the serum lipid esters and insulin sensitivity as measured by the hyperinsulinemic euglycemic clamp technique (TABLE 1). Again, there is the same fatty acid pattern in serum associated with insulin resistance characterized by high palmitic and palmitoleic acid, low LA, and a high proportion of DHLA.

What is (are) the reason(s) for the differences in serum lipid fatty acid composition in patients with insulin resistance or disorders related to insulin resistance and healthy people? Is this caused by different dietary habits or genetic differences or is it a consequence of the impaired insulin action? Against the latter interpretation speaks the fact that similar fatty acid patterns are seen in healthy subjects who later will develop diabetes[23] or coronary heart disease[24] in prospective studies. The same pattern is not seen in those who remain healthy during follow-up (TABLE 2).

If product/precursor ratios are used to calculate desaturase activities, it is obvious that insulin resistance and insulin resistance–related disorders are generally connected with a fatty acid pattern in serum indicating high activities of the Δ9 and Δ6 desaturases and a low activity of Δ5 desaturase (TABLE 3). Borkman and coworkers

TABLE 2. Serum cholesterol ester fatty acid composition of 50-year-old men who developed myocardial infarction (MI) during a 19-year follow-up period compared with those who remained healthy (mean; SD)

Fatty acid	Healthy ($n = 1593$)	MI ($n = 153$)
Myristic, 14:0	1.13; 0.25	1.19; 0.27**
Palmitic, 16:0	11.65; 0.97	11.94; 1.06***
Palmitoleic, 16:1 n-7	3.83; 1.36	4.11; 1.37*
Stearic, 18:0	1.15; 0.28	1.20; 0.32
Oleic, 18:1 n-9	19.38; 2.77	19.76; 2.62
Linoleic, 18:2 n-6	54.10; 5.24	52.90; 5.18**
Gamma linolenic, 18:3 n-6	0.70; 0.29	0.74; 0.31
Alpha linolenic, 18:3 n-3	0.66; 0.16	0.68; 0.17
Dihomo-gamma linolenic, 20:3 n-6	0.57; 0.13	0.60; 0.14**
Arachidonic, 20:4 n-6	4.77; 0.97	4.73; 1.01
Eicosapentaenoic, 20:5 n-3	1.35; 0.62	1.45; 0.82
Docosahexaenoic; 22:6 n-3	0.70; 0.21	0.72; 0.23

NOTE: *, **, *** = $p < 0.05$, 0.01, and 0.001, respectively, compared with men who developed MI. From reference 20.

TABLE 3. Insulin-resistant states are associated with the following fatty acid compositions in serum and skeletal muscle cell membrane

High 16:0 (palmitic acid)
Low 18:2 n-6 (linoleic acid)
High 16:1/16:0 ($\Delta 9$ desaturase)
High 18:3 n-6/18:2 n-6 ($\Delta 6$ desaturase)
Low 20:4 n-6/20:3 n-6 ($\Delta 5$ desaturase)

were the first to study the relationships between insulin sensitivity and fatty acid composition of the skeletal muscle phospholipids in humans.[21] They found positive associations between insulin sensitivity and the total number of long-chain polyunsaturated fatty acids in the skeletal muscle phospholipids, with a strong relation to the apparent $\Delta 5$ desaturase activity. Similar relationships were reported from a study of adult Pima Indians,[18] while we, in Uppsala, in a study of 70-year-old men,[22] found a significant and independent relation between the proportion of palmitic acid in the skeletal muscle membrane phospholipids and insulin sensitivity, but no associations between insulin sensitivity and the long-chain polyunsaturated fatty acids.

It is at present not possible to clearly explain the relationships between insulin sensitivity and the apparent desaturase activities. Can an increased $\Delta 5$ desaturase activity cause changes of the fatty acid proportions that directly contribute to an increased insulin sensitivity? Is the increased desaturase activity a consequence of an improved insulin sensitivity? Last, are the changes of the desaturase activities, as well as the improved insulin effect, parallel phenomena caused by a common factor?

TABLE 4. Calculated daily intake of energy (energy %) derived from fat on a diet rich in saturated fatty acids (SAT) and rapeseed oil (RO) (mean values)

	SAT	RO
Total fat	35.7	36.1
Saturated fatty acids	18.8	7.8
Monounsaturated fatty acids	10.7	16.2
Trans fatty acids	0.9	0.8
Polyunsaturated fatty acids	3.6	8.7
18:3 n-3	0.4	2.2
20:5–22:6 n-3	0.2	0.2

REGULATION OF DESATURASE ACTIVITIES BY DIETARY FATTY ACIDS

The desaturase activities are directly influenced by the fatty acid composition of the diet and the tissues. Changes of the proportions of saturated and unsaturated fatty acids in the diet affect the desaturase activities, as earlier shown by Laserre *et al.*[25] TABLE 4 shows the fatty acid composition of two diets given to healthy, mildly hyperlipidemic subjects during a randomized crossover study during two 24-day treatment periods under isoenergetic conditions. The diets were strictly controlled with identical nutrient content, differing only in fatty acid composition. One diet was rich in saturated fatty acids (butter-based); the other was based on rapeseed oil and contained much less saturated fatty acids, but more unsaturated fat, as oleic acid, LA, and (characteristically) a high proportion of ALA. TABLE 5 illustrates the fatty acid composition of the serum phospholipids at the end of the two treatment periods. As expected, the proportions of saturated fatty acids are lower and the content of oleic acid, LA, and ALA higher after the rapeseed oil–based diet. An important observation is that the proportions of AA, as well as of EPA and DHA, are identical after both diets despite much higher proportions of the precursor fatty acids, LA and ALA, respectively, in the rapeseed oil–based diet than in the butter-containing diet.

The explanation for similar proportions of long-chain polyunsaturated fatty acids in serum, in spite of largely diverging amounts of essential fatty acids in the diets, is apparent when the desaturase activities are calculated. During the unsaturated rapeseed oil–based diet, there were significant reductions of the $\Delta 9$ and $\Delta 6$ desaturase activities, while the $\Delta 5$ activity increased (TABLE 6). This means a reduced rate of desaturation of palmitic and stearic acid to their monounsaturated counterparts (palmitoleic and oleic acids, respectively) and a reduced rate of conversion of LA and ALA to their unsaturated metabolites. The fatty acid pattern after the rapeseed oil diet is very similar to that associated with an improved insulin sensitivity, while that after the diet rich in butter fat has a fatty acid composition identical to that usually seen in insulin-resistant states and disorders (TABLES 1 and 2). During these relatively short diet periods, there were, however, no changes of insulin sensitivity as measured by the clamp technique and the rate of insulin-induced glucose removal was not significantly different at the end of the treatments.

TABLE 5. Fatty acid composition (%) of serum phospholipids on a diet rich in saturated fatty acids (SAT) and rapeseed oil (RO)

Fatty acid	SAT	RO
16:0	32.8	30.8**
16:1 n-7	0.9	0.8*
18:0	15.1	14.9
18:1	12.1	13.7**
18:2 n-6	18.0	20.3**
18:3 n-3	0.3	0.5**
20:3 n-6	3.3	2.5*
20:4 n-6	7.2	7.1
20.5 n-3	1.8	1.7
22.5 n-3	1.1	0.9*
22:6 n-3	6.3	6.0

NOTE: * and ** = $p < 0.01$ and 0.001, respectively, compared with SAT diet.

TABLE 6. Apparent effects of diets enriched in butter fat (SAT) and rapeseed oil (RO) on desaturase activities as mirrored by serum cholesterol ester fatty acid composition

	SAT	RO	p value for difference between diets
"Δ9 desaturase", 16:1 n-7/16:0	+2%	−25%**	0.0001
"Δ6 desaturase", 18:3 n-6/18:2 n-6	−2%	−28%**	0.0001
"Δ5 desaturase", 20:4 n-6/20:3 n-6	−10%*	+19%**	0.0001

NOTE: * and ** = $p < 0.05$ and 0.001, respectively, for changes compared with admission values. Unpublished data.

RELATIONSHIPS BETWEEN CHANGES OF SERUM LIPID FATTY ACID COMPOSITION, DESATURASE ACTIVITIES, AND DIETARY FAT COMPOSITION

TABLE 7 includes data from a series of controlled dietary studies performed at our department during the last decade.[26–30] It summarizes the effects of variations in dietary fatty acid composition on the proportions of certain key fatty acids in the serum lipid esters (serum cholesterol or phospholipids) and on the apparent desaturase activities. The diets in all studies were tightly controlled, the fatty acid composition of the diets given to the participants was analyzed, and the studies were performed under isoenergetic conditions with randomized treatment periods of 3 weeks or more.

Some important conclusions from these studies are listed below:

TABLE 7. Effects of changes of dietary fat composition in controlled dietary intervention studies

Type of dietary intervention	Effects on fatty acid composition				Effects on desaturase activities		
	16:0	18:2 n-6	20:4 n-6	EPA + DHA	Δ9	Δ6	Δ5
SAT → MUFA (OO)	⇓	–	–	–	⇓	–	–
SAT → UFA (RO)	⇓	⇑	–	–	⇓	⇓	⇑
SAT → PUFA (n-6)	⇓	⇑	–	⇓	⇓	⇓	⇑
MUFA (OO) → PUFA (n-6)	⇓	⇑	–	⇓	⇓	⇓	↑
UFA (RO) → PUFA (n-6)	–	⇑	–	⇓	↓	↓	–
MUFA (OO) → UFA (RO)	–	⇑	–	–	–	–	–

NOTE: The table is based on data from references 26–30 and unpublished data. ⇑, increase; –, unchanged; ⇓, decrease; ↑, slight increase; ↓, slight decrease; SAT, diet with a high proportion of saturated fatty acids; MUFA (OO), diet with a high proportion of monounsaturated fatty acids (based on olive oil); UFA (RO), diet with a high proportion of both mono- and polyunsaturated fatty acids (based on rapeseed oil); PUFA (n-6), diet with a high proportion of polyunsaturated fatty acids (linoleic acid).

(1) In all studies where there is a shift from a saturated to a more unsaturated diet, there is a reduced ratio between 16:1 n-7 and 16:0, indicating a downregulation of the Δ9 desaturase. The activity of Δ9 desaturase is related to the proportion of saturated fatty acids, mainly palmitic acid (16:0), in the diet and varies inversely to the proportion of polyunsaturated fatty acids.

(2) When a diet with a high proportion of polyunsaturated fatty acids is substituted for a diet with much saturated fatty acids, there is a downregulation of Δ6 compatible with a product inhibition of the activity. Simultaneously, there is an apparent upregulation of the activity of the Δ5 desaturase. Δ6 and Δ5 seem to be regulated inversely, possibly due to the fact that the elongated product of Δ6 (DHLA) is the substrate for Δ5. If the proportion of DHLA is decreased, Δ5 desaturase has to be upregulated to maintain the AA level unchanged in the tissues.

(3) The proportion of AA in the serum lipid esters is kept unchanged despite large variations of the content of LA and of the ratio between LA and ALA in the diet and in serum. This indicates a very strict regulation and control of the content of AA in serum. The only situation when the proportion can be affected (decreased) by manipulating the composition of dietary fatty acids is by bypassing the desaturase and elongase system and introducing the long-chain polyunsaturated n-3 fatty acids (EPA and DHA) in the diet (or by directly increasing the amount of dietary AA).

(4) The proportions of EPA and DHA in the serum lipids are also efficiently regulated as long as there is no abundance of n-6 fatty acids in the diet, which inhibits the conversion of ALA to its metabolites. The proportion of EPA and DHA is identical in diets containing a high amount of saturated fatty acids and a diet containing a low amount of saturated fatty acids and a

high proportion of ALA, demonstrating how efficient a lack of n-3 fatty acids in the diet is compensated for by an increased endogenous production of the long-chain polyunsaturated metabolites. When there is a (too?) high proportion of LA in relation to ALA in the diet, the competition for the desaturases leads to a decreased formation of EPA and DHA.

DIET, SKELETAL MUSCLE FATTY ACID COMPOSITION, AND INSULIN SENSITIVITY

In a recent 3-month study, it was shown that insulin sensitivity was impaired when the proportion of saturated fat increased in the diet at the expense of monounsaturated fatty acids.[26] Supplementation of n-3 fatty acid to the diet did not affect insulin sensitivity, independent of the fatty acid composition of the background diet. When saturated fatty acids were substituted for oleic acid in the diet, the proportion of saturated fatty acids increased, while the content of oleic acid decreased in the serum lipids. There were no differences in the proportions of the long-chain polyunsaturated fatty acid metabolites in the serum lipids, or in the Δ6 or Δ5 desaturase activities, in spite of an improved insulin sensitivity. The addition of long-chain n-3 fatty acids, on the other hand, was associated with increased proportions of EPA and DHA in serum and reduced Δ6 and increased Δ5 activity, but without any effect on the insulin sensitivity.

The fatty acid composition of the skeletal muscle was studied at the end of the treatment periods in a subsample of the participants ($n = 32$). It was clearly shown that the fatty acid composition of the phospholipids in the skeletal muscle was affected in a similar way as in serum, with a reduction of saturated fatty acids and the Δ9 desaturase when switching from saturated to monounsaturated fatty acids, while the Δ6 and Δ5 desaturase activities did not change in spite of an improved insulin sensitivity. After supplementation with n-3 fatty acids, on the other hand, LA and AA decreased in the phospholipids, while EPA and DHA showed a large increase. At the same time, there were reductions of the Δ9 and Δ6 desaturases and an increase of Δ5, while insulin sensitivity was not affected (unpublished data).

PHYSICAL TRAINING AND SKELETAL MUSCLE FATTY ACID COMPOSITION

Physical activity and dietary fat quality do affect the composition of fatty acids in serum and skeletal muscle lipids independently, as recently shown in two studies.[31,32] Despite an identical fatty acid composition of the diet, subjects who were physically active showed a fatty acid pattern of the skeletal muscle phospholipids characterized by less palmitic acid and DHLA and an improved activity of Δ5 desaturase and of the elongase responsible for converting palmitic to stearic acid.[32] These results were corrected for differences in skeletal muscle fiber composition between trained and untrained subjects, which also influenced the fatty acid composition, with a higher proportion of long-chain n-3 fatty acids in subjects with a high proportion of type 1 slow-twitch fibers.

WHY ARE FATTY ACIDS DESATURATED AND ELONGATED IN THE BODY?

What is the aim of and mechanisms behind the metabolism (desaturation and elongation) of fatty acids in the body, including the responses to variations in dietary fat composition and physical activity? Some major reasons for modulation of these metabolic pathways seem obvious.

First, there is a need to control the degree of saturation of lipids in membranes throughout the body. The fatty acid composition, and the degree of saturation of the membranes, should be of great importance for fluidity and permeability of the membranes and for the efficiency of ion pumps, cell membrane receptor function, and transport of nutrients over the membranes, and it may be critical for the insulin-mediated glucose transport into the cell. The major saturated fatty acid in Western diets is palmitic acid, followed by stearic acid. The latter is less abundant in the diet and in the tissues and is also a better substrate for Δ9 desaturase than palmitic acid. There is no relationship between the proportions of stearic acid in the human tissues and insulin resistance (TABLE 1). Palmitic acid is not only the major fatty acid in membrane phospholipids in the body, but also a precursor of ceramide, an intracellular signal substance that may be critical in the development of insulin resistance.[33] It has also been suggested that palmitic, but not palmitoleic, acid has a lipotoxic effect on the pancreatic beta cells.[34] A major role for the Δ9 desaturase would seem to be to increase the unsaturation in membranes in response to increased dietary intake of saturated fat and reduce the availability of palmitic acid by converting palmitic to palmitoleic acid. Another way of reducing the pool of palmitic acid is to increase the rate of elongation to stearic acid (as seen in relation to physical activity) and a consequent desaturation to oleic acid.

The other main function for the system of desaturases and elongases would be to regulate the proportions of the long-chain polyunsaturated fatty acids in the body tissues. A major task seems to be to control the levels of the metabolically very potent eicosanoid precursors, specifically AA, but also EPA, and of the content of DHA, which is an essential component of the central nervous system with pronounced metabolic effects. The long-chain polyunsaturated fatty acids do not only modify the desaturase activities, but are also regulators of the expression of a series of other genes involved in lipogenesis and lipid oxidation.[9]

TENTATIVE CONCLUSIONS CONCERNING THE RELATIONSHIPS BETWEEN INSULIN SENSITIVITY, FATTY ACID COMPOSITION, AND FATTY ACID METABOLISM IN THE BODY

Experimental evidence in animals and humans indicates that the degree of saturation of the diet, and in the tissues in the body, is related to insulin action. Inclusion of more unsaturated fat in the diet and an increased degree of physical activity, both associated with an improved insulin sensitivity, are associated with changes of fatty acid metabolism favoring a reduced degree of saturation of the tissues with a lowered proportion of palmitic acid, which may be critical for good insulin sensitivity.

Changes of Δ6 and Δ5 desaturase activities in the body, as consequences of an increased content of polyunsaturated fatty acids in the diet or increased physical activity, have not been directly related to changes of insulin sensitivity. The strong relationships between insulin sensitivity and Δ5 desaturase activity seen in cross-sectional[18,21] and prospective[23] studies may be due to parallel effects by diet and/or physical activity on the fatty acid composition and insulin sensitivity and not due to a direct relationship between desaturase activity and insulin action.

The addition of n-3 fatty acids to human diets has not been shown to increase insulin sensitivity,[2,26] in spite of apparent changes of desaturase activities. The significant relationship between the proportions of n-3 fatty acids and insulin sensitivity seen in some studies may be due to simultaneous differences in fiber type in the skeletal muscles. It is also conceivable that supplementation with n-3 fatty acids might cause an improvement of insulin sensitivity in subjects with very low levels of n-3 fatty acids in the skeletal muscle due to genetic or dietary reasons.

ACKNOWLEDGMENTS

This work was supported by grants from the Swedish Council for Forestry and Agricultural Research.

REFERENCES

1. STORLIEN, L.H., L.A. BAUR, A.D. KRIKETOS *et al.* 1996. Dietary fats and insulin action. Diabetologia **39:** 621–631.
2. VESSBY, B. 2000. Dietary fat and insulin action in humans. Br. J. Nutr. **83**(suppl. 1): S91–S96.
3. HU, F.B., R.M. VAN DAM & S LIU. 2001. Diet and risk of type II diabetes: the role of types of fat and carbohydrate. Diabetologia **44:** 805–817.
4. NIKKARI, T., P. LUKKAINEN, P. PIETINEN *et al.* 1995. Fatty acid composition of serum lipid fractions in relation to gender and quality of dietary fat. Ann. Med. **27:** 491–498.
5. VAN STAVEREN, W.A., P. DEURENBERG, M. KATAN *et al.* 1986. Validity of the fatty acid composition of subcutaneous adipose tissue microbiopsies as an estimate of the diet of separate individuals. Am. J. Epidemiol. **123:** 455–463.
6. OZANNE, S.E., N.D. MARTENSZ, C.J. PETRY *et al.* 1998. Maternal low protein diet in rats programmes fatty acid desaturase activities in the offspring. Diabetologia **41:** 1337–1342.
7. KATAN, M.B., J.P. DESLYPERE, A.P.J.M. VAN BIRGELEN *et al.* 1997. Kinetics of the incorporation of dietary fatty acids into serum cholesteryl esters, erythrocyte membranes, and adipose tissue: an 18-month controlled study. J. Lipid Res. **38:** 2012–2022.
8. SIGUEL, E.N. & M. MACLURE. 1987. Relative activity of unsaturated fatty acid metabolic pathways in humans. Metabolism **36:** 664–669.
9. JUMP, D.B. & S.D. CLARKE. 1999. Regulation of gene expression by dietary fat. Annu. Rev. Nutr. **19:** 63–90.
10. KIM, Y.C. & J.M. NTAMBI. 1999. Regulation of stearoyl-CoA desaturase genes: role in cellular metabolism and preadipocyte differentiation. Biochem. Biophys. Res. Commun. **266:** 1–4.
11. ECK, M.G., J.O. WYNN, W.J. CARTER *et al.* 1979. Fatty acid desaturation in experimental diabetes. Diabetes **28:** 479–485.
12. MIMOUNI, V. & J.P. POISSON. 1992. Altered desaturase activities and fatty acid composition in liver microsomes of spontaneously diabetic Wistar BB rats. Biochim. Biophys. Acta **1123:** 296–302.
13. VON DOORMAL, J.J., F.A. MUSKIET, E. VAN BALLEGOIE *et al.* 1984. The plasma and erythrocyte fatty acid composition of poorly controlled, insulin-dependent (type I)

diabetic patients and the effect of improved metabolic control. Clin. Chim. Acta **144:** 203–212.
14. BASSI, A., A. AVOGARO, C. CREPALDI *et al.* 1996. Short-term diabetic ketosis alters n-6 polyunsaturated fatty acid content in plasma phospholipids. J. Clin. Endocrinol. Metab. **81:** 1650–1653.
15. LOIZOU, C.L., S.E. OZANNE & C.N. HALES. 1999. The effect of insulin on delta5 desaturation in hepG2 human hepatoma cells and L6 rat muscle myoblasts. Prostaglandins Leukotrienes Essent. Fatty Acids **61:** 89–95.
16. CLORE, J.N., L. LI & W.B. RIZZO. 2000. Effects of fructose and troglitazone on phospholipid fatty acid composition in rat skeletal muscle. Lipids **35:** 1281–1287.
17. SALOMAA, V., I. AHOLA, J. TUOMILEHTO *et al.* 1990. Fatty acid composition of serum cholesterol esters in different degree of glucose intolerance. Metabolism **39:** 1285–1291.
18. PAN, D.A., S. LILLIOJA, M.R. MILNER *et al.* 1995. Skeletal muscle membrane lipid composition is related to adiposity and insulin action. J. Clin. Invest. **96:** 2802–2808.
19. DESCI, T., G. CSABI, K. TOROK *et al.* 2000. Polyunsaturated fatty acids in plasma lipids of obese children with and without metabolic cardiovascular syndrome. Lipids **35:** 1179–1184.
20. ÖHRVALL, M., G. SUNDLÖF & B. VESSBY. 1996. Gamma, but not alpha, tocopherol levels in serum are reduced in coronary heart disease patients. J. Intern. Med. **239:** 111–117.
21. BORKMAN, M., L.H. STORLIEN, D.A. PAN *et al.* 1993. The relationship between insulin sensitivity and the fatty acid composition of skeletal-muscle phospholipids. N. Engl. J. Med. **328:** 238–244.
22. VESSBY, B., S. TENGBLAD & H. LITHELL. 1994. The insulin sensitivity is related to the fatty acid composition of the serum lipids and the skeletal muscle phospholipids in 70 year old men. Diabetologia **37:** 1044–1050.
23. VESSBY, B., A. ARO, E. SKARFORS *et al.* 1994. The risk to develop NIDDM is related to the fatty acid composition of the serum cholesterol esters. Diabetes **43:** 1353–1357.
24. ÖHRVALL, M., L. BERGLUND, I. SALMINEN *et al.* 1996. The serum cholesterol ester fatty acid composition, but not the serum concentration of alpha tocopherol predicts the development of myocardial infarction in 50-year-old men, 19 years follow up. Atherosclerosis **127:** 65–71.
25. LASERRE, M., F. MENDY, D. SPIELMANN *et al.* 1985. Effects of different dietary intakes of essential fatty acids on C 20:3 n-6 and C 20:4 n-6 serum levels in human adults. Lipids **20:** 227–233.
26. VESSBY, B., M. UUSITUPA, K. HERMANSEN *et al.* 2001. Substituting dietary for monounsaturated fat impairs insulin sensitivity in healthy men and women: the KANWU study. Diabetologia **44:** 312–319.
27. NYDAHL, M., I-B. GUSTAFSSON & B. VESSBY. 1994. Lipid lowering diets enriched with monounsaturated and polyunsaturated fatty acids, but low in saturated fatty acids have similar effects on serum lipid concentrations in hyperlipidemic patients. Am. J. Clin. Nutr. **59:** 115–122.
28. NYDAHL, M., I-B. GUSTAFSSON, M. ÖHRVALL *et al.* 1994. Similar serum lipoprotein cholesterol reductions in healthy subjects on diets enriched with rapeseed and with sunflower oil. Eur. J. Clin. Nutr. **48:** 128–137.
29. GUSTAFSSON, I-B., B. VESSBY, M. ÖHRVALL *et al.* 1994. A diet rich in monounsaturated rapeseed oil reduces the lipoprotein cholesterol concentration and increases the relative content of n-3 fatty acids in serum in hyperlipidemic subjects. Am. J. Clin. Nutr. **59:** 667–674.
30. NYDAHL, M., I-B. GUSTAFSSON, M. ÖHRVALL *et al.* 1996. Similar effects of rapeseed oil (canola oil) and olive oil in a lipid lowering diet for patients with hyperlipoproteinemia. J. Am. Coll. Nutr. **14:** 643–651.
31. ANDERSSON, A., A. SJÖDIN, R. OLSSON *et al.* 1998. Effect of physical exercise on phospholipid fatty acid composition in skeletal muscle. Am. J. Physiol. Endocrinol. Metab. **274**(37): E432–E438.
32. ANDERSSON, A., A. SJÖDIN, A. HEDMAN *et al.* 2000. Fatty acid profile of skeletal muscle phospholipids in trained and untrained young men. Am. J. Physiol. Endocrinol. Metab. **279:** E744–E751.

33. SCHMITZ-PEIFFER, C., D.L. CRAIG & T.J. BIDEN. 1999. Ceramide generation is sufficient to account for the inhibition of the insulin-stimulated PKB pathway in C2C12 skeletal muscle cells pretreated with palmitate. J. Biol. Chem. **274:** 24202–24210.
34. MAEDLER, K., G.A. SPINAS, D. DYNTAR *et al.* 2001. Distinct effects of saturated and monounsaturated fatty acids on beta-cell turnover and function. Diabetes **50:** 69–76.

Muscle Long-Chain Acyl CoA Esters and Insulin Resistance

G. J. COONEY, A. L. THOMPSON, S. M. FURLER, J. YE, AND E. W. KRAEGEN

Garvan Institute of Medical Research, St. Vincent's Hospital, Sydney, NSW 2010, Australia

ABSTRACT: A common observation in animal models and in humans is that accumulation of muscle triglyceride is associated with the development of insulin resistance. In animals, this is true of genetic models of obesity and nutritional models of insulin resistance generated by high-fat feeding, infusion of lipid, or infusion of glucose. Although there is a strong link between the accumulation of triglycerides (TG) in muscle and insulin resistance, it is unlikely that TG are directly involved in the generation of muscle insulin resistance. There are now other plausible mechanistic links between muscle lipid metabolites and insulin resistance, in addition to the classic substrate competition proposed by Randle's glucose–fatty acid cycle. The first step in fatty acid metabolism (oxidation or storage) is activation to the long-chain fatty acyl CoA (LCACoA). This review covers the evidence suggesting that cytosolic accumulation of this active form of lipid in muscle can lead to impaired insulin signaling, impaired enzyme activity, and insulin resistance, either directly or by conversion to other lipid intermediates that alter the activity of key kinases and phosphatases. Actions of fatty acids to bind specific nuclear transcription factors provide another mechanism whereby different lipids could influence metabolism.

KEYWORDS: insulin resistance; high-fat feeding; triglycerides; LCACoA; fatty acid metabolism

INTRODUCTION

Insulin resistance is found in many animal models that exhibit characteristics of human type 2 diabetes or obesity.[1] Accumulation of triglyceride in muscle appears to be an important contributing factor to this insulin resistance, and this review focuses on how increased lipid availability can be a major factor influencing muscle insulin action. Insulin-resistant states can be generated by dietary means, such as by high-fat feeding or high-fructose/sucrose feeding, both of which are commonly employed in rodent studies. For experimental purposes, insulin resistance can also be produced in muscle by more direct systemic delivery of nutrients. Examples are triglyceride/heparin infusions or glucose infusions to generate an overabundance of

Address for correspondence: Gregory J. Cooney, Garvan Institute of Medical Research, 384 Victoria Street, Darlinghurst, Sydney, NSW 2010, Australia. Voice: +61 2 9295 8209; fax: +61 2 9295 8201.
g.cooney@garvan.org.au

fatty acids or glucose supply to muscle, respectively. The aim of this review is to outline the general aspects of the association between lipid metabolism and insulin resistance and focus on the role played by long-chain acyl CoA derivatives (LCACoAs) in the interactions between lipid and glucose metabolism in muscle.

THE LINKS BETWEEN MUSCLE LIPIDS AND INSULIN RESISTANCE

There are now many studies confirming that increased fatty acid availability in muscle and liver is associated with insulin resistance.[2–5] Muscle triglyceride accumulation could occur because of increased availability and uptake of systemic free fatty acids (FFAs). The origin of the excess FFA supply would most likely result from lipoprotein lipase action on liver-derived VLDL-triglyceride or gut-derived chylomicrons in the postprandial state. Another possibility is that impaired muscle fatty acid oxidation (in the presence of unaltered uptake) leads to excess cytosolic lipid accumulation. The relative contribution of the different sources of excess lipid to muscle are not currently known in quantitative terms, and a combination of both factors may be a possibility. More detail is contained in recent reviews dealing with the uptake and fate of muscle lipids.[6,7]

Some of the major elements of the lipid supply hypothesis of muscle insulin resistance and the central role of LCACoAs are depicted in FIGURE 1. Muscle lipid

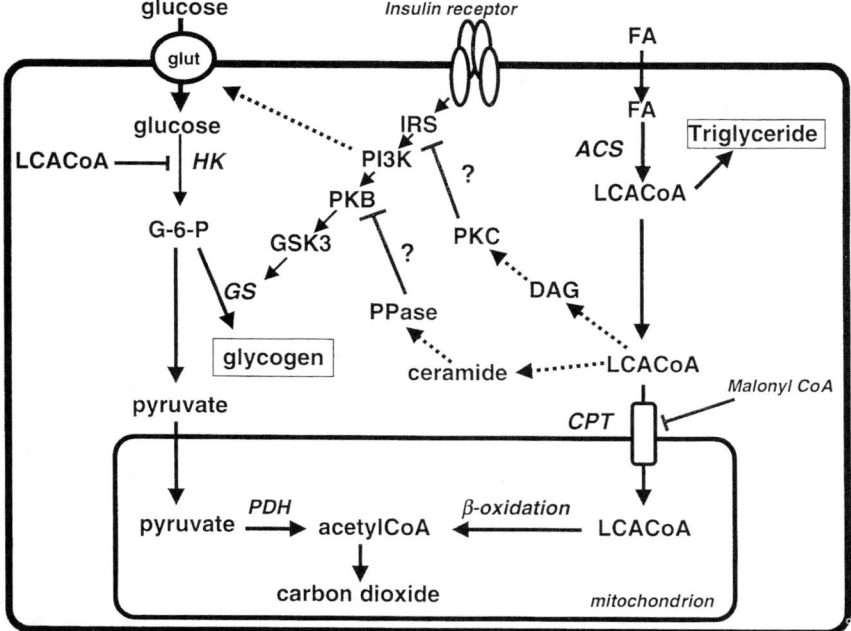

FIGURE 1. Some of the interactions, discussed in the text, whereby lipid metabolism may inhibit insulin action on glucose metabolism in muscle. LCACoA, long-chain fatty acyl CoAs; DAG, diacylglycerol; PKC, protein kinase C; IRS, insulin receptor substrate; PKB, protein kinase B; GSK3, glycogen synthase kinase 3.

accumulation can manifest itself as increased levels of either stored triglyceride or increased LCACoA. This metabolically active form of lipids may influence metabolism acutely by changing substrate availability or by altering key enzyme activities by allosteric regulation. Mitochondrial FFA oxidation may influence glucose oxidation via the classic glucose–fatty acid (Randle) cycle.[8] Changes in the level of specific fatty acids could result in production of other lipid species (e.g., specific diacylglycerols) that may alter insulin signaling pathways via protein kinase C activation. Ceramides can also be generated from increased lipid availability and have been shown to inhibit glycogen synthesis.[9] By blocking mitochondrial transfer of LCACoA, malonyl CoA can lead to accumulation of cytosolic LCACoAs and subsequent insulin resistance by the mechanisms mentioned previously. Some of these interactions will be considered in more detail in subsequent sections, but first we will consider the evidence that changes in LCACoA levels are an important indicator and possible mediator of insulin resistance.

INSULIN RESISTANCE AND INCREASED LCACoAs IN MUSCLE

Acute Systemic Fatty Acid Elevation

In the presence of acute FFA elevation, a reduction in insulin-mediated muscle glucose uptake occurs, although this takes several hours to manifest itself.[3,10–12] There is some evidence that any initial reduction in glycolysis, as might be predicted by the Randle cycle, is offset by an increased rate of glycogen synthesis.[12–14] However, after several hours, there is also a reduction in insulin-mediated glycogen synthesis and inhibition of insulin-stimulated muscle glucose uptake.[15–19] In the rat, 3–5 h of FFA elevation during a clamp produces significant whole body and muscle insulin resistance, and this is accompanied by a significant elevation of both muscle triglyceride and total LCACoAs.[10] This increase in LCACoAs is of a similar magnitude to that seen when insulin resistance is induced chronically by a high-fat diet.[20,21] Thus, it seems quite reasonable that similar mechanisms are operative in the induction of insulin resistance by acute FFA elevation as when insulin resistance is induced by a high dietary intake of lipids.

High Dietary Fat Intake

Much of the evidence for an association between muscle lipid accumulation and insulin resistance has come from animal dietary studies. Muscle insulin resistance appears to develop concomitantly with muscle triglyceride accumulation, leading to a significant association between muscle insulin resistance and triglyceride content after 3 weeks on the high-fat diet.[22,23] Similar to the situation with intralipid infusion, Kim et al. showed an initial enhancement of glycogen synthesis that offsets reduced glycolysis and explains the delayed effect of high-fat feeding on impairing insulin-mediated muscle glucose uptake.[24] However, it was also clear that glycogen synthesis was impaired by the time that there was impairment of insulin-mediated muscle glucose uptake.[22,24] Thus, the scenario of onset of diet-induced muscle insulin resistance seems similar to that induced by acute FFA elevation via triglyceride/ heparin, albeit over a longer time scale. In all studies of high-fat feeding where

LCACoAs have been measured, they have been found to be elevated.[20,21,25,26] Incubation of isolated soleus strips from rats with fatty acids for 4 h also impaired the ability of these muscles to take up and phosphorylate glucose in response to insulin. Different fatty acids have different effects on insulin action that correlate with the accumulation of LCACoA in the muscle strips during the incubation.[27]

Amelioration of Insulin Resistance and LCACoAs

It now seems clear that even short-term interventions that reduce muscle cytosolic lipid accumulation have a significant effect to lessen insulin resistance produced by high-fat feeding in rats. When chronic high-fat-fed rats are given a single low-fat, high-glucose meal as a replacement for one night, on the next day there is a highly significant amelioration of muscle insulin resistance, including effects on insulin-mediated glycogen synthesis.[20] Similar effects are produced by overnight fasting or by a 2-h bout of exercise on the day before assessment of insulin sensitivity.[20] A common feature in this reversal of insulin resistance is a decrease in the muscle LCACoA levels during the glucose clamp, suggesting a close association between insulin suppressibility of LCACoA and enhanced glucose metabolism. We have also established that the chronic activation/translocation of PKCθ and reduction in insulin's ability to phosphorylate protein kinase B (PKB), which are induced by chronic fat feeding, are both reversed by the single low-fat meal.[28]

Glucose Oversupply and LCACoAs

Chronic oversupply of glucose in the circulation produces a state of insulin resistance in muscle. It has been suggested that an increased glucose availability and oxidation might inhibit lipid oxidation[29] and set up a "glucose–fatty acid cycle in reverse".[30] Some possible mechanisms for this have been recently reviewed[31] and involve a possible buildup of lipid intermediates in the cytosol because of malonyl CoA inhibition of fatty acid oxidation. In the rat model of chronic glucose infusion, muscle becomes insulin-resistant, malonyl CoA increases, and presumably fatty acid oxidation is inhibited. In these animals, muscle triglyceride and LCACoAs are also significantly increased,[32,33] supporting the hypothesis that glucose- and lipid-induced insulin resistance in muscle may have several common features. Further, chronic glucose infusion leads to translocation and presumably activation of PKCε, which is another feature in common with fat-induced insulin resistance.[33] Thus, we believe there is evidence of a common scenario in the induction of muscle insulin resistance by either glucose or lipid oversupply. However, in the case of glucose oversupply, lipid metabolites are postulated to accumulate via reduced mitochondrial transfer of LCACoAs rather than via an excess muscle uptake of fatty acids.

Other Animal Models

Feeding rats a diet high in fructose produces increased circulating triglycerides and insulin resistance.[34] Preliminary investigations suggest that LCACoAs are also increased in muscle from fructose-fed rats (Ye and Iglesias, personal communication), and a recent report shows that LCACoAs are also elevated in muscle from insulin-resistant obese Zucker rats.[35]

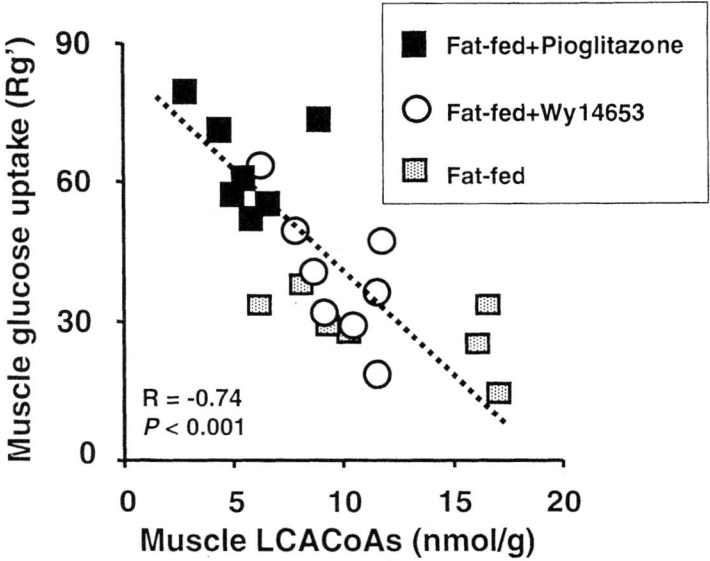

FIGURE 2. The correlation between LCACoA content and insulin-stimulated glucose uptake in muscle of fat-fed insulin-resistant rats, fat-fed rats treated with the PPARγ agonist pioglitazone, and fat-fed rats treated with the PPARα agonist WY-14653. (Adapted from reference 43.)

Pharmacological Manipulation of Muscle LCACoAs

PPAR agonists specifically target insulin resistance,[36–38] and the PPARγ receptor is principally (but not only) expressed in adipose tissue.[39] Activation of the PPARγ receptor by compounds such as pioglitazone, troglitazone, and rosiglitazone has been shown to have a key role in adipogenesis by increasing the gene expression of a number of key enzymes and/or transport proteins involved in adipose tissue metabolism.[36,38] Evidence suggests that these thiazolidinedione (TZD) compounds might principally affect muscle metabolism by acting in adipose tissue to favor retention of lipid.[40,41] Rosiglitazone, while enhancing insulin-mediated muscle glucose uptake and glycogen synthesis, also significantly reduces muscle triglyceride and DAG levels.[42] Recently, it has also been demonstrated that treatment of insulin-resistant fat-fed rats with the PPARγ agonist, pioglitazone, significantly reduces the LCACoA content of muscle and that LCACoA levels correlate with insulin-stimulated glucose uptake in the same muscles.[43] These decreases in LCACoA and DAG levels also coincide with reduced activation of the PKC isozymes that are associated with muscle lipid accumulation and insulin resistance in the high-fat-fed rat.[44]

PPARα is strongly expressed in liver, which is hence thought to be the principal target of action of PPARα agonists (e.g., fibrates, WY-14653). PPARα activation leads to expression of genes involved in β-oxidation of fatty acids, and it has been argued that the major action of the PPARα agonists is to reduce lipid by increasing the oxidation of fatty acids in liver. PPARα is also expressed in muscle, but its effects on lipid metabolism in this tissue are not clear at this stage. Although PPARα

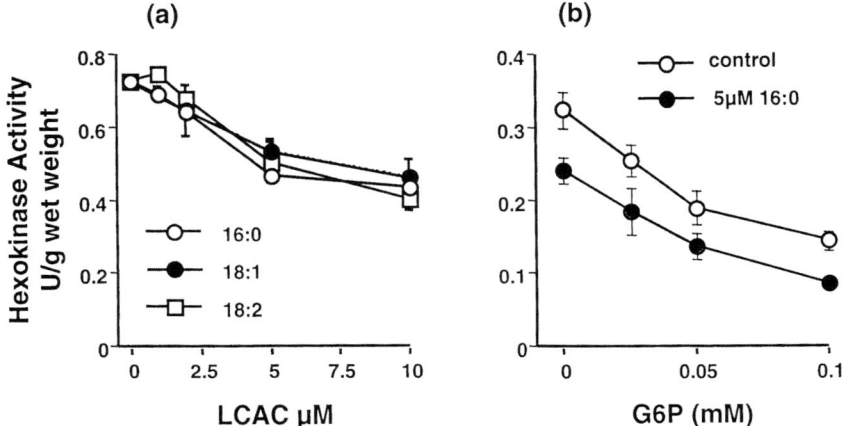

FIGURE 3. (a) The effect of increasing concentrations of LCACoA on the maximal activity of hexokinase in a crude homogenate of rat quadriceps muscle. (b) The additive effect of LCACoA and glucose-6-phosphate to inhibit hexokinase in a homogenate of human skeletal muscle. (Adapted from reference 45.)

agonists act via a different receptor to PPARγ agonists, one might predict that they should also lead to a reduction in muscle lipid accumulation and to amelioration of insulin resistance. This was recently demonstrated using the high-fat-fed rat model of insulin resistance.[43] In this study, treatment with the PPARα agonist, WY-14653, decreased LCACoA and triglyceride levels in muscle and increased insulin-stimulated glucose uptake in this tissue (FIG. 2).

EFFECTS OF LCACoAs ON INSULIN ACTION

Because of the consistent link between insulin action and LCACoAs in muscle, it is worth considering whether mechanisms involving actions of LCACoAs are responsible for the insulin resistance associated with increased lipid availability in muscle.

LCACoAs and Enzyme Regulation

The most directly relevant example of LCACoA inhibition of enzyme activity to muscle insulin resistance is the observation that LCACoAs can inhibit hexokinase, the first enzyme of glucose metabolism in muscle.[45] This inhibition was dependent on concentration and additive to the well-known inhibition of hexokinase by glucose-6-phosphate, suggesting that increased cytosolic LCACoAs could directly reduce glucose flux in muscle (FIG. 3). In liver, palmitoyl CoA can inhibit the activity of glycogen synthase by dissociation of enzyme subunits; however, it has never been confirmed that this also occurs in muscle.[46] LCACoAs also have been reported to influence activity of other liver enzymes, such as acyl CoA synthase,[47] glucokinase,[48] and glucose-6-phosphatase.[49] Palmitoyl CoA inhibits acetyl CoA carboxylase both directly[50] and indirectly.[51] Hormone-sensitive lipase, another key enzyme in lipid metabolism, is also regulated by LCACoAs directly and via LCACoA effects on AMP kinase.[52] While few of these effects have at this stage been related to the generation of insulin resistance in muscle, they do highlight the ability of LCACoAs to modulate the activity of enzymes involved in glucose and lipid metabolism.

LCACoAs, PKCs, and Signal Transduction

In insulin resistance, the phosphorylation state of many of the proteins involved in the insulin signaling pathway is reduced. These changes can be the result of reduced tyrosine phosphorylation of the initial step in the signaling pathway (insulin receptor) or interference with subsequent phosphorylation/dephosphorylation steps in the pathway. Some PKC subtypes can phosphorylate signaling proteins on serine or threonine residues and inhibit insulin signaling.[53] Several of the specific PKC subtypes are translocated and activated in muscle of insulin-resistant rats,[54] and increased intracellular availability of lipid species is a major stimulus for PKC activation.[54] There is some evidence that LCACoAs can directly modulate PKC activity, but this might also be indirect via LCACoAs driving the synthesis of specific diglycerides that are known activators of classical and novel classes of PKCs.[55] If this is the case, LCACoA content of muscle would be an indicator of the potential for increased DAG content and thus increased translocation and activation of the novel PKC subtypes.

LCACoAs and Ceramides

The sphingomyelin derivative, ceramide, is a signaling molecule that has been implicated in several physiological events, including inhibition of cell division and apoptosis. Ceramide levels are also elevated in insulin-responsive tissues of insulin-resistant animals,[56] and it has been postulated that agents that reduce glucose transport and inhibit insulin signaling (such as tumor necrosis factor [TNF]) may act by triggering the production of ceramide. Incubation of 3T3-L1 adipocytes with a membrane-permeable C2-ceramide inhibited insulin-stimulated glucose transport by 50% by reducing GLUT4 translocation and inhibiting the phosphorylation and activation of PKB (Akt).[57] This ceramide-induced inhibition of PKB occurs independently of any change in the ability of insulin to stimulate phosphorylation of IRS-1 or activate PI3-kinase and thus represents an independent pathway for the modification of insulin signaling. Recently, it was demonstrated[9] that incubation of C2C12 murine muscle cells with palmitate led to an increase in intracellular ceramide, a decrease in PKB phosphorylation, and a decreased ability of insulin to stimulate glycogen synthesis. The LCACoA palmitoyl CoA is one of the immediate precursors for the *de novo* generation of ceramide and this provides considerable evidence for the concept that changes in specific LCACoA species might have important roles in the generation of inhibitors (or stimulators) of specific signaling pathways.

LCACoAs and Membrane Trafficking

A major regulatory site for insulin-stimulated glucose uptake in muscle is the translocation of the GLUT4 glucose transporter from the endoplasmic reticulum to the plasma membrane. How this translocation of GLUT4 is achieved and what role defective translocation might play in fat diet–induced insulin resistance is still a matter of intense research effort. However, there is substantial evidence that LCACoAs are essential for either budding of vesicles from internal membranes or the fusion of vesicles with the plasma membrane.[58,59] The localization of acyl CoA synthase to GLUT4-containing vesicles suggests a role for reversible protein acylation in GLUT4 translocation,[60] and it is possible that the accumulation of a particular species of fatty acid or LCACoA may well interfere with translocation.

Fatty Acid and LCACoA Effects on Gene Regulation

It has been known for some time that increased lipid alters expression of specific genes, mostly related to pathways of lipid metabolism.[61,62] Recently, fatty acid–activated nuclear receptors have been identified that provide a mechanistic link between lipids and gene expression. The most important of these receptors are the peroxisome proliferator activated receptors (PPARs) and, in particular, PPARα and PPARγ, which are important for fat oxidation in the liver[63] and adipose tissue proliferation,[36,38] respectively. Different fatty acid types have different abilities to activate PPAR receptors,[64] and specific LCACoAs have been reported to activate or inhibit another type of transcription factor, hepatic nuclear factor-4α.[65] These effects on nuclear receptors provide another avenue by which the amount and activation states of fatty acids in a tissue might modulate gene expression and thus cellular metabolism.

IMPLICATIONS FOR HUMANS

A number of studies suggest that many of the interactions in animals discussed above may be operative in humans. Similarities in the time course for acute infusion of FFA inducing muscle insulin resistance in humans and rodents support this interpretation. There are also an increasing number of studies demonstrating lipid accumulation in muscle of insulin-resistant humans. These include studies based on muscle biopsy[66–68] as well as studies employing proton magnetic resonance spectroscopy.[69–72] The latter technique can distinguish signals from intramyocellular and extramyocellular lipid and thereby avoid problems associated with possible fat contamination of biopsied muscle tissue. A recent study using proton magnetic resonance spectroscopy reports that both acute infusion of fatty acids and 3 days of dietary manipulation can increase triglyceride content of muscle.[73] There is also evidence that lipids might accumulate in human muscle more by reduced fatty acid oxidation than by increased fatty acid uptake.[74,75] Reports of the measurement of LCACoAs in human muscle are few, but a recent study from our laboratory did show a negative correlation between insulin-mediated glucose disposal and the LCACoA content of muscle (obtained at the time of elective knee surgery) in a group of older men.[21] Regardless of possible differences in the way that muscle lipids might accumulate in humans and rodents, it is likely that subsequent lipid-induced biochemical[45] and signaling events are also operative in humans.

SUMMARY

There is ample evidence to support a hypothesis that a chronic oversupply of lipids is associated with lipid accumulation and the development of insulin resistance in muscle. There have also been considerable advances in understanding of possible mechanisms for lipid-induced muscle insulin resistance and the extent of alterations in insulin signaling and glucose metabolic parameters in models of lipid-induced insulin resistance. Although triglyceride content may predict the insulin sensitivity of muscle, changes in glucose utilization also appear to be closely related to the level of LCACoA in muscle. This important intermediate in the metabolism of fatty acids can have direct effects on glucose utilization by altering enzyme activity or by acting as an immediate precursor for other lipid intermediates that activate or inhibit key components of the insulin signaling pathway. The extent to which manipulation of LCACoA levels can affect glucose homeostasis remains to be determined.

REFERENCES

1. SHAFRIR, E. 1992. Animal models of non-insulin-dependent diabetes. Diabetes Metab. Rev. **8:** 179–208.
2. MCGARRY, J.D. 1992. What if Minkowski had been ageusic? An alternative angle on diabetes. Science **258:** 766–770.
3. BODEN, G. 1997. Role of fatty acids in the pathogenesis of insulin resistance and NIDDM. Diabetes **46:** 3–10.
4. KRAEGEN, E.W., D.G.P. CAREY & L.V. CAMPBELL. 1997. Effects of lipids on blood glucose regulation and insulin action. *In* Clinical Research in Diabetes and Obesity. Part 1: Methods, Assessment, and Metabolic Regulation, pp. 305–320. Humana Press. Totowa, New Jersey.
5. RANDLE, P.J. 1998. Regulatory interactions between lipids and carbohydrates: the glucose fatty acid cycle after 35 years. Diabetes Metab. Rev. **14:** 263–283.
6. CORTRIGHT, R.N., D.M. MUOIO & G.L. DOHM. 1997. Skeletal muscle lipid metabolism: a frontier for new insights into fuel homeostasis. Nutr. Biochem. **8:** 228–245.
7. RASMUSSEN, B.B. & R.R. WOLFE. 1999. Regulation of fatty acid oxidation in skeletal muscle. Annu. Rev. Nutr. **19:** 463–484.
8. RANDLE, P.J., P.B. GARLAND, C.N. HALES & E.A. NEWSHOLME. 1963. The glucose fatty-acid cycle: its role in insulin sensitivity and the metabolic disturbances of diabetes mellitus. Lancet **1:** 785–789.
9. SCHMITZ-PEIFFER, C., D.L. CRAIG & T.J. BIDEN. 1999. Ceramide generation is sufficient to account for the inhibition of the insulin-stimulated PKB pathway in C2C12 skeletal muscle cells pretreated with palmitate. J. Biol. Chem. **274:** 24202–24210.
10. CHALKLEY, S., M. HETTIARACHI, D.J. CHISHOLM & E.W. KRAEGEN. 1998. Five hour fatty acid elevation increases muscle lipids and impairs glycogen synthesis in the rat. Metabolism **47:** 1121–1126.
11. GRIFFIN, M.E., M.J. MARCUCCI, G.W. CLINE *et al.* 1999. Free fatty acid–induced insulin resistance is associated wth activation of protein kinase C theta and alterations in the insulin signaling cascade. Diabetes **48:** 1270–1274.
12. PARK, J.Y., C.H. KIM, S.K. HONG *et al.* 1998. Effects of FFA on insulin-stimulated glucose fluxes and muscle glycogen synthase activity in rats. Am. J. Physiol. **275:** E338–E344.
13. JENKINS, A.B., L.H. STORLIEN, D.J. CHISHOLM & E.W. KRAEGEN. 1988. Effects of non-esterified fatty acid availability on tissue-specific glucose utilization in rats *in vivo*. J. Clin. Invest. **82:** 293–299.
14. KRUSZYNSKA, Y.T., J.G. MCCORMACK & N. MCINTYRE. 1991. Effects of glycogen stores and non-esterified fatty acid availability on insulin-stimulated glucose metabolism and tissue pyruvate dehydrogenase activity in the rat. Diabetologia **34:** 205–211.

15. BODEN, G., F. JADALI, J. WHITE et al. 1991. Effects of fat on insulin stimulated carbohydrate metabolism in normal men. J. Clin. Invest. **88:** 960–966.
16. BODEN, G., X.H. CHEN, J. ROSNER & M. BARTON. 1995. Effects of a 48-h fat infusion on insulin secretion and glucose utilization. Diabetes **44:** 1239–1242.
17. KELLEY, D.E., M. MOKAN, J.A. SIMONEAU & L.J. MANDARINO. 1993. Interaction between glucose and free fatty acid metabolism in human skeletal muscle. J. Clin. Invest. **92:** 91–98.
18. KIM, J.K. & J.H. YOUN. 1997. Prolonged suppression of glucose metabolism causes insulin resistance in rat skeletal muscle. Am. J. Physiol. **272:** E288–E296.
19. JUCKER, B.M., A.J.M. RENNINGS, G.W. CLINE & G.I. SHULMAN. 1997. C-13 and P-31 NMR studies on the effects of increased plasma free fatty acids on intramuscular glucose metabolism in the awake rat. J. Biol. Chem. **272:** 10464–10473.
20. OAKES, N.D., K.S. BELL, S.M. FURLER et al. 1997. Diet-induced muscle insulin resistance in rats is ameliorated by acute dietary lipid withdrawal or a single bout of exercise—parallel relationship between insulin stimulation of glucose uptake and suppression of long-chain fatty acyl-CoA. Diabetes **46:** 2022–2028.
21. ELLIS, B.A., A. POYNTEN, A.J. LOWY et al. 2000. Long-chain acyl-CoA esters as indicators of lipid metabolism and insulin sensitivity in rat and human muscle. Am. J. Physiol. **279:** E554–E560.
22. KRAEGEN, E.W., P.W. CLARK, A.B. JENKINS et al. 1991. Development of muscle insulin resistance after liver insulin resistance in high-fat-fed rats. Diabetes **40:** 1397–1403.
23. STORLIEN, L.H., A.B. JENKINS, D.J. CHISHOLM et al. 1991. Influence of dietary fat composition on development of insulin resistance in rats. Diabetes **40:** 280–289.
24. KIM, J.K., J.K. WI & J.H. YOUN. 1996. Metabolic impairment precedes insulin resistance in skeletal muscle during high-fat feeding in rats. Diabetes **45:** 651–658.
25. OAKES, N.D., G.J. COONEY, S. CAMILLERI et al. 1997. Mechanisms of liver and muscle insulin resistance induced by chronic high-fat feeding. Diabetes **46:** 1768–1774.
26. CHEN, M.T., L.N. KAUFMAN, T. SPENNETTA & E. SHRAGO. 1992. Effects of high fat–feeding to rats on the interrelationship of body weight, plasma insulin, and fatty acyl-coenzyme-A esters in liver and skeletal muscle. Metabolism **41:** 564–569.
27. THOMPSON, A.L., M.Y-C. LIM-FRASER, E.W. KRAEGEN & G.J. COONEY. 2000. Specific fatty acids have distinct effects on insulin-stimulated glucose metabolism in soleus muscle *in vitro*. Am. J. Physiol. **279:** E577–E584.
28. BELL, K.S., M. LIM-FRASER, G. COONEY & E.W. KRAEGEN. 1999. Muscle protein kinase B phosphorylation is restored when fat-induced insulin resistance is reversed. Diabetologia **42**(suppl. 1): A174.
29. SIDOSSIS, L.S., C.A. STUART, G.I. SHULMAN et al. 1996. Glucose plus insulin regulate fat oxidation by controlling the rate of fatty acid entry into the mitochondria. J. Clin. Invest. **98:** 2244–2250.
30. WOLFE, R.R. 1998. Metabolic interactions between glucose and fatty acids in humans. Am. J. Clin. Nutr. **67:** S519–S526.
31. RUDERMAN, N.B., A.K. SAHA, D. VAVVAS & L.A. WITTERS. 1999. Malonyl-CoA, fuel sensing, and insulin resistance. Am. J. Physiol. **276:** E1–E18.
32. LAYBUTT, D.R., D.J. CHISHOLM & E.W. KRAEGEN. 1997. Specific adaptations in muscle and adipose tissue in response to chronic systemic glucose oversupply in rats. Am. J. Physiol. **273:** E1–E9.
33. LAYBUTT, D.R., C. SCHMITZ-PEIFFER, A.K. SAHA et al. 1999. Muscle lipid accumulation and protein kinase C activation in the insulin-resistant chronically glucose-infused rat. Am. J. Physiol. **277:** E1070–E1076.
34. BEZARRA, R.M., M. UENO, M.S. SILVA et al. 2000. A high fructose diet affects the early steps of insulin action in muscle and liver of rats. J. Nutr. **130:** 1531–1535.
35. FRANCH, J., J. KNUDSEN, B.A. ELLIS et al. 2002. Acyl coenzyme A binding protein expression is fibre type specific and elevated in muscles from the obese insulin resistant Zucker rat. Diabetes **51:** 152–158.
36. SPIEGELMAN, B.M. 1998. PPAR-gamma—adipogenic regulator and thiazolidinedione receptor. Diabetes **47:** 507–514.
37. SALTIEL, A.R. & J.M. OLEFSKY. 1996. Thiazolidinediones in the treatment of insulin resistance and type II diabetes. Diabetes **45:** 1661–1669.

38. SMITH, S.A. 1997. Peroxisome proliferator-activated receptors and the regulation of lipid oxidation and adipogenesis. Biochem. Soc. Trans. **25:** 1242–1248.
39. LEHMANN, J.M., L.B. MOORE, T.A. SMITHOLIVER *et al.* 1995. An antidiabetic thiazolidinedione is a high affinity ligand for peroxisome proliferator-activated receptor gamma (PPAR-gamma). J. Biol. Chem. **270:** 12953–12956.
40. OAKES, N.D., C.J. KENNEDY, A.B. JENKINS *et al.* 1994. A new antidiabetic agent, BRL 49653, reduces lipid availability and improves insulin action and glucoregulation in the rat. Diabetes **43:** 1203–1210.
41. OAKES, N.D., P.D. THALEN, S.M. JACINTO & B. LJUNG. 2001. Thiazolidinediones increase plasma-adipose tissue FFA exchange capacity and enhance insulin-mediated control of systemic FFA availability. Diabetes **50:** 1158–1165.
42. OAKES, N.D., S. CAMILLERI, S.M. FURLER *et al.* 1997. The insulin sensitizer, BRL 49653, reduces systemic fatty acid supply and utilization and tissue lipid availability in the rat. Metabolism **46:** 935–942.
43. YE, J.M., P.J. DOYLE, M.A. IGLESIAS *et al.* 2001. Peroxisome proliferator-activated receptor (PPAR)–alpha activation lowers muscle lipids and improves insulin sensitivity in high fat–fed rats: comparison with PPAR-gamma activation. Diabetes **50:** 411–417.
44. SCHMITZ-PEIFFER, C., N.D. OAKES, C.L. BROWNE *et al.* 1997. Reversal of chronic alterations of skeletal muscle protein kinase C from fat-fed rats by BRL-49653. Am. J. Physiol. **273:** E915–E921.
45. THOMPSON, A.L. & G.J. COONEY. 2000. Acyl CoA inhibition of hexokinase in rat and human skeletal muscle is a potential mechanism of lipid-induced insulin resistance. Diabetes **49:** 1761–1765.
46. WITITSUWANNAKUL, D. & K.H. KIM. 1977. Mechanism of palmitoyl coenzyme A inhibition of liver glycogen synthase. J. Biol. Chem. **252:** 7812–7817.
47. FAERGEMAN, N.J. & J. KNUDSEN. 1997. Role of long-chain fatty acyl-CoA esters in the regulation of metabolism and cell signalling. Biochem. J. **323:** 1–12.
48. TIPPETT, P.S. & K.E. NEET. 1982. An allosteric model for the inhibition of glucokinase by long acyl coenzyme A. J. Biol. Chem. **257:** 12846–12852.
49. FULCERI, R., A. GAMBERUCCI, H.M. SCOTT *et al.* 1995. Fatty acyl-CoA esters inhibit glucose-6-phosphatase in rat liver microsomes. Biochem. J. **307:** 391–397.
50. NIKAWA, J-I., T. TANABE, H. OGIWARA *et al.* 1979. Inhibitory effects of long-chain acyl coenzyme A analogues on liver acetyl coenzyme A carboxylase. FEBS Lett. **102:** 223–226.
51. CARLING, D., V.A. ZAMMIT & D.G. HARDIE. 1987. A common bicyclic protein kinase cascade inactivates the regulatory enzymes of fatty acid and cholesterol biosynthesis. FEBS Lett. **223:** 217–222.
52. JEPSON, C.A. & S.J. YEAMAN. 1992. Inhibition of hormone-sensitive lipase by intermediary lipid metabolites. FEBS Lett. **310:** 197–200.
53. DANIELSEN, A.G., F. LIU, Y. HOSOMI *et al.* 1995. Activation of protein kinase C alpha inhibits signalling by members of the insulin receptor family. J. Biol. Chem. **270:** 21600–21605.
54. SCHMITZ-PEIFFER, C., C.L. BROWNE, N.D. OAKES *et al.* 1997. Alterations in the expression and cellular localization of protein kinase C isozymes epsilon and theta are associated with insulin resistance in skeletal muscle of the high-fat-fed rat. Diabetes **46:** 169–178.
55. BRINDLEY, D.N. 1984. Intracellular translocation of phosphatide phosphohydrolase and its possible role in the control of glycerolipid synthesis. Prog. Lipid Res. **23:** 115–123.
56. TURINSKY, J., D.M. O'SULLIVAN & B.P. BAYLY. 1990. 1,2-Diacylglycerol and ceramide levels in insulin-resistant tissues of the rat *in vivo*. J. Biol. Chem. **265:** 16880–16885.
57. SUMMERS, S.A., L.A. GARZA, H. ZHOU & M.J. BIRNBAUM. 1998. Regulation of insulin-stimulated glucose transporter GLUT4 translocation and Akt kinase activity by ceramide. Mol. Cell. Biol. **18:** 5457–5464.
58. GLICK, B. & J. ROTHMAN. 1987. Possible role for fatty acyl-coenzyme A in intracellular protein transport. Nature **326:** 309–312.
59. PFANNER, N., B. GLICK, S. ARDEN & J. ROTHMAN. 1990. Fatty acylation promotes fusion of transport vesicles with Golgi cisternae. J. Cell Biol. **110:** 955–961.

60. SLEEMAN, M., N. DONEGAN, R. HELLER-HARRISON et al. 1998. Association of acyl-CoA synthetase-1 with GLUT4-containing vesicles. J. Biol. Chem. **273:** 3132–3135.
61. JUMP, D.B. & S.D. CLARKE. 1999. Regulation of gene expression by dietary fat. Annu. Rev. Nutr. **19:** 63–90.
62. RACLOT, T., R. GROSCOLAS, D. LANGIN & P. FERRE. 1997. Site-specific regulation of gene expression by n-3 polyunsaturated fatty acids in rat white adipose tissues. J. Lipid Res. **38:** 1963–1972.
63. LEMBERGER, T., B. DESVERGNE & W. WAHLI. 1996. Peroxisome proliferator-activated receptors: a nuclear receptor signaling pathway in lipid physiology. Annu. Rev. Cell Dev. Biol. **12:** 335–363.
64. KLIEWER, S.A., S.S. SUNDSETH, S.A. JONES et al. 1997. Fatty acids and eicosanoids regulate gene expression through direct interactions with peroxisome proliferator-activated receptors alpha and gamma. Proc. Natl. Acad. Sci. U.S.A. **94:** 4318–4323.
65. HERTZ, R., J. MAGENHELM, I. BERMAN & J. BAR-TANA. 1998. Fatty acyl-CoA thioesters are ligands of hepatic nuclear factor-4α. Nature **392:** 512–516.
66. FALHOLT, K., I. JENSEN, S. JENSEN et al. 1988. Carbohydrate and lipid metabolism of skeletal muscle in type 2 diabetic patients. Diabetic Med. **5:** 27–31.
67. PAN, D.A., S. LILLIOJA, A.D. KRIKETOS et al. 1997. Skeletal muscle triglyceride levels are inversely related to insulin action. Diabetes **46:** 983–988.
68. PHILLIPS, D.I.W., S. CADDY, V. ILIC et al. 1996. Intramuscular triglyceride and muscle insulin sensitivity—evidence for a relationship in nondiabetic subjects. Metabolism **45:** 947–950.
69. BOESCH, C., J. SLOTBOOM, H. HOPPELER & R. KREIS. 1997. In vivo determination of intra-myocellular lipids in human muscle by means of localized H-1-MR-spectroscopy. Magn. Reson. Med. **37:** 484–493.
70. SZCZEPANIAK, L.S., E.E. BABCOCK, F. SCHICK et al. 1999. Measurement of intracellular triglyceride stores by H spectroscopy: validation in vivo. Am. J. Physiol. **276:** E977–E989.
71. KRSSAK, M., K.F. PETERSEN, A. DRESNER et al. 1999. Intramyocellular lipid concentrations are correlated with insulin sensitivity in humans: a H-1 NMR spectroscopy study. Diabetologia **42:** 113–116.
72. JACOB, S., J. MACHANN, K. RETT et al. 1999. Association of increased intramyocellular lipid content with insulin resistance in lean nondiabetic offspring of type 2 diabetic subjects. Diabetes **48:** 1113–1119.
73. BACHMANN, O.P., D.K. DAHL, K. BRETCHEL et al. 2001. Effects of intravenous and dietary lipid challenge on intramyocellular lipid content and the relation with insulin sensitivity in humans. Diabetes **50:** 2579–2584.
74. GOODPASTER, B.H. & D.E. KELLEY. 1998. Role of muscle in triglyceride metabolism. Curr. Opin. Lipidol. **9:** 231–236.
75. TURPEINEN, A.K., T.O. TAKALA, P. NUUTILA et al. 1999. Impaired free fatty acid uptake in skeletal muscle, but not in myocardium in patients with impaired glucose tolerance—studies with PET and 14(R,S)-[F-18]fluoro-6-thia-heptadecanoic acid. Diabetes **48:** 1245–1250.

Eicosanoids and the Regulation of Cardiac Glucose Transport

OLAF DRANSFELD,[a] IRINI RAKATZI,[a] SHLOMO SASSON,[b] AND JÜRGEN ECKEL[a]

[a]*Department of Clinical Biochemistry and Pathobiochemistry, German Diabetes Research Institute, Düsseldorf, Germany*

[b]*Hebrew University of Jerusalem, Jerusalem, Israel*

ABSTRACT: Intact actin microfilaments are necessary for insulin-regulated GLUT4 translocation from intracellular pools to the plasma membrane. Products of the lipoxygenase (LO) pathway were shown to be implicated in the regulation of actin cytoskeleton rearrangement. The aim of this study was to examine the role of these LO products for cardiac insulin signaling and glucose uptake, GLUT4 translocation, and actin-based cytoskeleton structure. Exposure of cardiomyocytes to esculetin or NDGA, two structurally different LO inhibitors, induced a complete inhibition of insulin-stimulated glucose uptake, whereas control cells showed a threefold stimulation by insulin. Addition of 12(S)-HETE rendered the NDGA-treated cells insulin-sensitive. Early insulin signaling was not changed in cells exposed to LO inhibitors. Cell surface biotinylation of control cells showed a twofold increase of GLUT4 at the cell surface after insulin stimulation. In contrast, the LO inhibitors induced a complete inhibition of insulin-stimulated GLUT4 translocation. Labeling of the F-actin cytoskeleton revealed a prominent disassembly of actin fibers in cells exposed to the LO inhibitors. In conclusion, we show here that products of the LO reaction participate in the organization of the actin network in ventricular cardiomyocytes. Inhibition of LO blocks GLUT4 translocation without affecting insulin signaling events. These data suggest that products of the LO reaction participate in the regulation of glucose transport by contribution to a rearrangement of actin cytoskeletal elements.

KEYWORDS: insulin; glucose uptake; GLUT4; lipoxygenase (LO); actin

INTRODUCTION

At least three major mammalian lipoxygenases (LO) oxygenate carbon 5, 12, or 15 in arachidonic acid, thus termed 5-, 12-, and 15-LO, generating 5-, 12-, and 15-hydroxyeicosatetraenoic acids (HETEs).[1,2] Two forms of 12-LO were identified in mammalian cells: the platelet type and the leukocyte type.[3,4] So far, the biological function and significance of the different LO isoforms have remained unclear.[5] 12-LO

Address for correspondence: Prof. Dr. Jürgen Eckel, German Diabetes Research Institute, Auf'm Hennekamp 65, D-40225 Düsseldorf, Germany. Voice: +49 211 3382561; fax: +49 211 3382582.

eckel@uni-duesseldorf.de

was found to be expressed in the heart[1] and 12(S)-HETE represents the main LO metabolite in ventricular cardiomyocytes.[6]

Interestingly, 12(S)-HETE induces the phosphorylation of actin fibers, leading to an increased actin filament content and enhancement of actin polymerization.[7,8] Involvement of PKC in this process has been suggested.[9] Furthermore, eicosanoids may also bind directly to actin fibers.[10,11] So far, the involvement of LO products in the organization of the actin network in the cardiomyocyte and the potential functional implications have not been investigated. This is of special interest since 12(S)-HETE has been shown to play a role in the cardioprotective effect of ischemic preconditioning.[12–14]

Evidence has been provided for the involvement of actin cytoskeletal elements in insulin-regulated glucose transport and GLUT4 translocation. An intact actin network was shown to be essential for the insulin-mediated exocytosis of GLUT4 vesicles to the plasma membrane in both adipocytes and muscle cells.[15–18] Thus, factors involved in the rearrangement of the cytoskeleton may contribute to the regulation of glucose transport and may additionally modify the insulin sensitivity of this process. In the light of these considerations, we have now used adult cardiomyocytes[19–22] in order to assess the potential contribution of the LO pathway in the rearrangement of actin cytoskeletal elements and the functional implications for GLUT4 trafficking in these cells.

METHODS

Isolation and Culture of Ventricular Cardiomyocytes

Male Wistar rats weighing 260–310 g were used throughout the experiments. Ca^{2+}-tolerant myocytes were isolated by perfusion of the heart with collagenase and either used immediately or kept in primary culture as previously described.[23] The final cell suspension was washed three times with N-[2-hydroxyethyl]piperazine-N'-[2-ethane-sulfonic acid] (HEPES) buffer (composition: NaCl 130 mM, KCl 4.8 mM, KH_2PO_4 1.2 mM, HEPES 25 mM, glucose 5 mM, bovine serum albumin 20 g/L, pH 7.4, equilibrated with oxygen) and incubated in silicone-treated Erlenmeyer flasks in a rotating shaking water bath at 37°C. For overnight primary culture of cardiomyocytes, culture flasks were precoated with laminin (about 2 µg/cm^2) for at least 20 min before seeding.

Assay of 2-Deoxyglucose Uptake and 3-O-Methylglucose Transport

Determinations of 2-deoxyglucose (2-DOG) uptake were performed at 37°C in monolayer (2×10^5 cells per well) in serum-free culture medium containing 7.8 mM glucose as described previously.[24,25] Transport experiments were performed at 37°C in HEPES buffer containing $MgCl_2$ (1 mM) and $CaCl_2$ (1 mM). Carrier-mediated glucose transport was determined using a 10-s assay period and L-[^{14}C]glucose in order to correct for simple diffusion, as described in earlier reports from this laboratory.[19] Significance of reported differences was evaluated using the null hypothesis and t statistics for paired data.

Cell Surface Biotinylation and Immunoblotting

Cell surface biotinylation was adapted from previously published procedures.[26,27] About 10^6 cells in 100-mm culture plates were washed with ice-cold PBS and incubated with 0.5 mg/mL sulfo-NHS-LC biotin in PBS for 30 min at 4°C. The reaction was then stopped by rinsing the plates three times with 15 mM glycine in ice-cold PBS. The cells were then collected and solubilized for 30 min at 4°C. The supernatant was separated by centrifugation and mixed with 50 µL of streptavidin-agarose beads. The suspension was gently mixed overnight at 4°C and the beads were sedimented by centrifugation. After washing, the final pellet was resuspended in 50 µL of urea buffer (8.0 M urea, 2% (v/v) SDS, 100 mM Tris, pH 6.8) and incubated for 30 min at 37°C. The supernatant was mixed with Laemmli buffer and then frozen at −70°C until use. Protein content in the lysates was determined by the BCA protein assay using bovine serum albumin standard dissolved in the same buffer as used in the lysate preparation. Protein samples were separated by SDS-PAGE and immunoblotting was conducted as outlined in previous reports from this lab.[28,29]

Confocal Laser Microscopy

Cells on coverslips were washed twice with PBS. After fixation by 4% paraformaldehyde in PBS for 10 min at RT, the cells were covered with 0.1% Triton-X 100 in PBS for 10 min, washed with PBS, and covered with bovine serum albumin 1% in PBS for 1 h. For single staining of actin fibers, cells were incubated for 16 h at 6°C in a wet chamber with FITC-phalloidin (1:500). After two final washing steps, the cells were covered with fluorescence mounting medium and a coverslip. Immunostained cardiomyocytes on coverslips were analyzed using the Leica TCS-NT confocal laser scanning system with an argon-krypton laser on a Leica DM IRB inverted microscope.

RESULTS

Expression of 12-LO in adult cardiomyocytes was verified using Western blotting analysis. 12-LO was detected at a molecular mass of ~70–75 kDa. 12-LO was also visualized using confocal microscopy; however, the enzyme could not be detected colocalized to the actin network. In this study, we used two structurally different inhibitors of LO—NDGA and esculetin. In order to verify inhibition of LO, we used an ELISA specific for 12(S)-HETE, the product of the 12-LO and the 12/15-LO. Freshly isolated cardiomyocytes contain 40–50 ng of 12(S)-HETE per mg of protein. Both inhibitors reduced the cellular content of 12(S)-HETE by 70–80%.

Cardiomyocytes were then cultured overnight with or without the LO inhibitor, esculetin (100 µM), followed by an incubation in the absence or presence of insulin. As presented in FIGURE 1, insulin produced a 2- to 3-fold increase in 2-DOG uptake under control conditions. This effect was completely abrogated in the presence of esculetin, without affecting the basal transport rate. The role of LO for insulin action was studied in more detail using freshly isolated cardiomyocytes and the LO inhibitor, NDGA. Cells were incubated with 0.1% DMSO as control or increasing concentrations of the LO inhibitor, NDGA, followed by an incubation for 5 min with or without insulin. 3-O-Methylglucose transport was then determined and, in control

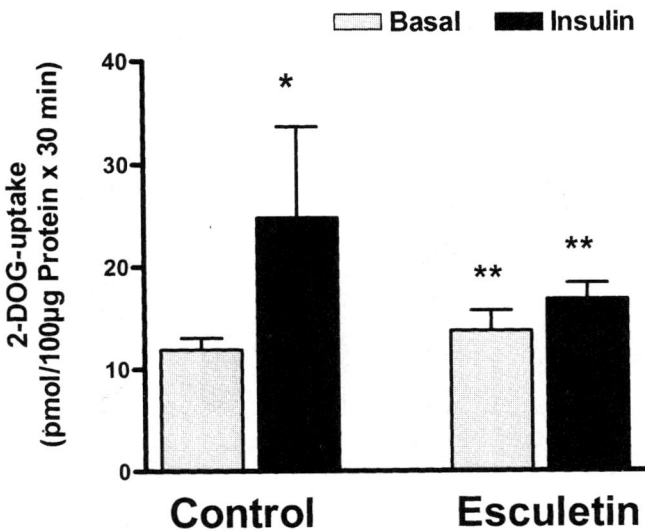

FIGURE 1. Effect of esculetin on glucose uptake in cultured ventricular cardiomyocytes. Ventricular cardiomyocytes were kept in culture for 16 h in the absence or presence of esculetin (100 µM). Basal uptake and insulin-stimulated uptake of 2-DOG were then determined as detailed in METHODS. Data are mean values ± SEM of four separate experiments. *Significantly different from basal control ($p < 0.05$); **not significantly different from basal ($p > 0.05$).

cells, a 4- to 5-fold increase in glucose transport in response to insulin could be detected (FIG. 2). In the NDGA-treated cells, basal glucose transport remained unchanged compared to control cells. Insulin-induced glucose transport was not altered in cells exposed to 0.5 µM NDGA. However, at 5 µM NDGA, the insulin effect was significantly reduced to about one-half, with a complete loss of insulin action on glucose transport at 50 µM (FIG. 2). In order to assess the specificity of this inhibitory effect, cells exposed to NDGA (50 µM) were additionally treated with exogenous 12(S)-HETE (3 µM). Under these conditions, insulin action on glucose transport could be restored to about 70% of that observed in control cells (FIG. 2).

Additional experiments were then performed to elucidate if the effect of the 12-LO inhibitors could be attributed to an inhibition of GLUT4 translocation. Cardiomyocytes were therefore treated with esculetin and insulin, followed by biotinylation of the cell surface glucose transporters, as described in METHODS. Cells were solubilized and proteins were analyzed by SDS-PAGE, transferred to PVDF membranes, and immunoblotted with a GLUT4-specific antiserum (FIG. 3). In control cells, insulin stimulation caused an increase of GLUT4 at the cell surface to 214±39% of basal ($n = 3$; $p < 0.05$). Exposure to esculetin did not modify the total cell content of GLUT4. However, after esculetin treatment, insulin was completely unable to increase the abundance of GLUT4 at the surface of the cardiomyocytes (FIG. 3). We thus conclude that the reduction of 12(S)-HETE in cardiac cells induces the complete loss of insulin-regulated GLUT4 translocation.

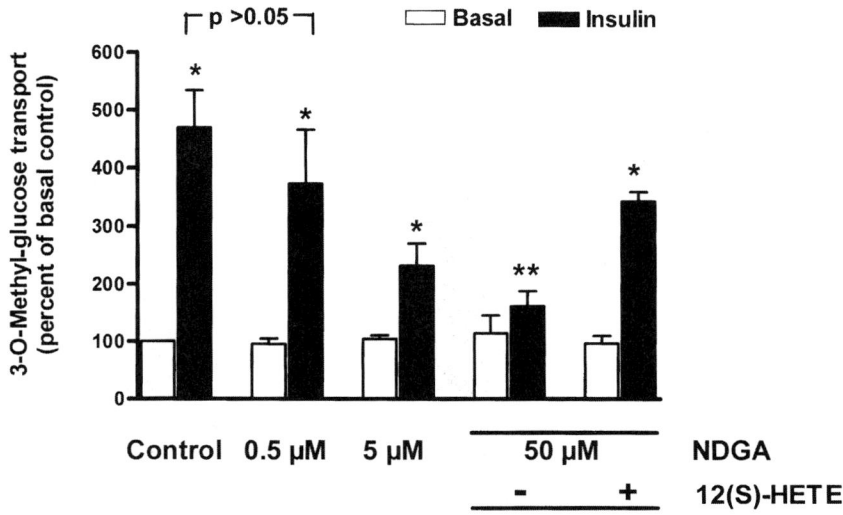

FIGURE 2. Effect of NDGA on glucose transport in freshly isolated cardiomyocytes. Freshly isolated cardiomyocytes were incubated for 1 h with increasing concentrations of NDGA (0.5–50 µM) or 0.1% DMSO as a control in HEPES buffer, followed by an incubation for 5 min with or without 100 nM insulin. Cells exposed to 50 µM NDGA were additionally treated with or without 12(S)-HETE (3 µM). 3-O-Methylglucose transport was then determined as outlined in METHODS. Results are means ± SEM of four experiments. *$p < 0.05$ compared with basal control; **not significantly different from basal control. Reproduced with permission from reference 28 (© the Biochemical Society).

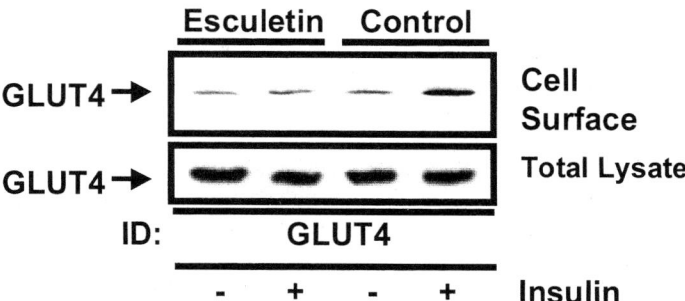

FIGURE 3. Immunodetection of GLUT4 in total lysates and cell surface biotinylated fractions of esculetin-treated cardiomyocytes. Cells were cultured for 16 h with esculetin (100 µM) and incubated further in the absence or presence of 1 µM insulin for 45 min. The cells were then treated with sulfo-NHS-LC biotin and solubilized as described in METHODS. The biotinylated lysates were processed for affinity purification or directly used for Western blotting. Total lysates or cell surface biotinylated fractions were analyzed by SDS-PAGE and immunoblotted with an anti-GLUT4 antibody. Representative blots from three independent experiments are shown. Reproduced with permission from reference 28 (© the Biochemical Society).

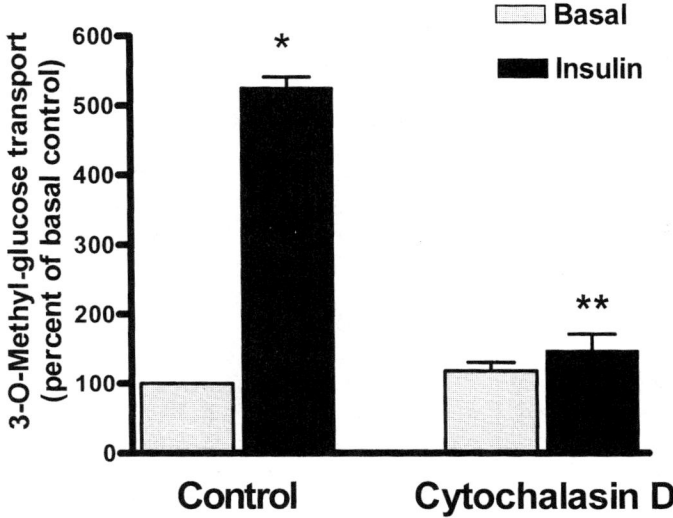

FIGURE 4. Effect of cytochalasin D on glucose transport. Freshly isolated cardiomyocytes were treated with cytochalasin D (1 µM) for 1 h, followed by a 5-min incubation with insulin (100 nM). 3-O-Methylglucose transport was then determined as described in FIGURE 2. Data are mean values ± SEM of four separate experiments. *Significantly different from basal ($p < 0.05$); **not significantly different from basal.

It may be argued that LO inhibition modifies insulin signaling pathways upstream of the GLUT4 translocation machinery. However, we did not detect any effect of NDGA treatment on the insulin-induced tyrosine phosphorylation of IRS-1 and -2, on the association of the p85 subunit of PI3-kinase with IRS-1, and on the insulin-dependent phosphorylation of the serine/threonine kinase Akt.

Actin filaments are critically involved in GLUT4 translocation in both muscle and adipose tissue.[15,30,31] To assess the involvement of actin network organization in GLUT4 translocation in cardiomyocytes, we used the actin fiber disrupter, cytochalasin D. Freshly isolated cardiomyocytes were incubated for 1 h with cytochalasin D (1 µM) and subsequently stimulated with insulin (FIG. 4). As can be seen from the data, the insulin response of glucose transport was completely abrogated in cells treated with cytochalasin D. Attempts were then made to correlate the effects of cytochalasin D and 12-LO inhibition with actin network organization using confocal microscopy. Cells were incubated with cytochalasin D or the LO inhibitors and were fixed, permeabilized, and stained for actin filaments by FITC-phalloidin. A dense actin network tightly separated into parallel fibers was detected in control cells. Exposure to LO inhibitors or cytochalasin D did not change the general actin filament pattern or the typical rod shape compared to untreated cells. In control cells, extremely parallel fibers can clearly be seen. In contrast, cells exposed to esculetin or NDGA revealed a prominent disassembly of actin fibers in a fashion similar to the effect induced by cytochalasin D. These data clearly indicate that LO inhibition leads to a marked disassembly of actin filaments most likely resulting in a loss of insulin-stimulated GLUT4 translocation.

DISCUSSION

We show here that inhibition of the LO pathway completely abrogates insulin-induced glucose uptake and GLUT4 translocation. The specificity of this effect is demonstrated by (i) the use of structurally different LO inhibitors that provide the same results with both freshly isolated and cultured cardiomyocytes, (ii) an unaltered basal rate of glucose transport, and (iii) the ability of 12(S)-HETE to restore insulin responsiveness in NDGA-treated cells. Insulin signaling pathways upstream of the glucose transporter remained undisturbed by LO inhibition, indicating that the LO products must be acting at the level of the GLUT4 translocation process itself.

Several studies suggest that 12(S)-HETE contributes to the regulation of actin network organization.[7,8,32] This may involve either binding to actin cytoskeletal elements[10,11] or the phosphorylation of major cytoskeletal proteins, like actin.[7,8] Using confocal microscopy, we show here that inhibition of the 12-LO pathway coupled to a strong reduction in the cellular content of 12(S)-HETE leads to a prominent disassembly of the actin network in ventricular cardiomyocytes in a fashion similar to the cytochalasin D–induced actin filament disassembly. Hypoxia or ischemia has been reported to stimulate the LO pathway, and the enhanced production of 12(S)-HETE has been suggested to be involved in the cardioprotection related to ischemic preconditioning.[13,14] Direct evidence for this assumption was recently reported when using 12-LO-deficient mice.[33] It may be speculated that the enhanced production of 12(S)-HETE observed in the ischemic heart may serve to retain proper actin network organization in the cardiomyocyte, providing the basis for undisturbed GLUT4 trafficking.

Most interestingly, it was recently reported that the 12/15-LO translocates from the cytosol to the plasma membrane and that this enzyme plays a major role in the local control of actin polymerization.[34] The rearrangement of the actin network at or near the plasma membrane is certainly of central importance for the docking and fusion of GLUT4 vesicles with the plasma membrane.[15,16] Recent studies in adipocytes and L6 myotubes revealed that insulin-regulated glucose uptake and GLUT4 translocation were markedly disturbed by inducing actin network disassembly using cytochalasin D.[16–18] Our data confirm this relationship for the heart and show that inhibition of the LO pathway produces a very similar effect. We therefore propose a model in which 12(S)-HETE may function to stabilize the actin network organization, thus contributing to the regulation of cardiac glucose transport. Additional work will be needed to study the relative contribution of different 12(S)-HETE levels to cytoskeletal rearrangement and the regulation of cardiac LO activity, including the possibility of insulin-mediated translocation of this enzyme to the plasma membrane. This work is currently under way.

ACKNOWLEDGMENTS

This work was supported by the Ministerium für Wissenschaft und Forschung des Landes Nordrhein-Westfalen, the Bundesministerium für Gesundheit, EU COST Action B5 and B17, EU BIOMED Concerted Action 3084, and Deutscher Akademischer Austauschdienst (DAAD). The secretarial assistance of Birgit Hurow is gratefully acknowledged.

REFERENCES

1. FREIRE-MOAR, J., A. ALAVI-NASSAB, M. NG et al. 1995. Cloning and characterization of a murine lipoxygenase. Biochim. Biophys. Acta **1254:** 112–116.
2. BRASH, A.R. 1999. Lipoxygenases: occurrence, functions, catalysis, and acquisition of substrate. J. Biol. Chem. **274:** 23679–23682.
3. YOSHIMOTO, T., H. SUZUKI, S. YAMAMOTO et al. 1990. Cloning and sequence analysis of the cDNA for arachidonate 12-lipoxygenase of porcine leukocytes. Proc. Natl. Acad. Sci. U.S.A. **87:** 2142–2146.
4. FUNK, C.D., L. FURCI & G.A. FITZGERALD. 1990. Molecular cloning, primary structure, and expression of the human platelet/erythroleukemia cell 12-lipoxygenase. Proc. Natl. Acad. Sci. U.S.A. **87:** 5638–5642.
5. KUHN, H. & B.J. THIELE. 1999. The diversity of the lipoxygenase family. FEBS Lett. **449:** 7–11.
6. BREITBART, E., Y. SOFER, A. SHAINBERG et al. 1996. Lipoxygenase activity in heart cells. FEBS Lett. **395:** 148–152.
7. RICE, R.L., D.G. TANG, M. HADDAD et al. 1998. 12(S)-Hydroxyeicosatetraenoic acid increases the actin microfilament content in B16a melanoma cells: a protein kinase–dependent process. Int. J. Cancer **77:** 271–278.
8. TANG, D.G., C.A. DIGLIO & K.V. HONN. 1994. Activation of microvascular endothelium by eicosanoid 12(S)-hydroxyeicosatetraenoic acid leads to enhanced tumor cell adhesion via up-regulation of surface expression of alpha V beta3 integrin: a post-transcriptional, protein kinase C–, and cytoskeleton-dependent process. Cancer Res. **54:** 1119–1129.
9. TANG, D.G. & K.V. HONN. 1997. Role of protein kinase C and phosphatases in 12(S)-HETE-induced tumor cell cytoskeletal reorganization. Adv. Exp. Med. Biol. **400A:** 349–361.
10. KANG, L.T. & J.Y. VANDERHOEK. 1998. Mono (S) hydroxy fatty acids: novel ligands for cytosolic actin. J. Lipid Res. **39:** 1476–1482.
11. KANG, L.T., T.M. PHILLIPS & J.Y. VANDERHOEK. 1999. Novel membrane target proteins for lipoxygenase-derived mono(S)hydroxy fatty acids. Biochim. Biophys. Acta **1438:** 388–398.
12. KUZUYA, T., T. HOSHIDA, Y. KIM et al. 1993. Free radical generation with arachidonate lipoxygenase reaction relates to reoxygenation induced myocardial cell injury. Cardiovasc. Res. **27:** 1056–1060.
13. MURPHY, E., W. GLASGOW, T. FRALIX et al. 1995. Role of lipoxygenase metabolites in ischemic preconditioning. Circ. Res. **76:** 457–467.
14. CHEN, W., W. GLASGOW, E. MURPHY et al. 1999. Lipoxygenase metabolism of arachidonic acid in ischemic preconditioning and PKC-induced protection in heart. Am. J. Physiol. **276:** H2094–H2101.
15. OMATA, W., H. SHIBATA, K. TAKATA et al. 2000. Actin filaments play a critical role in insulin-induced exocytotic recruitment, but not in endocytosis of GLUT4 in isolated rat adipocytes. Biochem. J. **346:** 321–328.
16. TSAKIRIDIS, T., P. TONG, B. MATTHEWS et al. 1999. Role of the actin cytoskeleton in insulin action. Microsc. Res. Tech. **47:** 79–92.
17. TSAKIRIDIS, T., Q. WANG, C. TAHA et al. 1997. Involvement of the actin network in insulin signalling. Soc. Gen. Physiol. Ser. **52:** 257–271.
18. TSAKIRIDIS, T., M. VRANIC & A. KLIP. 1994. Disassembly of the actin network inhibits insulin-dependent stimulation of glucose transport and prevents recruitment of glucose transporters to the plasma membrane. J. Biol. Chem. **267:** 29934–29942.
19. TILL, M., T. KOLTER & J. ECKEL. 1997. Molecular mechanisms of contraction-induced translocation of GLUT4 in isolated cardiomyocytes. Am. J. Cardiol. **80:** 85A–89A.
20. LIU, L.S., H. TANAKA, S. ISHII et al. 1998. The new antidiabetic drug MCC-555 acutely sensitizes insulin signaling in isolated cardiomyocytes. Endocrinology **139:** 4531–4539.
21. RAMRATH, S., H.J. TRITSCHLER & J. ECKEL. 1999. Stimulation of cardiac glucose transport by thioctic acid and insulin. Horm. Metab. Res. **31:** 632–635.

22. ECKEL, J., M. TILL & I. UPHUES. 2000. Cardiac insulin resistance is associated with an impaired recruitment of phosphatidylinositol 3-kinase to GLUT4-vesicles. Int. J. Obes. **24**(suppl. 2): S120–S121.
23. BÄHR, M., M. SPELLEKEN, M. BOCK et al. 1996. Acute and chronic effects of troglitazone (CS-045) on isolated rat ventricular cardiomyocytes. Diabetologia **39**: 766–774.
24. ECKEL, J., B. ASSKAMP & H. REINAUER. 1991. Induction of insulin resistance in primary cultured adult cardiac myocytes. Endocrinology **129**: 345–352.
25. BÄHR, M., M. VON HOLTEY et al. 1995. Direct stimulation of myocardial glucose transport and glucose transporter-1 (GLUT1) and GLUT4 protein expression by the sulfonylurea glimepiride. Endocrinology **136**: 2547–2553.
26. DRANSFELD, O., I. UPHUES, S. SASSON et al. 2000. Regulation of subcellular distribution in cardiomyocytes: Rab4A reduces basal glucose transport and augments insulin responsiveness. Exp. Clin. Endocrinol. Diabetes **107**: 26–36.
27. SASSON, S., N. KAISER, M. DAN-GOOR et al. 1997. Substrate autoregulation of glucose transport: hexose 6-phosphate mediates the cellular distribution of glucose transporters. Diabetologia **40**: 30–39.
28. DRANSFELD, O., I. RAKATZI, S. SASSON et al. 2001. Eicosanoids participate in the regulation of cardiac glucose transport by contribution to a rearrangement of actin cytoskeletal elements. Biochem. J. **359**: 47–54.
29. KESSLER, A., E. TOMAS, D. IMMLER et al. 2000. Rab11 is associated with GLUT4-containing vesicles and redistributes in response to insulin. Diabetologia **43**: 1518–1527.
30. WANG, Q., P.J. BILAN, T. TSAKIRIDIS et al. 1998. Actin filaments participate in the relocalization of phosphatidylinositol 3-kinase to glucose transporter–containing compartments and in the stimulation of glucose uptake in 3T3-L1 adipocytes. Biochem. J. **331**: 917–928.
31. STARKOPF, J., T.V. ANDREASEN et al. 1998. Lipid peroxidation, arachidonic acid, and products of the lipoxygenase pathway in ischemic preconditioning of the heart. Cardiovasc. Res. **37**: 66–75.
32. TANG, D.G., C.A. DIGLIO & K.V. HONN. 1993. 12(S)-HETE-induced microvascular endothelial cell retraction results from PKC-dependent rearrangement of cytoskeletal elements and alpha V beta 3 integrins. Prostaglandins **45**: 249–267.
33. GABEL, S.A., R.E. LONDON, C.D. FUNK et al. 2001. Leukocyte-type 12-lipoxygenase-deficient mice show impaired ischemic preconditioning–induced cardioprotection. Am. J. Physiol. Heart Circ. Physiol. **280**: H1963–H1969.
34. YAMADA, M. & A.D. PROIA. 2000. 8(S)-Hydroxyeicosatetraenoic acid is the lipoxygenase metabolite of arachidonic acid that regulates epithelial cell migration in the rat cornea. Cornea **19**(suppl. 3): S13–S20.

Regulation of Fat Metabolism in Skeletal Muscle

ASKER E. JEUKENDRUP

School of Sport and Exercise Sciences, University of Birmingham, United Kingdom

ABSTRACT: Regulation of carbohydrate and fat utilization by skeletal muscle at rest and during exercise has been the subject of investigation since the early 1960s when Randle *et al.* proposed the so-called glucose–fatty acid cycle to explain the reciprocal relationship between carbohydrate and fat metabolism. The suggested mechanisms were based on the premise that an increase in fatty acid (FA) availability would result in increased fat metabolism and inhibition of carbohydrate metabolism. Briefly, accumulation of acetyl-CoA would result in inhibition of pyruvate dehydrogenase (PDH), accumulation of citrate would inhibit phosphofructokinase (PFK), and accumulation of glucose-6-phosphate (G6P) would reduce hexokinase (HK) activity. Ultimately, this would inhibit carbohydrate metabolism with increasing availability and oxidation of FA. Although there is some evidence for the existence of the glucose-FA cycle at rest and during low-intensity exercise, it cannot explain substrate use at moderate to high exercise intensities. More recently, evidence has accumulated that increases in glycolytic flux may decrease fat metabolism. Potential sites of regulation are the transport of FA into the sarcoplasma, lipolysis of intramuscular triacylglycerol (IMTG) by hormone-sensitive lipase (HSL), and transport of FA across the mitochondrial membrane. There are several potential regulators of fat oxidation: first, malonyl-CoA concentration, which is formed from acetyl-CoA, catalyzed by the enzyme acetyl-CoA carboxylase (ACC), which in turn will inhibit carnitine palmitoyl transferase I (CPT I). Another possible mechanism is accumulation of acetyl-CoA that will result in acetylation of the carnitine pool, reducing the free carnitine concentration. This could theoretically reduce FA transport into the mitochondria. There is also some recent evidence that CPT I is inhibited by small reductions in pH that might be observed during exercise at high intensities. It is also possible that FA entry into the sarcolemma is regulated by translocation of FAT/CD36 in a similar manner to glucose transport by GLUT-4. Studies suggest that the regulatory mechanisms may be different at rest and during exercise and may change as the exercise intensity increases. Regulation of skeletal muscle fat metabolism is clearly multifactorial, and different mechanisms may dominate in different conditions.

KEYWORDS: exercise; fat; pyruvate dehydrogenase; phosphofructokinase; glucose-6-phosphate; carbohydrate metabolism

---

Address for correspondence: Dr. Asker Jeukendrup, Human Performance Laboratory, School of Sport and Exercise Sciences, University of Birmingham, Edgbaston B15 2TT, Birmingham, United Kingdom. Voice: +44-121-414-4124; +44-121-414-4121.
A.E.Jeukendrup@bham.ac.uk

INTRODUCTION

At rest and during exercise, skeletal muscle is the main site of oxidation of fatty acid (FA). In resting conditions and especially after fasting, FAs are the predominant fuel used by skeletal muscle. During low-intensity exercise, metabolism is elevated severalfold compared to resting conditions, and fat oxidation is increased. When the exercise intensity increases, fat oxidation increases further, until exercise intensities of about 65% VO_2max, after which a decline in the rate of fat oxidation is observed. In contrast to carbohydrate metabolism, which increases as a function of the aerobic work rate, fat oxidation is reduced at the high exercise intensities (FIG. 1). The changes in fat oxidation as a function of the exercise intensity have recently been described by Achten *et al.*[1] in a group of trained individuals. Given the amount of work done in this area, it is somewhat surprising that this is the first study to report changes in fat oxidation over a wide range of exercise intensities. Although large individual variation was observed, on average, maximal fat oxidation was reported at 64% VO_2max, after which fat oxidation decreased relatively rapidly. It was also concluded that fat oxidation was high over a wide range of intensities, but declined rapidly at high intensities (>80% VO_2max).[1] It is likely that this curve will shift to the right after (endurance) training and shift to the left after detraining. Research is currently being conducted to study the shape of this curve in response to training and other interventions.

This review will discuss potential mechanisms responsible for the upregulation of fat metabolism in the transition from rest to exercise and the downregulation of fat metabolism from low and moderate to high exercise intensities. Although factors

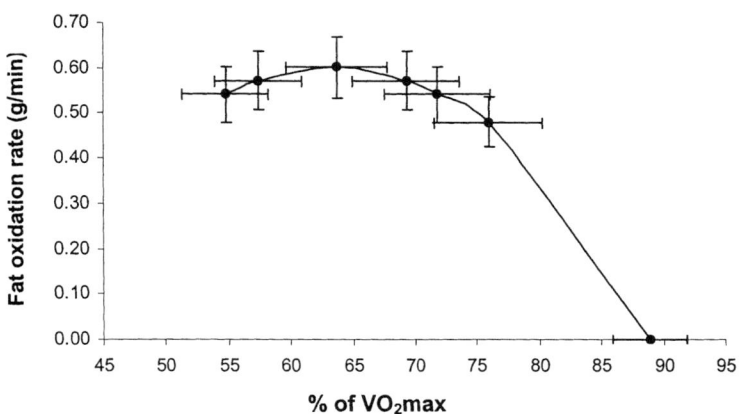

FIGURE 1. Fat oxidation as a function of exercise intensity in trained subjects. From low to moderate intensities, fat oxidation increased, peaked at 63% VO_2max, and decreased at high intensities. (From reference 1, reproduced with permission.)

outside skeletal muscle may influence fat oxidation (i.e., plasma FA concentration), this paper will mainly focus on the factors inside skeletal muscle that turn on skeletal muscle fat oxidation and inhibit fat oxidation. The data presented are based on human skeletal muscle where possible; however, when no information was available in humans, the results of rat studies were used.

FAT UTILIZATION AT REST AND DURING EXERCISE

After an overnight fast, most of the energy requirement at rest is covered by the oxidation of FAs derived from adipose tissue. Lipolysis in adipose tissue is mostly dependent on the concentrations of hormones (epinephrine to stimulate lipolysis and insulin to inhibit lipolysis). When exercise is initiated, the rate of lipolysis and the rate of FA release from adipose tissue are increased. During moderate-intensity exercise, lipolysis increases approximately threefold,[66] mainly because of an increased β-adrenergic stimulation. In addition, during moderate-intensity exercise, the blood flow to adipose tissue is doubled and the rate of reesterification is halved.[48,66] Also, blood flow in skeletal muscle is increased dramatically and therefore the delivery of FAs to the muscle is increased severalfold. During the first 15 min of exercise, plasma FA concentrations usually decrease because the rate of FA uptake by the muscle exceeds the rate of FA appearance from lipolysis. Thereafter, the rate of appearance is in excess of the utilization by muscle, and plasma FA concentrations increase. The rise in plasma FA concentration depends on the exercise intensity and the duration of exercise. During moderate exercise, FA concentrations may reach 1 mmol/L within 60 min of exercise; however, at higher exercise intensities, the rise in plasma FA is very small or may even be absent.

Romijn et al.[48] investigated fat metabolism at three different exercise intensities, 25%, 65%, and 85% VO_2max, and used stable isotopic tracers to measure maximum plasma FA oxidation and nonplasma FA oxidation. In their paper, the difference between total fat oxidation (measured by indirect calorimetry) and maximum plasma FA oxidation (determined by a ^2H-palmitate tracer) was referred to as intramuscular triacylglycerol (IMTG) oxidation. This assumes that plasma triacylglycerol (TG) and intermuscular fat (adipocytes in between muscle fibers) are not important fuels during exercise. However, some have suggested that these sources may contribute significantly to energy expenditure[23,24] and therefore it would be more correct to refer to this calculated fraction as TG oxidation (intramuscular and plasma) (FIG. 2). It must also be emphasized that the data by Romijn et al.[48] are imprecise. The data are based on calculations of the rate of disappearance of FA (Rd FA) and the assumption that all FAs taken up by the muscle are also oxidized. It has been shown that this is not the case,[53] and Rd FA thus overestimates plasma FA oxidation. In addition, all calculations are based on respiratory exchange ratios, and small changes in the ratio can have significant effects on calculated rates of TG oxidation. Furthermore, this study was conducted in well-trained individuals after an overnight fast, which may have forced the muscle to use more fat than would normally be the case. However, it is generally believed that the trends shown in this paper reflect the actual changes in substrate use that occur during exercise (FIG. 2).

The data from Romijn et al.[48] suggest that lipolysis increases from rest to exercise and from 25% to 65% VO_2max. However, at 85% VO_2max, there is no further

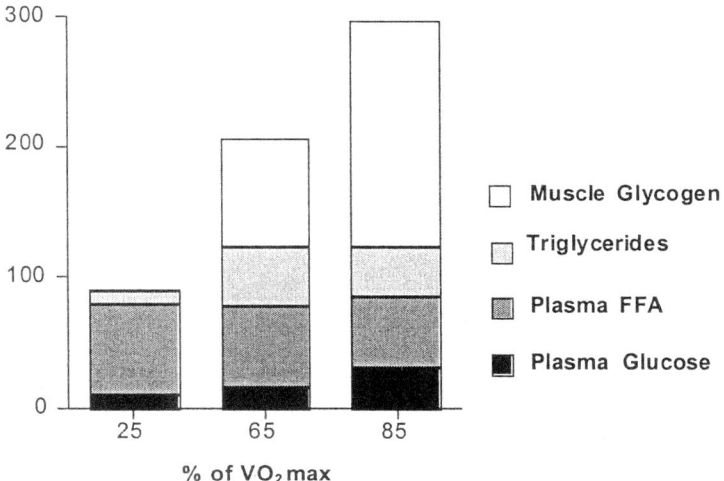

FIGURE 2. Energy expenditure increases with exercise intensity and there are substantial changes in substrate utilization. Fat oxidation increases at moderate intensities and decreases at high intensities. The contribution of plasma FA seems to decrease slightly with increasing intensity, whereas TG provides a more important substrate at 65% VO$_2$max. Plasma glucose and especially muscle glycogen become more important as the exercise intensity increases. Figure adapted from reference 48.

increase of lipolysis, and lipolytic rates are similar to those at 65% VO$_2$max. It is likely, however, that the FA availability at 85% VO$_2$max is reduced, as evidenced by 50% lower plasma FA concentrations after 30 min of exercise.[48] If we assume that blood flow to the working muscle is proportional to the exercise intensity, this must mean that, at 85% VO$_2$max, fewer FAs are delivered to the muscle, and this could contribute to the observed lower rates of fat oxidation. Indeed, Romijn *et al.*[48] reported a reduced Rd FA at 85% VO$_2$max.

At low intensities (25% VO$_2$max), most of the FAs oxidized are plasma-derived and TG oxidation is negligible (FIG. 2). However, when the exercise intensity is increased to 65% VO$_2$max, the contribution of plasma FAs is similar, but a significant increase in the contribution of TG can be observed that provides about half of the FAs used for total fat oxidation.[48] When the exercise intensity is further increased, fat oxidation decreases even though the rate of lipolysis is still high.[48] The blood flow to the adipose tissue may be decreased (due to sympathetic vasoconstriction) and this may result in a decreased removal of FAs from adipose tissue. It has also been suggested that, during high-intensity exercise, lactate accumulation may decrease lipolysis or increase the rate of reesterification of FAs.[6] However, a recent study using microdialysis probes in adipose tissue concluded that lactate did not impair lipolysis as no differences were found when lactate was infused or saline.[60]

Although the causes are still not entirely clear, plasma FA concentrations are usually low during intense exercise. However, this decreased availability of FA can only partially explain the reduced fat oxidation that is observed in these conditions.

EFFECT OF INCREASED AVAILABILITY OF FATTY ACIDS AT REST AND LOW/MODERATE-INTENSITY EXERCISE

One of the important factors determining substrate utilization at rest and during exercise is the availability of substrates. It has been repeatedly shown that fat oxidation can be increased when the availability of FA is increased.[8,64] It was originally thought that the classical glucose-FA cycle[45] could explain this reciprocal relationship between carbohydrate and fat metabolism (FIG. 3). This theory states that an increase in plasma FA concentration will result in an increased uptake of FAs. These FAs will undergo β-oxidation in the mitochondria where they will be broken down to acetyl-CoA. An increasing concentration of acetyl-CoA (or increased acetyl-CoA/CoA ratio) would inhibit pyruvate dehydrogenase (PDH), which is responsible for the breakdown of pyruvate to acetyl-CoA. Also, increased formation of acetyl-CoA from FAs would increase muscle citrate levels and, after diffusing into the sarcoplasm, these could inhibit phosphofructokinase (PFK), the rate-limiting enzyme in glycolysis. The effect of increased acetyl-CoA and citrate levels will therefore be a reduction in the rate of glycolysis. This, in turn, may cause accumulation of glucose-6-phosphate (G6P) in the muscle, which will inhibit hexokinase (HK) activity and thus reduce muscle glucose uptake.

Most of the evidence for the existence of the glucose-FA cycle is from isolated muscle experiments and *in vitro* studies of diaphragm or heart muscle.[14,44,45] There is relatively little information about skeletal muscle [47] and even less information in human skeletal muscle.

Odland *et al.*[35] investigated the metabolic effects of increasing FA availability by infusing TG (Intralipid®) plus heparin during low- and moderate-intensity exercise (40% and 65% VO$_2$max, respectively). The elevated FA levels resulted in a 4-fold increase in leg FA uptake and in a 23% reduction in glycogenolysis. The study provides therefore evidence for a shift from carbohydrate toward fat metabolism. However, in this study, the increased FA availability did not affect leg glucose uptake and the authors thus concluded that the regulation must take place at the level of glycogen phosphorylase or PDH (glycogen breakdown). Measurements of the active form of PDH (PDHa) were consistently lower with elevated FA. The PDH is phosphorylated (deactivated) by PDH kinase and dephosphorylated (activated) by PDH phosphatase.[46] The main regulator of the PDH phosphatase is Ca^{2+}, but it is unlikely that this is an important regulator of the transformation to PDHa because increases in Ca^{2+} during exercise would be similar with or without elevated FA concentrations. PDH kinase is inhibited by pyruvate and activated by high ATP, acetyl-CoA, and NADH levels at rest.[42] Odland *et al.*[35] showed that acetyl-CoA and ATP were similar with or without elevated plasma FA concentrations. A reduced pyruvate concentration and a higher NADH concentration in the presence of high FA availability have been suggested to relieve some of the inhibition of PDH kinase, resulting in a reduced activation of PDH.[56]

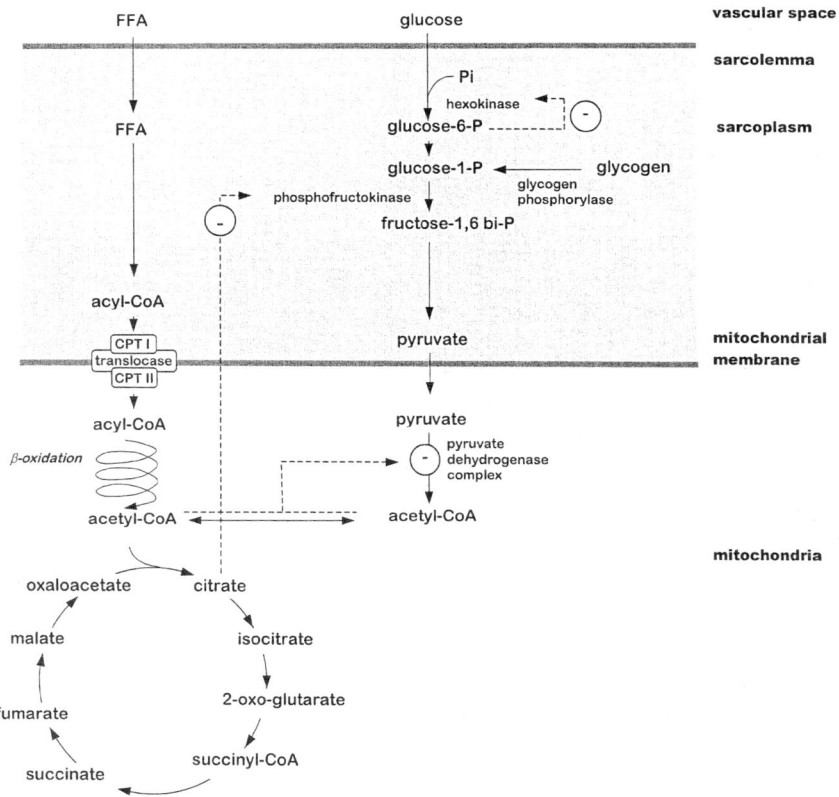

FIGURE 3. The glucose-FA or Randle cycle (adapted from reference 45) describes potential mechanisms involved in the interaction between fat and carbohydrate metabolism. In this model, FA availability seems to be the dominant factor. Briefly, increased FA availability would increase acetyl-CoA concentrations; accumulation of acetyl-CoA would result in inhibition of pyruvate dehydrogenase (PDH); accumulation of citrate would inhibit phosphofructokinase (PFK); and accumulation of glucose-6-phosphate (G6P) would reduce hexokinase (HK) activity. Ultimately, this would inhibit carbohydrate metabolism with increasing availability and oxidation of FA. Although there is some evidence for the existence of the glucose-FA cycle at rest and during low-intensity exercise, it cannot explain substrate use at moderate to high exercise intensities.

Thus, although the increased FA availability seemed to cause a shift in substrate metabolism at low and moderate exercise intensities, this did not seem to be through mechanisms as proposed by the glucose-FA cycle. Although a decreased activation of PDH was observed, this could not be explained by increases in acetyl-CoA concentrations as originally proposed. Increases in muscle citrate concentration were observed, but these may have been too small to significantly affect PFK activity *in vivo*.[41] Finally, no changes in glucose uptake were observed. It is therefore not entirely clear which factors cause the reduction in glycogenolysis in the presence of a high FA concentration at low and moderate intensities.

EFFECT OF INCREASED AVAILABILITY OF FATTY ACIDS DURING HIGH-INTENSITY EXERCISE

Dyck et al.[12,13] investigated a similar question at high exercise intensities (80%–85% VO_2max). A 45% reduction of glycogenolysis was observed after 15 min of cycling in the presence of an elevated plasma FA concentration (1.3–1.4 mmol/L). Muscle citrate and acetyl-CoA were unaffected, suggesting that regulation would take place at the level of phosphorylase. It was also observed that, with increased FA availability, there is a reduction of intramuscular free Pi and AMP accumulation during exercise, possibly because of a greater accumulation of mitochondrial NADH. This suggests a better match between ATP production and ATP breakdown. Since Pi and AMP are known to stimulate the enzyme, glycogen phosphorylase, it is possible that the reduction in Pi and AMP levels is at least partially responsible for the reduced muscle glycogen breakdown. No changes were observed in PDH, but glycogen phosphorylase was inhibited with increased FA availability.

In conclusion, at high-intensity exercise, increased FA availability results in a decreased breakdown of muscle glycogen. However, there is no evidence that this process is through the glucose-FA cycle. Allosteric regulators (Pi and AMP) seem to play a major role in the regulation of glycogen breakdown at the level of glycogen phosphorylase.

There is also an alternative explanation for reduced muscle glycogen breakdown after elevation of plasma FA concentrations in some studies.[8,64] When studying the plasma FA concentrations in more detail, it appears that in these studies plasma FAs were significantly elevated (by infusion of TG and injection of heparin) compared to the control condition. The plasma FA concentrations in the control condition, however, were very low (<0.2 mmol/L). It is conceivable that these FA levels are too low to provide the muscle with sufficient fat substrate. As a result of depriving the muscle of fat substrate, muscle glycogen breakdown may have been increased in the control condition. If this is true, the observed "sparing" of glycogen with the high FA concentrations was really caused by an increased breakdown of glycogen in the control condition. It is known that blocking lipolysis and reducing FA availability by giving nicotinic acid or a derivative will increase muscle glycogen breakdown during exercise.[19]

Another situation in which FA availability is reduced is during high-intensity exercise.[12,13,49,54] However, it has been shown that this decreased availability of FAs can only partially explain the reduced fat oxidation that is observed in these conditions. Romijn et al.[49] studied endurance-trained cyclists during high-intensity exercise (85% VO_2max). As observed in a previous study,[48] the plasma FA concentration and the rate of appearance of FA (Ra FA) were very low at this intensity.[49] When FA concentrations were restored to levels observed at moderate exercise intensities by infusing Intralipid® and heparin, fat oxidation was only slightly increased, but still lower than at moderate intensities.

This strongly suggests that FA oxidation is at least partly regulated at the muscular level. Further evidence for regulation of FA utilization at the muscular level is derived from tracer studies.[9,54] These studies investigated the oxidation of medium-chain fatty acids (MCFAs) versus long-chain fatty acids (LCFAs) when glycolytic flux is high. In the study by Coyle et al.[9] the glycolytic rate was increased by preexercise glucose feeding, whereas in the study by Sidossis et al.[54] this was achieved by increasing the exercise intensity from 40% to 80% VO_2max. Sidossis *et al.*[54] infused a ^{14}C-labeled MCFA (octanoate) and a ^{13}C-labeled LCFA (oleate). The percentage of the labeled LCFAs taken up and oxidized decreased at the high exercise intensity. There was, however, no difference in the oxidation of the MCFAs. Since these MCFAs are not as dependent on transport proteins or the specific MCFA transport proteins are not as heavily regulated, this suggests that the transport across membranes is involved in the regulation of FA oxidation and might at least partly explain the reduced fat oxidation at high exercise intensities. Sidossis *et al.*[54] argued that their results provided indirect evidence for regulation at the level of CPT I; however, regulation at the level of transport across the sarcolemma cannot be excluded, as will be discussed below.

Taken together, these studies suggest that FA availability (lipolysis and removal of FAs from the adipose tissue and blood flow to the working muscle) is a major regulatory factor. During low and moderate intensities, there is some evidence for the existence of the glucose-FA cycle. However, most of the regulation seems to be through deactivation of phosphorylase and possibly PDH. During high-intensity exercise, regulation seems to be solely through phosphorylase. Although FA availability plays an important role and fat oxidation may thus be partly regulated at the adipose tissue level, there is also evidence for regulation at the muscular level. The potential regulatory sites inside skeletal muscle will be discussed below in more detail.

POTENTIAL SITES OF REGULATION OF FAT METABOLISM IN SKELETAL MUSCLE

There are a few potential sites for the regulation of fat metabolism inside skeletal muscle (FIG. 4). These include (1) the transport of FAs from the vascular space across the sarcolemma into the sarcoplasm, (2) the release of FAs from IMTG under the influence of a hormone-sensitive lipase, and (3) the transport of FAs across the mitochondrial membrane, involving the enzyme, carnitine palmitoyl transferase I (CPT I). Once in the mitochondrial matrix, the fatty acyl-CoA is subjected to β-oxidation, a series of reactions that splits a two-carbon acetyl-CoA molecule of the multiple-carbon FA chain. This pathway is not believed to be a regulatory site as none of the enzymes seem to be acutely regulated.

TRANSPORT OF FATTY ACIDS INTO THE MUSCLE

For a long time it was believed that the transport of FAs into the muscle cell was a passive process. This was based on early observations that FA uptake increased linearly with FA concentration.[17] However, recently, specific carrier proteins have been identified in various tissues, including skeletal muscle (FIG. 4). In the sarcolemma, two proteins have thus far been identified that are involved in the transport of FAs across the membrane. These proteins are a specific plasma membrane FA-binding protein (FABPpm) and a FA translocase protein (FAT/CD36). A third protein has been identified (FA transport protein, FATP), but its transport role has been questioned because FATP content in the plasma membrane was inversely correlated

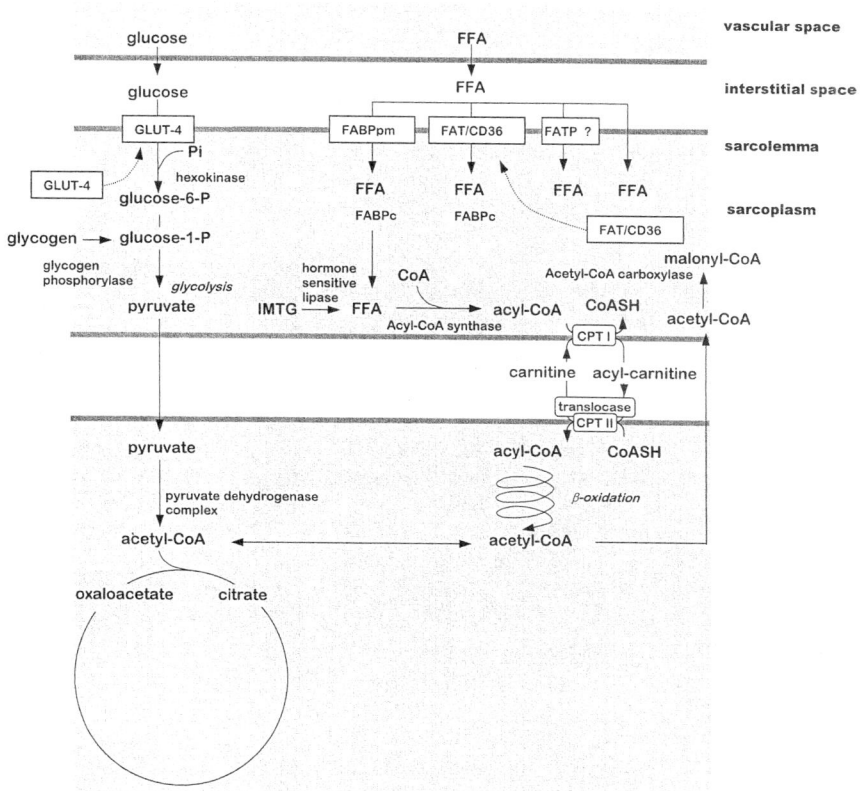

FIGURE 4. Transport of glucose and FAs from the vascular space to the mitochondria and their subsequent oxidation. Although it is possible that some FAs diffuse through the sarcolemma, the majority of the transport of FAs into the mitochondria involves transport proteins. Three candidates have been identified. FAT/CD36 translocates from its intracellular storage space to the membrane in the same manner that GLUT-4 translocates.

with LCFA uptake,[29] whereas positive correlations were found for FABPpm and FAT/CD36.

There is also physiological evidence for the existence of such transporters. In studies with isolated perfused hindlimbs, Turcotte et al.[61] showed that a saturation of FA transport occurred with increasing free palmitate concentration in the perfusion medium. Similar data were obtained in giant sarcolemmal vesicles of type I and type II muscle fibers.[4] These data make it highly likely that a large proportion of the LCFAs cross the membrane by transport rather than passive diffusion.

There is also evidence that this transport of FA into the muscle is regulated. After fasting, FABPpm is increased in slow-twitch oxidative muscle.[62] FABPpm also increases after exercise training.[25] After chronic stimulation (24 h/day, 7 days) of skeletal muscles, the FAT/CD36 protein was upregulated and this coincided with increased LCFA transport across the sarcolemma.[3]

More recently, Bonen et al.[5] demonstrated that FAT/CD36 can translocate from intracellular vesicles to the cell membrane in a similar manner as the GLUT-4 protein, indicating that FA transport can also be regulated acutely. It was demonstrated that muscle contraction increased plasma membrane FAT/CD36 and decreased the concentration of FAT/CD36 in the sarcoplasma. Along with a higher density of FAT/CD36 at the cell membrane, an increased LCFA transport into the cell was observed. It is currently not known what triggers the translocation of the FAT/CD36 to the cell membrane. However, it is tempting to speculate that similar factors that result in GLUT-4 translocation will also be responsible for the translocation of FAT/CD36. GLUT-4 translocation to the cell surface is stimulated by an AMP-activated protein kinase (AMPK)–dependent signaling pathway. Muscle contraction results in increased cyclic AMP levels, which in turn allosterically activate AMPK. Translocation of FAT/CD36 to the cell membrane is most likely under the control of phosphatidylinositol 3-kinase (PI-3 kinase) and AMPK (J. Luiken and J. Glatz, personal communication).

Although there is now clear evidence for an involvement of a transport mechanism for FA across the sarcolemma, its functional significance is still unclear. For example, it is not known whether there are any physiological conditions in which this transport limits fat oxidation.

In conclusion, FA transporters are likely to be responsible for most of the transport of FA across the sarcolemma, and these transporters can be regulated both acutely and chronically. At present, however, it is not known if there are any physiological situations in which this transport becomes limiting. We also do not know what the triggers are inside the muscle for the up- or downregulation of the transport proteins.

THE BREAKDOWN OF IMTG

IMTG stores, in trained muscle usually located adjacent to the mitochondria as lipid droplets[18](FIG. 5), have been recognized as an important energy source during exercise. However, although some studies reported IMTG breakdown after exercise measured by muscle biopsies, there are also a large number of studies in which no difference was found in muscle TG content before and after exercise.[23,26] However, there is additional evidence that IMTG stores provide an important fuel source. Studies in which muscle samples were investigated under a microscope, for example, revealed that the size of these lipid droplets decreased during exercise.[33] In addition, indirect measures using stable isotope measurements suggest an important role during exercise.[9,22,30,43,48,54,55] Studies using ^1H magnetic resonance spectroscopic techniques[2] also found utilization of IMTG during exercise.[10,11] Furthermore, in trained muscle the lipid droplets are located adjacent to the mitochondria,[18] whereas in untrained muscle they may not be associated with the mitochondria. This also suggests a functional role for IMTG.

The breakdown of IMTG is regulated by a lipase similar to that found in adipose tissue. Although lipases and their regulation have been extensively studied in a variety of tissues, little is known about the lipase responsible for the breakdown of TG in skeletal muscle. Langfort et al.[28] have now unequivocally shown that hormone-sensitive lipase (HSL), a neutral lipase, also exists in human skeletal muscle. HSL content correlates directly with muscle TG content in different muscle fiber types,

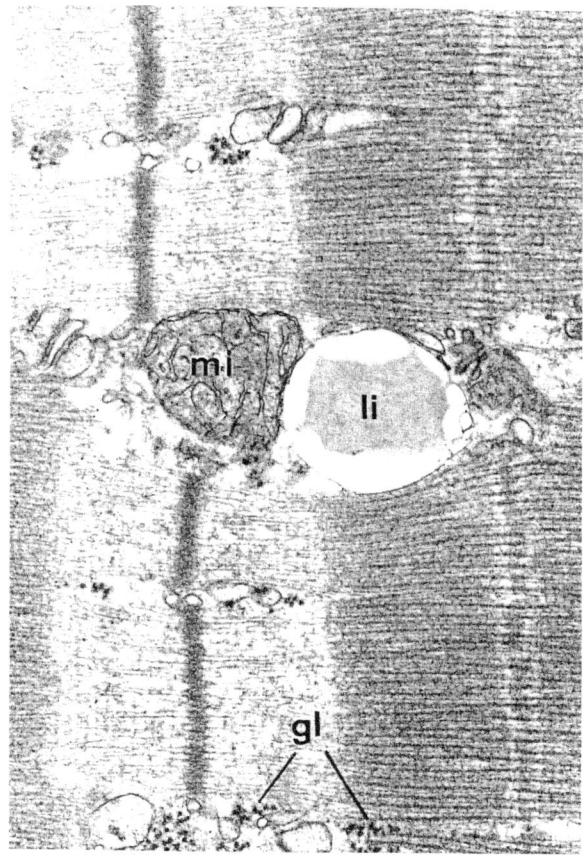

FIGURE 5. Lipid droplets in skeletal muscle. Terms: mi = mitochondria; li = lipid droplet; gl = glycogen. Courtesy of Hans Hoppeler, Bern, Switzerland. (From reference 21, reproduced with permission.)

being highest in slow-twitch oxidative fibers and lowest in fast-twitch glycolytic fibers.[28,40,57,58,63] It has also been shown that this HSL is activated by β-adrenergic stimulation (epinephrine) and that this activation is via the cyclic AMP (cAMP)–activated kinase.[28] In a subsequent study, Langfort et al.[27] showed that HSL is activated by muscle contraction independent of adrenergic stimulation. Although the activation of the enzyme probably involves phosphorylation, the mechanisms by which contraction stimulates HSL activity are largely unknown. Studies in rats and humans[27] have demonstrated a simultaneous activation of intramuscular HSL and glycogen phosphorylase during exercise. This suggests that similar triggers stimulate the two enzymes during exercise. It is well known that Ca^{2+} and metabolites related to the energy status of the cell (ADP, AMP, Pi) upregulate glycogen phosphorylase, but there are no studies to show such a relationship with HSL.

It is also important to note that at the very high exercise intensities, despite further increases in circulating epinephrine levels and possible increases in Ca^{2+}, ADP, AMP, and Pi, fat oxidation actually decreases. It is likely that other factors inside the muscle are responsible for reducing the activity of the enzyme at intensities above 70% VO_2max.

TRANSPORT OF FATTY ACIDS INTO THE MITOCHONDRIA

Another site that appears to play a role in the regulation of carbohydrate and fat metabolism in skeletal muscle involves the transport of LCFAs across the mitochondrial membrane. This process is depicted in FIGURE 4. FA in the sarcoplasm may be activated by the enzyme acyl-CoA synthethase or thiokinase to form an acyl-CoA complex (often referred to as an activated FA). This acyl-CoA complex can be used for the synthesis of IMTG or it can be bound to carnitine under the influence of the enzyme carnitine palmitoyl transferase I (CPT I), which is located at the outside of the outer mitochondrial membrane. The binding of carnitine with the activated FA is the first step in the transport of the FA into the mitochondria. As carnitine binds to the acyl-CoA moiety, free CoA is released. The acyl-carnitine complex is transported with a translocase and reconverted into acyl-CoA at the matrix side of the inner mitochondrial membrane by the enzyme carnitine palmitoyl transferase II (CPT II). The carnitine that is released diffuses back across the mitochondrial membrane into the cytoplasm and thus becomes available again for the transport of other FAs. Acyl carnitine crosses the inner membrane in a 1:1 exchange with a molecule of free carnitine.[63] Although it is often believed that short-chain fatty acids (SCFAs) and MCFAs can more freely diffuse into the mitochondrial matrix, carrier proteins with a specific maximum affinity for short- or medium-chain acyl-CoA transport at least some of these FAs.[15,50] It must be noted, however, that SCFAs and MCFAs represent only a small portion of the total FA oxidation.

It is generally believed that CPT I is the rate-limiting enzyme in the transport of FAs across the mitochondria and may be even rate-limiting for FA oxidation. There is substantial evidence that CPT I activity is influenced by numerous regulators. The importance of these regulators will be discussed in subsequent sections in relation to the changes that occur during exercise.

REGULATORS OF CPT I

FA transport into the mitochondria seems to be one of the most regulated steps in the fat oxidation process. Several regulators of CPT I activity have been proposed, including malonyl-CoA concentration, hydrogen ion accumulation in the sarcoplasm, and reduced free carnitine availability. The evidence for each of these mechanisms will be evaluated in the following sections.

Malonyl-CoA

A large number of *in vitro* studies have now established a role for malonyl-CoA in regulating the entry of LCFAs into the mitochondria in a variety of tissues, including skeletal muscle.[31,32,51,52] Malonyl-CoA is a potent inhibitor of CPT I and is thus a potential candidate for the regulation of fat metabolism. Malonyl-CoA is formed from acetyl-CoA, a reaction catalyzed by the enzyme acetyl-CoA carboxylase (ACC). Malonyl-CoA levels decrease in rodent skeletal muscle from rest to moderate-intensity exercise, when energy production from fat increases.[65] It is believed that the resting concentrations of malonyl-CoA are sufficiently high to inhibit CPT I, and a decrease in the malonyl-CoA concentration would therefore result in a relief of the inhibition of CPT I and increased LCFA transport into the mitochondria. It is also well known that acetyl-CoA concentration in the muscle increases rapidly at the onset of high-intensity exercise, which will stimulate the activity of ACC as it is the primary substrate for this enzyme. The resulting increased concentration of malonyl-CoA could possibly explain a reduced FA uptake into the mitochondria.[51,52]

Initial studies in humans,[34,36,37] though, show little or no evidence that malonyl-CoA is a very important regulator of FA metabolism. In a first *in vivo* study by Odland *et al.*,[36] malonyl-CoA was measured in human vastus lateralis muscle at rest and following 10 min of cycling at 40% VO_2max and 10 and 60 min at 65% VO_2max. Although fat oxidation was increased severalfold from rest to exercise, no significant changes in malonyl-CoA concentration were observed. These data suggest that malonyl-CoA is not an important regulator of fat metabolism in humans. In a follow-up study, Odland *et al.*[37] studied the changes in malonyl-CoA during exercise at different exercise intensities. It was hypothesized that malonyl-CoA would be increased at 90% VO_2max. However, such an increase was not observed despite a significant increase in its substrate, acetyl-CoA. Fat oxidation increased at 35% and 65% VO_2max, despite a lack of decrease in malonyl-CoA levels. This study suggests that a decrease in malonyl-CoA levels is not required in human skeletal muscle in order to increase LCFA uptake and oxidation. Furthermore, malonyl-CoA content does not increase during exercise at high-intensity exercise and does not contribute to the reduced rate of fat oxidation.[37] Although these studies seem to rule out a role for malonyl-CoA in the regulation of fat metabolism, it is interesting to note that the same authors also observed an increased sensitivity of CPT I to malonyl-CoA in trained compared to untrained muscle. This suggests that there may be a role for malonyl-CoA, but this role may not be as simple as suggested by *in vitro* studies.

There are also a few potential methodological limitations that could at least theoretically explain the lack of evidence for an important role of malonyl-CoA in human skeletal muscle. It is possible that the measurement of total muscle malonyl-CoA concentration does not reflect the local concentrations near CPT I. Obviously, we also have to be careful when extrapolating data from *in vitro* studies to *in vivo* situations and from rat skeletal muscle to human skeletal muscle. *In vitro* studies predict that CPT I activity is inhibited 85%–90% at all times, even when malonyl-CoA concentrations are very low. Finally, in humans, resting malonyl-CoA levels are theoretically high enough to completely inhibit CPT I, something that does obviously not occur.

Taken together, this information suggests that regulation of CPT I activity in human skeletal muscle is more complicated than regulation by malonyl-CoA concentration alone. Other factors may be interacting with malonyl-CoA and CPT I in regulating LCFA transport into the mitochondria and oxidation during exercise. One of the possible regulators that has recently been proposed is the accumulation of hydrogen ions.

Hydrogen Ion Accumulation

Another potential regulator of FA oxidation may be hydrogen ion accumulation. Starritt et al.[59] studied CPT I activity in isolated mitochondria from resting human skeletal muscle. It was found that small changes in pH from 7.0 to 6.8 inhibited CPT I activity by 50%.[59] Such changes in pH can be observed during exercise at 80% VO_2max in humans. Howlett et al.[20] reported muscle lactate levels of 38 and 108 mmol/kg dry muscle after only 10 min of cycling at 65% and 90% VO_2max, respectively. It can be calculated that corresponding pH values must have been 6.9 and 6.6 at these exercise intensities. It is also likely that the activity of the neutral lipase HSL is influenced by the acid environment created during high-intensity exercise.[39] Hydrogen ion accumulation in the muscle may therefore be responsible for a reduction in IMTG hydrolysis as well as a decreased FA uptake into the mitochondria. The relatively large decrease in pH during intense exercise (>80% VO_2max) could be responsible for the sharp decrease in fat oxidation. A small reduction in pH could explain the reduced fat oxidation from moderate- to high-intensity exercise.

It must be noted, though, that this hypothesis is based on *in vitro* studies and these findings have not been confirmed *in vivo*. Further studies are needed to investigate the role of pH on FA transport into the muscle and, ultimately, on fat oxidation.

Carnitine Availability

During low-intensity exercise, the flux through the PDH is lower than the flux through the TCA cycle.[7] This would result in minimal acetylation of the carnitine pool. Relatively low acetylcarnitine concentrations and high free carnitine concentrations have been observed in various studies.[7,16] With increasing exercise intensity, the flux through PDH may increase more rapidly than the flux through the TCA cycle. This would result in an accumulation of acetyl-CoA. In order to free up the CoA, the acetyl units are bound to free carnitine. This acetylation of the carnitine pool will result in a decrease of the free carnitine concentration.

Constantin-Teodosiu et al.[7] showed that, at very high intensities (90% VO_2max), a large percentage of the carnitine was bound to acetyl-CoA and the concentration of the free carnitine pool was reduced to very low levels. It is thus possible that the reduced rates of fat oxidation are caused by a reduced transport of FA into the mitochondria because the availability of free carnitine becomes rate-limiting. Although this is an attractive hypothesis, there is currently no direct evidence that this mechanism is important. It could be argued that the low levels of free carnitine observed at high exercise intensities could still support significant rates of FA transport across the mitochondrial membrane, especially since carnitine is recycled, not consumed, in this process. It is also likely that the majority of this carnitine is present in the cytosol[38] where it is needed. If free carnitine availability plays a role in the regulation of fat metabolism, it is likely that this will be limited to high-intensity exercise. At moderate intensities, the free carnitine availability decreases compared to rest and low-intensity exercise, but fat oxidation actually increases. There are currently no studies that have directly addressed this question, and further work is required before a role for free carnitine can be confirmed or ruled out.

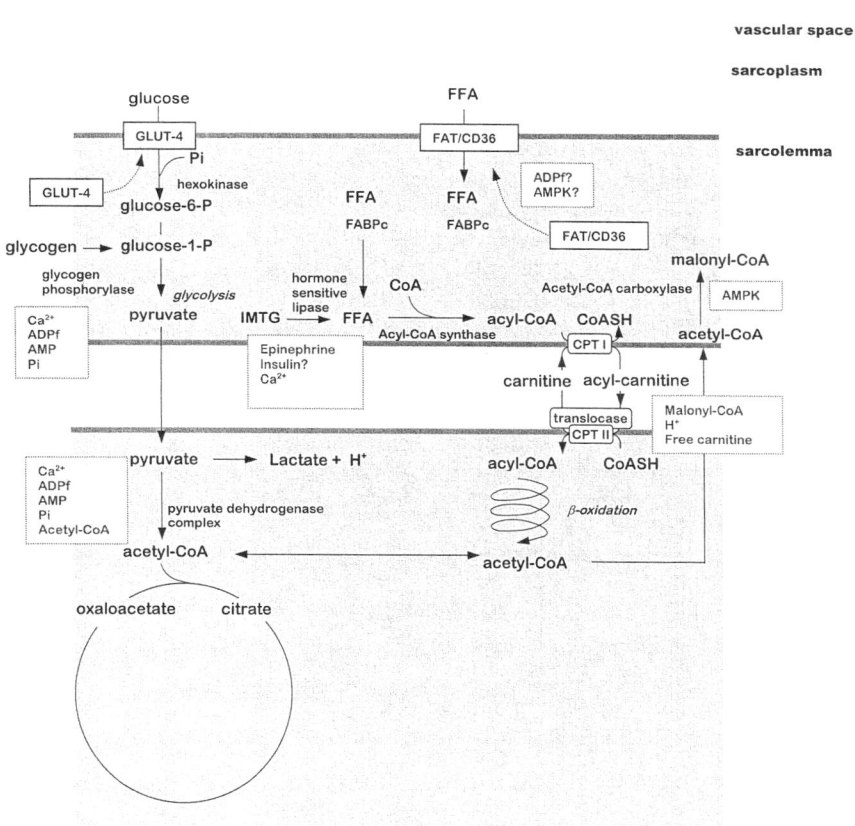

FIGURE 6. Potential regulators of the interaction between carbohydrate and fat metabolism. Three suggested sites of regulation are (1) transport of FA into the muscle controlled by transport proteins (especially the translocation of FA transporter, FAT/CD36), (2) hydrolysis of intramuscular triglycerides (IMTG) controlled by hormone-sensitive lipase (HSL), and (3) transport of FA into the mitochondria controlled by the carnitine palmitoyl transferase I (CPT I) enzyme. Transport of FA across the sarcolemma is regulated by translocation of FAT/CD36, and this translocation is triggered by yet unknown factors. It is possible that AMPK and free ADP play a role. Intramuscular lipolysis is regulated by HSL, which in turn is activated by muscle contraction (Ca^{2+}) and epinephrine. It is also likely that it is inhibited by the presence of insulin. The transport of FA into the mitochondria is regulated by the enzyme CPT I, and malonyl-CoA is known to inhibit this enzyme. It has also been suggested that a decreased pH will inhibit the CPT I, and a reduced free carnitine availability could also reduce FA transport. Regulation of carbohydrate metabolism seems more straightforward since there is a direct relationship between activation of these pathways and the aerobic work rate. Important triggers are related to the energy status of the cell (ADP, AMP, Pi, and AMPK), and the most regulated enzymes seem to be glycogen phosphorylase and pyruvate dehydrogenase (PDH).

SUMMARY

A summary of the potential mechanisms involved in the regulation of carbohydrate and fat metabolism is depicted in FIGURE 6. During low- and moderate-intensity exercise, lipolysis in adipose tissue is increased and this provides an important source of FA for the muscle. At high intensities, the delivery of FA from adipose tissue to the muscle and the use of FA by the muscle are reduced. There is clear evidence that the availability of FA is important for fat oxidation.

However, availability of FA alone cannot explain the changes observed during exercise at different intensities, and mechanisms inside skeletal muscle must be responsible. There are three potential sites of regulation of FA metabolism in skeletal muscle: transport of FA into the muscle, IMTG lipolysis, and transport of FA into the mitochondria. The increased fat oxidation from rest to low-intensity exercise and from low- to moderate-intensity exercise (65% VO_2max) may be mediated by similar factors that are responsible for increases in carbohydrate metabolism (epinephrine, Ca^{2+}, ADP, AMP, Pi, and AMPK). Other mediators may be responsible for the decrease in fat oxidation at high exercise intensities (>70% VO_2max): pH, acetyl-CoA, malonyl-CoA, or decreased free carnitine concentration. Although there have been suggestions that these mediators are involved in the regulation of FA uptake and oxidation, their roles are far from clear and there are many questions yet to be answered.

REFERENCES

1. ACHTEN, J., M. GLEESON & A.E. JEUKENDRUP. 2002. The relation between maximal fat oxidation and exercise intensity. Med. Sci. Sports Exercise **34:** 92–97.
2. BOESCH, C., J. SLOTBOOM, H. HOPPELER & R. KREIS. 1997. In vivo determination of intra-myocellular lipids in human skeletal muscle by means of localized ^{1}H-MR-spectroscopy. Magn. Reson. Med. **37:** 484–493.
3. BONEN, A., D.J. DYCK, A. IBRAHIMI & N.A. ABUMRAD. 1999. Muscle contractile activity increases fatty acid metabolism and transport and FAT/CD36. Am. J. Physiol. **276:** E642–E649.
4. BONEN, A., D.J. DYCK & J.J. LUIKEN. 1998. Skeletal muscle fatty acid transport and transporters. Adv. Exp. Med. Biol. **441:** 193–205.
5. BONEN, A., J.J. LUIKEN, Y. ARUMUGAM et al. 2000. Acute regulation of fatty acid uptake involves the cellular redistribution of fatty acid translocase. J. Biol. Chem. **275:** 14501–14508.
6. BOYD, A.E., III, S.R. GIAMBER, M. MAGER & H.E. LEBOVITZ. 1974. Lactate inhibition of lipolysis in exercising man. Metabolism **23:** 531–542.
7. CONSTANTIN-TEODOSIU, D., J.I. CARLIN, G. CEDERBLAD et al. 1991. Acetyl group accumulation and pyruvate dehydrogenase activity in human muscle during incremental exercise. Acta Physiol. Scand. **143:** 367–372.
8. COSTILL, D.L., E. COYLE, G. DALSKY et al. 1977. Effects of elevated plasma FFA and insulin on muscle glycogen usage during exercise. J. Appl. Physiol. **43:** 695–699.
9. COYLE, E.F., A.E. JEUKENDRUP, A.J.M. WAGENMAKERS & W.H.M. SARIS. 1997. Fatty acid oxidation is directly regulated by carbohydrate metabolism during exercise. Am. J. Physiol. **273:** E268–E275.
10. DECOMBAZ, J., M. FLEITH, H. HOPPELER et al. 2000. Effect of diet on the replenishment of intramyocellular lipids after exercise. Eur. J. Nutr. **39:** 244–247.
11. DECOMBAZ, J., B. SCHMITT, M. ITH et al. 2001. Postexercise fat intake repletes intra-myocellular lipids, but no faster in trained than in sedentary subjects. Am. J. Physiol. Regul. Integr. Comp. Physiol. **281:** R760–R769.

12. DYCK, D.J., S.A. PETERS, P.S. WENDLING et al. 1993. Regulation of muscle glycogen phosphorylase activity during intense aerobic cycling with elevated FFA. Am. J. Physiol. **265:** E116–E125.
13. DYCK, D.J., C.T. PUTMAN, G.J.F. HEIGENHAUSER et al. 1993. Regulation of fat-carbohydrate interaction in skeletal muscle during intense aerobic cycling. Am. J. Physiol. **265:** E852–E859.
14. GARLAND, P.B., E.A. NEWSHOLME & P.J. RANDLE. 1964. Regulation of glucose uptake by muscle: effect of fatty acids and ketone bodies, and of alloxan-diabetes and starvation, on pyruvate metabolism and on lactate/pyruvate and L-glycerol 3-phosphate/dihydroxyacetone phosphate concentration ratios in rat heart and rat diaphragm muscles. Biochem. J. **93:** 665–678.
15. GROOT, P.H.E. & W.C. HÜLSMANN. 1973. The activation and oxidation of octanoate and palmitate by rat skeletal muscle mitochondria. Biochim. Biophys. Acta **316:** 124–135.
16. HARRIS, R.C., C.V. FOSTER & E. HULTMAN. 1987. Acetylcarnitine formation during intense muscular contraction in humans. J. Appl. Physiol. **63:** 440–442.
17. HAVEL, R.J., A. NAIMARK & F. BORCHGREVINK. 1963. Turnover rate and oxidation of free fatty acids of blood plasma in man during prolonged exercise: studies during continuous infusion of palmitate-1-C^{14}. J. Clin. Invest. **42:** 1054–1063.
18. HOPPELER, H., H. HOWALD, K.E. CONLEY et al. 1985. Endurance training in humans: aerobic capacity and structure of skeletal muscle. J. Appl. Physiol. **59:** 320–327.
19. HOWLETT, K.F., L.L. SPRIET & M. HARGREAVES. 2001. Carbohydrate metabolism during exercise in females: effect of reduced fat availability. Metabolism **50:** 481–487.
20. HOWLETT, R.A., M.L. PAROLIN, D.J. DYCK et al. 1998. Regulation of skeletal muscle glycogen phosphorylase and PDH at varying exercise power outputs. Am. J. Physiol. **275:** R418–R425.
21. JEUKENDRUP, A.E. 1999. Dietary fat and physical performance. Curr. Opin. Clin. Nutr. Metab. Care **2:** 521–526.
22. JEUKENDRUP, A.E., W.H.M. SARIS & A.J.M. WAGENMAKERS. 1998. Fat metabolism during exercise: a review. Part I: Fatty acid mobilization and muscle metabolism. Int. J. Sports Med. **19:** 231–244.
23. KIENS, B., B. ESSEN-GUSTAVSSON, N.J. CHRISTENSEN & B. SALTIN. 1993. Skeletal muscle substrate utilization during submaximal exercise in man: effect of endurance training. J. Physiol. **469:** 459–478.
24. KIENS, B., B. ESSEN-GUSTAVSSON, P. GAD & H. LITHELL. 1987. Lipoprotein lipase activity and intramuscular triglyceride stores after long-term high-fat and high-carbohydrate diets in physically trained men. Clin. Physiol. **7:** 1–9.
25. KIENS, B., S. KRISTIANSEN, P. JENSEN et al. 1997. Membrane associated fatty acid binding protein (FABPpm) in human skeletal muscle is increased by endurance training. Biochem. Biophys. Res. Commun. **231:** 463–465.
26. KIENS, B. & E. RICHTER. 1998. Utilization of skeletal muscle triacylglycerol during postexercise recovery in humans. Am. J. Physiol. **275:** E332–E337.
27. LANGFORT, J., T. PLOUG, J. IHLEMANN et al. 2000. Stimulation of hormone-sensitive lipase activity by contractions in rat skeletal muscle. Biochem. J. **351:** 207–214.
28. LANGFORT, J., T. PLOUG, J. IHLEMANN et al. 1999. Expression of hormone-sensitive lipase and its regulation by adrenaline in skeletal muscle. Biochem. J. **340:** 459–465.
29. LUIKEN, J.J., L.P. TURCOTTE & A. BONEN. 1999. Protein-mediated palmitate uptake and expression of fatty acid transport proteins in heart giant vesicles. J. Lipid Res. **40:** 1007–1016.
30. MARTIN, W.H., III, G.P. DALSKY, B.F. HURLEY et al. 1993. Effect of endurance training on plasma free fatty acid turnover and oxidation during exercise. Am. J. Physiol. **265:** E708–E714.
31. MCGARRY, J.D., S.E. MILLS, C.S. LONG & D.W. FOSTER. 1983. Observations on the affinity for carnitine, and malonyl-CoA sensitivity, of carnitine palmitoyl transferase I in animal and human tissues. Biochem. J. **214:** 21–28.
32. MCGARRY, J.D., M.J. STARK & D.W. FOSTER. 1978. Hepatic malonyl-CoA levels of fed, fasted, and diabetic rats as measured using simple radioisotopic assay. J. Biol. Chem. **253:** 8291–8293.

33. OBERHOLZER, F., H. CLAASSEN, H. MOESCH & H. HOWALD. 1976. Ultrastrukturelle, biochemische, und energetische analyze einer extremen Dauerleistung (100 km Lauf). Schweiz. Z. Sportmed. **24:** 71–98.
34. ODLAND, L.M., G.J. HEIGENHAUSER & L.L. SPRIET. 2000. Effects of high fat provision on muscle PDH activation and malonyl-CoA content in moderate exercise. J. Appl. Physiol. **89:** 2352–2358.
35. ODLAND, L.M., G.J. HEIGENHAUSER, D. WONG et al. 1998. Effect of increased fat availability on fat carbohydrate interaction during prolonged exercise in men. Am. J. Physiol. **274:** R894–R902.
36. ODLAND, L.M., G.J.F. HEIGENHAUSER, G.D. LOPASCHUK & L.L. SPRIET. 1996. Human skeletal muscle malonyl-CoA at rest and during prolonged submaximal exercise. Am. J. Physiol. **270:** E541–E544.
37. ODLAND, L.M., R.A. HOWLETT, G.J. HEIGENHAUSER et al. 1998. Skeletal muscle malonyl-CoA content at the onset of exercise at varying power outputs in humans. Am. J. Physiol. **274:** E1080–E1085.
38. ORAM, J.F., J.I. WENGER & J.R. NEELY. 1975. Regulation of long chain fatty acid activation in heart muscle. J. Biol. Chem. **250:** 73–78.
39. OSCAI, L.B., D.A. ESSIG & W.K. PALMER. 1990. Lipase regulation of muscle triglyceride hydrolysis. J. Appl. Physiol. **69:** 1571–1577.
40. PETERS, S., D. DYCK, A. BONEN & L. SPRIET. 1998. Effects of epinephrine on lipid metabolism in resting skeletal muscle. Am. J. Physiol. **275:** E300–E309.
41. PETERS, S.J. & L.L. SPRIET. 1995. Skeletal muscle phosphofructokinase activity examined under physiological conditions in vitro. J. Appl. Physiol. **78:** 1853–1858.
42. PETTIT, F.H., J.W. PELLEY & L.J. REED. 1975. Regulation of pyruvate dehydrogenase kinase and phosphatase by acetyl-CoA/CoA and NADH/NAD ratios. Biochem. Biophys. Res. Commun. **65:** 575–582.
43. PHILLIPS, S.M., H.J. GREEN, M.A. TARNOPOLSKY et al. 1996. Effects of training duration on substrate turnover and oxidation during exercise. J. Appl. Physiol. **81:** 2182–2191.
44. RANDLE, P.J., E.A. NEWSHOLME & P.B. GARLAND. 1964. Regulation of glucose uptake by muscle: 8. Effects of fatty acids, ketone bodies, and pyruvate, and of alloxandiabetes and starvation, on the uptake and metabolic fate of glucose in rat heart and diaphragm muscles. Biochem. J. **93(3):** 652–665.
45. RANDLE, P.J., C.N. HALES, P.B. GARLAND & E.A. NEWSHOLME. 1963. The glucose–fatty acid cycle: its role in insulin sensitivity and the metabolic disturbances of diabetes mellitus. Lancet **1:** 785–789.
46. REED, L.J. 1981. Regulation of mammalian pyruvate dehydrogenase complex by a phosphorylation-dephosphorylation cycle. Curr. Top. Cell. Regul. **18:** 95–106.
47. RENNIE, M.J. & W.W. WINDER. 1976. A sparing effect of increased plasma fatty acids on muscle and liver glycogen content in the exercising rat. Biochem. J. **156:** 647–655.
48. ROMIJN, J.A., E.F. COYLE, L.S. SIDOSSIS et al. 1993. Regulation of endogenous fat and carbohydrate metabolism in relation to exercise intensity. Am. J. Physiol. **265:** E380–E391.
49. ROMIJN, J.A., E.F. COYLE, X-J. ZHANG et al. 1995. Fat oxidation is impaired somewhat during high intensity exercise by limited plasma FFA. J. Appl. Physiol. **79:** 1939–1945.
50. SAGGERSON, E.D. & C.A. CARPENTER. 1981. Carnitine palmitoyltransferase and carnitine octanoyltransferase activities in liver, kidney cortex, adipocyte, lactating mammary gland, skeletal muscle, and heart. FEBS Lett. **129:** 229–232.
51. SAHA, A., T. KUROWSKI & N. RUDERMAN. 1995. A malonyl-CoA fuel sensing mechanism in muscle: effects of insulin, glucose, and denervation. Am. J. Physiol. **267:** E95–E101.
52. SAHA, A.K., T.G. KUROWSKI & N.B. RUDERMAN. 1995. A malonyl-CoA mechanism in muscle: effect of insulin, glucose, and denervation. Am. J. Physiol. **269:** E283–289.
53. SIDOSSIS, L.S., A.R. COGGAN, A. GASTALDELLI & R.R. WOLFE. 1995. A new correction factor for use in tracer estimations of plasma fatty acid oxidation. Am. J. Physiol. **269:** E649–E656.
54. SIDOSSIS, L.S., A. GASTALDELLI, S. KLEIN & R.R. WOLFE. 1997. Regulation of plasma fatty acid oxidation during low- and high-intensity exercise. Am. J. Physiol. **272:** E1065–E1070.

55. SIDOSSIS, L.S., R.R. WOLFE & A.R. COGGAN. 1998. Regulation of fatty acid oxidation in untrained vs. trained men during exercise. Am. J. Physiol. **274:** E510–E515.
56. SPRIET, L.L. 1998. Regulation of fat/carbohydrate interaction in human skeletal muscle during exercise. Adv. Exp. Med. Biol. **441:** 249–261.
57. SPRIET, L.L., G.J.F. HEIGENHAUSER & N.L. JONES. 1986. Endogenous triacylglycerol utilization by rat skeletal muscle during tetanic stimulation. J. Appl. Physiol. **60:** 410–415.
58. STANKIEWICZ-CHOROSZUCHA, B. & J. GORSKI. 1978. Effect of beta-adrenergic blockade on intramuscular triglyceride utilization during exercise. Experientia **34:** 357–358.
59. STARRITT, E.C., R.A. HOWLETT, G.J. HEIGENHAUSER & L.L. SPRIET. 2000. Sensitivity of CPT I to malonyl-CoA in trained and untrained human skeletal muscle. Am. J. Physiol. Endocrinol. Metab. **278:** E462–E468.
60. TRUDEAU, F., S. BERNIER, E. DE GLISEZINSKI *et al.* 1999. Lack of antilipolytic effect of lactate in subcutaneous abdominal adipose tissue during exercise. J. Appl. Physiol. **86:** 1800–1804.
61. TURCOTTE, L.P., B. KIENS & E.A. RICHTER. 1991. Saturation kinetics of palmitate uptake in perfused skeletal muscle. FEBS Lett. **279:** 327–329.
62. TURCOTTE, L.P., A.K. SRIVASTAVA & J.L. CHIASSON. 1997. Fasting increases plasma membrane fatty acid–binding protein [FABP(PM)] in red skeletal muscle. Mol. Cell. Biochem. **166:** 153–158.
63. VAN DER VUSSE, G.J. & R.S. RENEMAN. 1996. Lipid metabolism in muscle. *In* Handbook of Physiology. Section 12—Exercise: Regulation, and Integration of Multiple Systems, pp. 952–994. Oxford University Press. London/New York.
64. VUKOVICH, M.D., D.L. COSTILL, M.S. HICKEY *et al.* 1993. Effect of fat emulsion infusion and fat feeding on muscle glycogen utilization during cycle exercise. J. Appl. Physiol. **75:** 1513–1518.
65. WINDER, W.W., J. AROGYASAMI, R.J. BARTON *et al.* 1989. Muscle malonyl-CoA decreases during exercise. J. Appl. Physiol. **67:** 2230–2233.
66. WOLFE, R.R., S. KLEIN, F. CARRARO & J-M. WEBER. 1990. Role of triglyceride–fatty acid cycle in controlling fat metabolism in humans during and after exercise. Am. J. Physiol. **258:** E382–E389.

The Sphingomyelin-Signaling Pathway in Skeletal Muscles and Its Role in Regulation of Glucose Uptake

JAN GÓRSKI, AGNIESZKA DOBRZYN, AND
MALGORZATA ZENDZIAN-PIOTROWSKA

Department of Physiology, Medical Academy of Bialystok, Bialystok, Poland

ABSTRACT: Sphingomyelin has been shown to be a source of bioactive compounds. This sphingolipid is located mostly in the outer layer of the plasma membrane and in the membranes of organelles. Sphingomyelin located in the plasma membrane is hydrolyzed into ceramide and phosphorylcholine. Ceramide is the principal second messenger in the sphingomyelin transmembrane signaling pathway. Products of ceramide metabolism, namely, sphingosine, sphingosine-1-phosphate, and ceramide-1-phosphate, also exert broad biological effects. The major effects of ceramide are induction of differentiation, inhibition of proliferation, regulation of inflammatory processes, and induction of apoptosis. There is also convincing evidence that ceramide counteracts insulin-stimulated glucose uptake. Ceramides are also present in skeletal muscles. We investigated ceramide metabolism in different skeletal muscle types of the rat at rest and after prolonged exercise of moderate intensity. Exercise reduced the total content of ceramide fatty acids and changed their composition in each muscle type. These data indicate that the sphingomyelin-signaling pathway functions in skeletal muscles and that its activity is downregulated during prolonged exercise. The content of ceramide in the muscles was inversely related to 2-deoxyglucose uptake by the muscles. This indicates that ceramide may be involved in regulation of glucose uptake by skeletal muscles *in vivo*.

KEYWORDS: sphingomyelin-signaling pathway; glucose uptake; skeletal muscle; bioactive compounds; ceramide

INTRODUCTION

There are two major fractions of phospholipids in the cell membranes, namely, glycerophospholipids and sphingolipids. A role of phospholipids in formation of the cell structure has been known for a long time. A role of glycerophospholipids in the transmembrane signal transduction was established over 20 years ago. The major glycerophospholipid-signaling pathway includes formation of diacylglycerol and inositol 1,4,5-trisphosphate from phosphatidylinositol 4,5-biphosphate in response to a variety of extracellular stimuli. Also, phosphatidylcholine was shown to be a source of diacylglycerol, phosphatidic acid, and arachidonic acid, a precursor of

Address for correspondence: Dr. Jan Górski, Department of Physiology, Medical Academy of Bialystok, 15-230 Bialystok, Poland. Voice: +48-85-7420330.
gorski@amb.edu.pl

bioactive compounds. Research of the last decade has established that sphingolipid-sphingomyelin is also involved in the transmembrane signal transduction. Its hydrolysis yields ceramide and phosphorylcholine. Ceramide is recognized as the second messenger in this pathway. Moreover, products of ceramide metabolism, that is, sphingosine, sphingosine-1-phosphate, and ceramide-1-phosphate, are also very active biologically. Ceramide is composed of a sphingoid base and a long-chain fatty acid joint in an amine bond. The major sphingoid base is C-18 and C-20 sphingosine. Ceramide is involved in induction of cell differentiation, inhibition of cell proliferation, regulation of inflammatory processes, and induction of apoptosis. There is also convincing evidence that ceramide inhibits insulin-mediated glucose uptake. The following contribution is focused on the function of the sphingomyelin-signaling pathway in skeletal muscles and on a role of ceramide in regulation of glucose uptake. It is based on the data from the literature and also includes unpublished results obtained by the authors.

OVERVIEW OF THE SPHINGOMYELIN-SIGNALING PATHWAY

The sphingomyelin transmembrane signaling pathway was described over a decade ago and it has been extensively studied ever since.[8,12,17] The major second messenger in this pathway is ceramide. In the process of signal transduction, ceramide is generated mostly from sphingomyelin located in the outer layer of the plasma membrane by the enzyme, neutral, Mg^{2+}-dependent sphingomyelinase, which is present there. The enzyme is activated by a receptor-mediated process. Certain amounts of ceramide are formed from sphingomyelin in endo/lysosomes by the enzyme, acidic sphingomyelinase. This enzyme is also activated in the receptor-mediated process (FIG. 1). Other sphingomyelinases have also been identified,

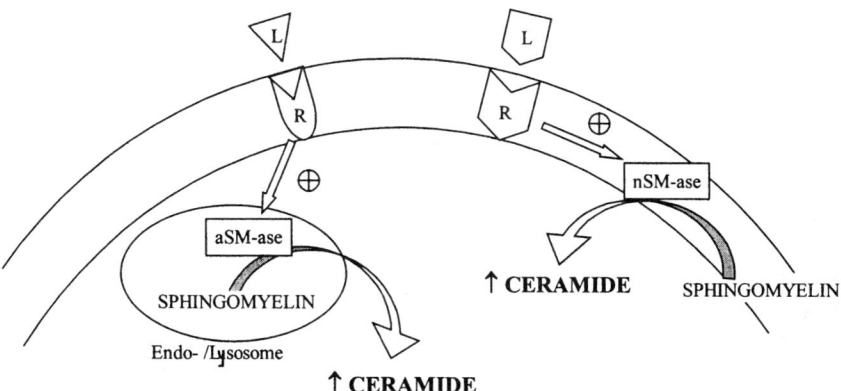

FIGURE 1. The sphingomyelin transmembrane signaling pathway. Sphingomyelin is hydrolyzed to ceramide and phosphocholine by the enzyme, neutral, Mg^{2+}-dependent sphingomyelinase, located in the plasma membrane (nSM-ase) and by endo/lysosomal acidic sphingomyelinase (aSM-ase). Activation of the enzymes is a receptor-mediated process. L, ligand; R, receptor.

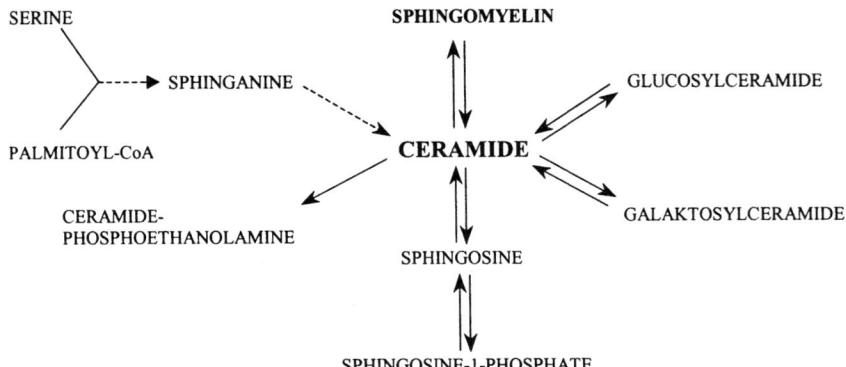

FIGURE 2. Ceramide metabolic pathways. Sphingomyelin is considered to be the major source of ceramide. Sphinganine is a key intermediate in the *de novo* synthesis pathway (other intermediates of this pathway are not shown). Ceramide is catabolized to sphingosine. It is important to note that most of the pathways can proceed in two directions.

namely, neutral, Mg^{2+}-independent sphingomyelinase; alkaline sphingomyelinase present in bile and in the digestive tract; and Zn^{2+}-dependent sphingomyelinase present in serum.[10,14] The biological role of the latter enzymes has not been fully elucidated. Ceramide is catabolized to sphingosine and a long-chain fatty acid in a reaction catalyzed by the enzyme, ceramidase. It can also be converted to sphingomyelin, glucosyl-ceramide, galactosyl-ceramide, ceramide-1-phosphate, and ceramide-phosphoethanolamine (FIG. 2).[8,17] However, the regulation in the content of ceramide in the cell is a very complex process since sphingomyelin is not the only source of the compound.

Ceramide is synthesized *de novo* from serine and palmitoyl-CoA. The first step in the synthesis is formation of 3-ketosphinganine in a reaction catalyzed by the enzyme, serine palmitoyl transferase. 3-Ketosphinganine is converted to sphinganine by the enzyme, ketosphinganine reductase. Sphinganine is acylated by the enzyme, ceramide synthase, and the resulting dihydroceramide is in turn converted to ceramide by the enzyme, dihydroceramide desaturase. Furthermore, not only sphingomyelin, but also sphingosine, glucosyl-ceramide, galactosyl-ceramide, and ceramide-1-phosphate can be converted back to ceramide (FIG. 2).[8,17] The quantitative contribution of these pathways for ceramide generation and removal in the regulation of the content of ceramide in the cell remains, however, to be established. At present, it is accepted that the major source of ceramide in the cell is sphingomyelin via the reaction catalyzed by membrane neutral, Mg^{2+}-dependent and acidic lysosomal sphingomyelinases. A substantial number of stimuli have been shown to increase the formation of ceramide. They have been grouped as follows: inducers of differentiation, inducers of apoptosis, damaging agents, and inflammatory cytokines. Examples of such stimuli are vitamin D_3, tumor necrosis factor α (TNFα), ionizing radiation, and interleukin-1β.[15] The biological effects of ceramide have been studied in different cell types, almost entirely *in vitro*, and were found to be diversified. The major effects include induction of cell differentiation, inhibition of cell proliferation, induction of apoptosis, and regulation of the inflammatory reactions.[12,14,17] There is also convincing evidence that ceramide is involved in the regulation of glucose uptake in different cell types, including myocytes (see below).

SPHINGOMYELIN-SIGNALING PATHWAY IN SKELETAL MUSCLES

Content and Composition of Ceramides and Sphingomyelins

There are very few data regarding the regulation of the sphingomyelin-signaling pathway in skeletal muscles. The presence of ceramide in skeletal muscles was first documented by Turinsky et al.[22] They showed that the content of ceramide in the soleus (a slow-twitch oxidative muscle) is lower than in the plantaris (a fast-twitch oxidative-glycolytic muscle). Denervation elevated the content of ceramide in both muscles. An important observation of this study was that the content of ceramide in the soleus and in plantaris was elevated in obese, insulin-resistant Zucker rats. The same group of authors examined the effect of insulin and exercise on the content of ceramide in calf muscles. They found that neither of these factors affected the content of ceramide in the muscles.[23] Unfortunately, the content of ceramide was determined in the calf muscles, so no information regarding particular muscle types was provided.

It should also be added that only relatively short-term, moderate-intensity exercise (stimulation of the sciatic nerve with single pulses for 25 min) was examined. The previous observation on elevation in the content of ceramide in denervated muscles was not confirmed in this study. Fasting and administration of endotoxin increased the content of ceramide in the muscles.[24] Further data on the content of ceramide in skeletal muscles *in vivo* came from our laboratory.[6] The content and composition of ceramide fatty acids along with that of sphingomyelin fatty acids and activity of neutral, Mg^{2+}-dependent sphingomyelinase were measured in different skeletal muscle types of the rat at rest and after prolonged exercise of moderate intensity on a treadmill. In this study, the total content of ceramide expressed as the content of ceramide fatty acids was about three times higher than that reported by Turinsky et al.,[22] and the difference was ascribed to different treatment of the rats (fed vs. fasted) and to different analytical procedures. The total content of ceramide fatty acids depended on the muscle type: it was 161.2 nmol/g in the soleus, 149.9 nmol/g in the red gastrocnemius, and 122.2 nmol/g in the white gastrocnemius (the muscles are composed predominantly of slow-twitch oxidative, fast-twitch oxidative-glycolytic, and fast-twitch glycolytic fibers, respectively[1,20]).

Twelve different ceramides were identified in each muscle in relation to the fatty acid residues (FIG. 3). They contained residues of the following acids: myristic (14:0), palmitic (16:0), palmitoleic (16:1), stearic (18:0), oleic (18:1), linoleic (18:2), linolenic (18:3), arachidonic (20:4), eicosapentaenoic (20:5), behenic (22:0), docosahexaenoic (22:6), and nervonic acid (24:1). The major saturated fatty acids were stearic and palmitic acids, and the major unsaturated fatty acid was oleic acid in each muscle type. Linoleic and docosahexaenoic acids were least represented. Twelve different sphingomyelins were also identified based on their fatty acid residues. They contained the same fatty acids as ceramides. As in the case of ceramides,

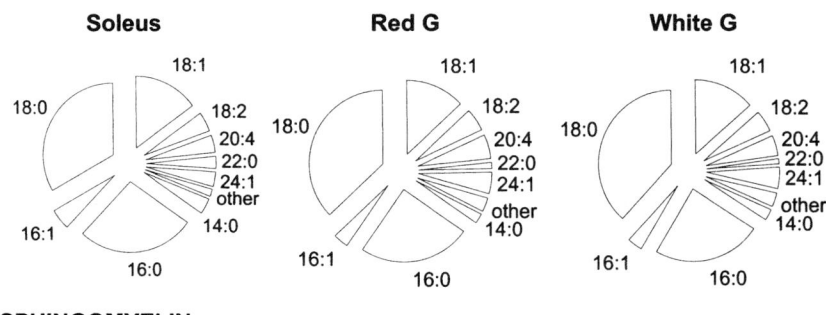
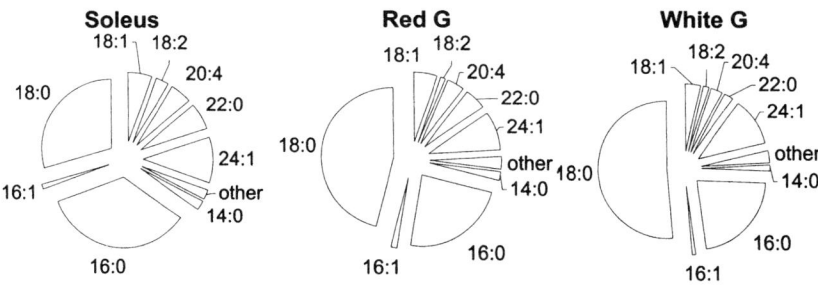

FIGURE 3. The percent composition of ceramides (**top**) and sphingomyelins (**bottom**) in different muscle types at rest, based on data published in our previous study.[6] The following acids are shown: myristic (14:0), palmitic (16:0), palmitoleic (16:1), stearic (18:0), oleic (18:1), linoleic (18:2), arachidonic (20:4), behenic (22:0), and nervonic (24:1). Others are linolenic (18:3), eicosapentaenoic (20:5), and docosahexaenoic (22:6) acids.

the major saturated fatty acids were palmitic and stearic acids; however, in contrast, the major unsaturated fatty acid was nervonic acid (FIG. 3).

The total content of sphingomyelin fatty acids in the soleus was higher than in either section of the gastrocnemius, with no difference between the latter two muscles. The activity of the neutral, Mg^{2+}-dependent sphingomyelinase also depended on the muscle type: it was highest in the soleus and lowest in the white gastrocnemius, with the value in the red gastrocnemius being in-between. The enzyme activity correlated negatively with the ratio, total content of sphingomyelin fatty acids/total content of ceramide fatty acids, which indicated that sphingomyelin was indeed the principal source of ceramide in the muscles. We have further shown that the ratio, individual sphingomyelin fatty acid/individual ceramide fatty acid, depended to a great extent on the fatty acid residue. This strongly indicates that susceptibility of sphingomyelins to the action of the sphingomyelinase is diversified and depends mostly on the fatty acid residue. Exercise considerably reduced the total content of ceramide fatty acids in each muscle. The total content of sphingomyelin fatty acids was also reduced, although to a much smaller degree than in the case of the ceramides. Exercise also produced substantial changes in the composition of acids in each compound. The activity of the sphingomyelinase was reduced after exercise in the soleus and in the red section of the gastrocnemius, while it remained stable in the white gastrocnemius. This suggests that inhibition of the sphingomyelin pathway contributes to the reduction in the content of ceramide during exercise in the two muscles with high oxidative capacity, that is, in the soleus and red gastrocnemius. The reason for the postexercise reduction in the content of ceramide in the white gastrocnemius remains obscure.[6]

Incorporation of ^{14}C-Palmitic Acid and ^3H-Oleic Acid into Skeletal Muscle Sphingomyelin and Ceramide

As mentioned above, ceramide may be generated not only from sphingomyelin, but also along several other routes. The aim of the following experiment was to examine incorporation of blood-borne ^{14}C-palmitic acid and ^3H-oleic acid into ceramide and sphingomyelin fractions in different muscle types of the rat at rest and after long-term exercise of moderate intensity. The general rationale behind this study was (a) to establish whether the blood-borne labeled fatty acids can be useful in the investigation of muscle ceramide *in vivo* and (b) to investigate further the regulation of the sphingomyelin-signaling pathway in the muscles using labeled fatty acids. If sphingomyelin were the main source of ceramide, we expected that changes in the radioactivity of ceramide containing a given acid would be inversely related to changes in the radioactivity of sphingomyelin containing the same label.

Methods

The experiments were carried out on male Wistar rats, 250–260 g of body weight, fed *ad libitum* a commercial pellet diet for rodents. They were divided into two groups: (1) control and (2) exercised until exhaustion. The rats in group 2 were made to run on a treadmill moving with a speed of 1200 m/h and set at $+10°$ incline. Exhaustion was regarded as the point at which rats refused further running and when placed on a table did not escape. The running time was 223 ± 22 min. The rats were anesthetized immediately after the run, and 10 µCi of ^{14}C-palmitic acid (specific activity, 55 mCi/mmol; DuPont) and 10 µCi of ^3H-oleic acid (specific activity, 5 Ci/mmol; DuPont) suspended in albumin were administered in the tail vein. Ten min after administration of the labels, samples of the soleus and red and white gastrocnemius were taken. The muscle samples were cleaned from any visible nerves, fat, and fascias and frozen in liquid nitrogen. Rats in the control group were anesthetized, the labels were given, and the muscle samples were taken 10 min later. The frozen muscles were pulverized in an aluminum mortar with a stainless pestle precooled in liquid nitrogen. The powder was transferred into tubes containing methanol at $-20°C$, and lipids were extracted with chloroform methanol. The methanol contained an antioxidant (butylated hydroxytoluene, Sigma, 30 mg/100 mL).[26] Sphingomyelin was isolated on silica plates (Kieselgel 60, 0.22 mm, Merck) using chloroform/methanol/acetic acid/water (50/37.5/3.5/2 v/v/v/v) as the developing solvent.[11] Ceramide was isolated using the plates as above according to reference 16 as modified by reference 6. First, they were developed to one-third of the plate's total length in chloroform/methanol/25% NH_3 (20/5/0.2 v/v/v). Next, they were dried and developed using a solvent composed of heptane, isopropyl ether, and acetic acid (60/40/3 v/v/v). Standards of sphingomyelin and ceramide (Sigma) were run along with the examined samples. The plates were dried and sprayed with a 0.5% solution of 3′,7′-dichlorofluorescein in absolute methanol; lipid bands were visualized under UV light. The bands corresponding to sphingomyelin and ceramide were scraped off the plates into the scintillation vials. A scintillation cocktail (Ultima Gold, Packard) was added, and the radioactivity was counted on a Tri-carb 1900 Packard counter. The results were evaluated statistically by means of Student's t test for unpaired data, with $n = 10$ for each mean.

Results and Discussion

^{14}C- and ^{3}H-radioactivity were present in sphingomyelin and ceramide moieties in each muscle, both at rest and after the run. It can be presumed that only a tiny amount of the acids could have been converted to other acids and incorporated into sphingomyelin and ceramide. Therefore, the radioactivity present in these compounds was represented almost entirely by palmitic and oleic acids, respectively. Exercise considerably increased incorporation of both acids into the sphingomyelin moiety and markedly reduced incorporation of the acids into the ceramide moiety in each muscle type (FIG. 4). Interestingly, sphingomyelin containing oleic acid represents only 5–10% and sphingomyelin containing palmitic acid represents as much as 18–35% (depending on the muscle type) of the total content of sphingomyelin fatty acids.[6] However, the percentage elevation in the incorporation of ^{3}H-oleate into sphingomyelin either is higher (the soleus and white gastrocnemius) or did not differ (the red gastrocnemius) from the incorporation of ^{14}C-palmitate. This indicates that oleate is incorporated into the sphingomyelin moiety in preference to palmitate. On the other hand, the postexercise radioactivity of ceramide-oleate is lower than the radioactivity of ceramide-palmitate, which in turn would indicate that less sphingomyelin-oleate is converted to ceramide. The ratio of radioactivity of sphingomyelin/ radioactivity of ceramide increased after exercise for both labels by ~10-fold in each muscle (FIG. 5). As mentioned above, labeled fatty acid could reach the ceramide moiety in two ways: either via sphingomyelin (i.e., on the sphingomyelin-signaling pathway) or on the *de novo* synthesis pathway. The latter pathway is considered to be a slow one.[12] We allowed only 10 min for the labels to incorporate and, thus, labeled sphingomyelin was the major source of the labeled ceramide. There are no data to suggest that the postexercise reduction in the content of ceramide was a consequence of accelerated catabolism of the compound. This would indicate that the exercise inhibited conversion of sphingomyelin into ceramide. Indeed, in the previous study, we reported reduced activity of neutral, Mg^{2+}-dependent sphingomyelinase in soleus and in red gastrocnemius, but not in the white section of the latter muscle after exercise.[6] This suggests that generation of ceramide from sphingomyelin was reduced in the first two muscles on the route operated by the enzyme. A reason for the reduction in radioactivity of ceramide after exercise in the white section of the gastrocnemius remains open. There is the possibility that acidic sphingomyelinase plays an important role in generation of ceramides in the white muscle. Its activity in skeletal muscles has not been measured to date. If it were the case, however, a reduction in the activity of this enzyme during prolonged exercise could be responsible for the reduction in formation of ceramide from sphingomyelin in the muscle.

Taken together, the data available indicate that the sphingomyelin-signaling pathway operates in skeletal muscles. Its activity is downregulated during prolonged

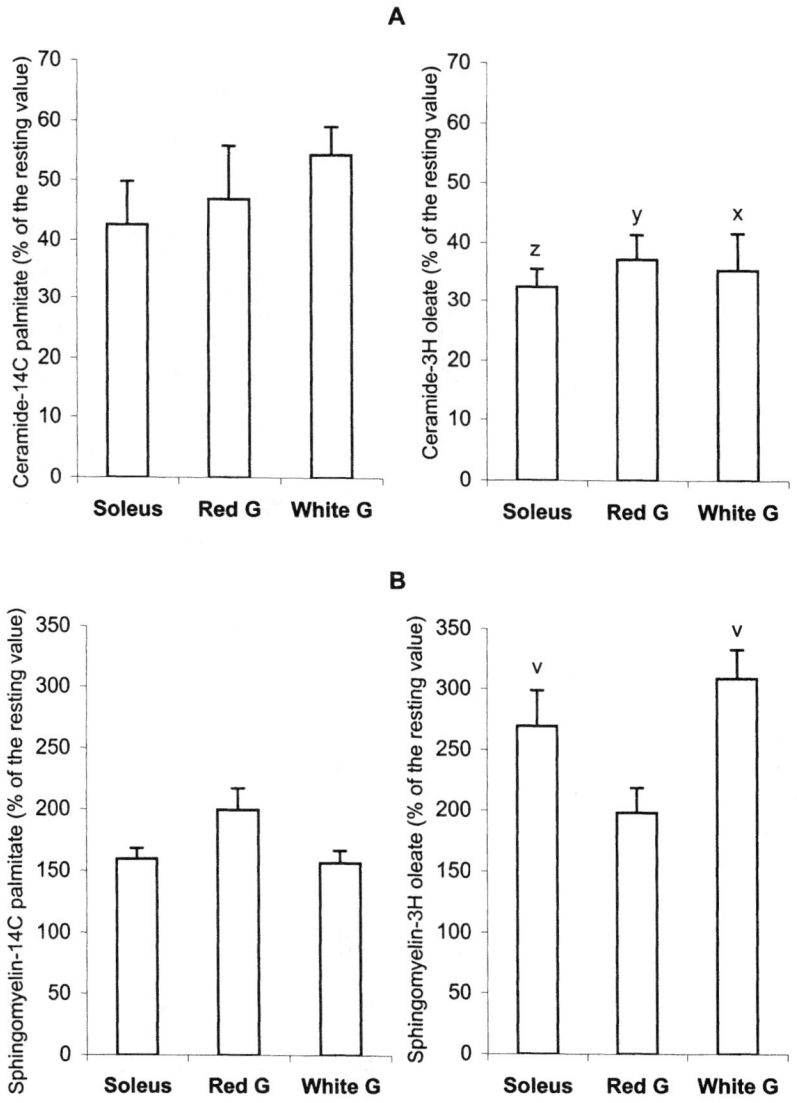

FIGURE 4. Effect of prolonged exercise on incorporation of blood-borne ^{14}C-palmitic and ^{3}H-oleic acids into sphingomyelin (**A**) and ceramide (**B**) in different muscle types. The mean value for each compound at rest was put at 100%, and the individual values after exercise are expressed as a percent of the mean. G, gastrocnemius. x, $p < 0.05$; y, $p < 0.02$; z, $p < 0.01$; v, $p < 0.001$ vs. the respective value for ^{14}C-palmitate.

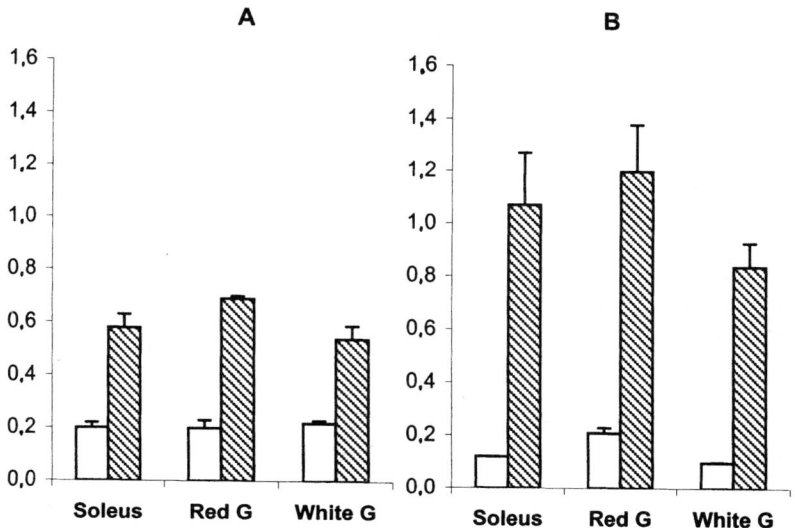

FIGURE 5. The ^{14}C-sphingomyelin/^{14}C-ceramide ratio (**A**) and ^{3}H-sphingomyelin/^{3}H-ceramide ratio (**B**) in different muscle types at rest (*open bars*) and after prolonged exercise (*hatched bars*). G, gastrocnemius. $p < 0.001$ for each postexercise value vs. the respective resting value.

exercise. This is reflected both by the reduction in the content of ceramide seen in the previous study[6] and by the reduction in the incorporation of the blood-borne labeled fatty acids into the muscle ceramide (the present data). There is good evidence that ceramide inhibits insulin-stimulated glucose entry into different cell types, including myocytes. The reduction in the content of ceramide during prolonged exercise would reduce the degree of inhibition and thus result in facilitation of glucose uptake.

CERAMIDE AND GLUCOSE UPTAKE

In most studies, the cell-permeable ceramide analogues containing 2 or 6 carbon fatty acids (C2 and C6 ceramides) and/or sphingomyelinase were used. Incubation of insulin-sensitive rat hepatoma Fao cells with bacterial sphingomyelinase resulted in time- and dose-dependent reduction in insulin-induced tyrosine phosphorylation of insulin receptor substrate-1 (IRS-1). A similar effect was observed after incubation with C2 and C6 ceramide analogues.[9] C2 and C6 ceramides and sphingomyelinase have also been shown to inhibit insulin-activated tyrosine phosphorylation of insulin receptor and IRS-1 in cultured 3T3-L1 adipocytes and myeloid 32D cells.[13] In another study on incubated 3T3-L1 adipocytes, C2 and C6 ceramides were shown to increase 2-deoxyglucose uptake in the absence of insulin. This was accompanied by an increased concentration of GLUT1 and GLUT4 in the plasma membranes.

Expression of GLUT1, but not GLUT4, was also elevated. However, the ceramides inhibited insulin-activated elevation in glucose uptake and translocation of GLUT1 and GLUT4 to the plasma membranes. C2 ceramide did not inhibit insulin-elevated phosphorylation of insulin receptor and IRS-1. This intriguing finding clearly indicates that (a) ceramide does affect glucose uptake and (b) its action on glucose uptake depends on the presence of insulin.[4,27]

Further data obtained in the same cell type confirmed the inhibitory action of C2 ceramide on insulin-stimulated transport of glucose and translocation of GLUT4 to the plasma membrane. These data also demonstrated the inhibitory action of C2 ceramide on the phosphorylation and activation of Akt kinase by insulin. Earlier data on the inhibitory action of ceramide on insulin-mediated tyrosine phosphorylation of IRS-1 were not confirmed in this study.[21] Detailed investigations of the role of ceramides in the regulation of glucose uptake in the absence of insulin were also performed on incubated rat epididymal adipocytes. It was found that sphingomyelinase remarkably increased glucose uptake by the cells. Concomitantly, changes in the distribution of GLUT4 occurred; namely, its content in the plasma membrane increased and the content in the microsomal fraction decreased. The distribution of GLUT1 remained unchanged. Sphingomyelinase did not affect the insulin receptor and IRS-1 tyrosine phosphorylation. These results show that the mechanism by which sphingomyelinase stimulates movement of GLUT4 to the plasma membrane is different from the mechanism of insulin action. Its details, however, remain to be elucidated.[5]

There are also convincing data showing that ceramide may be an important player in the regulation in glucose uptake by skeletal myocytes. In cultured skeletal muscle cells (line L6), C2 and C6 ceramides inhibit activation of the mitogen-activated protein kinase (MAPK) by insulin. They also activated protein phosphatase-2A and prevented the inhibitory action of insulin on this enzyme.[2,3] As in the case of adipocytes, sphingomyelinase stimulated basal glucose uptake in incubated skeletal muscle (m. soleus); however, unlike in the adipocytes, the enzyme potentiated glucose uptake induced by insulin. Sphingomyelinase did not influence either basal or insulin-stimulated activity of insulin receptor tyrosine kinase and phosphatidylinositol 3-kinase. This indicates that sphingomyelinase activates distal steps in the regulation of glucose uptake. Treatment by this enzyme markedly increased the content of ceramide in the muscle. However, unlike in the adipocytes, the ceramides had no effect on either basal or insulin-stimulated glucose uptake by the muscle.[25] Lack of ceramide action in the muscle was later ascribed to poor permeability of the myocyte membranes to the ceramides or, alternatively, to the inability to reach regulatory sites in the myocyte.[4] C2 ceramide inhibits insulin-stimulated glycogen synthesis and phosphorylation of protein kinase B and stimulates MAP kinase in incubated C2C12 myotubes. It has been postulated that ceramide mediates palmitate-induced insulin resistance in the myocytes.[19]

Finally, C2 ceramide was shown not to directly affect basal glucose uptake, but it abolished insulin-stimulated glucose uptake and glycogen synthesis in incubated L6 myotubes. This inhibition was caused not by the inhibition of phosphorylation of IRS-1 or activation of phosphoinositide 3-kinase, but by blockade of activation of protein kinase B.[7] The skeletal muscles constitute ~40% of the body weight and thus consume large amounts of glucose. Consumption of glucose by the muscles increases considerably during contractile activity. Skeletal muscles are insulin-sensitive. How-

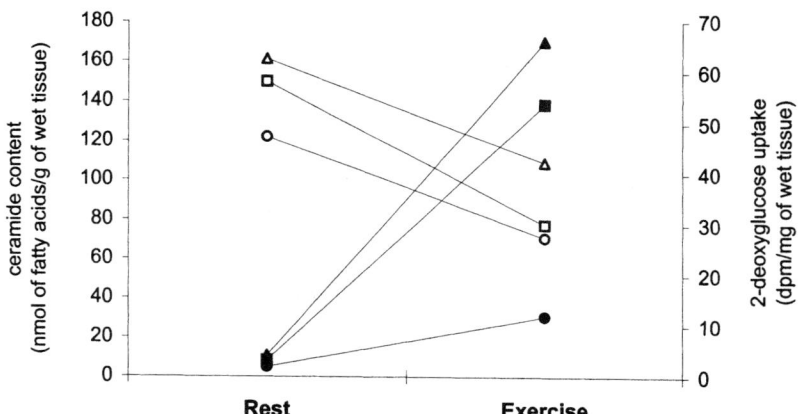

FIGURE 6. Relationship between the content of ceramide (*open figures*) and 2-deoxyglucose uptake (*closed figures*) in different skeletal muscle types (□■, soleus; ○●, red gastrocnemius; △▲, white gastrocnemius). Rats exercised on a treadmill (1200 m/h, +10° incline) until exhaustion. (From reference 6; reproduced with permission.)

ever, contractile activity itself increases glucose uptake, even in the absence of insulin.[18] We have provided indirect proof for the role of ceramide in the regulation in glucose uptake in skeletal muscles. In our study, we found that the postexercise elevation in glucose uptake by different skeletal muscle types in the rat was inversely related to the total content of ceramide in the muscles (FIG. 6).[6]

SUMMARY

Sphingomyelin has been shown to be a source of bioactive compounds. This sphingolipid is located mostly in the outer layer of the plasma membrane and in the membranes of organelles. Sphingomyelin located in the plasma membrane is hydrolyzed by the enzyme, neutral, Mg^{2+}-dependent sphingomyelinase, whereas sphingomyelin located in lysosomes is hydrolyzed by acidic sphingomyelinase. Both enzymes are activated in a receptor-mediated process. The products of hydrolysis are ceramide and phosphorylcholine. Ceramide is the principal second messenger in the sphingomyelin transmembrane signaling pathway. Products of ceramide metabolism, namely, sphingosine, sphingosine-1-phosphate, and ceramide-1-phosphate, also exert broad biological effects. The major effects of ceramide are induction of differentiation, inhibition of proliferation, regulation of inflammatory processes, and induction of apoptosis. There is also convincing evidence that ceramide counteracts insulin-stimulated glucose uptake.

Ceramides are also present in skeletal muscles. We investigated ceramide metabolism in different skeletal muscle types of the rat at rest and after prolonged exercise of moderate intensity. Twelve different ceramides and sphingomyelins were identified and quantified, based on fatty acid residues, in the soleus, red gastrocnemius, and white gastrocnemius (the muscles are composed mostly of the slow-twitch oxidative, fast-twitch oxidative-glycolytic, and fast-twitch glycolytic fibers, respectively). They contained the following acids: myristic, palmitic, palmitoleic, stearic, oleic, linoleic, linolenic, arachidonic, eicosapentaenoic, behenic, docosahexaenoic, and nervonic. Saturated fatty acids constitute nearly 70% of total ceramide fatty acids and almost 80% of sphingomyelin fatty acids. The major saturated fatty acids in ceramide and sphingomyelin moiety were palmitic and stearic acids. The major ceramide unsaturated fatty acid was oleic acid and the major sphingomyelin unsaturated fatty acid was nervonic acid. The total content of ceramide fatty acids and the activity of neutral, Mg^{2+}-dependent sphingomyelinase were highest in the soleus and lowest in the white gastrocnemius. The blood-borne ^{14}C-palmitic and ^{3}H-oleic acids were rapidly incorporated into ceramide and sphingomyelin moieties. Exercise reduced the total content of ceramide fatty acids and changed their composition in each muscle type. In addition, it reduced incorporation of the labeled acids into ceramide as well as decreased activity of the neutral, Mg^{2+}-dependent sphingomyelinase in the soleus and red gastrocnemius. These data indicate that the sphingomyelin-signaling pathway functions in skeletal muscles and that its activity is downregulated during prolonged exercise. The content of ceramide in the muscles was inversely related to 2-deoxyglucose uptake by the muscles. This indicates that ceramide may be involved in regulation of glucose uptake by skeletal muscles *in vivo*.

ACKNOWLEDGMENTS

This work was supported by the Polish Research Committee (KBN) (Grant No. 4/PO5B/022/15.

REFERENCES

1. ARIANO, M.A., R.B. ARMSTRONG & V.R. EDGERTON. 1979. Hindlimb muscle fibre population of five mammals. J. Histochem. Cytochem. **21:** 51–55.
2. BEGUM, N., L. RAGOLIA & M. SRINIVASAN. 1996. Effect of tumor necrosis factor-α on insulin-stimulated mitogen-activated protein kinase cascade in cultured rat skeletal muscle cells. Eur. J. Biochem. **238:** 214–220.
3. BEGUM, N. & L. RAGOLIA. 1996. Effect of tumor necrosis factor alpha on insulin action in cultured rat skeletal muscle cells. Endocrinology **137:** 2441–2446.
4. BRINDLEY, D.N., C.N. WANG, J. MEI *et al.* 1999. Tumor necrosis factor-α and ceramides in insulin resistance. Lipids **34:** S85–S88.
5. DAVID, T.S., P.A. ORTIZ, T.R. SMITH & J. TURINSKY. 1998. Sphingomyelinase has an insulin-like effect on glucose transporter translocation in adipocytes. Am. J. Physiol. **274:** R1446–R1453.
6. DOBRZYN, A. & J. GÓRSKI. 2002. Ceramides and sphingomyelins in skeletal muscles of the rat: content and composition—effect of prolonged exercise. Am. J. Physiol. (Endocrinol. Metab.) **282:** E277–E285.
7. HAJDUCH, E., A. BALENDRAN, I.H. BATTY *et al.* 2001. Ceramide impairs the insulin-dependent membrane recruitment of protein kinase B leading to a loss in downstream signalling in L6 skeletal muscle cells. Diabetologia **44:** 173–183.
8. HANNUN, Y.A. & C. LUBERTO. 2000. Ceramide in the eucaryotic stress response. Trends Cell Biol. **10:** 73–80.
9. KANETY, H., R. HEMI, M.Z. PAPA & A. KARASIK. 1996. Sphingomyelinase and ceramide suppress insulin-induced tyrosine phosphorylation of the insulin receptor substrate-1. J. Biol. Chem. **271:** 9895–9897.

10. LEVADE, T. & J-P. JAFFREZOU. 1999. Signalling sphingomyelinases: which, where, how, and why? Biochim. Biophys. Acta **1438:** 1–17.
11. MAHADEVAPPA, V.G. & B.J. HOLUB. 1987. Chromatographic analysis of phosphoinositides and their breakdown products in activated blood platelets/neutrophils. *In* Chromatography of Lipids in Biomedical Research and Clinical Diagnosis: Chromatography Library. Volume 37, pp. 225–265. Elsevier. Amsterdam/New York.
12. MATHIAS, S., L.A. PEÑA & R.N. KOESNICK. 1998. Signal transduction of stress via ceramide. Biochem. J. **335:** 465–480.
13. PERALDI, P., G.S. HOTAMISLIGIL, W.A. BUURMAN *et al.* 1996. Tumor necrosis factor (TNF)–α inhibits insulin signaling through stimulation of the p55 TNF receptor and activation of sphingomyelinase. J. Biol. Chem. **271:** 13018–13022.
14. PERRY, D.K. & Y.A. HANNUN. 1998. The role of ceramide in cell signaling. Biochim. Biophys. Acta **1436:** 233–243.
15. PFEILSCHIFTER, J. & A. HUWILER. 2000. Ceramides as key players in cellular response. News Physiol. Sci. **15:** 11–15.
16. PREVATI, M., L. BERTOLASO, M. TRAMARIN *et al.* 1996. Low nanogram range quantitation of diglycerides and ceramide by high-performance liquid chromatography. Anal. Biochem. **233:** 108–114.
17. RIBONI, L., P. VIANI, R. BASSI *et al.* 1997. The role of sphingolipids in the process of signal transduction. Prog. Lipid Res. **36:** 153–195.
18. RICHTER, E.A. 1996. Glucose utilization. *In* Handbook of Physiology. Section 12. Exercise: Regulation and Integration of Multiple Systems, pp. 912–951. Oxford University Press. London/New York.
19. SCHMITZ-PEIFFER, C., D.L. CRAIG & T.J. BIDEN. 1999. Ceramide generation is sufficient to account for the inhibition of the insulin-stimulated PKB pathway in C2C12 skeletal muscle cells pretreated with palmitate. J. Biol. Chem. **274:** 24202–24210.
20. SULLIVAN, T.E. & R.B. ARMSTRONG. 1978. Rat locomotory muscle fibre activity during trotting and galloping. J. Appl. Physiol. **44:** 358–363.
21. SUMMERS, S.A., L.A. GARZA, H. ZHOU & M.J. BIRNBAUM. 1998. Regulation of insulin-stimulated glucose transporter GLUT4 translocation and Akt kinase activity by ceramide. Mol. Cell. Biol. **18:** 5457–5464.
22. TURINSKY, J., D.M. O'SULLIVAN & B.P. BAYLY. 1990. 1,2-Diacylglycerol and ceramide levels in insulin-resistant tissues of the rat *in vivo*. J. Biol. Chem. **265:** 16880–16885.
23. TURINSKY, J., B.P. BAYLY & D.M. O'SULLIVAN. 1990. 1,2-Diacylglycerol and ceramide levels in rat skeletal muscle and liver *in vivo*: studies with insulin, exercise, muscle denervation, and vasopressin. J. Biol. Chem. **265:** 7933–7938.
24. TURINSKY, J., B.P. BAYLY & D.M. O'SULLIVAN. 1991. 1,2-Diacylglycerol and ceramide levels in rat liver and skeletal muscle *in vivo*. Am. J. Physiol. **261:** E620–E627.
25. TURINSKY, J., G.W. NAGEL, J.S. ELMENDORF *et al.* 1996. Sphingomyelinase stimulates 2-deoxyglucose uptake by skeletal muscle. Biochem. J. **313:** 215–222.
26. VAN DE VUSSE, G.J., T.H.M. ROEMAN & R.S. RENEMAN. 1980. Assessment of fatty acids in dog left ventriculum myocardium. Biochim. Biophys. Acta **617:** 347–352.
27. WANG, C.N., L. O'BRIEN & D.N. BRINDLEY. 1998. Effects of cell-permeable ceramides and tumor necrosis factor-α on insulin signaling and glucose uptake in 3T3-L1 adipocytes. Diabetes **47:** 24–31.

Genetic and Molecular Analyses of Complex Metabolic Disorders: Genetic Linkage

S. MENZEL

The Wellcome Trust Centre for Human Genetics, Oxford, United Kingdom

ABSTRACT: Wide efforts have taken place with complex metabolic disorders to emulate the success that linkage analysis has had in explaining the nature of monogenic metabolic diseases such as MODY (maturity-onset diabetes of the young) and FH (familial hypercholesterolemia). New linkage methods are being specifically developed and tested for complex disorders since some of the basic assumptions of traditional linkage analysis used with Mendelian traits are not valid. The nature of complex diseases precludes the use of extended families under the hypothesis that the same disease allele acts in most affected individuals throughout a pedigree. Rather, a multitude of genes and of rare and common alleles creates an apparently chaotic pattern of heterogeneity within and between families. Therefore, very simple family structures, in many studies even isolated sibling pairs, form the basis of efforts to compare the inheritance of disease with that of the chromosomal regions under investigation. Also, assumptions about how individual loci contribute to the overall disease inheritance used for the models applied in linkage computation have to be kept to a minimum. The overall effect of this, together with the potentially weak influence of many loci, is a heavy toll on the statistical power to detect individual contributing genes. This may be the reason why very few scans so far have yielded disease loci that meet genome-wide significance criteria. The confirmation of original loci in secondary studies has proven, as predicted, to be very difficult. Nevertheless, the overall emerging picture is very encouraging: one of the genome scans in type 2 diabetes has been carried through to the positional cloning of the underlying genetic variant, namely, the calpain 10–associated polymorphism in type 2 diabetes. Several other loci have been detected repeatedly throughout studies in various human racial groups, such as the chromosome 1q and 20q diabetes loci, and have become the target of collaborative fine-mapping efforts. Modifications to present methodology are in development with the goal to increase statistical power: examples are the use of intermediate traits with potentially increased genetic homogeneity, the investigation of admixed populations, and the study of linkage disequilibrium over wide genomic regions.

KEYWORDS: linkage analysis; genome; metabolic disorders; loci; mapping; trait(s); allele

Address for correspondence: S. Menzel, The Wellcome Trust Centre for Human Genetics, Oxford, United Kingdom. Voice: +44 (1865) 287 529; fax: +44 (1865) 287 533.
stephan@well.ox.ac.uk

LINKAGE ANALYSIS IN MONOGENIC METABOLIC DISEASES

Linkage analysis investigates whether, within groups of related individuals, diseases or other traits tend to be associated with alleles of a genetic marker. This intrafamilial association is usually present when the disease locus and marker reside on the same chromosome close enough to each other to restrain separation in recombination events (crossovers) during meiosis. Thus, when linkage can be established, its extent can serve as an approximation of the physical closeness between the marker involved and the causative genetic variant (e.g., the disease locus). This approach is especially powerful with rare Mendelian (i.e., monogenic) diseases, where extended families can be studied under the assumption that one allele (or two, if recessive) at the same disease locus is directly responsible for most of the cases.

In monogenic metabolic disorders, linkage with anonymous genetic markers has been very successful and has led to the identification of a series of (often surprising) genes and proteins that are key players in human energy metabolism, the endocrine system, and gene regulation.

Familial hypercholesterolemia (FH) was one of the first disease genes genetically mapped. The placement of the disease locus near the complement factor 3 gene was achieved using the first publicly available linkage analysis software (LIPED) and genetic markers based on protein polymorphisms in 1974.[1] The LDL receptor (LDLR) gene has since been identified as providing the causative variants. A recessive form of the disease recently has been positionally cloned (chromosome 1p36-p35), leading to the discovery of a novel LDLR modifier protein.[2]

Success stories of genome-wide linkage mapping efforts include also the identification of variants in the transcription factors, hepatic nuclear factors 1α[3] and 4α[4] (chromosome 12q24 and 20q12-q13.1, respectively), as the cause of MODY (maturity-onset diabetes of the young); a mutation in transcription initiation factor EIF2AK3 as the basis of Wolcott-Rallison syndrome;[5] and the discovery that mutations in the nuclear envelope protein lamin A (chromosome 1q21.2) cause familial partial lipodystrophy.[6] Recently, the gene for congenital generalized lipodystrophy, *BSCL1*, has been identified through positional cloning.[7]

Disease loci that have been mapped include the Alström syndrome locus on chromosome 2p13[8] and two loci for familial combined hyperlipidemia (*FCHL1*, chromosome 1q21-q23;[9] *FCHL2*, chromosome 11p[10]).

GENERAL APPROACH TO LINKAGE ANALYSIS IN COMMON METABOLIC DISEASES

In common diseases, allelic and locus heterogeneity, incomplete penetrance, and unknown mode of inheritance rule out the use of large families in the way that linkage is traditionally calculated.

The concept of an allele-sharing method in sibling pairs was established by Penrose in 1935[11] (for further bibliography, see Li: http://linkage.rockefeller.edu/bib/ibd/) and has since been developed to extract more complete information from small pedigrees.

In general, though, the power of linkage analysis in complex diseases is limited and only the strongest loci will be detected unless thousands of small families are studied.[12]

TECHNOLOGY

While genome-wide linkage mapping in extended families is possible in individual labs, genome scanning for multifactorial traits benefits from the involvement of "linkage factories" or specialized high-throughput core facilities, such as at the Marshfield Clinic (Michigan), the Centre National de Génotypage (Evry, France), or the Wellcome Trust Centre for Human Genetics (Oxford, United Kingdom). There, database-aided partially automated approaches combined with stringent genotype quality-control measures ensure that the data remain usable across large-scale genotyping projects.

Current high-throughput capable technologies are 96-head and 384-head pipetting stations and pipetting robots, 384-well thermocyclers, 96-capillary-array automated sequencers, laboratory information management systems (LIMS), and advanced genotyping software that automates trace analysis and continuously supplies quality-control monitoring of the data.

Widely used marker sets include the CHLC/Weber human genome-wide screening set version 10, 405 mostly tetranucleotide repeat markers (http://research.marshfieldclinic.org/genetics/sets/combo.html/), and the ABI LMSv.2 set (400 Généthon dinucleotide markers, http://www.appliedbiosystems.com/).

GENOME-WIDE LINKAGE SCANS IN COMMON METABOLIC DISEASES: THE EXAMPLE OF TYPE 2 DIABETES

The success of genome-wide mapping with autosomal-dominant forms of diabetes has prompted a series of scans with the common late-onset form (type 2 diabetes) of the disease.

The first scan to be published was carried out in 440 Texan, type 2 diabetic sib pairs of Mexican-American (i.e., with Caucasian, Native American, and African-American admixture) origin.[13] The results were encouraging since a locus was found that was supported by genome-wide significant evidence[14] for linkage. The locus on chromosome 2q, termed *NIDDM1*, was subsequently approached in a bold positional cloning endeavor. Through genetic detective work and an extended re-sequencing project, a high-risk genotype derived from variants of noncoding sequence across the calpain 10 gene was identified.[15] This genotype was also found to be associated with type 2 diabetes in several European populations. The mechanisms by which defects of calpain 10, a member of a family of calcium-regulated proteinases, cause type 2 diabetes remain so far unexplained. (See FIG. 1.)

All genetic linkage studies in metabolic diseases eventually aim at one overall goal: the better understanding of human intermediate and energy metabolism and its regulation. This final outcome will be valid across all human races and the study of a specific population merely serves as a tool to tease out gene effects that might be undetectable elsewhere. On the other hand, this sensitivity of linkage data to the specific population situation makes it very difficult to find again loci in other studies based on similar, but not identical, populations. This is illustrated by the failure of three other scans—two in other Mexican-Americans[16,17] and one in the Native American Pima population[18]—to detect the *NIDDM1* locus. Three other main loci, on chromosomes 10, 3, and 1, respectively, were found instead. Most other studies

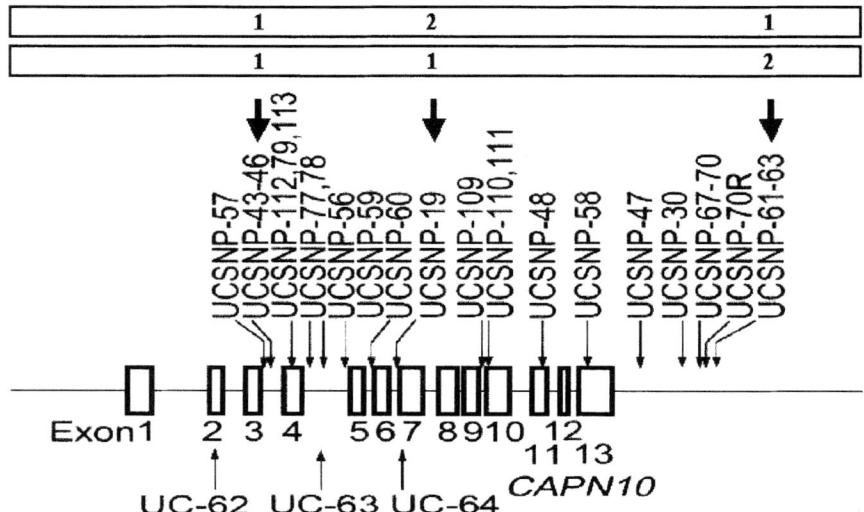

FIGURE 1. The first positionally cloned gene involved in a polygenic disease in humans is the calpain 10 gene on chromosome 2q (*NIDDM1*). A genotype made up of two haplotypes (1-2-1 and 1-1-2) of intronic SNPs confers risk to type 2 diabetes. Modified figure reprinted with permission from G. I. Bell.

for type 2 diabetes were carried out in Caucasian populations from the United States or Europe,[17,19–26] but one also involved African-American and Japanese-American families.[17]

One of the largest type 2 diabetes scans carried out so far, the FUSION (Finland–United States Investigation of Non-Insulin-Dependent Diabetes Mellitus Genetics) project,[25,27,28] aimed at garnering additional power (see below) through two mechanisms: the study of an isolated population (the Finnish) and the extensive inclusion of diabetes-related quantitative traits. Interestingly, the analysis of the dichotomous trait (diabetes[27]) yielded a set of loci different from those obtained with the continuous measurements.[28]

Another problem affecting the consistency of linkage results with common traits arises from the fact that studies are, in general, underpowered. Current sample sizes will allow the reliable identification of only major genes that take up a large proportion of the overall familial risk (e.g., sibling relative risk, λ_s). In practice, this means that studies need "luck", in addition to a high standard of data quality, to detect smaller loci, and other peaks of comparable size will always be found that are entirely due to spurious cosegregation of marker locus and trait. Thus, many efforts to confirm results in secondary studies are failing, even when great care is taken to sample the same population as in the original scan. Power-related problems and ways to circumvent them are impressively illustrated in a recent study on type 1 diabetes:[29] it is imperative to strive for international pooling of resources and joint analysis of data to help guide positional cloning efforts in multifactorial metabolic diseases.

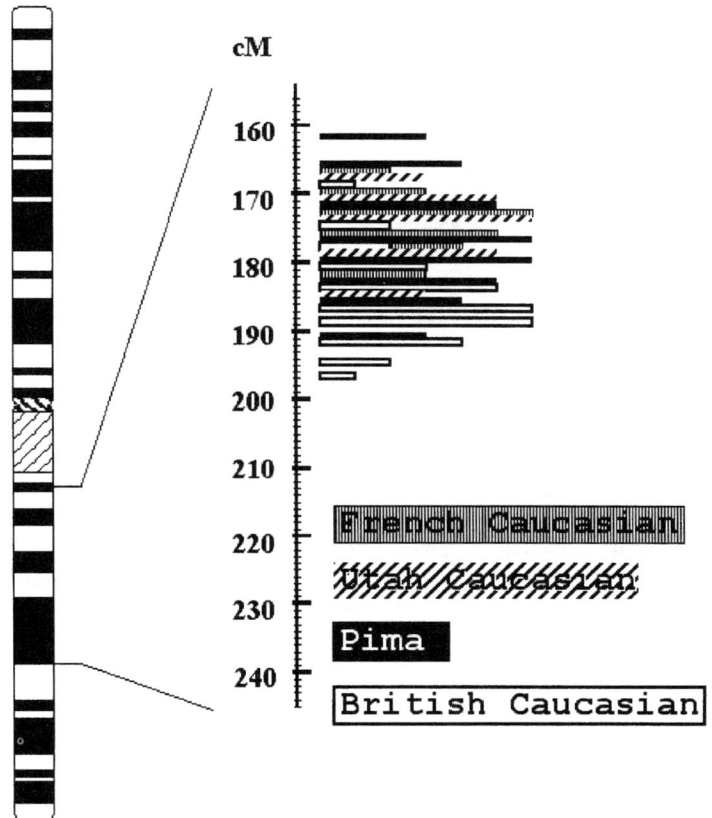

FIGURE 2. A type 2 diabetes locus detected on chromosome 1q shows linkage with the disease across several populations.

Already, by comparing results from several scans in type 2 diabetes, a few loci are emerging that seem to show consistency across studies. Interestingly, several scans have found peaks near MODY genes, that is, on chromosome 12q[19,22] and chromosome 20q.[19,24,25] This could occur by chance alone, or possibly mild changes in MODY genes might also contribute to the risk for developing type 2 diabetes. Alternatively, population samples used in these scans might contain families with monogenic forms of diabetes. Another type 2 diabetes locus, on chromosome 1q near *ApoA2*, occurs repeatedly in different populations[18,20,23,26] and can be considered to have been formally replicated (FIG. 2). Interestingly, it also overlaps with a diabetes locus from an experimental rat cross (GK rat diabetes model[30]) and with a region linked to *FCHL1*.[9]

GAIN OF STATISTICAL POWER THROUGH MODIFIED STUDY DESIGNS

Analysis of Quantitative Intermediate Traits

In place of the disease as a dichotomous trait, associated quantitative traits (QT) can be studied in genome-wide linkage analysis efforts (reviewed by Flint and Mott[31]). Underlying this strategy is the possibility that some of these traits might display increased heritability and that locus heterogeneity might be reduced, thus increasing the power to detect individual loci compared with studying the disease itself. Specific to metabolic disorders is the involvement of certain QT, such as measures of insulin resistance, with a whole series of associated diseases (the Metabolic Syndrome). Studying these continuous traits might therefore lead to the identification of common genetic origins for these diseases.

The ascertainment of unselected families suitable for QT studies is often easier than finding families with disease segregation. On the other hand, unselected population samples provide relatively low power, and schemes have been devised to use subject preselection to reduce the demand on sample sizes. Family collections can be "loaded" with disease-relevant alleles via ascertainment through one or more affected individuals, such as practiced in the FUSION study.[28] Another approach to solving the power problem is the selected study of relative pairs from tails of the trait distribution (extreme concordant or discordant pairs[32]). This ascertainment strategy has recently been applied to arterial hypertension.[33]

A trait that has recently come into focus is birth weight. It not only shows strong heritability and association with all components of the Metabolic Syndrome, but has also been demonstrated to be linked to monogenic[34] and complex forms[35] of diabetes through a genetic mechanism.

Study of Isolated Populations

Genetic studies in population isolates have been proposed (reviewed by Peltonen et al.[36]) to take advantage of the higher prevalence of certain diseases in specific populations, the enhanced power to detect recessive traits when significant inbreeding is present, a more uniform genetic and environmental background, the possibility of good genealogical records, and other characteristics. As Peltonen et al.[36] stress, the usefulness of certain isolates and sampling strategies for linkage studies always has to be reevaluated in the context of the specific trait to be studied. Population isolates also present advantages for the final identification of the causative genetic variant through association studies. They are expected to have undergone "population bottlenecks", that is, periods in their history when very few individuals provided the ancestry for later population expansions. These bottlenecks, in combination with genetic-drift effects, can lead to the presence of extended regions presenting linkage disequilibrium (LD) (see below), which improves the chance to detect a disease allele through association with the surrounding marker haplotype.

Linkage Disequilibrium (LD) Mapping

Systematic genome-wide searching for genes involved in complex metabolic disorders is presently possible only through linkage analysis using microsatellite markers. A potentially more powerful approach that also yields a finer map resolution is genome-wide LD mapping in a case control or family-based association study using haplotypes generated from single-nucleotide polymorphisms (SNPs).[37] The rationale behind this approach is to catch the decaying LD of causative gene variants with surrounding genomic regions through the genotyping of SNPs that are spaced densely enough to cover most of the human genome through their local LD relationships. Accordingly, the number of SNPs required is very large and depends on the extent and pattern of LD in the population under study. Whereas earlier simulations[38] estimated that LD is present over a few kilobases, more recent experimental studies[39] have indicated that LD typically extends across 60 kb in Northern European Caucasian populations and that genome-wide LD mapping could be possible with existing SNP resources. Suitable markers are identified in different populations through international collaborative efforts such as the SNP Consortium (http://snp.cshl.org/).

Investigation of Admixed Populations

The study of recently admixed populations is set to become the first practicable approach to genome-wide human LD mapping.[40] If the founder populations are sufficiently different in the rates of the disease under investigation, there is considerable power to detect the genes underlying these differences. If admixture took place only a few generations ago, widely spaced markers (from a 10-cM intermarker distance) can be used. This approach thus combines the statistical power of association studies with the broad mapping coverage of linkage studies.[40] A set of suitable population-specific markers will be obtained from current international SNP discovery programs. Since genetic resolution is poor with recently admixed families, additional resources will have to be collected for fine-mapping efforts.

CONCLUSIONS

Recent studies have shown both the problems and the great potential of linkage analysis as a tool to help uncover the natural history of complex metabolic diseases. The next years can be expected to yield exciting revelations about new genes and proteins implicated in disease causation and about new powerful approaches to the study of complex traits.

ACKNOWLEDGMENTS

The OMIM database (Online Mendelian Inheritance in Man, accessed through the NCBI Entrez server at http://www.ncbi.nlm.nih.gov:80/entrez/query.fcgi/db-OMIM/) has been used to prepare parts of this paper.

REFERENCES

1. OTT, J., et al. 1974. Linkage studies in a large kindred with familial hypercholesterolemia. Am. J. Hum. Genet. **26:** 598–603.

2. GARCIA, C.K., et al. 2001. Autosomal recessive hypercholesterolemia caused by mutations in a putative LDL receptor adaptor protein. Science **292:** 1394–1398.
3. YAMAGATA, K., et al. 1996. Mutations in the hepatocyte nuclear factor 1 alpha gene in maturity-onset diabetes of the young (MODY3). Nature **384:** 455–458.
4. YAMAGATA, K., et al. 1996. Mutations in the hepatocyte nuclear factor 4 alpha gene in maturity-onset diabetes of the young (MODY1). Nature **384:** 458–460.
5. DELEPINE, M., et al. 2000. EIF2AK3, encoding translation initiation factor 2-alpha kinase 3, is mutated in patients with Wolcott-Rallison syndrome. Nat. Genet. **25:** 406–409.
6. CAO, H. & R.A. HEGELE. 2000. Nuclear lamin A/C R482Q mutation in Canadian kindreds with Dunnigan-type familial partial lipodystrophy. Hum. Mol. Genet. **9:** 109–112.
7. MAGRE, J., et al. 2001. Identification of the gene altered in Berardinelli-Seip congenital lipodystrophy on chromosome 11q13. Nat. Genet. **28:** 365–370.
8. COLLIN, G.B., et al. 1997. Homozygosity mapping of Alström syndrome to chromosome 2p. Hum. Mol. Genet. **6:** 213–219.
9. PAJUKANTA, P., et al. 1998. Linkage of familial combined hyperlipidaemia to chromosome 1q21-q23. Nat. Genet. **18:** 369–373.
10. AOUIZERAT, B.E., et al. 1999. A genome scan for familial combined hyperlipidemia reveals evidence of linkage with a locus on chromosome 11. Am. J. Hum. Genet. **65:** 397–412.
11. PENROSE, L.S. 1935. The detection of autosomal linkage in data which consist of pairs of brothers and sisters of unspecified parentage. Ann. Eugen. (Lond.) **6:** 133–138.
12. RISCH, N. & K. MERIKANGAS. 1996. The future of genetic studies of complex human diseases. Science **273:** 1516–1527.
13. HANIS, C.L., et al. 1996. A genome-wide search for human non-insulin-dependent (type 2) diabetes genes reveals a major susceptibility locus on chromosome 2. Nat. Genet. **13:** 161–171.
14. LANDER, E. & L. KRUGLYAK. 1995. Genetic dissection of complex traits: guidelines for interpreting and reporting linkage results. Nat. Genet. **11:** 241–247.
15. HORIKAWA, Y., et al. 2000. Genetic variation in the gene encoding calpain-10 is associated with type 2 diabetes mellitus. Nat. Genet. **26:** 163–175.
16. DUGGIRALA, R., et al. 1999. Linkage of type 2 diabetes mellitus and of age at onset to a genetic location on chromosome 10q in Mexican Americans. Am. J. Hum. Genet. **64:** 1127–1140.
17. EHM, M.G., et al. 2000. Genome-wide search for type 2 diabetes susceptibility genes in four American populations. Am. J. Hum. Genet. **66:** 1871–1881.
18. HANSON, R.L., et al. 1998. An autosomal genomic scan for loci linked to type II diabetes mellitus and body-mass index in Pima Indians. Am. J. Hum. Genet. **63:** 1124–1132.
19. BOWDEN, D.W., et al. 1997. Linkage of genetic markers on human chromosomes 20 and 12 to NIDDM in Caucasian sib pairs with a history of diabetic nephropathy. Diabetes **46:** 882–886.
20. ELBEIN, S.C., et al. 1999. A genome-wide search for type 2 diabetes susceptibility genes in Utah Caucasians. Diabetes **48:** 1175–1182.
21. BEKTAS, A., et al. 1999. Evidence of a novel type 2 diabetes locus 50 cM centromeric to NIDDM2 on chromosome 12q. Diabetes **48:** 2246–2251.
22. MAHTANI, M.M., et al. 1996. Mapping of a gene for NIDDM associated with an insulin secretion defect by a genome scan in Finnish families. Nat. Genet. **14:** 90–95.
23. VIONNET, N., et al. 2000. Genomewide search for type 2 diabetes–susceptibility genes in French whites: evidence for a novel susceptibility locus for early-onset diabetes on chromosome 3q27-qter and independent replication of a type 2 diabetes locus on chromosome 1q21-q24. Am. J. Hum. Genet. **67:** 1470–1480.
24. JI, L., et al. 1997. New susceptibility locus for NIDDM is localized to human chromosome 20q. Diabetes **46:** 876–881.
25. GHOSH, S., et al. 1999. Type 2 diabetes: evidence for linkage on chromosome 20 in 716 Finnish affected sib pairs. Proc. Natl. Acad. Sci. U.S.A. **96:** 2198–2203.
26. WILTSHIRE, S., et al. 2001. A genomewide scan for loci predisposing to type 2 diabetes in a U.K. population (the Diabetes UK Warren 2 repository): analysis of 573 pedigrees

provides independent replication of a susceptibility locus on chromosome 1q. Am. J. Hum. Genet. **69:** 553–569.
27. GHOSH, S., *et al.* 2000. The Finland–United States Investigation of Non-Insulin-Dependent Diabetes Mellitus Genetics (FUSION) study. I. An autosomal genome scan for genes that predispose to type 2 diabetes. Am. J. Hum. Genet. **67:** 1174–1185.
28. WATANABE, R.M., *et al.* 2000. The Finland–United States Investigation of Non-Insulin-Dependent Diabetes Mellitus Genetics (FUSION) study. II. An autosomal genome scan for diabetes-related quantitative-trait loci. Am. J. Hum. Genet. **67:** 1186–1200.
29. COX, N.J., *et al.* 2001. Seven regions of the genome show evidence of linkage to type 1 diabetes in a consensus analysis of 767 multiplex families. Am. J. Hum. Genet. In press.
30. GAUGUIER, D., *et al.* 1996. Chromosomal mapping of genetic loci associated with non-insulin dependent diabetes in the GK rat. Nat. Genet. **12:** 38–43.
31. FLINT, J. & R. MOTT. 2001. Finding the molecular basis of quantitative traits: successes and pitfalls. Nat. Rev. Genet. **2:** 437–445.
32. RISCH, N. & H. ZHANG. 1995. Extreme discordant sib pairs for mapping quantitative trait loci in humans. Science **268:** 1584–1589.
33. XU, X., *et al.* 1999. An extreme-sib-pair genome scan for genes regulating blood pressure. Am. J. Hum. Genet. **64:** 1694–1701.
34. HATTERSLEY, A.T., *et al.* 1998. Mutations in the glucokinase gene of the fetus result in reduced birth weight. Nat. Genet. **19:** 268–270.
35. LINDSAY, R.S., *et al.* 2000. Type 2 diabetes and low birth weight: the role of paternal inheritance in the association of low birth weight and diabetes. Diabetes **49:** 445–449.
36. PELTONEN, L., *et al.* 2000. Use of population isolates for mapping complex traits. Nat. Rev. Genet. **1:** 182–190.
37. JORDE, L.B. 2000. Linkage disequilibrium and the search for complex disease genes. Genome Res. **10:** 1435–1444.
38. KRUGLYAK, L. 1999. Prospects for whole-genome linkage disequilibrium mapping of common disease genes. Nat. Genet. **22:** 139–144.
39. REICH, D.E., *et al.* 2001. Linkage disequilibrium in the human genome. Nature **411:** 199–204.
40. MCKEIGUE, P.M. 1997. Mapping genes underlying ethnic differences in disease risk by linkage disequilibrium in recently admixed populations. Am. J. Hum. Genet. **60:** 188–196.

Positional Cloning of an Obesity/Diabetes Susceptibility Gene(s) on Chromosome 11 in Pima Indians

LESLIE BAIER, PETER KOVACS, CHRISTOPHER WIEDRICH, KIMBERLY CRAY, AMY SCHEMIDT, GONG-QING SHEN, JEFFREY SUTHERLAND, PAMELA THUILLEZ, YUNHUA LI MULLER, MICHAEL TRAURIG, AND CLIFTON BOGARDUS

Phoenix Epidemiology and Clinical Research Branch, National Institute of Diabetes and Digestive and Kidney Diseases, National Institutes of Health, Phoenix, Arizona, USA

ABSTRACT: Prior results from our genomic scan in Pima Indians indicated an obesity locus in a region on chromosome 11q23-24 that was also linked to diabetes. Bivariate linkage analysis for the combined phenotype "diabesity" gave the strongest evidence for linkage (LOD = 5.2). Our aim is to positionally clone the gene(s) responsible for the linkage. Linkage disequilibrium mapping is being used to narrow the chromosomal region. Single nucleotide polymorphisms (SNPs) are being systematically identified and genotyped at 50-kb intervals across the region of linkage. To date, 455 SNPs have been genotyped in 1229 Pimas. A region containing a cluster of SNPs strongly associated with BMI and a second region, approximately 2 Mb telomeric, containing a cluster of SNPs associated with diabetes have been preliminarily identified.

KEYWORDS: type 2 diabetes; obesity; linkage disequilibrium mapping; single nucleotide polymorphisms

INTRODUCTION

The Pima Indians of Arizona have the world's highest reported incidence and prevalence of type 2 diabetes mellitus.[1] Their diabetes is prototypic of this disease and is characterized by obesity, insulin resistance, insulin secretory dysfunction, and increased rates of endogenous glucose production.[2] In prospective studies of prediabetic Pima Indians, insulin resistance and insulin secretory dysfunction are major predictors of the disease.[3] The Pima Indians also have a high prevalence of obesity. In 1988, more than 80% of Pima Indians between the ages of 20 and 55 years had a BMI > 27 kg/m^2 and the incidence of obesity has been steadily increasing.[4] Both type 2 diabetes and obesity in Pima Indians, as in other populations, have a substantial genetic basis.[5] Both diseases aggregate in families and both demonstrate a higher concordance rate among monozygotic twins as compared to dizygotic twins. Studies

Address for correspondence: Dr. Leslie Baier, NIDDK, National Institutes of Health, Phoenix, AZ 85016. Voice: 602-200-5240; fax: 602-200-5225.
lbaier@phx.niddk.nih.gov

in monozygotic twins in non–Native American populations estimate that the heritability for BMI ranges from 0.60 to 0.80.[6] Studies of monogenic forms of type 2 diabetes led to the identification of several diabetes genes with major effects in rare families. These include genes that encode insulin, insulin receptor, glucokinase, three hepatic nuclear transcription factors (HNF-1α, -4α, -1β), and another transcription factor, IPF-1.[7] A few genes that can cause severe, rare forms of obesity, such as leptin and its receptor and the MC4 receptor, have also been characterized.[8,9] However, the genes that cause the most common forms of type 2 diabetes and obesity remain unknown. Moreover, none of the genes identified to date appear to have a major role in the pathogenesis of type 2 diabetes or obesity in Pima Indians.

In an effort to determine the genetic basis for type 2 diabetes and obesity in Pima Indians, we have undertaken both a candidate gene approach as well as a genome scan approach. Candidate genes include genes that have a known physiologic role in glucose and/or lipid metabolism, as well as genes that are associated with type 2 diabetes or BMI in another population. To date, no variation in a physiologic candidate gene has been identified that has a major role in determining BMI or diabetic onset in Pima Indians.

We have recently completed a genomic scan in more than 1200 Pima Indians who had participated in a longitudinal study of diabetes.[10] Among 264 nuclear families containing 966 siblings, 516 autosomal markers with a median distance between adjacent markers of 6.4 cM were genotyped. Variance-component methods were used to test for linkage with an age-adjusted diabetes score and with BMI. In multipoint analyses, the strongest evidence for linkage with age-adjusted diabetes (LOD = 1.7) was on chromosome 11q, in a region that was also the most strongly linked with BMI (LOD = 3.6) (FIG. 1). Bivariate linkage analysis for the combined

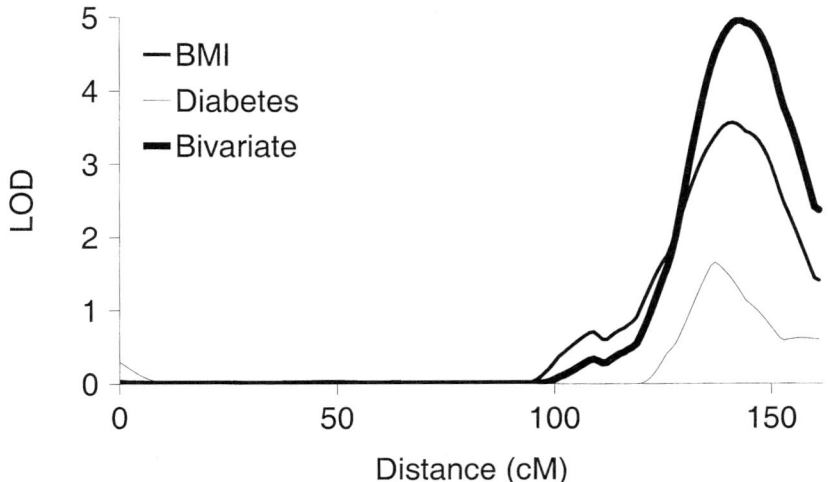

FIGURE 1. Multipoint linkage results for chromosome 11 for BMI and for diabetes, adjusted for age and sex by a cumulative incidence method. The bivariate analysis will test the null hypothesis of no linkage with either phenotype. (Adapted from reference 10.)

phenotype "diabesity" gave the strongest evidence for linkage (LOD = 5.2). The region of linkage is positioned at 11q23-24 (approximately 35 cM). Linkage of this region to type 2 diabetes and BMI has subsequently been replicated in a Mexican-American population and in the Framingham Population Study, respectively. The aim of the current study was to identify diabetes and/or obesity susceptibility genes on 11q23-24 that gave rise to the linkage.

METHODS

Linkage Disequilibrium Mapping of SNPs

Linkage disequilibrium analysis of SNPs is being used to narrow the chromosomal region along 11q23-24 harboring the susceptibility gene(s) for type 2 diabetes and BMI. Single nucleotide polymorphisms (SNPs) are being systematically identified and genotyped at 50-kb intervals across the 35-cM (~35-Mb) region of linkage.

SNP Identification/Verification

SNPs are systematically identified by *de novo* sequencing of 20 Pima Indian DNA samples or by searching public databases and verifying the database SNP in 20 Pima Indian DNA samples. For *de novo* sequencing, PCR primers are designed based on the published sequence or the sequence obtained by sequencing of bacterial artificial chromosome (BAC) ends that map to 11q23. PCR is typically performed with 60 ng of genomic DNA in buffer containing 1.5 mM $MgCl_2$, 0.25 mM dNTPs, and 0.75 U of AmpliTaq Gold (Perkin-Elmer) DNA polymerase. PCR conditions are typically 94°C for 10 min followed by 33 cycles of 94°C for 30 s, the optimal primer annealing temperature (55–60°C) for 30 s, and 72°C for 30 s, followed by a final extension at 72°C for 10 min. Sequencing of the PCR product is done with both the forward and reverse PCR primers. DNA cycle sequencing is carried out using Big Dye Terminator technology on a capillary DNA sequencer (model 3700, Applied Biosystems). SNPs that have a minor allele frequency of 20% or greater are considered "informative" and are genotyped in 1229 Pima Indian DNA samples.

Individual Genotyping of SNPs

Informative SNPs are genotyped in the Pima Indians used in our linkage study ($N = 1229$). Individual genotyping is done by either the Taqman allelic discrimination PCR technology, using a model 7700 ABI Prism instrument (PE Applied Biosystems), or the primer extension methodology of Pyrosequencing, using a PSQ 96 system (Pyrosequencing, Westborough, MA).

Genotyping of SNPs in Pooled DNA

Approximately 370 SNPs were genotyped at Sequenom (San Diego, CA) using their mass-array technology for pooled DNA. Each verified SNP was genotyped in pools of DNA representing the 1229 individuals from our initial linkage studies. The 1229 subjects were divided into quintiles (5 pools, ~220 subjects per pool) for the continuum of BMI. Ranges of BMI for the pools were as follows: pool 1, 20–30;

pool 2, 30–34; pool 3, 34–37; pool 4, 37–43; pool 5, 43–70. The same 1229 subjects were also divided into quintiles (~220 subjects per pool) for the "continuum" of diabetes. The diabetes pool 1 contained nondiabetic individuals between the ages of 31 and 70; pool 2 contained nondiabetic individuals between the ages of 12 and 31; pool 3 contained diabetic individuals with an age of onset of 40 years or greater; pool 4 contained diabetic individuals with an age of onset between 28 and 40 years; pool 5 contained diabetic individuals with an age of onset of less than 28 years. The sequences of 800 database SNPs were assayed in diabetes pool 1, and validated SNPs were genotyped in all 10 pools. The allele frequency for each SNP was determined in each pool. The frequencies were then analyzed by a chi-square test for trend.

Positional Candidate Genes

Positional physiologic candidate genes were selected based on a perceivable function in a physiologic pathway relevant to type 2 diabetes and/or obesity. For each positional candidate gene, all exons, putative promoter regions, and flanking regions were PCR amplified and sequenced, as above, in 20 Pima Indian DNA samples.

RESULTS

To date, we have sequenced 11 physiological candidate genes that map to chromosome 11q23-24. These genes encode a serotonin receptor, a dopamine receptor, three apolipoproteins, three zinc-finger proteins, two potassium-channel proteins, and a glucose-6-phosphate transferase. None of these physiologic candidate genes contain nucleotide variants that account for the linkage to diabetes and BMI on chromosome 11q23-24.[11]

Our goal for linkage disequilibrium mapping is to genotype an informative SNP every 50 kb across the 35-Mb region of linkage. This will require genotyping 700 informative SNPs. We initially selected 800 potential SNP sequences from public databases for SNP verification. Of these, 53% were polymorphic in Pima Indians. We also identified *de novo* 60 SNPs by sequencing DNA from 20 Pima Indians and comparing the sequences for variation. We identified a *de novo* SNP every 500–1000 base pairs, on average, depending upon the region of DNA being sequenced. For example, SNPs were much more common in intergenic regions of DNA as compared to the coding regions of genes.

To date, approximately 450 informative SNPs have been genotyped in 1229 Pima Indians. Of these 450 SNPs, 370 were genotyped in pooled DNA samples by the mass-array method of Sequenom, and 80 SNPs were individually genotyped by the methods of allelic discrimination or Pyrosequencing. The physical map position of each SNP, based upon the human genome sequence database (http://www.ncbi.nlm.nih.gov), was plotted against the significance of the difference in allelic frequencies among the pools.

Alleles that significantly differed among the BMI pools clustered in three regions, around 119×10^6, 130×10^6, and 139×10^6 base pairs (FIG. 2A). Similarly, alleles that were significantly different among the diabetes pools clustered in three regions, around 120×10^6, 130–134×10^6, and 142×10^6 base pairs (FIG. 2B). Two of these regions (120×10^6 and 139–142×10^6 base pairs) also contained SNPs that

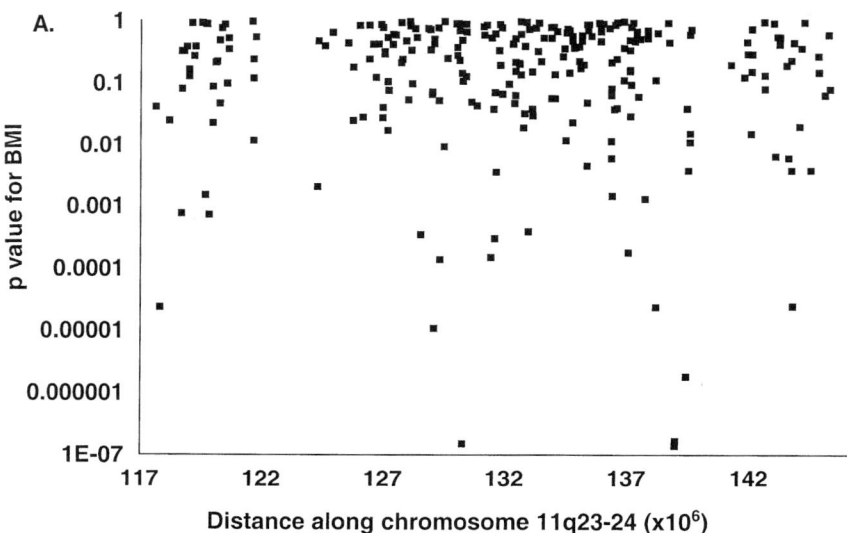

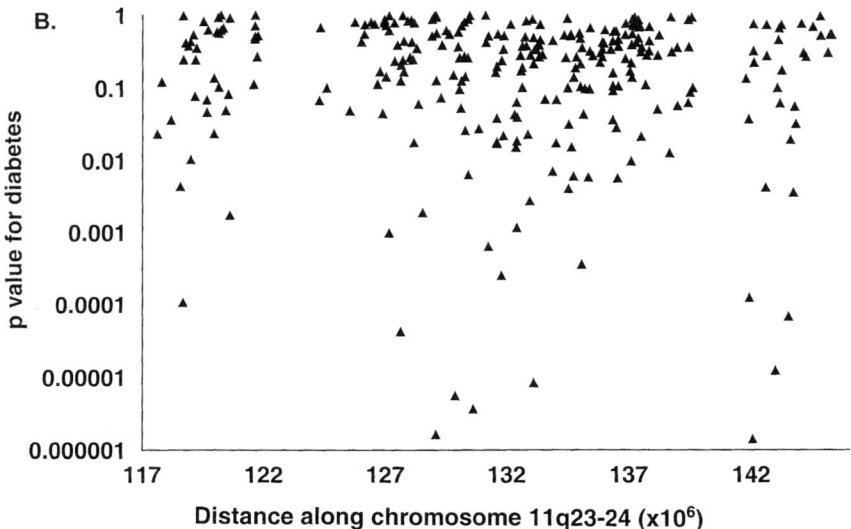

FIGURE 2. Plot of 370 SNPs genotyped in pools of DNA by the mass-array method of Sequenom (San Diego, CA). Each pool consisted of approximately 220 DNA samples. The allele frequency for each SNP was determined in each pool. The frequencies were then analyzed by a chi-square test for trend. **(A)** SNPs analyzed in 5 pools for the continuum of BMI. **(B)** SNPs analyzed in 5 pools for age of onset of diabetes.

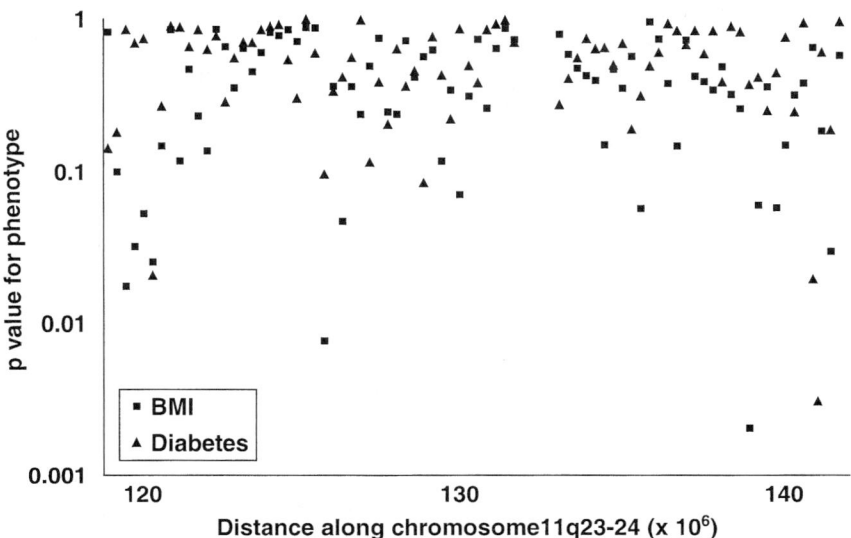

FIGURE 3. Plot of 80 SNPs genotyped in 1229 individual DNA samples and analyzed using an additive model for BMI and diabetes. Data were adjusted for age and sex.

were significantly associated with diabetes and BMI by individual genotyping (FIG. 3). We are currently filling in the gaps to complete a 50-kb map across the 35-Mb region.

DISCUSSION

The first example of using a positional cloning strategy, based on linkage analysis followed by linkage disequilibrium mapping, to successfully identify a susceptibility gene for a complex disease was the identification of SNP43 in the gene that encodes calpain-10.[12] A genome scan had identified linkage on chromosome 2q to type 2 diabetes in Mexican-Americans in Starr County, Texas.[13] The initial region of linkage encompassed a 12-cM region, and an interaction of the locus on chromosome 2 with a locus on chromosome 15 decreased the 1-LOD support interval from 12 to 7 cM. Linkage disequilibrium mapping was done by genotyping 133 SNPs, and haplotype analysis determined that SNP43, in combination with 2 other nearby SNPs, marks a heightened susceptibility to type 2 diabetes in Mexican-Americans in Starr County, Texas. The combination of 3 SNPs in the calpain-10 gene also conveys a risk in two northern European populations, whereas the single SNP43 is associated with increased insulin resistance in Pima Indians.[14]

A second success of using linkage disequilibrium mapping to identify a susceptibility locus for a complex disease is the identification of variants in the NOD2 gene that conferred a risk for Crohn's disease.[15] In this study, linkage analysis identified a 20 million base long region on chromosome 16 that appeared to harbor at least one risk gene. Subsequent mapping, using a database search and non-SNP markers, narrowed the region to a 160,000-nucleotide span, and finally 11 SNPs were genotyped for linkage disequilibrium mapping.

The most marked difference between the two successful examples of using linkage disequilibrium mapping to identify a complex disease susceptibility locus and our study of chromosome 11q23-24 in Pima Indians is the interval initially obtained in the genome scan. The region of linkage on chromosome 2q to type 2 diabetes in Mexican-Americans was narrowed to 7 cM and the region of linkage on chromosome 16 was narrowed to 160,000 base pairs before SNPs were systematically genotyped. Hence, it is perhaps not surprising that, in our 35-cM region, genotyping of more than 450 SNPs has not conclusively identified a susceptibility gene. It is likely that nearly double that number of SNPs must be systematically identified, mapped, and genotyped before we are certain of the precise location of the variant gene.

REFERENCES

1. KNOWLER, W.C., et al. 1978. Diabetes incidence and prevalence in Pima Indians: a 19-fold greater incidence than in Rochester, Minnesota. Am. J. Epidemiol. **108:** 497–505.
2. BOGARDUS, C. 1996. Metabolic abnormalities in the development of non-insulin dependent diabetes mellitus. *In* Diabetes Mellitus: A Fundamental and Clinical Text, pp. 459–467. Lippincott-Raven. Philadelphia.
3. LILLIOJA, S., et al. 1993. Insulin resistance and insulin secretory dysfunction as precursors of non-insulin-dependent diabetes mellitus: prospective studies of Pima Indians. N. Engl. J. Med. **329:** 1988–1992.
4. KNOWLER, W.C., et al. 1991. Obesity in the Pima Indians: its magnitude and relationship with diabetes. Am. J. Clin. Nutr. **53:** 1543S–1551S.
5. SAKUL, H., et al. 1997. Familiality of physical and metabolic characteristics that predict the development of non-insulin dependent diabetes mellitus in Pima Indians. Am. J. Hum. Genet. **60:** 651–656.
6. ALLISON, D.B., et al. 1996. The heritability of body mass index among an international sample of monozygotic twins reared apart. Int. J. Obes. Relat. Metab. Disord. **20:** 501–506.
7. ELBEIN, S.C. 1997. The genetics of human non-insulin dependent (type 2) diabetes mellitus. J. Nutr. **127:** 1891S–1896S.
8. MONTAGUE, C.T., et al. 1997. Congenital leptin deficiency is associated with severe early-onset obesity in humans. Nature **387:** 903–908.
9. VAISSE, C., et al. 1998. A frameshift mutation in human MC4R is associated with a dominant form of obesity. Nat. Genet. **20:** 113–114.
10. HANSON, R.L., et al. 1998. An autosomal genomic scan for loci linked to type 2 diabetes mellitus and body mass index in Pima Indians: an obesity-diabetes locus at 11q23-25. Am. J. Hum. Genet. **63:** 1130–1138.
11. JENKINSON, C.P., et al. 2000. Association of dopamine D2 receptor polymorphisms Ser311Cys and Taq1A with obesity or type 2 diabetes mellitus in Pima Indians. Int. J. Obes. **24:** 1233–1238.
12. HORIKAWA, Y., et al. 2000. Genetic variation in the gene encoding calpain-10 is associated with type 2 diabetes mellitus. Nat. Genet. **26:** 163–175.
13. HANIS, C.L., et al. 1996. A genome-wide search for human non-insulin-dependent (type 2) diabetes genes reveals a major susceptibility locus in chromosome 2. Nat. Genet. **13:** 161–166.
14. BAIER, L.J., et al. 2000. A calpain-10 gene polymorphism is associated with reduced muscle mRNA levels and insulin resistance. J. Clin. Invest. **106:** R69–R73.
15. HUGOT, J.P., et al. 2001. Association of NOD2 leucine-rich repeat variants with susceptibility to Crohn's disease. Nature **411:** 599–603.

Is a Pro12Ala Polymorphism of the PPARγ2 Gene Related to Obesity and Type 2 Diabetes Mellitus in the Czech Population?

DANIELA ŠRÁMKOVÁ,[a,b] MARIE KUNEŠOVÁ,[c] VOJTECH HAINER,[c] MARTIN HILL,[a] JOSEF VCELÁK,[a] AND BELA BENDLOVÁ[a]

[a]*Institute of Endocrinology, Prague, Czech Republic*

[b]*Department of Anthropology and Human Genetics, Natural Science Faculty, Charles University, Prague, Czech Republic*

[c]*Obesity Management Center, Third Department of Medicine, First Faculty of Medicine in Prague, Charles University, Prague, Czech Republic*

ABSTRACT: The peroxisome proliferator-activated receptors (PPARs) are members of the nuclear hormone receptor subfamily of transcription factors. PPARγ2 plays a key role in regulation of adipocyte differentiation and energy homeostasis. Recent studies provide evidence that the Pro12Ala polymorphism is linked to obesity and type 2 diabetes mellitus, but the results are controversial and depend on ethnicity. The aim of this study was to determine allele frequencies and to study the influence of the polymorphism on biochemical and anthropometric parameters in a Czech healthy adult population, in type 2 diabetics, and in a group of obese women. Results: The frequency of the Pro12Ala PPARγ2 gene polymorphism in Czech probands is similar to other central European populations. Frequency of the Pro12Ala substitution tends to be higher in obese women and diabetics compared with controls. The fasting insulin levels in the 12Ala carriers were significantly lower within the group of diabetics even after adjustment for age, BMI, and the length of diabetes duration. In obese women, higher WHR was found in subjects with the 12Ala allele. Conclusions: This study indicates that the substitution Pro12Ala is not associated with a decreased obesity or diabetes risk in the Czech population. However, the present data show that fasting insulin concentrations are lower in diabetics with the 12Ala allele than in those without it. This finding provides evidence that the polymorphism may influence glucose homeostasis.

KEYWORDS: obesity in humans; type 2 diabetes mellitus; PPARγ2 gene; Pro12Ala polymorphism

The peroxisome proliferator-activated receptor (PPAR) γ is a member of the nuclear hormone receptor family. This transcription factor has been identified as a functional receptor for the thiazolidinedione class of insulin-sensitizing drugs.[1] The predomi-

Address for correspondence: Mgr. Daniela Šrámková, Institute of Endocrinology, Prague, Czech Republic. Voice: +4202 24905301; fax: +4202 24905325.
Daniela.Sramkova@email.cz

nantly adipose isoform of PPARγ, PPARγ2, is expressed at a high level in adipose tissue,[2–4] where it modulates the expression of target genes implicated in adipocyte differentiation[5–8] and glucose homeostasis. The PPARγ2 is therefore a candidate gene for obesity and type 2 diabetes mellitus.

The Pro12Ala polymorphism[9] was found to lower the transactivation capacity of PPARγ.[10,11] This polymorphism occurs with extremely variable frequencies in populations of different ethnic origins.[9] It has been associated with lower body mass index (BMI) and higher insulin sensitivity in two large Finnish studies and in a group of second-generation Japanese-Americans, in which the 12Ala allele was also less frequent among subjects with type 2 diabetes than among normal controls.[11] A large Japanese study suggests that the 12Ala allele may protect from type 2 diabetes mellitus.[12] On the contrary, one study found a positive association with higher BMI in adult Caucasian subjects.[13] Some studies did not find an association with markers of adiposity and insulin resistance.[14–16]

The aim of this study was to determine allele frequencies of the polymorphism and to study the association of allele variants with biochemical and anthropometric parameters in a group of healthy Czech adults, in a group of type 2 diabetes mellitus patients, and in a group of obese women.

METHODS

Study Subjects

We studied the frequency of the Pro12Ala polymorphism in the Czech population in patients with type 2 diabetes mellitus ($n = 183$; age, 59.0±6.2 years; BMI, 30.1±4.8 kg/m^2), in a group of obese women ($n = 86$; age, 44.1±11.4 years; BMI, 37.5±5.8 kg/m^2), and in a group of healthy adult subjects ($n = 69$; age, 32.6±10.2 years; BMI, 23.9±3.7 kg/m^2). (See TABLE 1.)

Type 2 diabetes mellitus patients were diagnosed by criteria of the World Health Organization[17] at the Institute of Endocrinology, Prague. The group of obese women were recruited from the Prague Obesity Management Center. Control probands were volunteers between 20 and 60 years of age without serious health problems. All subjects gave their written informed consent to participate in the study.

Clinical and Anthropometric Characterization

Anthropometric data of the participants were obtained in the fasting state. Body weight, height, and waist and hip circumferences were measured, and the indexes WHR (waist-to-hip ratio) and BMI were calculated in all probands. In the control group, 11 anthropometric height measures, 14 circumferences, 12 width measures, and 14 skin folds were measured, and body composition (% fat mass, % muscles, % bone mass) was calculated using the ANTROPO[18] program. Sitting systolic and diastolic blood pressures were determined in a rest state.

After overnight fast, a venous blood sample was obtained for the determination of a number of biochemical parameters. Blood glucose level was measured by the glucose oxidase method (Beckman Glucose Analyzer 2), whereas glycosylated hemoglobin (HPLC BioRad, Czech Republic) or glycosylated proteins (spectro-

TABLE 1. Characteristics of the study subjects

Group	Number	Age (years) (± SD)	BMI (kg/m^2) (± SD)
Diabetics	183	59.0 ± 6.2	30.1 ± 4.8
Obese	86	44.1 ± 11.4	37.5 ± 5.8
Controls	69	32.6 ± 10.2	23.9 ± 3.7
Total	338		

photometric redox reaction using nitroblue tetrazolium as a sensitive redox indicator for the specific quantification of fructosamine in alkaline solution) were determined. Immunoreactive insulin (IRI) was assayed in probands not on insulin therapy using an immunoradiometric assay kit (Immunotech IRMA, Czech Republic). Proinsulin was analyzed using the ELISA kit (DRG Diagnostics, Germany). Serum levels of C-peptide were evaluated by the immunoradiometric assay kit (Immunotech IRMA, Czech Republic) and plasmatic glucagon levels were determined using the radio-immunoassay kit (Euro-Diagnostica AB, Sweden). Serum concentrations of total cholesterol (Merckotest, CHOD-PAP-Method), high-density lipoprotein (HDL) cholesterol (Merck System Cholesterin, CHOD-PAP-Method), and triglycerides (TG) (Merck System, GPO-PAP-Method) were measured using an automatic analyzer (Merck, Vitalab Eclipse). Low-density lipoprotein (LDL) cholesterol concentrations were calculated as [LDL = total cholesterol − (TG/5) − HDL]. Radioimmunoassays were used for the determination of growth hormone (Immunotech IRMA, Czech Republic), cortisol (our RIA method), SHBG (Immunotech kit, Czech Republic), and DHEA and DHEA-sulfate (Immunotech kit, Czech Republic). The status of thyroid hormones TSH, free T3, and free T4 was measured using the automatic analyzer Elecsys 2010 (Hitachi-Boehringer Mannheim, Germany). The whole subgroup of control probands underwent 3-h oGTT (oral glucose tolerance test) with 75 g of glucose and ivITT (intravenous insulin tolerance test) according to Young et al.[19]

Detection of the Pro12Ala Polymorphism of the PPARγ2 Using PCR and Hga I Restriction Endonuclease

DNA extracted from peripheral leukocytes was used to genotype for the two variants by the PCR-RFLP method. PCR amplification of the segment with the Pro12Ala polymorphism was carried out in a volume of 15 μL, containing 15 ng of genomic DNA, 0.5 μmol of each primer, 2 mM MgCl$_2$, 0.2 mM dNTPs (Takara), and 10× PCR buffer received together with Taq DNA polymerase (Top-Bio UNIS); 0.375 U of the enzyme was used.

The PCR conditions were denaturation at 94°C for 2 min followed by 40 cycles of denaturation for 40 s, annealing at 60°C for 40 s, extension at 72°C for 1 min, and final extension at 72°C for 10 min. The following primers[20] were used: forward primer 5′-GCCAATTCAAGCCCAGTC-3′ and reverse primer 5′-CGTCCCCAAT-AGCCGTATC-3′.

The substitution of alanine for proline creates an Hga I restriction site. RFLPs were detected after overnight digestion with 3 U of enzyme. The 12Pro allele gives one 306-bp fragment, whereas the 12Ala allele gives 220-bp and 86-bp fragments.

Statistical Analysis

The χ^2 test was used to assess differences in Pro12Ala frequencies between the group of type 2 diabetes patients, obese women, and controls. For evaluation of the differences in anthropometric and biochemical parameters between the 12Ala carriers and noncarriers, the nonparametric Mann-Whitney test was used. Since there were only four 12Ala homozygotes among all tested probands, they were added to 12Ala heterozygotes and included in the statistical analyses. For evaluation of the differences with adjustment for constant age, BMI, and (in diabetics) the length of diabetes duration, a two-factor ANOVA was used after power transformation of the original data to approximate a Gaussian distribution. The differences were considered statistically significant if $p < 0.05$. Two-tailed p values are reported. Odds ratio and 95% CI were calculated to evaluate the risk of type 2 diabetes and obesity for the 12Ala carriers. To assess insulin resistance and β-cell function, we used the homeostasis model assessment[21] (HOMA-R and HOMA-F, respectively). Moreover, disposition index[22] (DI) was calculated as follows: DI = HOMA-F · (G_0/I_0), where G_0 is fasting serum blood glucose and I_0 is fasting serum insulin.

The tests were done using statistic programs STATGRAPHIC Plus 3.0 (Manugistics, Rockville, MA) and NCSS 2000 (Statistical Solutions, Saugus, MA).

RESULTS

Frequency of the Pro12Ala Polymorphism in Czech Type 2 Diabetes Mellitus Patients, Obese Women, and Controls

Detection of the Pro12Ala polymorphism using Hga I restriction endonuclease is shown in FIGURE 1.

Taking all tested probands, the allele frequency of the Pro12Ala polymorphism in the PPARγ2 gene was similar to other central European populations.[23,24] However,

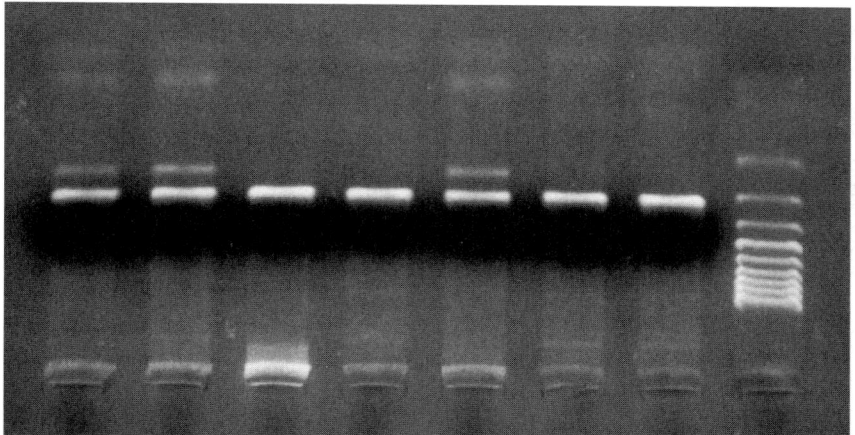

FIGURE 1. Detection of the Pro12Ala polymorphism in the PPARγ2 gene using PCR and Hga I restriction endonuclease.

TABLE 2. Frequency of the Pro12Ala polymorphism in the PPARγ2 gene in Czech type 2 diabetes mellitus patients, obese women, and controls

Group	Number	12Pro/12Ala	12Ala/12Ala	12Ala allele freq.
Diabetics	183	48 (26.23%)	2 (1.09%)	0.137
Obese	86	25 (29.07%)	1 (1.16%)	0.151
Controls	69	11 (15.94%)	1 (1.45%)	0.087
Total	338	84 (24.85%)	4 (1.18%)	0.130

12Ala tends to be higher in the group of diabetic patients (26.23% of heterozygotes and 1.09% of homozygotes) and obese women (29.07% of heterozygotes and 1.16% of homozygotes) compared with the control group (15.94% of heterozygotes and 1.45% of homozygotes). The difference between the group of diabetics and controls did not reach statistical significance ($\chi^2 = 2.92$, $p = 0.087$), whereas the difference between the obese and controls was stronger ($\chi^2 = 3.68$, $p = 0.055$). (See TABLE 2.)

Comparison of Phenotypic Features in Pro/Ala + Ala/Ala and Pro/Pro Genotypes in Type 2 Diabetes Mellitus Patients, Obese Women, and Controls

The Mann-Whitney nonparametric test revealed significantly lower fasting insulin and C-peptide levels among the 12Ala carriers compared with noncarriers in a group of diabetics (12.98±9.12 vs. 17.16±11.15, $p = 0.007$; and 0.90±0.91 vs. 0.91±0.43, $p = 0.037$; respectively). This association between the polymorphism and lower fasting insulin levels in diabetics was confirmed by ANOVA after adjustment for age, BMI, and duration of diabetes ($p = 0.004$). In a group of diabetics, among 12Ala carriers, lower fasting blood glucose levels were detected, although this association disappeared when corrected for age, BMI, and diabetes duration. In terms of anthropometric data, the only statistically significant difference between subjects carrying the 12Ala allele and subjects without it was found in obese women: higher WHR was found in 12Ala carriers (0.88±0.05 vs. 0.85±0.06, $p = 0.014$). No other differences were found in body composition (BMI, % fat mass, % muscles, % bone mass), nor were any significant differences found in plasmatic lipid levels or in other parameters tested (oGTT and ivITT process, steroid hormone concentrations, etc.), with the only exception being higher TSH levels among 12Ala carriers in comparison with noncarriers in a group of obese women (3.70±2.64 vs. 2.58±1.78, $p = 0.012$). Odds ratio, HOMA-R, HOMA-F, and DI were not significantly different between subjects with and without the 12Ala allele within any of the diabetic, obese, or control groups. (See TABLES 3 and 4.)

DISCUSSION

The aim of this study was to determine the frequency of the proline to alanine substitution in the human PPARγ2 gene in a Czech population and to investigate its impact on glucose and lipid metabolism and susceptibility for type 2 diabetes mellitus and obesity. According to allele frequencies, this study indicates that the substitution is not associated with a decreased risk of type 2 diabetes or obesity. However, the

TABLE 3. Some clinical and anthropometric characteristics of type 2 diabetes mellitus patients and controls in relation to the Pro12Ala polymorphism of the PPARγ2 gene (± SD)

Characteristic	Type 2 diabetics (n = 183)			Controls (n = 69)		
	Pro/Pro	Pro/Ala, Ala/Ala	p	Pro/Pro	Pro/Ala, Ala/Ala	p
Sex (M/F)	51/82	15/35		23/34	4/8	
WHR	0.92 ± 0.08	0.90 ± 0.09	0.098	0.78 ± 0.08	0.77 ± 0.06	0.647
Serum blood glucose fasting (mmol/L)	9.30 ± 3.42	8.19 ± 3.12	0.045	4.69 ± 0.48	4.91 ± 0.71	0.350
Serum insulin fasting (mIU/L)	17.16 ± 11.15	12.98 ± 9.12	0.007	7.51 ± 5.33	8.18 ± 7.05	0.883
Serum proinsulin fasting (pmol/L)	16.94 ± 16.58	21.35 ± 22.21	0.522	3.21 ± 1.58	3.22 ± 1.95	0.567
Glycosylated hemoglobin (%Hb)	9.51 ± 2.34	9.48 ± 2.79	0.569	6.28 ± 0.82	6.30 ± 0.96	0.855
Glycosylated proteins (mmol/L)	1.65 ± 0.30	1.71 ± 0.36	0.530	1.11 ± 0.14	1.10 ± 0.11	0.968
Serum C-peptide fasting (nmol/L)	0.91 ± 0.43	0.90 ± 0.91	0.037	0.65 ± 0.43	0.59 ± 0.30	0.657
Total cholesterol (mmol/L)	6.43 ± 1.30	6.57 ± 1.43	0.603	4.40 ± 0.98	4.67 ± 0.69	0.194
HDL cholesterol (mmol/L)	1.34 ± 0.39	1.44 ± 0.38	0.100	1.35 ± 0.30	1.36 ± 0.19	0.887
LDL cholesterol (mmol/L)	4.74 ± 1.06	4.90 ± 1.26	0.669	2.88 ± 0.93	3.11 ± 0.71	0.241
Triglycerides (mmol/L)	2.37 ± 2.16	2.01 ± 1.20	0.439	0.83 ± 0.39	1.00 ± 0.37	0.130
Growth hormone (mIU/L)	3.33 ± 3.98	3.60 ± 4.59	0.923	4.92 ± 7.79	17.94 ± 27.02	0.647
Cortisol (nmol/L)	529.61 ± 230.66	555.46 ± 199.30	0.490	661.68 ± 287.52	744.75 ± 362.47	0.590
SHBG (nmol/L)	30.40 ± 18.20	33.59 ± 21.40	0.163	53.33 ± 43.63	77.67 ± 61.33	0.082
DHEA-S (μmol/L)	3.63 ± 2.17	3.10 ± 1.63	0.297	6.37 ± 4.38	5.39 ± 2.57	0.457
DHEA (nmol/L)	2.86 ± 1.48	2.97 ± 2.13	0.746	22.06 ± 11.84	25.14 ± 17.08	0.912
TSH (mIU/L)	2.07 ± 2.29	1.98 ± 1.38	0.623	2.43 ± 1.37	2.40 ± 1.51	0.975
Free T3 (pmol/L)	5.68 ± 2.01	5.18 ± 2.04	0.069	5.42 ± 1.21	4.92 ± 1.03	0.199
Free T4 (pmol/L)	15.69 ± 3.00	15.44 ± 2.84	0.851	15.78 ± 2.42	15.43 ± 5.47	0.073

TABLE 4. Some clinical and anthropometric characteristics of obese women and controls in relation to the Pro12Ala polymorphism of the PPARγ2 gene (± SD)

Characteristic	Obese (n = 86)			Controls (n = 69)		
	Pro/Pro	Pro/Ala, Ala/Ala	p	Pro/Pro	Pro/Ala, Ala/Ala	p
Sex (M/F)	0/60	0/26		23/34	4/8	
WHR	0.85 ± 0.06	0.88 ± 0.05	0.014	0.78 ± 0.08	0.77 ± 0.06	0.647
Serum blood glucose fasting (mmol/L)	5.46 ± 2.00	5.66 ± 2.39	0.648	4.69 ± 0.48	4.91 ± 0.71	0.350
Serum insulin fasting (mIU/L)	13.17 ± 8.85	15.24 ± 15.29	0.922	7.51 ± 5.33	8.18 ± 7.05	0.883
Serum C-peptide fasting (nmol/L)	1.08 ± 0.46	1.09 ± 0.36	0.736	0.65 ± 0.43	0.59 ± 0.30	0.657
Total cholesterol (mmol/L)	5.37 ± 0.85	5.57 ± 0.95	0.278	4.40 ± 0.98	4.67 ± 0.69	0.194
HDL cholesterol (mmol/L)	1.27 ± 0.31	1.19 ± 0.32	0.276	1.35 ± 0.30	1.36 ± 0.19	0.887
Triglycerides (mmol/L)	1.90 ± 1.03	1.99 ± 1.18	0.680	0.83 ± 0.39	1.00 ± 0.37	0.130
Growth hormone (mIU/L)	2.45 ± 4.97	2.25 ± 5.90	0.750	4.92 ± 7.79	17.94 ± 27.02	0.647
SHBG (nmol/L)	49.87 ± 42.13	43.44 ± 26.53	0.753	53.33 ± 43.63	77.67 ± 61.33	0.082
DHEA-S (µmol/L)	4.87 ± 3.06	4.18 ± 2.30	0.524	6.37 ± 4.38	5.39 ± 2.57	0.457
DHEA (nmol/L)	7.18 ± 5.19	5.92 ± 3.97	0.327	22.06 ± 11.84	25.14 ± 17.08	0.912
TSH (mIU/L)	2.58 ± 1.78	3.70 ± 2.64	0.012	2.43 ± 1.37	2.40 ± 1.51	0.975

present data show that insulin concentrations are lower in diabetic subjects with the 12Ala allele than in those without it. This finding became even more evident after adjustment for age, BMI, and diabetes duration. These results may indicate that diabetics carrying the 12Ala allele are more insulin-sensitive than those without it, although this suggestion is not supported by significant difference in HOMA-R or DI. Recently, Mori et al.[25] reported that the 12Ala variant is associated with a lower level of insulin secretion in diabetic subjects and, consequently, with respect to other observations, he concluded that this can be explained by a reduced capacity of pancreas to secrete insulin. This is probably not the case in our study as fasting glucose levels tended to be lower among 12Ala carriers within the diabetic group. It is possible that genetic or environmental[26] factors causing diabetes interact with the PPARγ2 gene, leading to the differences in insulin sensitivity between subjects with and without Pro12Ala substitution in type 2 diabetes mellitus patients.

Increased insulin sensitivity might to some extent predispose subjects with the 12Ala variant to lipid accumulation under some environmental and genetic backgrounds and our study supports this idea: the 12Ala variant is carried by nearly 30% of individuals in the obese group compared with <16% of control individuals. Moreover, among the obese, the 12Ala carriers have higher WHR. However, we failed to detect any other differences in body composition between subjects with 12Ala and those without it.

In conclusion, the PPARγ2 Pro12Ala polymorphism was found to be significantly associated with lower fasting insulin levels among diabetic patients; thus, it is possible that it has influence on insulin sensitivity. Therefore, this polymorphism may have an important impact on glucose homeostasis.

ACKNOWLEDGMENTS

This study was supported by Grant IGA MH NB/5395-5.

REFERENCES

1. LEHMANN, J., L.B. MOORE, T.A. SMITH-OLIVER et al. 1995. An antidiabetic thiazolidinedione is a high affinity ligand for peroxisome proliferator-activated receptor γ (PPARγ). J. Biol. Chem. **270:** 12953–12956.
2. VIDAL-PUIG, A., R.V. CONSIDINE et al. 1997. Peroxisome proliferator-activated receptor gene expression in human tissues: effects of obesity, weight loss, and regulation by insulin and glucocorticoids. J. Clin. Invest. **99:** 2416–2422.
3. FAJAS, L., D. AUBOEUF, E. RASPE et al. 1997. Organization, promoter analysis, and expression of the human PPARγ gene. J. Biol. Chem. **272:** 18779–18789.
4. AUBOEUF, D., J. RIEUSSET, L. FAJAS et al. 1997. Tissue distribution and quantification of the expression of PPARs and LXRα in humans: no alterations in adipose tissue of obese and NIDDM patients. Diabetes **46:** 1319–1327.
5. SPIEGELMAN, B.M. 1998. PPAR-γ: adipogenic regulator and thiazolidinedione receptor. Diabetes **47:** 507–514.
6. TONTONOZ, P., E. HU & B.M. SPIEGELMAN. 1994. Stimulation of adipogenesis in fibroblasts by PPARγ2, a lipid-activated transcription factor. Cell **79:** 1147–1156.
7. FORMAN, B.M., P. TONTONOZ, J. CHEN et al. 1995. 15-Deoxy $\delta^{12,14}$-prostaglandin J2 is a ligand for the adipocyte determination factor PPARγ. Cell **83:** 803–812.

8. KLIEWER, S.A., J.M. LENHARD, T.M. WILLSON et al. 1995. A prostaglandin J_2 metabolite binds peroxisome proliferator-activated receptor γ and promotes adipocyte differentiation. Cell **83:** 813–819.
9. YEN, C.J., B.A. BEAMER, C. NEGRI et al. 1997. Molecular scanning of the human peroxisome proliferator activated receptor γ (hPPARγ) gene in diabetic Caucasians: identification of a Pro^{12}Ala PPARγ2 missense mutation. Biochem. Biophys. Res. Commun. **241:** 270–274.
10. MASUGI, J., Y. TAMORI et al. 2000. Inhibitory effect of a proline-to-alanine substitution at codon 12 of peroxisome proliferator-activated receptor-γ2 on thiazolidine-induced adipogenesis. Biochem. Biophys. Res. Commun. **268:** 178–182.
11. DEEB, S.S., L. FAJAS, M. NEMOTO et al. 1998. A Pro 12 Ala substitution in PPARγ2 associated with decreased receptor activity, lower body mass index, and improved insulin sensitivity. Nat. Genet. **20:** 284–287.
12. HARA, K., T. OKADA, K. TOBE et al. 2000. The Pro 12 Ala polymorphism in PPAR γ2 may confer resistance to type 2 diabetes. Biochem. Biophys. Res. Commun. **271:** 212–216.
13. BEAMER, B., C. YEN, R. ANDERSEN et al. 1998. Association of the Pro 12 Ala variant in the peroxisome proliferator activated receptor γ gene with obesity in two Caucasian populations. Diabetes **47:** 1806–1808.
14. MORI, Y., H. KIM-MOTOYAMA, T. KATAKURA et al. 1998. Effect of the Pro 12 Ala variant of the human peroxisome proliferator activated receptor γ gene on adiposity, fat distribution, and insulin sensitivity in Japanese men. Biochem. Biophys. Res. Commun. **251:** 195–198.
15. RINGEL, J., S. ENGELI, A. DISTLER et al. 1999. Pro 12 Ala PPARγ2 missense mutation of the peroxisome proliferator activated receptor γ and diabetes mellitus. Biochem. Biophys. Res. Commun. **254:** 450–453.
16. MANCINI, F., O. VACCARO, L. SABATINO et al. 1999. Pro 12 Ala substitution in the peroxisome proliferator-activated receptor-γ2 is not associated with type 2 diabetes. Diabetes **48:** 1466–1468.
17. WHO. 1985. Diabetes mellitus: report of a WHO study group. WHO Tech. Rep. Ser. **727:** 7–113.
18. BLÁHA, P. 1991. ANTROPO-ein Programm für automatische Beartbeitung anthropologischer Daten. Wiss. Z. Humboldt-Univ. Berl. **5:** 153–156.
19. YOUNG, R.P., J.A. CRITCHLEY, P.J. ANDERSON et al. 1996. The short insulin tolerance test: feasibility study using venous sampling. Diabet. Med. **13:** 429–433.
20. HAMANN, A., H. MÜNZBERG, P. BUTTRON et al. 1999. Missense variants in the human peroxisome proliferator-activated receptor-γ2 gene in lean and obese subjects. Eur. J. Endocrinol. **141:** 90–92.
21. MATTHEWS, D.R., J.P. HOSKER, A.S. RUDENSKI et al. 1985. Homeostasis model assessment: insulin resistance and beta-cell function from fasting plasma glucose and insulin concentrations in man. Diabetologia **28:** 412–419.
22. KAHN, S.E., R.L. PRIGEON, D.K. MCCULLOCH et al. 1993. Quantification of the relationship between insulin sensitivity and beta-cell function in human subjects: evidence for a hyperbolic function. Diabetes **42:** 1663–1672.
23. POIRIER, O., V. NICAUD, F. CAMBIEN et al. 2000. The Pro12Ala polymorphism in the peroxisome proliferator-activated receptor γ2 gene is not associated with postprandial responses to glucose or fat tolerance tests in young healthy subjects: the European Atherosclerosis Research Study II. J. Mol. Med. **78:** 346–351.
24. EVANS, D., J. DE HEER, C. HAGEMANN et al. 2001. Association between the P12A and c1431t polymorphisms in the peroxisome proliferator activated receptor gamma (PPAR gamma) gene and type 2 diabetes. Exp. Clin. Endocrinol. Diabetes **109:** 151–154.
25. MORI, H., H. IKEGAMI, Y. KAWAGUCHI et al. 2001. The Pro12→Ala substitution in PPAR-γ is associated with resistance to development of diabetes in the general population. Diabetes **50:** 891–894.
26. LUAN, J., P.O. BROWNE, A. HARDING et al. 2001. Evidence for gene-nutrient interaction at the PPARgamma locus. Diabetes **50:** 686–689.

A Custom-Built Insulin Resistance Gene Chip

KEN WALDER,[a] DAVID SEGAL,[a] SAM CHEHAB,[a] GUY AUGERT,[a,b]
DAVID CAMERON-SMITH,[c] MARK HARGREAVES,[c] AND GREG R. COLLIER[a,d]

[a]*Metabolic Research Unit, Deakin University, Geelong, VIC, Australia*

[b]*Merck-Lipha SA, Lyon, France*

[c]*Deakin University, Burwood, VIC, Australia*

[d]*Autogen Limited, South Melbourne, VIC, Australia*

ABSTRACT: *Objectives/Aim*—Microarray (gene chip) technology offers a powerful new tool for analyzing the expression of large numbers of genes in many experimental samples. The aim of this study was to design, construct, and use a gene chip to measure the expression levels of key genes in metabolic pathways related to insulin resistance. *Methods*—We selected genes that were implicated in the development of insulin resistance, including genes involved in insulin signaling; glucose uptake, oxidation, and storage; fat uptake, oxidation, and storage; cytoskeletal components; and transcription factors. The key regulatory genes in the pathways were identified, along with other recently identified candidate genes such as calpain-10. A total of 242 selected genes (including 32 internal control elements) were sequence-verified, purified, and arrayed on aldehyde-coated slides. *Results*—Where more than 1 clone containing the gene of interest was available, we chose those containing the genes in the 5′ orientation and an insert size of around 1.5 kb. Of the 262 clones purchased, 56 (21%) were found to contain sequences other than those expected. In addition, 2 (1%) did not grow under standard conditions and were assumed to be nonviable. In these cases, alternate clones containing the gene of interest were chosen as described above. The current version of the Insulin Resistance Gene Chip contains 210 genes of interest, plus 48 control elements. A full list of the genes is available at http://www.hbs.deakin.edu.au/mru/research/gene_chip_tech/genechip_three.htm/. *Conclusions*—The human Insulin Resistance Gene Chip that we have constructed will be a very useful tool for investigating variation in the expression of genes relevant to insulin resistance under various experimental conditions. Initially, the gene chip will be used in studies such as exercise interventions, fasting, euglycemic-hyperinsulinemic clamps, and administration of antidiabetic agents.

KEYWORDS: gene chip; microarray; insulin resistance

INTRODUCTION

Searching for individual genes, candidate by candidate, in a complex disease with polygenic or oligogenic origins, such as type 2 diabetes, is a daunting and near-

Address for correspondence: Ken Walder, Metabolic Research Unit, Deakin University, Geelong, VIC, Australia. Voice: +61-3-5227-2883; fax: +61-3-5227-2170.
walder@deakin.edu.au

impossible task. Candidate genes are normally selected from physiological pathways known to be important in the regulation of carbohydrate or fat metabolism and include a wide variety of genes. Hundreds of candidate gene association studies have been performed in recent decades identifying very few (if any) plausible, replicable relationships between sequence variants in individual genes and diabetes or its subphenotypes (e.g., fasting glucose and insulin concentrations, measures of insulin secretion and action). However, the accumulation of data detailing expression of these genes under a range of physiological conditions may contribute to our understanding of which genes are important in the response to dietary and exercise interventions and which genes are dysregulated during the development of insulin resistance and type 2 diabetes. Our group has previously used gene expression–based approaches to identify potential new therapeutic targets for obesity (beacon)[1] and diabetes (tanis).

Microarray (gene chip) technology provides a powerful technical means to establish such an expression database of both candidate genes and unknown transcripts.[2–6] Using gene chips, one can obtain comparative estimates of the level of gene expression of large numbers of genes (up to 20,000 per chip) in each sample. Gene chips comprise a solid substrate (usually an optically flat glass microscope slide) containing many "spots" of cDNA in an orderly array chemically coupled to its surface. Fluorescently labeled cDNA from experimental samples can be competitively hybridized to the gene chip against a reference cDNA sample, which is labeled with a different fluor. The relative intensity of each fluor at each cDNA spot gives a relative indication of the level of that particular RNA species in the experimental sample relative to the reference RNA. The ratio of fluorescence can be taken as a measure of the expression level of the gene corresponding to that spot in the experimental sample.[2–6]

To maximize the efficacy of such an approach, we have created a customized gene chip containing a range of insulin resistance candidate genes. The genes were chosen to represent key regulatory steps in various biochemical pathways, including glycolysis and fatty acid oxidation, as well as a number of transcription factors and cytoskeletal, signal transduction, and proteolytic proteins. This list of insulin resistance candidate genes is by no means complete; however, due to the versatility of the gene chip printing process, we are able to easily add genes to the chip as new candidates emerge. We anticipate that the insulin resistance gene chip will continue to grow over the coming years and will develop into a comprehensive tool for analysis of genes involved in the pathogenesis of insulin resistance and type 2 diabetes.

METHODS

Gene Chip Production

IMAGE cDNA clones corresponding to the chosen genes were purchased from Incyte Genomics (Palo Alto, CA). The clones were grown overnight in LB broth containing 100 µg/mL carbenicillin (Life Technologies, Rockville, MD), and the inserts were PCR-amplified using 5'-amino modified vector-specific primers (M13 reverse: 5' CAG GAA ACA GCT ATG AC 3'; 21M13 forward: 5' TGT AAA ACG ACG GCC AGT 3'). PCR amplifications were performed in a reaction volume of

20 μL containing 1× PCR buffer, 1.5 mM $MgCl_2$, 2 μmol of each primer, 0.2 mmol of each dNTP, 2.5 U of Taq DNA polymerase (QIAGEN, Mannheim, Germany), and 0.2 μL of overnight culture of each clone. PCR amplification of each clone was performed in a GeneAmp PCR System 9700 thermal cycler (PE Applied Biosystems, Sunnyvale, CA) for 35 cycles of denaturation at 94°C for 30 s, annealing at 56°C for 30 s, and extension at 72°C for 120 s. A final extension step was performed at 72°C for 5 min. Products were visualized by agarose gel electrophoresis and then excised and sequence-verified using a BigDye Terminator Cycle Sequencing Ready Reaction Kit (PE Applied Biosystems). The cDNA sequences corresponding to the genes of interest were then purified using an ArrayIT PCR Purification Kit (Telechem, Sunnyvale, CA) and resuspended in 20 μL of 1× Spotting Solution (Telechem) at a concentration of 0.5 mg/mL in a 384-well plate. The amplified, purified cDNA species were then arrayed onto SuperAldehyde Microarray Substrates (Telechem) using a ChipWriter Pro microarrayer (Virtek, Toronto, Canada) fitted with 32 Stealth SMP-03 pins (Telechem). Spotted DNAs were allowed to bond to the slide overnight and the slides were washed and blocked as recommended by the manufacturer (Telechem).

Cell Culture

HepG2 cells were maintained in DMEM containing 10% fetal calf serum (FCS) (Life Technologies). Prior to the experiments, the cells were washed and placed in DMEM with no FCS for 24 h, and then insulin (1 μM) was added to the cells for 6 h. The reference samples for gene chip experiments received no insulin treatment.

RNA Extraction

Total RNA was extracted from the cells by a two-step procedure using TRIzol Reagent (Life Technologies) followed by RNeasy columns (QIAGEN) according to the manufacturer's instructions.

Indirect Labeling of cDNA

Fluorescently labeled cDNA was prepared from 20 μg of total RNA using an indirect labeling method. cDNA synthesis was performed in a 30-μL reaction containing 5 μg of oligo-dT primer; 400 U of SuperscriptII (Life Technologies); 1× first-strand buffer; 0.01 M DTT; 0.5 mmol of each dATP, dCTP, and dGTP; 0.150 mM dTTP (Amersham, Buckinghamshire, United Kingdom); and 0.2 mM aminoallyl-dUTP (Sigma, St. Louis, MO). Synthesis was conducted in a GeneAmp PCR System 9700 (PE Applied Systems) at 42°C for 2 h. The reaction was stopped by addition of 5 μL of 0.5 M EDTA, and RNA was hydrolyzed by addition of 20 μL of 1 M NaOH at 70°C for 20 min. The reaction was neutralized with 25 μL of 1 M HEPES and the cDNA was purified using QIAGEN PCR purification kits according to the manufacturer's instructions and eluted in nuclease-free water. The cDNA was concentrated using Microcon30 spin columns (Millipore, Bedford, MA) and the volume retrieved was dried down under vacuum. The cDNA pellet was resuspended in 0.09 M sodium bicarbonate and coupled to Cy3 or Cy5 monofunctional NHS ester reactive dye (Amersham). The coupling reaction was conducted in the dark for 1 h.

The fluorescently labeled cDNA was purified using QIAGEN PCR purification columns and then combined and added to 10 µg of Human Cot1 DNA (Life Technologies). The cDNA was again concentrated to the required volume with Microcon30 spin columns (Millipore). The cDNA was hybridized in a 20-µL volume, containing the labeled cDNA, 5× SSC, 0.3% SDS, 4 µg of yeast tRNA, 8 µg of poly-dA, and 2.5× Denhart's solution. The cDNA was denatured at 98°C for 2 min and maintained at 60°C until required. Then, 18 µL of the hybridization solution was applied to coverslips and mounted onto the array slide. Hybridization was conducted in humid hybridization chambers in a hybridization oven at 65°C for 16–20 h.

Image Acquisition and Data Analysis

Fluorescent images of the microarrays were acquired using a ScanArray Lite confocal laser scanner (Gsi Lumonics, Boston, MA) and the images were analyzed using ImaGene software (BioDiscovery, Sunnyvale, CA).

RESULTS AND DISCUSSION

A number of candidate genes of interest were chosen primarily because of their key regulatory roles in selected metabolic processes and/or previous evidence of linkage and association with type 2 diabetes and its subphenotypes (TABLE 1). Some

TABLE 1. Examples of genes included on the Insulin Resistance Gene Chip in functional classes

Glucose metabolism	*Lipid metabolism*	*Transcription factors*
GLUT4	ACC	NOSs
HKII	LPL	NFKB1
PKCs	CD36	PDE5A
SNAP23,25,29	FABPs	PPARs
VAMP2	HADHA	STATs
GYS1	ACACB	SREBPs
	HSL	RXR
Cytoskeletal		RAR
DES	*Phosphatases*	C/EBPs
DAG1	PTPRE	
Nebulin	PTPRF	*Kinases*
Titin	PTP1B	JNKs
	PTPRK	MAPKs
Insulin signaling	PTPN11	MAPKKs
PIK3 subunits	PTPRA	ERKs
INSR		MEKs
IRSs	*Other*	
	CAPN10	
	CAP	
	RABs	

NOTE: The complete list is available from http://www.hbs.deakin.edu.au/mru/research/gene_chip_tech/genechip_three.htm/.

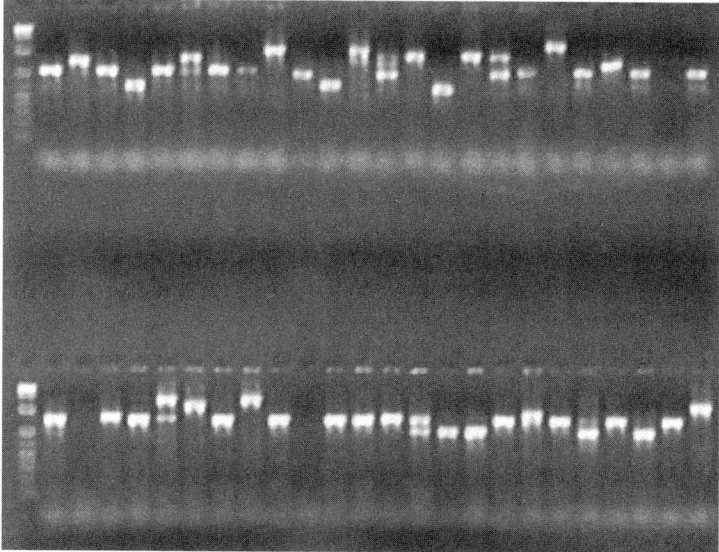

FIGURE 1. Representative agarose gel of amplified clones.

of the areas of particular interest included glucose and lipid metabolism, transcription factors, signal transduction, inflammation, and cytoskeletal proteins. The total number of genes currently on the Insulin Resistance Gene Chip is 250, which includes 48 internal control elements. The chip also contains 8 novel proprietary genes previously discovered in our laboratory.

A total of 262 clones were amplified and sequenced to produce this gene chip. Of these, 56 clones (21%) contained sequences different to that expected based on the annotation of that clone. In addition, 2 clones (1%) did not grow from the agar slabs provided and were assumed to be nonviable. In the case of 20 clones (8%), PCR amplification resulted in two distinct products, of which one corresponded with the expected sequence. FIGURE 1 shows a representative agarose gel of amplified products and includes examples of doublets and failures. The average size of the amplicons was 1.6 kb (range: 0.6–2.8 kb). Our finding that 21% of the clones obtained did not contain the annotated sequences is consistent with previous findings[7] and highlights an important consideration when designing projects such as this.

The purified PCR products were arrayed at a concentration of 100 ng/μL. A selection of the gene chips were stained for DNA using SYBRGold (Molecular Probes) and revealed a uniform printing pattern (FIG. 2). The median diameter of the features on the gene chip was 100 μm, with a preset spacing between the centers of adjacent features of 200 μm.

A series of experiments were conducted in parallel to assess the reproducibility of our gene chip system. These involved treatment of HepG2 hepatocytes with 1 μM insulin for 6 h. Experimental samples were labeled with Cy3 and reference samples were labeled with Cy5 and competitively hybridized to a gene chip. FIGURE 3 shows

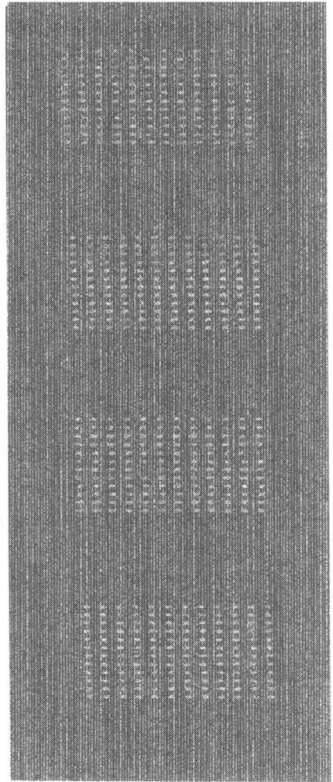

1 mm

FIGURE 2. SYBRGold stain of arrayed DNA on the Insulin Resistance Gene Chip.

a representative hybridization from this series of experiments. Initially, we conducted the same experiment four times and plotted the resulting hybridization intensities and ratios (experimental sample:reference sample). As shown in FIGURE 4, there was a very high degree of correlation between the results obtained in separate identical experiments, indicating a high degree of reproducibility in our system (correlation coefficient of 0.91).

To test for variance within hybridizations (i.e., across the slide), we printed the gene chip on six occasions on the same slide and conducted a hybridization with RNA obtained from insulin-treated HepG2 cells. Assessment of the variation across replicates revealed that the hybridization intensities were very consistent, with a mean coefficient of variation (CV) of 7%, while the ratios obtained exhibited a mean %CV of 18%.

FIGURE 3. Representative false color image of hybridization results using the Insulin Resistance Gene Chip.

The results of the replicated experiments were analyzed to determine which genes on the Insulin Resistance Gene Chip were affected by insulin treatment in HepG2 cells. Overall, 15 (7%) of the genes showed evidence of differential expression after insulin treatment (TABLE 2): 14 of these genes were upregulated after insulin treatment and 1 was downregulated. It should be noted that there are significant multiple testing issues with this type of analysis, and it could be argued that none of the genes listed in TABLE 2 would show significant differential expression after adequate correction for multiple testing. However, we present these findings in the context of the screen of 210 genes as those that show the best evidence of differential expression.

TABLE 2. Genes differentially expressed after treatment of HepG2 cells with insulin

Gene	Fold change	Nominal p value	Function
RAB4	1.52	0.002	GTPase
DAP3	1.37	0.002	Tumor suppressor
IL1RN	1.16	0.003	Inflammatory response
GOT1	1.17	0.005	TCA cycle; protein metabolism
PIK3R3	1.10	0.006	Insulin signaling
ATP5I	1.63	0.009	ATP synthesis

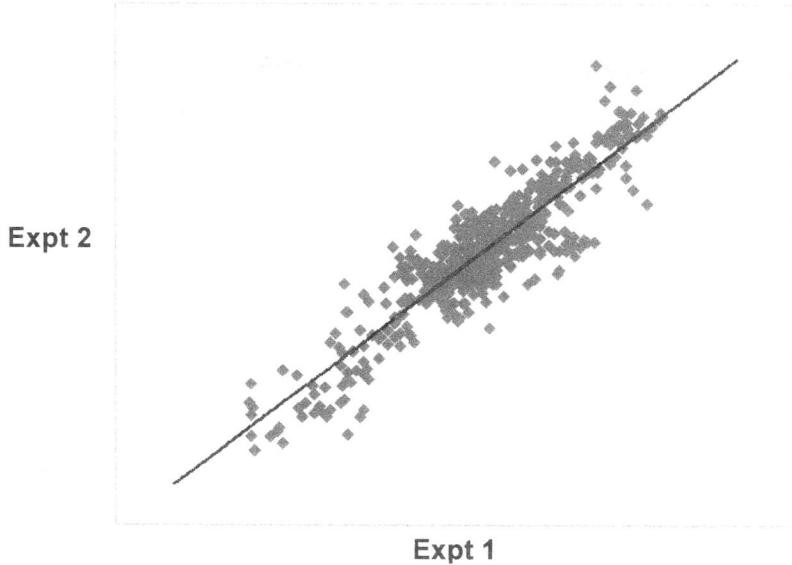

FIGURE 4. Reproducibility of repeated experiments using the Insulin Resistance Gene Chip ($r^2 = 0.83$).

The mean fold change in expression for all genes on the chip after insulin treatment was 1.05 (range: 0.82–1.63), indicating that overall there was little change in the expression of the selected genes. The genes identified as differentially expressed were from various functional groups that one would expect to be regulated by insulin treatment, such as fat and glucose oxidation, insulin signaling, and transcriptional control elements (TABLE 2). Note that the relatively small changes in gene expression caused by insulin treatment of HepG2 cells in this experiment are consistent with previous findings in our lab that glucose uptake and oxidation in these cells are only minimally affected by the addition of insulin (data not shown). We conclude that the HepG2 cells used in these experiments are relatively insulin-resistant.

CONCLUSIONS

The Insulin Resistance Gene Chip described here is a useful tool to profile the expression of genes in key metabolic pathways. We have shown that the results obtained using the gene chip are reproducible and reliable and that the system we are using is relatively robust. The Insulin Resistance Gene Chip will be used to profile changes in gene expression in tissues under a range of experimental conditions, such as type 2 diabetes, exercise training, dietary energy restriction, etc. These studies will increase our knowledge of the transcriptional changes associated with conditions associated with insulin resistance and should lead to greater understanding of the metabolic processes involved.

Furthermore, we intend to combine the Insulin Resistance Gene Chip with arrayed libraries of unknown clones. Using standard cluster analysis techniques, we will then be able to target unknown clones that demonstrate expression profiles similar or opposite to those of known key genes in insulin resistance. This may lead to the discovery of novel genes and pathways involved in the development of insulin resistance that represent potential new therapeutic targets for the treatment of type 2 diabetes.

REFERENCES

1. COLLIER, G.R., J.S. MCMILLAN, K. WINDMILL et al. 2000. Beacon: a novel gene involved in the regulation of energy balance. Diabetes **49:** 1766–1771.
2. SCHULZE, A. & J. DOWNWARD. 2001. Navigating gene expression using microarrays: a technology review. Nat. Cell Biol. **3:** E190–E195.
3. HEDGE, P., R. QI, K. ABERNATHY et al. 2000. A concise guide to cDNA microarray analysis. Biotechniques **29:** 548–556.
4. KHAN, J., M.L. BITTNER, Y. CHEN et al. 1999. DNA microarray technology: the anticipated impact on the study of human disease. Biochim. Biophys. Acta **1423:** M17–M28.
5. DEBOUCK, C. & P.N. GOODFELLOW. 1999. DNA microarrays in drug discovery and development. Nat. Genet. **21**(suppl.): 48–50.
6. BOWTELL, D.D. 1999. Options available—from start to finish—for obtaining expression data by microarray. Nat. Genet. **21**(suppl.): 25–32.
7. KNIGHT, J. 2001. When the chips are down. Nature **410:** 860–861.

Fatty Acid Regulation of Gene Expression

A Genomic Explanation for the Benefits of the Mediterranean Diet

STEVEN D. CLARKE,[a] DANIELA GASPERIKOVA,[b] CAROLANNE NELSON,[a] ALEXANDRE LAPILLONNE,[c] AND WILLIAM C. HEIRD[c]

[a]*Institute for Cellular and Molecular Biology and Division of Nutritional Sciences, University of Texas, Austin, Texas 78712, USA*

[b]*Diabetes and Nutrition Research Laboratory, Institute of Experimental Endocrinology, Slovak Academy of Sciences, SK-83306 Bratislava, Slovak Republic*

[c]*USDA/ARS Children's Nutrition Research Center, Baylor College of Medicine, Houston, Texas, USA*

ABSTRACT: The development of obesity and associated insulin resistance involves a multitude of gene products, including proteins involved in lipid synthesis and oxidation, thermogenesis, and cell differentiation. The genes encoding these proteins are in essence the blueprints that we have inherited from our parents. However, what determines the way in which blueprints are interpreted is largely dictated by a collection of environmental factors. The nutrients we consume are among the most influential of these environmental factors. During the early stages of evolutionary development, nutrients functioned as primitive hormonal signals that allowed the early organisms to turn on pathways of synthesis or storage during periods of nutrient deprivation or excess. As single-cell organisms evolved into complex life forms, nutrients continued to be environmental factors that interacted with hormonal signals to govern the expression of genes encoding proteins involved in energy metabolism, cell differentiation, and cell growth. Nutrients govern the tissue content and activity of different proteins by functioning as regulators of gene transcription, nuclear RNA processing, mRNA degradation, and mRNA translation, as well as functioning as posttranslational modifiers of proteins. One dietary constituent that has a strong influence on cell differentiation, growth, and metabolism is fat. The fatty acid component of dietary lipid not only influences hormonal signaling events by modifying membrane lipid composition, but fatty acids have a very strong direct influence on the molecular events that govern gene expression. In this review, we discuss the influence that (n-9), (n-6), and (n-3) fatty acids exert on gene expression in the liver and skeletal muscle and the impact this has on intra- and interorgan partitioning of metabolic fuels.

KEYWORDS: Mediterranean diet; obesity; insulin resistance; repartitioning agents; omega-3 fatty acids; HUFA

Address for correspondence: Steven D. Clarke, Ph.D., M. M. Love Chair of Nutritional, Cellular, and Molecular Sciences, 117 Gearing Building, University of Texas at Austin, Austin, TX 78712. Fax: 512-232-5864.
stevedclarke@mail.utexas.edu

INTRODUCTION

The development of obesity and associated insulin resistance involves a multitude of gene products, including proteins involved in lipid synthesis and oxidation, thermogenesis, and cell differentiation. The genes encoding these proteins are in essence the blueprints that we have inherited from our parents. However, what determines the way in which blueprints are interpreted is largely dictated by a collection of environmental factors. The nutrients we consume are among the most influential of these environmental factors. During the early stages of evolutionary development, nutrients functioned as primitive hormonal signals that allowed the early organisms to turn on pathways of synthesis or storage during periods of nutrient deprivation or excess. As single-cell organisms evolved into complex life forms, nutrients continued to be environmental factors that interacted with hormonal signals to govern the expression of genes encoding proteins involved in energy metabolism, cell differentiation, and cell growth. Nutrients govern the tissue content and activity of different proteins by functioning as regulators of gene transcription, nuclear RNA processing, mRNA degradation, and mRNA translation, as well as functioning as posttranslational modifiers of proteins. One dietary constituent that has a strong influence on cell differentiation, growth, and metabolism is fat. The fatty acid component of dietary lipid not only influences hormonal signaling events by modifying membrane lipid composition, but fatty acids have a very strong direct influence on the molecular events that govern gene expression. In the forthcoming

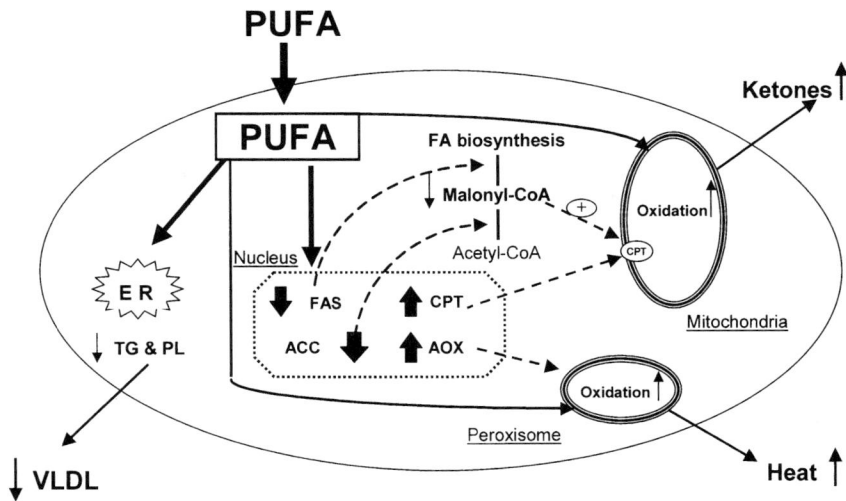

FIGURE 1. Polyunsaturated fatty acids (PUFA) as hepatic fatty acid repartitioning agents. This schematic describes the mechanisms by which dietary PUFA redirect fatty acids away from triglyceride storage and toward fatty acid oxidation. The data supporting the schematic have been derived from studies with the liver. ER, endoplasmic reticulum; TG, triglycerides; PL, phospholipid; VLDL, very-low-density lipoproteins; FAS, fatty acid synthase gene; ACC, acetyl-CoA carboxylase gene; CPT, liver-type carnitine palmitoyltransferase; AOX, peroxisomal acyl-CoA oxidase; FA, fatty acids.

review, we discuss the influence that (n-9), (n-6), and (n-3) fatty acids exert on gene expression in the liver and skeletal muscle and the impact this has on intra- and inter-organ partitioning of metabolic fuels (FIG. 1).

POLYUNSATURATED FATTY ACIDS AS FUEL REPARTITIONING AGENTS

In 1967, Nichaman et al.[1] reported that raising the linoleate content of the diet from 4% of calories to 18% reduced plasma triglyceride concentrations in normal and hyperlipemic subjects by 30% and 52%, respectively. Shortly after this, Allman and Gibson[2] discovered that the induction of hepatic lipogenesis associated with carbohydrate feeding could be inhibited by the inclusion of 18:2 (n-6) in the diet, but not by the addition of 16:0. Since the initial reports of Nichaman et al.[1] and Allman and Gibson,[2] numerous studies have demonstrated that the ingestion of fats rich in (n-6) and (n-3) fatty acids suppress hepatic *de novo* fatty acid biosynthesis, triglyceride production, and triglyceride secretion, while they enhance hepatic and skeletal muscle fatty acid oxidation.[3–16] These metabolic effects cannot be mimicked by dietary monounsaturated and saturated fatty acids nor by polyunsaturated fatty acids not derived from the (n-6) and (n-3) fatty acid families (e.g., 5,13-22:2).[15] Moreover, in order for 18:2 (n-6) and 18:3 (n-3) to influence gene expression and metabolism, they must be metabolized to highly unsaturated fatty acid (HUFA) products of the delta-6 desaturase pathway.[16]

In addition to regulating lipid metabolism, HUFAs may improve insulin sensitivity and increase nonoxidative glucose metabolism,[17,18] but this promising action of HUFA appears to be highly dependent on the amount of HUFA consumed. For example, we have recently discovered that rats fed a diet containing 40% of its energy as fish oil displayed a muscle gene expression pattern that would indicate glucose metabolism was severely impaired (e.g., a 20-fold increase in the expression of pyruvate dehydrogenase kinase 4). Interestingly, feeding rats a 40% olive oil diet resulted in a muscle gene expression pattern, suggesting that a diet rich in 18:1 (n-9) would minimally interfere with skeletal muscle glycolysis, appreciably increase glucose conversion to glycogen, and significantly reduce DNA damage (Nelson, Heird, and Clarke, unpublished data).

A key component of the HUFA repartitioning mechanism involves HUFA-dependent reduction in the hepatic activity of several glycolytic and lipogenic enzymes, including glucokinase, pyruvate kinase, acetyl-CoA carboxylase, fatty acid synthase, citrate lyase, malic enzyme, glucose-6-phosphate dehydrogenase, pyruvate dehydrogenase, and diacylglycerol acyl transferase.[2–15,19] Expression of these enzymes is very sensitive to the suppressive effects of HUFA. Diets that contain as little as 2–4% of their dietary energy as HUFA are sufficient to reduce hepatic fatty acid synthase activity by 25–40%, but maximum inhibition requires a diet that derives 10% of its energy from HUFA.[15]

Paralleling the HUFA-mediated decrease in hepatic lipogenesis is an increase in the rate of hepatic fatty acid oxidation and ketogenesis.[12–14] HUFA enhance fatty acid oxidation by coordinately increasing the expression of mitochondrial L-type carnitine palmitoyltransferase and peroxisomal acyl-CoA oxidase, by decreasing the synthesis of malonyl-CoA, and by decreasing the sensitivity of carnitine palmitoyltransferase to malonyl-CoA inhibition.[20] Carnitine palmitoyltransferase regulates the entry of fatty acids into the mitochondria, and acyl-CoA oxidase catalyzes the rate-limiting step in peroxisomal fatty acid oxidation. The contribution of hepatic peroxisomal fatty acid oxidation to overall fatty acid oxidation is unclear, but has been reported to be as high as 20%. Dietary HUFA readily induced the expression of both liver-type carnitine palmitoyltransferase and acyl-CoA oxidase. We have found that a diet providing as little as 7% kcal as HUFA will significantly increase the hepatic abundance of both transcripts (see reference 28; also Nelson, Heird, and Clarke, unpublished data).

In addition to the HUFA-dependent upregulation in hepatic capacity for fatty acid oxidation, metabolite regulation of fatty acid entry into the mitochondria is an important determinant of fatty acid oxidation.[21–23] In this respect, malonyl-CoA is a potent negative effector of carnitine palmitoyltransferase activity.[21–23] The concentration of malonyl-CoA is highly dependent on the rate of substrate flux through the fatty acid biosynthetic pathway.[21–23] Hepatic concentrations of malonyl-CoA increase 10- to 20-fold when fasted rats are fed a high-carbohydrate meal. Several years ago, we demonstrated that malonyl-CoA synthesis was significantly inhibited if 18:2 (n-6) supplemented the carbohydrate meal.[20] This, in fact, is one of the earliest metabolic changes detectable with HUFA feeding. Thus, the fatty acid biosynthetic pathway and its inhibition by dietary HUFA not only are important for governing the rate of conversion of dietary carbohydrate to fatty acids and triglycerides, but are also key determinants of the rate of fatty acid flux through the pathway of β-oxidation. This latter function may be most important in humans, where dietary HUFA are known to enhance fatty acid oxidation and lower hepatic triglyceride output, but where net fatty acid synthesis is reportedly very low.[24,25] In addition to suppressing the production of malonyl-CoA, HUFA also reduce the sensitivity of carnitine palmitoyltransferase to malonyl-CoA inhibition.[26] In this respect, 20- and 22-carbon (n-3) HUFA appear to be more potent than 18:2 (n-6).[26] Thus, dietary HUFA regulate hepatic lipid metabolism at three different levels: gene expression and enzyme abundance, malonyl-CoA synthesis, and membrane fluidity.[26,27]

Malonyl-CoA is also a key regulator of fatty acid oxidation in skeletal muscle.[22,23] Like the liver, the concentration of malonyl-CoA in soleus and gastrocnemius was highly dependent on nutritional state. For example, feeding fasted rats a high-carbohydrate diet increased skeletal muscle malonyl-CoA levels by 50–250% and, like the hepatic system, the rate of fatty acid oxidation decreased as the level of skeletal muscle malonyl-CoA increased.[22] However, unlike the liver, the production of malonyl-CoA in gastrocnemius muscle was not dependent on the amount of muscle-type acetyl-CoA carboxylase protein found in the muscle.[22] The expression of muscle-type acetyl-CoA carboxylase is also not affected by the type or amount of dietary HUFA (Nelson, Heird, and Clarke, unpublished data). Moreover, unlike the liver, muscle-type carnitine palmitoyltransferase and peroxisomal acyl-CoA oxidase gene expression are very resistant to upregulation by dietary HUFA. In fact to achieve a modest 50–100% increase in muscle-type carnitine palmitoyltransferase activity requires a diet to contain <35% HUFA kcal,[12–14] and even under these conditions peroxisomal acyl-CoA oxidase expression in gastrocnemius muscle remains virtually unchanged (Nelson, Heird, and Clarke, unpublished data). A close examination of the literature revealed that investigators have only employed high-HUFA diets or cells in culture to investigate the influence of fatty acid composition on muscle fatty acid oxidation.[12–14,28–30] Thus, while it is possible that low intakes of HUFA may increase mitochondrial membrane fluidity and reduce the malonyl-CoA sensitivity of muscle-type carnitine palmitoyltransferase, the impact of physiological intakes of HUFA on skeletal muscle fatty acid oxidation remains an open question. If increases in skeletal muscle fatty acid oxidation are to be seriously considered in formulating dietary recommendations for improvements in glucose sensitivity and for lowering blood triglycerides,[17,18,31] the influence of HUFA on skeletal muscle metabolism must be examined at lower, more meaningful dietary levels.

A GENOMIC MECHANISM FOR THE POLYUNSATURATED FATTY ACID REGULATION OF LIPID METABOLISM

Role of PPARα in PUFA Oxidation and Synthesis

HUFA inhibition of gene expression occurs too quickly (<90 min) to be explained simply by HUFA modifications of membrane-lipid environment-altered hormone receptor signaling.[7,30–33] Rather, the changes were more consistent with the idea that HUFA directly regulate the activity or abundance of a nuclear transcription factor. The 1990 discovery of a novel lipid-activated transcription factor, PPARα, provided the first evidence that the nucleus did, in fact, contain transcription factors that were dependent on fatty acid ligands.[34] PPARα is a member of the steroid receptor superfamily and, like other steroid receptors, it possesses a DNA-binding domain and a ligand-binding domain.[34–38] Interaction of PPARα with its DNA recognition site is markedly enhanced by ligands, such as the hypotriglyceridemic fibrate drugs, conjugated linoleic acid, and HUFA.[34–38] In general, PPARα activation leads to the induction of several genes encoding proteins involved in lipid transport, oxidation, and thermogenesis (FIG. 2).[29,39–42]

The importance of PPARα to overall glucose and fatty acid homeostasis was demonstrated by the observations that PPARα –/– mice lack the ability to increase rates of fatty acid oxidation during periods of food deprivation and they develop characteristics of adult-onset diabetes, including fatty livers, elevated blood triglycerides, and hyperglycemia.[43] The essentiality of PPARα to lipid oxidation was further underscored when hyperglycemia was found to suppress PPARα expression, induce PPARγ expression, increase β-cell and cardiomyocyte lipids, and accelerate cell death.[44,45]

The fact that PPARα may play a pivotal role in both fatty acid and glucose metabolism quickly led to the idea that PPARα was a "master switch" transcription factor that was targeted by HUFA to coordinately suppress genes encoding proteins of lipid synthesis and to induce genes encoding proteins of fatty acid oxidation and thermogenesis. More importantly, PPARα regulation by HUFA, particularly (n-3) HUFA and possibly conjugated linoleic acid, offered an explanation for the reported benefits of these fatty acids in protecting individuals from developing the detrimental characteristics of noninsulin-dependent diabetes.[18] This attractive hypothesis was strengthened by reports that potent pharmacological activators of PPARα modestly reduced lipogenic gene transcription.[33] However, PPARα does not appear to interact with HUFA-response sequences of lipogenic genes.[10] Moreover, HUFA

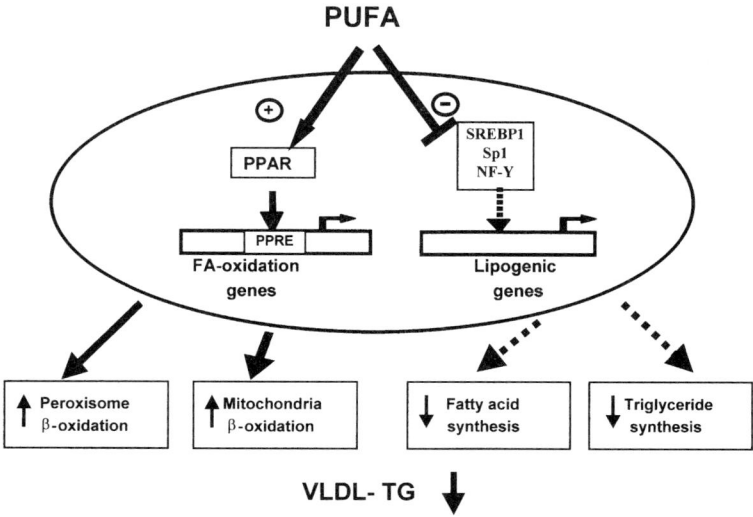

FIGURE 2. Nuclear mechanism for polyunsaturated fatty acid (PUFA) regulation of gene expression. FA, fatty acids; NF-Y, nuclear factor Y; PPAR, peroxisome proliferator-activated receptor; PPRE, peroxisome proliferator-activated receptor response element; Sp1, stimulatory protein 1; SREBP-1, sterol regulatory element binding protein-1; TG, triglycerides.

continue to suppress the transcription of hepatic lipogenic genes in PPARα −/− mice.[46] Thus, the HUFA inhibition of lipogenic gene expression is independent of PPARα and may simply reflect the increase in HUFA synthesis resulting from the PPARα-dependent induction of the delta-6 desaturase pathway (FIG. 2).[16,47]

HUFA Regulation of Transcription Factors in the Insulin Response Element and Lipogenic Gene Transcription

Dietary HUFA inhibit hepatic lipogenesis by suppressing the transcription of a number of hepatic enzymes involved in glucose metabolism and fatty acid biosynthesis, including glucokinase, pyruvate kinase, acetyl-CoA carboxylase, fatty acid synthase, stearoyl-CoA desaturase, and the delta-6 and delta-5 desaturases.[2–11,33] HUFA response sequences have been well characterized in only three genes: fatty acid synthase, S14, and L-type pyruvate kinase. The rat fatty acid synthase gene contains two independent HUFA regulatory sequences that are located between −118 and −43 and between −7250 and −7035 (Teran-Garcia and Clarke, unpublished data). Approximately 65% and 35% of the HUFA control of the fatty acid synthase gene can be attributed to the proximal and distal elements, respectively. Interestingly, the proximal HUFA response region of the fatty acid synthase gene has characteristics that are very similar to the HUFA response region of the S14 gene (−220 to −80), while the distal HUFA response region of the fatty acid synthase has similarities to the L-type pyruvate kinase HUFA response region (−160 to −97).[10]

The proximal HUFA response region of the fatty acid synthase gene imparts insulin responsiveness to the gene and contains DNA-binding sites for SREBP-1,

USF, Sp1, and NF-Y.[48–50] The nuclear abundance of USF is unaffected by dietary HUFA.[33] In contrast, HUFA rapidly reduce the nuclear content of hepatic SREBP-1 and this is associated with a decrease in the rate of fatty acid synthase and S14 gene transcription.[33] SREBPs are a family of transcription factors (i.e., SREBP-1a, -1c, and -2) that were first isolated as a result of their properties for binding to the sterol regulatory element. SREBP-2 is a regulator of genes encoding proteins involved in cholesterol metabolism.[48–53] SREBP-1 exists in two forms, 1a and 1c. SREBP-1a is the dominant form in cell lines and is a regulator of genes encoding proteins involved in both lipogenesis and cholesterogenesis. SREBP-1c constitutes 90% of the SREBP-1 found *in vivo* and is a determinant of lipogenic gene transcription.[54,55]

SREBP-1 is synthesized as a 125-kDa precursor protein that is anchored in the endoplasmic reticulum membrane.[48,53] Proteolytic release of the 68-kDa mature SREBP-1 occurs in the Golgi system, and movement of SREBP-1 from the endoplasmic reticulum to the Golgi requires the trafficking protein, SREBP cleavage-activating protein (SCAP).[53,56] Once released, mature SREBP-1 translocates to the nucleus and binds to both the classical sterol response element and/or to a palindrome CATG sequence. In the case of fatty acid synthase, SREBP-1 interacts with a CATG palindrome that also functions as an insulin response element.[48] Overexpression of mature SREBP-1a in liver is associated with high rates of fatty acid biosynthesis and the development of fatty liver.[57] In contrast, the ablation of the SREBP-1 gene results in low expression of lipogenic genes.[58] These observations led us to hypothesize that HUFA inhibit lipogenic gene transcription by impairing the proteolytic release of SREBP-1c and/or by suppressing SREBP-1c gene expression. Diets rich in 18:2 (n-6) or 20:5 and 22:6 (n-3) were found to reduce the hepatic nuclear and precursor content of mature SREBP-1 by 65% and 90% and by 60% and 75%, respectively.[33] The decrease in SREBP-1c was accompanied by a comparable decrease in the transcription rate of hepatic fatty acid synthase.[33] Unlike HUFA, saturated and monounsaturated fats had no effect on the nuclear content or precursor content of SREBP-1 or on lipogenic gene expression.[33] The HUFA-dependent reduction in hepatic content of SREBP-1 may explain how HUFA inhibit the transcription of several genes encoding proteins involved in hepatic glucose metabolism and fatty acid biosynthesis, including glucokinase, acetyl-CoA carboxylase, stearoyl-CoA desaturase, and the delta-6 and -5 desaturases.[59] Interestingly, the inhibition of lipogenic gene expression that reportedly occurs in adipose tissue with the ingestion of fish oil does not involve an SREBP-1-dependent mechanism.[51]

HUFA reduce the nuclear content of SREBP-1 by a two-phase mechanism. The first phase is a rapid (<60 min) inhibition of the proteolytic release process.[54,56,60] The second phase involves an adaptive (~48 h) reduction in the hepatic content of SREBP-1 mRNA, which is subsequently followed by a reduction in the amount of precursor SREBP-1 protein.[33] The mechanism by which HUFA acutely inhibit the proteolytic processes is unknown. However, nuclear run-on assays suggested that HUFA reduce the hepatic content of SREBP-1 mRNA by posttranscriptional mechanisms.[33] Using rat liver cells in primary culture, we have determined that HUFA reduced the half-life of SREBP-1c mRNA from 11 h to <5 h.[54] The mechanism by which HUFA control the half-life of SREBP-1 is unknown, but it may require the synthesis of a rapidly turning over HUFA-dependent protein.[54]

Recently, we found that feeding rats the PPARα-specific activator WY 14,643 induced hepatic delta-6 and delta-5 desaturase gene expression severalfold. The induction in delta-6 and delta-5 desaturase gene transcription was paralleled by an increase in the hepatic abundance of the respective transcripts and by a rise in enzymatic activity (Tang, Cho, and Clarke, unpublished data). This apparent increase in the rate of conversion of 18-carbon HUFA to 20- and 22-carbon products was paralleled by a 50% reduction in the expression of SREBP-1 and its target gene, fatty acid synthase.[33] Thus, it appears that nonlipid ligand activators of PPARα reduce hepatic lipogenic gene transcription by inducing activity of the delta-6 and delta-5 desaturase pathway. This suggests that PPARα activators (e.g., fibrates) lower hepatic triglyceride secretion and hence blood triglycerides in two ways: (i) by stimulating the hepatic oxidation of fatty acids and thereby diverting fatty acids away from triglyceride synthesis and (ii) by inducing the delta-6 desaturase pathway and the production of 20- and 22-carbon fatty acids, which in turn lower the SREBP-1 and thereby decrease lipogenic gene expression and hence the synthesis of malonyl-CoA for fatty acid biosynthesis and carnitine palmitoyltransferase regulation (FIGS. 1 and 2). It remains to be determined if the delta-6 and delta-5 desaturase pathway is enhanced by ligands for other PPARs, such as PPARγ and PPARδ. However, it is interesting that enrichment of skeletal muscle phospholipids with 20- and 22-carbon HUFA improves glucose uptake[17,18] and that PPARγ enhances skeletal muscle insulin sensitivity.[61]

SREBP-1c by itself possesses weak transactivating power, but the binding of SREBP-1c to its recognition sequence enhances the upstream DNA binding of NF-Y and Sp1 and this, in turn, amplifies the transactivating activities of the three transcription factors.[48] NF-Y is a heterotrimeric nuclear protein that reportedly plays a role in regulating chromatin structure by way of its interaction with histone acetyltransferases. The binding sites for NF-Y are essential for fatty acid synthase[49] and S14 promoter activity.[62] Mutations within the Y-box region of −104 to −99 of the S14 gene eliminated promoter activity by preventing NF-Y from interacting with upstream T3 (−2800 to −2500) and carbohydrate response (−1600 to −1400) regions.[62] Similarly mutating the Y-box motif of the rat fatty acid synthase gene eliminated 80% of the promoter activity (Teran-Garcia and Clarke, unpublished data). In contrast, eliminating the SREBP-1 site (−67 to −53) reduced fatty acid synthase promoter activity by only 40%. More importantly, only 25% of the HUFA inhibition of fatty acid synthase promoter activity was lost with the SREBP-1 site mutation, while 55% of the HUFA suppression fatty acid synthase promoter activity was lost with the NF-Y site mutation. Apparently, HUFA regulate fatty acid synthase gene transcription by governing the nuclear content of SREBP-1 and by interfering with the enhancer activity of NF-Y.

HUFA Regulation of the Carbohydrate Response Element and Lipogenic Gene Transcription

The insulin response region and its associated transcription factors (i.e., SREBP-1, NF-Y, and Sp1) are not the only nuclear factors regulated by HUFA. Transfection-reporter analyses indicate that HUFA exert a negative influence on the carbohydrate response element of the L-type pyruvate kinase[10,63] and fatty acid synthase genes (Teran-Garcia and Clarke, unpublished data). The nature of the transcription factors and the mechanism by which HUFA regulate them are not well defined. One hepatic protein that may be a HUFA target is HNF-4.[64] HNF-4 is a member of the steroid receptor superfamily. HNF-4 enhances the glucose/insulin induction of L-type pyruvate kinase transcription by binding as a homodimer to a direct repeat-1 motif.[63] Like PPARα, HNF-4 has a ligand-binding domain that interacts with acyl-CoA esters; unlike PPARα, fatty acyl-CoA binding to HNF-4 decreases its DNA binding activity.[63] This suggests that HUFA may exert part of its negative influence on gene transcription by reducing HNF-4 DNA-binding activity. Linker scanner mutations through the carbohydrate response region of the L-type pyruvate kinase promoter (i.e., −183 to −97) did, in fact, reveal that the HNF-4 recognition elements were essential for HUFA suppression of the promoter.[63] Recently, we found that sequences between −7242 and −7150 of the fatty acid synthase gene were required for glucose to induce fatty acid synthase gene transcription.[65,66] Subsequent studies have demonstrated that the −7242 to −7150 sequence contains DNA recognition sites for HNF-4 and a novel carbohydrate response factor.[66] Moreover, deleting this sequence eliminated 30–40% of the total HUFA suppression of the fatty acid synthase promoter (Teran-Garcia and Clarke, unpublished data). Thus, HUFA suppression of lipogenic gene expression may involve two mechanisms: (a) interference with the insulin regulation of transcription factor interaction with the insulin response element and (b) interference with the glucose signal governing the transcription factors affecting the carbohydrate response element.

HUFA Regulation of mRNA Processing

One of the first transcripts shown to be reduced by dietary HUFA was glucose-6-phosphate dehydrogenase mRNA. Interestingly, nuclear run-on assays revealed that the reduction in glucose-6-phosphate dehydrogenase mRNA was not due to an inhibition of gene transcription.[67] Recently, Salati and coworkers nicely demonstrated that HUFA govern the hepatic abundance of glucose-6-phosphate dehydrogenase mRNA by inhibiting the processing events involved in mRNA maturation.[67,68] HUFA regulation of transcript processing may be unique for glucose-6-phosphate dehydrogenase because transcription studies indicate that HUFA primarily exert their inhibitory influence by suppressing the rate of lipogenic gene transcription. Nevertheless, these results demonstrate that the dietary HUFA exert their influence of lipid synthesis and oxidation at a number of different regulatory steps, including transcription, mRNA processing, mRNA decay, and posttranslational modifications.

ARE OMEGA-3 FATTY ACIDS BETTER FUEL REPARTITIONERS THAN OMEGA-6?

To answer the relative potency question requires (i) a knowledge of the structural requirements for a HUFA to possess inhibitory activity and (ii) an appreciation for the fact that much of the work that has evaluated the effectiveness of omega-6 and omega-3 fatty acids has compared vegetable oils with marine oils. Such a comparison not only mixes the effects of (n-6) with (n-3) fatty acids, but also in essence compares delta-6 desaturase substrates [e.g., 18:2 (n-6)] with delta-6 desaturase products [e.g., 20:5 (n-3)].

To begin with, an inhibitory HUFA must consist of 18 carbons and contain at least two conjugated double bonds in the 9,12-position. The loss of one of the double bonds by hydroxylation renders the fatty acid inactive, but one of the double bonds may be in a *trans*-configuration.[15] A fatty acid may contain additional double bonds and still retain inhibitory potency (e.g., columbinic acid, 5t,9c,12c-18:3).[15] Most importantly, in order for a HUFA to influence fuel repartitioning, it must be a product of the delta-6 desaturase pathway.[16] These fatty acid structural requirements appear also to be instrumental in the HUFA regulation of skeletal muscle metabolism, but unequivocal data establishing these requirements remain to be acquired.

Delta-6 desaturase has a 2- to 3-fold greater affinity for 18:3 (n-3) than for 18:2 (n-6). Consequently, 18:3 (n-3) is more effective than 18:2 (n-6) as a regulator of lipid synthesis and oxidation. However, only small amounts of 18:2 (n-6) and 18:3 (n-3) undergo delta-6 desaturation (i.e., less than 1%).[69] Consequently, it is not surprising that dietary oils rich in fatty acid products of delta-6 desaturase are much more effective suppressors of hepatic lipogenesis and inducers of fatty acid oxidation than are the 18-carbon HUFA substrates (i.e., approximately 3-fold).[15] Moreover, when (n-6) and (n-3) HUFA are compared on an equal carbon basis, (n-6) and (n-3) HUFA appear to be equipotent regulators of lipid metabolism. For example, when 18:3 (n-6) and 20:4 (n-6) were compared with fish oil, these fatty acids were equipotent inhibitors of hepatic fatty acid biosynthesis.[15] Thus, the greater effectiveness of fish oil may have less to do with its (n-3) fatty acid content and more to do with the fact that fish oils are rich in 20- and 22-carbon HUFA, which bypass the regulated and required steps of desaturation and elongation. However, with this said, one cannot overlook the fact that 20:4 (n-6) may increase the production of eicosanoids and thereby offset the health benefits of HUFAs, for example, enhanced inflammatory responses. On the other hand, 20:5 (n-3) would give rise to eicosanoids with low bioactivity and, in this way, would provide benefits for both lipid metabolism and eicosanoid target diseases.

Finally, although (n-6) and (n-3) fatty acid products of delta-6 desaturase are modest inducers of fatty acid oxidation, both fatty acid groups are relatively weak ligand activators of PPARα. In contrast, eicosanoid products of (n-6) and (n-3) HUFA have 2- to 3-fold greater affinity for PPARα and hence are more effective activators of PPAR-responsive genes.[70] This raises the possibility that bioactive (n-6) or (n-3) fatty acid that governs lipid metabolism may be a prostaglandin, leukotriene, or hydroxylated fatty acid. While the answer to this question remains open, numerous studies using a wide array of eicosanoid synthesis inhibitors have failed to detect eicosanoid involvement in the regulation of hepatic lipid and glucose metabolism.[5,30] Prostaglandins have been implicated in the inhibition of lipogenic gene expression in 3T3-L1 adipocytes,[71] but their involvement in the *in vivo* regulation of lipogenesis remains to be established. In general, there are few data to support the argument that dietary HUFA exert their influence of hepatic lipogenesis by enhancing eicosanoid production. In fact, the signaling mechanism by which HUFA govern lipogenesis and lipogenic gene expression remains a mystery.

SUMMARY

A metabolic characteristic of noninsulin-dependent diabetes is the partitioning of fatty acids into triglycerides and away from fatty acid oxidation. Associated with this

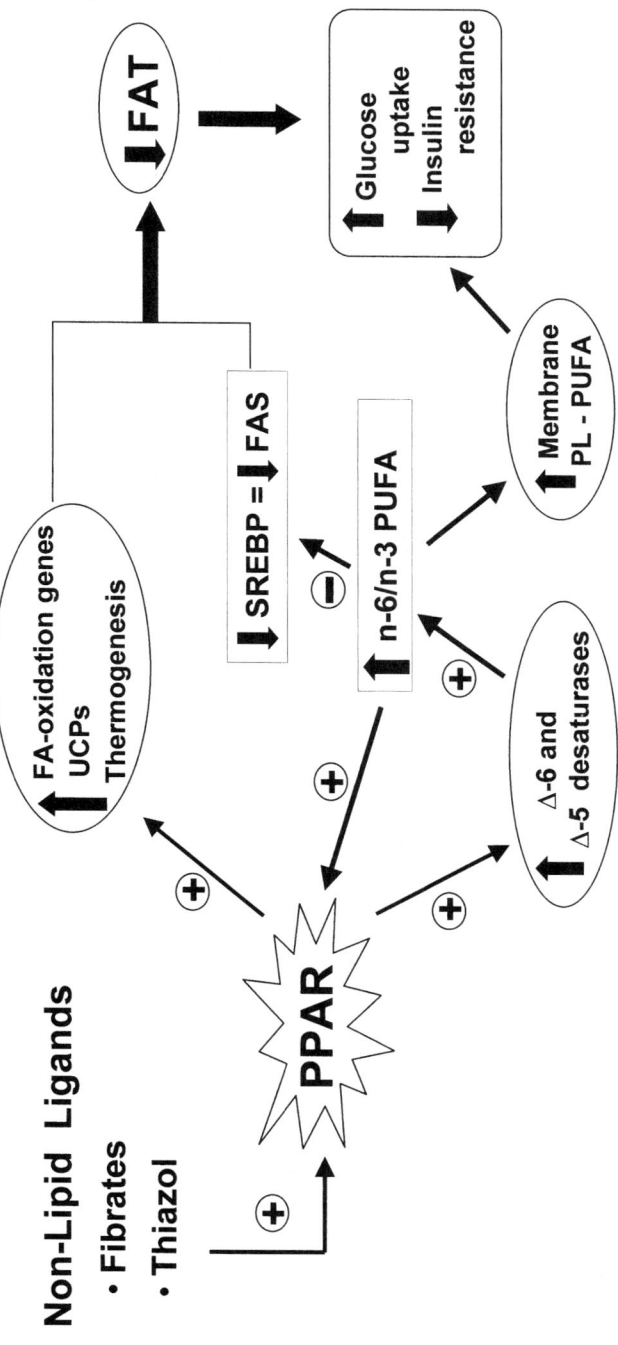

FIGURE 3. Schematic mechanism of polyunsaturated fatty acid (PUFA) regulation of fuel partitioning. Thiazol, thiazolidinediones; PPAR, peroxisome proliferator-activated receptor; UCPs, mitochondrial uncoupling proteins; SREBP, sterol regulatory element binding protein; FAS, fatty acid synthase; PL, phospholipids; +, activation or induction; −, suppression.

pattern of fuel partitioning is reduced insulin sensitivity, impaired skeletal muscle glucose utilization, and excessive triglyceride accumulation in select tissues (e.g., skeletal muscle, heart, and β cells) that may lead to accelerated rates of apoptosis (i.e., "lipotoxicity"). Using an integrated collection of human, animal, and cellular work, evidence was presented indicating that relatively low HUFA intakes [e.g., 2–5 g 20:5 and 22:6 (n-3) per day] lead to a repartitioning of hepatic fatty acids away from triglyceride synthesis and toward fatty acid oxidation (FIG. 3). HUFA intakes exert their influence on hepatic metabolism by modifying hepatic patterns of gene expression and by altering metabolic profiles that govern key regulatory enzymes of lipid and glucose metabolism. The repartitioning of metabolic fuels is most evident in the liver where HUFA (a) reduce fatty acid and triglyceride synthesis by inhibiting the nuclear actions of transcription factors involved in the control of lipogenic gene transcription; (b) lower malonyl-CoA synthesis and the sensitivity of carnitine palmitoyltransferase to malonyl-CoA inhibition and thereby enhance fatty acid entry into the mitochondria; and (c) increase the capacity for fatty acid oxidation by inducing the expression of genes encoding proteins of fatty acid oxidation. We suggest that fuel repartitioning of fatty acids occurs in skeletal muscle as well as liver, and this may may explain the positive association between skeletal muscle HUFA content and improved insulin sensitivity and glucose metabolism. However, studies to date have only examined the effects of high-HUFA intakes (e.g., <35% kcal); further, while these dietary studies have demonstrated an accelerated rate of muscle fatty acid oxidation and reduced muscle stores of triglyceride, recent gene expression profiles from our laboratory using microarray analyses indicate that high-HUFA intakes also severely interfere with skeletal muscle glucose utilization. Interestingly, high-olive-oil intakes had little negative impact on the expression of skeletal muscle glucose-metabolizing genes and, in fact, presented a gene pattern in gastrocnemius muscle that suggested the olive oil may facilitate glucose flux to glycogen and suppress skeletal muscle apoptosis. Collectively, the changes in gene expression profiles combined with observed adaptations in metabolic pathways has led to our speculation that a diet rich in olive oil combined with a balance of dietary fat derived from marine and vegetable lipids (e.g., 4 parts olive oil, 1 part fish oil, and 1 part vegetable oil) may reduce the risks associated with excessive tissue triglyceride accumulation (e.g., accelerated cellular apoptosis), which in turn may delay the onset of noninsulin-dependent diabetes. Unfortunately, data establishing the optimum dietary level of HUFA and the relative ratio of monounsaturated fatty acids to HUFA are lacking, but the use of RNA microarrays to establish skeletal muscle gene profiles has great potential for establishing future dietary recommendations.

ACKNOWLEDGMENTS

This work was supported by National Institutes of Health Grant Nos. DK-53872 (to S. D. Clarke) and HD-37133 (to W. C. Heird and S. D. Clarke) and by the sponsors of the M. M. Love Chair of Nutritional, Cellular, and Molecular Sciences.

REFERENCES

1. NICHAMAN, M.Z., R.E. OLSON & C.L. SWEELY. 1967. Metabolism of linoleic acid-1-14-C in normolipemic and hyperlipemic humans fed linoleate diets. Am. J. Clin. Nutr. **20**: 1070–1108.
2. ALLMAN, D.W. & D.W. GIBSON. 1969. Fatty acid synthesis during early linoleic acid deficiency in the mouse. J. Lipid Res. **6**: 51–62.
3. MUSCH, K., M.A. OJAKIAN & M.A. WILIAMS. 1974. Comparison of alpha-linolenate and oleate in lowering activity of lipogenic enzymes in rat liver: evidence for a greater effect of dietary linoleinate independent of food and carbohydrate intake. Biochim. Biophys. Acta **37**: 343–348.
4. CLARKE, S.D., D.R. ROMSOS & G.A. LEVEILLE. 1977. Differential effects of dietary methyl esters of long-chain saturated and polyunsaturated fatty acids on rat liver and adipose tissue lipogenesis. J. Nutr. **107**: 1170–1181.
5. FLICK, P.K., J. CHEN & P.R. VAGELOS. 1977. Effect of dietary linoleate on synthesis and degradation of fatty acid synthetase from rat liver. J. Biol. Chem. **252**: 4242–4247.
6. TOUSSANT, M.J., M.D. WILSON & S.D. CLARKE. 1981. Coordinate suppression of liver acetyl-CoA carboxylase and fatty acid synthetase by polyunsaturated fat. J. Nutr. **111**: 146–153.
7. CLARKE, S.D., M.K. ARMSTRONG & D.B. JUMP. 1990. Dietary polyunsaturated fats uniquely suppress rat liver fatty acid synthase and S14 mRNA content. J. Nutr. **120**: 225–232.
8. CLARKE, S.D. & S. ABRAHAM. 1992. Gene expression: nutrient control of pre- and posttranscriptional events. FASEB J. **6**: 3146–3152.
9. CLARKE, S.D. & D.B. JUMP. 1994. Dietary polyunsaturated fatty acid regulation of gene transcription. Annu. Rev. Nutr. **14**: 83–98.
10. JUMP, D.B. & S.D. CLARKE. 1999. Regulation of gene expression by dietary fat. Annu. Rev. Nutr. **19**: 63–90.
11. CLARKE, S.D. 2001. Polyunsaturated fatty acid regulation of gene transcription: a molecular mechanism to improve the metabolic syndrome. J. Nutr. **131**: 1129–1132.
12. IDE, T., H. KOBAYASHI, L. ASHAKUMARY et al. 2000. Comparative effects of perilla and fish oils on the activity and gene expression of fatty acid oxidation enzymes in rat liver. Biochim. Biophys. Acta **1485**: 23–35.
13. POWER, G.W. & E.A. NEWSHOLME. 1997. Dietary fatty acids influence the activity and metabolic control of mitochondrial carnitine palmitoyltransferase I in rat heart and skeletal muscle. J. Nutr. **127**: 2142–2150.
14. TAKADA, R., M. SAITOH & T. MORI. 1994. Dietary gamma-linolenic acid–enriched oil reduces body fat content and induces liver enzyme activities relating to fatty acid beta-oxidation in rats. J. Nutr. **124**: 469–474.
15. CLARKE, S.D. & D.B. JUMP. 1993. Regulation of hepatic gene expression by dietary gene expression by dietary fats: a unique role for polyunsaturated fatty acids. In Nutrition and Gene Expression. CRC Press. Boca Raton, FL.
16. NAKAMURA, M.T., H.P. CHO & S.D. CLARKE. 2000. Regulation of hepatic delta-6 desaturase expression and its role in the polyunsaturated fatty acid inhibition of fatty acid synthase gene expression in mice. J. Nutr. **130**: 1561–1565.
17. STORLIEN, L.H., E.W. KRAEGEN, D.J. CHISHOLM et al. 1987. Fish oil prevents insulin resistance induced by high-fat feeding in rats. Science **237**: 885–888.
18. BAUR, A., J. O'CONNOR, D.A. PAN & L.H. STORLIEN. 1999. Relationships between maternal risk of insulin resistance and the child's muscle membrane fatty acid composition. Diabetes **48**: 112–116.
19. ODIN, R.S., B.A. ADKINS-FINKE, W.L. BLAKE et al. 1987. Modification of fatty acid composition of membrane phospholipid in hepatocyte monolayer with n-3, n-6, and n-9 fatty acids and its relationship to triacylglycerol production. Biochim. Biophys. Acta **921**: 378–391.
20. WILSON, M.D., W.L. BLAKE, L.M. SALATI & S.D. CLARKE. 1990. Potency of polyunsaturated and saturated fats as short-term inhibitors of hepatic lipogenesis in rats. J. Nutr. **120**: 544–554.

21. MCGARRY, J.D. & N.F. BROWN. 1997. The mitochondrial carnitine palmitoyltransferase system: from concept to molecular analysis. Eur. J. Biochem. **244:** 1–8.
22. CHIEN, D., D. DEAN, A.K. SAHA *et al.* 2000. Malonyl-CoA content and fatty acid oxidation in rat muscle and liver *in vivo*. Am. J. Physiol. Metab. **279:** E259–E265.
23. ABU-ELHEIGA, L., M.M. MATZUK, A.H. ABO-HASHEMA & S.J. WAKIL. 2001. Continuous fatty acid oxidation and reduced fat storage in mice lacking acetyl-CoA carboxylase 2. Science **91:** 2613–2616.
24. HELLERSTEIN, M.K. 1995. Methods for measurement of fatty acid and cholesterol metabolism. Curr. Opin. Lipidol. **6:** 172–181.
25. COUET, C., J. DELARUE, P. RITZ *et al.* 1997. Effect of dietary fish oil on body fat mass and basal fat oxidation in healthy adults. Int. J. Obes. **21:** 637–643.
26. WONG, S.H., P.J. NESTEL, R.P. TRIMBLE *et al.* 1984. The adaptive effects of dietary fish and safflower oil on lipid and lipoprotein metabolism in perfused rat liver. Biochim. Biophys. Acta **792:** 103–109.
27. BROADWAY, N.M. & E.D. SAGGERSON. 1997. Effect of membrane environment on the activity and inhibitability by malonyl-CoA of the carnitine acyltransferase of hepatic microsomal membranes. Biochem. J. **322:** 435–440.
28. BAILLIE, R.A., R. TAKADA, M. NAKAMURA & S.D. CLARKE. 1999. Coordinate induction of peroxisomal acyl-CoA oxidase and UCP-3 by dietary fish oil: a mechanism for decreased body fat deposition. Prostaglandins Leukotrienes Essent. Fatty Acids **60:** 351–356.
29. BRANDT, J.M., F. DJOUADI & D.P. KELLY. 1998. Fatty acids activate transcription of the muscle carnitine palmitoyltransferase I gene in cardiac myocytes via the peroxisome proliferator-activated receptor alpha. J. Biol. Chem. **273:** 23786–23792.
30. CLARKE, S.D., M. TURINI, D.B. JUMP *et al.* 1998. Polyunsaturated fatty acid inhibition of fatty acid synthase transcription is independent of PPAR activation. Z. Ernaehrwiss. **37:** 14–20.
31. MALASANOS, T.H. & P.W. STACPOOLE. 1991. Biological effects of omega-3 fatty acids in diabetes mellitus. Diabetes Care **14:** 1160–1179.
32. CLARKE, S.D., D.R. ROMSOS & G.A. LEVEILLE. 1977. Influence of dietary fatty acids on liver and adipose tissue lipogenesis and on liver metabolites in meal-fed rats. J. Nutr. **107:** 1277–1287.
33. XU, J., M.T. NAKAMURA, H.P. CHO & S.D. CLARKE. 1999. Sterol regulatory element binding protein-1 expression is suppressed by dietary polyunsaturated fatty acids: a mechanism for the coordinate suppression of lipogenic genes by polyunsaturated fats. J. Biol. Chem. **274:** 23577–23583.
34. ISSEMANN, I. & S. GREEN. 1990. Activation of a member of the steroid hormone receptor superfamily by peroxisome proliferators. Nature **347:** 645–650.
35. KLIEWER, S.A., S.S. SUNDSETH, S.A. JONES *et al.* 1997. Fatty acids and eicosanoids regulate gene expression through direct interactions with peroxisome proliferator-activated receptors alpha and gamma. Proc. Natl. Acad. Sci. U.S.A. **94:** 4318–4323.
36. KREY, G., O. BRAISSANT, F. L'HORSET *et al.* 1997. Fatty acids, eicosanoids, and hypolipidemic agents identified as ligands of peroxisome proliferator-activated receptors by coactivator-dependent receptor ligand assay. Mol. Endocrinol. **11:** 779–791.
37. MOYA-CAMARENA, S.Y., J.P. VANDEN HEUVEL, S.G. BLANCHARD *et al.* 1999. Conjugated linoleic acid is a potent naturally occurring ligand and activator of PPAR-alpha. J. Lipid Res. **40:** 1426–1433.
38. KERSTEN, S., J. SEYDOUX, J. PETERS *et al.* 1999. Peroxisome proliferator-activated receptor alpha mediates the adaptive response to fasting. J. Clin. Invest. **103:** 1489–1498.
39. MASCARO, C., E. ACOSTA, J.A. ORTIZ & D. HARO. 1998. Control of human muscle-type carnitine palmitoyltransferase I gene transcription by peroxisome proliferator-activated receptor. J. Biol. Chem. **273:** 8560–8563.
40. RODRIGUEZ, J.C., B. GIL-GOMEZ, F.G. HEGRADT & D. HARO. 1994. Peroxisome proliferator-activated receptor mediates induction of the mitochondrial 3-hydroxy-3-methylglutaryl-CoA synthase gene by fatty acids. J. Biol. Chem. **269:** 18767–18772.
41. CLARKE, S.D., P. THUILLIER, R.A. BAILLIE & X. SHA. 1999. Peroxisome proliferator-activated receptors: a family of lipid-activated transcription factors. Am. J. Clin. Nutr. **70:** 566–577.

42. VARANASI, U., R. CHU, Q. HUANG et al. 1996. Identification of a peroxisome proliferator-responsive element upstream of the human peroxisomal fatty acyl coenzyme A oxidase gene. J. Biol. Chem. **271:** 2147–2155.
43. AOYAMA, T., J.M. PETERS, N. IRITANI et al. 1998. Altered constitutive expression of fatty acid–metabolizing enzymes in mice lacking the peroxisome proliferator-activated receptor alpha (PPARalpha). J. Biol. Chem. **273:** 5678–5684.
44. KAKUMA, T., Y. LEE, M. HIGA et al. 2000. Leptin, troglitazone, and the expression of sterol regulatory element binding proteins in liver and pancreatic islets. Proc. Natl. Acad. Sci. U.S.A. **97:** 8536–8541.
45. ZHOU, Y.T., P. GRAYBURN, A. KARIM et al. 2000. Lipotoxic heart disease in obese rats: implications for human obesity. Proc. Natl. Acad. Sci. U.S.A. **97:** 1784–1789.
46. REN, B., A.P. THELEN, J.M. PETERS et al. 1997. Polyunsaturated fatty acid suppression of hepatic fatty acid synthase and S14 gene expression does not require peroxisome proliferator-activated receptor alpha. J. Biol. Chem. **272:** 26827–26832.
47. MATSUI, H., K. OKUMURA, K. KAWAKAMI et al. 1997. Improved insulin sensitivity by bezafibrate in rats: relationship to fatty acid composition of skeletal-muscle triglycerides. Diabetes **46:** 348–353.
48. OSBORNE, T.F. 2000. Sterol regulatory element–binding proteins (SREBPs): key regulators of nutritional homeostasis and insulin action. J. Biol. Chem. **275:** 32379–32382.
49. MAGANA, M.M., S.H. KOO, H.C. TOWLE & T.F. OSBORNE. 2000. Different sterol regulatory element–binding protein-1 isoforms utilize distinct co-regulatory factors to activate the promoter for fatty acid synthase. J. Biol. Chem. **275:** 4726–4733.
50. XIONG, S., S.S. CHIRALA & S.J. WAKIL. 2000. Sterol regulation of human fatty acid synthase promoter I requires nuclear factor-Y– and Sp-1–binding sites. Proc. Natl. Acad. Sci. U.S.A. **97:** 3948–3953.
51. MATER, M.K., A.P. THELEN, D.A. PAN & D.B. JUMP. 1999. Sterol response element–binding protein 1c (SREBP1c) is involved in the polyunsaturated fatty acid suppression of hepatic S14 gene transcription. J. Biol. Chem. **274:** 32725–32744.
52. YAHAGI, N., H. SHIMANO, A.H. HASTY et al. 1999. A crucial role of sterol regulatory element–binding protein-1 in the regulation of lipogenic gene expression by polyunsaturated fatty acids. J. Biol. Chem. **274:** 35840–35844.
53. BROWN, M. & J.L. GOLDSTEIN. 1999. A proteolytic pathway that controls the cholesterol content of membranes, cells, and blood. Proc. Natl. Acad. Sci. U.S.A. **96:** 11041–11048.
54. XU, J., M. TERAN-GARCIA, J.H. PARK et al. 2001. Polyunsaturated fatty acids suppress hepatic sterol regulatory element–binding protein-1 expression by accelerating transcript decay. J. Biol. Chem. **276:** 9800–9807.
55. SHIMOMURA, I., Y. BASHMAKOV, S. IKEMOTO et al. 1999. Insulin selectively increases SREBP-1c mRNA in the livers of rats with streptozotocin-induced diabetes. Proc. Natl. Acad. Sci. U.S.A. **96:** 13656–13661.
56. HANNAH, V.C., J. OU, A. LUONG et al. 2001. Unsaturated fatty acids down-regulate SREBP isoforms 1a and 1c by two mechanisms in HEK-293 cells. J. Biol. Chem. **276:** 4365–4372.
57. SHIMOURA, I., R. HAMMER, J.A. RICHARDSON et al. 1998. Insulin resistance and diabetes mellitus in transgenic mice expressing nuclear SREBP-1c in adipose tissue: model for congenital generalized lipodystrophy. Genes Dev. **12:** 3182–3194.
58. SHIMANO, H., N. YAHAGI, M. AMEMIYA-KUFO et al. 1999. Sterol regulatory element–binding protein-1 as a key transcription factor for nutritional induction of lipogenic enzyme genes. J. Biol. Chem. **274:** 35832–35839.
59. SESSLER, A.M. & J.M. NTAMBI. 1998. Polyunsaturated fatty acid regulation of gene expression. J. Nutr. **128:** 923–926.
60. XU, J., M. GARCIA-TERAN, H.P. CHO & S.D. CLARKE. 2000. Polyunsaturated fatty acids (PUFA) regulate the action of sterol regulatory element binding protein-1 (SREBP-1) by inhibiting SREBP-1 nuclear localization and by enhancing SREBP-1 mRNA decay. FASEB J. **14:** A210.
61. KLETZIEN, R.F., L.A. FOELLMI, P.K.W. HARRIS et al. 1992. Adipocyte fatty acid–binding protein: regulation of gene expression *in vivo* and *in vitro* by an insulin-sensitizing agent. Mol. Pharmacol. **42:** 558–562.

62. JUMP, D.B., M.V. BADIN & A. THELEN. 1997. The CCAAT box binding factor, NF-Y, is required for thyroid hormone regulation of rat liver S14 gene transcription. J. Biol. Chem. **272:** 27778–27786.
63. LIIMATTA, M., H.C. TOWLE, S.D. CLARKE & D.B. JUMP. 1994. Dietary polyunsaturated fatty acids interfere with the insulin/glucose activation of L-type pyruvate kinase gene transcription. Mol. Endocrinol. **8:** 1147–1153.
64. HERTZ, R., J. MAGENHEIM, I. BERMAN & J. BAR-TANA. 1998. Fatty acyl-CoA thioesters are ligands of hepatic nuclear factor-4alpha. Nature **392:** 512–516.
65. RUFO, C., D. GASPERIKOVA, S.D. CLARKE et al. 1999. Identification of a novel enhancer sequence in the distal promoter of the rat fatty acid synthase gene. Biochem. Biophys. Res. Commun. **261:** 400–405.
66. RUFO, C., M. TERAN-GARCIA, M.T. NAKAMURA et al. 2001. Involvement of a unique carbohydrate-responsive factor in the glucose regulation of rat liver fatty-acid synthase gene transcription. J. Biol. Chem. **276:** 21969–21975.
67. STABILE, L.P., S.A. KLAUTKY, S.M. MINOR & L.M. SALATI. 1998. Polyunsaturated fatty acids inhibit the expression of the glucose-6-phosphate dehydrogenase gene in primary rat hepatocytes by a nuclear posttranscriptional mechanism. J. Lipid Res. **39:** 1951–1963.
68. AMIR-AHMADY, B. & L.M. SALATI. 2001. Regulation of the processing of glucose-6-phosphate dehydrogenase mRNA by nutritional status. J. Biol. Chem. **276:** 10514–10523.
69. PAWLOSKY, R.J., J.R. HIBBELN, J.A. NOVOTNY & N. SALEM, JR. 2001. Physiological compartmental analysis of alpha-linolenic acid metabolism in adult humans. J. Lipid Res. **42:** 1257–1265.
70. DEVCHAND, P.R., H. KELLER, J.M. PETERS et al. 1996. The PPARalpha-leukotriene B4 pathway to inflammation control. Nature **384:** 39–43.
71. MATER, M.K., D. PAN, W.G. BERGEN & D.B. JUMP. 1998. Arachidonic acid inhibits lipogenic gene expression in 3T3-L1 adipocytes through a prostanoid pathway. J. Lipid Res. **39:** 1327–1334.

Perinatal Supply and Metabolism of Long-Chain Polyunsaturated Fatty Acids

Importance for the Early Development of the Nervous System

ELVIRA LARQUE, HANS DEMMELMAIR, AND BERTHOLD KOLETZKO

*Division of Metabolism and Nutrition, Kinderklinik and Kinderpoliklinik,
Dr. von Hauner Children's Hospital, Ludwig-Maximilians-University of Munich,
D-80337 München, Germany*

ABSTRACT: The long-chain polyunsaturated fatty acids, arachidonic (AA) and docosahexaenoic acid (DHA), are essential structural lipid components of biomembranes. During pregnancy, long-chain polyunsaturated fatty acids (LC-PUFA) are preferentially transferred from mother to fetus across the placenta. This placental transfer is mediated by specific fatty acid binding and transfer proteins. After birth, preterm and full-term babies are capable of converting linoleic and α-linolenic acids into AA and DHA, respectively, as demonstrated by studies using stable isotopes, but the activity of this endogenous LC-PUFA synthesis is very low. Breast milk provides preformed LC-PUFA, and breast-fed infants have higher LC-PUFA levels in plasma and tissue phospholipids than infants fed conventional formulas. Supplementation of formulas with different sources of LC-PUFA can normalize LC-PUFA status in the recipient infants relative to reference groups fed human milk. Some, but not all, randomized, double-masked placebo-controlled clinical trials in preterm and healthy full-term infants demonstrated benefits of formula supplementation with DHA and AA for development of visual acuity up to 1 year of age and of complex neural and cognitive functions. From the available data, we conclude that LC-PUFA are conditionally essential substrates during early life that are related to the quality of growth and development. Therefore, a dietary supply during pregnancy, lactation, and early childhood that avoids the occurrence of LC-PUFA depletion is desirable, as was recently recommended by an expert consensus workshop of the Child Health Foundation.

KEYWORDS: arachidonic acid; docosahexaenoic acid; infant nutrition; dietary requirements; long-chain polyunsaturated fatty acids

---

Address for correspondence: Berthold Koletzko, M.D., Professor of Pediatrics, Division of Metabolism and Nutrition, Dr. von Hauner Children's Hospital, Ludwig-Maximilians-University of Munich, Lindwurmstraße 4, D-80337 München, Germany. Voice: +49-89-5160 3967; fax: +49-89-5160 3336.

Berthold.Koletzko@kk-i.med.uni-muenchen.de

INTRODUCTION

The potential of the early diet for modulation of the normal trajectory of brain development is of great interest. Long-chain polyunsaturated fatty acids (LC-PUFA), especially arachidonic acid (AA) and docosahexaenoic acid (DHA), are preferentially deposited in relatively high concentrations in developing neural cells and they modulate the structure, fluidity, and function of brain membranes.[1,2]

DHA acyl chains promote the function of the G protein–coupled system in membranes of photoreceptor cells and enhance the signaling pathways of metarhodopsin II.[3] The prenatal and postnatal accretion of LC-PUFA determine myelination and synaptogenesis during the postnatal brain growth spurt.[4] Recent studies have also provided evidence that DHA is involved in dopamine and serotonin metabolism.[2] In preterm babies, the availability of AA has been associated with weight at birth[5] and growth during the first year of life.[6]

This article discusses the potential roles of LC-PUFA during early human growth and development, under the perspective of preventive nutrition. The metabolism of essential fatty acids and their LC-PUFA metabolites, their transfer from mothers to their babies before and after birth, and differences between breast-fed and formula-fed infants are addressed. Available information on the influence of LC-PUFA on the development of visual and other neural functions in preterm and in healthy infants is reviewed.

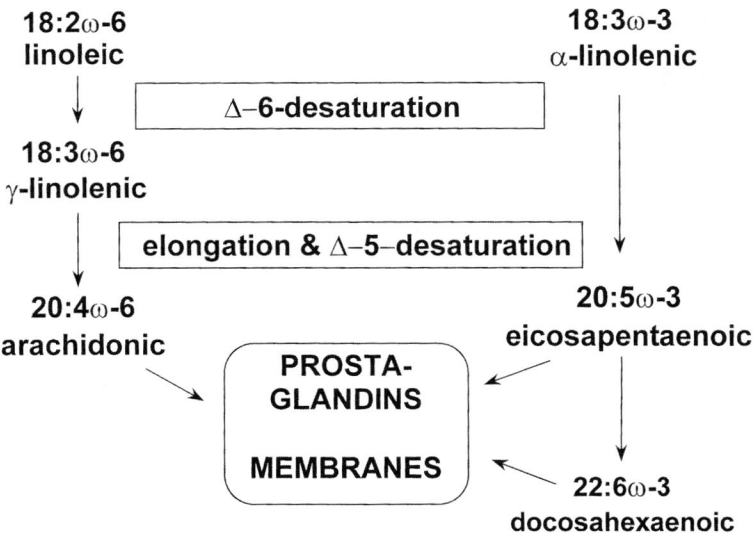

FIGURE 1. Simplified scheme of the conversion of the precursor essential fatty acids to their long-chain polyunsaturated metabolites.

LONG-CHAIN POLYUNSATURATED FATTY ACID METABOLISM

The liver is the major site for the metabolism of PUFA, both for synthesis of hepatic membrane phospholipids as well as for export and uptake by most other cells. AA and DHA can be synthesized by desaturation and elongation of the essential fatty acids, linoleic acid (LA, C18:2 n-6) and α-linolenic acid (ALA, C18:3 n-3), respectively. The metabolism of LA and ALA uses the same enzymes, resulting in competition between n-6 and n-3 fatty acids (FIG. 1).[7,8]

In the central nervous system, neurons appear unable to carry out fatty acid desaturation. By contrast, glial cells, astrocytes, and cerebral endothelium can elongate and desaturate precursors of LC-PUFA and accumulate DHA for maintaining a brain environment enriched in LC-PUFA.[9] However, the accumulation of preformed DHA and AA in the brain is far more efficient than the desaturation and elongation of the precursors.[10] In humans, the fetal and infant brain DHA content is relatively more affected by the diet than AA content, suggesting that endogenous metabolic regulation of AA content is more effective.[11]

SOURCES OF ESSENTIAL AND LONG-CHAIN POLYUNSATURATED FATTY ACIDS IN THE PERINATAL PERIOD

Fetal Essential Fatty Acid and LC-PUFA Supply

During pregnancy, the fetus is supplied with preformed LC-PUFA by placental transfer. The biochemical mechanisms involved in the underlying transport processes are not fully understood and it is unknown to which extent maternal dietary intakes can affect the LC-PUFA transfer from the mother to the fetus. Plasma lipids of mothers at birth contain higher levels of the essential fatty acid precursors, ALA and LA, than the cord blood lipids of their healthy term infants.[12] By contrast, percentage values for LC-PUFA are clearly and significantly higher in infants than in their mothers. These results point to a preferential and selective materno-fetal LC-PUFA transfer.[13]

Maternal triacylglycerols are hydrolyzed by lipoprotein lipase before they are transferred by the placenta.[14] Several proteins have been reported or proposed to be involved in the fatty acid movement across the placenta (TABLE 1).[15–18] The presence in placental tissue of plasma membrane fatty acid binding protein (FABPpm), fatty acid translocase (FAT), fatty acid transport protein (FATP), and recently a new fatty acid binding protein located exclusively on the maternal-facing membranes, named placental plasma membrane fatty acid binding protein (p-FABP$_{pm}$), has been reported. FABPpm, FAT, and FATP have not shown any specificity for particular types of free fatty acids.[19] However, pretreatment of human placental choriocarcinoma (Bewo) cells with antibodies against p-FABP$_{pm}$ inhibited most of the uptake of DHA (64%) and AA (68%), whereas oleic acid uptake was inhibited only 32% compared with the controls treated with preimmune serum.[20] Thus, p-FABP$_{pm}$ may be involved in the preferential uptake of LC-PUFA by these cells. p-FABP$_{pm}$ and classic FABPpm are both peripherally membrane-bound proteins of similar size (~40 kDa), but they differ in amino acid composition, pI value, and aspartate aminotransferase activity. Definitive evidence about the structure and function of p-FABP$_{pm}$ must await analysis of its complete amino acid and/or cDNA sequence.

TABLE 1. Plasma membrane fatty acid binding/transport proteins in the human placenta

Year	Research group	Name	Size (kDa)	Similarities	Proof of function	Reference
1985	Berk	FABPpm	43	MAspAT	Antibody inhibition, gene expression	15
1993	Abumrad	FAT	88	CD36	Inhibition by protein labeling	17
1994	Lodish	FATP	63	—	Gene expression	18
1994	Dutta-Roy	p-FABP$_{pm}$	40	—	Antibody inhibition	16

NOTE: Plasma membrane fatty acid binding protein = FABPpm; fatty acid translocase = FAT; fatty acid transport protein = FATP; placental FABPpm = p-FABP$_{pm}$.

Essential Fatty Acid and LC-PUFA Supply with Breast-Feeding

After birth, breast-fed infants receive appreciable amounts of preformed AA and DHA with human milk lipids that meet LC-PUFA accretion rates in membrane-rich tissues.[21] The fatty acid concentration of milk is related to the maternal diet, maternal plasma fatty acid composition, length of breast-feeding, and other factors. While LA values appear to be related to maternal dietary LA intake, AA values in milk are within a rather very narrow range.[1,21] By contrast, there are more than fourfold differences between the lowest and highest ALA and DHA values, respectively, which indicates a larger relative variability of n-3 than of n-6 fatty acid content in human milk.

Although milk LC-PUFA content is influenced by the maternal diet, studies investigating milk composition indicate some metabolic control of milk PUFA content.[1,21] We studied PUFA turnover in lactating women using LA uniformly labeled with the stable isotope ^{13}C.[22] The collection of milk and breath samples over a period of 5 days after the tracer application revealed that about 30% of milk LA is directly transferred from the diet, whereas about 11% of milk dihomo-γ-linolenic acid and 1.2% of milk AA originate from direct endogenous conversion of dietary LA.[22] Thus, the major portion of PUFA in human milk lipids is derived from maternal body stores and not directly from the maternal diet. This results in a relatively constant milk PUFA supply to the recipient infant, which might be of biological benefit (FIG. 2).

In a more recent study, we evaluated the contribution of dietary and endogenously synthesized AA to its milk secretion in 10 Mexican women on a habitual diet with a very low fat content.[23] The accumulated 72-h recovery of ^{13}C-LA in milk was 16.3±6.4% of the dose, but only 0.01% of the label was found as ^{13}C-AA. The AA stemming from conversion of dietary LA contributed only 1.1% to the total milk AA secreted. In this population, 70% of LA and almost 90% of AA secreted in milk were not derived from direct intestinal absorption. Thus, only a minor fraction of milk AA stemmed from conversion of LA, and maternal body stores are the major source of human milk LA and AA. We also studied supplementation of lactating women with n-3 LC-PUFA and found that this intervention can effectively prevent the postnatal decline of milk DHA without changing the percentage utilization of DHA.[24]

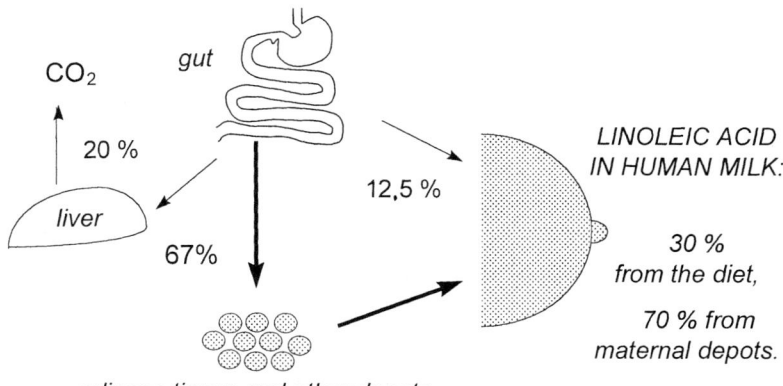

FIGURE 2. Schematic depiction of linoleic acid turnover in human lactation. Stable isotope studies with oral intake of ^{13}C-labeled linoleic acid indicate that 12.5% of dietary linoleic acid is transferred into milk, 20% is oxidized, and about 67% is deposited in maternal stores with slow turnover, such as adipose tissue.[22] Only 30% of the infant linoleic acid supply with human milk is derived directly from the maternal diet, but 70% originates from maternal body depots with slow turnover. Thus, breast-fed infants receive a relatively constant PUFA supply, even if there are short-term changes of maternal dietary intake.

Fatty Acid Status in Infants Fed Human Milk or Formula without and with LC-PUFA

In contrast to human milk, conventional milk formulas with fat derived from vegetable oils do not provide appreciable amounts of LC-PUFA. Infants fed such formulas depend on the utilization of body stores or on endogenous LC-PUFA synthesis for tissue deposition. A number of studies have evaluated the fatty acid composition of plasma and erythrocyte membrane lipid classes in full-term infants fed human milk or formula without LC-PUFA in order to estimate their LC-PUFA status.[25-34] A marked decrease in plasma and red blood cell AA and DHA in formula-fed as compared with breast-fed infants was observed in the absence of dietary supplementation at different ages. Moreover, not only plasma and red blood cell, but also tissue LC-PUFA contents appear to be affected. The proportion of DHA in the brain cortex of breast-fed infants was higher compared to those fed formula without LC-PUFA.[11]

Although infantile LC-PUFA synthesis has been demonstrated in newborns during the first week of life with refined stable isotope techniques,[8] a limited Δ6-desaturase activity and a high utilization of LC-PUFA for deposition, oxidation, and metabolic conversion to eicosanoids seem to result in an inadequacy of endogenous n-6 and n-3 LC-PUFA synthesis to prevent LC-PUFA depletion in infants fed conventional formulas without preformed LC-PUFA.[1,2,7,25]

Therefore, LC-PUFA supplementation of infant formulas has been proposed to achieve similar levels of LC-PUFA in the plasma and cells of formula-fed infants as found in breast-fed babies.[35] Carlson et al.[36] supplemented preterm formulas with marine oils rich in eicosapentaenoic acid (EPA) and DHA, which resulted in a reduction of AA plasma levels associated with lower growth rate in preterm babies. This effect was attributed to the presence of high EPA content in the supplemented formula, which might have inhibited AA synthesis or displaced AA from phospholipids. Further studies were carried out with other fat supplements, such as a fish oil with low EPA content, fractionated egg yolk phospholipids, evening primrose oil, and single cell oils from algae and fungi. The studies showed that a balanced supplementation with both DHA and AA can normalize LC-PUFA status in infants relative to reference groups fed human milk.[1,2] Incorporation of AA in formulas supplemented with DHA appears desirable to avoid subnormal plasma levels of AA.[1,2,27]

EFFECT OF DHA STATUS ON VISUAL FUNCTION IN PRETERM INFANTS

Since loss of DHA in brain due to dietary restrictions of n-3 fatty acids modifies neuronal function,[37] many studies have been carried out to assess whether improving LC-PUFA status affects visual and cognitive functions in preterm and full-term infants. These are difficult studies since neuronal processes are complex and multifactorial. Potentially confounding factors, such as birth weight, parental education and socioeconomic status, smoking, variability in the infant's DHA status at birth, different PUFA ratios among the formulas studied, sample size, and different test methodology may influence the results and obscure potential effects of dietary LC-PUFA supply.

In spite of these difficulties, five prospective, randomized clinical trials have been published reporting an effect of the addition of DHA to formula on the development of visual function in preterm infants. Uauy and coworkers[38] studied infants born at 27–33 weeks, with an average birth weight of about 1300 g. The infants were randomized to feeding formulas with corn oil (devoid of n-3 PUFA), soy oil (containing the n-3 precursor, ALA), and soy and marine oils (containing both ALA and n-3 LC-PUFA, including 0.35% of fatty acids as DHA) or were part of a nonrandomized reference group fed human milk. At 36 weeks postconceptional age (PCA), the group fed soy and marine oils had lower rod electroretinogram thresholds and higher amplitudes than the corn oil–fed group, similar to the human milk group. Soy oil alone showed a slightly better response than corn oil, but did not match the result of human milk or marine oil. Furthermore, at 57 weeks PCA, transient visual evoked potentials (VEPs) and Teller card tests showed visual acuity to be significantly better in infants fed human milk or marine oil formula than infants fed the other formulas.[39]

Carlson and coworkers[40] randomized infants born with an average gestational age of 29 weeks and a weight of 1100 g to conventional formula or formula enriched with marine oil (0.2% DHA), which were fed up to 79 weeks PCA. Teller card acuity was significantly better in marine oil–supplemented babies at 2 and 4 months of age (corrected for expected term birth).

In a further trial, Carlson and coworkers[41] studied infants with a mean gestational age of 28.5 weeks and a birth weight of 1100 g fed for a shorter time period with a marine oil–supplemented formula providing 0.2% DHA. In comparison to a control formula, they had an improved Teller card visual acuity at 2 months corrected age, but not thereafter. Interestingly, the subgroup of infants with chronic lung disease did not show an improvement of visual function with n-3 LC-PUFA.

Faldella and coworkers[42] studied flash VEPs at 52 weeks PCA in 58 preterm infants with an average gestational age of 31 weeks and a birth weight of 1500 g. The infants were fed breast milk or were randomized to a preterm formula supplemented with LC-PUFA (with 0.23% DHA) or a control formula. Breast milk and an LC-PUFA-supplemented formula resulted in similar VEP latencies, while the control formula group showed significantly delayed VEP latencies.

In a multicenter trial, O'Connor and coworkers[43] studied 470 preterm infants with birth weights of 750–1800 g and fed, in addition to varying amounts of human milk, with one of three formulas without LC-PUFA, a formula with 0.43% of fatty acids as AA and 0.27% DHA from fish and fungal oils, or a formula with 0.41% AA and 0.24% DHA from fish and egg oils. At term, the infants were switched to formulas with the same oil sources, but only 0.15–0.16% DHA. No adverse effects of any formula on growth or other side effects were found. Visual acuity measured by behavioral methods by different investigators in the various centers was not different. However, VEPs measured in three of the participating centers indicated an advantage of 3–4 cycles/degree at 6 months in the two groups fed enriched formulas relative to controls.

In summary, these trials support the efficacy of n-3 LC-PUFA intake on the early development of the visual system, which was not achieved to a similar extent with formulas providing the n-3 precursor PUFA, ALA. In a meta-analysis of the previously published results, SanGiovanni and coworkers concluded that DHA-supplemented formula versus DHA-free formula showed significant differences in visual resolution acuity at 2 and 4 months of age, with combined estimates of behaviorally based visual resolution acuity differences at these ages of 0.47 ± 0.14 and 0.28 ± 0.08 octaves, respectively, at 2 and 4 months. A 1-octave difference equals a reduction in the width of the stimulus elements by 50%. Similarly, a Cochrane review concluded that there is evidence that n-3 LC-PUFA supplementation of formula increases the early rate of visual maturation in preterm infants.[44]

EFFECT OF DHA STATUS ON VISUAL FUNCTION IN FULL-TERM INFANTS

For full-term infants, the published results are less clear. Makrides et al.[25] investigated full-term infants fed an experimental formula with 0.36% DHA (% weight/weight), 0.58% EPA, and 0.27% γ-linolenic acid (18:3 n-6), or a control formula without DHA, both at the ages of 16 and 30 weeks. Infants fed the formula with DHA exhibited significantly better VEP acuity than infants fed the control formula at both ages, while there was no difference of visual acuity between infants fed formula with DHA and breast-fed infants. Carlson et al.,[26] in another controlled clinical trial with full-term infants, randomized to formula either without or with a relatively low DHA content of only 0.1% and 0.43% AA, reported that, at the age of 2 months, infants fed the DHA-supplemented formula had a significantly better grating visual acuity measured according to the Teller acuity card procedure. However, there was no difference in visual acuity at later time points, despite a persistent significant difference of DHA levels in plasma and red blood cell lipids. The low concentration of DHA in this study might have caused the lack of a long-term effect in visual acuity.

In a multicenter trial, Auestad et al.[27] investigated infants receiving formula supplemented with DHA alone (0.23%) or with DHA and AA (0.12% and 0.43%, respectively), as well as infants fed formula without LC-PUFA and nonrandomized breast-fed reference infants. Using a somewhat different methodology at the four study sites, visual acuity was assessed with either an acuity card procedure or a VEP methodology from the ages of 2 to 12 months; the authors reported no group differences in visual functions. Since the study formula used was the same as the one used by Carlson et al.,[26] who found an effect on visual function at 2 months, it appears possible that the differences in results between the two studies are related to the methodologies applied.

Birch et al.[28] evaluated another infant formula with a higher DHA content either alone (0.35% DHA) or in combination with AA (0.36% DHA and 0.72% AA). They reported that children who received one of the two DHA-containing formulas had significantly better sweep VEP acuity at the ages of 6, 17, and 52 weeks; the provision of DHA improved visual function up to 1 year of age. By constrast, Jorgensen et al.[29] did not detect significant differences in swept steady-state VEP values of term infants, at the age of 4 months, fed infant formulas supplemented with DHA or not. However, these authors reported a nonsignificant improvement of visual acuity by the addition of DHA to the formula.

Hoffman et al.,[30] using similar supplemented formulas in healthy term infants as previously used by Birch et al.,[28] confirmed again more mature electroretinographic responses and VEPs in infants fed supplemented formulas than in infants fed conventional formulas in the first year of life. Makrides et al.[31] conducted a clinical trial with infants fed placebo, DHA-supplemented formula (0.35%) with only 0.1% EPA, DHA plus AA–supplemented formula (0.34% AA, 0.34% DHA), and breast-fed infants. There were no significant differences in measures of visual function at 16 or 34 weeks of age.

Auestad et al.[32] performed a double-masked, randomized, parallel trial in 239 term infants fed formulas without or with 0.46% AA and only 0.14% DHA for 1 year, and in 165 breast-fed infants weaned to formulas with and without AA plus DHA. No effects of this rather low dose of DHA on growth, visual acuity, information processing, general development, language, and temperament were detected.

In conclusion, some of the controlled trials in healthy term infants showed that DHA improved visual acuity during the first year of life, but others found no significant effect. None of the trials reported negative effects on visual acuity. Differences among the results of various studies may be due to differences in methodology and in strategies of supplementation. The dose and form of DHA supply may well be important causative factors for the differences between studies. Lauritzen et al.[45] recently reported a close dose-response relationship between DHA content in milk and VEP visual acuity in infants at 4 months of age.

EFFECTS OF LC-PUFA ON BEHAVIORAL DEVELOPMENT

While a large number of studies have evaluated visual acuity in infants, only few randomized studies have examined the effects of postnatal dietary LC-PUFA on neurodevelopment. Different tests have been used:[46] the Brunet-Lézine test and Bayley scales measure global neurodevelopment and provide indices of mental and motor skills relative to group norms. More specific tests of cognitive functions, information processing, and learning ability are the Fagan Test of Infant Intelligence (FTII), tests on problem solving and delayed-response tasks, and the MacArthur Communicative Development Inventory. At present, it remains unclear which tests are most sensitive to detect potential effects of LC-PUFA on nervous system development in infants.

In the aforementioned trial of O'Connor and coworkers,[43] preterm infants fed formulas with LC-PUFA showed advantages in novelty preference in the Fagan test at 6 months, but not at 9 months. At 12 months, there was no difference in the Bayley mental development index, but there was an improvement in the Bayley motor development index.

In term infants, Agostoni et al.[33] compared the psychomotor development of term infants randomly assigned to receive formulas with or without LC-PUFA (0.30% DHA and 0.44% AA). At the age of 4 months, Brunet-Lézine test results were significantly lower in infants fed formula without LC-PUFA than in those receiving the LC-PUFA-supplemented formula or human milk. The majority of the infants could be followed up to the age of 24 months, but at this time no differences were found with this test. However, the developmental quotients at the age of 2 years were still significantly and positively correlated to both AA and DHA contents of red blood cell phosphatidylcholine lipids at the age of 4 months.

Infant cognitive behavior was also assayed by Willats et al.[34] using a battery of problem-solving tests at the age of 10 months. In this study, infants at term were randomly assigned in a double-blind fashion in groups receiving during the first postnatal months a formula without or with both DHA and AA from egg phospholipids, providing 0.15–0.25% DHA and 0.30–0.40% AA, respectively. Infants who received preformed dietary LC-PUFA during early infancy achieved significantly more intentional solutions and better intention scores than infants whose formulas was devoid of LC-PUFA. Since higher problem-solving scores in infancy are related to higher childhood IQ scores, this study suggested a possible beneficial role of LC-PUFA supplementation on development.

By contrast, Makrides et al.[31] found no differences in Bayley developmental scores in term infants fed supplemented LC-PUFA formulas at the ages of 16 and 34 weeks of age. Also, Auestad et al.[32] reported the absence of significant effects of AA plus DHA formulas on multiple measures of general development, information processing, language, and temperament over the first 14 months after birth. Further studies are needed to clarify the effect of LC-PUFA availability, the dose and type of supplement, and other influencing factors on the performance in developmental tests in healthy full-term babies.

In relation to the maternal dietary habits, recent evidence describes that breast-fed infants whose mothers ate oily fish during pregnancy were more likely to achieve high-grade stereopsis than were children whose mothers did not eat oily fish (adjusted odds ratio: 1.57; 95% CI: 1.00–2.45).[47] Also, Jorgensen et al.[48] suggest a cause-effect relationship between infant milk DHA intake and visual acuity. Further research is needed to determine the functional benefits of the supplementation of LC-PUFA for pregnant and lactating women for infants and to define whether there is a minimum DHA requirement for mothers during pregnancy.

CONCLUSIONS

There is accumulating evidence for effects of perinatal PUFA supply and metabolism on early nervous system development, which is most fascinating. Although some studies could not detect effects of DHA provision on visual function in healthy infants, other trials have clearly documented that the addition of preformed DHA to infant formulas may improve visual acuity in preterm and term infants, with advantages reported up to the age of 1 year in some studies. Recommendations for the dietary requirements of n-6 and n-3 fatty acids for brain development are complex and involve both the amounts and balance of the precursors, LA and ALA, and the amounts and balance of AA and DHA. Variability of human milk fatty acid composition highlights the need for caution in using such data as the basis to define infant substrate requirements. The Child Health Foundation Expert Consensus Workshop[49] on the role of LC-PUFA in maternal and child health recently recommended that infant formulas for term infants should contain at least 0.2% of total fatty acids as DHA and 0.35% as AA. Since preterm infants are born with much less total body DHA and AA, they suggest that preterm formulas should include at least 0.35% DHA and 0.4% AA. It is possible that even higher levels might confer additional benefits. The relationship between dose and effect and the influence of other factors need to be further investigated since optimal dietary intakes for term and preterm infants remain to be defined.

ACKNOWLEDGMENTS

The work on perinatal lipid metabolism was financially supported in part by the Deutsche Forschungsgemeinschaft, Bonn, Germany, and by the Fifth Framework Research Programme, the European Commission, Brussels, Belgium. E. Larque is the recipient of a postdoctoral fellowship awarded by the Alexander von Humboldt Foundation, Bonn, Germany.

REFERENCES

1. KOLETZKO, B. & T. DECSI. 2001. The role of long chain polyunsaturated fatty acids for infant growth and development. *In* Preventive Nutrition. Vol. 2: Primary and Secondary Prevention. Chapter 14, pp. 237–252. Humana Press. Totowa, New Jersey.
2. INNIS, S. 2000. The role of dietary n-6 and n-3 fatty acids in the developing brain. Dev. Neurosci. **22:** 474–480.
3. LITMAN, B.J., S.L. NIU, A. POLOVA & D.C. MITCHELL. 2001. The role of docosahexaenoic acid containing phospholipids in modulating G protein–coupled signaling pathways: visual transduction. J. Mol. Neurosci. **16:** 237–242.
4. FERNSTROM, J.D. 1999. Effects of dietary polyunsaturated fatty acids on neuronal function. Lipids **32:** 161–169.
5. KOLETZKO, B. & M. BRAUN. 1991. Arachidonic acid and early human growth: is there a relation? Ann. Nutr. Metab. **35:** 128–131.
6. CARLSON, S.E., S.H. WERKMAN, J.M. PEEPLES *et al.* 1993. Arachidonic acid status correlates with first year growth in preterm infants. Proc. Natl. Acad. Sci. U.S.A. **90:** 1073–1077.
7. SAUERWALD, T.U., D.L. HACHEY, C.L. JENSEN *et al.* 1997. Intermediates in endogenous synthesis of C22:6ω3 and C20:4ω6 by term and preterm infants. Pediatr. Res. **41:** 183–187.

8. SZITANYI, P., B. KOLETZKO, A. MYDLILOVA & H. DEMMELMAIR. 1999. Metabolism of 13C-labeled linoleic acid in newborn infants during the first week of life. Pediatr. Res. **45:** 669–673.
9. MOORE, S.A., E. YODER, S. MURPHY et al. 1991. Astrocytes, not neurons, produce docosahexaenoic acid and arachidonic acid. J. Neurochem. **6:** 518–524.
10. GREINER, R.C., J. WINTER, P.W. NATHANIELSZ & J.T. BRENNA. 1997. Brain docosahexaenoate accretion in fetal baboons: bioequivalence of dietary alpha-linolenic and docosahexaenoic acids. Pediatr. Res. **42:** 826–834.
11. MAKRIDES, M., M.A. NEUMANN, R.W. BYARD et al. 1994. Fatty acid composition of brain, retina, and erythrocytes in breast- and formula-fed infants. Am. J. Clin. Nutr. **60:** 189–194.
12. BERGHAUS, T., H. DEMMELMAIR & B. KOLETZKO. 1998. Fatty acid composition of lipid classes in maternal and cord plasma at birth. Eur. J. Pediatr. **157:** 763–768.
13. KOLETZKO, B. & L. MÜLLER. 1990. Cis- and trans-isomeric fatty acids in plasma lipids of newborn infants and their mothers. Biol. Neonate **57:** 172–178.
14. SHAND, J.H. & R.C. NOBLE. 1979. The role of maternal triglycerides in the supply of lipids to the ovine fetus. Res. Vet. Sci. **26:** 117–119.
15. STREMMEL, W., G. STROHMEYER, F. BOCHARD et al. 1985. Isolation and partial characterization of a fatty acid binding protein in rat liver plasma membranes. Proc. Natl. Acad. Sci. U.S.A. **82:** 4–8.
16. CAMPBELL, F.M., M.J. GORDON & A.K. DUTTA-ROY. 1994. Plasma membrane fatty acid–binding protein (FABP pm) of the sheep. Biochim. Biophys. Acta **1214:** 187–192.
17. ABUMRAD, N.A., M. EL-MARABI, E. AMRI et al. 1993. Cloning of a rat adipocyte membrane protein implicated in binding or transport of long chain fatty acids that is induced during pre-adipocyte differentiation. J. Biol. Chem. **268:** 17665–17668.
18. SCHAFFER, J.E. & H.F. LODISH. 1994. Expression cloning and characterization of a novel adipocyte long chain fatty acid transport protein. Cell **79:** 427–436.
19. DUTTA-ROY, A.K. 2000. Transport mechanisms for long-chain polyunsaturated fatty acids in the human placenta. Am. J. Clin. Nutr. **71:** 315S–322S.
20. CAMPBELL, F.M., A.M. CLOHESSY, M.J. GORDON et al. 1997. Uptake of long chain fatty acids by human placental choriocarcinoma (BeWo) cells: role of plasma membrane fatty acid–binding protein. J. Lipid Res. **38:** 2558–2568.
21. RODRIGUEZ, P.M., B. KOLETZKO, C. KUNZ & R. JENSEN. 1999. Nutritional and biochemical properties of human milk: II. Lipids, micronutrients, and bioactive factors. Clin. Perinatol. **26:** 335–359.
22. DEMMELMAIR, H., M. BAUMHEUER, B. KOLETZKO et al. 1998. Metabolism of U^{13}C-labelled linoleic acid in lactating women. J. Lipid Res. **39:** 1389–1396.
23. DEL PRADO, M., S. VILLALPANDO, A. ELIZONDO et al. 2001. Contribution of dietary and newly formed arachidonic acid to human milk lipids in women eating a low fat diet. Am. J. Clin. Nutr. **74:** 90–95.
24. FIDLER, N., T. SAUERWALD, A. POHL et al. 2000. Docosahexaenoic acid transfer into human milk after dietary supplementation: a randomized clinical trial. J. Lipid Res. **41:** 1376–1383.
25. MAKRIDES, M., M. NEUMANN, K. SIMMER et al. 1995. Are long-chain polyunsaturated fatty acids essential nutrients in infancy? Lancet **345:** 1463–1468.
26. CARLSON, S.E., A.J. FORD, S.H. WERKMAN et al. 1996. Visual acuity and fatty acid status of term infants fed human milk and formulas with and without docosahexaenoate and arachidonate from egg yolk lecithin. Pediatr. Res. **39:** 882–888.
27. AUESTAD, N., M.B. MONTALTO, R.T. HALL et al. 1997. Visual acuity, eyrthrocyte fatty acid composition, and growth in term infants fed formulas with long chain polyunsaturated fatty acids for one year. Pediatr. Res. **41:** 1–10.
28. BIRCH, E.E., D.R. HOFFMAN, R. UAUY et al. 1998. Visual acuity and the essentiality of docosahexaenoic acid and arachidonic acid in the diet of term infants. Pediatr. Res. **44:** 201–209.
29. JORGENSEN, M.H., G. HOLMER, O. HERNELL & K. FLEISCHER-MICHAELSEN. 1998. Effect of formula supplemented with docosahexaenoic acid and gamma-linolenic acid on fatty acid status and visual acuity in term infants. J. Pediatr. Gastroenterol. Nutr. **26:** 412–421.

30. HOFFMAN, D.R., E.E. BIRCH, D.G. BIRCH *et al.* 2000. Impact of early dietary intake and blood lipid composition of long-chain polyunsaturated fatty acids on later visual development. J. Pedriatr. Gastroenterol. Nutr. **31:** 540–553.
31. MAKRIDES, M., M.A. NEUMAN, K. SIMMER & R.A. GIBSON. 2000. A critical appraisal of the role of dietary long-chain polyunsaturated fatty acids on neuronal indices of term infants: a randomized, controlled trial. Pediatrics **105:** 32–38.
32. AUESTAD, N., R. HALTER, R.T. HALL *et al.* 2001. Growth and development in term infants fed long-chain polyunsaturated fatty acids: a double-masked, randomized, parallel, prospective, multivariate study. Pediatrics **108:** 372–381.
33. AGOSTONI, C., S. TROJAN, R. BELLU *et al.* 1995. Neurodevelopmental quotient of healthy term infants at 4 months and feeding practice: the role of long-chain polyunsaturated fatty acids. Pediatr. Res. **38:** 262–266.
34. WILLATS, P., J.S. FORSYTH, M.K. DIMODUGNO *et al.* 1998. Effect of long-chain polyunsaturated fatty acids in infant formula on problem solving at 10 months of age. Lancet **352:** 688–691.
35. KOLETZKO, B., E. SCHMIDT, H.J. BREMER *et al.* 1989. Effects of dietary long-chain polyunsaturated fatty acids on the essential fatty acid status of premature infants. Eur. J. Pediatr. **148:** 669–675.
36. CARLSON, S.E., R.J. COOKE, S.H. WERKMAN & E.A. TOLLEY. 1992. First year growth of preterm infants fed standard compared to marine oil–n3 supplemented formulas. Lipids **27:** 901–907.
37. KOLETZKO, B., P.J. AGGETT, J.G. BINDELS *et al.* 1998. Growth, development, and differentiation: a functional food science approach. Br. J. Nutr. **80:** S5–S45.
38. UAUY, R., D.G. BIRCH, E.E. BIRCH *et al.* 1990. Effect of dietary omega-3 fatty acids on retinal function of very-low-birth-weight neonates. Pediatr. Res. **28:** 485–492.
39. BIRCH, D.G., E.E. BIRCH, D.R. HOFFMAN & R.D. MANY. 1992. Retinal development in very-low-birth weight infants fed diets differing in omega-3 fatty acids. Invest. Ophthalmol. Visual Sci. **33:** 2365–2376.
40. CARLSON, S.E., S.H. WERKMAN, P.G. RHODES & E.A. TOLLEY. 1993. Visual acuity development in healthy preterm infants: effect of marine-oil supplementation. Am. J. Clin. Nutr. **58:** 35–42.
41. CARLSON, S.E., S.H. WERKMAN & E.A. TOLLEY. 1996. Effect of long-chain n-3 fatty acid supplementation on visual acuity and growth of preterm infants with and without bronchopulmonary dysplasia. Am. J. Clin. Nutr. **63:** 687–697.
42. FALDELLA, G., M. GOVONI, R. ALESSANDRONI *et al.* 1996. Visual evoked potentials and dietary long chain polyunsaturated fatty acids in preterm infants. Arch. Dis. Child. Fetal Neonatal Ed. **75:** F108–F112.
43. O'CONNOR, D.L., R. HALL, D. ADAMKIN *et al.* 2001. Growth and development in preterm infants fed long-chain polyunsaturated fatty acids: a prospective, randomized controlled trial. Pediatrics **108:** 359–371.
44. SIMMER, K. 2001. Long chain polyunsaturated fatty acid supplementation in infants born at term (Cochrane Review). Cochrane Database Syst. Rev. **4:** CD000376.
45. LAURITZEN, L., H.S. HANSEN, M.H. JORGENSEN & K.F. MICHAELSEN. 2001. The essentiality of long chain n-3 fatty acids in relation to development and function of the brain and retina. Prog. Lipid Res. **40:** 1–94.
46. CARLSON, S.E. 2000. Behavioral methods used in the study of long-chain polyunsaturated fatty acid nutrition in primate infants. Am. J. Clin. Nutr. **71:** 268S–274S.
47. WILLIAMS, C., E.E. BIRCH, P.M. EMMETT & K. NORTHSTONE. 2001. Stereoacuity at age 3.5 y in children born full term is associated with prenatal and postnatal dietary factors: a report from a population-based cohort study. Am. J. Clin. Nutr. **73:** 316–322.
48. JORGENSEN, M.H., O. HERNELL, E. HUGHES & K.F. MICHAELSEN. 2001. Is there a relation between docosahexaenoic acid concentration in mother's milk and visual development in term infants? J. Pediatr. Gastroenterol. Nutr. **32:** 293–296.
49. KOLETZKO, B., A. AGOSTONI, S.E. CARLSON *et al.* 2001. Long-chain polyunsaturated fatty acids (LC-PUFA) and perinatal development. Acta Paediatr. **90:** 460–464.

The Responses of Serum and Adipose Fatty Acids to a One-Year Weight Reduction Regimen in Female Obese Monozygotic Twins

M. KUNEŠOVÁ,[a] S. PHINNEY,[b] V. HAINER,[a] E. TVRZICKÁ,[c] V. ŠTICH,[d] J. PARÍZKOVÁ,[a] A. ŽÁK,[c] AND A. STUNKARD[e]

[a]*Obesity Management Center, Third Department of Medicine, First Medical School, Charles University, Prague, Czech Republic*

[b]*Galileo Laboratories, Santa Clara, California 95054, USA*

[c]*Fourth Medical Department, First Medical School, Charles University, Prague, Czech Republic*

[d]*Third Medical School, Charles University, Prague, Czech Republic*

[e]*Department of Psychiatry, University of Pennsylvania, Philadelphia, Pennsylvania 19104, USA*

ABSTRACT: We have reported strong intrapair resemblances (IPRs) in serum phosphatidylcholine (PC) fatty acid composition within adult monozygotic twins living apart. This study assessed the contribution of genetic factors to changes in serum and adipose tissue fatty acids resulting from weight loss and followed by a subsequent year of weight maintenance. Eleven pairs of female obese monozygotic twins (age: 38.9 ± 1.8; BMI: 32.5 ± 0.9) were recruited for the study. Fasting serum and adipose tissue were obtained after 1 week of inpatient stabilization, after 1 month of inpatient very-low-calorie diet (VLCD), and again after 1 year of outpatient weight maintenance. Fatty acids in serum lipid fractions and adipose tissue were quantitated by gas chromatography. Using multiple regression adjusted for age and initial value, IPRs were determined for the changes induced by VLCD and by the year of weight maintenance. There were few IPRs in nonessential fatty acids. By contrast, there were numerous IPRs for essential fatty acids (EFA), especially in the n-3 family across the VLCD. Following the maintenance year, however, frequent IPRs for nonessential fatty acids were seen, particularly in serum PC, and strong IPRs were seen for 18:3 n-3 and 20:5 n-3 across multiple fractions. These results infer the existence of strong genetic factors determining both the nonessential and EFA compositions of tissue lipids in humans independent of diet. Of particular note were the consistent IPRs for n-3 fatty acids despite dietary stress, indicating that the conservation and distribution of this EFA family are subject to considerable genetic variance in humans.

KEYWORDS: obesity; twins; genetics; fatty acid composition; weight reduction; very-low-calorie diet; weight maintenance

Address for correspondence: M. Kunešová, M.D., Ph.D., Obesity Management Center, Third Department of Medicine, First Medical School, Charles University, U nemocnice 2, 128 08 Prague 2, Czech Republic. Voice: +420 2 24962921; fax: +420 2 24919780.
mkune@lf1.cuni.cz

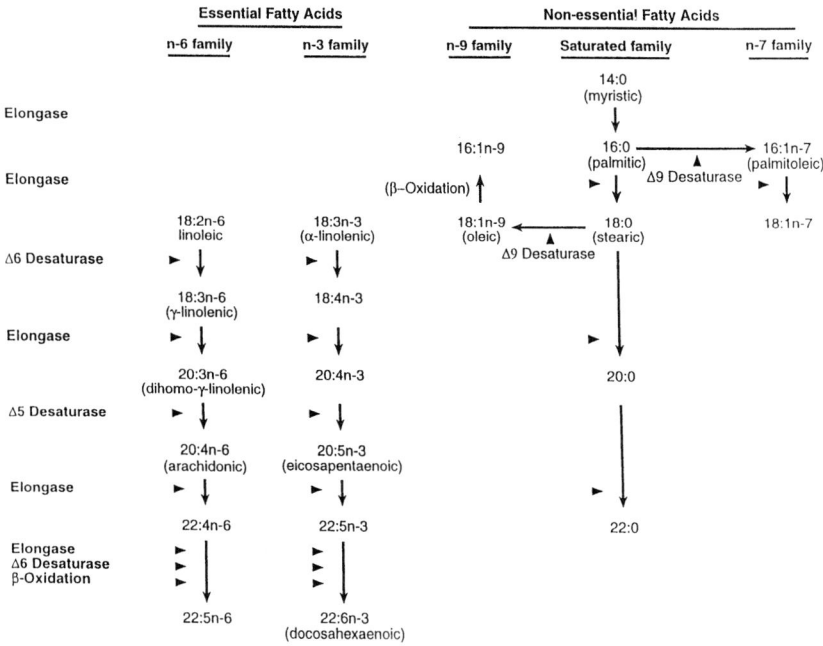

FIGURE 1. Nomenclature and metabolic pathways of long-chain fatty acids.

INTRODUCTION

Dietary fat is generally regarded as an energy-yielding substrate, with obesity being the state of its excess accumulation. However, dietary fat and its adipose tissue storage are also the source and reserves of essential fatty acids, which are required in humans for normal physiologic processes[1] that maintain health and well-being[2,3] (FIG. 1). There is general agreement that dietary fatty acid composition is the primary determinant of the mix of fatty acids stored in adipose triglyceride (TG),[4,5] but there is also evidence for other factors contributing to adipose TG fatty acid content. These can include endogenous *de novo* fatty acid production, as well as metabolic modification of dietary fatty acids by elongation and/or desaturation. There is also evidence for site-specific differences in adipose fatty acid composition within an individual, involving all classes of fatty acids, including the omega-6 (n-6) polyunsaturates.[6] This latter observation alerts us to the existence of selective mechanisms for the uptake and storage of specific fatty acids in adipocytes.

One mechanism contributing to the gradient between dietary fatty acid composition and that of adipose TG stores is selective release of specific fatty acids by adipose hormone-sensitive lipase.[7,8] Because fatty acid turnover through the adipocyte is greater than fatty acid oxidation,[9] this selectivity for fatty acids by hormone-sensitive lipase is magnified, particularly during periods of weight loss.[8,10–12] Interestingly, this selective mobilization of specific fatty acids distinguishes clearly between the n-6 and n-3 families of essential fatty acids.[8]

In addition to gradients in fatty acid composition between sites within a specific pool, such as adipose TG, there are also dramatic gradients between different lipid pools, including TG, cholesteryl esters (CE), and phospholipids (represented here by the phosphatidylcholine [PC] subfraction). Although there can be 20-fold gradients in the relative proportions of a specific fatty acid between different lipid pools (e.g., arachidonate between adipose TG and serum PC), not much is known of the mechanisms that establish such gradients. Because highly unsaturated fatty acids impart important physiological properties to membranes, there are clearly both defense and compensatory mechanisms to protect membrane properties against dietary stress.[13,14] Despite these obvious signs of homeostatic forces determining membrane composition, there is also evidence for genetically determined differences in membrane fatty acids within the same species, given identical diets.[15,16]

In prior reports, we have noted that obese adult monozygotic twins living apart are strongly concordant for the change of multiple phenotypic variables related to obesity (i.e., weight, BMI, and fat tissue),[17] and we have also noted remarkable similarities in the fatty acid composition of a number of lipid pools, particularly in the serum PC subfraction.[18] Expressed as intrapair resemblances (IPRs) determined by intraclass correlation, these similarities in serum PC fatty acids transcend similarity in diets. We have interpreted this as an indication that one's genotype is an important contributor to membrane composition independent of the diet. In addition, the assessment of fatty acid composition of adipose TG and serum CE in this cohort revealed strong correlations between the nonessential fatty acid palmitoleate (16:1 n-7) and multiple markers of obesity, suggesting a close (if not causal) relationship. To further assess the complex interrelationships between genotype and adiposity, we have performed a 13-month prospective study of 11 pairs of these twins, assessing IPRs for the changes in serum and adipose lipid fraction fatty acids across a period of weight loss and weight maintenance.

METHODS

Subjects

Eleven pairs of obese monozygotic premenopausal females, aged 38.9 ± 1.8 years (mean ± SE; range: 23–48) with a mean BMI of 32.5 ± 0.9 (range: 24.84–41.16), were recruited for the study through the media and gave their written informed consent to participate in this study. The study was approved by the Charles University Medical Ethics Committee.

The monozygosity of the twins was established on the basis of their physical appearance, the loci of the HLA antigen system, and DNA of the apo B gene. The initial physician evaluation included medical history and a physical examination. None of the subjects had a history of recent illness or severe cardiovascular or metabolic disease. Two pairs of twins were light smokers. None of the subjects had recently been on a diet or involved in a weight reduction regimen, and their body weight had been stable during the 6 months prior to recruitment. Each pair had been reared together, but (with the exception of the youngest pair) they had been living apart for at least 10 years at the time of recruitment. Twelve pairs were enrolled over a 15-week period, but 1 pair is not included in the calculations because of the loss of the serum in one twin after the 1-year follow-up.

Design of the Study

Four sets of twins at a time were admitted for an inpatient weight reduction regimen (FIG. 2). The regimen consisted of a 1-week eucaloric, inpatient baseline evaluation period, followed by a 4-week weight reduction period. At the end, 5 days of weight stabilization followed. During the 4-week weight reduction period, subjects received a very-low-calorie diet (VLCD, Redita®, Promil Nový Bydžov, Czech Republic), providing 1600 kJ/day, consisting of 37 g protein, 50 g carbohydrates, and 3 g fat. The reduction regimen also included light to moderate daily physical training totaling about 3 h per day.[17] Compliance to the diet was tested by daily semiquantitative measurement of urine ketone bodies and adherence to the physical training by pedometers.[19] During the last week of the inpatient stay, the subjects wrote a 1-week menu for the outpatient stay in consultation with a dietitian. The recommended energy intake was 5000–6000 kJ/day, with protein providing 14–17%, carbohydrates 53–56%, and fat less than or equal to 30%. To insure that the suggested menus met standards for the recommended daily vitamin and mineral intakes, the menus were evaluated by the PC database Výživa (KTS, Prague), which includes more then 2000 items and evaluates 14 nutrients, minerals, and vitamins. All these procedures were undertaken in the Obesity Unit, Fourth Department of Medicine, Charles University, Prague. After dismissal, the subjects were examined as outpatients every 3 months to monitor their weight, anthropometric characteristics, and food records. After 1 year, subjects were readmitted to the Obesity Unit for a 1-week inpatient reevaluation with a weight stabilization diet.

During the first and last weeks of the initial inpatient stay and also during the week after the 1-year follow-up, subjects remained weight-stable. At the end of each of these weeks of controlled eucaloric diet, fasting blood was drawn for lipid fraction fatty acid analysis; at the same time, adipose tissue was also obtained. The first two adipose tissue samples were obtained by making a small surgical incision, after using lidocaine as an anesthetic, on the abdomen about 20 mm to the left of the umbilicus

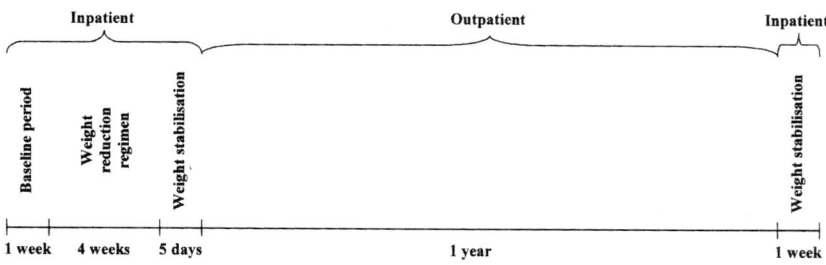

FIGURE 2. Study design.

and removing 300 mg of fat tissue; the 1-year sample was obtained by needle biopsy of subcutaneous fat tissue near the original location, yielding 50–100 mg.

Body Composition and Regional Tissue Distribution

Body density was determined by hydrostatic weighing, with the percentage of body fat estimated from body density using the method described by Parízková[20] and the equation of Brožek.[21] Pulmonary residual volume was determined using the helium dilution method.[22] Fat mass was obtained by multiplying the percentage of body fat by body weight. Anthropometric estimation of body fat was performed by measurement of 10 skinfolds according to Parízková[20] and of 4 skinfolds according to Durnin.[23] Waist and hip circumference and sagittal abdominal diameter at the L4/5 level were measured following the standardized procedure recommended at the Airlie Conference.[24]

Fatty Acid Composition Measurement

Fatty acid composition measurement was performed by gas chromatography after separation of individual serum fractions by thin-layer chromatography on silica gel. The detailed description of the method is given elsewhere.[18]

Statistical Methods

The changes of the fatty acid percentage from baseline to VLCD and from baseline to 1-year results were analyzed with a two-way ANOVA for repeated measures on one factor (time). The twins were considered nested within the pair, whereas the treatment effect was defined as a fixed variable. The second step was to estimate the genetic contribution to these changes by computing intraclass correlation coefficients for the changes across the three study points. The intraclass correlation coefficient was computed from the between pairs and the within pairs least squares.[25] Twin pairs concordant and discordant for dietary fat content were compared for intrapair differences in fatty acid composition in individual lipid classes. The analysis was performed with the Statgraphics package (Manugistic Inc., Rockville, MD) on a PC microcomputer. Data are expressed as means ± SE; $p = 0.05$ was taken as the threshold of statistical significance.

RESULTS

The characteristics of the subjects before the treatment, after VLCD, and after the 1-year follow-up are shown in TABLE 1. The subjects as a group succeeded in maintaining the weight loss achieved after the VLCD, despite the 1 year as outpatients. The predominant proportion of this sustained loss stemmed from reduced body fat stores, and the reduced waist and sagittal abdominal diameters indicate that abdominal fat (the site of abdominal biopsies) contributed to these fat losses.

TABLES 2–5 present the composition of fatty acids in serum PC, CE, and TG and in adipose tissue TG before the treatment, after the VLCD, and after the 1 year of outpatient follow-up. The values are given in mole %. Within the serum fractions (TABLES 2–4), the most consistent changes from baseline to 1 year were reductions

TABLE 1. Characteristics of the group ($n = 22$)

	Baseline	VLCD	One year
Age	38.9 ± 1.8		
Weight (kg)	89.33 ± 2.65	81.05 ± 2.49*	80.96 ± 2.53*
Weight loss (kg)		8.27 ± 0.32	8.38 ± 1.07
BMI (kg/m^2)	32.50 ± 0.88	29.51 ± 0.83*	29.39 ± 0.80*
Fat (kg)	41.07 ± 2.09	34.80 ± 1.80*	34.27 ± 1.60*
Fat loss (kg)		6.27 ± 0.43	6.80 ± 1.18
Sum of 10 skf (mm)	265 ± 12	196 ± 11*	172 ± 8*[a]
Sum of 4 skf (mm)	120 ± 5	86 ± 5*	76 ± 5*[a]
Waist (cm)	101.4 ± 2.6	90.3 ± 2.5*	91.4 ± 2.5*
SAD (cm)	27.3 ± 0.8	23.8 ± 0.9*	22.52 ± 0.9*
WHR	0.87 ± 0.02	0.82 ± 0.02*	0.82 ± 0.02*
WTR	1.51 ± 0.04	1.41 ± 0.03*	1.44 ± 0.04*
Centrality index[b]	1.09 ± 0.06	1.06 ± 0.08	1.25 ± 0.12

NOTE: Values are expressed as mean ± SE; *$p < 0.001$ in comparison with baseline values.
[a]$p < 0.05$ in comparison with values after VLCD.
[b]Subscapular/triceps ratio.

TABLE 2. Fatty acid composition in serum PC before and after VLCD and after 1-year follow-up in obese female identical twins (mol %) ($n = 22$)

Fatty acid	Baseline	VLCD	One year
14:0	0.64 ± 0.04	0.82 ± 0.12	0.28 ± 0.02*
16:0	39.72 ± 1.45	44.65 ± 1.32*	30.25 ± 0.69***
16:1 n-7	1.43 ± 0.07	1.77 ± 0.15*	0.61 ± 0.02***
18:0	20.16 ± 0.87	19.89 ± 0.70	14.61 ± 0.30***
18:1 n-9	16.56 ± 0.81	16.35 ± 0.53	11.30 ± 0.16***
18:1 n-7	3.42 ± 0.11	3.22 ± 0.16	2.52 ± 0.07***
18:2 n-6	12.13 ± 0.96	8.20 ± 0.89*	22.96 ± 0.77***
18:3 n-6	0.15 ± 0.04	0.14 ± 0.05	0.15 ± 0.02
18:3 n-3	0.30 ± 0.10	0.09 ± 0.02	0.22 ± 0.02
20:3 n-6	1.14 ± 0.15	0.53 ± 0.10**	2.85 ± 0.15***
20:4 n-6	3.31 ± 0.58	3.18 ± 0.76	9.58 ± 0.45***
20:5 n-3	0.25 ± 0.05	0.18 ± 0.04	0.90 ± 0.07***
22:5 n-3	0.18 ± 0.03	0.21 ± 0.05	0.70 ± 0.04***
22:6 n-3	0.62 ± 0.11	0.71 ± 0.19	3.11 ± 0.17***

NOTE: x ± SE; *$p < 0.05$, **$p < 0.01$, ***$p < 0.001$ in comparison with baseline values.

TABLE 3. Fatty acid composition in serum CE before and after VLCD and after 1-year follow-up in obese female identical twins (mol %) ($n = 22$)

Fatty acid	Baseline	VLCD	One year
14:0	1.15 ± 0.10	1.46 ± 0.12	1.71 ± 0.31
16:0	20.33 ± 0.81	23.48 ± 0.69**	14.80 ± 1.04***
16:1 n-7	4.25 ± 0.19	3.86 ± 0.14	2.79 ± 0.16***
18:0	7.76 ± 0.92	8.29 ± 0.79	3.49 ± 0.77***
18:1 n-9	23.81 ± 3.31	25.15 ± 0.89*	22.76 ± 0.56
18:1 n-7	1.99 ± 0.06	3.13 ± 0.77***	1.99 ± 0.08
18:2 n-6	34.80 ± 1.60	29.05 ± 1.63*	44.74 ± 2.24***
18:3 n-6	0.42 ± 0.04	0.19 ± 0.03*	0.70 ± 0.07**
18:3 n-3	0.48 ± 0.05	0.30 ± 0.06*	0.52 ± 0.03
20:3 n-6	0.40 ± 0.03	0.32 ± 0.07	0.78 ± 0.05***
20:4 n-6	3.78 ± 0.29	4.92 ± 0.59	4.75 ± 0.40
20:5 n-3	0.10 ± 0.04	0.17 ± 0.02	0.33 ± 0.04
22:5 n-3	0.08 ± 0.02	0.11 ± 0.03	0.15 ± 0.03
22:6 n-3	0.22 ± 0.03	0.31 ± 0.04	0.50 ± 0.05***

NOTE: x ± SE; *$p < 0.05$, **$p < 0.01$, ***$p < 0.001$ in comparison with baseline values.

TABLE 4. Fatty acid composition in serum TG before and after VLCD and after 1-year follow-up in obese female identical twins (mol %) ($n = 22$)

Fatty acid	Baseline	VLCD	One year
14:0	1.11 ± 0.11	0.70 ± 0.05	1.45 ± 0.15
16:0	27.14 ± 0.46	28.81 ± 0.46	25.65 ± 0.53
16:1 n-7	5.10 ± 0.23	3.95 ± 0.13	3.57 ± 0.15**
18:0	3.38 ± 0.11	3.27 ± 0.11	3.36 ± 0.12
18:1 n-9	44.46 ± 0.68	45.38 ± 0.87	43.46 ± 0.49
18:1 n-7	4.21 ± 0.07	3.94 ± 0.10*	3.53 ± 0.06**
18:2 n-6	12.07 ± 0.72	12.15 ± 1.01	15.31 ± 0.47*
18:3 n-6	0.21 ± 0.05	0.17 ± 0.04	0.30 ± 0.04
18:3 n-3	0.76 ± 0.12	0.50 ± 0.07	0.95 ± 0.05
20:3 n-6	0.13 ± 0.02	0.08 ± 0.01*	0.27 ± 0.02**
20:4 n-6	0.74 ± 0.08	0.87 ± 0.13	1.07 ± 0.07
20:5 n-3	0.08 ± 0.02	0.08 ± 0.02	0.20 ± 0.02
22:5 n-3	0.21 ± 0.03	0.23 ± 0.03	0.35 ± 0.03*
22:6 n-3	0.39 ± 0.06	0.38 ± 0.05	0.53 ± 0.05

NOTE: x ± SE; *$p < 0.05$, **$p < 0.001$ in comparison with baseline values.

TABLE 5. Fatty acid composition in adipose tissue TG before and after VLCD and after 1-year follow-up in obese female identical twins (mol %) (n = 22)

Fatty acid	Baseline	VLCD	One year
14:0	2.03 ± 0.09	2.09 ± 0.08	1.16 ± 0.13***
16:0	21.61 ± 0.34	21.67 ± 0.37	19.32 ± 0.49***
16:1 n-7	5.16 ± 0.19	4.97 ± 0.26	3.14 ± 0.28***
18:0	4.42 ± 0.18	4.56 ± 0.22	5.22 ± 0.30*
18:1 n-9	49.66 ± 0.32	50.15 ± 0.38	54.12 ± 0.56**
18:1 n-7	4.69 ± 0.09	4.70 ± 0.08	3.93 ± 0.13**
18:2 n-6	10.57 ± 0.32	10.17 ± 0.51	11.07 ± 0.67
18:3 n-6	0.03 ± 0.008	0.03 ± 0.01	0.09 ± 0.02*
18:3 n-3	0.88 ± 0.05	0.78 ± 0.05	0.86 ± 0.06
20:3 n-6	0.17 ± 0.01	0.17 ± 0.01	0.23 ± 0.03
20:4 n-6	0.40 ± 0.09	0.35 ± 0.04	0.33 ± 0.02
20:5 n-3	0.04 ± 0.01	0.01 ± 0.01	0.10 ± 0.03*
22:5 n-3	0.15 ± 0.02	0.15 ± 0.01	0.17 ± 0.01
22:6 n-3	0.20 ± 0.04	0.20 ± 0.04	0.24 ± 0.02

NOTE: $x \pm$ SE; $*p < 0.05$, $**p < 0.01$, $***p < 0.001$ in comparison with baseline values.

in 16:1 n-7 and increases in 18:2 n-6, the latter being most likely due to an increased proportion (but not increased total amount) of dietary fat coming from vegetable oil sources. The lack of rise in adipose TG 18:2 n-6 over the outpatient year (TABLE 5) is consistent with the subjects' constraint of dietary fat intake as part of their successful weight maintenance. The reduced proportion of 16:1 n-7 seen in the serum fractions was also reflected in adipose TG, indicating significant mobilization over the course of the outpatient year. The uniform decline in 18:3 n-3 following the VLCD in all four lipid fractions (significant only in the serum CE) is consistent with prior published reports of adipose composition changes with weight loss[10-12] and appears to fully recover to baseline values after the maintenance year.

TABLE 6 provides a summary of the IPRs for the changes in individual fatty acids occurring across the VLCD and also the changes between baseline and the end of 1 year of weight maintenance.

To distinguish the role of dietary factors and the role of metabolic processing, we evaluated twin pairs concordant and discordant in fat intake with regard to their resemblance in the change of fatty acid composition after the 1 year of weight maintenance. We did not find any significant difference in fatty acid composition in twin pairs concordant and discordant in fat intake, except for TG 14:0 ($p < 0.05$).

DISCUSSION

This study observed the effects of weight loss and sustained weight maintenance on the distribution of serum and adipose tissue fatty acids within and between pairs

TABLE 6. Summary of significant IPRs in the change of serum lipids and adipose tissue TG after VLCD and after 1-year weight maintenance

Fatty acid	Change after VLCD (in comparison with basal value)				Change after 1-year treatment (in comparison with basal value)			
	PC	CE	Serum TG	Adipose TG	PC	CE	Serum TG	Adipose TG
14:0	0.50(*)		0.62**		0.45(*)	0.47*	0.74**	
16:0			0.54*		0.94***			
16:1 n-7					0.47*	0.51*	0.52*	0.48*
18:0	0.55*		0.54*		0.47*			
18:1 n-9	0.64**		0.55*	0.52*	0.83***		0.62**	
18:2 n-6			0.48*		0.46*			
18:3 n-6		0.71**				0.61**		
20:3 n-6								0.59**
20:4 n-6								
18:3 n-3	0.87***		0.41(*)		0.84***	0.71**		
20:5 n-3			0.45(*)		0.44(*)	0.43(*)	0.50*	0.52*
22:6 n-3	0.56*				0.44(*)		0.42(*)	0.47*

NOTE: (*)$p < 0.08$, *$p < 0.05$, **$p < 0.01$, ***$p < 0.001$.

of identical twins. The fatty acid composition of the various lipid fractions for the subject group across the 13-month study was influenced by the rapid mobilization of adipose stores during the VLCD and then by a marked change in dietary habits in the subsequent year of successful weight maintenance. Despite these dramatic changes in the lipid fraction fatty acids induced by dietary stress and change, there was a consistent pattern of IPRs across both the 1 month of VLCD and the 12-month maintenance period. This pattern of IPRs was most consistent for the n-3 precursor 18:3 n-3 and also for its metabolic product 20:5 n-3. Of note is the fact that 18:3 n-3 in the serum fractions underwent considerable change over the course of the study, with the IPRs persisting across these changes.

IPRs for a change in response to diet reflect the degree to which the specific fatty acid within a specific pool stays constant within an individual genotype independent of diet. Thus, a pattern of IPRs for an individual fatty acid across one or more lipid fractions over time indicates that there is genetic control over the metabolism of that specific fatty acid and that one or more factors influencing this metabolic control vary among genotypes in the subject group. If a fatty acid's content in a specific lipid pool were determined solely by diet, then (for sets of monozygotic twins subjected to the same diet) any IPRs resulting from diet similarities within twin pairs should be reduced or lost across treatment. Alternatively, if the content of a nutrient in a particular pool was tightly regulated to the same level for all individuals in a species, then IPRs would be minor or insignificant to begin with (i.e., the consistency of values within pairs would be matched by the consistency of values between pairs). This conclusion is also supported by our observation that, with one exception, there was a lack of significant difference between concordant twin pairs and those discordant in the change of fat intake during the 1-year follow-up. The only exception was the change in TG 14:0, potentially because of a type-1 error.

Seen from this perspective, our data support the conclusion that individual genotypes defend "fingerprint" fatty acid profiles for selected fatty acids in selected lipid pools. Clearly, these fingerprints can be altered by dietary stress (e.g., weight loss, change in type or amount of dietary fat); however, despite such influences, each genotype maintains its unique pattern. We have previously reported strong IPRs for the individual fatty acids in serum and adipose lipid fractions in these subjects at baseline.[18] This report extends that observation to a dynamic dietary state over time, showing in the context of change for the group as a whole that the twin pairs defend the uniqueness of their individual fatty acid fingerprints in multiple lipid fractions.

Of particular note in TABLE 6 is the relatively consistent pattern of IPRs for the n-3 fatty acids across the VLCD and then across the year of maintenance. This consistency of IPRs for the n-3 fatty acids was most pronounced for 18:3 n-3, occurring despite quite profound diet-induced changes in the group mean values for this family of essential fatty acids (TABLES 2–5). By comparison, however, there was a much more limited pattern of IPRs for the n-6 essential fatty acid family, including its precursor 18:2 n-6 and its product 20:4 n-6 (FIG. 1). This observation does not necessarily imply that the contents of arachidonate and its precursor are less well regulated in lipid pools than is the n-3 family. Rather, our data are consistent with there being relative homogeneity in the control of the n-6 fatty acid composition across all genotypes within this group of subjects, contrasted to the n-3 family being characterized by a greater genetic heterogeneity in the control mechanisms.

This conclusion is also consistent with a prior report by Phinney *et al.*[6] of remarkably uniform differences in all of the n-6 fatty acids between central and peripheral adipose tissue depots in a group of women, contrasted with no significant gradients for the n-3 fatty acids between adipose tissue locations. Taking this information in total, it yields a pattern suggesting that the regulation of the n-6 fatty acids is species-specific, whereas the n-3 fatty acids appear to be just as tightly regulated, but in a genotype-specific manner.

One limitation of our study in assessing IPRs for the n-6 fatty acids was the selection of a relatively homogeneous group of obese female subjects. One of us has previously reported variations in the arachidonate content of lipid fraction fatty acids associated with obesity in humans[26] and in animals.[15,16] By limiting our subject pool to obese twin pairs, we may have reduced the range of genetically mediated variation in n-6 fatty acid metabolism, thereby limiting our ability to discern between-pair variation for this class of fatty acids.

The prior report of variations in serum phospholipid 20:4 n-6 between lean and obese humans also noted a marked increase in the arachidonate content of phospholipids associated with VLCD.[26] That effect of VLCD was not observed in the present study (TABLE 2), possibly due to the 5 days of weight stabilization prior to obtaining the blood sample after the VLCD. If so, it indicates a very rapid response of the proportion of arachidonate in the serum PC fraction to changes in energy balance.

The mechanisms through which such a response might occur or through which the gradients between lipid fractions (TABLES 2–5) are established remain poorly understood. There is even less known about how a specific genotype could maintain a fingerprint profile of an essential fatty acid class, such as the n-3 family. Clearly, this is not limited to control of fatty acid desaturases as the most significant IPRs observed for this family were for the precursor 18:3 n-3. This points us toward mechanisms of selective uptake and retention into the various lipid pools. It is now well established that there is a fatty acid–specific selectivity for hormone-sensitive lipase[7,8] that is consistently seen across species.[10–12] Postprandial selective uptake of fatty acids into adipose tissue in humans was also shown.[27] Our results imply that there is considerable genetic diversity for this or another similar enzyme within humans, resulting in multiple variants for the genotype of n-3 fatty acid retention.

The metabolic implications of our observation are also in need of further exploration. Highly unsaturated fatty acids (HUFAs) are the substrates for eicosanoid synthesis and, thus, variations in precursor concentration could affect the variety of cellular regulatory processes effected by eicosanoids. Both prospective intervention[28,29] and epidemiologic studies[30,31] have demonstrated strong relationships between n-3 fatty acid intake and coronary disease risk and stroke;[32] thus, it is credible that individual genotype variations in n-3 retention could also be a factor in coronary risk independent of diet. HUFAs are also regulators of gene expression,[33] providing not only a functional role for the observed IPRs, but also a potential self-regulatory pathway (if the genes affected were involved in selectivity of fatty acid uptake).

A final point worthy of note is the consistent reduction in 16:1 n-7 in all of the lipid fractions across the course of the study. In some lipid fractions such as serum CE and TG, this may be the result of displacement by the increasing proportion of 18:2 n-6 due to changing dietary fat sources. However, this cannot explain the decline in 16:1 n-7 in adipose TG as the proportions of other major fatty acids did not change enough to have this effect. We have previously reported that serum CE and adipose TG 16:1 n-7 at baseline for the complete cohort of these subjects correlates positively with multiple measures of adiposity.[18] In this context and in view of the observations that human lipogenesis may occur in adipose tissue[34,35] and that overtly lipogenic animals have marked elevations of 16:1 in serum CE and adipose TG,[15] it is possible that the changes in 16:1 n-7 seen in this study are a biomarker for reduced lipogenesis associated with successful weight maintenance.

ACKNOWLEDGMENTS

We express our gratitude to Vera Lánská and Lubomír Poušek for statistical management and graphical documentation of the study. The project was supported by Charles University Grant No. GA UK 203030/54 91/99 and by Grant No. NB/70313 from the Ministry of Health.

REFERENCES

1. HOLMAN, R.T. 1971. Essential fatty acids deficiency. Prog. Chem. Fats Other Lipids **9:** 275–348.
2. RIELLA, M.C., J.W. BROCIAC, M. WELLS & B.H. SCHRIBNER. 1975. Essential fatty acid deficiency in human adults during total parenteral nutrition. Ann. Intern. Med. **83:** 786–789.
3. HOLMAN, R.T., S.B. JOHNSON & T.F. HATCH. 1982. A case of human linoleic acid deficiency involving neurological abnormalities. Am. J. Clin. Nutr. **35:** 617–623.

4. HIRSCH, J. 1962. Composition of adipose tissue. *In* Adipose Tissue as an Organ, pp. 79–123. Thomas. Springfield, IL.
5. BERRY, E.M., J. HIRSCH, J. MOST *et al.* 1985. The relationship of dietary fat to plasma lipid levels as studies by factor analysis of adipose tissue fatty acid composition in free-living population of middle-aged American men. Am. J. Clin. Nutr. **44:** 220–231.
6. PHINNEY, S.D., J.S. STERN, A.B. TANG *et al.* 1994. Human subcutaneous adipose tissue shows site-specific differences in fatty acid composition, including omega-6 polyunsaturates. Am. J. Clin. Nutr. **60:** 725–729.
7. RACLOT, T. & R. GROSCOLAS. 1993. Differential mobilization of white adipose tissue fatty acids according to chain length, unsaturation, and positional isomerism. J. Lipid Res. **34:** 1515–1526.
8. RACLOT, T., E. MIOSKOWSKI, A.C. BACH & R. GROSCOLAS. 1995. Selectivity of fatty acid mobilization: a general feature of adipose tissue. Am. J. Physiol. **296:** R1060–R1067.
9. WOLFE, R.R., S. KLEIN, F. CARRARO & J.M. WEBER. 1994. Role of triglyceride–fatty acid cycle in controlling fat metabolism in humans during and after exercise. Am. J. Physiol. **258:** E382–E389.
10. PHINNEY, S.D., A.B. TANG, S.B. JOHNSON & R.T. HOLMAN. 1990. Reduced adipose 18:3ω3 with weight loss by very low calorie dieting. Lipids **25:** 798–806.
11. HUDGINS, L.C. & J. HIRSCH. 1991. Changes in abdominal and gluteal adipose tissue fatty acid compositions in obese subjects after weight gain and weight loss. Am. J. Clin. Nutr. **53:** 1372–1377.
12. TANG, A.B., K.Y. NISHIMURA & S.D. PHINNEY. 1993. Preferential reduction in adipose tissue 18:3ω3 during very low calorie dieting despite supplementation with 18:3ω3. Lipids **28:** 987–993.
13. HAZEL, J. & E.E. WILLIAMS. 1990. The role of alterations in membrane lipid composition in enabling physiological adaptation of organisms to their physical environment. Prog. Lipid Res. **29:** 167–227.
14. MOUSSA, M., J. GARCIA, J. GHISOLFI *et al.* 1996. Dietary essential fatty acid deficiency affects tissues of rats. J. Nutr. **126:** 3040–3045.
15. PHINNEY, S.D., A.B. TANG, D.C. THURMOND *et al.* 1993. Abnormal polyunsaturated lipid metabolism in the obese Zucker rat, with partial metabolic correction by gamma-linolenic acid administration. Metabolism **42:** 1127–1140.
16. PHINNEY, S.D., J.S. FISLER, A.B. TANG & C. WARDEN. 1994. Liver fatty acid composition correlates with body fat and gender in multigenic mouse model of obesity. Am. J. Clin. Nutr. **60:** 61–67.
17. HAINER, V., A.J. STUNKARD, M. KUNEŠOVÁ *et al.* 2000. Intrapair resemblance in very low calorie diet–induced weight loss in female obese identical twins. Int. J. Obes. Relat. Metab. Disord. **24:** 1051–1057.
18. KUNEŠOVÁ, M., V. HAINER, E. TVRZICKÁ *et al.* 2002. Assessment of dietary and genetic factors influencing serum and adipose fatty acid composition in obese female identical twins. Lipids. In press.
19. HAINER, V., A.J. STUNKARD, M. KUNEŠOVÁ *et al.* 2001. A twin study of weight loss and metabolic efficiency. Int. J. Obes. Relat. Metab. Disord. **25:** 533–537.
20. PAŘÍZKOVÁ, J. 1997. Body Fat and Physical Fitness. Nijhoff. The Hague.
21. BROŽEK, J., F. GRANDE, J.T. ANDERSON & A. KEYS. 1967. Densitometric analysis of body composition: revision of some quantitative assumptions. Ann. N.Y. Acad. Sci. **110:** 113–140.
22. MENEELY, G.R. & N.L. KALTREIDER. 1949. The volume of the lung determined by helium dilution: description of the method and comparison with other procedures. J. Clin. Invest. **28:** 129–139.
23. DURNIN, J.V. & A.G. WOMMERSLEY. 1974. Body fat assessed from total body density and its estimation from skinfold thickness: measurements on 481 men and women aged from 16 to 72 years. Br. J. Nutr. **32:** 77–97.
24. LOHMAN, T., A. ROCHE & R. MARTOREL, Eds. 1989. Standardization of Anthropometric Measurements. Human Kinetics Pub. Champaign, IL.
25. WINER, B.J. 1971. Statistical Principles in Experimental Design. McGraw–Hill. New York.
26. PHINNEY, S.D., P.G. DAVIS, S.B. JOHNSON & R.T. HOLMAN. 1991. Obesity and weight loss alter polyunsaturated lipid metabolism in humans. Am. J. Clin. Nutr. **53:** 831–838.

27. SUMMERS, L.K.M., C. BARNES, B.A. FIELDING et al. 2000. Uptake of individual fatty acids into adipose tissue in relation to their presence in the diet. Am. J. Clin. Nutr. **71:** 1470–1477.
28. BURR, M.L., A.M. FEHILY, J.F. GILBERT et al. 1989. Effect of change in fat, fish, and fibre intakes on death and reinfarction: diet and reinfarction trial (DART). Lancet **2:** 757–761.
29. DE LORGERIL, M., S. RENAUD et al. 1994. Mediterranean alpha-linolenic acid–rich diet in secondary prevention of coronary heart disease. Lancet **343:** 1454–1459.
30. KROMHOUT, D., E.B. BOSCHIETER & C.L. COULANDER. 1985. The inverse relation between fish consumption and 20-year mortality from coronary heart disease. N. Engl. J. Med. **312:** 1205–1209.
31. SISCOVICK, D.S., T.E. RAGHUNATHAN, I. KING et al. 1995. Dietary intake and cell membrane levels of long-chain n-3 polyunsaturated fatty acids and the risk of primary cardiac arrest. JAMA **274:** 1363–1367
32. ISO, H., K.M. REXRODE, M.J. STAMPFER et al. 2001. Intake of fish and omega-3 fatty acids and risk of stroke in women. JAMA **285:** 304–312.
33. CLARKE S.D. & D.B. JUMP. 1994. Dietary polyunsaturated fatty acid regulation of gene transcription. Annu. Rev. Nutr. **14:** 83–98.
34. AARSLAND, A., D. CHIRKES & R.R. WOLFE. 1997. Hepatic and whole-body fat synthesis in humans during carbohydrate overfeeding. Am. J. Clin. Nutr. **65:** 1774–1782.
35. LAMMERT, O., N. GRUNNERT, P. FABER et al. 2000. Effect of isoenergetic overfeeding of either carbohydrate or fat in young men. Br. J. Nutr. **84:** 233–245.

Dietary Fat as a Risk Factor for Type 2 Diabetes

BARBARA V. HOWARD

MedStar Research Institute, Washington, District of Columbia, USA

> ABSTRACT: Too much information is missing for a definitive recommendation to be made concerning the role of diet in the development of diabetes. Although studies in certain population subgroups show a positive correlation between amount of dietary fat intake and occurrence of type 2 diabetes, these cannot be considered definitive, in part because they are confounded by many other variables that might relate to the development of diabetes. Furthermore, dietary fat intake has not been correlated with certainty to diabetes risk factors. With regard to diabetes risk, type of dietary fat consumed may be more important than total dietary fat intake. Several studies suggest that diets higher in saturated fat may pose a higher diabetes risk than those higher in unsaturated fat. Although there is a general consensus that reduced consumption of saturated fat and cholesterol reduces CVD risk, there are no definite nutrition recommendations to prevent diabetes. The most important recommendation may be for health-care providers to carefully assess diabetes risk in their patients and institute dietary changes on an individual basis as needed to achieve and maintain body weight.
>
> KEYWORDS: dietary fat; type 2 diabetes; carbohydrate; cholesterol

As with other chronic diseases, there are limited long-term data concerning the effect of diet on the incidence of diabetes, and these have been analyses of epidemiologic studies. Cross-country comparisons[1,2] and migrant studies[3] uniformly show associations between higher prevalence of diabetes and higher proportion of fat intake; however, these observations are greatly confounded by the many other variables differing in these comparisons that might relate to the development of diabetes. A few longitudinal cohort studies have analyzed the relationship between baseline diet intake and the incidence of diabetes.[4,5] In the Zutphen Study of nondiabetic men, incidence of diabetes was associated with increased intake of saturated fat and cholesterol.[5] In a prospective analysis of the diet of Pima Indian women, higher incidence of diabetes was associated with a Western diet compared to a traditional diet; the latter contained less fat.[6] In the San Luis Valley Study of Mexican-Americans, incidence of diabetes was associated with higher fat intake.[4] In a study of Japanese-Americans, diets high in fat and saturated fat predicted the development of diabetes.[7]

Address for correspondence: Barbara V. Howard, Ph.D., MedStar Research Institute, 108 Irving Street NW, Washington, D.C. 20010. Voice: 202-877-6530; fax: 202-877-0845.
barbara.v.howard@medstar.net

On the other hand, a recent analysis of the Nurses' Health Study[8] did not show a relation between dietary fat intake and diabetes incidence. Epidemiologic analyses are, however, limited by the survey instrument, that is, the inability of self-reported questionnaires to accurately assess diet intake, and by the interrelatedness of the various dietary variables, especially dietary fat and total calories, and dietary fat and the amount of fruits, vegetables, and grains. At this point, there are no randomized clinical trials assessing the effects of varying dietary fat intake on diabetes incidence.

We are thus left to draw conclusions from relatively short-term studies in diabetic individuals that examine the relationship between dietary fat and various metabolic parameters and risk factors for diabetes. The many studies of lipid and lipoprotein levels in individuals with diabetes are in firm agreement that increasing saturated fat intake in diabetic, as in nondiabetic, individuals increases LDL cholesterol and that lowering saturated fat (and cholesterol) intake reduces LDL and total cholesterol levels.

Conversely, other types of fats may have beneficial effects on LDL concentrations. Polyunsaturated fats clearly lower LDL levels. Monounsaturated fats may be neutral in isolated feeding studies, but other studies have shown decreases in LDL with diets high in olive oil.[9] Lowering the proportion of any kind of fat in the diet decreases HDL concentrations in individuals with diabetes. Thus, replacing saturated fat with mono- or polyunsaturated fat rather than carbohydrate may result in more favorable HDL concentrations.[10] Substitution of dietary fat with carbohydrates may lead to increases in triglycerides in diabetic individuals and this triglyceride-raising effect of carbohydrate may be greater in individuals with a predisposition to hypertriglyceridemia. Triglyceride elevations may be attenuated by the type of carbohydrate ingested. For example, high-carbohydrate diets enriched in fiber, especially soluble fiber, cause clinically meaningless triglyceride elevations.[11] A low-fat diet with complex carbohydrates may also have less effect on HDL because HDL decreases with triglyceride elevations.

Also being examined are the effects of varying dietary fat on glycemia and insulin resistance. In type 1 diabetes, insulin requirements increase as dietary fat is substituted for carbohydrates because the carbohydrates, not fat, require insulin for disposal. Individuals with type 2 diabetes who have had the disease for a long time and who have greatly diminished β cell function also might be expected to show greater postprandial glycemia as the proportion of carbohydrate in the diet is increased. Effects of lowering fat and increasing carbohydrate on glycemia, however, are highly dependent on the type of carbohydrate. Carbohydrates containing soluble fiber and those with low glycemic indices often have no measurable effect on glucose control, especially in type 2 diabetics who have not had the disease very long. These have recently been shown to be associated with lower incidence of diabetes in the Nurses' Health Study.[8] However, a recent randomized study followed the effects of counseling and lower dietary fat intake and, in 162 individuals with glucose intolerance, adherence to the dietary recommendations resulted in a sustained improvement in glycemia.[12]

The effect of dietary fat content on insulin resistance has been a subject of debate for a number of years. Although animal studies almost uniformly show increases in insulin resistance accompanying high-fat diets,[13–15] this increase has never been confirmed in humans. When controlled randomized comparisons of high-fat versus high-carbohydrate diets using careful methods for evaluating insulin action are reviewed, the data show no differences in insulin-mediated glucose disposal in diets varying in content from 20% to 60% fat.[16–18] All of these studies, however, were relatively short-term (2–3 months) and, thus, the question of long-term variation in dietary fat and carbohydrate content on insulin action remains unanswered. Percent fat, however, was shown to correlate with insulin resistance in the IRAS study of whites, blacks, and Hispanics.[19] A recent report has suggested that the type rather than amount of fat may influence insulin resistance. A 3-month diet high in saturated fat led to higher insulin resistance than a similar diet high in monounsaturated fat.[20]

Recommendations for dietary fat intake in individuals with diabetes are influenced by the potential impact of the diet composition on body weight. Most people who have type 2 diabetes are overweight and it is universally agreed that weight loss is the primary goal of dietary therapy. There is great controversy concerning whether dietary fat intake influences weight gain or loss. The few studies available in individuals with diabetes show that short-term changes in dietary fat and carbohydrate content of isocaloric diets have no influence on basal or daily energy expenditure.[21] However, because many fat-containing foods are calorie-dense and taste good, there is ample reason for concern that consumption of high-fat foods could lead to increased calorie intake. Successful weight-loss regimens in individuals with diabetes generally have employed diets with low to moderate fat content. Further, all of these regimens were generally accompanied by exercise. Thus, there are no long-term data on the effects of *ad libitum* consumption of high-fat vs. high-carbohydrate diets on weight loss in diabetic individuals.

Other metabolic issues should be considered when formulating dietary fat recommendations in individuals with diabetes, although at present there are few data that shed light on these issues.[22] One of these issues is the potential effect of diet composition on postprandial lipoproteins. At least one study shows that high-carbohydrate diets increase postprandial lipoprotein levels in diabetic individuals,[23] but other studies show decreases in remnant particles on high-carbohydrate diets.[24] Another study suggests that leptin may be decreased in diabetic individuals on a low-carbohydrate diet,[25] although this needs further exploration. The possible role of dietary fat on thrombotic factors also needs to be explored because diabetes is associated with increased thrombotic potential.[26]

In conclusion, there is general consensus that individuals at high risk for diabetes need to decrease total calorie consumption and consume a diet low in saturated fat and cholesterol. How to balance the remaining fat and carbohydrate is still an open question. The National Heart, Lung, and Blood Institute and the American Heart Association both recommend that diets contain no more than 30% of total fat, with less than 10% saturated fat. It would be reasonable to extend this recommendation to individuals at risk for diabetes. One approach might be to advocate a slightly higher protein intake (perhaps 20% of total calories) in individuals who have diabetes, but no renal disease (there is no evidence that this is detrimental except in diabetic individuals with renal dysfunction). Carbohydrate intake (regardless of the presence or absence of diabetes) should stress consumption of fruits, vegetables, and whole grain, high-fiber, complex carbohydrates. The American Diabetes Association has called for flexibility in the amount of fat recommended for individuals with diabetes.[27] In view of the lack of evidence that one approach is clearly better than the other, this flexibility can permit tailoring of palatable, low-calorie diets based on cultural backgrounds and individual tastes. For example, the traditional diets of many ethnic groups were low in fat and high in carbohydrate; these groups might be more successful in making dietary changes geared toward those of their traditional lifestyles. On the other hand, individuals who prefer higher-fat diets might actually be more successful in maintaining and losing weight by not drastically removing dietary fat because they may be more likely to substitute the calorie-dense, "no-fat" commercial substitutes.

As stated above, individuals may vary in their response to changes in dietary carbohydrates, especially those with genetic hypertriglyceridemia or pancreatic insufficiency. Therefore, health-care providers must monitor both glucose and triglyceride levels as any dietary changes are instituted.

REFERENCES

1. FESKENS, E.J., S.M. VIRTANEN, L. RASANEN et al. 1995. Dietary factors determining diabetes and impaired glucose tolerance: a 20-year follow-up of the Finnish and Dutch cohorts of the Seven Countries Study. Diabetes Care **18:** 1104–1112.
2. RAVUSSIN, E., M.E. VALENCIA, J. ESPARZA et al. 1994. Effects of a traditional lifestyle on obesity in Pima Indians. Diabetes Care **17:** 1067–1074.
3. FUJIMOTO, W.Y. 1996. Overview of non-insulin-dependent diabetes mellitus (NIDDM) in different population groups. Diabetic Med. **13**(9, suppl. 6): S7–S10.
4. MARSHALL, J.A., R.F. HAMMAN & J. BAXTER. 1991. High-fat, low-carbohydrate diet and the etiology of non-insulin-dependent diabetes mellitus: the San Luis Valley Diabetes Study. Am. J. Epidemiol. **134:** 590–603.
5. FESKENS, E.J.M. & D. KROMHOUT. 1990. Habitual dietary intake and glucose tolerance in euglycemic men: the Zutphen Study. Int. J. Epidemiol. **19:** 953–959.
6. WILLIAMS, D.E., W.C. KNOWLER, C.J. SMITH et al. 2001. The effect of Indian or Anglo dietary preference on the incidence of diabetes in Pima Indians. Diabetes Care **24:** 811–816.
7. FUJIMOTO, W.Y., R.W. BERGSTROM, E.J. BOYKO et al. 2000. Type 2 diabetes and the metabolic syndrome in Japanese Americans. Diabetes Res. Clin. Pract. **50**(suppl. 2): S73–S76.
8. SALMERON, J., F.B. HU, J.E. MANSON et al. 2001. Dietary fat intake and risk of type 2 diabetes in women. Am. J. Clin. Nutr. **73:** 1019–1026.
9. KRIS-ETHERTON, P.M., J. DERR, D.C. MITCHELL et al. 1993. The role of fatty acid saturation on plasma lipids, lipoproteins, and apolipoproteins: I. Effects of whole food diets high in cocoa butter, olive oil, soybean oil, dairy butter, and milk chocolate on the plasma lipids of young men. Metabolism **42:** 121–129.
10. GUMBINER, B., C.C. LOW & P.D. REAVEN. 1998. Effects of a monounsaturated fatty acid–enriched hypocaloric diet on cardiovascular risk factors in obese patients with type 2 diabetes. Diabetes Care **21:** 9–15.
11. RICARDI, G. & A.A. RIVELLESE. 1991. Effects of dietary fiber and carbohydrate on glucose and lipoprotein metabolism in diabetic patients. Diabetes Care **14:** 1115–1125.
12. SWINBURN, B.A., P.A. METCALF & S.J. LEY. 2001. Long-term (5-year) effects of a reduced-fat diet intervention in individuals with glucose intolerance. Diabetes Care **24:** 619–624.
13. KRAEGEN, E.W., D.E. JAMES, L.H. STORLIEN et al. 1986. In vivo insulin resistance in individual peripheral tissues of the high fat fed rat: assessment of euglycaemic clamp plus deoxyglucose administration. Diabetologia **29:** 192–198.
14. STORLIEN, L.H., D.E. JAMES, K.M. BURLEIGH et al. 1986. Fat-feeding causes widespread in vivo insulin resistance, decreased energy expenditure, and obesity in rats. Am. J. Physiol. **251:** E576–E583.
15. STORLIEN, L.H., A.B. JENKINS, D.J. CHISHOLM et al. 1991. Influence of dietary fat composition on development of insulin resistance in rats: relationship to muscle triglyceride and w-3 fatty acids in muscle phospholipid. Diabetes **40:** 280–289.

16. BORKMAN, M., L.V. CAMPBELL, D.J. CHISHOLM et al. 1991. Comparison of the effects on insulin sensitivity of high carbohydrate and high fat diets in normal subjects. J. Clin. Endocrinol. Metab. **72:** 432–437.
17. SWINBURN, B.A., V.L. BOYCE, R.N. BERGMAN et al. 1991. Deterioration in carbohydrate metabolism and lipoprotein changes induced by modern, high fat diet in Pima Indians and Caucasians. J. Clin. Endocrinol. Metab. **73:** 156–165.
18. GARG, A., S.M. GRUNDY & R.H. UNGER. 1992. Comparison of effects of high and low carbohydrate diets on plasma lipoproteins and insulin sensitivity in patients with mild NIDDM. Diabetes **41:** 1278–1285.
19. MAYER-DAVIS, E.J., J.H. MONACO, H.M. HOEN et al. 1997. Dietary fat and insulin sensitivity in a triethnic population: the role of obesity—the Insulin Resistance Atherosclerosis Study (IRAS). Am. J. Clin. Nutr. **65:** 79–87.
20. VESSBY, B., M. UUSITUPA, K. HERMANSEN et al. 2001. Substituting dietary saturated fat for monounsaturated fat impairs insulin sensitivity in healthy men and women: the KANWU study. Diabetologia **44:** 312–319.
21. ABBOTT, W.G., B.V. HOWARD, G. RUOTOLO et al. 1990. Energy expenditure in humans: effects of dietary fat and carbohydrate. Am. J. Physiol. **258**(2, part 1)**:** E347–E351.
22. GRUNDY, S.M. 1991. Dietary therapy in diabetes mellitus: is there a single best diet? Diabetes Care **14:** 796–801.
23. PERROTTI, N., D. SANTORO, S. GENOVESE et al. 1984. Effect of digestible carbohydrates on glucose control in insulin-dependent diabetic subjects. Diabetes Care **7:** 354–359.
24. GEORGOPOULOS, A., J.P. BANTLE, M. NOUTSOU et al. 1998. Differences in the metabolism of postprandial lipoproteins after a high-monounsaturated-fat versus a high-carbohydrate diet in patients with type 1 diabetes mellitus. Arterioscler. Thromb. Vasc. Biol. **18:** 773–782.
25. JENKINS, A.B., T.P. MARKOVIC, A. FLEURY & L.V. CAMPBELL. 1997. Carbohydrate intake and short-term regulation of leptin in humans. Diabetologia **40:** 348–351.
26. OSTERMAN, H. & J. VAN DE LOO. 1986. Factors of the hemostatic system in diabetic patients: a survey of controlled study. Hemostasis **16:** 386–416.
27. AMERICAN DIABETES ASSOCIATION. 1998. Nutrition recommendations and principles for people with diabetes mellitus [position statement]. Diabetes Care **21**(suppl. 1)**:** S32–S35.

Type of Dietary Fat and Insulin Resistance

ANGELA A. RIVELLESE, CLAUDIA DE NATALE, AND STEFANIA LILLI

Department of Clinical and Experimental Medicine, Federico II University Medical School, Napoli, Italy

ABSTRACT: Animal studies have already shown the possibility to modulate insulin action by changing not only the amount of total fat, but also the type of fat. In these studies, saturated fat significantly increased insulin resistance, long- and short-chain ω_3 fatty acids significantly improved it, whereas the effects of monounsaturated and ω_6 polyunsaturated fatty acids ranged somewhere in between the two. A recent multicenter study (the Kanwu study) on humans has shown that shifting from a diet rich in saturated fatty acids to one rich in monounsaturated fat improved insulin sensitivity in healthy people, while a moderate ω_3 supplementation did not affect it; this second finding confirms previous results in type 2 diabetic patients with hypertriglyceridemia. There are also other aspects of the metabolic syndrome that can be influenced by the different type of dietary fat, particularly blood pressure and lipid metabolism. With respect to blood pressure, the majority of studies show that ω_3 fatty acids are able to reduce blood pressure in hypertensive patients, but not in normotensive individuals; this result has been confirmed also by the Kanwu study, where no changes in blood pressure were seen after ω_3 supplementation in healthy people. On the other hand, in this study, the change from saturated to monounsaturated fatty acids was able to significantly reduce diastolic blood pressure. As to the lipid abnormalities more frequently present in the metabolic syndrome (i.e., hypertriglyceridemia and low HDL cholesterol), the main effects are related to ω_3 fatty acids, which surely reduce triglyceride levels, but at the same time increase LDL cholesterol. In conclusion, there is so far sound evidence in humans that the quality of dietary fat is able to influence insulin resistance and some of the related metabolic abnormalities.

KEYWORDS: dietary fat; insulin resistance; insulin sensitivity; metabolic syndrome; cholesterol

INTRODUCTION

Insulin resistance is the pathogenic link underlying the different metabolic abnormalities making up the so-called metabolic syndrome. It can be modulated by different environmental factors, mainly dietary habits. The influence of diet on insulin sensitivity is mediated by not only its energy content, but also its composition. Since insulin resistance is pathogenically linked to other metabolic abnormalities and, directly or through these abnormalities, to cardiovascular diseases, its modulation by

---

Address for correspondence: Angela A. Rivellese, Department of Clinical and Experimental Medicine, Federico II University Medical School, Via S. Pansini 5, 80131 Napoli, Italy. Voice: +39 081 746 2117; fax: +39 081 546 6152.

rivelles@unina.it

TABLE 1. Effect of diet composition on insulin sensitivity

Author	Fat content	Participants	Duration (days)	Outcome
Chen et al.[2]	55% vs. 0%	Healthy	4	↓
Borkman et al.[1]	37% vs. 22%	Healthy	21	=
Parillo et al.[5]	40% vs. 20%	Type 2 DM	14	↓
Garg et al.[3]	50% vs. 25%	Type 2 DM	21	=
Hughes et al.[4]	30% vs. 20%	IGT	84	=

diet implies also a possible change in other related alterations and, more important, a possible prevention of cardiovascular diseases.

Regarding the effect of diet composition on insulin resistance, initially attempts have been made to influence insulin action by changing the total amount of nutrients, in particular the amount of total fat and carbohydrates. TABLE 1 summarizes some of the main studies related to this topic.[1–5] Only in the case of a sharp change in the amount of fat (from 55% to 0%) and, thus, in that of CHO did insulin sensitivity decrease with the very high fat diet. With moderate changes, which are more feasible in normal living conditions, the increase in the total amount of fat does not impair insulin action in humans, but may even induce a slight improvement in type 2 diabetic patients, at least when the increase in fat is mainly due to monounsaturated fat.[5]

More recently, attention has been given also to the possibility to influence insulin sensitivity by changing the type of nutrients, in particular the type of fat. The importance of this topic has been raised by animal studies, which have shown that (1) diets rich in saturated fat impair insulin action and, thus, significantly increase insulin resistance; (2) long and short ω_3 fatty acids significantly improve insulin action, reverting completely the negative effects of saturated fat; (3) monounsaturated and ω_6 polyunsaturated fatty acids have less negative effect on insulin sensitivity than saturated fat.[6] Moreover, in animals, there is also a strong significant correlation between the amount of long-chain ω_3 fatty acids present in the phospholipids of muscle membranes and insulin action.[6] This relationship, together with other data, strongly suggests that the fatty acid composition of muscle cell membranes is of great importance in the regulation of insulin action;[7] therefore, it is possible to hypothesize that dietary fatty acids influence insulin sensitivity through a modification in the fatty acid composition of cell membranes.

This report, along with other studies on animals,[7] clearly shows that the type of dietary fat has a strong influence on insulin sensitivity—hence the need to examine our current knowledge on this topic in humans.

TYPE OF FAT AND INSULIN SENSITIVITY IN HUMANS

Some epidemiological studies have found that the different types of dietary fatty acids are associated with insulin sensitivity.[8] Among these, the Normative Aging Study,[9] performed in more than 600 normal people, found that total and saturated fat are directly and significantly related to both fasting and postprandial insulin levels, taken as indexes of insulin resistance, also after adjusting for body mass index (BMI).

On the other hand, in the Insulin Resistance Atherosclerosis Study,[10] performed in three different ethnic populations and including also subjects with impaired glucose tolerance and type 2 diabetic patients, there was a significant inverse correlation between insulin sensitivity, on the one hand, and total fat, oleic acid, and ω_6 polyunsaturated fatty acids (PUFA) on the other, but the correlation with saturated fat was not statistically significant. Furthermore, in this particular study, all the relationships between insulin sensitivity and dietary fat seemed largely mediated by the effect of dietary fat on adiposity since they were no longer significant after adjusting for BMI.

Other epidemiological studies have evaluated the relationship between insulin sensitivity and dietary fatty acids, taking as index of dietary fatty acid intake a more objective measurement than dietary questionnaires, such as the amount of the different fatty acids present in the phospholipids of serum or cell membranes. These studies have shown that the relationship may differ according to the type of population studied. As a matter of fact, in a group of Swedish elderly men, there was a significant inverse correlation between the amount of saturated fat (in particular palmitic acid) in skeletal muscle cell membranes and their insulin sensitivity.[11] Instead, in a group of Australian men, there was a significant direct correlation between the amount of long-chain PUFA (both ω_6 and ω_3) and insulin sensitivity.[12] These differences might be due to different dietary habits and different genetic background.

Therefore, trying to summarize the data deriving from epidemiological studies, a possible conclusion could be that the association between different types of dietary fat and insulin sensitivity is not very consistent and, in any case, not as strong as one would expect from studies in animals. Furthermore, associations do not necessarily imply a casual role, which can be proven only by intervention studies.

The majority of these studies evaluated the effects of long-chain ω_3 fatty acids on insulin resistance, in particular in type 2 diabetic patients (TABLE 2).[13–17] On the whole, they show rather clearly that ω_3 fatty acids do not have any significant effect on the insulin resistance of these subjects. However, all these studies have been performed for periods of time too short to allow a real change in the composition of cell membrane phospholipids. Of course, if dietary fat is able to modulate insulin action through modifications in the composition of cell membranes, the duration of experiments is of crucial importance. To this regard, two studies have been performed in humans with such a duration as to allow a change in the composition of cell membranes. One was performed in a small group of type 2 diabetic patients with hypertriglyceridemia[18] and the other one in a large group of healthy individuals.[19] In the first study, which lasted 6 months, type 2 diabetic patients with hypertriglyceridemia were randomly assigned to two groups, one receiving fish oil supplementation (2.5 g/day for the first 2 months and 1.7 g/day for the following 4 months) and the other placebo. After a period of 6 months, there was a really significant change in the composition of phospholipid fatty acids, at least for erythrocytes, which were significantly enriched in long-chain ω_3 fatty acids only in the group receiving fish oil (TABLE 3). Nonetheless, insulin-mediated glucose utilization, evaluated by the clamp technique at the baseline and at the end of the study period in the two groups of patients, was practically unchanged (TABLE 3).[18]

The other study related to this topic lasted 3 months and was performed in 162 healthy individuals recruited from five different centers (Kuopio, Århus, Naples, Wallongog, Uppsala). In this case, the aim was to evaluate whether the level of insulin sensitivity in healthy people could be modulated by the different types of

TABLE 2. Controlled studies of ω_3 fatty acid dietary supplementation on insulin sensitivity in type 2 diabetic patients

Study	Subjects (n)	Fish oil dose (g)	Design	Duration (weeks)	Insulin sensitivity
Borkman et al.[13]	10	3	Crossover	3	Unchanged
Annuzzi et al.[14]	8	3	Crossover	2	Unchanged
Boberg et al.[15]	11	3	Crossover	8	Unchanged
McManus et al.[16]	11	3	Crossover	12	Unchanged
Luo et al.[17]	12	6	Crossover	8	Unchanged

TABLE 3. Effects of a moderate ω_3 fatty acid supplementation on the phospholipid fatty acid composition of erythrocytes and insulin-mediated glucose disposal in type 2 diabetic patients (M ± SEM)

	ω_3 supplementation		Placebo	
	Baseline	After 6 months	Baseline	After 6 months
EPA (%)	0.7 ± 0.2	2.2 ± 0.3*	0.7 ± 0.1	0.6 ± 0.1
DHA (%)	5.1 ± 0.3	6.7 ± 0.2*	5.0 ± 0.4	4.7 ± 0.3
M (mg/kg/min)	4.04 ± 0.82	3.96 ± 0.50	3.5 ± 0.62	4.09 ± 0.49

NOTE: EPA, eicosapentaenoic acid; DHA, docosahexaenoic acid; *$p < 0.001$ vs. baseline.

dietary fatty acids, not only long-chain ω_3 fatty acids, but also saturated and monounsaturated fatty acids (MUFA). To this aim, subjects were randomized to receive two types of isoenergetic diet, one rich in saturated fat and the other in monounsaturated fat. Within each diet, a second randomization was performed for supplementation with long-chain ω_3 fatty acids (3.6 g/day of fish oil in forms of capsules) or placebo. The results of this study clearly show that insulin sensitivity improves on the high MUFA diet and that this improvement is much more relevant when the total amount of fat consumed by participants is not very high (<37% of total energy content of the diet).[19] On the other hand, also in these healthy subjects, a moderate long-chain ω_3 fatty acid supplementation is not able to modify insulin sensitivity.[20]

TYPE OF FAT AND METABOLIC ABNORMALITIES LINKED TO INSULIN RESISTANCE

Insulin resistance is strictly related to other metabolic abnormalities, among which the most important ones are lipid abnormalities, characterized mainly by high triglyceride levels and low HDL cholesterol. Since the quality of dietary fat influences insulin resistance, it is of clinical importance to see whether these changes are followed by changes also in the other metabolic abnormalities. With respect to the triglyceride metabolism, long-chain ω_3 fatty acids are most powerful in reducing triglyceride levels in humans.[21] Even if they have no effect on insulin resistance, the reduction of triglycerides by long-chain ω_3 fatty acids has been found in hypertriglyceridemic individuals as well as in type 2 diabetic patients[18] and has been largely confirmed also in healthy people by the Kanwu study.[20] However, it is important to underline that this hypotriglyceridemic effect is very often accompanied by an increase in LDL cholesterol, not only in hyperlipidemic individuals, but also in healthy people with very normal lipid levels.[20]

In relation to blood pressure, it is first of all important to remember that some epidemiological studies have found significant associations between the different types of dietary fatty acids and blood pressure levels, indicating that the higher the consumption of saturated fat the higher the blood pressure, while the higher the intake of MUFA the lower the blood pressure.[22,23] Intervention studies in humans comparing the effects of saturated, monounsaturated, and polyunsaturated ω_6 fat on blood pressure are few and give no consistent results. Instead, studies with long-chain ω_3 fatty acids seem to suggest that these are able to reduce blood pressure, but only in hypertensive people and patients with vascular disease, with a dose-related effect.[24]

The multicenter Kanwu study, previously described, has shown that a moderate shift from saturated to monounsaturated fat is able to significantly reduce diastolic blood pressure. On the other hand, a moderate supplementation of long-chain ω_3 fatty acids for 3 months has absolutely no influence on blood pressure in perfectly normotensive people.

CONCLUSIONS

Summarizing the results of the intervention studies performed in humans, it is certainly possible to conclude that the quality of dietary fat is able to influence insulin sensitivity as well as other abnormalities linked to insulin resistance (TABLE 4). However, the effects are not always interrelated, suggesting different underlying mechanisms of action.

In particular, saturated fat increases insulin resistance, while monounsaturated fat improves insulin sensitivity. Long-chain ω_3 fatty acids have no effect on insulin sensitivity in humans, at least with a moderate consumption, which in any case is the only possible intake, considering the potentially negative effects of higher amounts. The effects of ω_6 polyunsaturated fat have not yet been studied with large intervention trials, but it is possible that their effects are similar to those induced by monounsaturated fat.

TABLE 4. Effects of different types of dietary fat on insulin resistance and some of the related metabolic abnormalities

	IR	Plasma tg	LDL chol	HDL chol	LDL size	BP
SAFA	↑	–	↑↑	–	–	↑
MUFA	↓[a]	–	↓	–↑	–	↓
PUFA ω_6	–	↓	↓	–↓	–	–
PUFA ω_3	–	↓↓	↑	–	–	↓ (H), – (N)

NOTE: (H) = hypertensive; (N) = normotensive.
[a]Especially when intake of total fat is not very high.

The most consistent change in plasma triglycerides is due to long-chain ω_3 fatty acids, which, however, induce negative effects on LDL cholesterol also in healthy people.

The quality of fat does not seem to influence HDL cholesterol nor LDL size, at least when changes in the type of dietary fat are not so extreme.

Finally, saturated fat increases and monounsaturated fat decreases blood pressure, while the effects of long-chain ω_3 fatty acids on blood pressure seem to be limited only to hypertensive subjects.

All these data underline the need to implement a reduction in the consumption of foods rich in saturated fat in favor of foods and vegetable oils rich in polyunsaturated fat, particularly monounsaturated ones, not only for their positive effects on LDL cholesterol, but also for their beneficial influence on insulin resistance and some of the related metabolic abnormalities.

REFERENCES

1. BORKMAN, M., L.V. CAMPBELL, D.J. CHISHOLM & L.H. STORLIEN. 1991. Comparison of the effects on insulin sensitivity of high carbohydrate and high fat diets in normal subjects. J. Clin. Endocrinol. Metab. **72:** 432–437.
2. CHEN, M., R. BERGMAN & D. PORTE. 1988. Insulin resistance and beta-cell dysfunction in aging: the importance of dietary carbohydrate. J. Clin. Endocrinol. Metab. **67:** 951–957.
3. GARG, A., S.M. GRUNDY & R.H. UNGER. 1992. Comparison of the effects of high and low carbohydrate diets on plasma lipoproteins and insulin sensitivity in patients with mild NIDDM. Diabetes **41:** 1278–1285.
4. HUGHES, V.A., M.A. FIATARONE, R.A. FIELDING et al. 1995. Long-term effects of a high-carbohydrate diet and exercise on insulin action in older subjects with impaired glucose tolerance. Am. J. Clin. Nutr. **62:** 426–433.
5. PARILLO, M., A.A. RIVELLESE, A.V. CIARDULLO et al. 1992. A high monounsaturated fat/low carbohydrate diet improves peripheral insulin sensitivity in non-insulin dependent diabetic patients. Metabolism **41:** 1373–1378.
6. STORLIEN, L.H., A.B. JENKINS, D.J. CHISHOLM et al. 1991. Influence of dietary fat composition on development of insulin resistance in rat: relationship to muscle triglyceride and omega-3 fatty acids in muscle phospholipid. Diabetes **40:** 280–289.
7. STORLIEN, L.H., L.A. BAUR, A.D. KRIKETOS et al. 1996. Dietary fats and insulin action. Diabetologia **39:** 621–631.
8. HU, F.B., R.M. VAN DAM & S. LIU. 2001. Diet and risk of type II diabetes: the role of types of fat and carbohydrate. Diabetologia **44:** 805–817.
9. PARKER, D.R., S.T. WEISS, T. TROISI et al. 1993. Relationship of dietary saturated fatty acids and body habitus to serum insulin concentrations: the Normative Aging Study. Am. J. Clin. Nutr. **58:** 129–136.
10. MAYER-DAVIS, E.J., J.H. MONACO, H.M. HOEN et al. 1997. Dietary fat and insulin sensitivity in a triethnic population: the role of obesity—the Insulin Resistance Atherosclerosis Study (IRAS). Am. J. Clin. Nutr. **65:** 79–87.
11. VESSBY, B., S. TENGBLAD & H. LITHELL. 1994. Insulin sensitivity is related to the fatty acid composition of serum lipids and skeletal muscle phospholipids in 70-year-old men. Diabetologia **37:** 1044–1050.
12. BORKMAN, M., L.H. STORLIEN, D.A. PAN et al. 1993. The relation between insulin sensitivity and fatty-acid composition of skeletal-muscle phospholipids. N. Engl. J. Med. **328:** 238–244.
13. BORKMAN, M., D.J. CHISHOLM, S. FURLER et al. 1989. Effects of fish oil supplementation on glucose and lipid metabolism in NIDDM. Diabetes **38:** 1314–1319.

14. ANNUZZI, G., A.A. RIVELLESE, B. CAPALDO et al. 1991. A controlled study on the effects of ω-3 fatty acids on lipid and glucose metabolism in non-insulin dependent diabetic patients. Atherosclerosis **87:** 65–73.
15. BOBERG, M., T. POLLARE, A. SIEGBAHN & B. VESSBY. 1992. Supplementation with ω-3 fatty acids reduces triglycerides, but increases PAI-1 in non-insulin dependent diabetes mellitus. Eur. J. Clin. Invest. **22:** 645–650.
16. MCMANUS, R.M., J. JUMPSON, D.T. FINEGOOD et al. 1996. A comparison of the effects of ω-3 fatty acids from linseed oil and fish oil in well controlled type II diabetes. Diabetes Care **19:** 463–476.
17. LUO, J., S.W. RIZKALLA, H. VIDAL et al. 1998. Moderate intake of ω-3 fatty acids for 2 months has no detrimental effect on glucose metabolism and could ameliorate the lipid profile in type II diabetic men: results of a controlled study. Diabetes Care **21:** 717–724.
18. RIVELLESE, A.A., A. MAFFETTONE, C. IOVINE et al. 1996. Long-term effects of fish oil on insulin resistance and plasma lipoproteins in NIDDM patients with hypertriglyceridemia. Diabetes Care **19:** 1207–1213.
19. VESSBY, B., M. UNSITUPA, K. HERMAN et al. 2001. Substituting dietary saturated for monounsaturated fat impairs insulin sensitivity in healthy men and women: the Kanwu study. Diabetologia **44:** 312–319.
20. RIVELLESE, A.A. 2000. Effects of ω-3 fatty acids on carbohydrate and lipid metabolism in healthy people: the Kanwu study. Diabetologia **43:** A3.
21. HARRIS, W.S. 1997. ω-3 fatty acids and serum lipoproteins: human studies. Am. J. Clin. Nutr. **65**(suppl.): 1645s–1654s.
22. STAMLER, J., A.W. CAGGIULA & G.A. GRANDITS. 1997. Relation of body mass and alcohol, nutrient, fiber, and caffeine intakes to blood pressure in the special intervention and usual care groups in the Multiple Risk Factor Intervention Trial. Am. J. Clin. Nutr. **65**(suppl.): 338s–365s.
23. TREVISAN, M., V. KROGH, J. FREUDENHEIM et al. 1990. Consumption of olive oil, butter, and vegetable oils and coronary heart disease risk factors. JAMA **263:** 688–692.
24. MORRIS, M.C., F. SACKS & B. ROSNER. 1993. Does fish oil lower blood pressure? A meta-analysis of controlled trials. Circulation **88:** 523–533.

Treatment of Hypertriglyceridemia with Fenofibrate, Fatty Acid Composition of Plasma and LDL, and Their Relations to Parameters of Lipoperoxidation of LDL

M. ZEMAN, A. ŽÁK, M. VECKA, E. TVRZICKÁ, S. ROMANIV, AND M. KONÁRKOVÁ

Fourth Department of Medicine, First Faculty of Medicine, Charles University, Prague, Czech Republic

> ABSTRACT: The purpose of this study was to determine oxidation and oxidability of VLDL and LDL in connection with changes in their composition and content of FA in LDL after treatment with fenofibrate in patients with HTG.
>
> KEYWORDS: hypertriglyceridemia; fenofibrate; lipoperoxidation; fatty acids; LDL; VLDL

Based on contemporary knowledge, hypertriglyceridemia (HTG) is considered to be an independent risk factor (RF) of CHD.[1] There are several mechanisms by which HTG contributes to the risk of athero- and thrombogenesis (postprandial HTG, phenotype B of LDL particle size, low HDL-C, insulin resistance, and procoagulation state). Moreover, in experimental studies, HTG caused endothelial dysfunction connected with increased generation of O^{2-} and decrease of NO.[2] Oxidative modification of LDL is considered the key process of the early stage of atherosclerosis.[3] Two of the sources of higher oxidative stress influencing oxidative modification of lipoproteins (LP) are hypercholesterolemia[4] and HTG.[5–7] We have previously described increased lipoperoxidation of LDL and also VLDL in patients with dyslipidemia[8] and increased resistance to oxidation of LDL in patients with hypercholesterolemia after treatment with pravastatin.[9] However, other authors presented different results.[10]

The purpose of this study was to determine oxidation and oxidability of VLDL and LDL in connection with changes in their composition and content of fatty acids in LDL after treatment with fenofibrate in patients with HTG.

Address for correspondence: M. Zeman, Fourth Department of Medicine, First Faculty of Medicine, Charles University, Prague, Czech Republic. Voice: +420 2 2496 2092; fax: +420 2 2492 3524.

zemanm@vfn.cz

MATERIALS AND METHODS

The studied group consisted of 45 persons (28/17 M/F) with primary HTG [LP phenotype IV/II B (26/19)] consecutively recruited from the outpatient lipid clinic. They were advised to follow the AHA step-1 diet[11] at least 6 weeks before the study and to maintain it throughout the trial. Subjects with atherosclerotic vascular disease, diabetes mellitus (DM), obesity [body mass index (BMI) > 35 kg/m^2], severe renal or liver disorders, and endocrinopathies were excluded from the study. The average age of the studied group was 53.1 years (range: 33–68 years) and BMI was 27.9 ± 1.1 (mean ± SEM, kg/m^2). The patients were given micronized fenofibrate (Lipanthyl 200 M®, Fournier, France) at a dose of 200 mg per day for 6 weeks.

Before and after treatment, the following parameters were determined: total cholesterol (TC), triglycerides (TG), phospholipids (PL), and apolipoproteins (apo) in plasma, VLDL, and LDL; the composition of fatty acids (FA) in lipid classes [phosphatidylcholine (PC), TG, cholesteryl esters (CE)] in both plasma and LDL; and markers of lipoperoxidation in VLDL and LDL, which were isolated by two-step sequential ultracentrifugation.[12] Blood samples were collected after overnight fast. Concentrations of TC, TG, and PL were determined by enzymatic-colorimetric methods, HDL$_2$-C, and HDL$_3$-C according to Kostner *et al.*[13] Levels of apo B and A-I were determined by Laurell rocket immunoelectrophoresis using specific standards and antibodies (Behring Werke AG Marburg, Boehringer Mannheim, Germany). Content of FA in LDL CE, TG, and PC was determined by capillary GC as described previously.[14] Parameters of LDL and VLDL oxidation were measured by the method of Cu^{2+}-mediated kinetics of conjugated dienes.[15] Statistical evaluation was performed by means of the BMDP Statistical Software.

RESULTS AND DISCUSSION

The concentrations of plasma lipids, apolipoproteins, and selected FA in CE, PC, and TG in both plasma and LDL; the composition of VLDL and LDL; and the parameters of their lipoperoxidation before and after treatment are summarized in TABLE 1. As expected, we observed a significant decrease of plasma TC concentration (by 13%), TG (by 34%), and apo B (by 14%) and an increase in HDL-C (by 19%), HDL$_2$-C (by 23%), and HDL$_3$-C (by 11%) (all $p < 0.05$). After treatment, LDL-TG concentration was decreased (by 25%, $p < 0.05$). The separated ultracentrifugation fraction ($d < 1.063$ g/mL) contained also a subfraction of IDL ($1.006 < d < 1.019$ g/mL). Thus, the LDL-TG decrease after treatment can reflect the decreases in IDL that are atherogenous per se and in some studies predicted progression of the coronary atherosclerosis.[16–18] There are only few data concerning the influence of fibrate treatment on VLDL composition. In this work, the treatment resulted in decreased concentration of VLDL-C (by 28%), VLDL-TG (by 26%), and VLDL-PL (by 27%) (all $p < 0.05$). Changes in VLDL composition in our study implicated a smaller proportion of buoyant VLDL particles that are known to be precursors of small dense LDL.[19] Hypertriglyceridemic VLDL are atherogenous per se.[20] After treatment of HTG, we found significant prolongation of the lag phase in both LDL (+33%, $p < 0.05$) and VLDL (+16%, $p < 0.01$). Determination of lag phase in used assays reflects resistance of VLDL and LDL to oxidation.[15] Some authors,[21]

TABLE 1. Plasma lipids, lipoproteins, plasma and LDL fatty acid composition, and lipoperoxidation of VLDL and LDL

	Before treatment	After treatment[c]
TC (mmol/L)	7.12 ± 0.35[a]	6.19 ± 0.36*
TG (mmol/L)	5.74 ± 0.97[b]	3.78 ± 0.64*
HDL-C (mmol/L)	0.86 ± 0.06	1.02 ± 0.06*
HDL_3-C (mmol/L)	0.65 ± 0.05	0.72 ± 0.04*
HDL_2-C (mmol/L)	0.21 ± 0.08	0.26 ± 0.03*
Apo B (g/L)	1.16 ± 0.12	1.00 ± 0.06*
Apo A-I (g/L)	1.38 ± 0.07	1.53 ± 0.06*
LDL-C (mmol/L)	3.50 ± 0.40	3.38 ± 0.32
LDL-TG (mmol/L)	1.05 ± 0.18	0.68 ± 0.06*
LDL-PL (mmol/L)	1.17 ± 0.06	1.14 ± 0.09
LDL-apo B (g/L)	0.93 ± 0.11	0.88 ± 0.09
VLDL-C (mmol/L)	1.71 ± 0.29	1.23 ± 0.25*
VLDL-TG (mmol/L)	3.44 ± 0.68	2.54 ± 0.52
VLDL-PL (mmol/L)	1.00 ± 0.17	0.70 ± 0.13*
VLDL-apo B (mg/dL)	19.37 ± 4.91	20.05 ± 4.46
LDL-basal absorbance (A_{234nm})	0.331 ± 0.070	0.278 ± 0.019
LDL-lag phase (min)	78.83 ± 7.01	104.96 ± 9.98*
LDL-slope (δA_{234nm}/min)	0.019 ± 0.001	0.022 ± 0.005
VLDL-basal absorbance (A_{234nm})	0.580 ± 0.107	0.430 ± 0.074
VLDL-lag phase (min)	198.06 ± 16.27	228.62 ± 15.64**
VLDL-slope (δA_{234nm}/min)	0.0092 ± 0.0009	0.0068 ± 0.0015
Plasma 18:2 n-6 in PC	20.17 ± 0.80[a]	17.01 ± 1.55*
Plasma 18:2 n-6 in TG	14.17 ± 1.11	12.28 ± 0.64*
LDL 14:0 in CE	0.76 ± 0.07[b]	1.22 ± 0.06**
LDL 14:0 in TG	1.39 ± 0.22[d]	2.10 ± 0.35*
LDL 18:0 in TG	4.38 ± 0.26	5.06 ± 0.17*
LDL 18:2 n-6 in CE	51.91 ± 1.30	44.22 ± 2.27*
LDL 18:2 n-6 in TG	14.79 ± 1.23	12.22 ± 0.78*
LDL 16:1 n-7 in CE	27.72 ± 0.37	28.69 ± 0.33*
LDL 18:1 n-9 in PC	1.56 ± 0.08	2.27 ± 0.05*
LDL ΣMFA in PC	14.75 ± 0.49	15.64 ± 0.51*

[a]Mean ± SEM.
[b]Mann-Whitney test: *$p < 0.05$; **$p < 0.01$.
[c]Two hundred mg micronized fenofibrate per day (6 weeks).
[d]Only relevant and significant changes of FA are shown.

but not all,[22] proved negative correlations between the extent of atherosclerotic lesions and lag phase. Treatment of HTG with fenofibrate led also to decreased basal absorbances (BA) in both LDL (−16%) and VLDL (−25%), but the differences did not reach statistical significance. The decrease in LP oxidability after treatment with statins and fibrates can be explained by the decrease of the substrate concentration, changes in FA profile, or direct antioxidative effects of some hypolipidemics.[23,24] The significant prolongation of the LDL lag phase after treatment of HLP with fibrates was observed after application of clofibrate,[6] bezafibrate,[7] and gemfibrozil,[7] but it was not proved in other studies.[25,26] In the literature, there are only rare studies aimed at VLDL lipoperoxidation with inconsistent findings. In patients with well-compensated type 2 DM, Rabini et al. found decreased resistance of isolated VLDL and HDL to oxidation in comparison with controls. These patients were characterized by lower proportion of saturated FA, with compensatory higher proportion of unsaturated FA in VLDL and HDL.[27] Prolongation of the lag phase of a mixture of VLDL and LDL with a decrease in the rate of oxidation was found in rats also after fenofibrate treatment.[28] On the other hand, De Man et al.[10] observed decreased resistance of VLDL and LDL in HTG patients after treatment with bezafibrate. Moreover, LP of HTG patients were more resistant to oxidation in comparison with controls.

Treatment with fenofibrate led to significant decrease of linoleic acid (LA, 18:2 n-6) in PC and TG of plasma and in CE and TG of LDL; increase of palmitoleic acid (POA, 16:1 n-7) in LDL CE and of oleic acid (OA, 18:1 n-9) in LDL PC; significantly higher concentration of total monoenic FA (MFA) in LDL PC and CE; and significant increase of the proportion of myristic acid (14:0) in CE and of myristic and stearic acid (18:0) in LDL TG. It is generally accepted that MFA and saturated FA increase LDL resistance to oxidation, while polyunsaturated FA (PUFA) revealed the opposite effects.[29–31] The decrease in PUFA, especially in LA, may be connected with their increased oxidative degradation in peroxisomes that is induced by fibrate treatment.[32]

In conclusion, the treatment of HTG patients with micronized fenofibrate led to significant changes in LDL and VLDL composition that were connected with a decrease in oxidation and oxidability of both LDL and VLDL and were accompanied by changes in the composition of FA in plasma and LDL.

SUMMARY

The aim of the study was to evaluate the influence of treatment of primary hypertriglyceridemia (HTG) with fenofibrate on fatty acid (FA) composition of the main lipid classes in plasma and LDL [phosphatidylcholine (PC), TG, cholesteryl esters (CE)] and on lipoperoxidation in VLDL and LDL, isolated by preparative ultracentrifugation. Six-week fenofibrate treatment of 45 patients with HTG led to significant decrease in LDL-TG concentration (−25%, $p < 0.05$) and also in VLDL-C (−28%), VLDL-TG (−26%), and VLDL-PL (−27%) (all $p < 0.05$). Concomitantly, there was a significant decrease of linoleic acid (18:2 n-6) content in PC and TG plasma and in CE and TG LDL, and an increase of the proportion of palmitoleic acid (16:1 n-7) content in LDL CE and of oleic acid (18:1 n-9) content in LDL PC. These changes were connected with a significantly higher content of total monoenic FA in LDL PC and CE and a significant increase of the proportion of myristic acid (14:0) in CE and myristic and stearic acid (18:0) in LDL TG. At the same time, we have observed significant prolongation of the lag phase of both LDL (+33%, $p < 0.05$) and VLDL (+16%, $p < 0.01$) particles.

In conclusion, the treatment of HTG patients with micronized fenofibrate led to significant changes in LDL and VLDL composition that were connected with a decrease in oxidation and oxidability of both LDL and VLDL and were accompanied by changes in the composition of FA in plasma and LDL.

ACKNOWLEDGMENTS

This study was supported by Research Project No. J13/98 1111 0000 2-1 from the Ministry of Education, Czech Republic.

REFERENCES

1. HOKANSON, J.E. & M.A. AUSTIN. 1996. Plasma triglyceride level is a risk factor for cardiovascular disease independent of high-density lipoprotein cholesterol level: a meta-analysis of population-based prospective studies. J. Cardiovasc. Risk **3:** 213–219.
2. KUSTERER, K., T. POHL, H-P. FORTMEYER et al. 1999. Chronic selective hypertriglyceridemia impairs endothelium-dependent vasodilatation in rats. Cardiovasc. Res. **42:** 783–793.
3. WITZTUM, J.L. 1993. Role of oxidised low-density lipoprotein in atherogenesis. Br. Heart J. **69**(suppl.): S12–S18.
4. OHARA, Y., T.E. PETERSON & D.G. HARRISON. 1993. Hypercholesterolaemia increases endothelial superoxide anion production. J. Clin. Invest. **91:** 2546–2551.
5. CHAPMAN, M.J. & E. BRUCKERT. 1996. The atherogenic role of triglycerides and small dense low density lipoproteins: impact of ciprofibrate therapy. Atherosclerosis **124**(suppl.): S21–S28.
6. DE GRAAF, J., J.C.M. HENDRIKS, P.N.M. DEMACKER et al. 1993. Identification of multiple dense LDL subfractions with enhanced susceptibility to in vitro oxidation among hypertriglyceridemic subjects: normalization after clofibrate treatment. Arterioscler. Thromb. **13:** 712–719.
7. YOSHIDA, H., T. ISHIKAWA, M. AYAORI et al. 1998. Beneficial effect of gemfibrozil on the chemical composition and oxidative susceptibility of low-density lipoprotein: a randomized, double-blind, placebo-controlled study. Atherosclerosis **139:** 179–187.
8. ŽÁK, A., M. ZEMAN, E. TVRZICKÁ et al. 2000. Fatty acid composition and parameters of lipoperoxidation of VLDL and LDL in dyslipidemia [in Czech]. Cas. Lek. Cesk. **139:** 18–23.
9. ZEMAN, M., A. ŽÁK, E. TVRZICKÁ et al. 1999. Influence of hypolipidemic therapy on the composition of VLDL and LDL, fatty acids, and parameters of lipoperoxidation in the patients with hypercholesterolemia [in Czech]. Cas. Lek. Cesk. **138:** 628–631.
10. DE MAN, F.H., I.J. JONKERS, E. SCHWEDHELM et al. 2000. Normal oxidative stress and enhanced lipoprotein resistance to in vitro oxidation in hypertriglyceridemia: effects of bezafibrate therapy. Arterioscler. Thromb. Vasc. Biol. **20:** 2434–2440.
11. THE EXPERT PANEL. 1988. Report of the National Cholesterol Education Program Expert Panel on detection, evaluation, and treatment of high cholesterol in adults. Arch. Intern. Med. **148:** 36–69.
12. SCHUMACKER, V.N. & D.I. PUPPIONE. 1986. Sequential flotation ultracentrifugation. In Methods in Enzymology. Vol. 128. Academic Press. Orlando.
13. KOSTNER, G.M., E. MOLINARI & P. PICHLER. 1985. Evaluation of a new HDL2/HDL3 quantitation method based on precipitation with polyethyleneglycol. Clin. Chim. Acta **148:** 139–147.

14. TVRZICKÁ, E., E. CVRCKOVÁ, B. MÁCA & M. JIRÁSKOVÁ. 1994. Changes in the liver, kidney, and heart fatty acid composition following ibuprofen administration in mice. J. Chromatogr. **B656:** 51–57.
15. PUHL, H., G. WAEG & H. ESTERBAUER. 1994. Methods to determine oxidation of low density lipoproteins. *In* Oxygen Radicals in Biological Systems: Methods in Enzymology. Vol. 233, pp. 425–441. Academic Press. San Diego.
16. KRAUSS, R.M., F.T. LINDGREN, P.T. WILLIAMS *et al.* 1987. Intermediate-density lipoproteins and progression of coronary heart disease. Lancet **II:** 62–66.
17. WATTS, G.F., S. MANDALIA, J.N.H. BRUNT *et al.* 1993. Independent association between plasma lipoprotein subfraction levels and the course of coronary artery disease in the St. Thomas' Atherosclerosis Regression Study (STARS). Metabolism **42:** 1461–1467.
18. ILLINGWORTH, D.R. 2000. Management of hypercholesterolemia. Med. Clin. North Am. **84:** 23–42.
19. PACKARD, C.J. & J. SHEPHERD. 1997. Lipoprotein heterogeneity and apolipoprotein B metabolism. Arterioscler. Thromb. Vasc. Biol. **17:** 3542–3556.
20. WHITMAN, S.C., C.G. SAWYER, D.B. MILLER *et al.* 1998. Oxidized type IV hypertriglyceridemic VLDL-remnants cause greater macrophage cholesterol ester accumulation than oxidized LDL. J. Lipid Res. **39:** 1008–1020.
21. SALONEN, R., K. NYYSSONEN, E. PORKKALA *et al.* 1995. Kuopio Atherosclerosis Prevention Study (KAPS): a population-based primary preventive trial of the effect of LDL lowering on atherosclerotic progression in carotid and femoral arteries. Circulation **92:** 1758–1764.
22. FRUEBIS, J., D.A. BIRD, J. PATTISON & W. PALINSKI. 1997. Extent of antioxidant protection of plasma LDL is not a predictor of the antiatherogenic effect of antioxidants. J. Lipid Res. **38:** 2455–2464.
23. AVIRAM, M., G. DANKNER, U. COGAN *et al.* 1992. Lovastatin inhibits low-density lipoprotein oxidation and alters its fluidity and uptake by macrophages: *in vitro* and *in vivo* studies. Metabolism **41:** 229–235.
24. HOFFMAN, R., G.J. BROOK & M. AVIRAM. 1995. Hypolipidemic drugs reduce lipoprotein susceptibility to undergo lipid peroxidation: *in vitro* and *ex vivo* studies. Atherosclerosis **93:** 105–113.
25. BREDIE, S.J.H., T.W.A. DE BRUIN, P.N.M. DEMACKER *et al.* 1995. Comparison of gemfibrozil versus simvastatin in familial combined hyperlipidemia and effects on apolipoprotein-B containing lipoproteins, low-density lipoprotein subfraction profile, and low-density lipoprotein oxidizability. Am. J. Cardiol. **75:** 348–353.
26. STALENHOEF, A.F.H., *et al.* 2000. The effect of concentrated n-3 fatty acids versus gemfibrozil on plasma lipoproteins, low density lipoprotein heterogeneity, and oxidability in patients with hypertriglyceridemia. Atherosclerosis **153:** 129–138.
27. RABINI, R.A., M. TESEI, T. GALEAZZI *et al.* 1999. Increased susceptibility to peroxidation of VLDL from non-insulin-dependent diabetic patients: a possible correlation with fatty acid composition. Mol. Cell. Biochem. **199:** 63–67.
28. CHAPUT, E., D. MAUBROU-SANCHEZ, F.D. BELLAMY *et al.* 1999. Fenofibrate protects lipoproteins from lipid peroxidation: synergistic interaction with α-tocopherol. Lipids **34:** 497–502.
29. LOUHERANTA, A.M., E.K. PORKKALA-SARATAHO, M.K. NYYSSONEN *et al.* 1996. Linoleic acid intake and susceptibility of very low density and low density lipoproteins to oxidation in men. Am. J. Clin. Nutr. **63:** 698–703.
30. REAVEN, P.D., B.J. GRASSE & D.L. TRIBBLE. 1994. Effects of linoleate-enriched and oleate-enriched diets in combination with α-tocopherol on the susceptibility of LDL and LDL subfractions to oxidative modification in humans. Arterioscler. Thromb. **14:** 557–566.
31. CROFT, K.D., *et al.* 1995. Oxidation of low-density lipoproteins: effect of antioxidant content, fatty acid composition, and intrinsic phospholipase activity on susceptibility to metal ion–induced oxidation. Biochim. Biophys. Acta **1254:** 250–256.
32. SCHOONJANS, K., *et al.* 1997. Peroxisome proliferator activated receptors, orphans with ligands, and functions. Curr. Opin. Lipidol. **8:** 159–166.

Effect of Iron Depletion on Cardiovascular Risk Factors

Studies in Carbohydrate-Intolerant Patients

FRANCESCO S. FACCHINI[a,b,c] AND KAMI L. SAYLOR[b]

[a]*Department of Medicine, Division of Nephrology, San Francisco General Hospital and University of California, San Francisco, California, USA*

[b]*Department of Medicine, Kaiser Foundation Hospitals, and Permanente Medical Group, Incorporated, Oakland, California, USA*

[c]*Zentrum für Schlafmedizin und Stoffwechselstörungen, Dortmund, Germany*

> ABSTRACT: Controversy surrounds the role of iron (Fe) in atherosclerosis (ASCVD), mainly due to the inaccuracy of assessing body Fe stores with serum ferritin and transferrin saturation. Quantitative phlebotomy was used to test whether or not (a) Fe stores are increased in individuals at high risk for ASCVD and (b) Fe depletion to near-deficiency (NID) levels is associated with reduction of risk factors for ASCVD. Thirty-one carbohydrate-intolerant subjects completed the study. Fe stores were within normal limits (1.5 ± 0.1 g). At NID, a significant increase of HDL-cholesterol ($p < 0.001$) and reductions of blood pressure ($p < 0.001$), total and LDL-cholesterol ($p < 0.001$), triglyceride ($p < 0.001$), fibrinogen ($p < 0.001$) and glucose and insulin responses to oral glucose loading ($p < 0.001$) were noted, while homocysteine plasma concentration remained unchanged. These effects were largely reversed by a 6-month period of Fe repletion with reinstitution of Fe sufficiency. Thus, although individuals at high risk for ASCVD are not Fe-overloaded, they seem to benefit, metabolically and hemodynamically, from lowering of body Fe to levels commonly seen in premenopausal females.
>
> KEYWORDS: iron; atherosclerosis; hypertension; glucose intolerance; insulin; insulin resistance; diabetes

INTRODUCTION

In 1981, Sullivan proposed that the gender difference in the incidence of atherosclerotic cardiovascular disease (ASCVD) might be related to Fe accumulation in men as opposed to premenopausal females.[1] Since then, many epidemiological studies tested such a hypothesis with conflicting results.[2,3] Methodological bias is likely at the origin of such controversy as dietary Fe intake, iron saturation, and serum ferritin concentration (SFC) are all indirect and unreliable measurements of

---

Address for correspondence: Dr. Francesco S. Facchini, Box 1341, UCSF, San Francisco, CA 94080-1341. Fax: 510-420-1988.
 FSTE2000@yahoo.com

Fe stores. Dietary Fe content bears little relation with absorption since Fe bioavailability is either markedly reduced or enhanced by so many factors contained in food.[4] Transferrin saturation (Fe saturation) has a day-to-day variability of about 30%[5] and SFC becomes insensitive in the range of 70–250 µg/L, where only 21% of its variance relates to the size of body Fe stores.[6] Therefore, relying upon such methodologies cannot resolve whether or not Fe stores are increased in individuals at risk for ASCVD. The most compelling evidence against the hypothesis that body Fe progressively increases with aging is the fact that intestinal absorption of Fe is tightly regulated and, as body Fe increases, absorption rapidly decreases.[6,7] Such negative-feedback inhibition seems very efficient, preventing further accumulation of Fe at SFC of ~60 µg/L (corresponding to a total body Fe pool of ~0.5–1.5 g).[8] Accordingly, autopsy studies in normal subjects confirmed that, after the fourth decade of life, there is no further increase in liver Fe beyond these levels.[8,9] These findings thus argue against the possibility that Fe accumulation occurs in aging individuals and raise the possibility that the protection seen in fertile women from ASCVD may be related to a state of menstruation-induced Fe deficiency or near-iron deficiency[10] (NID) rather than overload. Support for this hypothesis comes from studies showing failure of estrogen replacement therapy in protecting postmenopausal women from ASCVD,[11] as well as from other studies where risk of ASCVD before the age of natural menopause was enhanced by isolated hysterectomy rather than oophorectomy.[12]

As carbohydrate (CHO) intolerance, dyslipidemia, and hypertension tend to cluster in the same individuals,[13] synergistically increasing risk of ASCVD,[14–16] a key question would thus be whether or not Fe depletion, to a degree similar to that seen in fertile women, might provide protection against ASCVD by ameliorating such cluster of risk factors. Two small-scale trials partially attempted to address this question in diabetic individuals by lowering body Fe with deferoxamine.[17,18] SFC was lowered from a "high" to a "high-normal" level, Fe loss was not quantitated, and opposite results were found, leaving the foregoing question unsettled.

The current work originates from the following hypothesis: if there is a beneficial effect secondary to low Fe status, it should best be noted at NID, e.g., that state where Fe depletion is maximal and yet compatible with lack of anemia. Furthermore, it would be important to show an eventual beneficial effect in individuals who are at high risk for ASCVD rather than in an unselected sample of patients. Upon these premises, a study was initiated to address the following questions:

(1) Are body Fe stores truly increased in patients at high risk for ASCVD?
(2) Is SFC a valid surrogate measure of body Fe stores in such patients?
(3) Is lowering of Fe stores to NID associated with better ASCVD risk profile?

METHODS

The study was approved by the local ethical committees, experimental procedures were in accordance with institutional guidelines, and informed consent obtained.

Patients with either recent type-2 diabetes or impaired glucose tolerance, none of whom were insulin-treated, with no prior history of cardiovascular, retinal, or renal complications were invited to participate. None of the patients had any acute, chronic inflammatory or neoplastic disease, and fasting glycemia was <7.8 mM on three repeated occasions. Genetic hemochromatosis (GH) was excluded by a serum iron saturation < 50% and absent C282Y/H63D.[19] After overnight fasting, a 3-h OGTT with hourly measurement of plasma glucose and insulin[20] was performed. Fasting total cholesterol, triglyceride, HDL-cholesterol, HbA1c, fructosamine, fibrinogen, Fe, transferrin, SFC, and homocysteine were all measured by routine automated techniques. LDL-cholesterol was estimated with Friedewald's formula.[21] Twenty-four-hour, ambulatory blood pressure was measured hourly with a portable monitoring device (model TM 2421, A&D Engineering, Milpitas, CA). For analysis, all readings from each monitoring period were averaged for heart rate and systolic and diastolic blood pressure. Quantitative phlebotomy was used to measure body Fe stores.[22] Half-liter phlebotomies were performed monthly or bimonthly after topical anesthesia with 0.5–1.0 mL of 1% lidocaine. Fe indexes and a CBC were measured before every phlebotomy to exclude anemia. Hematocrit (Hct) was not allowed to decrease more than 5% (from baseline values) by delaying the next phlebotomy as necessary. Phlebotomies were then discontinued at NID defined as serum ferritin ≤ 30 μg/L, iron saturation ≤ 15%, and mean corpuscular cell volume (MCV) ≤ 82 fl. Body iron stores (mg) were estimated with the following formula:

$$(\text{baseline Hct} + \text{NID Hct}/2) \times \text{blood volume removed} \times 1 \text{ mg/mL}.$$

Baseline measurements were repeated at 30–45 days after discontinuation of phlebotomies. A preliminary study[23] indicated that, within such time span, Hct entirely returns to baseline levels. The potential confounding effect of changes of blood viscosity on blood pressure regulation were therefore avoided.

Six months later, patients were studied again (Fe replenishment control phase). Throughout the study, an effort was made to avoid changes in medication and lifestyle habits.

Statistical Analyses

Results are expressed as means ± SEM. Frequency distributions were estimated for each variable. Mean values at NID were compared with baseline and 6-month post-NID results by paired (two-tailed) Student's t test or the Wilcoxon-matched pairs test for normally distributed and nonparametric correlated variables, respectively. Nonparametric variables were serum fructosamine, ferritin, triglyceride, and post-OGTT insulin concentrations. Associations among serum indexes of Fe status and size of Fe stores were estimated by calculating Pearson's correlation coefficients. Statistical calculation was performed with a commercial software (Statsoft-Inc., Tulsa, OK) for the MacIntosh (iMAC, Apple Computers, Cupertino, CA).

RESULTS

Demographic and clinical variables of the 31 study patients are shown in TABLE 1. No changes in body weight were observed. Mean weight was 90 ± 4 kg at baseline, 90 ± 4 kg at NID, and 88 ± 4 kg at 6 months after NID ($p = \text{NS}$).

Phlebotomy lowered both SFC and Fe saturation to target values (TABLE 2), while MCV fell from 88 ± 1 fl to 81 ± 1 fl ($p < 0.001$). A cumulative, average blood volume

TABLE 1. Demographic and clinical characteristics

N (M/F)	31 (21/10)
Age (years)	51 ± 4
Weight (kg)	90 ± 4
BMI (kg/m^2)	27 ± 2
Type-2 DM (%)	70 (21/31)
HTN (%)	78 (24/31)
Both (%)	52 (16/31)
Duration of type-2 DM (years)	4 ± 1
Duration of HTN (years)	9 ± 2
Smoker status (%)	12 (4/31)
Type-2 DM on OHA (%)	82 (18/22)
HTN on ≥2 drugs (%)	75 (18/24)

NOTE: Type-2 DM = type-2 diabetes mellitus; HTN = essential hypertension; OHA = oral hypoglycemic agent.

TABLE 2. Size of body Fe stores (mg of Fe removed) and serum parameters of Fe status at baseline, at near-iron deficiency (NID), and 6 months after phlebotomies were discontinued

Variable	Baseline	NID	6 months	p
Fe stores	1535 ± 122	112 ± 8*	360 ± 64*	<0.001
Ferritin (µg/L)	272 ± 32	14 ± 1	45 ± 8	<0.001
Iron (µmol/L)	17 ± 0.9	8 ± 0.5	15 ± 1	<0.001
TIBC (g/L)	3.2 ± 0.08	4.2 ± 0.08	3.7 ± 0.08	<0.001
Fe-sat	38 ± 2	10 ± 1	23 ± 2	<0.001
MCV (fl)	88 ± 1	81 ± 1	85 ± 1	<0.001
Hct (%)	42 ± 1	41 ± 1	42 ± 1	NS

NOTE: p values—baseline vs. NID. TIBC = transferrin; Fe-sat = iron saturation; MCV = mean corpuscular volume of red blood cells; Hct = hematocrit; * = estimated from SFC (references 24 and 25).

of 3646 ± 155 mL was removed, corresponding to 1.5 ± 0.1 g of Fe. Stored Fe poorly correlated with SFC ($r = 0.4$), explaining only about 20% of the SFC variance. This information is shown in TABLE 3. From TABLE 3, it can also be seen that neither Fe saturation nor MCV correlated with Fe removed. At NID, statistically significant hemodynamic and metabolic changes occurred (TABLES 4 and 5; FIGURE 1).

Ambulatory blood pressure recordings were an average of 20 ± 2 readings/session-patient vs. 21 ± 2 vs. 18 ± 1 at baseline, NID, and 6 months after NID, respectively. When average ambulatory blood pressure readings were considered (TABLE 4), both SBP and DBP were 10% lower at NID than at baseline ($p < 0.001$). Six months later, SBP returned to baseline values, while DBP was still lower ($p < 0.02$) than at baseline. TABLE 5 and FIGURE 1 show that chronic (HbA1c, fructosamine) and acute

TABLE 3. Correlation coefficients among Fe stores and serum parameters of Fe status

Variable	r	p
Ferritin	0.40	<0.05
TIBC	−0.36	<0.05
Iron	0.08	NS
Fe-saturation	0.12	NS
MCV	0.07	NS
Hct	0.14	NS

NOTE: r = Pearson's correlation coefficients; TIBC = transferrin; Fe-saturation = serum iron/TIBC; MCV = mean corpuscular volume; Hct = hematocrit.

TABLE 4. Hemodynamic changes at NID and 6 months later

Variable	Baseline	NID	6 months
SBP (mmHg)	150 ± 4	148 ± 4	156 ± 5
DBP (mmHg)	85 ± 2	83 ± 2	87 ± 3
AMBU SBP	160 ± 4	146 ± 5*	156 ± 4
AMBU DBP	91 ± 3	80 ± 3*	87 ± 3†
AMBU HR (BPM)	79 ± 2	74 ± 2†	81 ± 2

NOTE: Significances—*$p < 0.001$; †$p < 0.02$. SBP: systolic blood pressure; DBP: diastolic blood pressure; AMBU: mean ambulatory blood pressure and heart rate (HR) recordings were calculated from an average of 20 ± 2 vs. 21 ± 2 vs. 18 ± 1 readings/session-patient at baseline, at NID, and 6 months after NID, respectively.

TABLE 5. Metabolic changes at NID and 6 months later

Variable	Baseline	NID	6 months
FAGlu (mmol/L)	8.7 ± 0.5	7.5 ± 0.4*	8.8 ± 0.5
FAIns (pmol/L)	230 ± 27	165 ± 20†	181 ± 19
HbA1c (%)	8.0 ± 0.3	6.9 ± 0.2*	7.7 ± 0.4
Fructos. (µmol/L)	300 ± 12	268 ± 8*	292 ± 13
TC (mmol/L)	5.6 ± 0.2	4.9 ± 0.2*	5.3 ± 0.2
TG (mmol/L)	1.87 ± 0.1	1.54 ± 0.1*	1.86 ± 0.2
LDL-C (mmol/L)	3.46 ± 0.2	2.80 ± 0.2*	3.30 ± 0.2
HDL-C (mmol/L)	1.09 ± 0.05	1.29 ± 0.05*	1.20 ± 0.05†
TC/HDL	5.2 ± 0.2	4.0 ± 0.2*	4.7 ± 0.2‡
Homocysteine (µmol/L)	9 ± 1	9 ± 1	9 ± 1
Fibrinogen (g/L)	3.4 ± 0.1	2.9 ± 0.1*	3.1 ± 0.1

NOTE: Significances—*$p < 0.001$; †$p < 0.01$; ‡$p < 0.02$. FAGlu = fasting glucose; FAIns = fasting insulin; Fructos. = fructosamine; TC = total cholesterol; TG = triglycerides; LDL-C = low-density lipoprotein cholesterol; HDL-C = high-density lipoprotein cholesterol.

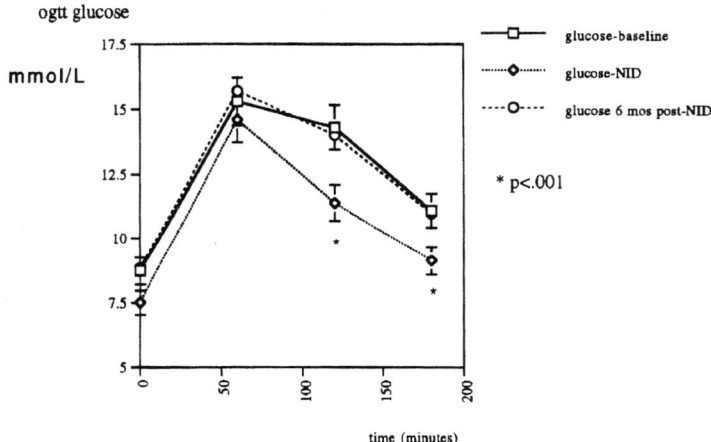

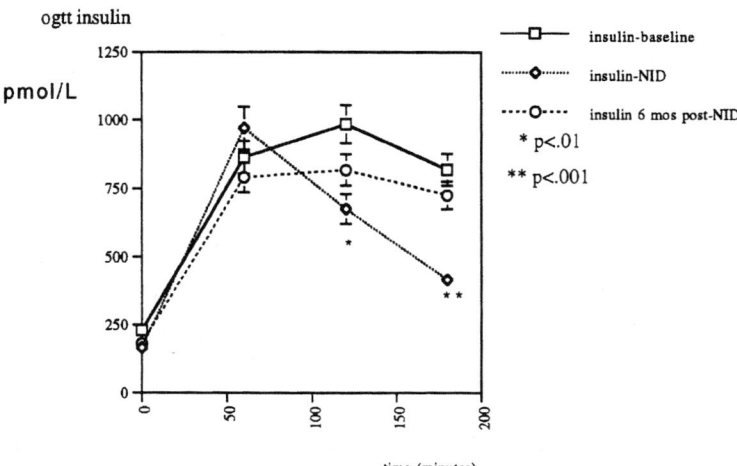

FIGURE 1. (Top) Plasma glucose concentrations during a glucose tolerance test with measurements at baseline and 60, 120, and 180 min. Paired values at baseline, NID, and 6 months post-NID are illustrated. **(Bottom)** Plasma insulin concentrations during a glucose tolerance test with measurements at baseline and 60, 120, and 180 min. Paired values at baseline, NID, and 6 months post-NID are illustrated.

(OGTT glucose and insulin) glucoregulatory indexes were 10–30% lower at NID than at baseline ($p < 0.001$ to $p < 0.01$). When comparing baseline with NID, the reduction of plasma insulin values was more pronounced at 120 min (from 984 ± 72 to 675 ± 50 pmol/L; $p < 0.01$) and 180 min (from 818 ± 70 to 415 ± 31 pmol/L; $p < 0.001$) than at peak levels (60 min) (862 ± 62 vs. 969 ± 72 pmol/L; $p = NS$). Reductions of glucose and insulin concentrations were paralleled by changes in total cholesterol (–12%; $p < 0.001$), triglyceride (–20%; $p < 0.001$), LDL-cholesterol (–20%; $p < 0.001$), TC/HDL (–24%; $p < 0.001$), and fibrinogen (–15%; $p < 0.001$), while plasma HDL-cholesterol was increased (+18%; $p < 0.001$). Homocysteine concentration was unchanged. Six months (199 ± 6 days) after interruption of phlebotomy, most ASCVD risk factors returned to baseline values (TABLE 5), with the exception of fasting insulin (–21%; $p < 0.02$), TC/HDL (–10%; $p < 0.02$), and HDL-C (+10%; $p < 0.01$).

DISCUSSION

To the aim of quantifying body Fe stores, quantitative phlebotomy remains the standard of reference.[22,24,25] In the current investigation, 31 CHO-intolerant patients underwent quantitative phlebotomy and their body Fe stores were normal. Former liver biopsy studies came to the same conclusion,[26] while Moirand et al. reported Fe stores of 2.5 g in individuals with clinical evidence of insulin resistance.[27] The much greater SFC (>500 µg/L vs. 272 µg/L) and prevalence of males (M:F = 12:1 vs. 2:1) probably accounted for the additional 1 g of Fe found in that study. Moreover, those patients were selected on the basis of increased SFC, while selection in the present study strictly relied on different criteria. Therefore, rather than discordant, the present results integrate those from Moirand et al. and indicate an important inter-relationship among Fe and CHO-intolerance, independent of hemochromatosis, and detectable even at lower values of the storage Fe frequency distribution curve. Thus, CHO-intolerant patients have normal or mildly elevated body Fe stores and use of SFC would have overestimated their iron burden of 40–70%.[24,25] Hallberg et al., by directly measuring body Fe in normal subjects, came to similar conclusions and suggested that factors other than Fe storage may become more important determinants of SFC above 70 µg/L.[6] In this context, it was demonstrated that insulin-resistant individuals often present elevation of inflammatory indexes, such as fibrinogen, WBC count, and C-reactive protein.[28] Phagocyte-mediated, Fe-catalyzed oxidant damage to the endothelium is presumably involved in atherogenesis, and endothelial ferritin synthesis is an important cytoprotective antioxidant stratagem.[29] Thus, there is the possibility that the hyperferritinemia noted in individuals with the Metabolic Syndrome X[30,31] reflects ongoing oxidant-induced endovascular inflammation rather than increased body Fe stores. Other abnormalities commonly found in individuals with the Metabolic Syndrome X (or insulin resistance) are hyperfibrinogenemia,[28,32] hyper-LDL cholesterolemia,[16] hypertriglyceridemia, hypertension, and low HDL-cholesterolemia, as well as relative hyperglycemia and marked hyperinsulinemia after CHO intake.[13] These individuals are at greater risk of developing type-2 diabetes and all together, or in various combinations, these abnormalities make them prone to ASCVD. In the current investigation, all of these abnormalities, which are known to cluster[13,16,28,30,31] in insulin-resistant individuals with the exception of plasma homocysteine concentration,[33] improved with depletion of Fe stores to NID. In particular, both plasma glucose and insulin responses to oral glucose loading were about 30% lower than at baseline. This finding indicates a state of reduced insulin resistance[34] and that one of NID's most important effects was to enhance insulin-stimulated glucose disposal. Considerable evidence, *in vitro* and *in vivo*, supports this notion. Potashnick *et al.* demonstrated upregulation of GLUT-1 in cultured myocytes incubated with iron chelators with a 2-fold increase of glucose uptake.[35] Glucose transport also doubled in rats[36] and veal calves[37] made Fe-deficient and anemic by Fe-poor diets, while venesection to NID increased ~50% glucose disposal in lean, normal humans.[38] Thus, regardless of the method used to induce Fe depletion (bleeding, Fe-chelation, or selective Fe dietary exclusion) the same effect was observed: a 50–200% increase of glucose metabolism. The fact that this effect occurred with chelators and with normal hematocrits, as in the present study, indicates that Fe, rather than anemia, is the glucose tolerance (GT) factor. However, Fe depletion might enhance glucose metabolism indirectly, for example, by stimulating norepinephrine turnover.[39] Regardless of the specific mechanism involved, the present study demonstrates that the insulin-sparing effect of NID also occurs in CHO-intolerant humans and likely influenced the significant improvement of all Syndrome X–related ASCVD risk factors. In fact, similar improvements are known to happen with any other insulin-sparing agent, for example, physical activity, calorie restriction, metformin, or thiazolidinediones. However, such changes were largely abolished after 6 months of Fe repletion, despite the fact that storage Fe remained below baseline levels. Thus, for individuals at high risk for ASCVD, NID seems a more favorable metabolic state than Fe sufficiency. This view is congruous with the cardiovascular protection of premenopausal, menstruating females who are often near-iron-deficient[10] as well. The fact that hysterectomy eliminated such protection,[12] while recent controlled trials of estrogen replacement had neutral or negative impact on ASCVD,[11] lends further credence to the notion that menstrual bleeding, not estrogen, is the protective factor. Obviously, phlebotomy removes many elements besides Fe. Could substrates other than Fe have modified glucose homeostasis? Such possibility appears unlikely as lowering of no other metabolite or ion is known to enhance GT. On the contrary, depletion of trace elements such as Zn,[40] Cu,[41] and Cr[42] should worsen, not enhance, GT, indicating the possibility that dietary Fe depletion may be more effective than phlebotomy.

In summary, Fe stores were normal in individuals at high risk for ASCVD. Secondary to the insulin-sparing effect of NID, risk factors for ASCVD were improved at NID and worsened again by iron sufficiency. It is concluded that depletion of Fe to levels similar to those of premenopausal females has beneficial effect on most cardiovascular risk factors of CHO-intolerant individuals.

REFERENCES

1. SULLIVAN, J. 1981. Iron and the sex difference in heart disease risk. Lancet **2**: 1293–1294.
2. DANESH, J. & P. APPLEBY. 1999. Coronary heart disease and iron status: meta-analysis of prospective studies. Circulation **99**: 852–854.
3. DEVALK, B. & J.J.M. MARX. 1999. Iron, atherosclerosis, and ischemic heart disease. Arch. Intern. Med. **159**: 1542–1548.
4. LYNCH, S.R. 1997. Interaction of iron with other nutrients. Nutr. Rev. **55**: 102–110.

5. BOREL, M.J., S.M. SMITH, J. DERR et al. 1991. Day-to-day variation in iron status parameters in healthy men and women. Am. J. Clin. Nutr. **54:** 729–735.
6. HALLBERG, L., L. HULTEN & E. GRAMATKOVSKI. 1997. Iron absorption from the whole diet in men: how effective is the regulation of iron absorption? Am. J. Clin. Nutr. **66:** 347–356.
7. HULTEN, L., E. GRAMATKOVSKI, A. GLEERUP et al. 1995. Iron absorption from the whole diet: relation to meal composition, iron requirements, and iron stores. Eur. J. Clin. Nutr. **49:** 794–808.
8. CHARLTON, R.W., D.M. HAWKINS, W.O. MAVOR et al. 1970. Hepatic storage concentrations in different population groups. Am. J. Clin. Nutr. **23:** 358–371.
9. STURGEON, P. & A. SHODEN. 1971. Total liver storage iron in normal populations of the USA. Am. J. Clin. Nutr. **24:** 469–474.
10. HALLBERG, L., C. BENGTSSON, L. LAPIDUS et al. 1993. Screening for iron deficiency: an analysis based on bone-marrow examination and ferritin determination in a population of women. Br. J. Haematol. **85:** 787–798.
11. MOSCA, L., P. COLLINS, D.M. HERRINGTON et al. 2001. Hormone replacement therapy and cardiovascular disease. Circulation **104:** 499–503.
12. CENTERWALL, B.S. 1981. Premenopausal hysterectomy and cardiovascular disease. Am. J. Obstet. Gynecol. **139:** 58–61.
13. REAVEN, G.M. 1988. Role of insulin resistance in human disease. Diabetes **37:** 1595–1607.
14. FULLER, J.H., M.J. SHIPLEY, G. ROSE et al. 1980. Coronary-heart-disease risk and impaired glucose tolerance: the Whitehall study. Lancet **28**(1[8183]): 1373–1376.
15. HAFFNER, S.M., S. LEHTO, T. RONNEMAA et al. 1998. Mortality from coronary heart disease in subjects with type 2 diabetes and in nondiabetic subjects with and without prior myocardial infarction. N. Engl. J. Med. **339:** 229–234.
16. FACCHINI, F.S., N.W. HUA, F. ABBASI et al. 2001. Insulin resistance as a predictor of age-related diseases. J. Clin. Endocrinol. Metab. **86:** 3574–3578.
17. CUTLER, P. 1989. Deferoxamine treatment in high-ferritin diabetes. Diabetes **38:** 1207–1210.
18. REDMON, J.B., K. PYZDROWSKY & R.P. ROBERTSON. 1993. No effect of deferoxamine therapy on glucose homeostasis and insulin secretion in individuals with NIDDM and elevated serum ferritin. Diabetes **42:** 544–549.
19. BEUTLER, E., T. GELBART, C. WEST et al. 1996. Mutation analysis in genetic hemochromatosis. Blood Cells Mol. Dis. **22:** 187–194.
20. HALES, C.H. & P.J. RANDLE. 1963. Immunoassay of insulin with insulin-antibody precipitate. Biochem. J. **88:** 137–146.
21. FRIEDEWALD, W.T., R.I. LEVY & D.S. FREDRICKSON. 1972. Estimation of plasma low-density lipoprotein cholesterol concentration without use of the preparative ultracentrifuge. Clin. Chem. **18:** 499–509.
22. HASKINS, D., A.R. STEVENS, S. FINCH et al. 1952. Iron stores in man as measured by phlebotomy. J. Clin. Invest. **31:** 543–547.
23. FACCHINI, F.S. 1998. Effect of phlebotomy on plasma glucose and insulin concentrations. Diabetes Care **21:** 2190.
24. FINCH, C.A., J.D. COOK, R.F. LABBE et al. 1977. Effect of blood donation on iron stores as evaluated by serum ferritin. Blood **50:** 441–447.
25. BASSETT, M.L., J.W. HALLIDAY, R.A. FERRIS et al. 1984. Diagnosis of hemochromatosis in young subjects: predictive accuracy of biochemical screening tests. Gastroenterology **87:** 628–633.
26. DINNEEN, S.F., J.D. SILVERBERG, K.P. BATTS et al. 1994. Liver iron stores in patients with non-insulin dependent diabetes mellitus. Mayo Clin. Proc. **69:** 13–15.
27. MOIRAND, R., A.M. MORTAJI, O. LOREAL et al. 1997. A new syndrome of iron overload with normal transferrin saturation. Lancet **349:** 95–97.
28. FESTA, A., R. D'AGOSTINO, G. HOWARD et al. 2000. Chronic subclinical inflammation as part of the insulin resistant syndrome: the Insulin Resistance Atherosclerosis Study (IRAS). Circulation **102:** 42–47.
29. BALLA, G., H.S. JACOB, J. BALLA et al. 1992. Ferritin: a cytoprotective antioxidant stratagem of endothelium. J. Biol. Chem. **267:** 18148–18153.

30. TUOMAINEN, T.P., K. NYYSSONEN, R. SALONEN et al. 1997. Body Fe stores are associated with insulin and glucose concentrations. Diabetes Care **20:** 426–428.
31. FERNANDEZ-REAL, J.M., W. RICART-ENGEL, E. ARROYO et al. 1998. Serum ferritin as a component of the insulin resistance syndrome. Diabetes Care **21:** 62–68.
32. AGEWALL, S. 1999. Insulin sensitivity and hemostatic factors in men at high and low cardiovascular risk. J. Intern. Med. **246:** 489–495.
33. ABBASI, F., F.S. FACCHINI, M.H. HUMPHREYS et al. 1999. Plasma homocysteine concentrations are not related to differences in insulin-mediated glucose disposal. Atherosclerosis **146:** 175–178.
34. MYKKÄNEN, L., D.J. ZACCARO, C.N. HALES et al. 1999. The relation of proinsulin and insulin to insulin sensitivity and acute insulin response in subjects with newly diagnosed type II diabetes: the IRA study. Diabetologia **42:** 1060–1066.
35. POTASHNICK, R., N. KOZLOVSKY, S. BEN-EZRA et al. 1995. Regulation of glucose transport and GLUT-1 expression by chelators in muscle cells in culture. Am. J. Physiol. **269:** E1052–E1058.
36. BOREL, M.J., J.L. BEARD & P.A. FARREL. 1993. Hepatic glucose production and insulin sensitivity and responsiveness in Fe deficient rats. Am. J. Physiol. **264:** E380–E390.
37. HOSTETTLER-ALLEN, R., L. TAPPY & J.W. BLUM. 1993. Enhanced insulin-dependent glucose utilization in iron-deficient veal calves. J. Nutr. **123:** 1656–1667.
38. HUA, N.W., R.A. STOOHS & F.S. FACCHINI. 2001. Low iron stores and enhanced insulin sensitivity in lacto-ovo vegetarians. Br. J. Nutr. **86:** 515–519.
39. BEARD, J. & B. TOBIN. 1987. Feed efficiency and norepinephrine turnover in iron deficiency. Proc. Soc. Exp. Biol. Med. **184:** 337–344.
40. FAURE, P., A. ROUSSEL, C. COUDRAY et al. 1992. Zinc and insulin sensitivity. Biol. Trace Elem. Res. **32:** 305–310.
41. HASSEL, C.A., J.A. MARCHELLO & K.Y. LEI. 1983. Impaired glucose tolerance in copper deficient rats. J. Nutr. **113:** 1081–1083.
42. ANDERSON, R.A., N. CHENG, N.A. BRYDEN et al. 1997. Elevated intakes of supplemental chromium improve glucose and insulin levels in individuals with type 2 diabetes. Diabetes **46:** 1786–1791.

Erythrocyte Membrane Ion Transport in Offspring of Hypertensive Parents

Effect of Acute Hyperinsulinemia and Relation to Insulin Action

GABRIELA SUCHÁNKOVÁ, ZUZANA VLASÁKOVÁ, JOSEF ZICHA, MARTINA VOKURKOVÁ, ZDENA DOBEŠOVÁ, AND TEREZIE PELIKÁNOVÁ

Institute for Clinical and Experimental Medicine, and Institute of Physiology, Czech Academy of Sciences, Prague, Czech Republic

ABSTRACT: Some patients with essential hypertension exhibit insulin resistance (IR) and several red blood cell (RBC) ion transport abnormalities. The aims of the study were to assess RBC ion transport acitivities under basal conditions, to test *in vivo* the effect of acute hyperinsulinemia, and to evaluate the relationship to IR in the offspring of hypertensive parents ($n = 12$; OHP) and healthy controls ($n = 14$; C). Activities of the Na^+-K^+ pump, Na^+-K^+ cotransport, Na^+-Li^+ countertransport (SLC), and Na^+, Rb^+, and Li^+ leaks (passive membrane permeability) were measured before and after a hyperinsulinemic (75 µU/mL) euglycemic clamp (HIC) and compared to those found under isoinsulinemic isovolumic conditions in OHP and C. An insulin action was calculated as glucose disposal and insulin sensitivity index (M/I) after HIC. OHP were characterized by lower M/I (0.12 ± 0.07 vs. 0.20 ± 0.09 mg/kg/min/µU/mL; $p < 0.05$) and elevated SLC and Li^+ and Rb^+ leaks ($p < 0.05$) compared with C. Although acute hyperinsulinemia did not modify significantly any ion transport parameter studied, negative correlation was observed between insulin action and membrane cation leaks. Glucose disposal correlated with an Li^+ leak in C ($r = -0.736$; $p < 0.01$) and all subjects ($r = -0.424$; $p < 0.05$) after HIC and in OHP with an Na^+ leak ($r = -0.727$; $p < 0.05$) before HIC. In conclusion, OHP displayed higher insulin resistance, enhanced activity of SLC, and augmented Li^+ and Rb^+ leaks. Acute hyperinsulinemia did not modify any ion transport parameter studied, although negative correlation was observed between insulin action and membrane leaks.

KEYWORDS: hypertension; insulin sensitivity; passive membrane permeability; sodium-lithium countertransport

---

Address for correspondence: Gabriela Suchánková, M.D., Diabetes Center, Institute for Clinical and Experimental Medicine, Vídenská 1958/9, 140 21 Prague, Czech Republic. Voice: +420-2-61362150; fax: +420-2-61362820.
 gasu@medicon.cz

INTRODUCTION

In the past decades, abnormalities of red blood cell (RBC) ion transport were reported in patients with essential hypertension and their normotensive offspring.[1] The increased activity of Na^+-Li^+ countertransport (SLC) is believed to be a genetic marker of essential hypertension[1,2] and an increased risk of cardiovascular morbidity.[3–5] Within a hypertensive population, elevated countertransport activity is associated with nonmodulating hypertension, alterations in body mass index, and changes in lipid and uric acid levels.[6] These metabolic abnormalities are a part of the syndrome associated with the resistance to insulin-stimulated glucose disposal.[7] On the other hand, insulin resistance has been linked to both microalbuminuria and SLC, which characterize the hypertensive population.[8,9] It is known that environmental factors can also affect SLC.[10,11] It has been suggested that SLC is a mode of operation of the cell membrane sodium-hydrogen exchanger,[12] and patients with essential hypertension have indeed higher activity of the Na^+-H^+ antiport.[13]

In the present study, we thus determined the effect of acute hyperinsulinemia on RBC ion transport and the relationship between insulin-mediated glucose disposal and RBC ion transport in offspring of hypertensive parents (OHP) and healthy controls (C).

METHODS

Subjects

The group of offspring of hypertensive parents (OHP) consisted of 12 young lean males with a positive family history of hypertension (one or both parents) and a negative family history of diabetes mellitus, morbid obesity, and ischemic heart disease. They were not taking any drugs. One or both parents had to suffer from hypertension requiring antihypertensive medication. The OHP were matched according to age, body mass index (BMI), and waist-to-hip ratio (WHR) to a group of 14 healthy controls without a family history of hypertension, diabetes mellitus, and ischemic heart disease, who were not taking any drugs that could affect glucose metabolism. Both groups consisted of men; women were excluded from insulin-sensitivity testing to eliminate potential variations in glucose disposal related to ovarian function. All subjects had a normal OGTT using the criteria recently defined by the Expert Committee on the Diagnosis and Classification of Diabetes Mellitus.[14] Physical health was assessed by routine clinical examination. Written informed consent was obtained from each participant in this study after the purpose, nature, and potential risks of the study had been explained. Clinical characteristics of the subjects are summarized in TABLE 1.

Procedures

Subjects were examined on two separate occasions: a hyperinsulinemic euglycemic clamp (HIC) test was carried out first, which was followed by saline infusion as a second test within the next month.

The subjects were admitted to the research laboratory after 12 h of fasting. The one-step HIC study, taking 5 h to complete, was conducted as previously

TABLE 1. Characteristics of the offspring of hypertensive parents (OHP; $n = 12$) and healthy controls (C; $n = 14$)

	OHP	C	p
Age (years)	27.5 ± 3.6	26.3 ± 0.6	ns
BMI (kg/m^2)	25.6 ± 3.4	24.5 ± 1.5	ns
WHR	0.89 ± 0.07	0.84 ± 0.06	ns
SBP (mmHg)	129.5 ± 6.03	115.5 ± 3.9	***
DBP (mmHg)	79.6 ± 4.9	67.9 ± 4.8	***
TG (mmol/L)	2.53 ± 2.27	1.26 ± 0.5	*
CHOL (mmol/L)	5.11 ± 0.9	3.87 ± 0.7	**
HDL-C (mmol/L)	1.12 ± 0.3	1.01 ± 0.1	ns
LDL-C (mmol/L)	2.68 ± 0.9	2.08 ± 0.56	ns
HbA$_1$c (%)	5.12 ± 0.2	5.08 ± 0.2	ns

NOTE: Data are means ± SD. $*p < 0.05$, $**p < 0.01$, $***p < 0.001$; ns = not significant. SBP and DBP, systolic and diastolic blood pressure; BMI, body mass index; WHR, waist-to-hip ratio; TG, serum triglycerides; CHOL, total serum cholesterol; HDL-C, high-density lipoprotein cholesterol; LDL-C, low-density lipoprotein cholesterol; HbA$_1$c, glycosylated hemoglobin.

described.[15] Briefly, a Teflon cannula (Venflon; Viggo, Helsingborg, Sweden) was inserted into an antecubital vein for the infusion of all test substances. A second cannula was inserted retrogradely into a wrist vein for blood sampling, and the hand was placed in a heated (65°C) box in order to achieve venous blood arterialization. A stepwise primed-continuous insulin infusion (1 mU/kg/min of Actrapid HM; NovoNordisk, Copenhagen, Denmark) was administered to acutely raise and maintain the plasma concentrations of insulin at ~75 µU/mL; plasma glucose was clamped at 5 mmol/L with a coefficient of variation up to 5% by continuous infusion of 15% glucose. Decreases in serum potassium concentrations during the insulin infusion were prevented by adding potassium chloride (30 mmol of 7.5% KCl/L of glucose). Blood samples for plasma immunoreactive insulin (IRI) determination were taken before (0 min) and at 100, 120, 180, 200, 280, and 300 min of the clamp study. Blood sampling for measurement of RBC ion transport was performed before and after HIC. During and at the end of the study, a urine sample was collected to measure glucosuria. After the end of insulin infusion, the patients consumed a meal, and glucose infusion continued for another 20 min. The individuals remained in our research department until their serum glucose had returned to a normal level. The time- and volume load–controlled studies with saline infusion (4 mmol/L) and tap water drinking were performed using an identical protocol of blood sampling.

Laboratory Measurements

Blood glucose was determined immediately using the glucose oxidase method (Beckman Glucose Analyzer; Beckman Instruments, Fullerton, CA). IRI was measured by an immunoradioassay using an IMMUNOTECH Insulin IRMA kit (IMMUNOTECH a.s., Prague, Czech Republic). C-peptide was determined by an

IMMUNOTECH C-peptide IRMA kit (IMMUNOTECH a.s.). Total serum cholesterol (CHOL), high-density lipoprotein cholesterol (HDL-C), and serum triglycerides (TG) were measured by an enzymatic method using CHOD-PAP tests (Hoffmann-LaRoche, Basel, Switzerland). Low-density lipoprotein cholesterol (LDL-C) was calculated using a modified version of the formula of Friedewald[16] as total cholesterol minus [0.16 × (TG + HDL-C)]. HbA$_1$c concentrations in blood were measured by ion exchange high-performance liquid chromatography (HLPC) using a Bio-Rad Hemoglobin A$_1$c Column Test (Bio-Rad Laboratories, München, Germany).

Measurement of Erythrocyte Ion Transport

Erythrocyte ion transport activities were measured according to the methods described earlier.[17–19] Briefly, triplicates of hematocrit and hemoglobin content, as well as Na$^+$, K$^+$, Li$^+$, and Rb$^+$ contents, were determined in fresh erythrocytes obtained from heparinized blood. After centrifugation (10 min, 4000g, 4°C), plasma, buffy coat, and the uppermost red blood cells were aspirated. The remaining erythrocytes were washed three times with 137 mmol/L choline chloride medium [in mmol/L: glucose, 5; morpholinopropanesulfonic acid (MOPS)–tris(hydroxymethyl)aminomethane (TRIS)–phosphate (P)[10]]. We studied ouabain-sensitive net Na$^+$ extrusion and Rb$^+$ (K$^+$) uptake mediated by the Na$^+$-K$^+$ pump, bumetanide-sensitive net Na$^+$ and Rb$^+$ (K$^+$) transport mediated by the Na$^+$-K$^+$-Cl$^-$ cotransport system, as well as cation leaks reflecting passive membrane permeability (defined as residual Na$^+$ and Rb$^+$ fluxes resistant to ouabain and bumetanide). Net Na$^+$ movements and unidirectional Rb$^+$ (K$^+$) fluxes were assessed at intracellular Na$^+$ and extracellular Rb$^+$ (K$^+$) concentrations that were close to *in vivo* values. Na$^+$ and Rb$^+$ (K$^+$) fluxes susceptible to the inhibition by 0.2 mmol/L ouabain (Sigma, St. Louis, MO) and 10 µmol/L bumetanide (Leo Pharmaceutical Products, Ballerup, Denmark) were measured in erythrocytes incubated in saline medium (in mmol/L: NaCl, 145; glucose, 5; phosphoric acid, 2.5; MOPS-TRIS, 10; titrated to pH 7.4 at 37°C with TRIS, 315 mOsm/L) containing 5 mmol/L RbCl. The incubation was started by the addition of 100 µL of 50% erythrocyte suspension to 1.75 mL of prewarmed incubation medium. Uptake of Li$^+$ for measurement of Na$^+$-Li$^+$ countertransport (SLC) was initiated by adding 200 µL of 50% erythrocyte suspension to 1.75 mL of incubation medium (in mmol/L: MgCl$_2$, 70; sucrose, 85; RbCl, 5; glucose, 5; MOPS-TRIS-P, 10; LiCl, 5) containing 0.2 mmol/L ouabain. Phloretin (dissolved in ethanol) was added to the incubation medium in a final concentration of 0.1 mmol/L.

Red blood cells were incubated for 60 min at 37°C. Cell sediments were washed three times with 2 mL of ice-cold choline chloride (137 mmol/L) and hemolyzed with 1.25 mL of 6% *n*-butanol containing 0.1% CsCl.

Cation concentrations were determined by an atomic absorption spectrophotometer. The RBC cation contents were calculated from hemoglobin and cation concentrations in the lysates on the basis of the mean cellular hemoglobin content as determined in the fresh blood. All measurements were done in duplicates.

Data Analyses

Insulin sensitivity was assessed as the mean glucose infusion rate (M, mg/kg/min) and as the insulin sensitivity index (M/I) in the 280- to 300-min time period of HIC. M/I was calculated by dividing M (the average of three measurements after HIC) by the average of two IRI concentrations during the same interval.[15]

We assumed that, during euglycemic hyperinsulinemia at the achieved level of plasma insulin, hepatic glucose production was completely suppressed.[20] The glucose infusion rate then reflects total insulin-stimulated glucose metabolism during steady-state hyperinsulinemia.

Statistical Analyses

Data in the text and FIGURES 1–3 are presented as mean ± SD; RBC ion transport data are presented as mean ± SEM. Student's two-tailed t test for unpaired data was used to compare data between OHP and C. When the data were not normally distributed, the Mann-Whitney rank-sum test for unpaired data was used. Analysis of variance with repeated measures and grouping factor was applied for comparison of RBC ion transport activities between OHP and C and also for determining the effect of acute hyperinsulinemia on RBC transport. Significant differences between basal and insulin clamp periods within each group were tested by one-way ANOVA for repeated measures. Standard equations were used to calculate correlation coefficients.

RESULTS

The clinical and biochemical characteristics of the two groups are presented in TABLE 1. Systolic and diastolic blood pressure were within the normal range, but significantly higher in OHP ($p < 0.001$). Plasma cholesterol and TG were significantly elevated in OHP compared to C ($p < 0.05$). There were no differences in HDL-C levels. LDL-C levels were slightly (but not significantly) higher in subjects with a family history of hypertension. The levels of HbA$_1$c were equal in both groups.

Hyperinsulinemic Euglycemic Clamp (HIC)

During the one-step 5-h HIC, similar steady-state plasma insulin levels were maintained in OHP and C (75.6 ± 3.1 vs. 73.15 ± 5.76 µU/mL). No significant differences in insulin-mediated glucose disposal (M) were observed between OHP and C (10.44 ± 2.52 vs. 12.69 ± 3.10 mg/kg/min) after HIC, but the insulin sensitivity index (M/I) was lower in OHP ($p < 0.05$) than in C (FIG. 1).

RBC Ion Transport

As shown in TABLE 2, OHP displayed enhanced activity of SLC and augmented Li$^+$ and Rb$^+$ leaks ($p < 0.05$) compared to C. No significant differences between OHP and C were found in other ion transport parameters studied. RBC Na$^+$ content (Na^+_i) did not significantly differ between the two groups under basal conditions nor after HIC or saline infusion. Hyperinsulinemia lasting 5 h did not cause significant changes in any investigated RBC ion transport in OHP or C. By contrast, saline infusion was accompanied by a significant ($p < 0.05$) decrease of Rb$^+$ leak in OHP and an increase of Na$^+$-K$^+$ pump activity in C (TABLE 2). Glucose disposal (M) was inversely related to Na$^+$ leak in OHP before HIC ($p < 0.05$) (FIG. 2) as well as to Li$^+$ leak in C ($p < 0.01$) (FIG. 3) and all subjects ($r = -0.425$, $n = 26$, $p < 0.05$) after HIC.

TABLE 2. Basal RBC ion transport activities (T0) and their changes after hyperinsulinemic euglycemic clamp (ΔHIC) or saline infusion (ΔSaline) in offspring of hypertensive parents (OHP; $n = 12$) and healthy controls (C; $n = 14$)

Activity (mmol/h/L cells)	T0		ΔHIC		ΔSaline	
	OHP	C	OHP	C	OHP	C
Na^+-K^+ pump	1.509 ± 0.073	1.584 ± 0.052	0.052 ± 0.075	0.124 ± 0.092	0.004 ± 0.075	0.249 ± 0.101**
Na^+-K^+ cotransport	-0.159 ± 0.031	-0.135 ± 0.036	0.027 ± 0.088	0.060 ± 0.089	0.027 ± 0.063	-0.004 ± 0.093
Na^+-Li^+ countertransport	0.080 ± 0.004	0.068 ± 0.003*	0.002 ± 0.004	0.002 ± 0.005	0.005 ± 0.004	-0.003 ± 0.003
Na^+ leak	1.684 ± 0.076	1.834 ± 0.076	-0.147 ± 0.187	-0.003 ± 0.085	-0.100 ± 0.120	-0.007 ± 0.197
Rb^+ leak	0.160 ± 0.014	0.120 ± 0.007*	-0.001 ± 0.008	-0.013 ± 0.015	-0.054 ± 0.019**	0.010 ± 0.008
Li^+ leak	0.106 ± 0.004	0.093 ± 0.003*	-0.004 ± 0.005	-0.004 ± 0.008	-0.010 ± 0.006	0.002 ± 0.005
Na^+_i	6.265 ± 0.171	6.687 ± 0.173	0.217 ± 0.271	0.011 ± 0.161	-0.023 ± 0.106	-0.023 ± 0.152

NOTE: Data are expressed as means ± SEM. *$p < 0.05$, OHP vs. C; **$p < 0.05$.

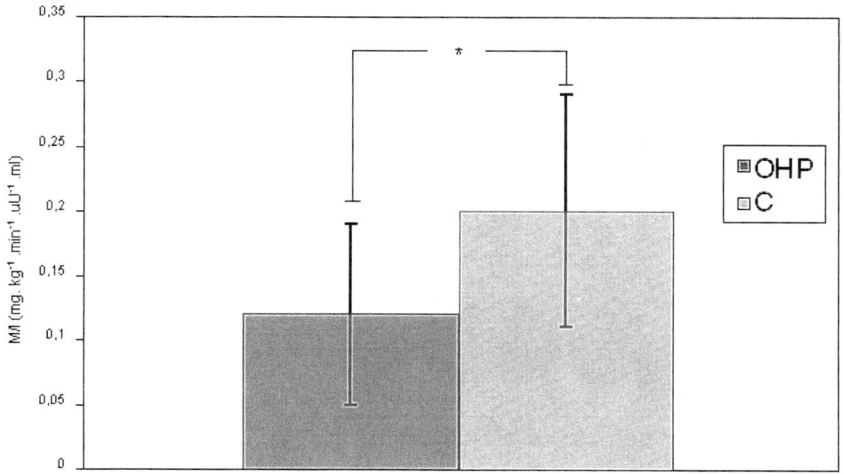

FIGURE 1. Insulin sensitivity index (M/I) in offspring of hypertensive parents (OHP; $n = 12$) and controls (C; $n = 14$) during 280–300 min of hyperinsulinemic euglycemic clamp. Data are means ± SD. Statistical significance: $*p < 0.05$.

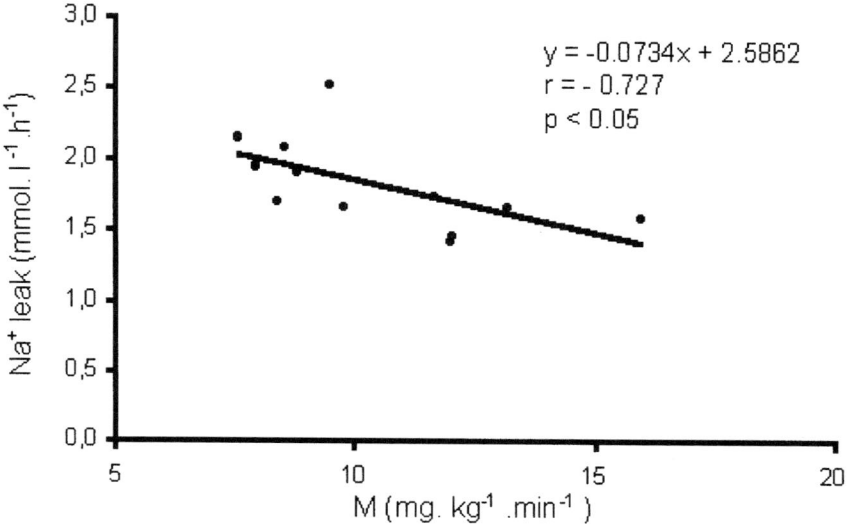

FIGURE 2. Relationship between glucose disposal (M) and Na$^+$ leak in offspring of hypertensive parents (OHP; $n = 12$) before hyperinsulinemic euglycemic clamp.

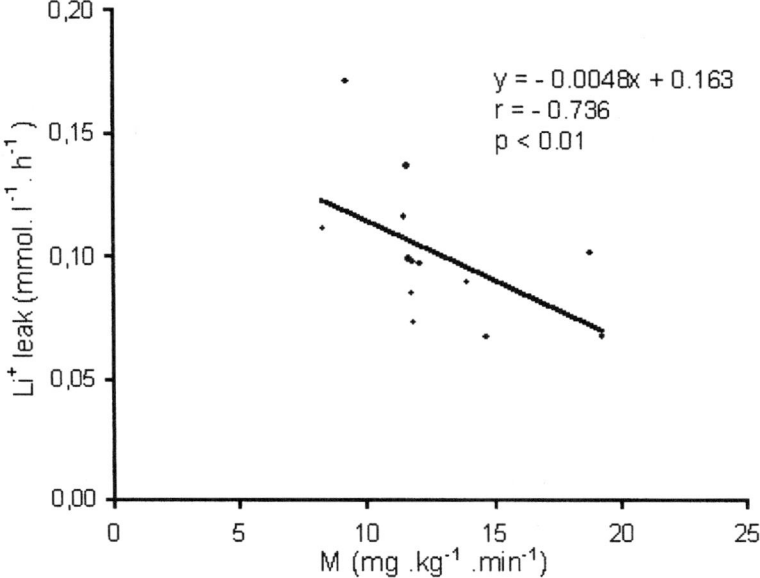

FIGURE 3. Relationship between glucose disposal (M) and Li^+ leak in healthy controls (C; $n = 14$) after hyperinsulinemic euglycemic clamp.

No other type of RBC ion transport was significantly associated with glucose disposal.

DISCUSSION

The results of our study demonstrate that the offspring of hypertensive parents, who were characterized by higher systolic and diastolic blood pressure, increased plasma levels of total cholesterol and triglycerides, and lower insulin sensitivity, also exhibited higher activity of sodium-lithium countertransport and augmented passive membrane permeability for Li^+ and Rb^+ (cation leaks) compared to healthy controls.

Since the original report by Canessa et al.,[1] several studies confirmed an increase in erythrocyte SLC activity in essential hypertension. Moreover, normotensive first-degree relatives of hypertensive patients have higher values of SLC than individuals without a family history of hypertension.[1,21] Earlier studies have indicated that the level of SLC activity is a subject of modification by both genetic and environmental factors. It has been estimated that as much as 80–90% of the individual variation in SLC activity is accounted for by heritability.[22] It is also known that body weight positively correlates with SLC[23] and that changes of SLC activity occurring over 2.5 years correlate with changes in serum TG and the degree of obesity, independently of the baseline levels.[24] Raised SLC has also been associated with decreased sensitivity to exogenous insulin in patients with essential hypertension,[25] as well as in hypertensive or microalbuminuric type 1 diabetic patients.[26,27]

Our present findings of increased SLC in OHP, in accord with previous studies, can be explained by the presence of factors associated with increased SLC, that is, higher systolic and diastolic blood pressure (although still within the normal range), elevated levels of serum cholesterol and TG, and greater insulin resistance found in OHP compared to healthy subjects. As far as we are concerned, our data provide the first evidence of an increased passive membrane permeability for Li^+ and Rb^+ in insulin-resistant subjects.

Our data also show reduced insulin sensitivity in the offspring of hypertensive parents when assessed as an insulin sensitivity index (M/I). This is consistent with earlier reports on hyperinsulinemia and insulin resistance in hypertensive or borderline hypertensive patients.[20,28]

We have found significant negative correlations between glucose disposal (M) and Na^+ leak in OHP under basal conditions, as well as between M and Li^+ leak in C and all subjects after HIC. Herlitz et al.[29] disclosed a significant negative relation between erythrocyte SLC and glucose disposal rate during a hyperinsulinemic clamp. This is in agreement with the observation that SLC was associated with peripheral insulin resistance in hypertensive men and women.[25] Although this was not confirmed by other authors,[30] glucose disposal in our study did not significantly correlate with SLC in both OHP and C. Unfortunately, the studies mentioned above did not investigate correlations between glucose disposal and other types of RBC ion transport because they were aimed at SLC only. The altered membrane lipid composition with resulting changes in membrane microviscosity might be the common factor underlying abnormal transmembrane cationic transport in hypertension[31] and diabetes.[32] Borkman et al.[33] described the significant correlation between decreased insulin sensitivity and decreased concentrations of polyunsaturated fatty acids in skeletal muscle phospholipids.

The effect of acute hyperinsulinemia on RBC ion transport activities was evaluated after a 5-h HIC. We did not demonstrate any effect of hyperinsulinemia on ion transport in both OHP and C. Trends of RBC ion transport changes in OHP were not significant during HIC compared with those under time- and volume load–controlled saline infusion; the effect of hypokalemia was prevented by infusion of KCl during HIC.[34]

In conclusion, we found that the offspring of hypertensive parents displayed higher insulin resistance, enhanced activity of Na^+-Li^+ countertransport, and augmented Li^+ and Rb^+ leaks in comparison to healthy controls. However, acute hyperinsulinemia did not modify any type of RBC ion transport studied, although a negative correlation was observed between insulin action and passive membrane permeability for Li^+ and Na^+.

ACKNOWLEDGMENTS

We thank D. Lapešová and D. Šišáková for their assistance during the clinical studies. We are also very grateful to J. Skibová and V. Lánská, who gave statistical advice. This work was supported by Grant No. 6682-3 awarded by the Internal Grant Agency of the Ministry of Health, Czech Republic.

REFERENCES

1. CANESSA, M., et al. 1980. Increased sodium-lithium countertransport in red cells of patients with essential hypertension. N. Engl. J. Med. **302:** 772–776.
2. REBBECK, T.R., et al. 1993. Sodium-lithium countertransport genotype and the probability of hypertension in adults. Hypertension **22:** 560–568.
3. YAP, L., et al. 1989. Is there increased cardiovascular risk in essential hypertensive patients with abnormal kinetics of red blood cell sodium-lithium countertransport? J. Hypertens. **7:** 667–673.
4. CARR, S., et al. 1990. Elevated sodium-lithium countertransport: a familial marker of hyperlipidaemia and hypertension? J. Hypertens. **8:** 139–146.
5. KRZESINSKI, J.M., et al. 1993. Red blood cell Na-Li countertransport, hypertensive heredity, and cardiovascular risk in young adults. Am. J. Hypertens. **6:** 314–316.
6. SEMPLICINI, A. 1993. The NaLi countertransport in hypertension. In Ionic Transport in Hypertension: New Perspectives, pp. 90–117. CRC Press. Boca Raton, FL.
7. DEFRONZO, R. & E. FERRANNINI. 1991. A multifaceted syndrome responsible for NIDDM, obesity, hypertension, dyslipidemia, and atherosclerotic cardiovascular disease. Diabetes Care **14:** 173–194.
8. CANESSA, M., et al. 1993. Red blood cell sodium-proton exchange in hypertensive blacks with insulin-resistant glucose disposal. Hypertension **22:** 204–213.
9. BIANCHI, S., et al. 1994. Elevated serum insulin levels in patients with essential hypertension and microalbuminuria. Hypertension **23:** 681–687.
10. ADRAGNA, A., et al. 1985. Effects of exercise on cation transport in human red cells. Hypertension **7:** 132–139.
11. WINCOUR, P.H., et al. 1992. Serum triglyceride and insulin levels are associated with erythrocyte sodium-lithium countertransport activity in normoglycaemic individuals. Clin. Chim. Acta **208:** 193–203.
12. CANESSA, M., et al. 1988. Genetic differences in lithium-sodium exchange and regulation of the sodium-hydrogen exchanger in essential hypertension. J. Cardiovasc. Pharmacol. **12**(suppl. 3): S92–S98.
13. IVES, H.E. 1989. Ion transport defects and hypertension: where is the link? Hypertension **14:** 587–590.
14. EXPERT COMMITTEE ON THE DIAGNOSIS AND CLASSIFICATION OF DIABETES MELLITUS. 1999. Report of the Expert Committee on the Diagnosis and Classification of Diabetes Mellitus. Diabetes Care **22**(suppl. 1): S5–S19.
15. DEFRONZO, R.A., et al. 1979. Glucose clamp technique: a method for quantifying insulin secretion and resistance. Am. J. Physiol. **237:** E214–E223.
16. FRIEDEWALD, W.T., et al. 1972. Estimation of the concentration of low density lipoprotein cholesterol in plasma without use of the preparative ultracentrifuge. Clin. Chem. **18:** 499–502.
17. DUHM, J. & B.O. GOBEL. 1983. Sodium-lithium exchange and sodium-potassium cotransport in human erythrocytes. Part 1: Evaluation of a simple uptake test to assess the activity of the two transport systems. Hypertension **4:** 468–476.
18. KUNEŠ, J., et al. 1994. Erythrocyte ion transport alterations in hypertriglyceridaemic rats. Clin. Sci. **86:** 11–13.
19. ZICHA, J., et al. 1997. Relationship of red blood cell ion transport alterations and serum lipid abnormalities in Lyon genetically hypertensive rats. Can. J. Physiol. Pharmacol. **75:** 1123–1128.
20. FERRANNINI, E., et al. 1987. Insulin resistance in essential hypertension. N. Engl. J. Med. **317:** 350–357.
21. CARR, S., et al. 1990. Increase in glomerular filtration rate in patients with insulin-dependent diabetes and elevated erythrocyte sodium-lithium countertransport. N. Engl. J. Med. **322:** 500–505.
22. BOERWINKLE, E., et al. 1986. Analysis of the distribution of erythrocyte sodium-lithium countertransport in a sample representative of the general population. Genet. Epidemiol. **3:** 365–378.
23. TREVISAN, M., et al. 1983. Abnormal red blood cell ion transport and hypertension: The People's Gas Company Study. Hypertension **3:** 363–367.

24. HUNT, S., et al. 1986. Associations of three erythrocyte cation transport systems with plasma lipids in Utah subjects. Hypertension **8:** 39–46.
25. DORIA, A., et al. 1991. Insulin resistance is associated with high sodium-lithium countertransport in essential hypertension. Am. J. Physiol. **261:** E684–E691.
26. TREVISAN, R., et al. 1992. Clustering of risk factors in hypertensive insulin-dependent diabetics with high sodium-lithium countertransport. Kidney Int. **41:** 855–861.
27. DE LOPES, F.J. 1992. Sodium-lithium countertransport activity and insulin resistance in normotensive IDDM patients. Diabetes **41:** 610–615.
28. JULIUS, S., et al. 1998. "White coat" versus "sustained" borderline hypertension in Tecumseh, Michigan. J. Hypertens. **16:** 617–623.
29. HERLITZ, H., et al. 1996. Relationship between sodium-lithium countertransport and insulin sensitivity in mild hypertension. J. Int. Med. **239:** 235–240.
30. MATTIASSON, I., et al. 1998. Insulin sensitivity, sodium-lithium countertransport, and platelet free calcium concentrations in normotensive men with a family history of hypertension. J. Hum. Hypertens. **12:** 259–264.
31. CORROCHER, R., et al. 1992. Effects induced by olive-rich diet on erythrocyte membrane lipids and sodium-potassium transports in postmenopausal hypertensive women. J. Endocrinol. Invest. **15:** 369–376.
32. LIJNEN, P., et al. 1994. Influence of cholesterol lowering on plasma membrane lipids and cationic transport systems. J. Hypertens. **12:** 59–64.
33. BORKMAN, M., et al. 1993. The relation between insulin sensitivity and the fatty-acid composition of skeletal-muscle phospholipids. N. Engl. J. Med. **328:** 238–244.
34. BEUCKELMANN, D., et al. 1984. Exogenous factors influencing the human erythrocyte sodium-lithium countertransport system. Eur. J. Clin. Invest. **14:** 392–397.

Increased Fat Intake, Impaired Fat Oxidation, and Failure of Fat Cell Proliferation Result in Ectopic Fat Storage, Insulin Resistance, and Type 2 Diabetes Mellitus

ERIC RAVUSSIN AND STEVEN R. SMITH

Pennington Biomedical Research Center, Baton Rouge, Louisiana 70808-4124, USA

ABSTRACT: It is widely accepted that increasing adiposity is associated with insulin resistance and increased risk of type 2 diabetes. The predominant paradigm used to explain this link is the portal/visceral hypothesis. This hypothesis proposes that increased adiposity, particularly in the visceral depots, leads to increased free fatty acid flux and inhibition of insulin action via Randle's effect in insulin-sensitive tissues. Recent data do not entirely support this hypothesis. As such, two new paradigms have emerged that may explain the established links between adiposity and disease. (A) Three lines of evidence support the *ectopic fat storage syndrome*. First, failure to develop adequate adipose tissue mass in either mice or humans, also known as *lipodystrophy*, results in severe insulin resistance and diabetes. This is thought to be the result of ectopic storage of lipid into liver, skeletal muscle, and the pancreatic insulin-secreting beta cell. Second, most obese patients also shunt lipid into the skeletal muscle, the liver, and probably the beta cell. The importance of this finding is exemplified by several studies demonstrating that the degree of lipid infiltration into skeletal muscle and liver correlates highly with insulin resistance. Third, increased fat cell size is highly associated with insulin resistance and the development of diabetes. Increased fat cell size may represent the failure of the adipose tissue mass to expand and thus to accommodate an increased energy influx. Taken together, these three observations support the *acquired lipodystrophy* hypothesis as a link between adiposity and insulin resistance. (B) The *endocrine paradigm* developed in parallel with the ectopic fat storage syndrome hypothesis. Adipose tissue secretes a variety of endocrine hormones, such as leptin, interleukin-6, angiotensin II, adiponectin (also called ACRP30 and adipoQ), and resistin. From this viewpoint, adipose tissue plays a critical role as an endocrine gland, secreting numerous factors with potent effects on the metabolism of distant tissues. These two new paradigms provide a framework to advance our understanding of the pathophysiology of the insulin-resistance syndrome.

KEYWORDS: acquired lipodystrophy; ectopic fat storage syndrome; endocrine paradigm; adiponectin; resistin

---

Address for correspondence: Eric Ravussin, Pennington Biomedical Research Center, 6400 Perkins Road, Baton Rouge, LA 70808-4124. Voice: 225-763-3186; fax: 225-763-3030.
ravusse@pbrc.edu

It is estimated that approximately 250 million people by the year 2020 will be affected by type 2 diabetes mellitus worldwide.[1] Although the exact causes of the disease are not known, it is clear that insulin resistance plays a major role in its development. Insulin resistance is present many years before the onset of the disease and is a major predictor of its development.[2,3] Besides the failure of peripheral tissues such as skeletal muscle and liver to respond to physiologic doses of insulin, it is now well recognized that the beta cell eventually fails to secrete adequate amounts of insulin in response to elevated plasma glucose concentration, leading to overt diabetes.

THE IMPACT OF BODY FAT ON INSULIN SENSITIVITY

About 10 years ago, the concept of syndrome X was introduced by Reaven[4] to explain the clustering of cardiovascular risk factors, such as hypertension, obesity, insulin resistance, and high triglyceride and low HDL cholesterol concentrations. In 1998, the WHO proposed a unifying definition and renamed it the *metabolic syndrome* instead of the *insulin resistance syndrome* because it had not been established that insulin resistance was the cause of all components present in the syndrome.

Numerous cross-sectional studies have shown an association between obesity and type 2 diabetes. In the third National Health Examination Survey (NHANES III) data, the prevalence of reported diabetes was 2.9 times higher in overweight than in non-overweight persons.[5] Prospective studies confirmed this association. For example, Knowler *et al.* showed that, in Pima Indians, the likelihood of developing diabetes increases steeply with increasing body weight.[6] The reason for the association between obesity and diabetes is at least in part attributed to the increase in insulin resistance[7] as insulin resistance is a clear predisposing factor for the development of type 2 diabetes.[4,8,9]

In obesity, experimental evidence has confirmed the presence of insulin resistance in adipose tissue, muscle,[10] and liver.[11] As a consequence, when obese subjects develop diabetes, there is a weakened inhibition of splanchnic glucose output with simultaneous insulin increase and elevated concentrations of free fatty acids (FFAs).[4] However, the insulin resistance that accompanies obesity is related to a decrease in total body glucose disposal and is thought to occur predominantly in the skeletal muscle.[12]

It has been clearly demonstrated that both an increase in fatness and a preferential upper-body accumulation of fat are independently related to insulin resistance.[13] Obese women with a greater proportion of upper-body fat (measured by body circumferences) tend to be more insulin-resistant, hyperinsulinemic, glucose-intolerant, and dyslipidemic than obese women with a greater proportion of lower-body fat.[14] When imaging techniques, such as magnetic resonance imaging (MRI) and computed tomography (CT), were used, visceral-fat accumulation was found to be specifically associated with the metabolic alterations of obesity in men and women.[15–18] Combined with Randle's hypothesis,[19–21] these observations led to the portal hypothesis, which states that the complications of obesity are attributable to increases in visceral adipose tissue with an associated rise in portal vein plasma FFA concentrations.[22]

PROBLEMS WITH THE PORTAL HYPOTHESIS

In contrast with the portal hypothesis, associations between the amount of subcutaneous fat on the trunk and insulin resistance have been recently reported in obese nondiabetic men[23,24] and in men with type 2 diabetes.[20,25,26] Subcutaneous fat, which does not drain into the portal vein, must cause insulin resistance by a nonportal mechanism. Similarly, insulin resistance in obese women is best related to overall elevated fat mass or to an increase in truncal subcutaneous fat mass as measured by the sum of skinfolds[27] or by whole body MRI.[18] Moreover, insulin resistance is predicted independently by an increased truncal subcutaneous fat mass and an increased amount of visceral fat.[17,27]

The observation that total fat mass and subcutaneous fat mass are important to the insulin resistance syndrome, combined with the growing experimental evidence that does not support the Randle/portal hypothesis, calls for a change in the scientific paradigm.[28] The two emerging paradigms are (i) the "ectopic fat storage syndrome" and (ii) a view of the adipocyte as an endocrine organ.

ECTOPIC FAT STORAGE AND INSULIN RESISTANCE

Skeletal Muscle Fat Content in Obesity and Type 2 Diabetes

The content of triglyceride within skeletal muscle is increased in obesity and in type 2 diabetes mellitus.[24] This is of pathophysiological importance. Muscle triglyceride content is a strong predictor of insulin resistance, both in animals and in humans.[29] Muscle biopsies are one method to assess triglyceride content,[30] but there are also noninvasive techniques, such as CT and spectral nuclear magnetic resonance (sNMR).

Kelley *et al.* reported that obesity affects the distribution of CT attenuation within skeletal muscle.[31] Obese subjects have an accumulation of muscle with attenuation values more than two standard deviations below the distribution range of lean muscle. They also found associations between skeletal muscle CT characteristics and adiposity, insulin resistance, oxidative enzymatic capacity of muscle, and maximal aerobic capacity.[32,33] More specifically, in obese individuals with a BMI > 30 kg/m^2, muscle attenuation was a stronger body composition correlate of insulin resistance than either visceral adiposity or overall adiposity.[24] In a weight-loss intervention study, the area of skeletal muscle displaying low CT attenuation was reduced with weight loss (67 ± 5 to 55 ± 4 cm^2), reflecting a reduction in skeletal muscle fat content in parallel with an improvement in insulin sensitivity.[34] In another study by the same authors, although subcutaneous adipose tissue was not significantly related to insulin resistance in obesity, subfascial thigh adipose tissue was.[32] Recent studies using ^{1}H-NMR have closely shown that intramyocellular triglyceride content is a strong determinant of *in vivo* insulin resistance in humans.[35–37]

Examples of Ectopic Fat Storage and Insulin Resistance

Lipodystrophy in humans is an acquired or hereditary syndrome characterized by a decrease in adipose tissue mass and diabetes mellitus. The insufficient adipose tissue mass leads to excess energy storage as triglyceride in the liver and skeletal muscle and to insulin resistance and diabetes.[38,39] Several animal models support this sequence of events. Recent transgenic animal models that block adipose tissue development revealed lipid infiltration of skeletal muscle and fatty liver along with insulin resistance, glucose intolerance, and diabetes.[40–42] Similarly, a mutation of the *lipin* gene results in a fatless mouse with a similar metabolic phenotype.[43,44] The transplantation of adipose tissue back into lipoatrophic animals reverses the elevated glucose levels.[45] Conversely, surgical removal of adipose tissue in normal hamsters produces a strikingly similar lipodystrophic syndrome.[46] These latter two experiments demonstrate that the inadequate adipose tissue mass leads to ectopic fat storage and is the cause of the metabolic disturbances. In other words, too little fat is as deleterious as too much fat and predisposes to the development of insulin resistance and ultimately type 2 diabetes.

Obesity Is Another Ectopic Fat Storage Syndrome

Positive energy balance in our obesogenic environment (see FIG. 1) produces a pattern similar to lipodystrophy in humans: excess lipid storage as triglyceride in liver[47] and skeletal muscle,[32,48] followed by insulin resistance, glucose intolerance, and diabetes. In contrast to lipodystrophic diabetes, adipose tissue stores are adequate or even large in these patients. These observations suggest that the adipose tissue mass is inadequate to sequester dietary lipid away from liver, skeletal muscle, and pancreas. Conversely, weight loss decreases fat cell size at the same time that the metabolic syndrome improves. Recently, hepatic fat content measured by sNMR was also shown to be associated with insulin resistance.[47] Thiazolidinediones, through the activation of PPAR-γ, promote differentiation of new fat cells in subcutaneous, but not intraperitoneal, adipose tissue.[49] *In vivo*, this translates into a gain in subcutaneous, but not intraperitoneal, adipose tissue[50] and a decrease in lipid infiltration in skeletal muscle. The above observations suggest that drug therapies might be designed to target ectopic fat storage as a strategy to improve insulin action and prevent or reverse diabetes.

WHY IS LIPID STORED IN THE "WRONG" PLACES?

Failure of New Fat Cell Formation

Enlarged adipocytes correlate better with insulin resistance than any other measures of adiposity.[51–53] These enlarged fat cells are resistant to insulin-stimulated glucose uptake,[54] but are unlikely to directly cause systemic insulin resistance since adipose tissue takes up only a minimal portion of ingested or infused glucose. Recent results from studies in Pima Indians seem to favor the role of an impairment in proliferation and differentiation capacity of adipocytes as a precipitating factor for the development of type 2 diabetes.[53,55] Paolisso first showed that the predictive effect of high plasma FFA concentration on diabetes disappeared after adjustment for abdominal fat cell size.[55] More recently, Weyer *et al.*[53] reported that enlarged abdominal adipocytes predict the development of type 2 diabetes independently of insulin resistance and insulin secretion.

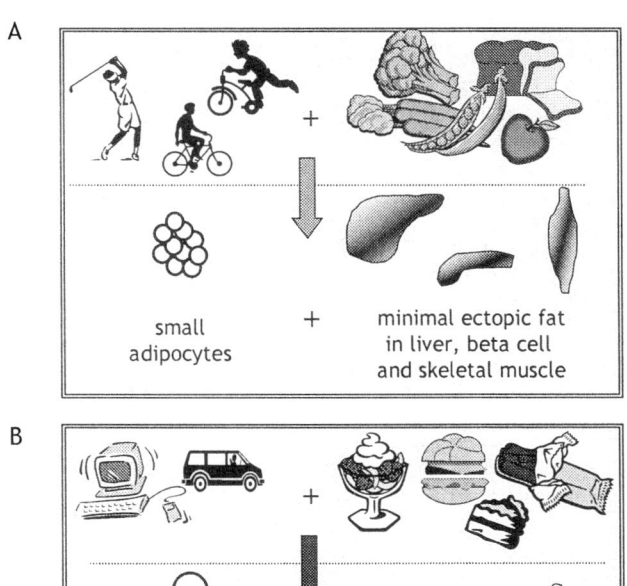

FIGURE 1. Energy balance and ectopic fat storage. (**A**) When energy intake and energy expenditure are in balance, fat intake equals fat oxidation and small adipocytes predominate. (**B**) When energy intake is greater than energy expenditure, excess dietary fat causes adipocyte hypertrophy and fat storage in ectopic sites, such as liver, skeletal muscle, and the pancreatic beta cell. Ectopic (intracellular) lipid decreases insulin action, leading to hyperinsulinemia and predisposing to the development of type 2 diabetes mellitus.

Why are the fat cells large in patients with obesity and the metabolic syndrome? As discussed above, the evidence supports the view that large fat cells are indicative of a failure of proliferation and/or differentiation of adipocytes (FIG. 2). Adipocytes develop from mesenchymal stem cells via a complex cascade of transcriptional and nontranscriptional events.[56] When proliferation is stopped, mesenchymal stem cells are able to differentiate along several mutually exclusive pathways.[57] A mesenchymal stem cell is able to differentiate into a chondrocyte, osteocyte, tenocyte, or adipocyte, depending on the paracrine, endocrine, and metabolic milieu[58] (FIG. 3). The control mechanisms for adipocyte differentiation are complex, but the transcriptional and extracellular signals necessary for stem cell differentiation into a preadipocyte and from the preadipocyte into a mature fat cell are being elucidated.[59] The mouse cell line 3T3-L1 is the most commonly studied cell *in vitro*. These cells are

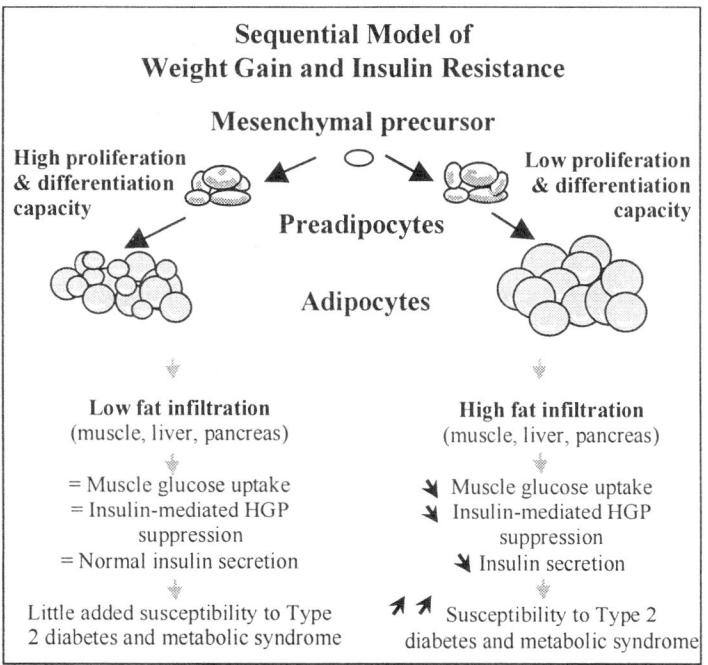

FIGURE 2. The sequential model of weight gain and insulin resistance. Adipocytes are derived from pleuripotent stem cell precursors. Individuals with a low capacity for proliferation and/or differentiation of mesenchymal precursors into mature fat-storing adipocytes are susceptible to hypertrophy of the existing adipocytes under conditions of energy and/or fat excess, so common in our modern "obesogenic" environment. Adipocyte hypertrophy leads to ectopic fat storage, insulin resistance, and eventually type 2 diabetes mellitus. HGP: hepatic glucose production.

characteristic of the preadipocyte *in vivo* and can only differentiate into mature lipid-storing adipocytes.

As recently reviewed by Nadler and Attie, the regulation of adipocyte differentiation involves growth arrest and the coordinate activation/inactivation of nuclear transcription factors.[60] These transcription factors then regulate the transcription, translation, and/or activation of a variety of genes necessary for lipid storage and insulin sensitivity. ADD/SREB-1; C/EBP-α, -β, and -δ; and PPAR-γ are a few of the known transcription factors.[56,57,59,61–63] These transcription factors are regulated in response to extracellular signals, such as prostaglandins, cytokines (LIF,[64] TNF-α[65]), and hormones, including corticosteroids and insulin. Defects in any one of these steps are potentially important in the failure of proliferation or differentiation of adipocytes. As an example, single nucleotide substitutions in the PPAR-γ gene are associated with risk for the development of diabetes.[66]

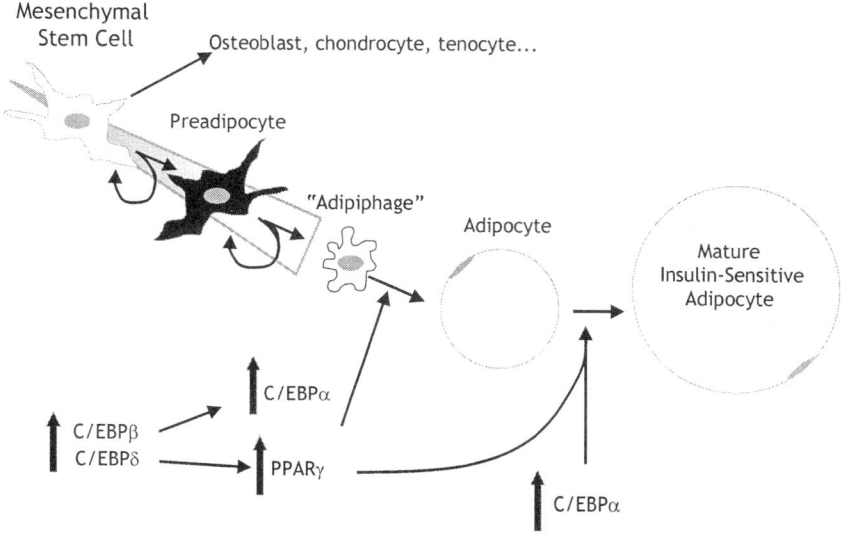

FIGURE 3. Adipocyte proliferation and differentiation. Adipocytes are derived from pleuripotent stem cell precursors: the mesenchymal stem cells (MSC). The MSC can differentiate into an osteoblast, a chondroblast, a myocyte, and other cell types. These cells are thought to be self-replicating and capable of differentiating into a mature adipocyte. An intermediate cell type, the "adipiphage", has been proposed.[116] Transcription factors, such as SREBP/ADD, C/EBP-α and -β, PPAR-γ, and the RXR receptor are important components of the differentiation cascade. Several other regulatory systems include the autocrine/paracrine growth factors, TNF-α, and the Wnt family of proteins.[117] Additional factors that are secreted from the adipocyte and act as autocrine/paracrine regulators of adipocyte proliferation and differentiation are listed in FIGURE 4.

Impaired Fat Oxidation

An alternative explanation for the increased intracellular lipid is that whole body fat oxidation is impaired, leading to ectopic accumulation of intracellular lipid. This view holds that the machinery to oxidize fat is not sufficient to match the dietary fat load or is not activated in a timely fashion by a signal (or signals) necessary to oxidize fat. It is also possible that mitochondrial fatty acid availability or transport is decreased. If fat oxidation is decreased, fat accumulation, in the form of increased intracellular lipid, might follow. Consistent with the impaired fat oxidation hypothesis, McGarry and colleagues demonstrated that inhibition of fat oxidation in rodents increased intracellular lipid and decreased insulin action *in vivo*, and the two were highly correlated ($r = 0.96$).[67] In humans, an increased respiratory quotient (RQ), indicative of decreased postabsorptive fat oxidation, predicts weight gain[68–70] and is associated with deterioration of insulin sensitivity.

Recent studies have shown a large degree of interindividual variability in the capacity to oxidize excess dietary fat.[71] Such results are consistent with the inadequate fat oxidizing machinery hypothesis proposed above as a low aerobic capacity was highly correlated with the ability to oxidize excess dietary fat. Fasting insulin values were also related to the capacity to oxidize fat.

Fat oxidation in muscle can be modulated by a number of factors. Chronic exercise increases fat oxidation in muscle by increasing LPL, carnitine palmitoyltransferase I (CPT-I),[72] and the number of mitochondria.[73] Collectively, these processes enhance the availability of FFA, their transport into mitochondria, and their rate of oxidation mostly by muscle. Physical fitness is correlated with the concentrations of enzymes needed to oxidize fatty acids.[74] As mentioned above, intramuscular lipid stores are associated with maximal aerobic capacity,[32] and physical fitness is associated with the ability to oxidize dietary fat.[71] Furthermore, exercise decreases intramuscular lipid stores.[75]

Other neural-endocrine mechanisms may regulate fat oxidation. When healthy young men were treated with a synthetic beta-3 agonist, there was a decrease in 24-h RQ,[76] indicating an increase in fat oxidation. In addition, infusions of beta-adrenergic agonists not only increase energy expenditure, but also increase fat oxidation[77,78] and improve insulin sensitivity. As such, it is possible that the sympathetic nervous system (SNS), acting through the beta-3 adrenoreceptor or other adrenoreceptors, regulates fat oxidation. Importantly, a low SNS activity has been linked to weight gain[79] and decreased fat oxidation,[80] again suggesting the importance of the SNS in the regulation of fat oxidation, body weight, and insulin sensitivity.

Endocrine hormones secreted from the adipocyte or from the neuroendocrine systems may also regulate fat oxidation. For example, adiponectin, leptin, and CRH have all been shown to increase fat oxidation.[81–83] The endocrine hypothesis will be discussed in greater detail below.

The allosteric regulators of CPT-I (malonyl CoA) and AMP kinase (ATP) provide minute-to-minute regulation of substrate utilization. In addition to defects in these pathways, individuals with increased intracellular lipid may have inadequate machinery to oxidize an increased flux of fatty acids. In other words, the metabolic signals may be sufficient, but the capacity of the system may be inadequate. Regulation of fuel oxidation may also occur over a longer period through transcriptional or translational regulation of enzymes for fatty acid oxidation. For example, fatty acid transport by CPT-I is regulated not only by malonyl CoA, but also by the concentration of CPT-I. CPT-I, in turn, is regulated in a tissue-specific fashion by the nuclear transcription factors, PPAR-α and PPAR-γ.[84–86] Importantly, these two nuclear transcription factors, in concert with the coactivator PGC-1,[87] regulate transcription of the entire repertoire of genes involved in fatty acid oxidation, such as ACO.[88,89]

AN ALTERNATE PARADIGM: THE ENDOCRINE HYPOTHESIS

In this model, the adipocytes respond to various stimuli to integrate metabolic, hormonal, and neural stimuli by releasing hormones (FIG. 4).[81,90–111] The molecular revolution brought to light many adipocyte-secreted factors, some of which are secreted into the blood stream, such as IL-6 and leptin,[90] whereas others, such as TNFα, exert their effects in an autocrine-paracrine fashion.[90] Large fat cells,

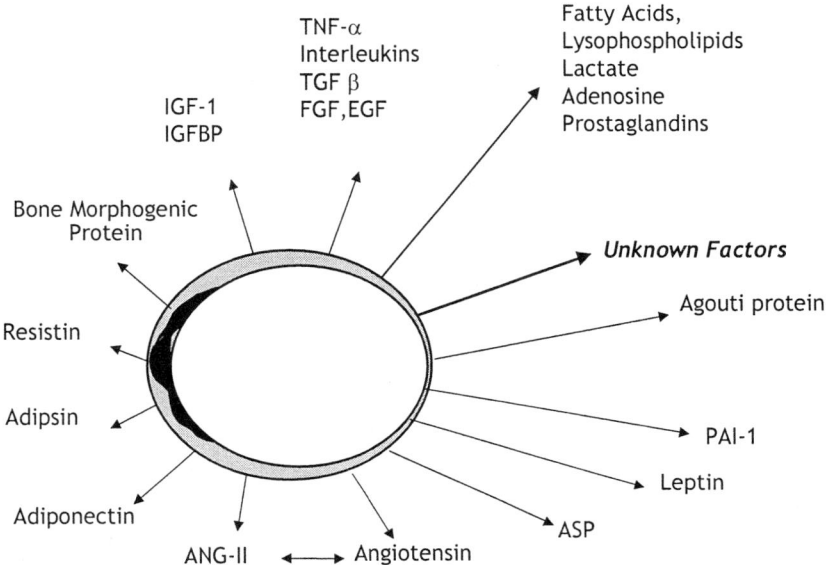

FIGURE 4. Proteins secreted by the adipocyte. The list of proteins and factors secreted by the adipocyte is growing rapidly. Several of these proteins/factors act as endocrine hormones, for example, leptin and IL-6, whereas others act locally, for example, TNF-α and the growth factors. The endocrine proteins/factors are proposed to influence distant tissues, such as the liver, skeletal muscle, or brain, to modulate insulin action, that is, the *endocrine hypothesis*. Paracrine/autocrine factors may also modulate insulin action if they promote or inhibit adipocyte proliferation and/or differentiation. (See references 81 and 90–111.)

evidence of increased energy influx, are likely to secrete an entirely different pattern of hormones than small adipocytes. Other autocrine, paracrine, or endocrine factors remain undiscovered.

An alternate hypothesis to that of the endocrine/paracrine effect of large fat cells on the regulation of insulin sensitivity is that enlarged adipocytes may not have a pathophysiological significance by themselves, but are instead only a manifestation (end organ response) of other unknown pathogenic factors. These factors may lead independently to enlarged adipocytes, insulin resistance, and other traits of the metabolic syndrome (FIG. 5). Glucocorticoids, growth hormones, IGF-1, and sex steroids (modulated by the neuroendocrine systems) might represent common factors since they all have an impact on both glucose homeostasis and adipocyte size.[112] Neuropeptides, such as NPY and AGRP, may also influence adipocyte function and endocrine secretion.[113–115]

CONCLUSIONS

Our obesogenic environment provides constant opportunities for excess energy intake and a sedentary lifestyle. Recent experimental data reveal several flaws in the portal/Randle's hypothesis. Two alternate paradigms are emerging. The first hypoth-

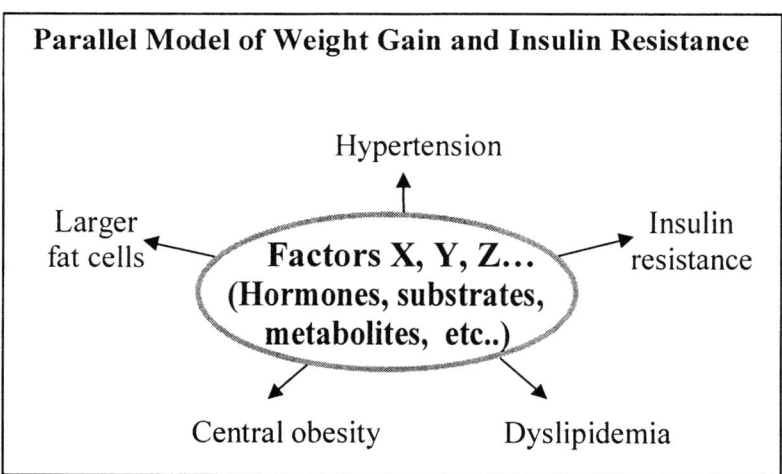

FIGURE 5. The parallel model of weight gain and insulin resistance. From this viewpoint, unknown factor(s) may be responsible for all of the observed components of the syndrome: for example, insulin resistance, dyslipidemia, central obesity, hypertrophy of adipocytes, and hypertension. The factor(s) could be endocrine, metabolic, or neurologic in origin. (See text for details.)

esis, the *ectopic fat storage syndrome*, states that, if fat mass cannot expand through the proliferation and differentiation of new adipocytes, then adipocyte hypertrophy results, and excess dietary fat leads to ectopic storage of fat in the liver, skeletal muscle, and the beta cell. Alternately, the ectopic fat storage syndrome could also be due to defects in the oxidation of fat, leading to intracellular accumulation of lipids. A second paradigm, the *endocrine hypothesis*, is based on the growing number of factors that are released from adipocytes. These hormones have potent effects on metabolism and insulin sensitivity in many insulin-sensitive tissues. In the next decade, these two paradigms probably will constitute a new framework for the study of the links between our obesogenic environment and the risk of developing diabetes. It may also lead to the development of new pharmacological strategies to treat insulin resistance and possibly reduce the galloping increased incidence of type 2 diabetes mellitus.

REFERENCES

1. O'RAHILLY, S. 1997. Science, medicine, and the future: non-insulin dependent diabetes mellitus—the gathering storm. Br. Med. J. **314**(7085): 955–959.
2. WARRAM, J.H., B.C. MARTIN, A.S. KROLEWSKI *et al.* 1990. Slow glucose removal rate and hyperinsulinemia precede the development of type II diabetes in the offspring of diabetic parents. Ann. Intern. Med. **113**(12): 909–915.
3. LILLIOJA, S., D.M. MOTT, B.V. HOWARD *et al.* 1988. Impaired glucose tolerance as a disorder of insulin action: longitudinal and cross-sectional studies in Pima Indians. N. Engl. J. Med. **318**(19): 1217–1225.

4. REAVEN, G.M. 1988. Banting lecture: role of insulin resistance in human disease. Diabetes **37**(12): 1595–1607.
5. HARRIS, M.I., K.M. FLEGAL, C.C. COWIE et al. 1998. Prevalence of diabetes, impaired fasting glucose, and impaired glucose tolerance in U.S. adults: the Third National Health and Nutrition Examination Survey, 1988–1994 [see comments]. Diabetes Care **21**(4): 518–524.
6. KNOWLER, W.C., D.J. PETTITT, P.J. SAVAGE & P.H. BENNETT. 1981. Diabetes incidence in Pima Indians: contributions of obesity and parental diabetes. Am. J. Epidemiol. **113**(2): 144–156.
7. OLEFSKY, J.M. 1981. Lilly lecture 1980: insulin resistance and insulin action—an *in vitro* and *in vivo* perspective. Diabetes **30**(2): 148–162.
8. MARTIN, B.C., J.H. WARRAM, A.S. KROLEWSKI et al. 1992. Role of glucose and insulin resistance in development of type 2 diabetes mellitus: results of a 25-year follow-up study [see comments]. Lancet **340**(8825): 925–929.
9. LILLIOJA, S., D.M. MOTT, M. SPRAUL et al. 1993. Insulin resistance and insulin secretory dysfunction as precursors of non-insulin-dependent diabetes mellitus: prospective studies of Pima Indians. N. Engl. J. Med. **329**(27): 1988–1992.
10. RABINOWITZ, D. & K.L. ZIERLER. 1962. Forearm metabolism in obesity and its response to intraarterial insulin: characterization of insulin resistance and evidence for adaptive hyperinsulinism. J. Clin. Invest. **41**: 2173–2179.
11. FELIG, P., J. WAHREN, R. HENDLER & J. BRUNDIN. 1974. Splanchnic glucose and amino acid metabolism in obesity. J. Clin. Invest. **53**: 582–590.
12. CARO, J.F., L.G. DOHM, W.J. PORIES & M.K. SINHA. 1989. Cellular alterations in liver, skeletal muscle, and adipose tissue responsible for insulin resistance in obesity and type II diabetes. Diabetes Metab. Rev. **5**(8): 665–689.
13. CLAUSEN, J.O., K. BORCH-JOHNSEN, H. IBSEN et al. 1996. Insulin sensitivity index, acute insulin response, and glucose effectiveness in a population-based sample of 380 young healthy Caucasians: analysis of the impact of gender, body fat, physical fitness, and life-style factors. J. Clin. Invest. **98**(5): 1195–1209.
14. KISSEBAH, A.H. & A.N. PEIRIS. 1989. Biology of regional body fat distribution: relationship to non-insulin-dependent diabetes mellitus. Diabetes Metab. Rev. **5**(2): 83–109.
15. DESPRES, J.P., S. LEMIEUX, B. LAMARCHE et al. 1995. The insulin resistance–dyslipidemic syndrome: contribution of visceral obesity and therapeutic implications. Int. J. Obes. Relat. Metab. Disord. **19**(suppl. 1): S76–S86.
16. BANERJI, M.A., R.L. CHAIKEN, D. GORDON et al. 1995. Does intra-abdominal adipose tissue in black men determine whether NIDDM is insulin-resistant or insulin-sensitive? Diabetes **44**(2): 141–146.
17. ALBU, J.B., L. MURPHY, D.H. FRAGER et al. 1997. Visceral fat and race-dependent health risks in obese nondiabetic premenopausal women. Diabetes **46**(3): 456–462.
18. ALBU, J.B., A.J. KOVERA & J.A. JOHNSON. 2000. Fat distribution and health in obesity. Ann. N.Y. Acad. Sci. **904**: 491–501.
19. RANDLE, P.J., P.B. GARLAND, C.N. HALES et al. 1963. The glucose fatty acid cycle: its role in insulin sensitivity and metabolic disturbances of diabetes mellitus. Lancet **i**: 7285–7289.
20. KELLEY, D.E. & L.J. MANDARINO. 2000. Fuel selection in human skeletal muscle in insulin resistance: a reexamination. Diabetes **49**(5): 677–683.
21. BERGMAN, R.N. & M. ADER. 2000. Free fatty acids and pathogenesis of type 2 diabetes mellitus. Trends Endocrinol. Metab. **11**(9): 351–356.
22. BJORNTORP, P. 1990. "Portal" adipose tissue as a generator of risk factors for cardiovascular disease and diabetes. Arteriosclerosis **10**(4): 493–496.
23. ABATE, N., A. GARG, R.M. PESHOCK et al. 1995. Relationships of generalized and regional adiposity to insulin sensitivity in men. J. Clin. Invest. **96**(1): 88–98.
24. GOODPASTER, B.H., F.L. THAETE, J.A. SIMONEAU & D.E. KELLEY. 1997. Subcutaneous abdominal fat and thigh muscle composition predict insulin sensitivity independently of visceral fat. Diabetes **46**(10): 1579–1585.
25. SMITH, S.R., J.C. LOVEJOY, F. GREENWAY et al. 2001. Contributions of total body fat, abdominal subcutaneous adipose tissue compartments, and visceral adipose tissue to the metabolic complications of obesity. Metabolism **50**(4): 425–435.

26. ABATE, N., A. GARG, R.M. PESHOCK et al. 1996. Relationship of generalized and regional adiposity to insulin sensitivity in men with NIDDM. Diabetes **45**(12): 1684–1693.
27. MARCUS, M.A., L. MURPHY, F.X. PI-SUNYER & J.B. ALBU. 1999. Insulin sensitivity and serum triglyceride level in obese white and black women: relationship to visceral and truncal subcutaneous fat. Metabolism **48**(2): 194–199.
28. KUHN, T.F. 1962. The Structure of Scientific Revolutions. Second edition. University of Chicago Press. Chicago.
29. GOODPASTER, B.H. & D.E. KELLEY. 1998. Role of muscle in triglyceride metabolism. Curr. Opin. Lipidol. **9**(3): 231–236.
30. PAN, D.A., S. LILLIOJA, A.D. KRIKETOS et al. 1997. Skeletal muscle triglyceride levels are inversely related to insulin action. Diabetes **46**(6): 983–988.
31. KELLEY, D.E., B.S. SLASKY & J. JANOSKY. 1991. Skeletal muscle density: effects of obesity and non-insulin-dependent diabetes mellitus. Am. J. Clin. Nutr. **54**(3): 509–515.
32. GOODPASTER, B.H., F.L. THAETE & D.E. KELLEY. 2000. Thigh adipose tissue distribution is associated with insulin resistance in obesity and in type 2 diabetes mellitus. Am. J. Clin. Nutr. **71**(4): 885–892.
33. KELLEY, D.E., B. GOODPASTER, R.R. WING & J.A. SIMONEAU. 1999. Skeletal muscle fatty acid metabolism in association with insulin resistance, obesity, and weight loss. Am. J. Physiol. **277**(6, part 1): E1130–E1141.
34. GOODPASTER, B.H., D.E. KELLEY, R.R. WING et al. 1999. Effects of weight loss on regional fat distribution and insulin sensitivity in obesity. Diabetes **48**(4): 839–847.
35. SZCZEPANIAK, L.S., E.E. BABCOCK, F. SCHICK et al. 1999. Measurement of intracellular triglyceride stores by H spectroscopy: validation in vivo. Am. J. Physiol. **276**(5, part 1): E977–E989.
36. PERSEGHIN, G., P. SCIFO, F. DE COBELLI et al. 1999. Intramyocellular triglyceride content is a determinant of in vivo insulin resistance in humans: a ^{1}H-^{13}C nuclear magnetic resonance spectroscopy assessment in offspring of type 2 diabetic parents. Diabetes **48**(8): 1600–1606.
37. KRSSAK, M., K. FALK PETERSEN, A. DRESNER et al. 1999. Intramyocellular lipid concentrations are correlated with insulin sensitivity in humans: a ^{1}H NMR spectroscopy study. Diabetologia **42**(1): 113–116.
38. ROBBINS, D.C., E. DANFORTH, JR., E.S. HORTON et al. 1979. The effect of diet on thermogenesis in acquired lipodystrophy. Metabolism **28**(9): 908–916.
39. ROBBINS, D.C., E.S. HORTON, O. TULP & E.A. SIMS. 1982. Familial partial lipodystrophy: complications of obesity in the non-obese? Metabolism **31**(5): 445–452.
40. REITMAN, M.L., M.M. MASON, J. MOITRA et al. 1999. Transgenic mice lacking white fat: models for understanding human lipoatrophic diabetes. Ann. N.Y. Acad. Sci. **892**: 289–296.
41. SHIMOMURA, I., R.E. HAMMER, S. IKEMOTO et al. 1999. Leptin reverses insulin resistance and diabetes mellitus in mice with congenital lipodystrophy. Nature **401**(6748): 73–76.
42. KIM, J.K., O. GAVRILOVA, Y. CHEN et al. 2000. Mechanism of insulin resistance in A-ZIP/F-1 fatless mice. J. Biol. Chem. **275**(12): 8456–8460.
43. REUE, K., P. XU, X.P. WANG & B.G. SLAVIN. 2000. Adipose tissue deficiency, glucose intolerance, and increased atherosclerosis result from mutation in the mouse fatty liver dystrophy (fld) gene. J. Lipid Res. **41**(7): 1067–1076.
44. PETERFY, M., J. PHAN, P. XU & K. REUE. 2001. Lipodystrophy in the fld mouse results from mutation of a new gene encoding a nuclear protein, lipin. Nat. Genet. **27**(1): 121–124.
45. GAVRILOVA, O., B. MARCUS-SAMUELS, D. GRAHAM et al. 2000. Surgical implantation of adipose tissue reverses diabetes in lipoatrophic mice. J. Clin. Invest. **105**(3): 271–278.
46. WEBER, R.V., M.C. BUCKLEY, S.K. FRIED & J.G. KRAL. 2000. Subcutaneous lipectomy causes a metabolic syndrome in hamsters. Am. J. Physiol. Regul. Integr. Comp. Physiol. **279**(3): R936–R943.
47. RYYSY, L., A.M. HAKKINEN, T. GOTO et al. 2000. Hepatic fat content and insulin action on free fatty acids and glucose metabolism rather than insulin absorption are associated with insulin requirements during insulin therapy in type 2 diabetic patients. Diabetes **49**(5): 749–758.

48. SHULMAN, G.I. 2000. Cellular mechanisms of insulin resistance. J. Clin. Invest. **106**(2): 171–176.
49. ADAMS, M., C.T. MONTAGUE, J.B. PRINS *et al.* 1997. Activators of peroxisome proliferator-activated receptor gamma have depot-specific effects on human preadipocyte differentiation. J. Clin. Invest. **100**(12): 3149–3153.
50. AKAZAWA, S., F. SUN, M. ITO *et al.* 2000. Efficacy of troglitazone on body fat distribution in type 2 diabetes [in process citation]. Diabetes Care **23**(8): 1067–1071.
51. BJORNTORP, P., P. BERCHTOLD & G. TIBBLIN. 1971. Insulin secretion in relation to adipose tissue in men. Diabetes **20**(2): 65–70.
52. SCHNEIDER, B.S., I.M. FAUST, R. HEMMES & J. HIRSCH. 1981. Effects of altered adipose tissue morphology on plasma insulin levels in the rat. Am. J. Physiol. **240**(4): E358–E362.
53. WEYER, C., J.E. FOLEY, C. BOGARDUS *et al.* 2000. Enlarged subcutaneous abdominal adipocyte size, but not obesity itself, predicts type II diabetes independent of insulin resistance. Diabetologia **43**: 1498–1506.
54. CZECH, M.P. 1976. Cellular basis of insulin insensitivity in large rat adipocytes. J. Clin. Invest. **57**(6): 1523–1532.
55. PAOLISSO, G., P.A. TATARANNI, J.E. FOLEY *et al.* 1995. A high concentration of fasting plasma non-esterified fatty acids is a risk factor for the development of NIDDM. Diabetologia **38**(10): 1213–1217.
56. RANGWALA, S.M. & M.A. LAZAR. 2000. Transcriptional control of adipogenesis [in process citation]. Annu. Rev. Nutr. **20**: 535–559.
57. MORRISON, R.F. & S.R. FARMER. 1999. Insights into the transcriptional control of adipocyte differentiation. J. Cell. Biochem. Suppl. **32/33**: 59–67.
58. PITTENGER, M.F., A.M. MACKAY, S.C. BECK *et al.* 1999. Multilineage potential of adult human mesenchymal stem cells. Science **284**(5411): 143–147.
59. ROSEN, E.D., C.J. WALKEY, P. PUIGSERVER & B.M. SPIEGELMAN. 2000. Transcriptional regulation of adipogenesis. Genes Dev. **14**(11): 1293–1307.
60. NADLER, S.T. & A.D. ATTIE. 2001. Please pass the chips: genomic insights into obesity and diabetes. J. Nutr. **131**(8): 2078–2081.
61. ROSEN, E.D. & B.M. SPIEGELMAN. 2000. Molecular regulation of adipogenesis. Annu. Rev. Cell Dev. Biol. **16**: 145–171.
62. WOLF, G. 1999. The molecular mechanism of the stimulation of adipocyte differentiation by a glucocorticoid. Nutr. Rev. **57**(10): 324–326.
63. AUWERX, J. 1999. PPARγ, the ultimate thrifty gene. Diabetologia **42**(9): 1033–1049.
64. AUBERT, J., S. DESSOLIN, N. BELMONTE *et al.* 1999. Leukemia inhibitory factor and its receptor promote adipocyte differentiation via the mitogen-activated protein kinase cascade. J. Biol. Chem. **274**(35): 24965–24972.
65. STEPHENS, J.M., M. BUTTS, R. STONE *et al.* 1993. Regulation of transcription factor mRNA accumulation during 3T3-L1 preadipocyte differentiation by antagonists of adipogenesis. Mol. Cell. Biochem. **123**(1/2): 63–71.
66. ALTSHULER, D., J.N. HIRSCHHORN, M. KLANNEMARK *et al.* 2000. The common PPAR-gamma Pro12Ala polymorphism is associated with decreased risk of type 2 diabetes [in process citation]. Nat. Genet. **26**(1): 76–80.
67. DOBBINS, R.L., L.S. SZCZEPANIAK, B. BENTLEY *et al.* 2001. Prolonged inhibition of muscle carnitine palmitoyltransferase-1 promotes intramyocellular lipid accumulation and insulin resistance in rats. Diabetes **50**(1): 123–130.
68. ZURLO, F., S. LILLIOJA, A. ESPOSITO–DEL PUENTE *et al.* 1990. Low ratio of fat to carbohydrate oxidation as predictor of weight gain: study of 24-h RQ. Am. J. Physiol. **259**(5, part 1): E650–E657.
69. SEIDELL, J.C., D.C. MULLER, J.D. SORKIN & R. ANDRES. 1992. Fasting respiratory exchange ratio and resting metabolic rate as predictors of weight gain: the Baltimore Longitudinal Study on Aging. Int. J. Obes. Relat. Metab. Disord. **16**(9): 667–674.
70. VALTUENA, S., J. SALAS-SALVADO & P.G. LORDA. 1997. The respiratory quotient as a prognostic factor in weight-loss rebound. Int. J. Obes. Relat. Metab. Disord. **21**(9): 811–817.
71. SMITH, S.R., L. DE JONGE, J.J. ZACHWIEJA *et al.* 2000. Fat and carbohydrate balances during adaptation to a high-fat diet. Am. J. Clin. Nutr. **71**(2): 450–457.

72. BERTHON, P.M., R.A. HOWLETT, G.J. HEIGENHAUSER & L.L. SPRIET. 1998. Human skeletal muscle carnitine palmitoyltransferase I activity determined in isolated intact mitochondria. J. Appl. Physiol. **85**(1): 148–153.
73. HOLLOSZY, J.O. & E.F. COYLE. 1984. Adaptations of skeletal muscle to endurance exercise and their metabolic consequences. J. Appl. Physiol. **56**(4): 831–838.
74. RABEN, A., E. MYGIND & A. ASTRUP. 1998. Lower activity of oxidative key enzymes and smaller fiber areas in skeletal muscle of postobese women. Am. J. Physiol. **275**(3, part 1): E487–E494.
75. BRECHTEL, K., A.M. NIESS, J. MACHANN et al. 2001. Utilisation of intramyocellular lipids (IMCLs) during exercise as assessed by proton magnetic resonance spectroscopy (^{1}H-MRS). Horm. Metab. Res. **33**: 63–66.
76. WEYER, C., P.A. TATARANNI, S. SNITKER et al. 1998. Increase in insulin action and fat oxidation after treatment with CL 316,243, a highly selective beta3-adrenoceptor agonist in humans. Diabetes **47**(10): 1555–1561.
77. ENOCKSSON, S., M. SHIMIZU, F. LONNQVIST et al. 1995. Demonstration of an *in vivo* functional beta3-adrenoceptor in man. J. Clin. Invest. **95**(5): 2239–2245.
78. SCHIFFELERS, S.L., V.J. VAN HARMELEN, H.A. DE GRAUW et al. 1999. Dobutamine as selective beta(1)-adrenoceptor agonist in *in vivo* studies on human thermogenesis and lipid utilization. J. Appl. Physiol. **87**(3): 977–981.
79. TATARANNI, P.A., J.B. YOUNG, C. BOGARDUS & E. RAVUSSIN. 1997. A low sympathoadrenal activity is associated with body weight gain and development of central adiposity in Pima Indian men. Obes. Res. **5**(4): 341–347.
80. SNITKER, S., P.A. TATARANNI & E. RAVUSSIN. 1998. Respiratory quotient is inversely associated with muscle sympathetic nerve activity. J. Clin. Endocrinol. Metab. **83**(11): 3977–3979.
81. YAMAUCHI, T., J. KAMON, H. WAKI et al. 2001. The fat-derived hormone adiponectin reverses insulin resistance associated with both lipoatrophy and obesity. Nat. Med. **7**(8): 941–946.
82. SMITH, S.R., L. DE JONGE, M. PELLYMOUNTER et al. 2001. Peripheral administration of human corticotropin-releasing hormone: a novel method to increase energy expenditure and fat oxidation in man. J. Clin. Endocrinol. Metab. **86**(5): 1991–1998.
83. MUOIO, D.M., G.L. DOHM, F.T. FIEDOREK, JR. et al. 1997. Leptin directly alters lipid partitioning in skeletal muscle. Diabetes **46**(8): 1360–1363. [1997. Erratum. Diabetes **46**(10): 1663.]
84. MASCARO, C., E. ACOSTA, J.A. ORTIZ et al. 1998. Control of human muscle-type carnitine palmitoyltransferase I gene transcription by peroxisome proliferator-activated receptor. J. Biol. Chem. **273**(15): 8560–8563.
85. LAPSYS, N.M., A.D. KRIKETOS, M. LIM-FRASER et al. 2000. Expression of genes involved in lipid metabolism correlate with peroxisome proliferator-activated receptor gamma expression in human skeletal muscle [in process citation]. J. Clin. Endocrinol. Metab. **85**(11): 4293–4297.
86. KLIEWER, S.A., S.S. SUNDSETH, S.A. JONES et al. 1997. Fatty acids and eicosanoids regulate gene expression through direct interactions with peroxisome proliferator-activated receptors alpha and gamma. Proc. Natl. Acad. Sci. U.S.A. **94**(9): 4318–4323.
87. VEGA, R.B., J.M. HUSS & D.P. KELLY. 2000. The coactivator PGC-1 cooperates with peroxisome proliferator-activated receptor alpha in transcriptional control of nuclear genes encoding mitochondrial fatty acid oxidation enzymes. Mol. Cell. Biol. **20**(5): 1868–1876.
88. LEHMAN, J.J., P.M. BARGER, A. KOVACS et al. 2000. Peroxisome proliferator-activated receptor gamma coactivator-1 promotes cardiac mitochondrial biogenesis [in process citation]. J. Clin. Invest. **106**(7): 847–856.
89. KNUTTI, D., A. KAUL & A. KRALLI. 2000. A tissue-specific coactivator of steroid receptors, identified in a functional genetic screen. Mol. Cell. Biol. **20**(7): 2411–2422.
90. MOHAMED-ALI, V., S. GOODRICK, A. RAWESH et al. 1997. Subcutaneous adipose tissue releases interleukin-6, but not tumor necrosis factor-alpha, *in vivo*. J. Clin. Endocrinol. Metab. **82**(12): 4196–4200.
91. BERG, A.H., T.P. COMBS, X. DU et al. The adipocyte-secreted protein Acrp30 enhances hepatic insulin action. Nat. Med. **7**(8): 947–953.

92. LaNoue, K.F. & L.F. Martin. 1994. Abnormal A1 adenosine receptor function in genetic obesity. FASEB J. **8**(1): 72–80.
93. Weyer, C., T. Funahashi, S. Tanaka et al. 2001. Hypoadiponectinemia in obesity and type 2 diabetes: close association with insulin resistance and hyperinsulinemia. J. Clin. Endocrinol. Metab. **86**(5): 1930–1935.
94. Cianflone, K., M. Maslowska & A. Sniderman. 1995. The acylation stimulating protein–adipsin system. Int. J. Obes. Relat. Metab. Disord. **19**(suppl. 1): S34–S38.
95. Xue, B., N. Moustaid, W.O. Wilkison & M.B. Zemel. 1998. The agouti gene product inhibits lipolysis in human adipocytes via a Ca^{2+}-dependent mechanism. FASEB J. **12**(13): 1391–1396.
96. Ailhaud, G., A. Fukamizu, F. Massiera et al. 2000. Angiotensinogen, angiotensin II, and adipose tissue development. Int. J. Obes. Relat. Metab. Disord. **24**(suppl. 4): S33–S35.
97. Witthuhn, B.A. & D.A. Bernlohr. 2001. Upregulation of bone morphogenetic protein GDF-3/Vgr-2 expression in adipose tissue of FABP4/aP2 null mice. Cytokine **14**(3): 129–135.
98. Serrero, G. & N. Lepak. 1996. Endocrine and paracrine negative regulators of adipose differentiation. Int. J. Obes. Relat. Metab. Disord. **20**(suppl. 3): S58–S64.
99. Tabata, Y., M. Miyao, T. Inamoto et al. 2000. De novo formation of adipose tissue by controlled release of basic fibroblast growth factor. Tissue Eng. **6**(3): 279–289.
100. Ramsay, T.G., M.E. White & C.K. Wolverton. 1989. Insulin-like growth factor 1 induction of differentiation of porcine preadipocytes. J. Anim. Sci. **67**(9): 2452–2459.
101. Wabitsch, M., E. Heinze, K.M. Debatin & W.F. Blum. 2000. IGF-I- and IGFBP-3-expression in cultured human preadipocytes and adipocytes. Horm. Metab. Res. **32**(11/12): 555–559.
102. Bruun, J.M., S.B. Pedersen & B. Richelsen. 2000. Interleukin-8 production in human adipose tissue: inhibitory effects of anti-diabetic compounds, the thiazolidinedione ciglitazone, and the biguanide metformin. Horm. Metab. Res. **32**(11/12): 537–541.
103. DiGirolamo, M., F. Newby & J. Lovejoy. 1992. Lactate production in adipose tissue: a regulated function with extra-adipose implications. FASEB J. **6**(7): 2405–2412.
104. Zhang, Y., R. Proenca, M. Maffei et al. 1994. Positional cloning of the mouse obese gene and its human homologue. Nature **372**(6505): 425–432.
105. Pages, C., P. Valet, O. Jeanneton et al. 1999. Alpha2-adrenergic receptor–mediated release of lysophosphatidic acid by adipocytes: a paracrine signal for preadipocyte growth. Lipids **34**(suppl.): S79.
106. Valet, P., C. Pages, O. Jeanneton et al. 1998. Alpha2-adrenergic receptor–mediated release of lysophosphatidic acid by adipocytes: a paracrine signal for preadipocyte growth. J. Clin. Invest. **101**(7): 1431–1438.
107. Lundgren, C.H., S. Brown, T. Nordt et al. 1996. Elaboration of type-1 plasminogen activator inhibitor from adipocytes: a potential pathogenetic link between obesity and cardiovascular disease. Circulation **93**(1): 106–110.
108. Sul, H.S., C. Smas, B. Mei & L. Zhou. 2000. Function of pref-1 as an inhibitor of adipocyte differentiation. Int. J. Obes. Relat. Metab. Disord. **24**(suppl. 4): S15–S19.
109. Darimont, C., G. Vassaux, G. Ailhaud & R. Negrel. 1994. Differentiation of preadipose cells: paracrine role of prostacyclin upon stimulation of adipose cells by angiotensin-II. Endocrinology **135**(5): 2030–2036.
110. Berger, A. 2001. Resistin: a new hormone that links obesity with type 2 diabetes. Br. Med. J. **322**(7280): 193.
111. Lofgren, P., V. van Harmelen, S. Reynisdottir et al. 2000. Secretion of tumor necrosis factor-alpha shows a strong relationship to insulin-stimulated glucose transport in human adipose tissue. Diabetes **49**(5): 688–692.
112. Rajkumar, K., T. Modric & L. Murphy. 1999. Impaired adipogenesis in insulin-like growth factor binding protein-1 transgenic mice. J. Endocrinol. **162**(3): 457–465.
113. Margareto, J., M. Aguado, J.A. Oses-Prieto et al. 2000. A new NPY-antagonist strongly stimulates apoptosis and lipolysis on white adipocytes in an obesity model. Life Sci. **68**(1): 99–107.
114. Labelle, M., Y. Boulanger, A. Fournier et al. 1997. Tissue-specific regulation of fat cell lipolysis by NPY in 6-OHDA-treated rats. Peptides **18**(6): 801–808.

115. BOSTON, B.A. 1999. The role of melanocortins in adipocyte function. Ann. N.Y. Acad. Sci. **885:** 75–84.
116. COUSIN, B., O. MUNOZ, M. ANDRE *et al.* 1999. A role for preadipocytes as macrophage-like cells. FASEB J. **13**(2): 305–312.
117. ROSS, S.E., N. HEMATI, K.A. LONGO *et al.* 2000. Inhibition of adipogenesis by Wnt signaling. Science **289**(5481): 950–953.

Leptin Signaling, Adiposity, and Energy Balance

ERIC JÉQUIER

Institute of Physiology, University of Lausanne, CH-1005 Lausanne, Switzerland

ABSTRACT: A chronic minor imbalance between energy intake and energy expenditure may lead to obesity. Both lean and obese subjects eventually reach energy balance and their body weight regulation implies that the adipose tissue mass is "sensed", leading to appropriate responses of energy intake and energy expenditure. The cloning of the *ob* gene and the identification of its encoded protein, leptin, have provided a system signaling the amount of adipose energy stores to the brain. Leptin, a hormone secreted by fat cells, acts in rodents via hypothalamic receptors to inhibit feeding and increase thermogenesis. A feedback regulatory loop with three distinct steps has been identified: (1) a sensor (leptin production by adipose cells) monitors the size of the adipose tissue mass; (2) hypothalamic centers receive and integrate the intensity of the leptin signal through leptin receptors (LRb); (3) effector systems, including the sympathetic nervous system, control the two main determinants of energy balance—energy intake and energy expenditure. While this feedback regulatory loop is well established in rodents, there are many unsolved questions about its applicability to body weight regulation in humans. The rate of leptin production is related to adiposity, but a large portion of the interindividual variability in plasma leptin concentration is independent of body fatness. Gender is an important factor determining plasma leptin, with women having markedly higher leptin concentrations than men for any given degree of fat mass. The *ob* mRNA expression is also upregulated by glucocorticoids, whereas stimulation of the sympathetic nervous system results in its inhibition. Furthermore, leptin is not a satiety factor in humans because changes in food intake do not induce short-term increases in plasma leptin levels. After its binding to LRb in the hypothalamus, leptin stimulates a specific signaling cascade that results in the inhibition of several orexigenic neuropeptides, while stimulating several anorexigenic peptides. The orexigenic neuropeptides that are downregulated by leptin are NPY (neuropeptide Y), MCH (melanin-concentrating hormone), orexins, and AGRP (agouti-related peptide). The anorexigenic neuropeptides that are upregulated by leptin are α-MSH (α-melanocyte-stimulating hormone), which acts on MC4R (melanocortin-4 receptor); CART (cocaine and amphetamine-regulated transcript); and CRH (corticotropin-releasing-hormone). Obese humans have high plasma leptin concentrations related to the size of adipose tissue, but this elevated leptin signal does not induce the expected responses (i.e., a reduction in food intake and an increase in energy expenditure). This suggests that obese humans are resistant to the effects of endogenous leptin. This resistance is also shown by the lack of effect of exogenous leptin administration to induce weight loss in obese patients. The mechanisms that

---

Address for correspondence: Prof. E. Jéquier, Institut de Physiologie, 7 rue du Bugnon, 1005 Lausanne, Switzerland. Fax: +41-21-692-55-95.
Eric.Jequier@iphysiol.unil.ch

Ann. N.Y. Acad. Sci. 967: 379–388 (2002). © 2002 New York Academy of Sciences.

may account for leptin resistance in human obesity include a limitation of the blood-brain-barrier transport system for leptin and an inhibition of the leptin signaling pathways in leptin-responsive hypothalamic neurons. During periods of energy deficit, the fall in leptin plasma levels exceeds the rate at which fat stores are decreased. Reduction of the leptin signal induces several neuro-endocrine responses that tend to limit weight loss, such as hunger, food-seeking behavior, and suppression of plasma thyroid hormone levels. Conversely, it is unlikely that leptin has evolved to prevent obesity when plenty of palatable foods are available because the elevated plasma leptin levels resulting from the increased adipose tissue mass do not prevent the development of obesity. In conclusion, in humans, the leptin signaling system appears to be mainly involved in maintenance of adequate energy stores for survival during periods of energy deficit. Its role in the etiology of human obesity is only demonstrated in the very rare situations of absence of the leptin signal (mutations of the leptin gene or of the leptin receptor gene), which produces an internal perception of starvation and results in a chronic stimulation of excessive food intake.

KEYWORDS: obesity; body weight; food intake; energy expenditure; starvation

ENERGY BALANCE AND BODY WEIGHT REGULATION

The prevalence of obesity is increasing worldwide, which indicates that the primary cause of obesity lies in environmental and behavioral changes rather than in genetic modifications. Among the environmental influences that disrupt body weight regulation, two factors play a major role: the passive overconsumption of energy-dense, high-fat diets and the decline in physical activity. It is important to emphasize that a minor imbalance between energy intake and energy expenditure may lead to severe obesity: if energy intake exceeds energy expenditure by 5% every day, this results in a gain of 5 kg fat mass over 1 year. A gain of 5 kg fat mass per year leads to morbid obesity over several years, yet a difference of 5% between energy intake and energy expenditure is hardly measurable with available techniques. This simple calculation illustrates why it is so difficult to assess the etiology of obesity in humans.

Obesity results in an increase in the number and the size of adipocytes. Yet, changes in expression of genes that control adipose cell differentiation and development are not primarily responsible for the gain in body weight. Cellular events can cause obesity only if they affect energy balance.[1] The critical issue in addressing the problem of alterations in body weight regulation is not intake or expenditure taken separately, but the adjustment of one to the other under *ad libitum* food intake conditions.[2] Phenomena controlling food intake exert a greater influence on the energy balance than do the small changes in metabolic rates that occur during overfeeding or underfeeding. In addition, daily variations in food intake are large (coefficient of variation of ±23%),[3] whereas daily variations in energy expenditure are small (about ±2% in subjects spending several days in a respiration chamber[4]). Daily variations in energy expenditure mainly depend on lifestyle (i.e., enhanced physical activity with the practice of sports), but most adult individuals with a tendency to weight gain have a sedentary lifestyle.

It has been argued that slight differences in energy expenditure (such as a low resting metabolic rate or a high metabolic efficiency) are important for the development of obesity.[5] Such a concept does not consider that a low metabolic rate is often matched by a low energy intake. A similar reasoning applies to conditions of enhanced energy expenditure: the extensive search for drugs capable of stimulation energy expenditure to treat obesity has been a failure.[6] These drugs could contribute to body weight loss only if food intake remains unchanged; this is unlikely because food intake is stimulated when energy expenditure increases, such as after a period of enhanced physical activity.

The concept of "a low metabolic rate at the origin of weight gain"[5] does not take into account the fact that the gain in body weight over a 4-year follow-up period is accompanied by an enhancement of resting energy expenditure, a process that normalizes the low metabolic rate. A gain of 5–10 kg is sufficient to normalize the metabolic rate and this shows that a low metabolic rate (or a high metabolic efficiency) can explain up to 10 kg weight gain, but greater increases in body weight, which characterize the development of obesity,[7] must result from excessive food intake and/or low physical activity.

If the precise mechanisms responsible for the development of human obesity are difficult to demonstrate, it is, however, possible to understand the reasons for the maintenance of adiposity in obese subjects. Most obese individuals reach energy balance; their body weight and composition remain relatively constant during prolonged periods of time. Energy balance is reached when the enlarged body weight induces an increase in energy expenditure commensurate with the level of energy intake. In addition, the increased fat mass leads to a stimulation of fat oxidation up to a level matching fat intake. Thus, changes in body composition during the development of obesity elicit mechanisms that modify energy expenditure and nutrient oxidation until energy and nutrient balances (i.e., a stable body weight and a constant body composition) are reached.

The new insights in the molecular mechanisms involved in energy homeostasis have enhanced our understanding of how energy balance is maintained for prolonged periods of time. The discovery of leptin, a hormone secreted by fat cells that is involved in the maintenance of energy stores, has been a great advance.[8] This review describes recent developments in this field.

LEPTIN: AN IMPORTANT MEDIATOR IN A FEEDBACK REGULATORY LOOP BETWEEN ADIPOSE TISSUE AND THE HYPOTHALAMUS

The cloning of the *ob* gene and the identification of its encoded protein leptin[8] have provided a system signaling the amount of adipose energy stores to the brain.[9] Leptin acts via hypothalamic receptors to inhibit feeding and increase thermogenesis, resulting in a decrease in body weight.

A feedback regulatory loop with three distinct steps has been identified: (1) a sensor (leptin production by adipose cells) monitors the level of energy stores (the size of the adipose tissue mass); (2) hypothalamic centers receive and integrate the intensity of the leptin signal through leptin receptors; (3) effector systems, including the sympathetic nervous system, control the two main determinants of energy balance— energy intake and energy expenditure.

This feedback regulatory loop seems ideal to maintain a constant body weight. Leptin production is increased following weight gain,[9] whereas leptin plasma levels decrease following weight loss.[10] Chronic administration of leptin to *ob/ob* mice (these obese mice lack leptin because of a mutation of the *ob* gene) causes the animals to lose weight and to maintain their weight loss.[11–13] Leptin administration also reduces food intake and body weight in diet-induced obese (DIO) mice, but these animals need higher leptin doses than *ob/ob* mice, showing a certain degree of resistance to the effects of leptin.[11]

Although this feedback regulatory loop is well established in rodents, there are many unsolved questions about its applicability to body weight regulation in humans. Serum leptin concentration is related to the size of adipose tissue mass in humans.[9,14] The increase in serum leptin concentration involves both the number of adipose cells and the induction of *ob* mRNA per cell. Small fat cells express less *ob* mRNA than large fat cells,[14] the latter being present in obese subjects with hypertrophy and hyperplasia of adipose tissue. When small fat cells fill with lipid, there is a threshold size that causes the stimulation of *ob* gene expression. The mechanisms that produce adequate transcription factors when fat cell size increases may involve metabolites of triacylglycerol, such as diacylglycerol and free fatty acids.[15] Cell stretching may also be a signal for stimulation of *ob* gene expression[16] as an increased tension exogenously applied on cells can induce signaling.[17]

The rate of leptin production is related to adiposity, but a large portion of the interindividual variability in plasma leptin concentration is independent of body fatness. Gender is an important factor determining plasma leptin, with women having markedly higher leptin concentrations than men for any given degree of fat mass.[18] Plasma leptin in women increases during the luteal phase of the menstrual cycle, suggesting that sex hormones play a role in the regulation of leptin secretion by adipose cells.[19,20] The *ob* mRNA expression is also upregulated by glucocorticoids.[21] By contrast, stimulation of the sympathetic nervous system or the increase in circulating epinephrine results in inhibition of *ob* mRNA expression through increased intracellular cAMP levels in adipose cells.[21]

DOES LEPTIN ACT AS A SATIATION OR A SATIETY SIGNAL?

During a meal, the ingestion of food induces a suppression of hunger that leads to the termination of eating. This process is referred to as satiation and involves mechanisms that determine the meal size. Satiation is followed by a period of variable duration that is characterized by the absence of hunger: this is referred to as satiety. Termination of the period of satiety coincides with the resurgence of the feeling of hunger, leading to consumption of the following meal. The mechanisms that promote satiation are different from those that induce satiety.

Leptin could play a role in food intake regulation in rodents by acting as a satiety factor. *Ob* gene expression is stimulated after mice have started eating and remains elevated during several hours thereafter. Furthermore, leptin administration to mice acutely reduces food intake.[11] By contrast, fasting decreased leptin plasma levels, and refeeding fasted rats restored *ob* mRNA within 4 hours to levels of fed animals.[10] In rats, insulin is an important stimulus for *ob* gene expression: injection of insulin into fasted rats doubled mRNA in adipose tissue cells within 4 hours.[10]

Leptin is neither a satiation nor a satiety factor in humans because acute changes in food intake do not induce short-term increases in plasma leptin levels. The postprandial rise in serum insulin concentration is not associated with a stimulation of leptin secretion.[9] Leptin secretion exhibits a circadian rhythm with a nocturnal rise over daytime secretion, but these changes in leptin plasma concentrations are not influenced by meal ingestion and meal-induced increases in circulating insulin concentration. Leptin plasma concentrations are low around noon and begin to rise towards 3 PM; they reach maximal values during the night. The nocturnal increase in plasma leptin concentration precedes the early morning rise of ACTH and cortisol, showing that the nocturnal rise in plasma leptin does not result from induction of *ob* gene expression by cortisol. Since leptin secretion is not related to meal intake, leptin cannot be involved in the control of meal size.

HYPOTHALAMIC NEURAL CIRCUITS CONTROLLED BY LEPTIN

Leptin binds to leptin receptors on two populations of hypothalamic neurons. One population of neurons synthesize and release two orexigenic (feeding inducing) neuropeptides: the neuropeptide Y (NPY) and the agouti-related peptide (AGRP). Leptin reduces the expression of these two neuropeptides. Another population of neurons express genes encoding the anorexigenic peptide α-melanocyte-stimulating hormone (α-MSH), which is derived from pro-opiomelanocortin (POMC). The latter neurons also express cocaine and amphetamine-related transcript (CART). Leptin induces the expression of these two anorexigenic peptides (α-MSH and CART). Thus, the leptin-induced inhibition of food intake results from both the suppression of orexigenic and the induction of anorexigenic neuropeptides. The presence of these redundant pathways to control a key behavior (i.e., induction of feeding) may explain why recent investigations on mutant mice deficient for NPY show that these animals maintain a normal body weight and adiposity in spite of the lack of this important orexigenic neuropeptide.[22] Treatment of these NPY-deficient mice with leptin for 5 days significantly reduced their food intake and body weight, indicating that leptin can suppress feeding via signaling pathways independent of NPY.

The neuropeptide α-MSH is an agonist of the melanocortin-4 receptor (MC4R), which reduces food intake when activated, while AGRP is an antagonist of this receptor. About 5% of severe human obesity is due to a mutation of the MC4R gene.[23] The α-MSH pathway leading to stimulation of MC4R is important for the regulation of human energy balance. This is illustrated by the fact that mutations in the POMC gene, which prevents the production of α-MSH, produce severe human obesity.[24] MC4R and POMC mutations result in a syndrome of hyperphagic obesity in humans, indicating that the central melanocortin pathway is mainly involved in the control of food intake.

OBESITY AND LEPTIN RESISTANCE

Rodents with dietary-induced obesity (DIO) are hyperleptinemic and they become resistant to the effects of exogenous leptin administration. These obese rodents do respond, however, to administration of low leptin doses by the intracerebroventricular route, showing that a defect in the blood-brain barrier plays a role in these animals.[25,26] Peripherally administered leptin to DIO mice inhibited food intake after 4 days of exposure to a high-fat diet, but the mice became resistant to peripheral leptin administration after 16 days of the high-fat diet.[26] By contrast, a leptin dose (0.1 µg) that was 4000 times smaller than the latter, given directly into the CNS through icv cannula, inhibited food intake and decreased body weight in these leptin-resistant DIO mice.

Similar to rodents with DIO, human obese subjects have high plasma leptin concentrations that do not induce the expected responses (i.e., a reduction in food intake and an increase in energy expenditure). This suggests that obese humans are resistant to the effects of endogenous leptin.[9,27] Leptin resistance is also demonstrated in obese patients by the lack of effect of exogenous leptin administration to induce weight loss.[28] The mechanisms that may account for leptin resistance in human obesity include a limitation of the blood-brain-barrier leptin transport system,[29,30] a mutated leptin receptor in the hypothalamus,[31] or an inhibition of the leptin signaling mechanisms in leptin-responsive neurons of the hypothalamus.[32–34]

Leptin resistance might result from a defect in the transporter system of leptin through the blood-brain barrier.[25,29] This transporter system is mediated by the short form of the leptin receptor (LRa), a protein expressed in the choroid plexus that acts to transport leptin into the cerebrospinal fluid (CSF).[35] This leptin transport system is saturable[25,29] and its efficiency (measured as the CSF/plasma leptin concentration ratio) is lower in obese than in lean individuals. The CSF leptin concentration is probably similar to the hypothalamic interstitial leptin concentration and, thus, a transport defect may explain why obese persons do not have the expected responses to hyperleptinemia (i.e., a reduction in food intake and an increase in resting energy expenditure). There is a threshold plasma leptin concentration (about 25 ng/mL) above which the transport of leptin into the CSF does not increase anymore in spite of high values of plasma leptin concentration. Hence, in patients with severe obesity, the large increase in leptin plasma levels may not be adequately translated in regulatory signals to the hypothalamus to control the excessive weight gain.

After its transport through the blood-brain barrier mediated by the LRa, leptin binds and activates the long form of the leptin receptor (LRb) in the hypothalamus.[36] Leptin resistance in human obesity very rarely results from a mutation of the LRb,[31] and most obese patients have an unaltered LRb.[37] Another mechanism that might explain leptin resistance in obese humans is a downregulation of leptin transporters induced by hyperleptinemia, as demonstrated in *db/db* mice.[38,39]

Several mechanisms of inhibition of the leptin signaling pathways in leptin-responsive neurons of the hypothalamus have been recently described.[32–34] Leptin administration to normal mice induces hypothalamic signaling through the STAT3 pathway. After 15 weeks on a high-fat diet, peripherally administered leptin was unable to activate hypothalamic STAT3 signaling, whereas icv leptin administration induced STAT3 activation, but of a smaller magnitude than in control mice.[40] These results suggest the existence of two sites of leptin resistance in these DIO mice: a defect in access of leptin to the hypothalamic sites of action and an intracellular signaling defect upstream of STAT3 activation.

LRb signaling is initiated by leptin binding to LRb, leading to phosphorylation of two tyrosine residues through Jak 2 activation. This mediates distinct signals, such as STAT3 phosphorylation, which activates transcription of the SOCS3 (suppressor of cytokine signaling) gene. After long-time stimulation, the SOCS3-translated protein binds to phosphorylated Tyr of LRb and thereby inhibits the various LRb-mediated signals.[33,34]

The consequence of these recent findings is an incentive for pharmaceutical developments to create small leptin analogues with increased capacity to cross the blood-brain barrier. On the other hand, this approach can be limited by factors that inhibit leptin signaling. The understanding of the precise molecular mechanisms involved in the inhibition of leptin signaling in hypothalamic neurons is an important goal for the development of new drugs that may enhance leptin's effects for treating human obesity.

DIRECT ACTIONS OF LEPTIN ON PERIPHERAL TARGET CELLS

The LRb is not only expressed in hypothalamic cells, but also in tissues including skeletal muscle, liver, adipose tissue, and pancreatic β cells. In tissues such as muscle, leptin stimulates lipid oxidation, which would promote insulin sensitivity.[41,42] Unger[43] has proposed that an important physiologic role of leptin is to prevent ectopic lipogenesis through its antilipogenic effect. The *fa/fa* ZDF rats, in which all leptin receptors are mutated, have a large increase in the triglyceride content of their pancreatic islets, which is associated with impairment of β-cell function. In these animals, hyperphagia leads to hyperinsulinemia, which upregulates transcription factors leading to hypertrophy of adipocytes. Furthermore, secretion of insulin-like growth factor I (IGF-1) and other adipocyte products induces adipocyte hyperplasia. In nonadipocytes, the hyperphagia-induced hyperinsulinemia results in deposition of triacylglycerol, providing an excess of fatty acid substrate for damaging pathways of nonoxidative metabolism of fatty acids, such as ceramide synthesis.[44] This ectopic deposition of triacylglycerol in β cells and myocardium causes functional impairment, which may lead to diabetes and cardiomyopathy. Thus, the chronic lack of leptin action in the *fa/fa* ZDF rats enhances lipogenesis in nonadipose tissues, predisposing these tissues to genetic lipotoxicity.

Diet-induced lipotoxicity develops at a much slower rate than genetic lipotoxicity. In young DIO rodents, the increasing hyperleptinemia exerts a protective action by stimulating lipid oxidation in nonadipose tissues. Later, when obesity has developed, leptin resistance appears not only in hypothalamic cells, but also in various peripheral nonadipose tissues. In obese humans, cardiomyopathy related to lipotoxicity is not yet recognized as a clinical entity, but recent evidence shows that lipid excess can occur in the myocardium of obese humans.[44] Lipotoxicity in human islet cells has not yet been clearly demonstrated as a case of diabetes, but increased neutral lipids have been recently demonstrated in β cells of older humans.[45] Furthermore, lipopenic drugs, such as troglitazone, lower the islet lipid content in obese prediabetic rats[46] and improve glucose tolerance.

LEPTIN IS NOT AN ANTIOBESITY HORMONE

During evolution, there was no pressure against obesity, and the thrifty gene hypothesis argues that transient obesity may be a survival asset.[47] It is unlikely that leptin has evolved to prevent obesity during exposure to nutritional excess since obese individuals become resistant to leptin action. Hyperleptinemia is not an efficient signal to the CNS due to a limitation of leptin transport across the blood-brain barrier or to negative feedback signals that inhibit leptin signaling.[48] Leptin resistance may have arisen during evolution to facilitate energy storage in times of plenty in order to avoid the consequences of starvation when food availability was limited.[1]

Whereas leptin secretion in humans is not acutely modulated by single meal ingestion, a more chronic stimulus of food intake, such as a decrease or an increase over several days, does affect leptin secretion. For instance, fasting over several days, which induced a decrease of 10% in body weight, was associated with 53% reduction in serum leptin.[9] Thus, a reduced energy intake, which is accompanied by a low fasting plasma insulin concentration, induces a large decrease in leptin secretion[49,50] unrelated to the small reduction in adipose tissue mass. The decline in leptin levels could contribute to the reduction in energy expenditure resulting from weight loss. During starvation, the fasting-induced decrease in leptin secretion may lead to energy conservation by decreasing thyroid hormone–induced thermogenesis; in addition, the lowering of plasma leptin levels favors mobilization of energy stores by increasing secretion of glucocorticoids.[51–53] Hence, a sharp decline in leptin plasma levels plays a major role in the body's short-term adaptation to starvation. In humans, leptin is considered more as a hormone involved in the survival of people exposed to situations of famine rather than a hormone that can limit the excessive weight gain of people exposed to an environment of plenty.

REFERENCES

1. SPIEGELMAN, B.M. & J.S. FLIER. 2001. Obesity and the regulation of energy balance. Cell **104:** 531–543.
2. FLATT, J.P. 1997. How not to approach the obesity problem. Obes. Res. **5:** 632–633.
3. BINGHAM, S.A., C. GILL, A. WELCH et al. 1995. Comparison of dietary assessment methods in nutritional epidemiology: weighed records v. 24 h recalls, food-frequency questionnaires, and estimated-diet records. Br. J. Nutr. **74:** 141–143.
4. JÉQUIER, E. & Y. SCHUTZ. 1983. Long-term measurements of energy expenditure in humans using a respiration chamber. Am. J. Clin. Nutr. **38:** 989–998.
5. RAVUSSIN, E., S. LILLIOJA, W.C. KNOWLER et al. 1988. Reduced rate of energy expenditure as a risk factor for body-weight gain. N. Engl. J. Med. **318:** 467–472.
6. WEYER, C., J.F. GAUTIER & E.J. DANFORTH. 1999. Development of beta 3-adrenoceptor agonists for the treatment of obesity and diabetes—an update. Diabetes Metab. **25:** 11–21.
7. JÉQUIER, E. & Y. SCHUTZ. 1985. New evidence for a thermogenic defect in human obesity. Int. J. Obes. **9:** 1–7.
8. ZHANG, Y., R. PROENCA, M. MAFFEI et al. 1994. Positional cloning of the mouse *ob* gene and its human homologue. Nature **372:** 425–432.
9. CONSIDINE, R.V., M.K. SINHA, M.L. HEIMAN et al. 1996. Serum immunoreactive-leptin concentrations in normal-weight and obese humans. N. Engl. J. Med. **334:** 292–295.
10. SALADIN, R., P. DEVOS, M. GUERRO-MILLO et al. 1995. Transient increase in obese gene expression after food intake or insulin administration. Nature **377:** 527–529.
11. CAMPFIELD, L.A., F.J. SMITH, Y. GUISEZ et al. 1995. Recombinant mouse OB protein: evidence for a peripheral signal linking adiposity and central neural networks. Science **269:** 546–549.
12. HALAAS, J.L., K.S. GAJIWALA, M. MAFFEI et al. 1995. Weight-reducing effects of the plasma protein encoded by the obese gene. Science **269:** 543–546.

13. PELLEYMOUNTER, M.A., M.J. CULLEN, M.B. BAKER et al. 1995. Effects of the *obese* gene product on body weight regulation in *ob/ob* mice. Science **269:** 540–543.
14. HAMILTON, B.S., D. PAGLIA, A.Y.M. KWAN & M. DEITEL. 1995. Increased obese mRNA expression in omental fat cells from massively obese humans. Nat. Med. **1:** 953–956.
15. LISCOVITCH, M. & L.C. CANTLEY. 1994. Lipid second messengers. Cell **77:** 329–334.
16. AILHAUD, G., P. GRIMALDI & R. NÉGREL. 1992. Cellular and molecular aspects of adipose tissue development. Annu. Rev. Nutr. **12:** 207–233.
17. LI, X. & R.B. NICKLAS. 1995. Mitotic forces control a cell-cycle checkpoint. Nature **377:** 630–632.
18. SAAD, M.F., S. DAMANI, R.L. GINGERICH et al. 1997. Sexual dimorphism in plasma leptin concentration. J. Clin. Endocrinol. Metab. **82:** 579–584.
19. HAFFNER, S.M., H. MIETTINEN, P. KARHAPAA et al. 1997. Leptin concentrations, sex hormones, and cortisol in nondiabetic men. J. Clin. Endocrinol. Metab. **82:** 1807–1809.
20. THOMAS, T., B. BURGUERA, L.J. MELTON III et al. 2000. Relationship of serum leptin levels with body composition and sex steroid and insulin levels in men and women. Metabolism **49:** 1278–1284.
21. SLIEKER, L.J., K.W. SLOOP, P.L. SURFACE et al. 1996. Regulation of expression of *ob* mRNA and protein by glucocorticoids and cAMP. J. Biol. Chem. **271:** 5301–5304.
22. ERICKSON, J.C., K.E. CLEGG & R.D. PALMITER. 1996. Sensitivity to leptin and susceptibility to seizures of mice lacking neuropeptideY. Nature **381:** 415–421.
23. FAROOQI, I.S., G.S. YEO, J.M. KEOGH et al. 2000. Dominant and recessive inheritance of morbid obesity associated with melanocortin 4 receptor deficiency. J. Clin. Invest. **106:** 271–279.
24. KRUDE, H., H. BIEBERMANN, W. LUCK et al. 1998. Severe early-onset obesity, adrenal insufficiency, and red hair pigmentation caused by POMC mutations in humans. Nat. Genet. **19:** 155–157.
25. BANKS, W.A., A.J. KASTIN, W. HUANG et al. 1996. Leptin enters the brain by a saturable system independent of insulin. Peptides **17:** 305–311.
26. VAN HEEK, M., D.S. COMPTON, C.F. FRANCE et al. 1997. Diet-induced obese mice develop peripheral, but not central, resistance to leptin. J. Clin. Invest. **99:** 385–390.
27. MAFFEI, M., J. HALAAS, E. RAVUSSIN et al. 1995. Leptin levels in human and rodent: measurement of plasma leptin and *ob* RNA in obese and weight-reduced subjects. Nat. Med. **1:** 1155–1161.
28. HEYMSFIELD, S.B., A.S. GREENBERG, K. FUJIOKA et al. 1999. Recombinant leptin for weight loss in obese and lean adults: a randomized, controlled, dose-escalation trial. JAMA **282:** 1568–1575.
29. CARO, J., J.W. KOLACZYNSKI, M.R. NYCE et al. 1996. Decreased cerebrospinal-fluid/serum leptin ratio in obesity: a possible mechanism for leptin resistance. Lancet **348:** 159–161.
30. SCHWARTZ, M.W., E. PESKIND, M. RASKIND et al. 1996. Cerebrospinal fluid leptin levels: relationship to plasma levels and to adiposity in humans. Nat. Med. **2:** 589–592.
31. CLÉMENT, K., C. VAISSE, N. LAHLOU et al. 1998. A mutation in the human leptin receptor gene causes obesity and pituitary dysfunction. Nature **392:** 398–401.
32. VAISSE, C., J.L. HALAAS, C.M. HORVATH et al. 1996. Leptin activation of Stat3 in the hypothalamus of wild-type and ob/ob mice, but not db/db mice. Nat. Genet. **14:** 95–97.
33. BJORBAEK, C., H.J. LAVERY, S.H. BATES et al. 2000. SOCS3 mediates feedback inhibition of the leptin receptor viaTyr985. J. Biol. Chem. **275:** 40649–40657.
34. BJORBAEK, C., R.M. BUCHHOLZ, S.M. DAVIS et al. 2001. Divergent roles of SHP-2 in ERK activation by leptin receptors. J. Biol. Chem. **276:** 4747–4755.
35. LEE, G.H., R. PROENCA, J.M. MONTEZ et al. 1996. Abnormal splicing of the leptin receptor in diabetic mice. Nature **379:** 632–635.
36. ELMQUIST, J.K., C.F. ELIAS & C.B. SAPER. 1999. From lesions to leptin: hypothalamic control of food intake and body weight. Neuron **22:** 221–232.
37. CONSIDINE, R.V., E.L. CONSIDINE, C.J. WILLIAMS et al. 1996. The hypothalamic leptin receptor in humans. Diabetes **19:** 992–994.
38. LYNN, R.B., G.U. CAO, R.V. CONSIDINE et al. 1996. Autoradiographic localization of leptin binding in the choroid plexus of *ob/ob* and *db/db* mice. Biochem. Biophys. Res. Commun. **219:** 884.

39. MALIK, K.F. & W.S. YOUNG. 1996. Localization of binding sites in the central nervous system for leptin (OB protein) in normal, obese (*ob/ob*), and diabetic (*db/db*) C57BL/6J mice. Endocrinology **137:** 1497–1500.
40. EL-HASCHIMI, K., D.D. PIERROZ, S.M. HILEMAN *et al.* 2000. Two defects contribute to hypothalamic leptin resistance in mice with diet-induced obesity. J. Clin. Invest. **105:** 1827–1832.
41. MUOIO, D.M., G.L. DOHN, F.T. FIEDOREK, JR. *et al.* 1997. Leptin directly alters lipid partitioning in skeletal muscle. Diabetes **48:** 1360–1363.
42. SHIMABUKURO, M., K. KOYAMA, G. CHEN *et al.* 1997. Direct antidiabetic effect of leptin through triglyceride depletion of tissues. Proc. Natl. Acad. Sci. U.S.A. **94:** 4637–4641.
43. UNGER, R.H. 2000. Leptin physiology: a second look. Regul. Pept. **92:** 87–95.
44. UNGER, R.H. & L. ORCI. 2001. Diseases of liporegulation: new perspective on obesity and related disorders. FASEB J. **15:** 312–321.
45. CNOP, M., A. GRUPPING, A. HOORENS *et al.* 2000. Endocytosis of low-density lipoprotein by human pancreatic beta cells and uptake in lipid-storing vesicles, which increase with age. Am. J. Pathol. **156:** 237–244.
46. HIGA, M., Y.T. ZHOU, M. RAVAZZOLA *et al.* 1999. Troglitazone prevents mitochondrial alterations, beta cell destruction, and diabetes in obese prediabetic rats. Proc. Natl. Acad. Sci. U.S.A. **96:** 11513–11518.
47. NEEL, J.V. 1999. The "thrifty genotype" in 1998. Nutr. Rev. **57:** S2–S9.
48. BJORBAEK, C., K. EL-HASCHIMI, J.D. FRANTZ & J.S. FLIER. 1999. The role of SOCS-3 in leptin signaling and leptin resistance. J. Biol. Chem. **274:** 30059–30065.
49. BODEN, G. 1997. Role of fatty acids in the pathogenesis of insulin resistance and NIDDM. Diabetes **46:** 3–10.
50. KOLACZYNSKI, J.W., R.V. CONSIDINE, J. OHANNESIAN *et al.* 1996. Responses of leptin to short-term fasting and refeeding in humans. Diabetes **45:** 1511–1515.
51. AHIMA, R.S., D. PRABAKARAN, C. MANTZOROS *et al.* 1996. Role of leptin in the neuroendocrine response to fasting. Nature **382:** 250–252.
52. LEGRADI, G., C. EMERSON, R. AHIMA *et al.* 1997. Leptin prevents fasting-induced suppression of prothyrotropin-releasing hormone messenger ribonucleic acid in neurons of the hypothalamic paraventricular nucleus. Endocrinology **138:** 2569–2576.
53. YU, W.H., M. KIMURA, A. WALCZEWSKA *et al.* 1997. Role of leptin in hypothalamic-pituitary function. Proc. Natl. Acad. Sci. U.S.A. **94:** 1023–1028.

Neuroimaging and Obesity

Mapping the Brain Responses to Hunger and Satiation in Humans Using Positron Emission Tomography

ANGELO DEL PARIGI,[a] JEAN-FRANCOIS GAUTIER,[a] KEWEI CHEN,[b] ARLINE D. SALBE,[a] ERIC RAVUSSIN,[a] ERIC REIMAN,[b] AND P. ANTONIO TATARANNI[a]

[a]*Clinical Diabetes and Nutrition Section, National Institute of Diabetes and Digestive and Kidney Diseases, National Institutes of Health, Phoenix, Arizona 85016, USA*

[b]*Good Samaritan Regional Medical Center, Phoenix, Arizona 85012, USA*

ABSTRACT: The hypothalamus has a major role in the control of food intake. However, neurotracing studies have shown that the hypothalamus receives input from several other regions of the brain that are likely to modulate its activity. Of particular interest to the understanding of human eating behavior is the possible involvement of the cortex. Using positron emission tomography (PET), we generated functional brain maps of the neuroanatomical correlates of hunger (after a 36-h fast) and satiation (after oral administration of a liquid formula meal) in lean and obese subjects. Results in lean individuals indicate that the neuroanatomical correlates of hunger form a complex network of brain regions including the hypothalamus, thalamus, and several limbic/paralimbic areas such as the insula, hippocampal/parahippocampal formation, and the orbitofrontal cortex. Satiation was associated with preferentially increased neuronal activity in the prefrontal cortex. Our studies also indicate that the brain responses to hunger/satiation in the hypothalamus, limbic/paralimbic areas (commonly associated with the regulation of emotion), and prefrontal cortex (thought to be involved in the inhibition of inappropriate response tendencies) might be different in obese and lean individuals. In conclusion, neuroimaging of the human brain is proving to be an important tool for understanding the complexity of brain involvement in the regulation of eating behavior. PET studies might help to unravel the neuropathophysiology underlying human obesity.

KEYWORDS: brain; humans; hunger; hypothalamus; mapping; neuroimaging; obesity; positron emission tomography (PET); satiation

INTRODUCTION

The critical role of the brain in the maintenance of the body's homeostasis has long been recognized. However, unprecedented advances in understanding the physiology of energy and body weight regulation have been made with the discovery of

Address for correspondence: P. Antonio Tataranni, Clinical Diabetes and Nutrition Section, NIH-NIDDK, 4212 North 16th Street, Room 541, Phoenix, AZ 85016. Voice: 602-200-5301; fax: 602-200-5335.
antoniot@mail.nih.gov

leptin,[1] a hormone that is secreted by adipose tissue and binds to specific receptors in the brain, primarily in the hypothalamus.[2] Leptin exerts its anorectic effects by controlling the electrical activity and neuropeptidergic content and release of a complex network of neurons localized in several hypothalamic nuclei.[3] Leptin is also thought to modulate short-term signals for the regulation of food intake (e.g., nutrients, insulin, gut peptides).[4]

The discovery of leptin irrefutably established that body fat content is under homeostatic control.[5,6] However, while we continue to refine the understanding of these control mechanisms, we cannot lose sight of the fact that animals and humans seldom eat in response to acute changes in energy balance, which makes eating patterns generally idiosyncratic. Eating is not a simple, stereotypic behavior, but requires a complex set of tasks to be carried out by the central and peripheral nervous system to coordinate the initiation of a meal episode, procurement of food, consumption of the procured food, and termination of the meal.[7] Most of these tasks are not part of the innate drive to eat, but are behaviors learned after weaning.[8] Experimental evidence from neurotracing studies indicates that the hypothalamus receives neural inputs from the amygdala, hippocampal complex, insula, and several polymodal association areas of the prefrontal cortex.[7,9] Therefore, the hypothalamus is likely to be only one of several brain centers involved in the regulation of eating behavior, all of which integrate central and peripheral signals before in turn transmitting impulses to sets of yet unidentified efferent sites. Thus, it is vital to the understanding of the neurophysiology of eating behavior that the global participation of the brain in these processes be elucidated. Of particular interest to the understanding of human eating behavior is the possible involvement of cortical regions. We submit that noninvasive functional neuroimaging represents one possible approach to achieve this goal.

FUNCTIONAL IMAGING OF THE HUMAN BRAIN

During the past two decades, there has been a remarkable development of noninvasive brain imaging techniques. Together, brain imaging techniques and measures of local neuronal activity have permitted researchers to investigate regions of the brain that are functionally related to a variety of sensory and cognitive tasks.[10] Functional neuroimaging techniques include positron emission tomography (PET), functional magnetic resonance imaging (fMRI), magnetoencephalography (MEG), electroencephalography (EEG), and several other methods. A detailed description of the principles and specific applications of each of these techniques is beyond the scope of this manuscript and the reader is referred to reviews on this topic.[11–13]

Measurement of local neuronal activity by PET requires the use of certain positron-emitting radiotracers (e.g., ^{15}O-water for the measurement of cerebral blood flow [CBF] and ^{18}F-fluorodeoxyglucose for the measurement of the cerebral metabolic rate for glucose). A positron-emitting radioisotope, which is typically generated using a cyclotron, has an unstable nucleus. For the measurement of regional CBF, ^{15}O-water is administered intravenously and distributed to tissues throughout the body. Since this tracer readily diffuses across the blood-brain barrier, it is suitable for the measurement of CBF, which in turn is an established marker of local neuronal activity.[14] The unstable nucleus of ^{15}O decays with a half-life of approximately 2 min by emitting positrons that travel a short distance from the nucleus before colliding with an electron in the surrounding tissue. This results in an annihilation of the two particles and generation of two γ rays at 180° from one another. The site of the annihilation is recorded as the site of the signal by one pair of the collinear detectors encircling the subject's head. Because the annihilation typically takes place 1–6 mm away from the emitting nuclei, this limits the spatial resolution of PET. One of the great advantages of ^{15}O-water is its 2-min radioactive half-life, which makes it possible to acquire multiple images during a single scanning session and thus makes it possible for each person to serve as his or her own control. Subtraction of an experimental PET image from a baseline image improves a researcher's ability to detect and localize state-dependent changes in regional CBF. Automated algorithms permit researchers to convert each person's PET images (or each person's "subtraction images") into the coordinates of a standard brain atlas, average data from different subjects, and produce statistical parametric maps of state-dependent changes in regional CBF, further increasing a researcher's ability to distinguish subtle changes in regional brain activity from noise.[14] When superimposed onto an MRI scan of the same subject, these maps allow precise identification of the neuroanatomical location of the change in neuronal activity (FIG. 1).

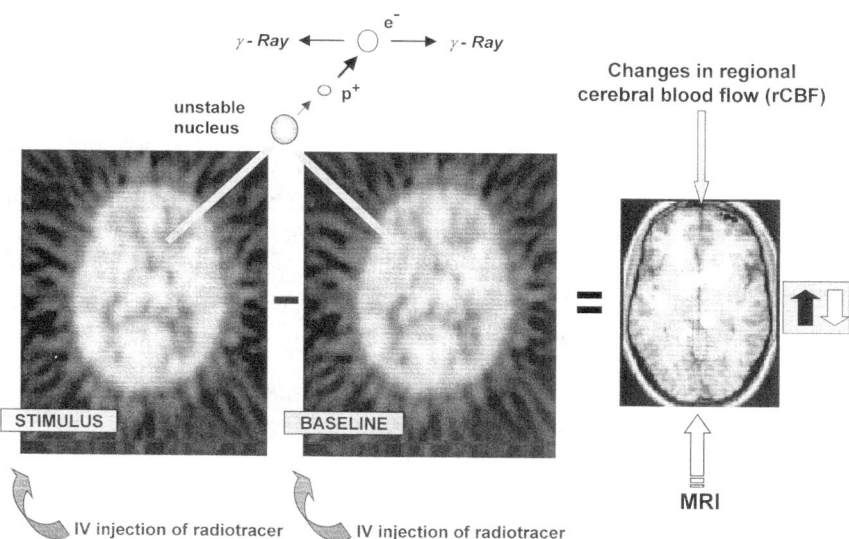

FIGURE 1. PET imaging requires peripheral injection of radiotracers with unstable nuclei, which emit positrons (p^+). The p^+ travels a certain distance before it collides with an electron (e^-) and is annihilated. This results in the emission of γ rays that travel in opposite directions and are detected by one pair of the collinear detectors encircling the head of the individual, resulting in the gray scale brain images presented in the left and central portion of the figure. Subtraction image averaging (stimulus minus baseline) and statistical parametric mapping are then used to detect and localize state-dependent changes in regional CBF (rCBF), shown in gray scale in the brain image presented in the right side of the figure. When superimposed onto an MRI scan of the same subject, these maps allow precise identification of the neuroanatomical location of the change in neuronal activity.

TABLE 1. Population characteristics

	Obese ($n = 23$)	Lean ($n = 21$)	Obese vs. lean p
Gender (M/F)	11/12	11/10	ns
Age (years)	29 ± 6 [18–39]	33 ± 9 [19–49]	0.05
Weight (kg)	112.4 ± 12.5 [93.4–132.9]	67.2 ± 9.9 [52.3–85.3]	<0.0001
Height (m)	1.68 ± 0.07 [1.57–1.81]	1.71 ± 0.10 [1.58–1.88]	ns
BMI (kg/m^2)	39.6 ± 3.8 [34.3–47.8]	22.9 ± 2.2 [19.2–25.6]	<0.0001
Body fat (%)	39 ± 5 [27–51]	22 ± 7 [8–33]	<0.0001
REE (kcal/day)	1812 ± 310 [1410–2835]	1366 ± 194 [1091–1800]	<0.0001

NOTE: Mean ± SD [range]. Group differences were tested by χ^2 statistics (gender) and Student's t test using the procedures of the SAS Institute (Version 6, Cary, NC).

NEUROANATOMICAL CORRELATES OF HUNGER AND SATIATION

We have used PET and ^{15}O-water to study the brain response to hunger (after a 36-h fast) and satiation (in response to a liquid meal providing 50% of the resting energy requirements) in lean and obese men and women (TABLE 1).[15–18] The general experimental design has been described in detail elsewhere[15,19] and is schematically represented in FIGURE 2. An example of the brain responses to hunger and satiation in men and women is shown in FIGURE 3. We employed this approach to answer three main questions:

(1) Can the neuroanatomical correlates of hunger and satiation be imaged in humans?
(2) Are there selective differences in the brain response to a meal between lean and obese individuals?
(3) What is the pathophysiological relevance, if any, of these differences?

As for the first question, our results demonstrated that the administration of a meal to hungry individuals was associated with increased neuronal activity in the prefrontal cortex (generally involved in inhibition of inappropriate response tendencies[20]) and decreased neuronal activity in the hypothalamus, thalamus, several limbic/paralimbic areas (generally involved in affect and motivation), basal ganglia, temporal cortex, and cerebellum. Among the limbic/paralimbic areas, we observed decreased activity in response to the meal in the insular cortex (a visceral sensory area also involved in monitoring distressing and potentially dangerous internal sensations[20]), the anterior cingulate (selectively involved in response to noxious stimuli[21]), and the orbitofrontal cortex (previously shown to respond to hunger in nonhuman primates[22]). Some of these findings were recently replicated in a study of the changes in brain activity related to eating solid food.[23] We postulated that the

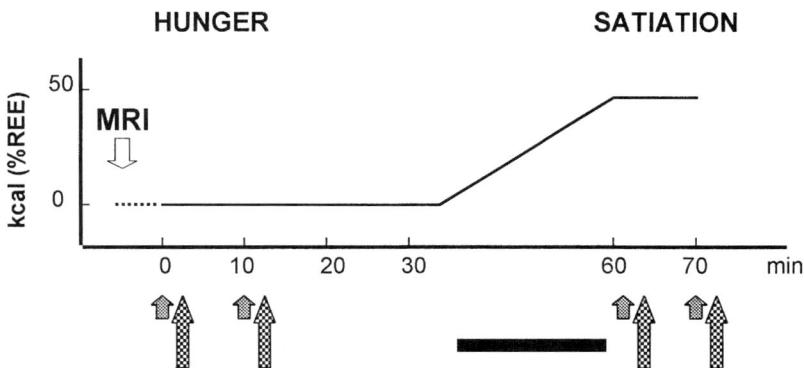

FIGURE 2. Schematic representation of the experimental design. Following an MRI scan to obtain the neuroanatomical detail, individuals undergo four consecutive PET scans (checkered arrows) to measure regional cerebral blood flow (rCBF) in response to hunger (after a 36-h fast) and satiation (after administration of a liquid meal delivering 50% of the daily energy requirements) (black bar). The gray arrows preceding the PET scans indicate administration of 2 mL of drinking water. This is done to cancel out the possible confounding effects of swallowing and the tactile stimulation of the tongue.

activation of the prefrontal cortex is an important component of the central response aimed at promoting the termination of a feeding episode. We further postulated that the hypothalamus, thalamus, limbic/paralimbic areas, and basal ganglia represent a central orexigenic network. Because the prefrontal cortex has efferent inhibitory projections to this central orexigenic network (especially to the lateral hypothalamus), we concluded that the prefrontal cortex exerts inhibiting effects on eating by suppressing the neuronal activity of these brain areas.

As for the second question, we observed that obese individuals respond to satiation with greater activation of the prefrontal cortex and greater deactivation of some of the limbic/paralimbic areas compared to lean subjects (FIG. 4).[16] These responses were generally consistent in men and women.[17,18] In men only, deactivation of the hypothalamus, thalamus, and anterior cingulate was attenuated in obese compared to lean subjects.[16] A similar observation was reported in lean and obese individuals in response to the ingestion of glucose.[24] The explanation for the greater postmeal activation of the prefrontal areas in obese compared to lean subjects remains unclear. It is possible that the greater neuronal activity of the inhibiting areas (prefrontal cortex) may simply be proportional to the neuronal activity of the inhibited areas (limbic/paralimbic areas, etc.). Alternatively, differences in macronutrient and hormonal postprandial excursions between obese and lean individuals may underlie the observed differences in brain response. We found significant associations between postprandial changes in insulin (FIG. 5), glucose, and FFA and postprandial changes in neuronal activity in several areas of the brain, which suggests that hormones and metabolites might act as modulators of postprandial neuronal events. In some instances, the correlations between postprandial changes in hormones/metabolites and neuronal activity were in opposite directions in obese and lean individuals.[16,17]

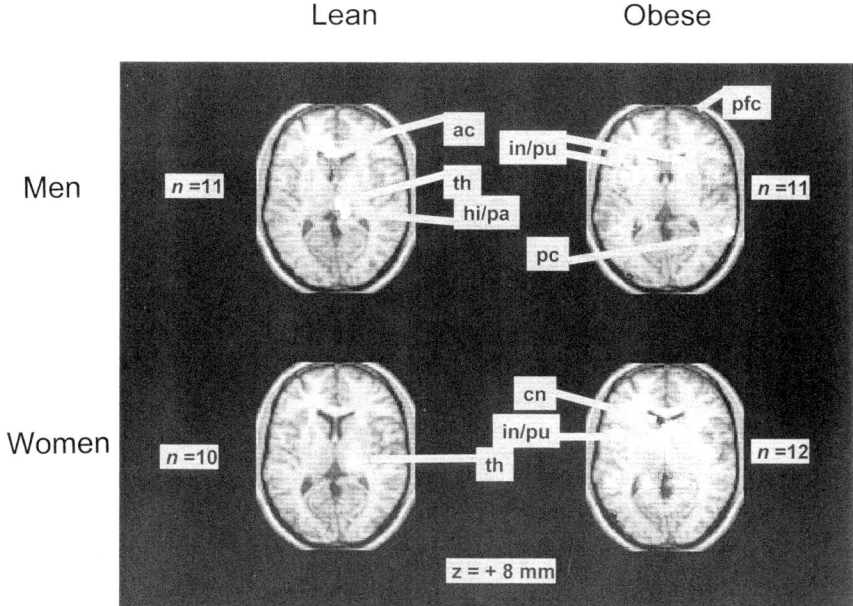

FIGURE 3. Images of brain responses to hunger and satiation in lean and obese men and women at +8 mm from a horizontal plane between the anterior and posterior commissures (coordinates of the Talairach and Tournoux brain atlas). The right hemisphere in each section is on the reader's right. Brain regions with significant increase in rCBF in response to satiation are shown in black; brain regions with significant decrease in rCBF in response to satiation are shown in light gray. Images were generated using PET and MRI data. Gray-coded images are superimposed onto the MRI template. The figure is intended for visual inspection only of several areas of the brain, including anterior cingulate (ac), insula/putamen (in/pu), thalamus (th), hippocampus/parahippocampal gyrus (hi/pa), caudate nucleus (cn), parietal cortex (pc), and prefrontal cortex (pfc).

Whether or not the observed differences in the brain response to a meal in obese and lean individuals have any pathophysiological relevance remains to be proven. Prader-Willi syndrome (PWS) is the most common form of inherited obesity (1 in 16,000 individuals) and is characterized by severe hyperphagia, gross overweight, mild mental retardation, and altered sexual maturation.[25] By comparing the meal-induced brain response in individuals with PWS, whose neurophysiology is likely to be more homogeneous by virtue of their genetic disorder, with the brain response of individuals with obesity of unknown origin, we hope to identify with more certainty areas of the brain closely linked to the hyperphagia that leads to obesity. By studying meal-induced brain responses in postobese individuals who have successfully achieved and maintained a near-normal body weight by lifestyle changes,[26] we intend to establish more precisely the etiologic relevance of the differences uncovered between lean and obese individuals. Functional similarity between postobese (i.e., people with a proven predisposition to obesity) and obese subjects would strongly suggest a causative role. Ultimately, longitudinal studies in individuals at high risk

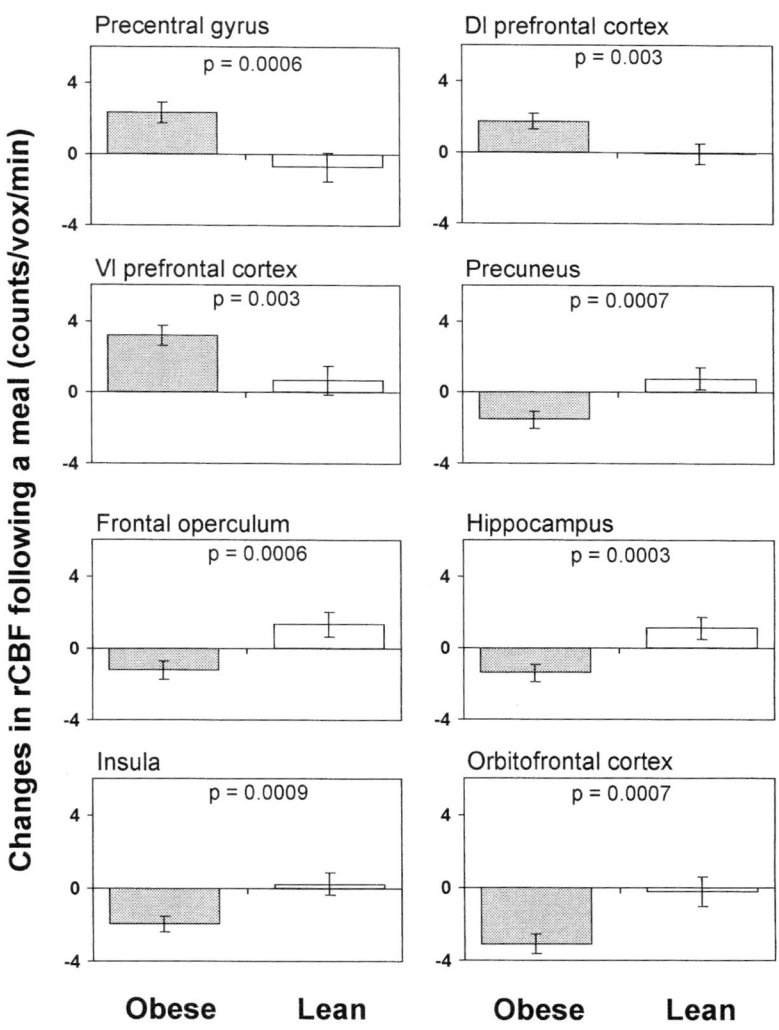

FIGURE 4. Areas of the brain with largest meal-induced differences in rCBF in 23 obese and 21 lean individuals (all $p < 0.005$). The location of the maximal change in rCBF in each of the eight brain regions was identified using statistical parametric mapping. Bar graphs represent the mean ± SD of the relative change in PET counts (normalized for changes in whole brain blood flow) in response to a meal in the voxel with the largest change in rCBF in each of the eight regions.

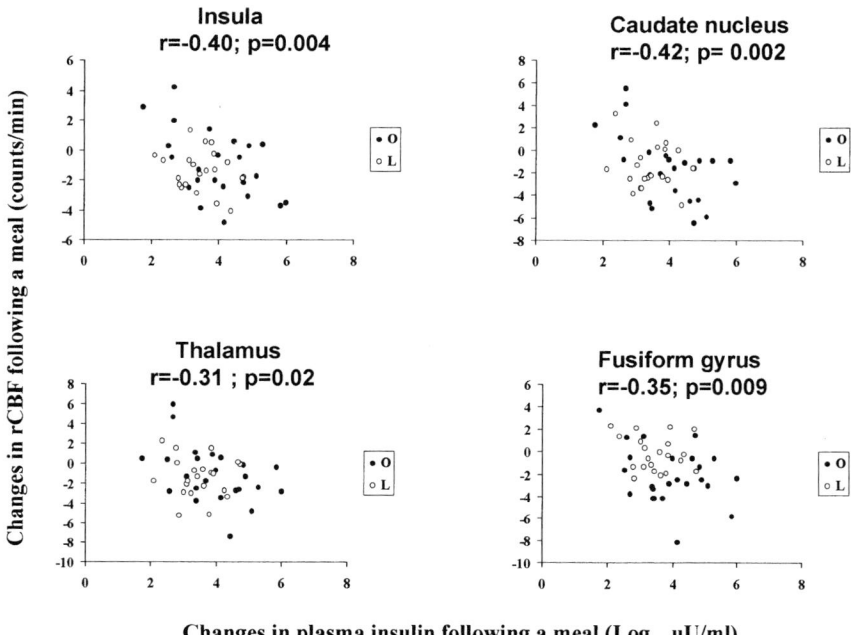

FIGURE 5. Correlations between changes in plasma insulin concentrations and rCBF in response to a liquid meal in 23 obese (filled circles) and 21 lean (open circles) individuals in four regions of the brain. In each of these regions, rCBF data were extracted from the voxels identified by statistical parametric mapping as showing the maximal degree of correlation between rCBF and plasma insulin concentrations.

for the development of obesity and studies designed to uncover the neurochemistry responsible for the observed differences in brain hemodynamics will need to be undertaken to confirm any hypothesis generated by the above approaches.

CONCLUSIONS

Studies of the neuroanatomical correlates of hunger and satiation indicate that functional neuroimaging is a viable approach to describe the complexity of the human brain response to a meal. According to results from these studies, a complex network of brain regions participates in the regulation of hunger and satiation. Glucose, insulin, and FFA may function as modulators of postprandial neuronal events in some of these regions. The regional cerebral blood flow changes in response to a meal in the prefrontal cortex and some limbic/paralimbic areas of the brain are different in lean and obese individuals. This indicates that obesity may be associated with abnormal neuronal activity in certain regions of the brain, some of which may have a role in the pathophysiology of the disease. Additional studies, however, are needed to uncover the neurochemical causes for the observed differences in brain activity.

REFERENCES

1. ZHANG, Y., R. PROENCA, M. MAFFEI et al. 1994. Positional cloning of the mouse obese gene and its human homologue. Nature **372:** 425–432.
2. TARTAGLIA, L.A., M. DEMBSKI, X. WENG et al. 1995. Identification and expression cloning of a leptin receptor, OB-R. Cell **83:** 1263–1271.
3. COWLEY, M.A., J.L. SMART, M. RUBINSTEIN et al. 2001. Leptin activates anorexigenic POMC neurons through a neural network in the arcuate nucleus. Nature **411:** 480–484.
4. FRIEDMAN, J.M. 2000. Obesity in the new millennium. Nature **404:** 632–634.
5. FRIEDMAN, J.M. 1999. Leptin and the regulation of body weight. Harvey Lect. **95:** 107–136.
6. SCHWARTZ, M.W., S.C. WOODS, D.J. PORTE et al. 2000. Central nervous system control of food intake. Nature **404:** 661–671.
7. BERTHOUD, H.R. 2000. An overview of neural pathways and networks involved in the control of food intake and selection. In Neural and Metabolic Control of Macronutrient Intake, pp. 361–387. CRC Press. Washington, D.C.
8. DREWNOWSKI, A. & J.M. SHULTZ. 2001. Impact of aging on eating behaviors, food choices, nutrition, and health status. J. Nutr. Health Aging **5:** 75–79.
9. DEFALCO, J., M. TOMISHIMA, H. LIU et al. 2001. Virus-assisted mapping of neural inputs to a feeding center in the hypothalamus. Science **291:** 2608–2613.
10. RAICHLE, M.E. 1998. Imaging the mind. Semin. Nucl. Med. **28:** 278–289.
11. GEORGE, J.S., C.J. AINE, J.C. MOSHER et al. 1995. Mapping function in the human brain with magnetoencephalography, anatomical magnetic resonance imaging, and functional magnetic resonance imaging. J. Clin. Neurophysiol. **12:** 406–431.
12. AINE, C.J. 1995. A conceptual overview and critique of functional neuroimaging techniques in humans: I. MRI/fMRI and PET. Crit. Rev. Neurobiol. **9:** 229–309.
13. KOSSLYN, S.M. 1999. If neuroimaging is the answer, what is the question? Philos. Trans. R. Soc. Lond. B Biol. Sci. **354:** 1283–1294.
14. ACTON, P.D. & K.J. FRISTON. 1998. Statistical parametric mapping in functional neuroimaging: beyond PET and fMRI activation studies. Eur. J. Nucl. Med. **25:** 663–667.
15. TATARANNI, P.A., J-F. GAUTIER, K. CHEN et al. 1999. Neuroanatomical correlates of hunger and satiation in humans using positron emission tomography. Proc. Natl. Acad. Sci. U.S.A. **96:** 4569–4574.
16. GAUTIER, J-F., K. CHEN, A.D. SALBE et al. 2000. Differential brain responses to satiation in obese and lean men. Diabetes **49:** 838–846.
17. GAUTIER, J-F., A. DEL PARIGI, K. CHEN et al. 2001. Effect of satiation on brain activity in obese and lean women. Obes. Res. **9:** 676–684.
18. DEL PARIGI, A., K. CHEN, J-F. GAUTIER et al. 2002. Sex differences in the human brain's response to hunger and satiation. Am. J. Clin. Nutr. In press.
19. GAUTIER, J-F., K. CHEN, A. UECKER et al. 1999. Regions of the human brain affected during a liquid-meal taste perception in the fasting state: a positron emission tomography study. Am. J. Clin. Nutr. **70:** 806–810.
20. REIMAN, E.M. 1997. The application of positron emission tomography to the study of normal and pathologic emotions. J. Clin. Psychiatry **58:** 4–12.
21. CRAIG, A.D., E.M. REIMAN, A. EVANS et al. 1996. Functional imaging of an illusion of pain. Nature **384:** 258–260.
22. ROLLS, E.T. 1996. The orbitofrontal cortex. Philos. Trans. R. Soc. Lond. B Biol. Sci. **351:** 1433–1443.
23. SMALL, D.M., R.J. ZATORRE, A. DAGHER et al. 2001. Changes in brain activity related to eating chocolate: from pleasure to aversion. Brain **124:** 1720–1733.
24. MATSUDA, M., Y. LIU, S. MAHANKALI et al. 1999. Altered hypothalamic function in response to glucose ingestion in obese humans. Diabetes **48:** 1801–1806.
25. GREENSWAG, L.R. & R.A. ALEXANDER, Eds. 1995. Management of Prader-Willi Syndrome. Springer-Verlag. New York/Berlin.
26. KLEM, M.L., R.R. WING, M.T. MCGUIRE et al. 1997. A descriptive study of individuals successful at long-term maintenance of substantial weight loss. Am. J. Clin. Nutr. **66:** 239–246.

Uncoupling Proteins, Leptin, and Obesity

An Updated Review

JEAN-PAUL GIACOBINO

Département de Biochimie Médicale, C. M. U., 1211 Geneva 4, Switzerland

ABSTRACT: The hypothesis that the novel uncoupling protein UCP3 is thermogenic and/or thermoregulatory is discussed. *In vitro*, *ex vivo*, and *in vivo* models are presented. The β_3-adrenoceptors are crucial for the appearance of UCP1-expressing cells in the white adipose tissue. These cells might differ from classical brown adipocytes. Besides its well-known effect on brown adipose tissue UCP1, leptin might stimulate indirectly muscle UCP3 expression.

KEYWORDS: uncoupling protein; β_3-adrenoceptor; leptin; obesity; brown and white adipose tissue; muscle

We are going to discuss three of the so-called "obesity genes" that code for proteins connected by an interesting network of interactions. These genes are the uncoupling proteins (UCPs), among which we will focus on UCP3 and UCP1, the β_3-adrenoceptor, and leptin.

THE UNCOUPLING PROTEINS WITH PARTICULAR EMPHASIS ON MUSCLE UCP3

The uncoupling protein 1 (UCP1) is an uncoupler of oxidative phosphorylation. It dissipates the proton gradient created by the respiratory chain, thereby producing heat instead of ATP. It is expressed in rodents exclusively in brown adipose tissue (BAT). Since there is no BAT in adult humans,[1] it was generally admitted that there was no thermogenic protein in humans.

In 1997, two novel members of the mitochondrial carrier family were discovered that displayed a high (≥55%) amino acid sequence identity to UCP1. Based on these identities, the two novel proteins were considered as possible uncoupling proteins and were called UCP2[2] and UCP3.[3] UCP2 was found to be highly expressed not only in rodents, but also in humans in most tissues studied.[2] UCP3 was also highly expressed in both rodents and humans, mainly in skeletal muscle.[3]

Address for correspondence: Jean-Paul Giacobino, Département de Biochimie Médicale, C. M. U., 1, rue Michel-Servet, 1211 Geneva 4, Switzerland. Voice: 41-22 702 54 86; fax: 41-22 702 55 02.

jean-paul.giacobino@medecine.unige.ch

Since the discovery of UCP2 and UCP3, many experimental approaches have been developed to test the hypothesis that these proteins might be uncouplers of oxidative phosphorylation and, therefore, participate in heat dissipation. For the discussion on the possible similarities with the gold standard of the uncoupling proteins, UCP1, we will focus on UCP3.

Data Supporting or Not the Notion that UCP3 Is an Uncoupling Protein

C_2C_{12} myoblasts, which do not express UCP3, were transfected either with a vector containing human UCP3 or with an empty vector. The mitochondrial membrane potential was measured by the uptake of a potential-sensitive fluorescent dye, TMRE. The fluorescence was found to be lower in the transfected cells, indicating that UCP3 uncouples oxidative phosphorylations.[4]

Using heterologous yeast cell expression systems, UCP3 was shown to decrease mitochondrial membrane potential, increase whole cell basal O_2 consumption, increase heat production, and increase slightly mitochondria state 4 respiration.[5–7] Therefore, in *in vitro* studies, there was a unanimity to support the notion that UCP3 is an uncoupling protein. It was argued, however, that the incorporation of a large amount of a heterologous protein might disorganize the mitochondria inner membrane and thereby induce a state of uncoupling. The results obtained with transfected cells should thus be considered with caution.

UCP3 knockout (KO) mice were created by two different groups and phenotyped.[8,9] Muscle mitochondria were prepared and *ex vivo* studies performed. Maximal O_2 consumption was similar in wild-type (WT) and UCP3KO animals. State 4 respiration, however, was 20% lower in the UCP3KO as compared to the WT animals, which revealed a state of overcoupling.[8] *In vivo*, no change in any of the energy balance parameters tested could be observed.[8,9]

Mice overexpressing UCP3 in their skeletal muscle (UCP3 Tg) were also created.[10] The overexpression was massive, that is, about 50-fold the level in WT animals. In muscle mitochondria isolated from UCP3 Tg mice, an uncoupling was observed, characterized by a 25% increase in state 4 respiration and by a decrease from 3.4 to 2.4 of the respiratory control as compared with WT animals. *In vivo* studies revealed, in UCP3 Tg as compared to WT mice, a 33–54% increase in food intake and a 24% lower body weight, revealing a strong decrease in food efficiency.[10]

In conclusion, mitochondria isolated from UCP3KO mouse muscle were overcoupled, whereas those isolated from UCP3 Tg were uncoupled, suggesting that UCP3 is an uncoupling protein. The results of the *in vivo* studies showed that UCP3 might also be thermoregulatory if highly overexpressed.

Modulation Studies of UCP3 in Vivo

From the very beginning, it was observed that muscle UCP3 mRNA expression varied considerably with changes in metabolic conditions. Increases up to 10-fold or more of UCP3 mRNA levels were observed upon fasting and diabetes, as well as β_3-agonist or T3 administration. In many of these instances, the stimulation of UCP3 mRNA expression seemed to be mediated by an increase in circulating and/or intracellular fatty acid level.[5–7] These observations suggested the so-called FFA theory that proposed that UCP3 expression in muscle is positively modulated by fatty acids.

This hypothesis was confirmed by direct experiments in rodents and humans. The experiments consisted of increasing circulating FFA by intralipid plus heparin administration in rodents[11] or by 5-h triglyceride infusion in humans.[12] Under both conditions, a dramatic increase in muscle UCP3 mRNA expression was observed. The effects of fatty acids on muscle UCP3 might be mediated by an interaction with PPARs. Only thyroid hormones seem to control directly UCP3 expression in muscle.

The FFA theory has largely contributed to feed the controversy about UCP3. It suggested the hypothesis that UCP3 might not be an uncoupling protein, but rather involved in the fatty acid degradation pathway.

THE β_3-ADRENOCEPTOR (β_3AR) AND ITS SPECIFIC INTERACTION WITH UCP1

The β_3AR is a 7-transmembrane receptor bound to adenylate cyclase. It is expressed mainly in rodent BAT and white adipose tissue (WAT), where it is the predominant subtype. In BAT, the stimulation of the β_3AR mediates the thermogenic effects of catecholamines. Two types of thermogenesis occur in BAT: cold- and diet-induced thermogenesis.

In both cases, the signaling is a stimulation of the sympathetic nervous system (SNS). Norepinephrine released by the SNS nerve endings, via the β_3AR, stimulates lipolysis, fatty acid β-oxidation, and (due to the presence of UCP1) thermogenesis. Norepinephrine also induces UCP1 expression.

In WAT, stimulation of the β_3AR mediates the lipolytic effects of catecholamines during fasting.

Therefore, the stimulation of the β_3AR would, by activating thermogenesis in BAT and lipolysis in WAT, prevent the development of obesity.

With this in mind, we created a β_3AR KO (β_3KO) mouse. The phenotype of this mutant was rather mild.[13] The mice did not become overtly obese, but showed a tendency to accumulate fat.

Our β_3KO mice being of a mixed genetic background, we performed a backcross and got pure C57BL/6J and 129Sv/ev mice. In these pure backgrounds, we studied the infiltration of WAT by brown adipocyte-like cells expressing UCP1.

The expression of UCP1 mRNA was measured in pooled parametrial and periovarian WAT and in interscapular BAT of C57BL/6J WT and β_3KO mice. In WAT of WT mice, the basal level of UCP1 mRNA at 24°C was found to be increased after 2 or 10 days of cold exposure. The effects of a lack of β_3AR were striking. The UCP1 mRNA expression levels in WAT of β_3KO mice were strongly depressed as compared to those measured in WT mice kept at 24°C or exposed to 6°C for 2 or 10 days. An effect of cold exposure on UCP1 mRNA expression was also seen in WAT of β_3KO mice, but only after 10 days.

In BAT of C57BL/6J mice, 2- or 10-day cold exposure increased UCP1 mRNA expression to the same extent in WT and β_3KO mice. The β_3KO did not affect UCP1 mRNA expression under all conditions studied.

The expression level of UCP1 mRNA in WAT relative to that in BAT, assessed on the same Northern blot, was about 3% and 11% in mice kept at 24°C or cold-exposed for 10 days, respectively.[14]

LEPTIN

Leptin, secreted by the white adipocytes, acts as a satiety signal via the ventromedial hypothalamus. It also stimulates the SNS innervating the BAT. Thus, leptin controls body weight, not only by decreasing food intake, but also by increasing energy expenditure. In rodents, the β_3AR mediates the effects of leptin on BAT thermogenesis.[15] The stimulatory effect of leptin on UCP1 expression in BAT is well established. The possible effect of leptin on UCP3 expression in muscle is still a matter of debate.[5-7]

In collaboration with the group of Françoise Jeanrenaud, we studied mice with intracerebroventricular (ICV) infusion of leptin. Since this treatment decreases food intake, we used pair-fed animals as controls, We found that pair feeding strongly decreased muscle UCP3 expression and that ICV infusion of leptin prevented this decrease.[16]

The stimulatory effect of leptin on muscle UCP3 expression is probably not mediated by a stimulation of the SNS. An interesting hypothesis is that it might be mediated by thyroid hormones. Indeed, pair feeding was found to decrease circulating T3 level, and the ICV infusion of leptin prevented this decrease. It could be hypothesized that it was the maintenance in T3 level in pair-fed animals that mediated the effect of leptin on muscle UCP3. In fact, in hypothyroid animals, leptin did not prevent anymore the fall in muscle UCP3 induced by pair feeding.[17] These data support the hypothesis that the effect of leptin on muscle UCP3 might be indirect, that is, mediated by changes in circulating T3.

CONCLUSIONS

The role of muscle UCP3 as a thermogenic and/or thermoregulatory protein remains to be established.

The β_3-adrenoceptor mediates the recruitment of UCP1-expressing cells in WAT.

Muscle UCP3 and recruitment of UCP1-expressing cells in WAT might represent two interesting targets for antiobesity drugs.

Leptin might mediate thermogenesis via β_3-adrenoceptor activation of UCP1 expression in BAT and by acting indirectly on muscle UCP3. Its possible role in recruitment should be studied.

REFERENCES

1. CUNNINGHAM, S., P. LESLIE, D. HOPWOOD et al. 1985. The characterization and energetic potential of brown adipose tissue in man. Clin. Sci. **69:** 343–348.
2. FLEURY, C., M. NEVEROVA, S. COLLINS et al. 1997. Uncoupling protein-2: a novel gene linked to obesity and hyperinsulinemia. Nat. Genet. **15:** 269–272.
3. BOSS, O., S. SAMEC, A. PAOLONI-GIACOBINO et al. 1997. Uncoupling protein-3: a new member of the mitochondrial carrier family with tissue-specific expression. FEBS Lett. **408:** 39–42.
4. BOSS, O., S. SAMEC, F. KUHNE et al. 1998. Uncoupling protein-3 expression in rodent skeletal muscle is modulated by food intake, but not by changes in environmental temperature. J. Biol. Chem. **273:** 5–8.

5. Boss, O., P. Muzzin & J-P. Giacobino. 1998. The uncoupling proteins: a review. Eur. J. Endocrinol. **139:** 1–9.
6. Muzzin, P., O. Boss & J-P. Giacobino. 1999. Uncoupling protein 3: its possible biological role and mode of regulation in rodents and humans. J. Bioenerg. Biomembr. **31:** 467–473.
7. Boss, O., T. Hagen & B.B. Lowell. 2000. Uncoupling proteins 2 and 3: potential regulators of mitochondrial energy metabolism. Diabetes **49:** 143–156.
8. Vidal-Puig, A.J., D. Grujic, C.Y. Zhang et al. 2000. Energy metabolism in uncoupling protein 3 gene knockout mice. J. Biol. Chem. **275:** 16258–16266.
9. Gong, D.W., S. Monemdjou, O. Gavrilova et al. 2000. Lack of obesity and normal response to fasting and thyroid hormone in mice lacking uncoupling protein-3. J. Biol. Chem. **275:** 16251–16257.
10. Clapham, J., J. Arch, H. Chapman et al. 2000. Mice overexpressing human uncoupling protein-3 in skeletal muscle are hyperphagic and lean. Nature **406:** 415–418.
11. Weigle, D.S., L.E. Selfridge, M.W. Schwartz et al. 1998. Elevated free fatty acids induce uncoupling protein 3 expression in muscle: a potential explanation for the effect of fasting. Diabetes **47:** 298–302.
12. Khalfallah, Y., S. Fages, M. Laville et al. 2000. Regulation of uncoupling protein-2 and uncoupling protein-3 mRNA expression during lipid infusion in human skeletal muscle and subcutaneous adipose tissue. Diabetes **49:** 25–31.
13. Revelli, J.P., F. Preitner, S. Samec et al. 1997. Targeted gene disruption reveals a leptin-independent role for the mouse β_3-adrenoceptor in the regulation of body composition. J. Clin. Invest. **100:** 1098–1106.
14. Jimenez, M., G. Barbatelli, R. Allevi et al. 2001. Targeted disruption of the β_3-adrenoceptor in mice suppresses the occurrence of "ectopic" brown adipocytes in white adipose tissue: importance of the genetic background. Submitted.
15. Giacobino, J-P. 1996. Role of the β_3-adrenoceptor in the control of leptin expression. Horm. Metab. Res. **28:** 633–637.
16. Cusin, I., K.E. Zakrzewska, O. Boss et al. 1998. Chronic central leptin infusion enhances insulin-stimulated glucose metabolism and favors the expression of uncoupling proteins. Diabetes **47:** 1014–1019.
17. Cusin, I., J. Rouru, T. Visser et al. 2000. Involvement of thyroid hormones in the effect of intracerebroventricular leptin infusion on uncoupling protein-3 expression in rat muscle. Diabetes **49:** 1101–1105.

New Approaches to Gene Discovery with Animal Models of Obesity and Diabetes

GREG COLLIER,[a,b] KEN WALDER,[a] ANDREA DE SILVA,[a] JANETTE TENNE-BROWN,[a] ANDREW SANIGORSKI,[a] DAVID SEGAL,[a] LAKSHMI KANTHAM,[a] AND GUY AUGERT[c]

[a]*Metabolic Research Unit, School of Health Sciences, Deakin University, Geelong, Australia*

[b]*Autogen Limited, Melbourne, Victoria, Australia*

[c]*Merck-Lipha, Lyon, France*

> ABSTRACT: DNA-based approaches to the discovery of genes contributing to the development of type 2 diabetes have not been very successful despite substantial investments of time and money. The multiple gene-gene and gene-environment interactions that influence the development of type 2 diabetes mean that DNA approaches are not the ideal tool for defining the etiology of this complex disease. Gene expression–based technologies may prove to be a more rewarding strategy to identify diabetes candidate genes. There are a number of RNA-based technologies available to identify genes that are differentially expressed in various tissues in type 2 diabetes. These include differential display polymerase chain reaction (ddPCR), suppression subtractive hybridization (SSH), and cDNA microarrays. The power of new technologies to detect differential gene expression is ideally suited to studies utilizing appropriate animal models of human disease. We have shown that the gene expression approach, in combination with an excellent animal model such as the Israeli sand rat (*Psammomys obesus*), can provide novel genes and pathways that may be important in the disease process and provide novel therapeutic approaches. This paper will describe a new gene discovery, beacon, a novel gene linked with energy intake. As the functional characterization of novel genes discovered in our laboratory using this approach continues, it is anticipated that we will soon be able to compile a definitive list of genes that are important in the development of obesity and type 2 diabetes.
>
> KEYWORDS: animal model of obesity and diabetes; gene discovery; RNA approaches to gene discovery

DNA-based approaches for the discovery of genes that contribute to the development of type 2 diabetes have not been very successful despite substantial investment of time and money. The etiology of a complex disease such as type 2 diabetes is not easily solved using current DNA-based approaches due to the multiple gene-gene and gene-environment interactions. Large numbers and/or variable combinations of

Address for correspondence: Greg Collier, Metabolic Research Unit, School of Health Sciences, Deakin University, Geelong, Australia. Voice: (613) 5227 2547; fax: (613) 5227 2170.
barbedwa@deakin.edu.au

small gene defects leading to the final disease state also complicate these analyses and make identification of important genes difficult. However, there are alternative strategies in the field of gene discovery in type 2 diabetes, such as RNA- (gene expression) or proteomics-based approaches. Gene expression–based technologies may prove to be a more rewarding approach to identify diabetes candidate genes and, in combination with appropriate animal models, will be a powerful tool in understanding the underlying mechanisms of human polygenic diseases such as diabetes.

A number of RNA-based technologies are available to identify genes that are differentially expressed in various tissues. These include differential display polymerase chain reaction (ddPCR), suppression subtractive hybridization (SSH), and cDNA microarrays. Both ddPCR and SSH have been successfully used to identify novel genes involved in energy metabolism. For example, in our laboratory, ddPCR was used to identify beacon, a novel polypeptide involved in the regulation of energy balance, which is differentially expressed in the hypothalamus of obese/diabetic and lean/nondiabetic Israeli sand rats (*Psammomys obesus*).[1] Recently, Steppan and colleagues[2] used the SSH technique to identify genes differentially expressed in adipocytes exposed to rosiglitazone, a PPARγ agonist used to treat type 2 diabetes. One of the identified genes, resistin, is a soluble molecule that may provide a link between obesity and type 2 diabetes.

Representational difference analysis (RDA) is a powerful RNA-based technique that allows the comparison of two populations of mRNA and obtains clones of genes that are expressed in one population and not in the other. RDA has the advantage of analyzing gene differences in the 5′-portion of cDNAs, allowing alternatively spliced variants and importantly coding regions to be detected. Unfortunately, several rounds of subtraction are necessary to isolate rare transcripts. To overcome the technical limitations of traditional subtractive methods, a new method has been developed called suppression subtractive hybridization (SSH).[3] The key advantage of SSH is its ability, via second-order hybridization kinetics, to exponentially amplify and equalize both rare and abundant differentially expressed transcripts while suppressing sequences that are common in both populations. The generated cDNAs can then be directly inserted into a variety of cloning strategies such as a T/A cloning vector. After picking, either the clones can be screened using cDNA dot blots to eliminate false positives and then sequenced or alternatively clones can be used to generate subtracted cDNA libraries for future microarray analysis. Generating cDNA libraries using SSH is relatively quick and, more importantly, it is species- and tissue-specific.

Techniques such as SSH and ddPCR can be used to generate target genes using standard laboratory equipment and at a reasonable cost. The major disadvantages of these methods are that they are labor-intensive and they identify only small numbers of differentially expressed genes. A further limitation is the binary nature of the comparisons using these techniques, which makes them unsuited to complex experiments such as time-course or dose-response studies. However, cDNA microarray experiments can provide gene expression data for thousands of genes and from large numbers of experimental samples. This technology is ideally suited to complex, multivariate analyses that generate detailed expression profiles, which in turn have the potential to increase our understanding of the underlying mechanisms associated with a disease. Current analysis techniques tend to focus on clustering algorithms that identify genes exhibiting similar expression patterns.[4] Recent advances in the analysis of microarray data have led to the construction of models to identify groups of genes involved in selected physiological processes.[5] Current progress in this field suggests that cDNA microarray technology will facilitate the identification of key genes and pathways involved in the pathogenesis of type 2 diabetes.

The power of these new technologies to detect differential gene expression is ideally suited to studies utilizing appropriate animal models of human disease because very few polygenic, outbred rodent models of obesity and type 2 diabetes exist. One such model, *Psammomys obesus* (Israeli sand rat) has been studied extensively in our laboratory.[1,6–12] *Psammomys obesus* are gerbil-like rodents found in the desert areas of the Middle East and North Africa. They remain lean and free from diabetes in their native habitat, subsisting on a diet composed mainly of salt bush (*Atriplex halimus*).[13] When taken into the laboratory, however, and allowed free access to standard rodent chow, varying degrees of obesity, insulin resistance, and type 2 diabetes develop.[6,14] Adult *Psammomys obesus* have a wide range of body weight and body fat content that forms a continuous distribution. It is the heterogeneous response to a relatively energy-dense diet that makes *Psammomys obesus* more analogous to the pattern of human obesity and type 2 diabetes than homogenous single-gene animal models.

A number of metabolic disturbances have been identified in obese, diabetic *Psammomys obesus* relative to their lean littermates: hyperglycemia, insulin resistance, hyperphagia, obesity, and dyslipidemia.[6–8,12] Hepatic insulin resistance[15] and a defective insulin receptor signaling pathway have been reported in diabetic *Psammomys obesus*.[16,17] These animals also have hyperproinsulinemia, reduced pancreatic insulin storage capacity, and beta cell apoptosis.[18,19] Increased expression of protein kinase (PKC epsilon) in skeletal muscle of *Psammomys obesus* has recently been reported. This overexpression of PKC epsilon may be causally related to the development of insulin resistance in these animals, possibly by increasing the degradation of insulin receptors.[20] Elevated levels of leptin concentrations in obese diabetic animals have been reported,[10] with resistance to the effects of peripheral (intraperitoneal) leptin administration in obese, but not lean, animals.[11] As a species, *Psammomys obesus* appear to be relatively insensitive to the effects of leptin administration and it is possible that the leptin resistance exhibited in this study may be contributing to the development of obesity.

The body weight distribution in *Psammomys obesus* approximates a normal distribution and closely resembles that observed in human populations. Animals above the 75th percentile for body weight have increased body fat content and a greater risk of developing diabetes. Increased visceral fat content was also associated with elevated blood glucose and plasma insulin concentrations.[12] Cross-sectional analysis of the animal population of *Psammomys obesus* reveals heterogenous distributions of blood glucose, plasma insulin, and body weight.[6,12] It is this aspect of the development of diabetes and obesity in *Psammomys obesus* that is of most importance because these distributions are almost identical to the patterns observed in cross-sectional studies of human populations, including the inverted U-shaped relationship between blood glucose and insulin concentrations termed "Starling's curve of the pancreas".[21,22] Moreover, we have demonstrated that genetic factors account for 51% of the variation in body weight and 23–26% of the variation in blood glucose and plasma insulin concentrations in *Psammomys obesus*.[12] Recently, during the sequential development of insulin resistance and diabetes in *Psammomys obesus*, it has been shown that various disturbances occur in plasma lipid profile and lipoprotein composition, as well as in liver cholesterol metabolism.[23] Thus, *Psammomys obesus* represents an excellent animal model of obesity and type 2 diabetes that exhibits a phenotypic pattern closely resembling that observed in human population studies.

In our laboratory, the use of modern technologies to detect differential gene expression, in combination with an excellent animal model such as *Psammomys obesus*, has resulted in the identification of novel genes important in the development of type 2 diabetes. Using ddPCR, the beacon gene was identified as a gene product overexpressed in the hypothalamus of obese, diabetic animals. Further studies in a larger group of animals demonstrated that the beacon gene was expressed in the hypothalamus in direct proportion to the body fat content.[1] Beacon is expressed ubiquitously throughout the body and encodes a small protein of 73 amino acids. The human beacon gene consists of 2194 nucleotides arranged into 5 exons and 4 introns and has been mapped to chromosome 19. The *Psammomys obesus* beacon gene is composed of 4 exons, has a shorter 5′-untranslated region, and consequently lacks the first exon present in the human gene[1] (Genbank Accession: AF318186).

Candidate gene identification is only the first step in determining the gene's importance in the development of diabetes. It is necessary to determine the physiological function of the protein produced by the novel gene and to validate the potential of this protein as a therapeutic target. Recent years have seen the addition of various new tools to the existing classical biochemical techniques for understanding the role of proteins of unknown function. The research tools and strategies for functional studies that have been applied in our own and various other laboratories are outlined in FIGURE 1 and will be discussed here in the context of functional analysis of the beacon protein.

Bioinformatic tools can assist in the recognition of certain structural elements and consensus sequence motifs present in the putative amino acid sequence. For example, it is possible to predict the regions corresponding to transmembrane domains, hydrophilic and hydrophobic regions, and sequence motifs for posttranslational modifications such as phosphorylation, glycosylation, isoprenylation, and geranylation. When present, the consensus targeting sequences for secretion or those that direct the newly synthesized proteins to different subcellular compartments such as mitochondria, peroxisomes, endoplasmic reticulum, and nucleus can also be detected using a number of commercially or publicly available software tools. It is rare, however, that the bioinformatics of a protein will provide sufficient clues to indicate the function of a protein. Implementation of hypothesis-based experimental strategies are often necessary to arrive at a deeper understanding of the precise function of a gene product.

In our laboratory, a research strategy is designed for each selected target based on the information obtained from bioinformatics and the physiological context in which the gene is discovered. The functional studies are directed to understand the function at a molecular level with isolated proteins or at a cellular level by examining the altered biochemical or metabolic functions of cells as a consequence of inhibition or overexpression of the gene of interest. Where possible, we apply *in vivo* studies either by direct administration of the protein or by gene delivery into the animals using viral expression vectors.

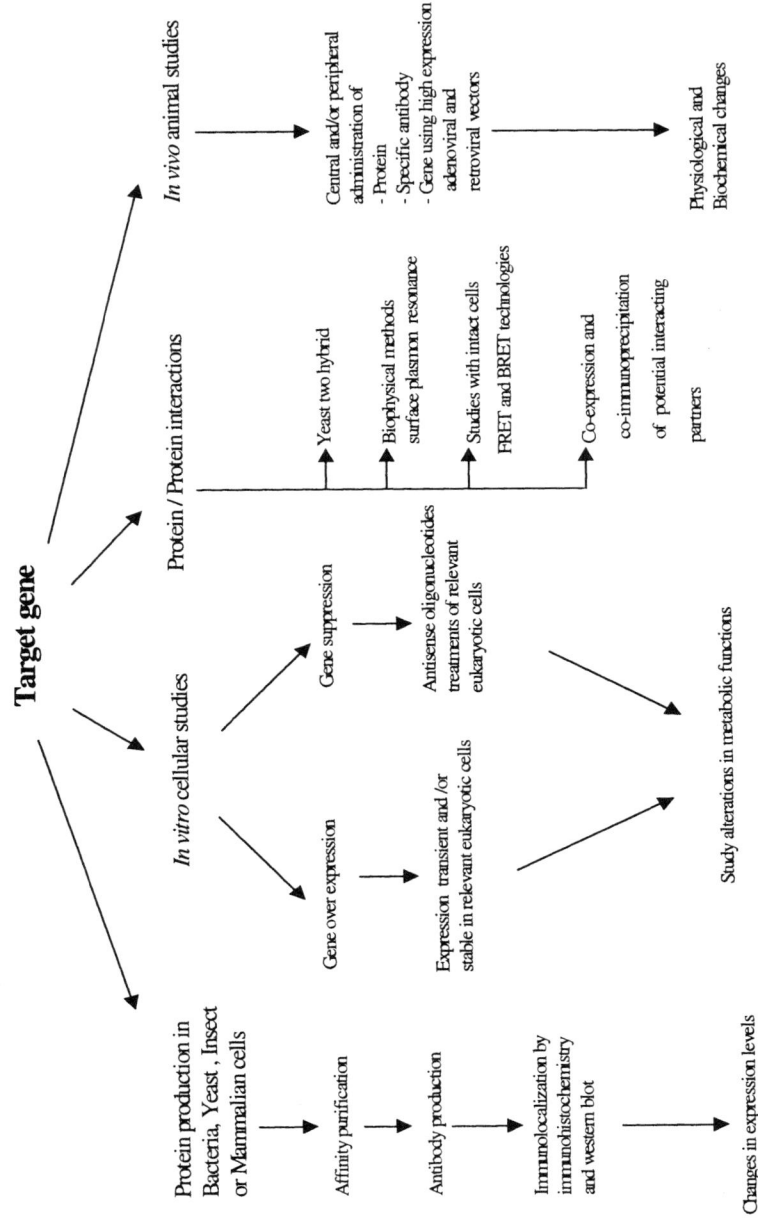

FIGURE 1. Tools for validation of functional role of target gene.

It is helpful to have the purified protein available not only as a reagent for *in vivo* and *in vitro* experiments, but also for raising antibodies. It is relatively more simple to genetically engineer bacterial strains and produce recombinant proteins than to isolate native proteins from tissue sources. Beacon was predicted to be a soluble protein with no obvious targeting signals and transmembrane domains. We have cloned beacon into a bacterial expression vector designed for expression of GST fusion (Glutathione S-transferase) proteins with a built-in cleavage site for subsequent removal of the GST tag with thrombin. The expressed GST-beacon fusion protein was separated from the rest of the proteins in the bacterial lysate by affinity binding of the GST tag to Glutathione Sepharose beads. Under optimized conditions of expression and purification, we have been able to produce 25 mg of pure homogeneous beacon protein per liter of culture.

Monoclonal and polyclonal antibodies make valuable tools for several functional studies such as immunoblotting, immunohistochemistry, and immunoprecipitation. Preparation of monoclonal antibodies is labor-intensive; however, once the clones are established, new stocks can be easily prepared. A monoclonal antibody specific for beacon was produced and used in immunohistochemical localization studies in brain sections. Beacon was highly expressed in a region of the hypothalamus called the retrochiasmatic nucleus,[1] an area implicated in the regulation of energy balance.

Work involved in producing polyclonal antibodies is less than for monoclonal antibodies and, as multiple epitopes are recognized, is in general more effective in recognizing antigens present in native or denatured conformations and for immunoprecipitation of antigens from dilute test samples. It is known that small and highly conserved proteins make weak antigens and therefore we chose to immunize rabbits with beacon conjugated to a much larger diphtheria toxoid protein. The conjugated beacon aggregates produced a good immune response. To obtain pure beacon-specific antibodies, we bound beacon protein to chemically activated "Aminolink" solid beads and passed the sera through the matrix followed by elution of bound beacon-specific antibodies. The sera purified on the affinity matrix were highly enriched and beacon-specific. On Western blots, beacon was detectable in several of the *Psammomys obesus* tissues such as brain, liver, kidney, skeletal muscle, and adipose tissue. The functional role of beacon in peripheral tissues is yet to be assessed.

A variety of eukaryotic plasmid and viral (retroviral or adenoviral) expression vectors were employed to overexpress genes of interest in insulin responsive cell lines such as differentiated 3T3-L1 (adipocyte), C2C12 or L6 (muscle), HepG2 or H4IIE (hepatocyte), or min6 (pancreatic). As an alternative, carefully designed antisense oligonucleotides are used to inhibit endogenous expression of genes of interest. Studies can be conducted in which the gene of interest is overexpressed or inhibited, and the cell's response to insulin, such as changes in glucose uptake, glycogen synthesis, or lipogenesis, is examined. This information can reveal the nature of metabolic pathways in which the gene of interest is involved. When the studies in cell culture models are completed, our plan is to apply the same vectors and antisense oligonucleotides that produce an effect for *in vivo* studies in live animals and validate the observed effects.

Almost all proteins fulfill their functional role via interaction with other proteins. Identification of interacting proteins and any knowledge of their functional properties can also provide clues for positioning the candidate protein in a defined signal trans-

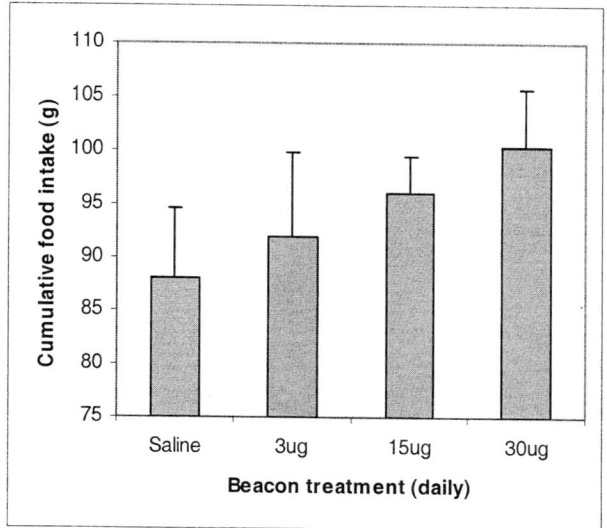

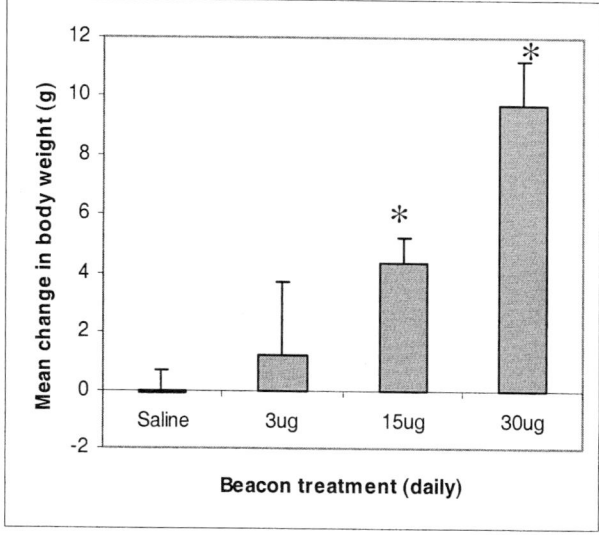

FIGURE 2. Animals were treated with 3, 15, or 30 μg beacon per day for 7 days. Cumulative food intake (**A**) and change in body weight (**B**) are shown. *$p < 0.05$ vs. saline group.

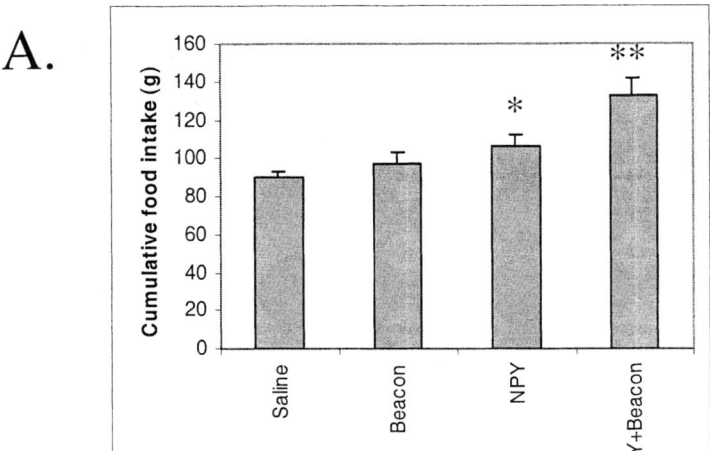

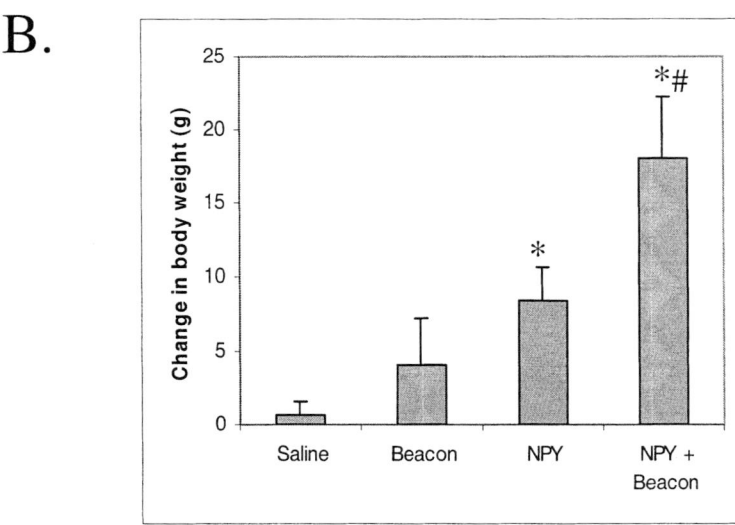

FIGURE 3. Animals were treated with 15 µg of beacon or 15 µg of both beacon and NPY per day for 7 days. Cumulative food intake (**A**) and change in body weight (**B**) are shown. $*p < 0.05$ vs. saline group; $^{\#}p < 0.05$ vs. NPY group; $**p < 0.05$ vs. NPY, beacon, and saline groups.

duction pathway. Protein-protein interactions can be identified using coexpression technologies such as the yeast two-hybrid system[24–27] or biophysical techniques such as Fluorescence and Bioluminescence Resonance Energy Transfer[28,29] or Biacore.[30] All of these techniques aim to identify interactions between two proteins. Using the yeast two-hybrid method, we have identified a novel kinase that interacts with beacon. The events of the intracellular kinase cascade play a significant role in the signaling pathway downstream to binding of insulin to its receptor. The novel kinase identified in our laboratory may thus provide a novel candidate pathway in the pathogenesis of obesity and type 2 diabetes. Importantly, this pathway now provides a target for the development of chemical interventions to decrease beacon action, which can then be developed further as a potential therapeutic agent in the treatment of obesity.

The physiological function of beacon in the regulation of energy balance was confirmed by direct administration *in vivo* in *Psammomys obesus*. Intracerebroventricular (ICV) administration of beacon for 7 days resulted in a dose-dependent increase in food intake and body weight gain (FIG. 2) and in a 2-fold increase in hypothalamic expression of NPY.[1] In addition, coadministration of beacon (15 μg/day) and NPY (15 μg/day) had a synergistic orexigenic effect, resulting in a dramatic increase in food intake and body weight gain that was significantly greater than the sum of the responses seen when the peptides were administered separately (FIG. 3).[1] The substantial body weight gain after ICV administration of beacon was due to increased fat accumulation.[31] Indirect calorimetry was used to investigate effects of ICV beacon administration on energy expenditure, physical activity, and substrate utilization. Beacon treatment had no effect on these parameters in *Psammomys obesus* and did not affect feed efficiency.[31] Therefore, the actions of beacon in the hypothalamus result in increased food intake, which in turn causes accumulation of body fat leading to excessive body weight gain. The observation of dramatic physiological effects of beacon *in vivo* strongly supports the hypothesis that beacon plays a major role in the regulation of energy balance in *Psammomys obesus*. Thus, beacon represents an interesting new target with great potential for the design of therapeutic agents for obesity and type 2 diabetes.

In summary, with the combination of an excellent animal model for the study of diabetes in *Psammomys obesus* and RNA-based technologies such as ddPCR, we discovered a novel gene, beacon. After establishing that the beacon gene was expressed in the hypothalamus in direct proportion to body fat content, it was necessary to determine the physiological function of the gene product. Using a raft of technologies, the functional role of the gene protein was established and interaction with other proteins examined. Confirmation of the function of the novel protein occurred by direct application *in vivo* in an animal model. Finally, with details of the pathway of beacon action defined, it is hoped that this information will be used for high-throughput screening in the quest for a new therapeutic approach to obesity and diabetes.

REFERENCES

1. COLLIER, G.R., J. MCMILLAN, K. WINDMILL *et al.* 2000. Beacon, a novel gene involved in the regulation of energy balance. Diabetes **49**: 1766–1771.

2. STEPPAN, C.M., S.T. BAILEY, S. BHAT *et al.* 2001. The hormone resistin links obesity to diabetes. Nature **409:** 307–312.
3. DIATCHENKO, L., Y-F.C. LAU, A.P. CAMPBELL *et al.* 1996. Suppression subtractive hybridization: a method for generating differentially regulated or tissue-specific cDNA probes and libraries. Proc. Natl. Acad. Sci. U.S.A. **93:** 6025–6030.
4. EISEN, M.B., P.T. SPELLMAN, P.O. BROWN *et al.* 1998. Cluster analysis and display of genome wide expression patterns. Proc. Natl. Acad. Sci. U.S.A. **95:** 14863–14868.
5. KIM, S., E.R. DOUGHERTY & Y. CHEN. 2000. Multivariate measurement of gene expression relationships. Genomics **67:** 201–209.
6. BARNETT, M., G.R. COLLIER, F.M. COLLIER *et al.* 1994. A cross-sectional and short-term longitudinal characterization of NIDDM in *Psammomys obesus*. Diabetologia **37:** 671–676.
7. HABITO, R.C., M. BARNETT, A. YAMAMOTO *et al.* 1995. Basal glucose turnover in *Psammomys obesus*: an animal model of type 2 (non-insulin-dependent) diabetes mellitus. Acta Diabetol. **32:** 187–192.
8. COLLIER, G.R., K. WALDER, P. LEWANDOWSKI *et al.* 1997. Leptin and the development of obesity and diabetes in *Psammomys obesus*. Obes. Res. **5**(5)**:** 455–458.
9. COLLIER, G.R., A. DE SILVA, A. SANIGORSKI *et al.* 1997. Development of obesity and insulin resistance in the Israeli sand rat (*Psammomys obesus*): does leptin play a role? Ann. N.Y. Acad. Sci. **827:** 50–63.
10. WALDER, K., M. WILLET, P. ZIMMET *et al.* 1997. *Ob* (obese) gene expression and leptin levels in *Psammomys obesus*. Biochim. Biophys. Acta **1354:** 272–278.
11. WALDER, K., P. LEWANDOWSKI, G. MORTON *et al.* 1999. Leptin resistance in a polygenic, hyperleptinemic animal model of obesity and NIDDM *Psammomys obesus*. Int. J. Obes. **2:** 83–89.
12. WALDER, K.R., R.P. FAHEY, G.J. MORTON *et al.* 2000. Characterization of obesity phenotypes in *Psammomys obesus* (Israeli sand rats). Int. J. Exp. Diabetes Res. **1:** 177–184.
13. SHAFRIR, E. & A. GUTMAN. 1993. *Psammomys obesus* of the Jerusalem colony: a model for nutritionally induced, non-insulin-dependent diabetes. J. Basic Clin. Physiol. Pharmacol. **4:** 83–99.
14. KALDERON, B., A. GUTMAN & E. LEVY. 1996. Characterization of stages in the development of obesity-diabetes syndrome in sand rat (*Psammomys obesus*). Diabetes **45:** 717–724.
15. ZIV, E., R. KALMAN, K. HERSHKOP *et al.* 1996. Insulin resistance in the NIDDM model *Psammomys obesus* in the normoglycemic, normoinsulinemic state. Diabetologia **39:** 1269–1275.
16. KANETY, H., S. MOSHE & E. SHAFRIR. 1994. Hyperinsulinemia induces a reversible impairment in insulin receptor function leading to diabetes in the sand rat model of non-insulin-dependent diabetes mellitus. Proc. Natl. Acad. Sci. U.S.A. **91:** 1853–1857.
17. SHAFRIR, E. & E. ZIV. 1998. Cellular mechanism of nutritionally induced insulin resistance: the desert rodent *Psammomys obesus* and other animals in which insulin resistance leads to a detrimental outcome. J. Basic Clin. Physiol. Pharmacol. **9:** 347–385.
18. GADOT, M., G. LEIBOWITZ & E. SHAFRIR. 1994. Hyperproinsulinemia and insulin deficiency in the diabetic *Psammomys obesus*. Endocrinology **135:** 610–616.
19. BAR-ON, H., R. BEN-SASSON, E. ZIV *et al.* 1999. Irreversibility of nutritionally induced NIDDM in *Psammomys obesus* is related to beta-cell apoptosis. Pancreas **18:** 259–265.
20. IKEDA, Y., G.S. OLSEN, E. ZIV *et al.* 2001. Cellular mechanism of nutritionally induced insulin resistance in *Psammomys obesus*: overexpression of protein kinase C epsilon in skeletal muscle precedes the onset of hyperinsulinemia and hyperglycemia. Diabetes **50**(3)**:** 584–592.
21. ZIMMET, P., S. WHITEHOUSE & J. KISS. 1979. Ethnic variability in the plasma insulin response to oral glucose in Polynesian and Micronesian subjects. Diabetes **28:** 624–628.
22. DEFRONZO, R.A. 1988. The triumvirate B-cell, muscle, and liver: a collusion responsible for NIDDM. Diabetes **37:** 667–688.
23. ZOLTOWSKA, M., E. ZIV, E. DEVLIN *et al.* 2001. Circulating lipoproteins and hepatic sterol metabolism in *Psammomys obesus* prone to obesity. Atherosclerosis **157**(1)**:** 85–96.
24. FIELDS, S. & O. SONG. 1989. A novel genetic system to detect protein-protein interactions. Nature **340:** 245–246.

25. GYURIS, J., E. GOLEMIS, H. CHATKOU *et al.* 1993. Cdi1, a human G1 and S phase protein phosphatase that associates with Cdk2. Cell **75:** 791–803.
26. DURFEE, T., K. BECHERERN & P.L. CHEN. 1993. The retinoblastoma protein associates with the protein phosphatase type 1 catalytic subunit. Gene Dev. **7:** 555–569.
27. VIDAL, M., R.K. BRACHMANN, A. FATTAEY *et al.* 1996. Reverse two-hybrid and one-hybrid systems to detect dissociation of protein-protein and DNA-protein interactions. Proc. Natl. Acad. Sci. U.S.A. **93:** 10315–10320.
28. DAMELIN, M. & P. SILVER. 2000. Mapping interactions between nuclear transport factors in living cells reveals pathways through the nuclear pore complex. Mol. Cell **5:** 133–140.
29. ANGERS, S., A. SALAHPOUR, E. JOLY *et al.* 2000. Detection of B_2-adrenergic receptor dimerization in living cells using Bioluminescence Resonance Energy Transfer (BRET). Proc. Natl. Acad. Sci. U.S.A. **97:** 3684–3689.
30. NAGATA, K. & H. HANDA, Eds. 2000. Real-Time Analysis of Biomolecular Interactions. Springer-Verlag. Berlin/New York/Tokyo.
31. WALDER, K.R., J.S. MCMILLAN, S. LEE *et al.* 2001. Effects of beacon administration on energy expenditure and substrate utilization in *Psammomys obesus* (Israeli sand rats). Int. J. Obes. In press.

High-Fat High-Energy Feeding Impairs Fasting Glucose and Increases Fasting Insulin Levels in the Göttingen Minipig

Results from a Pilot Study

MARIANNE OLHOLM LARSEN,[a,b] BIDDA ROLIN,[a] MICHAEL WILKEN,[c] RICHARD DAVID CARR,[a] AND OVE SVENDSEN[b]

[a]*Department of Pharmacological Research I, Novo Nordisk A/S, Bagsvaerd, Denmark*

[b]*Department of Pharmacology and Pathobiology, The Royal Veterinary and Agricultural University, Copenhagen, Denmark*

[c]*Department of Assay and Cell Technology, Novo Nordisk A/S, Bagsvaerd, Denmark*

ABSTRACT: High-fat diet and obesity are known to be of major importance for development of type 2 diabetes in humans. High-fat feeding can induce syndromes of glucose intolerance and/or insulin resistance in several species, and the Göttingen minipig might be a useful model for studying the effect of dietary high-fat intake and obesity on glucose homeostasis and the susceptibility to diabetes. The present study was designed as a pilot study to investigate the effects of obesity caused by high-fat high-energy feeding on oral and intravenous glucose tolerance. Male Göttingen minipigs were fed a control diet (CD) or a high-fat high-energy diet (HFD) for 3 months. Body weight (32.6 ± 2.4 kg vs. 24.9 ± 0.5 kg, $p < 0.001$) and total (13.2 ± 3.2% vs. 6.1 ± 0.5%, $p = 0.002$) and truncal (11.0 ± 3.9% vs. 1.8 ± 1.1%, $p = 0.001$) fat percent were increased significantly, whereas relative lean body mass was decreased (84.8 ± 3.3% vs. 91.9 ± 0.5%, $p = 0.002$) in the HFD group compared to CD. Fasting plasma glucose (4.3 ± 0.4 mM vs. 3.6 ± 0.3 mM, $p = 0.023$) and insulin (80 ± 23 pM vs. 23 ± 21 pM, $p = 0.012$) were increased in the HFD group compared to CD, but oral glucose tolerance was not significantly changed. Insulin responses to intravenous glucose were increased (6741 ± 2538 vs. 3938 ± 771 pM × min, $p = 0.050$), while glucose clearance was not changed by HFD vs. CD, thus indicating insulin resistance. In conclusion, changes in body weight and composition, resulting in minor abnormalities in glucose tolerance and insulin sensitivity, characterized by slight hyperglycemia and compensatory hyperinsulinemia, can be induced in the male Göttingen minipig by high-fat high-energy feeding for 3 months. This approach seems to be an interesting and promising method for establishment of a nonrodent model of insulin resistance or type 2 diabetes.

KEYWORDS: Göttingen minipigs; dietary fat; obesity; glucose tolerance; insulin sensitivity; type 2 diabetes; high-fat diet; male; lipids

Address for correspondence: Marianne Olholm Larsen, Department of Pharmacological Research I, Novo Nordisk A/S, 6A1.117 Novo Allé, 2880 Bagsvaerd, Denmark. Voice: +45 4442 7675; fax: +45 4442 7488.
mmla@novonordisk.com

INTRODUCTION

Type 2 diabetes is a metabolic disorder with multiple etiologies. Defects in important mechanisms of insulin secretion and/or insulin action are involved in the development of the disease. The specific reasons for development of these disorders are not yet defined,[1] although high-energy intake and obesity are known to be of major importance. Obese subjects have several abnormalities in their glucose homeostasis, including increased hepatic glucose output[2] and gluconeogenesis,[3] insulin resistance,[4,5] and abnormal insulin secretion.[5–7] Furthermore, the triglyceride (TG) content in peripheral tissues plays an important role in regulation of both insulin sensitivity and secretion,[8] and intracellular lipids in the β-cells are suspected to affect the destruction of β-cells seen in diabetes.[9] In mice,[10,11] rats,[12,13] and dogs,[14] the feeding of high-fat diets has proven useful for the establishment of glucose intolerance and insulin-resistant syndromes. Feeding of high-fat diets to miniature pigs is known to induce slight insulin resistance[15] and increase total cholesterol (TC) levels considerably.[16–18] Thus, the Göttingen minipig might be useful, as a supplement to rodent models, for studying the effect of dietary high-fat intake and obesity on glucose homeostasis. The present study was designed as a pilot study to investigate the effects of obesity caused by high-fat feeding on glucose tolerance. The experimental diets used have previously been shown to induce slight insulin resistance in female Göttingen minipigs.[15]

MATERIALS AND METHODS

Animals

Male Göttingen minipigs, 7–8 months of age, were obtained from the barrier unit at Ellegaard Göttingen Minipigs ApS (Dalmose, Denmark). Animals were randomized to two groups with similar body weight ($n = 5$ in control group; $n = 4$ in high-fat group). Animals were housed in single pens under controlled conditions (temperature between 18°C and 22°C, relative air humidity 30–70%, with 4 air changes per hour) and fed twice daily on a restricted schedule with either control diet (CD) or high-fat high-energy diet (HFD). Body weights were recorded weekly and food ratios were adjusted to keep the control group on the normal weight curve[19] according to their age (225 g per meal) and to gradually increase the weight in the high-fat group to 150% of normal weight for their age (300 g per meal). The type of study was approved by the Animal Experiments Inspectorate, Ministry of Justice, Denmark. Animals were carefully trained so that all experiments could be performed in conscious, unstressed animals in their usual pens.

Experimental Diets

The composition of both diets is summarized in TABLE 1. Analysis of diets was performed at Steins Laboratorium A/S (Brorup, Denmark).

TABLE 1. Characteristics of experimental diets (CD = control diet; HFD = high-fat high-energy diet)

	CD	HFD
Protein (wt %)	18	14
Carbohydrate (wt %)	58	55
Fat (wt %)	5	20
Energy (MJ/kg)	9.1	12.1
Energy intake (MJ/kg)	4.1	7.3

Scan for Body Composition

A dual-energy X-ray absorptiometry scanning (DXA)–scanning (QDR-1000 W, Hologic, Zaventem, Belgium) for body composition was performed before and after the first 3 months of diet feeding in anesthetized animals [zolazepam 0.83 mg/kg, tiletamine 0.83 mg/kg (Zoletil® 50 vet., Boehringer Ingelheim, Copenhagen, Denmark), xylazine 0.90 mg/kg (Rompun® vet. [20 mg/mL], Bayer, Lyngby, Denmark), ketamine 0.83 mg/kg (Ketaminol® vet. [100 mg/mL], Rosco, Taastrup, Denmark), and methadone 0.20 mg/kg (Metadon "DAK" [10 mg/mL], Nycomed, Roskilde, Denmark); im]. Analysis of body composition was performed using the Whole Body Analysis protocol in the Hologic software package.

Implantation of Central Venous Catheters

Two central venous catheters (Certo 455, B. Braun Melsungen AG, Melsungen, Germany) were inserted surgically under general anesthesia induced as described above and maintained with isoflurane (1–3%) (Forene®, Abbot, Gentofte, Denmark) in 100% oxygen. Postsurgical analgesia was maintained by injection of 0.03 mg/kg buprenorfine [Anorfin® (0.3 mg/mL) GEA, Frederiksberg, Denmark] and 4 mg/kg carprofen [Rimadyl® vet. (50 mg/mL), Pfizer, Ballerup, Denmark] im before the end of anesthesia and for 3 days postsurgery by injection of 4 mg/kg carporfen once daily im.

Oral Glucose Tolerance Test

An oral glucose tolerance test (OGTT) was performed after 3 months of diet feeding. After an 18-h overnight fast, animals were offered a mixed meal glucose tolerance test of 25 g SDS (SDS, Essex, United Kingdom) minipig fodder and 2 g/kg glucose (500 g/L, SAD, Copenhagen, Denmark). The meal was eaten from a bowl under supervision. Blood samples were obtained from the catheters at −15, −5, 0, 15, 30, 45, 60, 90, 120, 150, and 180 min relative to the glucose load.

Frequent Sampling Intravenous Glucose Tolerance Test

An intravenous glucose tolerance test (IVGTT) was performed in conscious animals after 3 months of diet feeding. After an 18-h overnight fast, animals were dosed intravenously through the catheters with 0.3 g/kg glucose (500 g/L, SAD,

Copenhagen, Denmark) over 30 s at $t = 0$. At $t = 20$, animals were dosed intravenously through the catheters with Actrapid® (0.05 IU/kg) (Novo Nordisk A/S, Bagsvaerd, Denmark). One-mL blood samples were obtained from the catheters at −30, −20, −10, 0, 2, 3, 4, 5, 6, 8, 10, 12, 14, 16, 19, 22, 23, 24, 25, 27, 30, 35, 40, 45, 50, 55, 60, 65, 70, 80, 90, 100, 110, 120, 140, 160, and 180 min relative to the glucose load.

Clinical Chemistry

Blood samples were obtained in fasting animals and plasma levels of fructosamine, FFA, TG, and TC were determined.

Handling and Analysis of Blood Samples

Blood samples (2 mL for samples obtained during OGTT and 1 mL for samples obtained during IVGTT) were immediately transferred to vials containing EDTA (1.6 mg/mL final concentration) and 500 kIU aprotinin/mL full blood (Trasylol®, 10,000 kIU/mL, Bayer, Lyngby, Denmark) and were kept on ice until centrifugation. Samples were centrifuged (4°C, 10 min, 3500 rpm) and plasma separated and stored at −20°C until analysis. Plasma glucose was analyzed using the immobilized glucose oxidase method, with 10 μL of plasma in 0.5 mL of buffer (EBIO plus autoanalyzer and solution, Eppendorf, Hamburg, Germany). Plasma insulin was analyzed in a two-site immunometric assay with monoclonal antibodies as catching and detecting antibodies (catching antibody HUI-018 raised against the A-chain of human insulin; detecting antibody OXI-005 raised against the B-chain of bovine insulin)[20] and using purified porcine insulin for calibration of the assay. Analyses for TG, FFA, TC, and fructosamine were performed using a COBAS MIRA plus autoanalyzer (Roche Diagnostic Systems, Basel, Switzerland).

Evaluation

Effects of diet feeding on body composition were evaluated based on changes in absolute and relative lean and fat mass and relative truncal fat mass. Effects of diet feeding on glucose tolerance were evaluated based on changes in fasting concentrations of glucose and insulin and area under the plasma concentration curve (AUC) for glucose and insulin during OGTT and IVGTT. Glucose disappearance rate (KG) during IVGTT was calculated as the slope of the least-squares regression line relating the natural logarithm of the glucose concentration to time[21,22] between 8 and 19 min after the glucose bolus. Furthermore, insulin sensitivity (S_I) and glucose effectiveness (S_G) calculated by minimal modeling[21,23] were evaluated during IVGTT. Data are presented as mean ± SD.

Calculations and statistical evaluation by paired or unpaired two-tailed Student's t test were performed using Excel (2000) and GraphPad Prism version 3.00 for Windows (GraphPad software, San Diego, CA). Calculation of insulin sensitivity was done using IS-Ciba (Ciba-Geigy, Basel, Switzerland). p values of 0.05 or less were considered significant.

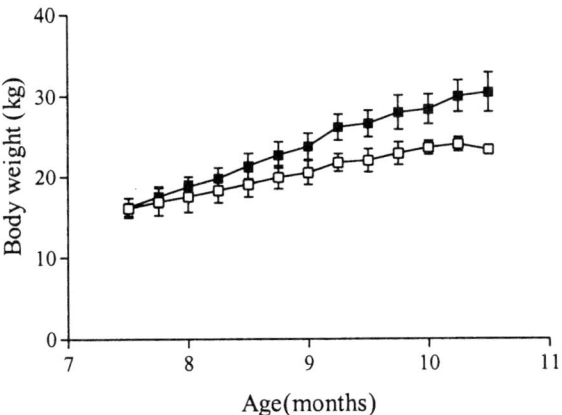

FIGURE 1. Body weights in male Göttingen minipigs during feeding of control (□) ($n = 5$) or high-fat high-energy (■) ($n = 4$) diet. Data are means ± SD.

RESULTS

Body Weights

Body weights in the two experimental groups were similar at the start of diet feeding (16.1 ± 1.2 kg in control group vs. 16.2 ± 1.3 kg in high-fat group, $p = 0.878$), and the development in body weights is shown in FIGURE 1. After 3 months, body weight in the HFD group was significantly higher (32.6 ± 2.4 kg vs. 24.9 ± 0.5 kg, $p < 0.001$) compared to CD.

DXA Scan for Body Composition

Results from the DXA scans are shown in FIGURE 2. No differences were found before the start of diet feeding in truncal or total fat percentage or total lean percentage. In the CD group, no significant changes were found in truncal or total body fat percentage or total lean percentage after 3 months of diet feeding. In the HFD group, both truncal ($p = 0.015$) and total ($p = 0.022$) fat percentage had increased significantly after 3 months of diet feeding, whereas relative total lean mass had decreased ($p = 0.022$), although absolute total lean mass had increased (from 14.9 ± 1.1 kg to 27.4 ± 1.5 kg, $p < 0.001$). Furthermore, both truncal ($p = 0.001$) and total ($p = 0.002$) fat percentage were significantly higher in the HFD group, whereas total lean percentage was significantly lower ($p = 0.002$) and absolute total lean mass was significantly higher (27.4 ± 1.5 kg vs. 22.9 ± 0.5 kg, $p < 0.001$) compared to CD after 3 months of diet feeding.

Oral Glucose Tolerance

A significant increase in fasting plasma glucose (4.3 ± 0.4 mM vs. 3.6 ± 0.3 mM, $p = 0.023$) and insulin (80 ± 23 pM vs. 23 ± 21 pM, $p = 0.012$) was seen after HFD,

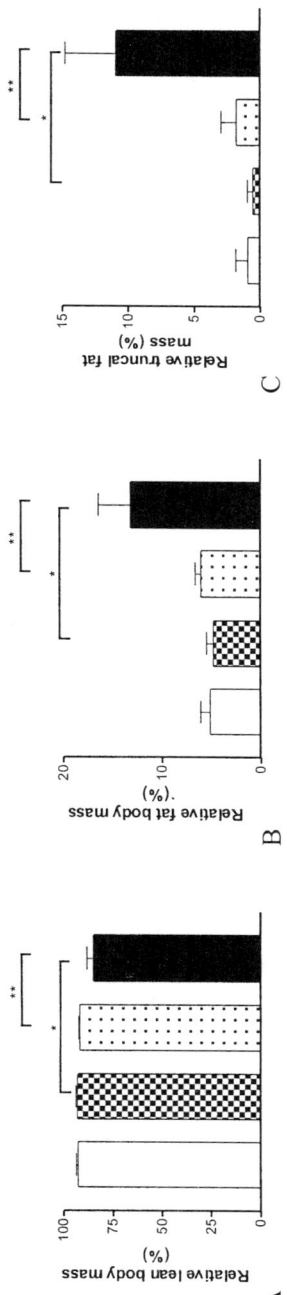

FIGURE 2. Body composition in male Göttingen minipigs determined by DXA scanning before [control diet: white bars ($n = 5$); high-fat high-energy diet: dark hatched bars ($n = 4$)] and after (control diet: light hatched bars; high-fat high-energy diet: black bars) 3 months of diet feeding: **(A)** relative lean body mass; **(B)** relative fat body mass; **(C)** relative truncal fat mass. Data are means ± SD; *$p < 0.05$, **$p < 0.01$ (Student's t test).

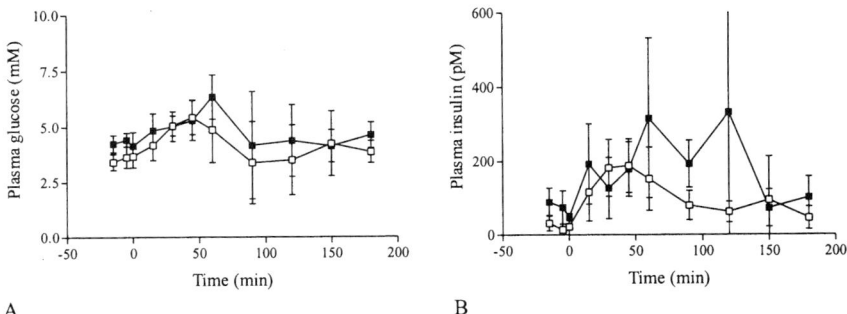

FIGURE 3. Plasma levels of glucose (**A**) and insulin (**B**) during OGTT in male Göttingen minipigs after 3 months of control (□) ($n = 5$) or high-fat high-energy feeding (■) ($n = 4$). Data are means ± SD.

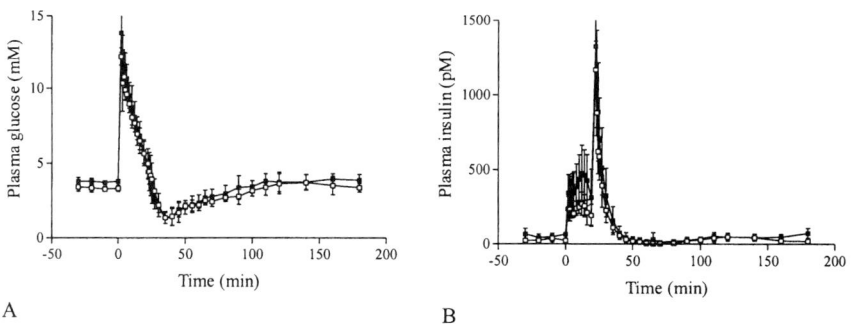

FIGURE 4. Plasma levels of glucose (**A**) and insulin (**B**) during IVGTT in male Göttingen minipigs after 3 months of control (□) ($n = 5$) or high-fat high-energy feeding (■) ($n = 4$). Data are means ± SD.

whereas no changes were seen in AUC glucose (923 ± 98 vs. 803 ± 141 mM × min, $p = 0.192$) or AUC insulin (35,523 ± 21,969 vs. 18,958 ± 3952 pM × min, $p = 0.137$) (FIG. 3).

Intravenous Glucose Tolerance

In parallel with the OGTT, a significant increase was observed in fasting plasma glucose (3.8 ± 0.1 mM vs. 3.4 ± 0.2 mM, $p = 0.008$) and insulin (55 ± 21 pM vs. 27 ± 12 pM, $p = 0.038$) after HFD. AUC glucose was not significantly changed by HFD (807 ± 54 vs. 749 ± 61 mM × min, $p = 0.180$). AUC insulin from 2–19 min, representing the endogenous insulin response to glucose, was significantly higher in the HFD group (6741 ± 2538 vs. 3938 ± 771 pM × min, $p = 0.05$) (FIG. 4). KG did not differ between the groups (4.2 ± 0.7%/min in CD group vs. 4.8 ± 0.5%/min in HFD group, $p = 0.151$). Due to failure of data to be processed in the model, S_I and

S_G could only be determined in 4 of the 5 animals in the CD group and in 3 of the 4 animals in the HFD group. S_I was not significantly different in the two experimental groups after 3 months of diet feeding (8.0 E-04 ± 1.1 E-03 in CD group vs. 6.5 E-04 ± 8.6 E-04 min^{-1}/pM in HFD group, $p = 0.851$). S_G was 4.1 E-02 ± 3.5 E-02 min^{-1} for the CD group and 4.4 E-02 ± 4.6 E-02 min^{-1} for the HFD group.

Plasma Lipids and Fructosamine

HFD significantly increased plasma levels of TC (1.80 ± 0.23 mM vs. 0.87 ± 0.03 mM, $p = 0.003$), whereas levels of fructosamine (216 ± 8 μM vs. 225 ± 9 μM, $p = 0.494$), FFA (0.34 ± 0.14 mM vs. 0.17 ± 0.08 mM, $p = 0.317$), and TG (0.72 ± 0.06 mM vs. 1.03 ± 0.35 mM, $p = 0.473$) were not changed.

DISCUSSION

In this study, 3 months of high-fat high-energy feeding resulted in an increase in body weight of approximately 50% compared to normal values, with an increase in truncal fat from around 2% to 11% and total fat from 6% to 13%, whereas total relative lean mass was decreased from 93% to 85%, although absolute lean mass was increased. Since the normal weight curve for minipigs[19] is based on animals fed a restricted diet, the body weight obtained after HFD in the present study might not represent substantial overweight. Nevertheless, the changes induced by HFD resulted in increased levels of fasting plasma glucose and insulin. This is in accordance with observations in mice[10,11,24–26] and rats,[12] whereas a previous study of female minipigs did not show increased fasting glucose and insulin levels.[15] The pigs in the present study did not show significant changes in glucose tolerance, in accordance with results from female Göttingen minipigs fed a high-fat diet,[15] but in contrast to what has been reported in rats[12,13] and mice.[11,25] Furthermore, the insulin secretory response to glucose was increased in the present study, as has also been found in female Göttingen minipigs,[15] but is in contrast to observations in mice.[10,26] Some of these differences could be attributable to strains showing different susceptibility towards the effects of high-fat high-energy feeding as has been demonstrated in both rats[12] and mice.[11,25] Alternatively, a longer period of HFD feeding might possibly induce more pronounced effects on glucose tolerance in pigs. The use of minimal modeling for estimation of insulin sensitivity and glucose effectiveness has been validated in both man[27–29] and dog,[30] but its precision has been questioned by others.[31] Although the values of both S_I and S_G derived from the minimal model in the present study are in the range previously found for both dogs[30,32] and humans,[33] and the minimal model has previously been used in pigs,[34] the use of the model in this species has not, to our knowledge, been validated formally. After 3 months of HFD, results from minimal modeling did not indicate changes in insulin sensitivity, although the coexistence of fasting hyperglycemia and hyperinsulinemia and the increase in endogenous insulin response during IVGTT, without a simultaneous decrease in AUC glucose and increase in glucose disappearance rate during the same test, do indicate decreased insulin sensitivity after HFD in accordance with findings in pigs,[15] dogs,[14] rats,[12] and mice.[10,11] In contrast to what has been reported in mice,[26] the insulinotropic effect of glucose was not impaired in the high-fat-fed pigs, as has also recently been reported.[15] The lack of changes in TG levels during high-fat feeding is in accordance with previous findings in male minipigs,[17,18] whereas high-fat feeding in female minipigs has been reported to increase levels of TG.[15,18] The increase in TC levels induced by HFD in the present study is much lower than what has previously been reported after using cholesterol-enriched diets,[16–18] whereas similar moderate increases have been reported after feeding high-fat diets with relatively low cholesterol content.[15,35] The diet used in the present study was not enriched with cholesterol and this probably explains the difference. In conclusion, these studies have shown that changes in body weight and composition, resulting in minor abnormalities in glucose tolerance, insulin sensitivity, and TC levels, can be induced in the male Göttingen minipig by HFD feeding for 3 months, aiming at body weights of approximately 150% of normal weight. Abnormalities are characterized by slight hyperglycemia and compensatory hyperinsulinemia, indicating some degree of insulin resistance.

High-fat high-energy feeding of Göttingen minipigs seems to be an interesting and promising approach for establishment of a large animal model of insulin resistance or type 2 diabetes. The advantage of this model, compared to chemical induction of diabetes, is that the pathogenesis involved resembles the situation in humans and includes the possibility of inducing insulin resistance, mimicking the characteristics of the human disease. It is possible that a longer period of fat feeding would cause glucose tolerance to deteriorate further. It is also an interesting aspect to combine the moderate abnormalities in glucose tolerance and reduced insulin sensitivity caused by HFD with a small amount of chemically induced β-cell damage, thereby decreasing glucose tolerance further.

REFERENCES

1. WHO. 1999. Diagnosis and classification of diabetes mellitus. *In* Definition Diagnosis and Classification of Diabetes Mellitus and Its Complications, pp. 1–59. World Health Organization. Geneva.
2. HENRY, R.R., *et al.* 1986. Effects of weight loss on mechanisms of hyperglycemia in obese non-insulin-dependent diabetes mellitus. Diabetes **35:** 990–998.
3. GASTALDELLI, A., *et al.* 2000. Influence of obesity and type 2 diabetes on gluconeogenesis and glucose output in humans—a quantitative study. Diabetes **49:** 1367–1373.
4. KELLEY, D.E., *et al.* 1996. The effect of non-insulin-dependent diabetes mellitus and obesity on glucose transport and phosphorylation in skeletal muscle. J. Clin. Invest. **97:** 2705–2713.
5. WARD, W.K., *et al.* 1984. Diminished b cell secretory capacity in patients with non-insulin dependent diabetes mellitus. J. Clin. Invest. **74:** 1318–1328.
6. POLONSKY, K.S., *et al.* 1988. Twenty-four-hour profiles and pulsatile patterns of insulin secretion in normal and obese subjects. J. Clin. Invest. **81:** 442–448.
7. O'MEARA, N.M., *et al.* 1993. Lack of control by glucose of ultradian insulin secretory oscillations in impaired glucose tolerance and in non-insulin-dependent diabetes mellitus. J. Clin. Invest. **92:** 262–271.
8. KOYAMA, K., *et al.* 1997. Tissue triglycerides, insulin resistance, and insulin production: implications for hyperinsulinemia of obesity. Am. J. Physiol. Endocrinol. Metab. **273:** E708–E713.
9. MCGARRY, J.D. & R.L. DOBBINS. 1999. Fatty acids, lipotoxicity, and insulin secretion. Diabetologia **42:** 128–138.
10. AHREN, B., *et al.* 1997. Dissociated insulinotropic sensitivity to glucose and carbachol in high-fat diet–induced insulin resistance in C57BL/6J mice. Metabolism **46:** 97–106.

11. SURWIT, R.S., et al. 1988. Diet-induced type II diabetes in C57BL/6J mice. Diabetes **37:** 1163–1167.
12. LEVIN, B.E., et al. 1997. Selective breeding for diet-induced obesity and resistance in Sprague-Dawley rats. Am. J. Physiol. Endocrinol. Metab. **272:** R725–R730.
13. OAKES, N.D., et al. 1997. Mechanisms of liver and muscle insulin resistance induced by chronic high-fat feeding. Diabetes **46:** 1768–1774.
14. BERGMAN, R.N., et al. 2001. Central role of the adipocyte in the metabolic syndrome. J. Invest. Med. **49:** 119–126.
15. JOHANSEN, T., et al. 2001. The obese Göttingen minipig as a model of the metabolic syndrome: dietary effects on obesity, insulin sensitivity, and growth hormone profile. Comp. Med. **51:** 150–155.
16. JACOBSSON, L. 1986. Comparison of experimental hypercholesterolemia and atherosclerosis in Gottingen mini-pigs and Swedish domestic swine. Atherosclerosis **59:** 205–213.
17. DIXON, J.L., et al. 1999. Dyslipidemia and vascular dysfunction in diabetic pigs fed an atherogenic diet. Arterioscler. Thromb. Vasc. Biol. **19:** 2981–2992.
18. JACOBSSON, L. 1989. Comparison of experimental hypercholesterolemia and atherosclerosis in male and female mini-pigs of the Göttingen strain. Artery **16:** 105–117.
19. BOLLEN, P. & L. ELLEGAARD. 1996. Developments in breeding Göttingen minipigs. *In* Advances in Swine in Biomedical Research, pp. 59–66. Plenum. New York.
20. ANDERSEN, L., et al. 1993. Enzyme immunoassay for intact human insulin in serum or plasma. Clin. Chem. **39:** 578–582.
21. BERGMAN, R.N., et al. 1979. Quantitative estimation of insulin sensitivity. Am. J. Physiol. **236:** E667–E677.
22. KAHN, S.E., et al. 1994. The contribution of insulin-dependent and insulin-independent glucose uptake to intravenous glucose tolerance in healthy human subjects. Diabetes **43:** 587–592.
23. TOFFOLO, G., et al. 1980. Quantitative estimation of beta cell sensitivity to glucose in the intact organism: a minimal model of insulin kinetics in the dog. Diabetes **29:** 979–990.
24. AHREN, B. 1999. Plasma leptin and insulin in C57BL/6J mice on a high-fat diet: relation to subsequent changes in body weight. Acta Physiol. Scand. **165:** 233–240.
25. AHREN, B. & A.J.W. SCHEURINK. 1998. Marked hyperleptinemia after high-fat diet associated with severe glucose intolerance in mice. Eur. J. Endocrinol. **13:** 461–467.
26. SIMONSSON, E. & B. AHREN. 1998. Potentiated beta-cell response to non-glucose stimuli in insulin-resistant C57BL/6J mice. Eur. J. Pharmacol. **350:** 243–250.
27. SAAD, M.F., et al. 1994. A comparison between the minimal model and the glucose clamp in the assessment of insulin sensitivity across the spectrum of glucose tolerance: Insulin Resistance Atherosclerosis Study. Diabetes **43:** 1114–1121.
28. COATES, P.A., et al. 1995. Comparison of estimates of insulin sensitivity from minimal model analysis of the insulin-modified frequently sampled intravenous glucose tolerance test and the isoglycemic hyperinsulinemic clamp in subjects with NIDDM. Diabetes **44:** 631–635.
29. BEST, J.D., et al. 1990. Practical application of methods for *in vivo* assessment of insulin secretion and action. Horm. Metab. Res. Suppl. **24:** 60–66.
30. FINEGOOD, D.T., G. PACINI & R.N. BERGMAN. 1984. The insulin sensitivity index: correlation in dogs between values determined from the intravenous glucose tolerance test and the euglycemic glucose clamp. Diabetes **33:** 362–368.
31. FOLEY, J.E., et al. 1985. Estimates of *in vivo* insulin action in humans: comparison of the insulin clamp and the minimal model techniques. Horm. Metab. Res. **17:** 406–409.
32. ADER, M., et al. 1985. Importance of glucose per se to intravenous glucose tolerance: comparison of the minimal-model prediction with direct measurements. Diabetes **34:** 1092–1103.
33. BERGMAN, R.N. 1989. Toward physiological understanding of glucose tolerance: minimal-model approach. Diabetes **38:** 1512–1527.
34. BEHME, M.T. 1996. Dietary fish oil enhances insulin sensitivity in miniature pigs. J. Nutr. **126:** 1549–1553.
35. POND, W.G., H.J. MERSMANN & J.T. YEN. 1985. Effects of obesity per se on plasma lipid and aortic response to diet in swine. Proc. Soc. Exp. Biol. Med. **179:** 90–95.

Comparison of the Extrapancreatic Action of BRX-220 and Pioglitazone in the High-Fat Diet–Induced Insulin Resistance

ELENA ŠEBÖKOVÁ,[a] MARIA KÜRTHY,[b] T. MOGYOROSI,[b] KAROLY NAGY,[b] EDITA DEMCÁKOVÁ,[a] JOZEF UKROPEC,[a] LASZLO KORANYI,[b] AND IWAR KLIMEŠ[a]

[a]*Diabetes and Nutrition Research Laboratory, Institute of Experimental Endocrinology, Slovak Academy of Sciences, SK-83306 Bratislava, Slovak Republic*

[b]*Biorex R&D Company, Veszprém, Hungary*

ABSTRACT: A new Biorex molecule, BRX-220, has been shown to be effective in animal models of diabetic neuro- and retinopathy. Recent *in vitro* studies showed that it might also have an insulin-sensitizing action. Therefore, the effect of BRX-220 on insulin sensitivity was compared with the action of pioglitazone (PGZ) in high fat (HF) diet–induced insulin resistance (IR) of rats. *Methods*—Male Wistar rats were fed for 3 weeks a standard chow (PD) or the HF (70-cal%) diet. The HF-fed rats were also given daily BRX-220 (20 mg/kg BW) or PGZ (6 mg/kg BW) by gavage. *In vivo* insulin action was assessed by the euglycemic hyperinsulinemic clamp. Glucose, insulin, FFA, triglyceride (TG), and glycerol levels in blood were also measured, as well as tissue TG content. *Results*—Increased levels of fed TG in circulation after HF diet (PD: 2.0 ± 0.2 vs. HF: 5.0 ± 0.8 mmol/L) were partially corrected by BRX-220 (HF+BRX: 3.8 ± 0.3) and normalized by PGZ (HF+PGZ: 2.6 ± 0.3). Both molecules prevented the increase in fed serum FFA levels after HF diet (PD: 0.5 ± 0.06; HF: 1.8 ± 0.2 mmol/L), with a more pronounced effect of PGZ (HF+BRX: 1.2 ± 0.1; HF+PGZ: 0.7 ± 0.06). Tissue TG levels increased significantly in response to HF feeding in both liver (HF: 16 ± 3.0; PD: 6.4 ± 1.1 μmol/g) and skeletal muscle (HF: 7.7 ± 1.2; PD: 2.4 ± 0.7). This increase was completely normalized by both agents in the liver (HF+BRX: 8.8 ± 0.8; HF+PGZ: 8.8 ± 1.0), and only partially in the skeletal muscles. HF diet–induced *in vivo* IR (PD: 25.4 ± 0.5; HF: 15.7 ± 0.5 mg/kg/min) was significantly reduced by BRX-220 (HF+BRX: 18.7 ± 0.3) and PGZ (HF+PGZ: 22.8 ± 0.4) treatment. *Conclusions*—(1) Subchronic administration of BRX-220 leads to an improvement of *in vivo* insulin action. (2) This insulin-sensitizing effect is, however, not as pronounced as that of PGZ. (3) It is accompanied by a decrease of circulating TG and FFA levels in the postprandial state and (4) by lower TG content in liver and skeletal muscle.

KEYWORDS: BRX-220; pioglitazone; HF diet; lipids; insulin action

Address for correspondence: Iwar Klimeš, Diabetes and Nutrition Research Laboratory, Institute of Experimental Endocrinology, Slovak Academy of Sciences, Vlárska 3, SK-83306 Bratislava, Slovak Republic. Voice: +421-2-5477 2687; fax: +421-2-5477 2678.
ueeniwar@savba.sk

INTRODUCTION

Insulin resistance (IR) is not only a substantial etiological factor in development of the metabolic syndrome X and a risk factor for cardiovascular complications,[1,2] but also a target for drug development. Although treatment has previously been limited to metformin or nonpharmacological measures, efforts are now being made to discover new drugs active against IR, so-called insulin sensitizers.[3] Some drugs sensitize peripheral tissues to the action of insulin. For instance, biguanides facilitate translocation of the glucose transporters to the membrane in the presence of insulin.[4] Other compounds such as vanadate[5] or IGF-1[6] mimic some peripheral action of insulin. Finally, blockade of FFA oxidation by specific inhibitors[7] can limit IR[8] as well. There are a number of other molecules under investigation with potential extrapancreatic action, which were originally developed as new insulin secretagogues, for example, A-4166,[9] insulin sensitizers,[10] or agents for treatment of chronic diabetic complications.

A good example for the latter is bimoclomol, which has been tested in the treatment of diabetic neuropathy[11] or retinopathy.[12] However, recently, the first indirect evidence[13] was provided that a bimoclomol derivative might improve also the peripheral glucose utilization. Very recently, Biorex Research Laboratories have developed BRX-220, which was supposed to have an even more pronounced insulin-sensitizing action.[14]

Indeed, recent *in vivo* studies, in particular the measurement of the whole body insulin-induced glucose utilization using the euglycemic hyperinsulinemic clamp technique, showed that a long-term administration of these compounds to rats with HF diet–induced IR improved the reduced *in vivo* insulin action when compared to control rats.[15,16]

Thiazolidinediones are a new class of orally active antidiabetic agents that reduce IR in the periphery and in the liver, resulting in increased insulin-dependent glucose disposal and reduced hepatic glucose output.[17] The precise mechanism of action of these drugs is not yet completely understood. However, there is clear evidence demonstrating that thiazolidinediones, as potent highly selective agonists for peroxisome proliferator-activated nuclear receptors (PPAR) gamma, modulate the transcription of a number of insulin-responsive genes involved in glucose and lipid metabolism.[18]

Thus, in this study, the *in vivo* action of BRX-220 was compared with that of an established insulin-sensitizing agent, pioglitazone (PGZ), in the HF diet–induced IR of rats.

MATERIALS AND METHODS

Animals and Diets

Fifteen-week-old male Wistar Charles River rats (AnLab, Prague, Czech Republic) were used. The treated animals were fed for 3 weeks an IR-inducing HF diet (70-cal% fat).[19] For the whole study period, animals received BRX-220 (20 mg/kg/BW) or PGZ (6 mg/kg/BW) suspended in 0.5% methylcellulose in water once daily by gavage. This approach created the following groups of animals: PD, HF, HF + BRX, and HF + PGZ.

Euglycemic Clamp Studies

After 3 weeks of feeding the above diets, rats were anesthetized and fitted with chronic artery and jugular cannula as described by Koopmans et al.[20] After surgery, feeding continued and studies were conducted 72 h after catheter implantation in the unrestrained sedentary conscious state. Food was removed 16 h before the study. Euglycemic hyperinsulinemic clamps were performed according to Kraegen et al.[21] as described in detail earlier.[9] Blood samples were obtained for glucose and insulin determination in all clamp studies at 15-min intervals.

Analytical Methods

Serum glucose concentrations were measured with the aid of the Beckman Glucose Analyzer (Fullerton, CA). TG levels in serum or in tissue lipid extracts were measured using a specific, commercially available enzymatic set (TG-DST-P, Australia). For measurement of FFA and glycerol in serum, the enzymatic kit from Randox Laboratories (Ardmore, United Kingdom) was used. Serum insulin levels were measured using the rat insulin RIA kit from Linco (U.S.A.).

Statistical Analyses

Results are expressed as the mean ± SEM. Differences between groups were analyzed using analysis of variance (ANOVA) with the Bonferroni post-hoc test[22] at the overall significance threshold of $\alpha = 0.05$.

RESULTS

Animal Characteristics and Serum Parameters

The initial body weights between the individual groups of rats did not differ. The body weights increased also in a similar way during the 3-week feeding period

TABLE 1. Effect of BRX-220 and PGZ treatment on body weight gain, food consumption, serum glucose and insulin levels, and insulin sensitivity index GIR in HF diet–fed animals

	PD	HF	HF + BRX	HF + PGZ
Body weight increase [g]	93 ± 8.0^a	93 ± 4.4^a	84 ± 4.8^a	88 ± 5.9^a
Food consum. [g/day/kg/BW]	74.0 ± 1.0^a	53.0 ± 1.5^b	51.2 ± 1.5^b	49.0 ± 0.9^b
Glucose [mmol/L]	4.9 ± 0.2^a	4.9 ± 0.1^a	4.8 ± 0.1^a	4.8 ± 0.1^a
Insulin [mU/L]	16.4 ± 2.5^a	52.5 ± 5.2^b	50.8 ± 4.1^b	39.1 ± 2.4^b
GIR [mg/kg/min]	25.4 ± 0.5^a	15.7 ± 0.5^b	18.7 ± 0.3^c	22.8 ± 0.4^d

NOTE: Data are expressed as mean ± SEM of 8 rats. Values without a common superscript (a, b, c, d) within the rows are significantly different at $p < 0.05$. PD = control lab chow diet; HF = high-fat diet; HF + BRX = high-fat diet plus BRX-220 (20 mg/kg/day) given by gavage, HF + PGZ = high-fat diet plus PGZ (6 mg/kg/day) given by gavage. Fasted serum glucose and/or insulin values were obtained from blood withdrawn after 3 weeks of dietary treatment of rats that had not yet undergone the vascular surgery. GIR = glucose infusion rate.

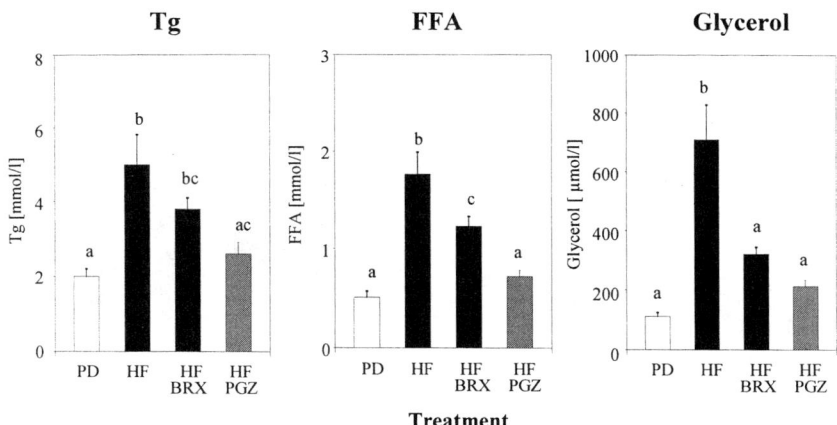

FIGURE 1. Effect of BRX-220 and PGZ treatment on lipid parameters in serum obtained from rats in the fed state. Bars represent mean ± SEM of 8 rats. Values without a common superscript (a, b, c) are significantly different at $p < 0.05$. PD = control lab chow diet; HF = high-fat diet; HF + BRX = high-fat diet plus BRX-220 (20 mg/kg/day) given by gavage; HF + PGZ = high-fat diet plus PGZ (6 mg/kg/day) given by gavage.

(TABLE 1). The food consumption was lower in all HF diet–fed groups, which was expected due to the higher caloric content of this diet in comparison to the PD-fed rats (TABLE 1).

Feeding rats the HF diet led to an increase in fasting insulinemia with no change in fasting plasma glucose levels, indicating a decrease of insulin action (TABLE 1). Administration of the BRX-220 and/or PGZ compound to the HF diet–fed animals, however, failed to normalize fasting serum insulin levels (TABLE 1).

The HF feeding led to a statistically significant increase in fed TG, FFA, and glycerol levels in circulation. Treatment of rats with both molecules lowered significantly serum FFA and glycerol levels, with a more pronounced effect in the PGZ-treated animals. However, only PGZ was able to normalize the increased levels of circulating TG in the HF diet–fed animals (FIG. 1).

Tissue Parameters

Feeding animals the HF diet for 3 weeks led to a marked increase in tissue TG in both the liver and skeletal muscle. Application of PGZ to rats fed the HF diet prevented the accumulation of TG in both tissues. On the contrary, in skeletal muscle, a significant effect of the 3-week treatment was seen in the PGZ group only. Nevertheless, a tendency to a decrease was seen also in the HF diet–fed animals given the BRX-220 compound (FIG. 2).

Euglycemic Hyperinsulinemic Clamp Data

The HF diet–fed control rats had a low (15.7 ± 0.5 mg/kg/min) exogenous glucose infusion rate (GIR) required to maintain euglycemia during hyperinsulinemia

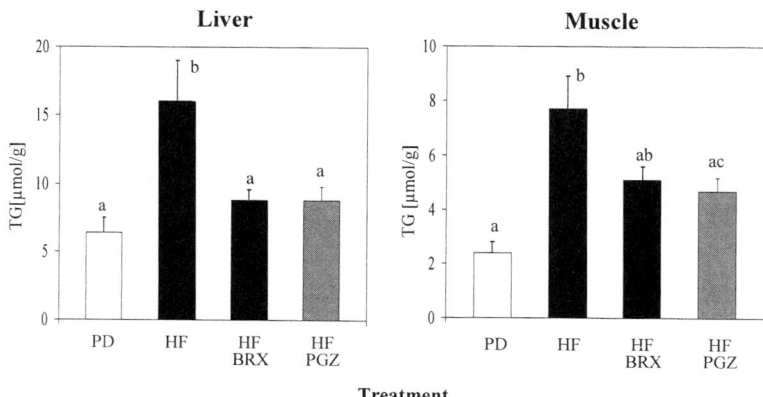

FIGURE 2. Effect of BRX-220 and PGZ treatment on TG content in liver and skeletal muscle. Bars represent mean ± SEM of 8 rats. Values without a common superscript (a, b, c) are significantly different at $p < 0.05$. PD = control lab chow diet; HF = high-fat diet; HF + BRX = high-fat diet plus BRX-220 (20 mg/kg/day) given by gavage; HF + PGZ = high-fat diet plus PGZ (6 mg/kg/day) given by gavage.

(TABLE 1). In fact, their *in vivo* insulin-induced glucose utilization was reduced by close to 60% when compared to the GIR of the control rats with normal insulin action at a comparable steady-state plasma insulin level. A long-term, once-a-day oral administration of both the BRX-220 and the PGZ compounds was accompanied by a clear improvement of insulin action, as reflected by a higher GIR (18.7 ± 0.3 and 22.8 ± 0.4 mg/kg/min, respectively) required to maintain euglycemia. Nevertheless, none of the tested compounds was capable of restoring a normal *in vivo* insulin action (TABLE 1) under the dose and experimental setting used.

DISCUSSION

Our study has shown that the newly developed molecule, BRX-220, is able to improve the *in vivo* insulin-induced whole body glucose utilization in the HF diet–induced IR. Indeed, the BRX-220-treated insulin-resistant rats required a higher GIR during a hyperinsulinemic clamp in order to maintain euglycemia when compared to the GIR of the HF diet–fed animals. However, the effect of this compound was not as pronounced when compared with the effect of PGZ, the well-established insulin-sensitizer.

Although body weight alterations are generally important for changes in the *in vivo* insulin action, we did not observe any significant difference in body weight increments in the HF diet–fed rats whether supplemented or not with the BRX-220 or PGZ compounds.

Feeding rats for several weeks a diet containing 70-cal% of fat (about 40-wt% of saturated fatty acids) results in widespread IR with major effects in skeletal muscle,[23,24] as assessed by measurement of the glucose metabolic index Rg′ in insulin target tissues after a single bolus of ^{3}H-deoxyglucose during the insulin clamp.[21] This type of IR is usually accompanied by raised TG availability, that is, by increased plasma TG and/or by their accumulation in skeletal muscles.[9,25] Moreover, an increased concentration of FFA in circulation is also present. In contrast, insulin action improves by lowering plasma and/or intramuscular TG and FFA in circulation (e.g., by benfluorex[26] or the thiazolidinedione derivative ciglitazone[27]).

Our findings, namely, that beneficial effects of both tested compounds are associated with a significant decrease of TG, FFA, and glycerol in circulation, are in harmony with the aforementioned information from the literature. They also provide another evidence for the important role of these lipids for the development of IR.

The mechanism by which BRX-220 lowers FFA and TG is not clear. PGZ acts as a potent PPAR-activating agent, but we have no direct evidence that BRX-220 acts in the same way. The insulin-sensitizing effect of BRX-220 may come from its ability to suppress the accumulation of TG in muscle and liver to an extent comparable with that of PGZ. In searching for potential mechanisms responsible for the improvement of insulin action after BRX-220 and PGZ treatment, we have clearly demonstrated that an improvement of *in vivo* insulin sensitivity after both treatments is associated with lowering of TG availability not only in the circulation, but also in liver and more importantly in the skeletal muscles. This is in harmony with recent data obtained in various animal models[28] of IR and in humans with type 2 diabetes.[29,30]

In summary, a good deal of evidence has now accumulated to show that BRX-220, originally developed for the treatment of diabetic complications, has also some beneficial effect on the whole body insulin sensitivity. The latter is accompanied also by an improvement of lipid metabolism. Further studies at the cellular and molecular levels are needed to shed more light on the mechanism of action of this potentially interesting compound.

ACKNOWLEDGMENTS

The skillful technical assistance of Alica Mitková and Katarína Sušienková is greatly appreciated. This study was supported in part by Research Grant No. 2/7210/21 from the Grant Agency of the Slovak Academy of Sciences VEGA and the COST B17 Program.

REFERENCES

1. DEFRONZO, R.A., R.C. BONADONNA & E. FERRANNINI. 1997. Pathogenesis of NIDDM. *In* International Textbook of Diabetes Mellitus, pp. 635–712. Wiley. New York/Chichester.
2. HAFFNER, S.M. 1997. Progress in population analyses of the insulin resistance syndrome. Ann. N.Y. Acad. Sci. **827:** 1–12.
3. NILSSON, P., K.F. ERIKSSON & L. GROOP. 1998. New drugs against insulin resistance. Lakartidningen **95:** 2832–2834.
4. WIERNSPERGER, N.F. 1999. Membrane physiology as a basis for the cellular effects of metformin in insulin resistance and diabetes. Diabete Metab. (Paris) **25:** 110–127.
5. FANTUS, I.G. & E. TSIANI. 1998. Multifunctional actions of vanadium compounds on insulin signaling pathways: evidence for preferential enhancement of metabolic versus mitogenic effects. Mol. Cell. Biochem. **182:** 109–119.

6. LE MARCHAND–BRUSTEL, Y. 1999. Molecular mechanisms of insulin action in normal and insulin-resistant states. Exp. Clin. Endocrinol. Diabetes **107:** 126–132.
7. FOLEY, J.E., et al. 1997. Pharmacological strategies for reduction of lipid availability. Ann. N.Y. Acad. Sci. **827:** 231–245.
8. VIALETTES, B. & P. SILVESTRE. 1992. Pharmacological approach in the treatment of insulin resistance. Horm. Res. **38:** 51–56.
9. KLIMEŠ, I., et al. 1998. The effect of the new oral hypoglycemic agent A-4166 on glucose turnover in the high fat diet–induced and/or in the hereditary insulin resistance of rats. Arch. Physiol. Biochem. **106:** 325–332.
10. GRANBERRY, M.C., E.F. SCHNEIDER & V.A. FONSECA. 1998. The role of troglitazone in treating the insulin resistance syndrome. Pharmacotherapy **18:** 973–987.
11. BIRO, K., et al. 1997. Bimoclomol (BRLP-42) ameliorates peripheral neuropathy in streptozotocin-induced diabetic rats. Brain Res. Bull. **44:** 259–263.
12. BIRO, K., et al. 1998. Bimoclomol improves early electro-physiological signs of retinopathy in diabetic rats. Neuroreport **9:** 2029–2033.
13. BIOREX R&D. 1998. BRX-SCR-DB-057 study report. Veszprém, Hungary.
14. NAGY, K. & L. KORANYI. 1999. Personal communication.
15. KLIMEŠ, I., et al. 2000. The new Biorex compound increases insulin mediated glucose disposal in high fat diet–induced insulin resistance. XV. Conference of the Hungarian Diabetes Association (Tihany, Hungary).
16. KLIMEŠ, I., et al. 1999. Evaluation of the extrapancreatic effects of new biologically active substances (BRX 220 and BRX 221): Part A. Study report. Biorex R&D. Veszprém, Hungary.
17. SALTIEL, A.R. & J.M. OLEFSKY. 1996. Thiazolidinediones in the treatment of insulin resistance and type II diabetes. Diabetes **45:** 1661–1669.
18. MURAKAMI, K., et al. 1998. A novel insulin sensitizer acts a co-ligand for peroxisome proliferator-activated receptor α and γ: effect of PPARα activation on normal lipid metabolism in liver of Zucker fatty rats. Diabetes **47:** 1841–1847.
19. STORLIEN, L.H., et al. 1991. Influence of dietary fat composition on development of insulin resistance in rats. Diabetes **40:** 280–289.
20. KOOPMANS, S.J., et al. 1992. In vivo insulin responsiveness for glucose uptake at eu- and hyperglycemic levels in diabetic rats. Biochim. Biophys. Acta **1115:** 230–238.
21. KRAEGEN, E.W., et al. 1985. Dose response curve for in vivo insulin sensitivity in individual tissues in rats. Am. J. Physiol. **248:** E353–E362.
22. AFIFI, A.A. & S.P. AZEN. 1972. Statistical Analysis: A Computer Oriented Approach. Academic Press. New York/London.
23. STORLIEN, L.H., et al. 1987. Fish oil prevents insulin resistance induced by high fat feeding in rats. Science **237:** 885–888.
24. OAKES, N.D., et al. 1997. Mechanisms of liver and muscle insulin resistance induced by chronic high fat feeding. Diabetes **46:** 1768–1774.
25. STORLIEN, L.H., et al. 1993. High fat diet–induced insulin resistance: lessons and implications from animal studies. Ann. N.Y. Acad. Sci. **683:** 82–90.
26. STORLIEN, L.H., et al. 1993. "Syndromes of insulin resistance" in the rat: induction by diet and amelioration with benfluorex. Diabetes **42:** 457–462.
27. KRAEGEN, E.W., et al. 1989. A potent in vivo effect of ciglitazone on muscle insulin resistance induced by high fat feeding of rats. Metabolism **38:** 1089–1093.
28. KRAEGEN, E.W., G.J. COONEY, J. YE & A.L. THOMPSON. 2001. Triglycerides, fatty acids, and insulin resistance–hyperinsulinemia. Exp. Clin. Endocrinol. Diabetes **109:** 516–526.
29. KELLEY, D.E. & B.H. GOODPASTER. 2001. Skeletal muscle triglycerides: an aspect of regional adiposity and insulin resistance. Diabetes Care **24:** 933–941.
30. VIGH, L. 1997. Bimoclomol: a non-toxic, hydroxylamine derivative with stress protein-inducing activity and cytoprotective effects. Nat. Med. **3:** 1150–1154.

Opposing Effects of Fatty Acids and Acyl-CoA Esters on Conformation and Cofactor Recruitment of Peroxisome Proliferator-Activated Receptors

CLAUS JØRGENSEN,[a] ANNE-M. KROGSDAM,[a] IRINA KRATCHMAROVA,[a] TIMOTHY M. WILLSON,[b] JENS KNUDSEN,[a] SUSANNE MANDRUP,[a] AND KARSTEN KRISTIANSEN[a]

[a]*Department of Biochemistry and Molecular Biology, University of Southern Denmark, DK-5230 Odense, Denmark*

[b]*GlaxoSmithKline, Nuclear Receptor Discovery Research, Research Triangle Park, North Carolina 27709-3398, USA*

ABSTRACT: The peroxisome proliferator-activated receptors (PPARs) bind and are activated by a variety of fatty acids and derivatives thereof. Agonist binding enhances PPAR-mediated transactivation via release of corepressors and recruitment of coactivator complexes. Recently, we and others have reported that acyl-CoA esters act as PPAR antagonists *in vitro*. Here, we show that the binding of the nonhydrolyzable acyl-CoA analogue, S-hexadecyl-CoA, differentially affected conformation and coactivator recruitment of the individual PPAR subtypes. In protease protection assays, S-hexadecyl CoA increased the sensitivity of PPARα and PPARδ towards chymotrypsin, whereas the action of chymotrypsin on PPARγ was only marginally affected, suggesting distinct subtype-dependent differences in the effects of S-hexadecyl-CoA on conformation of the PPARs. In keeping with these findings, S-hexadecyl-CoA abrogated ligand-induced recruitment of coactivators to PPARα and PPARδ, whereas coactivator recruitment to PPARγ was unaffected by S-hexadecyl-CoA.

KEYWORDS: peroxisome proliferator-activated receptor (PPAR); fatty acids; acyl-CoA esters; conformation; coactivator recruitment

INTRODUCTION

Peroxisome proliferator-activated receptors (PPARs) are ligand-activated transcription factors that play pivotal roles in lipid homeostasis, cellular proliferation, and differentiation (for recent reviews, see references 1 and 2). The PPAR family comprises three PPAR subtypes: PPARα, PPARγ, and PPARδ (also called PPARβ, NUC1, or FAAR). The importance of PPARα and PPARγ in cellular regulation is

---

Address for correspondence: Karsten Kristiansen, Department of Biochemistry and Molecular Biology, University of Southern Denmark, Campusvej 55, DK-5230 Odense M, Denmark. Voice: +45-6550-2408; fax: +45-6550-2467.
kak@bmb.sdu.dk

well established and, recently, important information on the biological significance of PPARδ has emerged.[3–5] The PPARs bind and are activated by a large variety of fatty acids and metabolites derived from polyunsaturated fatty acids. Binding of these ligands to the PPARs enhances PPAR-mediated transactivation via recruitment of different coactivator complexes.[6,7] Uptake and intracellular transport of fatty acids are facilitated by numerous membrane-bound transporters and intracellular fatty acid–binding proteins, respectively. It is generally appreciated that these proteins must play a role in modulating PPAR-mediated transactivation, but very few studies have addressed this question.[8–10] Activation of fatty acids to their corresponding acyl-CoA derivatives occurs rapidly following uptake and it has been suggested that this process also may promote uptake.[11] While fatty acid–binding proteins may bind acyl-CoA esters with moderate affinity, the main intracellular acyl-CoA pool former and transporter is the acyl-CoA-binding protein (ACBP), which binds acyl-CoA esters with very high affinity.[12] It was initially observed that overexpression of an acyl-CoA synthetase blunted fatty acid–induced PPARα activation and it was suggested that this simply reflected the reduced abundance of the activating free fatty acids.[13] However, recently, we and others presented evidence that acyl-CoA esters directly bind to PPARα and PPARγ and affect DNA binding, conformation, and cofactor recruitment in a manner suggesting that acyl-CoA esters may, in fact, function as PPAR antagonists.[14,15] Thus, we showed that a nonhydrolyzable acyl-CoA analogue, S-hexadecyl-CoA, prevented agonist-induced recruitment of SRC-1 to PPARα, but enhanced interaction with the corepressor NCoR.[14] In this report, we present a comparison of the effects of S-hexadecyl-CoA on PPARα, PPARδ, and PPARγ conformation and cofactor recruitment. We show that the acyl-CoA analogue differentially affects conformation and cofactor interaction of the individual PPAR subtypes. The conformation of PPARα and PPARδ is profoundly affected by the acyl-CoA analogue and, in keeping with this, the acyl-CoA analogue prevented agonist-induced recruitment of coactivators. In contrast, the acyl-CoA analogue only marginally affected PPARγ conformation as determined by protease protection experiments and, accordingly, the acyl-CoA analogue only had minor effects on PPARγ cofactor recruitment.

MATERIALS AND METHODS

Plasmids

The plasmid encoding SRC-1 (pBKCMV-hSRC1) was a kind gift from B. W. O'Malley.[16] pCMV-T7-PBP was kindly provided by J. K. Reddy.[17] pTLI-mPPAR-ΔAB was kindly provided by M. Leid.[18] pGEX-5X-1-mPPARα-LBD (aa: 166–469) has been described previously.[19] pGEX-5X-1-mPPARγ(DF) and mPPARδ(DF) were subcloned from pGBT9-mPPARγ(DF) and pGBT9-mPPARδ(DF), respectively.[20]

Ligands

Wy14643 was from Biomol. S-Hexadecyl-CoA (a nonhydrolyzable palmitoyl-CoA analogue) was synthesized as described by Rosendal et al.[21] Rosiglitazone and fenofibric acid were kindly provided by Novo Nordisk A/S. L165041 was kindly supplied by Merck Research Laboratories. GW0072[22] and GW9662[23] were synthesized at GlaxoSmithKline.

Differential Protease Sensitivity Assay

In vitro transcription/translation, protease protection, gel electrophoresis, and visualization were conducted as previously described.[14]

GST Pull-Downs

GST fusion proteins were captured on glutathione-Sepharose beads as described.[19] The pull-down experiments were performed essentially as described in reference 14, with the exceptions that binding was performed in buffer A containing 100 mM NaCl and washing was performed by washing once in 150 µL buffer A and twice in 150 µL buffer A without milk powder. The bound proteins were eluted by boiling in SDS-PAGE sample buffer, resolved by electrophoresis on a 10% SDS-polyacrylamide gel, and visualized using the phosphorimager. Quantitation was done using the ImageQuant v.5.0 software (Molecular Dynamics).

RESULTS AND DISCUSSION

Differential protease sensitivity assays have been widely used for analyses of conformational changes induced by the binding of agonists or antagonists to nuclear receptors.[18,24–26] We employed this assay to compare the effects of known PPAR subtype-selective ligands and the nonhydrolyzable acyl-CoA analogue, S-hexadecyl-CoA, on the conformation of PPARα, PPARδ, and PPARγ. FIGURE 1 shows that the PPARα-selective agonist, fenofibric acid, decreased the sensitivity of PPARα towards chymotrypsin, leading to the preservation of a diagnostic protease-resistant fragment. As reported previously, S-hexadecyl-CoA markedly increased the sensitivity of PPARα to chymotrypsin digestion.[14] A similar pattern was observed for PPARδ. Here, the PPARδ-selective agonist, L165041, significantly increased the resistance of PPARδ to digestion with chymotrypsin, whereas addition of S-hexadecyl-CoA strongly increased the protease sensitivity (FIG. 2). As expected, the PPARγ-selective agonist, rosiglitazone, decreased the sensitivity towards chymotrypsin, resulting in the preservation of two protease-resistant fragments of 32 and 22 kDa, respectively. However, interestingly and contrasting the results obtained with PPARα and PPARδ, addition of S-hexadecyl-CoA had only a marginal effect on the sensitivity of PPARγ to chymotrypsin digestion (FIG. 3), suggesting that PPARγ in this respect differs from PPARα and PPARδ. Alignment of the sequences of the ligand-binding domains of the PPAR subtypes suggested that PPARα and PPARδ are more closely related to each other than to PPARγ, and a single amino acid substitution in the ligand-binding domain of PPARα has been shown to generate a PPARδ phenotype.[27] However, a detailed comparison of the topology of the ligand-binding pockets of the three PPAR subtypes now indicates that PPARα and PPARγ, in fact, are related more closely to each other than to PPARδ, in which the pocket is more narrow in the region adjacent to the AF-2 helix.[28] Therefore, our finding that S-hexadecyl-CoA selectively affects the conformation of PPARα and PPARδ might suggest that S-hexadecyl-CoA binding involves regions and/or residues outside of the usual ligand-binding pocket.

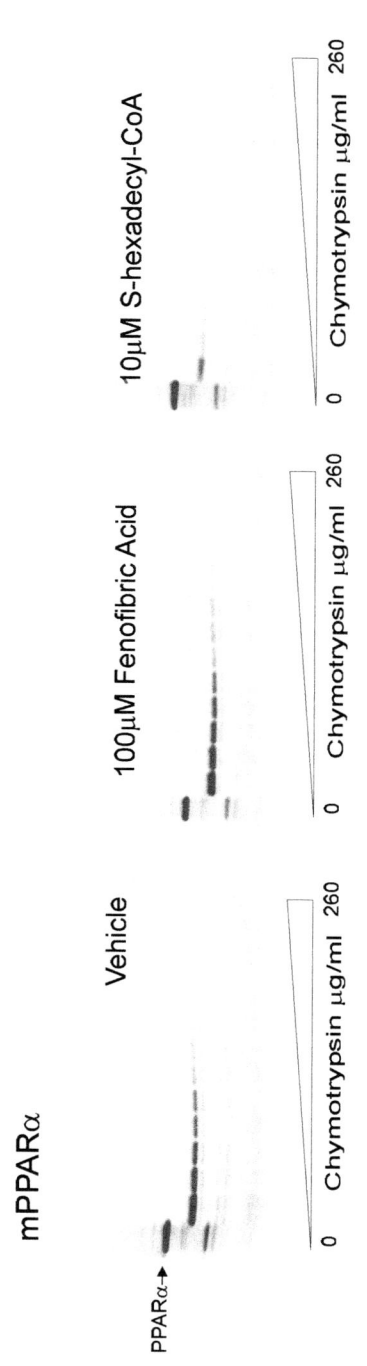

FIGURE 1. Opposing effects of fenofibric acid and S-hexadecyl-CoA on PPARα conformation. *In vitro* transcribed/translated mPPARα ΔA/B was incubated with vehicle, fenofibric acid (100 μM), or S-hexadecyl-CoA (10 μM) and subsequently subjected to chymotrypsin digestion. Fragments were separated by SDS gel electrophoresis.

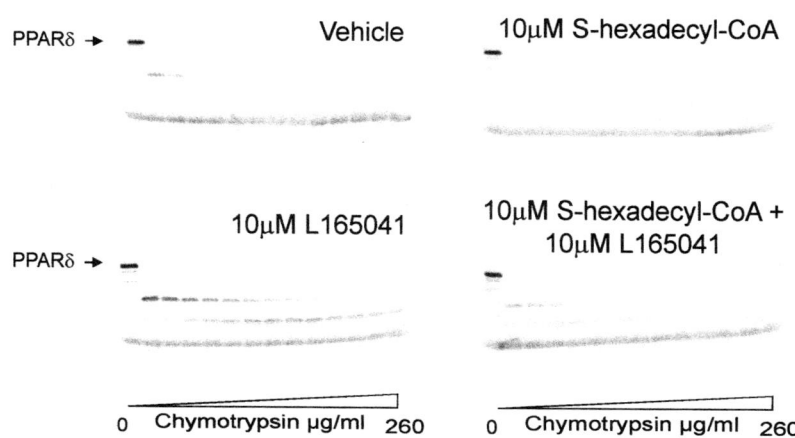

FIGURE 2. Opposing effects of L165041 and S-hexadecyl-CoA on PPARδ conformation. *In vitro* transcribed/translated mPPARδ A/F was incubated with vehicle, L165041 (10 μM), or S-hexadecyl-CoA (10 μM) followed by chymotrypsin digestion. Fragments were separated by SDS gel electrophoresis.

A number of PPARγ antagonists have been synthesized and characterized. We compared the effect of two of those on the conformation of PPARγ in differential protease sensitivity assays (FIG. 3). GW0072 is a PPARγ modulator with unique binding properties. Unlike other known PPARγ ligands, which bind to the ligand-binding domain of PPARγ with the polar head group oriented towards AF-2 helix H12, GW0072 binds in the opposite orientation with no contact to the AF-2 helix.[22] We have argued that this orientation may also be adopted by acyl-CoA esters bound to the PPARα ligand-binding pocket[14] and, hence, we speculated whether GW0072 would confer increased protease sensitivity to PPARγ analogously with the increased protease sensitivity observed upon binding of S-hexadecyl-CoA to PPARα and PPARδ. However, this appeared not to be the case. If anything, GW0072 slightly decreased the protease sensitivity of PPARγ, with a modestly enhanced preservation of both the 32- and 22-kDa fragment. GW9662 is a functional PPARγ antagonist, which binds covalently to a cysteine residue in helix 3 in PPARγ.[23] This ligand also decreased the sensitivity of PPARγ to chymotrypsin; in contrast, though, to the other PPARγ ligands, GW9662 only enhanced protection of the 32-kDa fragment. In conclusion, addition of S-hexadecyl-CoA did not alter the sensitivity of PPARγ to chymotrypsin and, thus, if S-hexadecyl-CoA does bind to PPARγ, as would be expected from competition experiments using normal acyl-CoA esters,[15] the effect of binding on conformation is distinct from that observed for PPARα and PPARδ.

Binding of antagonists to nuclear receptors diminishes or prevents agonist-induced recruitment of coactivators. Using GST pull-downs, we examined the effect

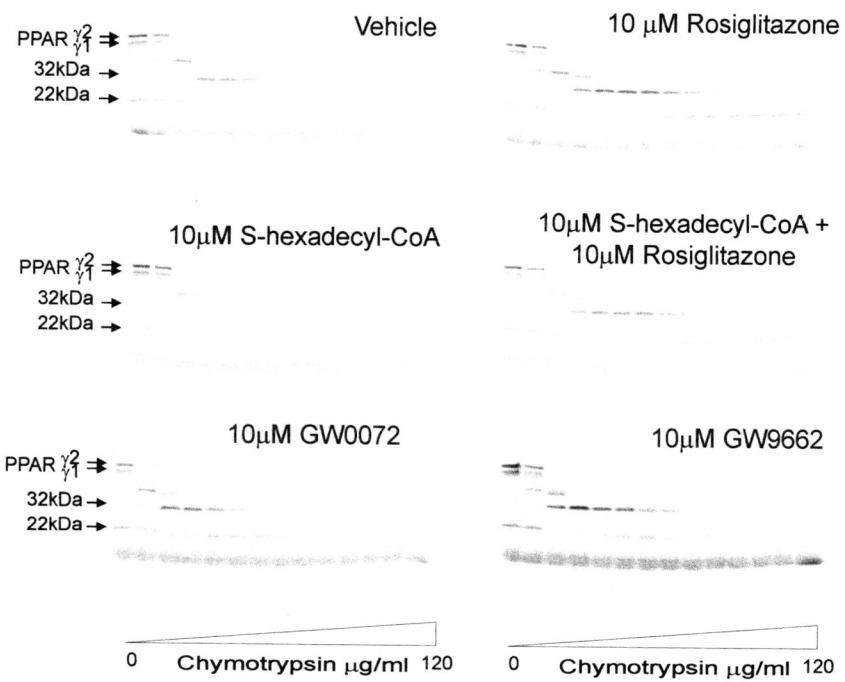

FIGURE 3. The effects of *S*-hexadecyl-CoA and known PPARγ ligands on PPARγ conformation. *In vitro* transcribed/translated mPPARγ was incubated with vehicle, rosiglitazone (10 μM), *S*-hexadecyl-CoA (10 μM), GW0072 (10 μM), or GW9662 (10 μM). Chymotrypsin was added, and digestion was allowed to proceed for 20 min. Fragments were separated by SDS gel electrophoresis.

of *S*-hexadecyl-CoA on recruitment of the coactivators PBP and SRC-1 to the three PPARs. FIGURE 4 shows that addition of ligands, as expected, increased recruitment of PBP and SRC-1 to each PPAR subtype. However, *S*-hexadecyl-CoA differentially affected PPAR-coactivator interaction. Most importantly, the interactions between PPARγ and PBP or SRC-1 in the absence or presence of rosiglitazone were only modestly affected, if affected at all, by *S*-hexadecyl-CoA. In contrast, *S*-hexadecyl-CoA clearly influenced interaction of coactivators with PPARα and PPARδ. Of note, it appeared that the effect of *S*-hexadecyl-CoA was dependent on PPAR subtype and coactivator. Thus, the basal interaction between PPARα and PBP was unaffected by *S*-hexadecyl-CoA, and the ligand-induced recruitment was only partially abolished. In contrast, *S*-hexadecyl-CoA diminished the basal interaction between PPARα and SRC-1 and reduced SRC-1 recruitment to the same level, even in the presence of the strong PPARα agonist, Wy14643. Interaction of PBP or SRC-1 with PPARδ responded similarly to the addition of *S*-hexadecyl-CoA. No effect was observed on the basal level of coactivator recruitment, and ligand-induced recruitment was completely reversed in both cases by the addition of *S*-hexadecyl-CoA.

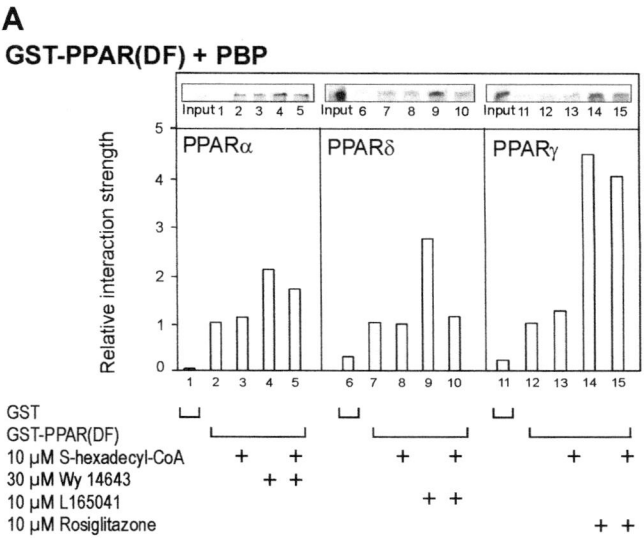

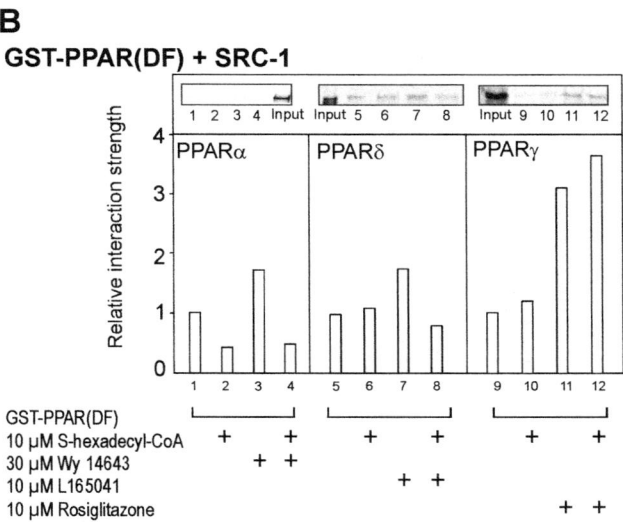

FIGURE 4. S-Hexadecyl-CoA antagonizes ligand-induced coactivator binding to PPARα and PPARδ. Bacterially expressed GST-PPARα, -PPARδ, and -PPARγ (ligand-binding domains), immobilized on glutathione-Sepharose, were incubated with *in vitro* transcribed/translated ^{35}S-labeled PBP (**A**) or SRC-1 (**B**) in the presence of vehicle, Wy14643, L165041, rosiglitazone, or S-hexadecyl-CoA. Bound proteins were recovered by boiling in SDS sample buffer, separated by SDS gel electrophoresis, and quantified by phosphorimaging. Interaction in the presence of agonists or acyl-CoA was normalized to the level of interaction observed in the absence of ligands, which was set equal to 1.

Taken together, our results indicate that acyl-CoA esters *in vivo* might antagonize ligand-induced recruitment of coactivators to PPARα and PPARδ, but not PPARγ. Thus, fatty acid–dependent induction of PPAR α- and PPARδ-mediated transactivation would be expected to be controlled by the levels of and the ratio between fatty acids and their corresponding CoA esters, whereas PPARγ-dependent transactivation would be relatively unaffected by the accumulation of acyl-CoA esters.

ACKNOWLEDGMENTS

We thank J. K. Reddy, M. Leid, and B. W. O'Malley for the kind gifts of plasmids. We thank Novo Nordisk A/S for the generous supply of rosiglitazone and fenofibric acid, and Merck Research Laboratories for L165041. This work was conducted within the Center for Experimental BioInformatics and supported by the Danish Biotechnology Program, the Danish Cancer Society, the Danish Natural Science Research Council, and the Novo Nordisk Foundation.

REFERENCES

1. KLIEWER, S.A., H.E. XU, M.H. LAMBERT *et al.* 2001. Peroxisome proliferator-activated receptors: from genes to physiology. Recent Prog. Horm. Res. **56:** 239–263.
2. HUANG, E.Y., J. ZHANG, E.A. MISKA *et al.* 2000. Nuclear receptor corepressors partner with class II histone deacetylases in a Sin3-independent repression pathway. Genes Dev. **14:** 45–54.
3. OLIVER, W.R.J., J.L. SHENK, M.R. SNAITH *et al.* 2001. A selective peroxisome proliferator-activated receptor δ agonist promotes reverse cholesterol transport. Proc. Natl. Acad. Sci. U.S.A. **98:** 5306–5311.
4. PETERS, J.M., S.S. LEE, W. LI *et al.* 2000. Growth, adipose, brain, and skin alterations resulting from targeted disruption of the mouse peroxisome proliferator-activated receptor β (δ). Mol. Cell. Biol. **20:** 5119–5128.
5. WESTERGAARD, M., J. HENNINGSEN, M.L. SVENDSEN *et al.* 2001. Modulation of keratinocyte gene expression and differentiation by PPAR-selective ligands and tetradecylthioacetic acid. J. Invest. Dermatol. **116:** 702–711.
6. YANG, W., C. RACHEZ & L.P. FREEDMAN. 2000. Discrete roles for peroxisome proliferator-activated receptor γ and retinoid X receptor in recruiting nuclear receptor coactivators. Mol. Cell. Biol. **20:** 8008–8017.
7. KODERA, Y., K. TAKEYAMA, A. MURAYAMA *et al.* 2000. Ligand type-specific interactions of peroxisome proliferator-activated receptor γ with transcriptional coactivators. J. Biol. Chem. **275:** 33201–33204.
8. MANDRUP, S., R.V. SORENSEN, T. HELLEDIE *et al.* 1998. Inhibition of 3T3-L1 adipocyte differentiation by expression of acyl-CoA-binding protein antisense RNA. J. Biol. Chem. **273:** 23897–23903.
9. WOLFRUM, C., C.M. BORRMANN, T. BORCHERS *et al.* 2001. Fatty acids and hypolipidemic drugs regulate peroxisome proliferator-activated receptors alpha- and gamma-mediated gene expression via liver fatty acid binding protein: a signaling path to the nucleus. Proc. Natl. Acad. Sci. U.S.A. **98:** 2323–2328.
10. HELLEDIE, T., M. ANTONIUS, R.V. SORENSEN *et al.* 2000. Lipid-binding proteins modulate ligand-dependent trans-activation by peroxisome proliferator-activated receptors and localize to the nucleus as well as the cytoplasm. J. Lipid Res. **41:** 1740–1751.
11. SCHAFFER, J.E. & H.F. LODISH. 1994. Expression cloning and characterization of a novel adipocyte long chain fatty acid transport protein. Cell **79:** 427–436.
12. FAERGEMAN, N.J. & J. KNUDSEN. 1997. Role of long-chain fatty acyl-CoA esters in the regulation of metabolism and in cell signalling. Biochem. J. **323**(part 1): 1–12.

13. HERTZ, R., I. BERMAN & J. BAR-TANA. 1994. Transcriptional activation by amphipathic carboxylic peroxisomal proliferators is induced by the free acid rather than the acyl-CoA derivative. Eur. J. Biochem. **221:** 611–615.
14. ELHOLM, M., I. DAM, C. JORGENSEN *et al.* 2001. Acyl-CoA esters antagonize the effects of ligands on peroxisome proliferator-activated receptor α conformation, DNA binding, and interaction with co-factors. J. Biol. Chem. **276:** 21410–21416.
15. MURAKAMI, K., T. IDE, T. NAKAZAWA *et al.* 2001. Fatty-acyl-CoA thioesters inhibit recruitment of steroid receptor co-activator 1 to α and γ isoforms of peroxisome-proliferator-activated receptors by competing with agonists. Biochem. J. **353:** 231–238.
16. ONATE, S.A., S.Y. TSAI, M.J. TSAI *et al.* 1995. Sequence and characterization of a coactivator for the steroid hormone receptor superfamily. Science **270:** 1354–1357.
17. ZHU, Y., C. QI, S. JAIN *et al.* 1997. Isolation and characterization of PBP, a protein that interacts with peroxisome proliferator-activated receptor. J. Biol. Chem. **272:** 25500–25506.
18. DOWELL, P., V.J. PETERSON, T.M. ZABRISKIE *et al.* 1997. Ligand-induced peroxisome proliferator-activated receptor alpha conformational change. J. Biol. Chem. **272:** 2013–2020.
19. KUSSMANN-GERBER, S., I. KRATCHMAROVA, S. MANDRUP *et al.* 1999. Microaffinity columns for analysis of protein-protein interactions. Anal. Biochem. **271:** 102–105.
20. KROGSDAM, A-M., C.A.F. NIELSEN, S. NEVE *et al.* 2001. NCoR-dependent repression of PPAR-δ-mediated transactivation. Biochem. J. **363:** 157–165.
21. ROSENDAL, J., P. ERTBJERG & J. KNUDSEN. 1993. Characterization of ligand binding to acyl-CoA-binding protein. Biochem. J. **290:** 321–326.
22. OBERFIELD, J.L. & J.L. COLLINS. 1999. A peroxisome proliferator-activated receptor γ ligand inhibits adipocyte differentiation. Proc. Natl. Acad. Sci. U.S.A. **96:** 6102–6106.
23. HUANG, J.T., J.S. WELCH, M. RICOTE *et al.* 1999. Interleukin-4-dependent production of PPAR-γ ligands in macrophages by 12/15-lipoxygenase. Nature **400:** 378–382.
24. BERGER, J., M.D. LEIBOWITZ, T.W. DOEBBER *et al.* 1999. Novel peroxisome proliferator-activated receptor (PPAR) γ and PPARδ ligands produce distinct biological effects. J. Biol. Chem. **274:** 6718–6725.
25. BERGER, J., P. BAILEY, C. BISWAS *et al.* 1996. Thiazolidinediones produce a conformational change in peroxisomal proliferator-activated receptor-γ: binding and activation correlate with antidiabetic actions in db/db mice. Endocrinology **137:** 4189–4195.
26. ALLAN, G.F., X. LENG, S.Y. TSAI *et al.* 1992. Hormone and antihormone induce distinct conformational changes which are central to steroid receptor activation. J. Biol. Chem. **267:** 19513–19520.
27. TAKADA, I., R.T. YU, H.E. XU *et al.* 2000. Alteration of a single amino acid in peroxisome proliferator-activated receptor-α (PPARα) generates a PPAR δ phenotype. Mol. Endocrinol. **14:** 733–740.
28. XU, H.E., M.H. LAMBERT, V.G. MONTANA *et al.* 2001. Structural determinants of ligand binding selectivity between the peroxisome proliferator-activated receptors. Proc. Natl. Acad. Sci. U.S.A. **98:** 13919–13924.

Relationship between Insulin Resistance and Muscle Triglyceride Content in Nonobese and Obese Experimental Models of Insulin Resistance Syndrome

JANA DIVIŠOVÁ, LUDMILA KAZDOVÁ, MIRIAM HUBOVÁ, AND ELEN MESCHIŠVILI

Institute for Clinical and Experimental Medicine, Prague, Czech Republic

ABSTRACT: In nonobese hereditary hypertriglyceridemic (HTg) rats and obese HTg Koletsky SHROB/Kol rats, muscle triglyceride accumulation and its relationship to glucose homeostasis and tissue sensitivity to insulin action were examined. Soleus muscle and diaphragm triglyceride contents were markedly increased in HTg rats as compared with controls and were further increased in obese animals. On the other hand, glucose intolerance and impairment of insulin-stimulated *in vitro* glycogen synthesis were of a similar degree in nonobese as well as obese animals. Results indicate that insulin resistance did not increase proportionally to muscle triglyceride content.

KEYWORDS: insulin resistance; triglycerides; skeletal muscle; insulin sensitivity

INTRODUCTION

Hypertriglyceridemia, hyperinsulinemia, impaired glucose tolerance, and hypertension are all linked to tissue resistance to insulin action and, with these, there is a marked increase in risk for type 2 diabetes mellitus (DM 2) and cardiovascular disease.[1,2] It has been shown that not only circulating, but also skeletal muscle triglycerides are abnormally high within the context of insulin resistance.[3,4] The mechanisms underlying the association between muscle triglyceride content and the cluster of metabolic disturbances characterizing the insulin-resistant state have been extensively investigated. The strong relationship between muscle triglyceride content and whole body insulin sensitivity has been observed.[4] Recently, direct evidence of a significant negative correlation between intramyocellular lipid concentration and insulin sensitivity was found by Krssak *et al.* in nondiabetic subjects using ^{1}H nuclear magnetic resonance spectroscopy (NMR) and the hyperinsulinemic-euglycemic clamp test.[5] There are also inconsistent data available in the literature concerning mutual links between muscle triglyceride content, total adiposity, and

---

Address for correspondence: Jana Divišová, Department of Metabolic Research, Institute for Clinical and Experimental Medicine, Vídenská 1958/9, 140 21 Prague, Czech Republic. Voice: +420-2-61083256; fax: +420-2-61083490.
    jadi@medicon.cz

insulin resistance. In the study of Krssak et al.,[5] muscle triglyceride content was not related to body mass index, in contrast to Manco et al.,[6] although both studies found correlations between muscle triglyceride content and insulin sensitivity.

The aim of the present study is to compare muscle triglyceride content in nonobese and obese experimental models of insulin resistance syndrome and to investigate whether it is related to glucose homeostasis and tissue sensitivity to insulin action.

METHODS

Animals

The experiments were carried out in adult male nonobese hereditary hypertriglyceridemic (HTg) and obese hypertriglyceridemic Koletsky SHROB/Kol (Kol) nonfasting rats. The models used, that is, nonobese HTg and obese Kol rat strains, manifest most symptoms of the metabolic syndrome: tissue resistance to insulin action, hypertension, hyperinsulinemia, and impaired glucose tolerance.[7–9] Normotriglyceridemic Wistar rats and Koletsky SHROB/Kol lean siblings were used as controls. All animals were fed a standard laboratory diet *ad libitum*.

Analysis

Plasma triglyceride levels were measured in nonfasting animals. The oral glucose tolerance test (OGTT) was performed after overnight fasting (14 h). Blood for glycemia determination was drawn from the tail at intervals of 0, 30, 60, and 120 min after intragastric glucose administration was given to conscious rats (3 g/kg BW, 30% aqueous solution). Commercially available analytical kits were employed to determine plasma glucose and serum triglyceride concentrations (Lachema, Czech Republic). Following decapitation in the nonfasting state, diaphragm and soleus muscles were removed. For muscle triglyceride analysis, each sample was made free of visible contamination from fat. Tissue lipids were extracted according to Folch,[10] and triglycerides were determined using an analytical kit (Lachema, Czech Republic).

In vitro insulin sensitivity was determined in isolated diaphragmatic muscle by measuring the effect of insulin on ^{14}C-U glucose incorporation into glycogen. After decapitation, isolated diaphragms were incubated in Krebs-Ringer bicarbonate buffer, at pH 7.4, containing 0.1 µCi/mL of ^{14}C-U glucose, 5 mmol/L of unlabeled glucose, and 2.5 mg/mL of bovine serum albumin (Armour, Fraction V), with or without 250 µU/mL insulin. Tissues were incubated for 2 h at 37°C; atmosphere was 95% O_2 + 5% CO_2. Glycogen from the diaphragmatic muscles was extracted, and radioactivity was measured as described before.[11]

Statistical Analysis

Results are given as means ± SEM. Intergroup differences were evaluated using ANOVA and the Tukey-Kramer multiple comparisons test. Differences were considered statistically significant at the level of $p < 0.05$.

TABLE 1. Body weight and serum triglyceride levels in nonobese HTg rats and Kol obese rats as compared with control Wistar and Kol lean rats, respectively

	Wistar	HTg	Kol lean	Kol obese
Body weight (g)	321 ± 18	312 ± 13	310 ± 17	618 ± 52***
Serum triglycerides (mmol/L)	1.10 ± 0.13	2.61 ± 0.07***	1.08 ± 0.05	2.53 ± 0.29***

NOTE: Data are given as means ± SEM, n = 4–11 animals per group; ***$p < 0.001$ vs. corresponding control group.

RESULTS

The body weights of nonobese HTg, Kol, and Wistar groups did not differ significantly, in contrast to the morbidly obese Kol group (TABLE 1). Triglyceridemia was elevated more than twofold in both HTg and Kol obese animals in comparison with the corresponding control groups (TABLE 1).

The soleus muscle triglyceride content (FIG. 1A) was higher in HTg rats as compared with Wistar controls and it was further increased in association with obesity in Kol obese animals (to nearly threefold of the control value). The situation was similar in diaphragm (FIG. 1B). The OGTT demonstrated impaired glucose tolerance in HTg and Kol obese rats (FIG. 2). After the intragastric administration of glucose, serum glucose levels were invariably higher in HTg than in Wistar rats and in Kol obese than in Kol lean rats at all evaluated time intervals. This finding is distinct when calculating the test results as the area under the curve (AUC). FIGURE 3 shows the

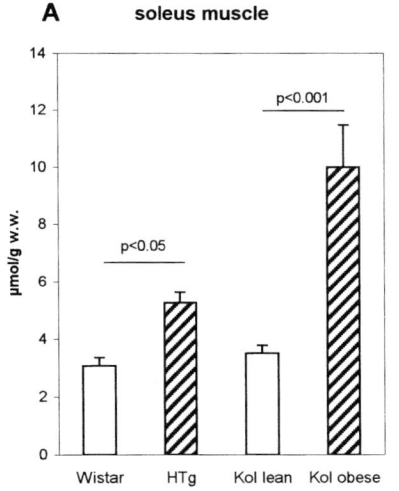
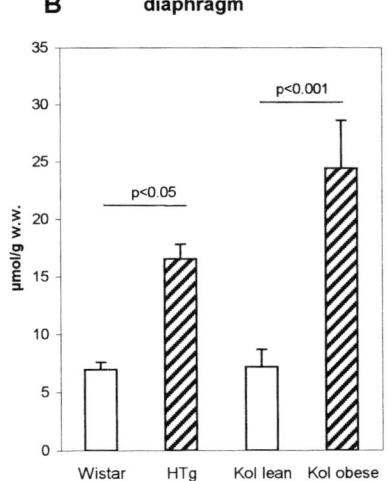

FIGURE 1. Triglyceride content in soleus muscle (**A**) and diaphragm (**B**) of nonobese HTg rats and Kol obese rats as compared with control Wistar and Kol lean rats, respectively. Data: means ± SEM; n = 5–7 animals per group.

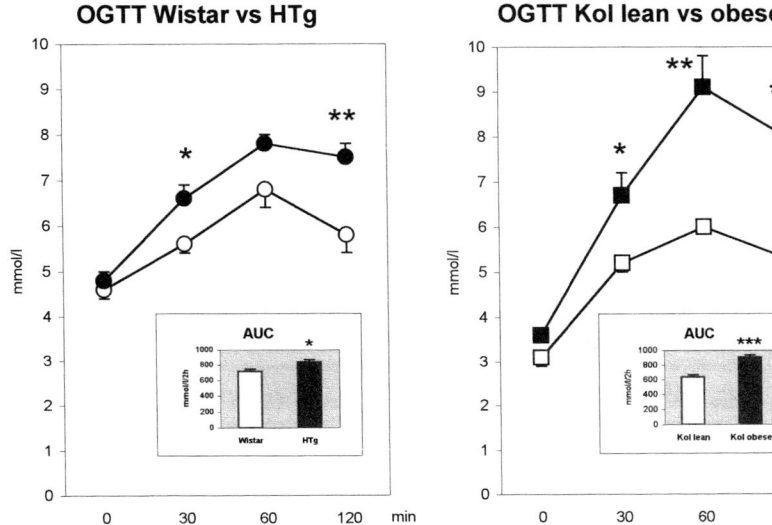

FIGURE 2. The oral glucose tolerance test (OGTT) in nonobese HTg rats and Kol obese rats (*black*) as compared with control Wistar and Kol lean rats (*white*), respectively. Glycemia was determined at time intervals of 0, 30, 60, and 120 min after intragastric administration of glucose (3 g/kg BW) after an overnight fast (14 h). AUC is the area under the curve of the OGTT. Data: means ± SEM, $n = 6–11$ animals per group; ***$p < 0.001$, **$p < 0.01$, *$p < 0.05$ vs. controls.

sensitivity of skeletal muscle to insulin action as evaluated by *in vitro* ^{14}C-U glucose incorporation into diaphragmatic glycogen. Glucose incorporation under the basal conditions (without insulin) was lower in both nonobese and obese experimental groups than in controls by approximately 30% (FIG. 3). Moreover, measurement of insulin-stimulated glucose incorporation showed tissue resistance to insulin action in HTg and Kol obese animals, being almost 60% of the control value.

It is noteworthy that glucose intolerance and deterioration of tissue insulin sensitivity evaluated by OGTT (AUC: HTg 839 ± 29 vs. Kol obese 906 ± 56 mmol/L/120 min, NS) and by insulin-stimulated *in vitro* incorporation of ^{14}C-U glucose into diaphragmatic glycogen (HTg 150 ± 38 vs. Kol obese 138 ± 28 mmol glucose/g tissue, NS), respectively, were of similar degree in HTg and Kol obese animals.

DISCUSSION

In our study, we evaluated the relationship between muscle triglyceride content and the degree of insulin resistance in nonobese and obese experimental models of insulin resistance. Skeletal muscles contain considerable amounts of intracellular triglycerides. These are important as immediately available energy sources and might thus interfere with glucose metabolism and contribute to insulin resistance. Abnormally elevated muscle triglyceride content was frequently observed in associ-

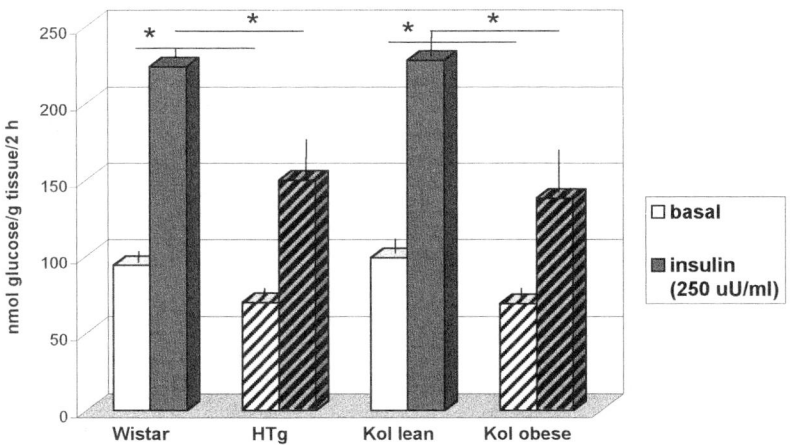

FIGURE 3. Basal and insulin-stimulated *in vitro* ^{14}C-U glucose incorporation into diaphragmatic glycogen in nonobese HTg rats and Kol obese rats as compared with control Wistar and Kol lean rats, respectively. Data: means ± SEM, $n = 5–9$ animals per group; $*p < 0.05$.

ation with insulin resistance in both nonobese[3,12] and obese[4] nondiabetic subjects. It is not clear if the accumulation of intracellular triglycerides is a consequence of metabolic derangements of the insulin resistance syndrome or if increased muscle triglyceride stores might be responsible for a reduced muscle insulin sensitivity.[13]

Our results show skeletal muscle triglyceride content was elevated in both insulin-resistant experimental models. It is of interest that triglyceride concentration, despite being the same in serum, was almost twofold higher in musculus soleus and diaphragm of Kol obese rats as compared with nonobese HTg rats. Similarly, Manco *et al.* reported approximately twofold higher muscle triglyceride content in obese than in nonobese subjects using muscle biopsy.[6] They found a positive correlation between muscle triglycerides and body mass index (BMI). On the other hand, Krssak *et al.* reported no relationship between BMI and muscle triglyceride content as assessed by ^{1}H NMR in normal-weight nondiabetic subjects.[5] In another study, muscle triglycerides from muscle biopsy were not related to any measurement of adiposity in obese Pima Indians.[4] The different findings may be explained by the diverse range of BMI (18–38 vs. 21–27 kg/m^2), by different techniques used (lipid extraction from whole muscle, or muscle biopsy might include extracellular triglycerides), or by different muscles that have been analyzed.[12]

Interestingly, in our study, a degree of insulin resistance did not increase proportionally to muscle triglyceride content. Insulin sensitivity—as assessed by OGTT and *in vitro* ^{14}C-U glucose incorporation into diaphragmatic glycogen—was of a similar degree in nonobese as well as obese experimental animals, in spite of markedly higher muscle triglyceride content associated with obesity. The results of Manco *et al.* seem to disagree with our findings.[6] These investigators reported negative linear correlation between muscle triglyceride content and insulin sensitivity in nonobese and obese subjects. However, looking at the figure demonstrating this dependence, we can see no further deterioration of insulin sensitivity in the range of muscle triglyceride levels, that is, 9–15 μmol/g w/w. Thus, to understand exactly the relationship between muscle triglyceride accumulation and insulin sensitivity, especially when associated with obesity, detailed investigation is required.

The results obtained in this study showed that muscle triglyceride accumulation was markedly increased in association with insulin resistance in both nonobese and obese models of the metabolic syndrome. Obesity leads to further elevation of muscle triglyceride content; however, it was not accompanied with further deterioration of glucose homeostasis and tissue sensitivity to insulin action.

ACKNOWLEDGMENTS

This study was supported by Grant Nos. 6367-3 and 6678-3 from the Internal Grant Agency of the Ministry of Health of the Czech Republic.

REFERENCES

1. REAVEN, G.M. 1988. Banting lecture 1988: role of insulin resistance in human disease. Diabetes **37:** 1595–1607.
2. MYKKÄNEN, L., S.M. HAFFNER, T. RÖNNEMANAA et al. 1997. Low insulin sensitivity is associated with clustering of cardiovascular disease risk factors. Am. J. Epidemiol. **146:** 315–321.
3. GOODPASTER, B.H., F.L. THEATE, J-A. SIMONEAU & D.E. KELLEY. 1997. Subcutaneous abdominal fat and thigh muscle composition predict insulin sensitivity independently of visceral fat. Diabetes **46:** 1579–1585.
4. PAN, D.A., S. LILLIOJA, A.D. KRIKETOS et al. 1997. Skeletal muscle triglyceride levels are inversely related to insulin action. Diabetes **46:** 983–988.
5. KRSSAK, M., K.F. PETERSEN, A. DRESNER et al. 1999. Intramyocellular lipid concentrations are correlated with insulin sensitivity in humans: a ^{1}H NMR spectroscopy study. Diabetologia **42:** 113–116.
6. MANCO, M., G. MINGRONE, A.V. GRECO et al. 2000. Insulin resistance directly correlates with increased saturated fatty acids in skeletal muscle triglycerides. Metabolism **49:** 220–224.
7. VRÁNA, A., L. KAZDOVÁ, Z. DOBEŠOVÁ et al. 1993. Triglyceridemia, glucoregulation, and blood pressure in various rat strains. Ann. N.Y. Acad. Sci. **683:** 57–68.
8. DIVIŠOVÁ, J., L. KAZDOVÁ, A. VRÁNA et al. 1999. The hereditary hypertriglyceridemic rat—an experimental model for study of the pathogenesis of insulin resistance and its cardiovascular complications. Cor Vasa **41:** K121–K122.
9. KOLETSKY, S. 1975. New type of spontaneously hypertensive rats with hyperlipemia and endocrine gland defects. Am. J. Pathol. **80:** 194–197.
10. FOLCH, J., M. LEES & G.H.S. STANLEY. 1957. A simple method for the isolation and purification of total lipids from animal tissues. *In* Methods in Enzymology. Volume 3, pp. 497–511. Academic Press. New York.
11. VRÁNA, A. & L. KAZDOVÁ. 1970. Insulin sensitivity of rat adipose tissue and of diaphragm *in vitro*: effect of the type of dietary carbohydrate (starch-sucrose). Life Sci. **9:** 257–265.
12. PERSEGHIN, G., P. SCIFO, F. DE COBELLI et al. 1999. Intramyocellular triglyceride content is a determinant of *in vivo* insulin resistance in humans: a ^{1}H-^{13}C nuclear magnetic resonance spectroscopy assessment in offspring of type 2 diabetic parents. Diabetes **48:** 1600–1606.
13. KRAEGEN, E.W., G.J. COONEY, J. YE & A.L. THOMPSON. 2001. Triglycerides, fatty acids, and insulin resistance–hyperinsulinemia. Exp. Clin. Endocrinol. Diabetes **109:** S516–S526.

Insulin Resistance in the Hereditary Hypertriglyceridemic Rat Is Associated with an Impairment of Δ-6 Desaturase Expression in Liver

D. GAŠPERÍKOVÁ, E. DEMCÁKOVÁ, J. UKROPEC, I. KLIMEŠ, AND E. ŠEBÖKOVÁ

Diabetes and Nutrition Research Laboratory, Institute of Experimental Endocrinology, Slovak Academy of Sciences, SK-833 06 Bratislava, Slovak Republic

ABSTRACT: Our previous studies have shown that insulin resistance (IR) in the hereditary hypertriglyceridemic (hHTg) rat is accompanied by a specific fatty acid (FA) profile in insulin target tissues, possibly due to a defect in the desaturation pathway. Increased dietary intake of n-3 polyunsaturated fatty acids (PUFAs) was shown to shape FA composition and to improve insulin sensitivity in this animal strain. Thus, the aim of this study is twofold: (1) to evaluate a defect in the FA desaturation by direct measurement of enzyme activity and gene expression for Δ-6 desaturase (Δ-6 D) in liver of hHTg rats and (2) to investigate the effect of dietary n-3 PUFAs on hepatic Δ-6 D in relation to tissue FA composition. Male Wistar or hHTg rats were fed *ad libitum* for 21 days either the basal or fish oil (FO)–supplemented diets. Triglyceride (Tg) levels in serum and tissue lipid extracts were measured with the aid of a commercially available enzymatic set. Hepatic activity of the Δ-6 D was determined radiometrically in a microsomal fraction using 1-^{14}C-linoleic acid as a substrate. The Δ-6 D mRNA levels were measured using the Northern blot technique. Tissue FA composition was determined by gas chromatography in the total phospholipid fraction after TLC separation. Increased levels of Tg in hHTg rat circulation were accompanied by raised accumulation of Tg in skeletal muscles. FO feeding lowered the concentration of Tg in serum and prevented their accumulation in skeletal muscles of hHTg rats. A pronounced decrease in the hepatic Δ-6 D activity in hHTg rats (by about 80%) was not further diminished by FO feeding. On the other hand, the activity of Δ-6 D in liver of control rats was reduced by about 40% after FO supplementation. These changes were paralleled by a decrease in the Δ-6 D index as calculated from the liver phospholipid FA profile. In particular, an increase in the amount of 18:2 n-6 and a decrease in arachidonic acid and PUFA n-6 metabolites were found. The results indicate that a decrease of insulin action in hHTg rats is accompanied by an impairment of the hepatic Δ-6 D activity already at the gene level, which is not further affected by n-3 PUFA supplementation.

KEYWORDS: Δ-6 desaturase; hHTg rat; insulin resistance; fatty acid composition

Address for correspondence: Daniela Gašperíková, Ph.D., Diabetes and Nutrition Research Laboratory, Institute of Experimental Endocrinology, Slovak Academy of Sciences, Vlárska 3, SK-833 06 Bratislava, Slovak Republic. Voice/fax: +4212-5477-26-87.
ueengasp@savba.sk

INTRODUCTION

There is an increasing body of evidence demonstrating that in both rats[1] and humans[2] insulin resistance is related to a specific fatty acid (FA) profile in the insulin target tissues. Several studies have shown that FA composition of the skeletal muscle cell membrane phospholipids is closely related to insulin sensitivity.[3,4] An increased proportion of saturated fatty acids and a decreased concentration of long-chain polyunsaturated fatty acids (PUFAs) that are vital components of the membrane phospholipids seem to play a critical role in insulin resistance.[5,6] The availability of long-chain PUFAs is greatly dependent upon the rate of linoleic acid (18:2 n-6) metabolism, especially at the level of Δ-6 desaturase (Δ-6 D), with a prominent site of action in the liver.[7]

Beneficial effects of dietary marine fish oil rich in long-chain n-3 PUFAs have been demonstrated in prevention of insulin resistance and/or abnormal lipid profile in rodents and humans as well.[3,8] There is a substantial body of evidence demonstrating that the favorable effect of marine fish oil (FO) may be at least partly explained by an increase of FA oxidation, suppression of hepatic FA synthesis and secretion, and induction of enzymes involved in FA elongation and desaturation. These changes are associated with alterations in the genes involved in FA metabolism already at the transcription level.[9]

The hereditary hypertriglyceridemic (hHTg) rat is a nonobese animal model of hypertriglyceridemia and insulin resistance.[10] The impaired insulin action in these animals is featured by deviant lipid metabolism leading to skeletal muscle Tg accumulation[10] and changes in tissue FA composition, suggesting a defect in the FA desaturation pathway.[11] Therefore, the aim of this study was to evaluate a defect in the FA desaturation by a direct measurement of enzyme activity and gene expression for Δ-6 D in liver of hHTg rats in relation to the FA spectrum of tissue phospholipids. In addition, the effect of dietary PUFAs from the FO, known to modulate the activity of Δ-6 D already at the transcriptional level,[12] has been evaluated as well.

MATERIALS AND METHODS

Animals

All the experiments reported here were approved by the Institute of Experimental Endocrinology Animal House Ethics Committee. Adult male Wistar Charles River rats (AnLab, Prague, Czech Republic) and hHTg and insulin-resistant rats of this Institute's colony were kept in a temperature ($22 \pm 2°C$)– and light (12-h light/dark cycle, lights off at 1800 h)–controlled room. The rats (4–6 animals per group) were fed for 21 days either the commercially available standard laboratory chow (ST1 Velaz, Prague, Czech Republic) or the FO-supplemented diet, in which 10 wt % was replaced by FO, rich in n-3 PUFAs (EPAX 5500, Pronova Biocare, Norway). This approach created the following groups of animals: C, C-FO, hHTg, and hHTg-FO. The rats were sacrificed by decapitation in the fed state. Blood and selected tissues were immediately dissected and either used fresh for measurement of Δ-6 D activity or frozen in liquid nitrogen and stored at −70°C for later analysis.

Biochemical Analyses

Tg levels in serum and tissue lipid extracts were measured by a spectrophotometric method using a specific, commercially available enzymatic set (Tri-glyceridy DST-P, DOT-Diagnostics, Prague, Czech Republic).

Δ-6 D enzymatic activity was determined radiometrically in a microsomal fraction prepared from fresh liver using 1-^{14}C-linoleic acid (NEN, Boston, MA) as a substrate, as described in detail elsewhere.[13] Total lipids from the reaction mixture were saponified in methanolic KOH, and FAs (extracted with hexane after acidification) were converted to methyl esters using BF_3. Thin-layer chromatography on silica gel 60 plates (Merck, Darmstadt, Germany) was used to separate the radioactive products. The radioactivity of the spots was measured by a liquid scintillation counter (Beckman, Fullerton, CA). Enzyme activity was expressed as pmol of γ-linolenic acid formed from linoleic acid per min per mg of microsomal proteins.

Δ-6 Desaturase mRNA

The relative abundance of Δ-6 D mRNA was determined by Northern blot. Briefly, the RNA extracted according to Chomczynski and Sacchi[14] was size-fractionated by agarose gel electrophoresis and by capillary blotting transferred to membrane (HYBOND C-extra, Amersham, United Kingdom). Abundance of the transcript of interest was determined by hybridization with the specific cDNA probe (kindly donated by Steven Clarke,[15] University of Texas, Austin, TX) using Quick-Hyb (Stratagene, La Jolla, CA). Probes were radiolabeled using PRIME-IT RmT (Stratagene, La Jolla, CA) and [α-^{32}P]-dCTP (Amersham, Buckinghamshire, United Kingdom). Results were corrected to G3PDH mRNA level.

Fatty Acid Analyses

Lipids from the rat liver and skeletal muscle were extracted with chloroform-methanol-water (1:1:0.9). Total phospholipids were separated from neutral lipids using TLC, and FA methyl esters were estimated by GC.[11]

Statistical Evaluation

Results are expressed as mean ± SEM ($n = 4$–6). Differences between groups were analyzed using analysis of variance (ANOVA) followed by the appropriate post-hoc test at the overall significance threshold of $p < 0.05$.

RESULTS

Animal Characteristics

Although the differences in final body weight between the strains were not significant, the body weight gain was significantly lower in both groups of hHTg rats (TABLE 1). Increased levels of Tg in hHTg rat circulation were accompanied by raised accumulation of Tg in skeletal muscle, without any effect found in the liver. Raised dietary intake of FO lowered the concentration of Tg in serum and prevented their accumulation in skeletal muscle, but not in liver of hHTg rats.

TABLE 1. Animal characteristics

	C	C-FO	hHTg	hHTg-FO
Final BW (g)	389 ± 10[a]	378 ± 11.6[a]	328 ± 15[a]	329 ± 14[a]
BW gain (g)	126 ± 4[a]	118 ± 8.7[a]	10 ± 2[b]	11 ± 7[b]
Serum Tg (mmol/L)	1.5 ± 0.2[a]	0.8 ± 0.1[a]	2.5 ± 0.3[b]	1.1 ± 0.1[a]
Liver Tg (nmol/mg)	5.2 ± 0.5[a]	5.9 ± 1.1[a]	5.8 ± 0.9[a]	5.2 ± 0.9[a]
Muscle Tg (nmol/mg)	3.9 ± 0.4[a]	4.4 ± 0.4[a]	6.1 ± 0.5[b]	5.4 ± 0.7[a,b]

NOTE: Given values represent the mean ± SEM. Values without a common superscript (a, b) are significantly different ($p < 0.05$). C = controls (Wistar rats) fed a standard lab chow; hHTg = hereditary hypertriglyceridemic rats fed a standard lab chow; C-FO = Wistar rats fed a marine fish oil–supplemented diet; hHTg-FO = hHTg rats fed a marine fish oil–supplemented diet.

TABLE 2. Effect of marine fish oil (FO) supplementation on fatty acid composition in liver phospholipids of control (C) and hereditary hypertriglyceridemic (hHTg) rats

Wt %	C	C-FO	hHTg	hHTg-FO
18:2 n-6	12.2 ± 0.47[a]	9.4 ± 0.38[b]	17.5 ± 0.19[c]	10.7 ± 0.28[d]
18:3 n-6	0.08 ± 0.004[a]	0.04 ± 0.003[b]	0.09 ± 0.005[a]	0.06 ± 0.006[c]
20:3 n-6	1.57 ± 0.02[a]	1.46 ± 0.10[a,b]	1.35 ± 0.06[a,b]	1.25 ± 0.02[b]
20:4 n-6	26.2 ± 0.62[a]	18.3 ± 0.44[b]	21.8 ± 0.04[c]	17.9 ± 0.20[b]
Total PUFA n-6	40.6 ± 0.88[a]	29.8 ± 0.69[b,c]	41.1 ± 0.11[a]	30.4 ± 0.46[c]
Total PUFA n-6M	28.4 ± 0.59[a]	20.4 ± 0.49[b]	23.7 ± 0.09[c]	19.7 ± 0.19[b]
18:3 n-6/18:2 n-6 Δ-6 DI (× 1000)	6.55 ± 0.26[a]	3.68 ± 0.16[b]	5.25 ± 0.22[c]	5.8 ± 0.43[a,c]

NOTE: Given values represent the mean ± SEM. Values without a common superscript (a, b, c, d) are significantly different ($p < 0.05$). C = controls (Wistar rats) fed a standard lab chow; hHTg = hereditary hypertriglyceridemic rats fed a standard lab chow; C-FO = Wistar rats fed a marine fish oil–supplemented diet; hHTg-FO = hHTg rats fed a marine fish oil–supplemented diet; PUFA n-6 = n-6 polyunsaturated fatty acids; PUFA n-6M = metabolites of n-6 polyunsaturated fatty acids; Δ-6 DI = Δ-6 desaturase index.

Fatty Acid Composition

Major FAs of the n-6 series in the total phospholipid fraction are listed in TABLE 2. Thus, in the liver of hHTg rats, linoleic (18:2 n-6) acid was significantly elevated, while the content of dihomo-γ-linoleic (20:3 n-6) and arachidonic (20:4 n-6) acid, the main products of the Δ-6 D reaction, was decreased. Although the total n-6 PUFA levels did not differ between the control and hHTg rats, n-6 PUFA metabolites were significantly lowered in the total phospholipid fraction in liver of hHTg rats.

Feeding the rats an FO-supplemented diet resulted in a notable decrease of linoleic (18:2 n-6) acid in the total phospholipid fraction in liver of both hHTg and control rats, with a more pronounced effect being found in hHTg animals. Also, the amount of arachidonic acid (20:4 n-6) was significantly suppressed in both rat strains after FO supplementation, resulting in a marked decrease in n-6 metabolites of the PUFAs.

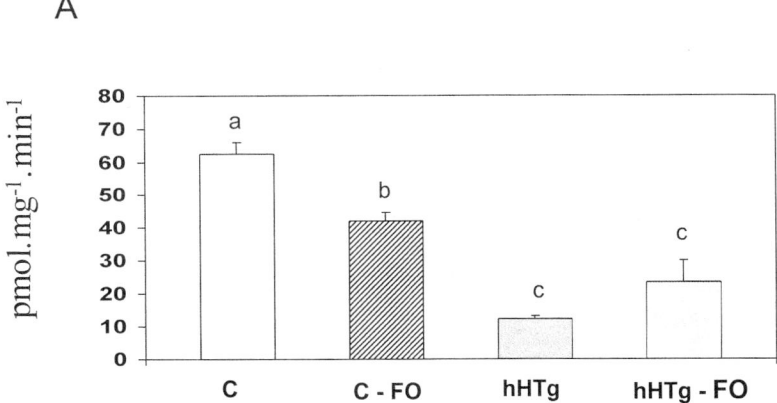
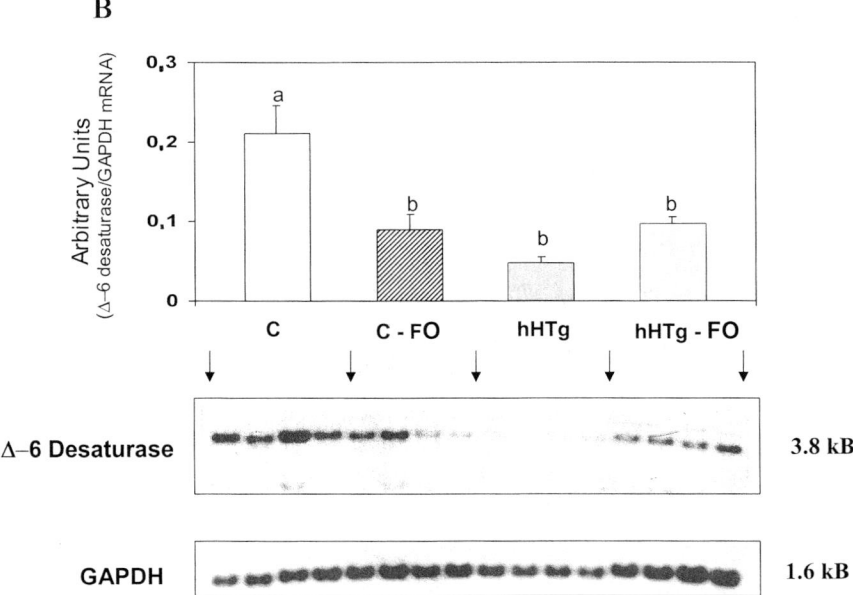

FIGURE 1. Effect of fish oil (FO) supplementation on Δ-6 desaturase activity **(A)** and mRNA levels **(B)** in control and hHTg rats. Bars represent mean ± SEM of 4 rats. Values without a common superscript are significantly different ($p < 0.05$). C = controls (Wistar rats) fed a standard lab chow; hHTg = hereditary hypertriglyceridemic rats fed a standard lab chow; C-FO = Wistar rats fed a marine FO-supplemented diet; hHTg-FO = hHTg rats fed a marine FO-supplemented diet; G3PDH = glyceraldehyde-3-phosphate dehydrogenase.

In accordance with the aforementioned changes in the spectrum of FAs in liver phospholipids, the index of Δ-6 D (Δ-6 DI, calculated as the ratio of 18:3 n-6 to 18:2 n-6), which is an indirect measure of enzyme activity, was significantly reduced in hHTg rats in comparison to controls. Feeding rats the FO-supplemented diet resulted in a striking suppression of the Δ-6 DI in liver of control rats. However, it did not further reduce this parameter in the hHTg animals.

Δ-6 Desaturase Activity and mRNA

Direct measurement of enzyme activity for Δ-6 D in liver microsomes from hHTg rats confirmed the results on FA composition, implying a defect in the Δ-6 D enzyme. As shown in FIGURE 1A, the hHTg rats had a significantly lower activity of Δ-6 D in liver when compared to control rats. FO supplementation into the diet led to a striking reduction of the enzyme activity in control rats. However, it had, again, no effect in insulin-resistant hHTg animals (FIG. 1A).

Northern analysis revealed that the decrease in enzyme activity in liver of hHTg rats was paralleled by a significant reduction in the hepatic abundance of Δ-6 D mRNA (FIG. 1B). Also, the FO supplementation led to a significant reduction of both activity and mRNA abundance for Δ-6 D in liver of control rats. On the other hand, n-3 PUFAs did not further suppress the Δ-6 D mRNA abundance in insulin-resistant hHTg rats, which corresponds with the activity results.

DISCUSSION

In harmony with our previous findings,[11] our results showed that the insulin resistance of hHTg rats is accompanied by changes in FA composition in liver phospholipids. In particular, higher amounts of 18:2 n-6 and lower amounts of 20:4 n-6 and PUFA n-6M were found. In accordance with these changes, Δ-6 DI was significantly reduced in hHTg rats. The aforementioned changes were adequately reflected by direct measurement of the enzyme activity for Δ-6 D, which was dramatically decreased by about 80% as compared with control animals. Northern analysis indicated clearly that the decrease in hepatic Δ-6 D activity is paralleled by a comparable reduction in the hepatic abundance of Δ-6 D mRNA. Thus, it appears that the activity of Δ-6 D in liver of hHTg rats is largely regulated by pretranslational events.

These data are in harmony with previous studies showing that Δ-6 D activity is reduced both in experimental diabetes and in human type 1 or type 2 diabetes mellitus.[16,17] In particular, lower activity and mRNA levels for Δ-6 D were observed in animal models of streptozotocin-induced diabetes[18,19] and in spontaneously diabetic rats.[20] Both models are characterized by insulin deficiency. Thus, the aforementioned changes can be reversed by acute insulin treatment. On the contrary, the hHTg rats have rather normal or slightly higher circulating insulin levels after an overnight fast. Therefore, a decrease in the activity and gene expression for Δ-6 D in liver gives the impression that this could be another feature of insulin resistance[21] in this animal line.

Several studies,[22,23] including our earlier data,[13] indicate that animals fed diets supplemented with PUFAs of the n-6 and n-3 families have reduced Δ-6 D enzymatic activity. This is in agreement with our recent results on lower activity of Δ-6 D in control rats fed the FO-supplemented diet. Moreover, we found that raised dietary FO intake suppressed the hepatic abundance of mRNA for Δ-6 D in control Wistar rats. The latter observation is consistent with our earlier findings[24] and is supported also by other investigators[25] demonstrating that dietary PUFAs coordinately suppress the expression of a wide array of lipogenic and glycolytic enzymes. While the mechanism(s) of this suppression remains yet under investigation, recent evidence suggests that n-6 and n-3 PUFAs inhibit the hepatic transcription of lipogenic genes by reducing the mRNA and protein abundance of the hepatic sterol response element binding protein-1.[25,26]

However, the intrinsic decrease of Δ-6 D activity and mRNA in the liver of hHTg rats fed a standard lab chow was not further changed by raised dietary intake of marine fish oil. Unfortunately, to the best of our knowledge, our data cannot be compared to any appropriate research setting. Therefore, it is suggested that the activity of Δ-6 D in liver of hHTg rats cannot be further influenced anymore (i.e., lowered) by raised dietary fish oil supplementation. Further studies are needed, though, to shed more light on the aforementioned findings.

In conclusion, the endogenous synthesis of long-chain PUFAs in the liver of hHTg rats seems to be reduced due to a decrease in the enzyme activity and gene expression for Δ-6 desaturase. Thus, an impaired regulation of the Δ-6 desaturation process might participate in the pathogenesis of insulin resistance in this animal model of the insulin resistance syndrome.[21]

ACKNOWLEDGMENTS

We are grateful to Steven D. Clarke, University of Texas at Austin, for generously providing us with the cDNA probe for Δ-6 desaturase. Thanks are also due to Pavol Bohov for GC analyses of the liver FA composition. The skillful technical assistance of Alica Mitková, Silvia Kuklová, and Katarína Sušienková is greatly appreciated. This work was supported in part by Slovak Academy of Sciences VEGA Grant No. 2/7210/21 and by a research grant from the Slovak Diabetes Society.

REFERENCES

1. WILKES, J.J., A. BOHEN & R.C. BELL. 1998. A modified high-fat diet induces insulin resistance in rat skeletal muscle, but not adipocytes. Am. J. Physiol. **275:** E679–E686.
2. VESSBY, B., S. TENGBLAD & H. LITHELL. 1994. Insulin sensitivity is related to the fatty acid composition of serum lipids and skeletal muscle phospholipids in 70-year-old men. Diabetologia **37:** 1044–1050.
3. STORLIEN, L.H., et al. 1996. Dietary fats and insulin action. Diabetologia **39:** 621–631.
4. PARK, K.S., et al. 2000. Induction of insulin resistance in human skeletal muscle cells by downregulation of glycogen synthase protein expression. Metabolism **49:** 962–968.
5. BAUR, L.A., et al. 1998. The fatty acid composition of skeletal muscle membrane phospholipid: its relationship with the type of feeding and plasma glucose levels in young children. Metabolism **47:** 106–112.
6. MANN, J.L. 2000. Can dietary intervention produce long-term reduction in insulin resistance? Br. J. Nutr. **83**(suppl. 1): S169–S172.
7. SPRECHER, H. 1981. Biochemistry of essential fatty acids. Prog. Lipid Res. **20:** 13–22.
8. GAŠPERÍKOVÁ, D., et al. 1997. Glucose transport and insulin signaling in rat muscle and adipose tissue: effect of lipid availability. Ann. N.Y. Acad. Sci. **827:** 144–157.

9. CLARKE, S.D. 2001. Polyunsaturated fatty acid regulation of gene transcription: a molecular mechanism to improve the metabolic syndrome. J. Nutr. **131**(4): 1129–1132.
10. VRÁNA, A. & L. KAZDOVÁ. 1990. The hereditary hypertriglyceridemic nonobese rat: an experimental model of human hypertriglyceridemia. Transplant. Proc. **22**: 2579.
11. BOHOV, P., et al. 1997. Fatty acid composition in fractions of structural and storage lipids in liver and skeletal muscle of hereditary hypertriglyceridemic rats. Ann. N.Y. Acad. Sci. **827**: 494–509.
12. NAKAMURA, M.T., H.P. CHO & S.D. CLARKE. 2000. Regulation of hepatic Δ-6 desaturase expression and its role in the polyunsaturated fatty acid inhibition of fatty acid synthase gene expression in mice. J. Nutr. **130**: 1561–1565.
13. GARG, M.L., et al. 1998. Δ-6 desaturase activity in liver microsomes of rats fed diets enriched with cholesterol and/or ω-3 fatty acids. Biochem. J. **249**: 351–356.
14. CHOMCZYNSKI, P. & N. SACCHI. 1987. Single-step method of RNA isolation by acid guanidinium thiocyanate–phenol–chloroform extraction. Anal. Biochem. **162**: 156–159.
15. CHO, H.P., M.T. NAKAMURA & S.D. CLARKE. 1999. Cloning, expression, and nutritional regulation of the mammalian delta 6 desaturase. J. Biol. Chem. **274**: 471–477.
16. HORROBIN, D.F. 1988. The roles of essential fatty acids in the development of diabetic neuropathy and other complications of diabetes mellitus. Prostaglandins Leukotrienes Med. **31**: 181–197.
17. MOHAN, I.K. & U.N. DAS. 2000. Effect of L-arginine–nitric oxide system on the metabolism of essential fatty acids in chemical-induced diabetes mellitus. Prostaglandins Leukotrienes Essent. Fatty Acids **62**: 35–46.
18. BRENNER, R.R. 2000. Nutritional and hormonal factors influencing desaturation of essential fatty acids. Prog. Lipid Res. **20**: 41–48.
19. RIMOLDI, O.J., G.S. FINARELLI & R.R. BRENNER. 2001. Effects of diabetes and insulin on hepatic delta 6 desaturase gene expression. Biochem. Biophys. Res. Commun. **4**: 323–326.
20. BROWN, J.E., R.M. LINDSAY & R.A. RIEMERSMA. 2000. Linoleic acid metabolism in the spontaneously diabetic rat: delta-6 desaturase activity vs. product/precursor ratios. Lipids **35**: 1319–1323.
21. KLIMEŠ, I., et al. 1995. The hereditary hypertriglyceridemic rat, a new model of the insulin resistance syndrome. *In* Lessons from Animal Diabetes, pp. 271–283. Smith-Gordon. United Kingdom.
22. CHRISTIANSEN, E.N., J.S. LUND, T. RORTVEIT & A.C. RUSTAN. 1991. Effect of dietary n-3 and n-6 fatty acids on fatty acid desaturation in rat liver. Biochim. Biophys. Acta **1082**: 57–62.
23. PELUFFO, R.O., A.M. NERVI & R.R. BRENNER. 1976. Linoleic acid desaturation activity of liver microsomes of essential fatty acid deficient and sufficient rats. Biochim. Biophys. Acta **441**: 25–31.
24. ŠEBÖKOVÁ, E., et al. 1996. Regulation of gene expression for lipogenic enzymes in the liver and adipose tissue of hereditary hypertriglyceridemic, insulin-resistant rats: effect of dietary sucrose and marine fish oil. Biochim. Biophys. Acta **1303**: 56–62.
25. XU, J., et al. 1999. Sterol regulatory element binding protein-1 expression is suppressed by dietary polyunsaturated fatty acids: a mechanism for the coordinate suppression of lipogenic genes by polyunsaturated fats. J. Biol. Chem. **274**: 23577–23583.
26. CLARKE, S.D., et al. 2002. Fatty acid regulation of gene expression: a genomic explanation for the benefits of the Mediterranean diet. Ann. N.Y. Acad. Sci. This volume.

Heart Remodeling in the Hereditary Hypertriglyceridemic Rat

Effect of Captopril and Nitric Oxide Deficiency

F. SIMKO,[a] I. LUPTAK,[a] J. MATUSKOVA,[a] P. BABAL,[b] O. PECHANOVA,[c] I. BERNATOVA,[c] AND I. HULIN[a]

[a]*Department of Pathophysiology and* [b]*Department of Pathology, Comenius University, School of Medicine, Bratislava, Slovak Republic*

[c]*Institute of Normal and Pathological Physiology, Slovak Academy of Sciences, Bratislava, Slovak Republic*

ABSTRACT: *Aim*—The hereditary hypertriglyceridemic (hHTg) rat is characterized by insulin resistance, hypertension, and hypertriglyceridemia. Thus, we investigated whether (a) remodeling of the heart left ventricle (LV) is present under the given hypertensive situation and (b) whether this potential alteration could be influenced by an inhibition of the angiotensin converting enzyme (ACE) and/or by a blockade of nitric oxide production. *Methods*—Five groups of rats were investigated: control Wistar (C) rats, hHTg rats, hHTg rats given captopril (100 mg/kg/day) (hHTg + CAP) or N^G-nitro-L-arginine methyl ester (L-NAME, 40 mg/kg/day) (hHTg + L-NAME), and hHTg rats given the combination of both drugs (hHTg + CAP + L-NAME) for 28 days. Systolic blood pressure (SBP) was measured by tail-cuff plethysmography each week. After cervical dislocation, the relative weights of the left and right ventricles (LV/BW, RV/BW) were obtained, the LV nucleic acid concentrations were analyzed, and the fibrosis amount was quantified with aid of a semiquantitative histological technique. *Results*—In the hHTg group, the increased SBP (141.7 ± 4.4 vs. 117.2 ± 3.1 mmHg in controls) was linked to hypertrophy of the LV (1.63 ± 0.05 vs. 1.30 ± 0.03 g/kg in controls) with only a minimum of fibrosis. DNA concentration in the LV was decreased (0.45 ± 0.03 vs. 0.69 ± 0.04 mg/g w.w. in controls) in the hHTg group. Captopril normalized SBP and decreased the LV/BW (1.44 ± 0.04 g/kg). Chronic administration of L-NAME to the hHTg rats additionally enhanced (189.3 ± 5.9 mmHg) the already raised SBP, stimulated fibrosis development, and increased DNA concentration (0.54 ± 0.02 mg/g w.w.) in the LV compared to hHTg group, yet without additional weight increase of the LV. The combined treatment of the hHTg rats with CAP and L-NAME resulted in normal SBP and the development of LV hypertrophy, and fibrosis was substantially reduced. *Conclusions*—(a) The heart of hHTg rats carries signs of LV hypertrophy with minimal fibrosis. (b) Nevertheless, LV fibrosis was increased in the hHTg + L-NAME group. (c) Captopril normalized SBP and decreased the extent of LV hypertrophy in both the nontreated hHTg and the hHTg + L-NAME groups and (d) substantially reduced the develop-

Address for correspondence: Prof. Fedor Simko, M.D., Ph.D., Department of Pathophysiology, Faculty of Medicine, Comenius University, Sasinkova 4, 813 72 Bratislava, Slovak Republic. Voice: 4212 59357 276; fax: 4212 59357 601.
simko@fmed.uniba.sk

ment of LV fibrosis in the hHTg + L-NAME group. LVH in hHTg rats may be induced by sympathoadrenal system activation, circulating volume enlargement, and impairment of nitric oxide (NO) production rather than by activation of the renin-angiotensin-aldosterone system.

KEYWORDS: hereditary hypertriglyceridemia (hHTg); heart remodeling; rat; captopril

INTRODUCTION

Hereditary hypertriglyceridemic (hHTg) rats are characterized by hypertension, hyperinsulinemia, and insulin resistance.[1] Despite numerous metabolic studies performed on this model of hypertension, no attention has been devoted to the potential heart muscle alterations. Pressure hemodynamic overload can be expected to induce the adaptive growth of the heart, which could be modified by the specific neurohormonal and metabolic disorders. Thus, we investigated whether hypertrophic growth of the heart is present in hHTG rats and how the character of this potential remodeling could be altered by the angiotensin converting enzyme (ACE) inhibitor, captopril, and by blockade of nitric oxide (NO) synthase activity by N^G-nitro-L-arginine methyl ester (L-NAME).

MATERIALS AND METHODS

Animals and Treatment

The experiments were performed on male Wistar and hHTG rats, 12 weeks old, which were fed a standard pellet mixture. Five groups of rats were investigated: control Wistar (C) rats ($n = 10$), hHTg rats ($n = 7$), hHTg rats given captopril (100 mg/kg/day) (hHTg + CAP) ($n = 7$) or L-NAME (40 mg/kg/day) (hHTg + L-NAME) ($n = 8$), and hHTg rats given the combination of both drugs (hHTg + CAP + L-NAME) ($n = 8$) for 28 days.

Systolic blood pressure (SBP) was measured by a noninvasive method of tail-cuff plethysmography at the beginning of the experiment and at the end at 4 weeks. After 4 weeks, the animals were killed; the body weight (BW), heart weight (HW), left ventricle weight (LVW), and right ventricle weight (RVW) were determined; and the ratios, LVW/BW and RVW/BW, were calculated.

Determination of Nucleic Acid Concentration

RNA concentration was determined using the guanidinium thiocyanate–phenol–chloroform extraction method.[2] DNA concentration was analyzed according to Sambrook et al.[3]

Morphometric Determination of Fibrosis

The samples of myocardium from the central parts of the left ventricle were oriented perpendicularly to the sectioning plane. The tissues were routinely processed in paraffin after 24-h fixation in 10% phosphate-buffered formalin, and serial 5-μm-

TABLE 1. Effect of captopril and/or L-NAME treatment on body weight (BW), left ventricle weight (LVW), right ventricle weight (RVW), right ventricle to body weight ratio (RVW/BW), and fibrosis in the left ventricle myocardium in hHTg rats

	Wistar	hHTg	hHTg + CAP	hHTg + L-NAME	hHTg + L-NAME + CAP
BW [g]	316.11 ± 4.39	337.86 ± 5.86	319.86 ± 6.76	301.38 ± 7.73**	299.00 ± 8.68**
RVW [mg]	139.89 ± 5.44	187.17 ± 10.49^^	187.67 ± 10.22^^	135.38 ± 5.5***	144.86 ± 8.15***
LVW [mg]	410.56 ± 8.17	559.33 ± 23.27^^^	456.67 ± 9.51**	525.38 ± 20.96^^^	461.00 ± 24.26**
RVW/BW	0.44 ± 0.02	0.55 ± 0.03^	0.59 ± 0.03^^^	0.45 ± 0.02*	0.48 ± 0.02*
Fibrosis in the LV myocardium	−	±	±	+++	+

NOTE: Values are means ± SEM (^$p < 0.05$, ^^$p < 0.01$, ^^^$p < 0.001$ compared to Wistar; *$p < 0.05$, **$p < 0.01$, ***$p < 0.001$ compared to hHTg).

thick sections were stained with hematoxylin and eosin and by van Gieson's staining for collagen. The amount of fibrosis was evaluated semiquantitatively by help of an optical grid as % of the section area: (−) <1%; (±) 1–2%; (+) 2–5%; (++) 5–7%; (+++) >7%.

Statistical Analysis

The results were expressed as mean ± SEM. Values were considered to differ significantly if the p value was less than 0.05. For analysis, one-way ANOVA, Bonferroni test, was used.

RESULTS

In hHTg rats, SBP was increased by 20.8% when compared to Wistar controls ($p < 0.001$) and captopril treatment normalized it. After 4 weeks of L-NAME treatment, SBP increased by ~34% ($p < 0.001$) compared to untreated hHTg rats. Simultaneous treatment with L-NAME and captopril did not influence SBP in hHTg rats (FIG. 1).

The LV/BW ratio was 25% higher ($p < 0.001$) in the hHTg group when compared to the Wistar controls. Captopril reduced the relative weight of the LV in the hHTg rats by 12% ($p < 0.05$). L-NAME did not induce additional growth of the LV, but addition of captopril to L-NAME reduced the LV/BW ratio by 11.5% ($p < 0.05$) compared to the hHTg + L-NAME group (FIG. 1). BW, LVW, RVW, and RVW/BW are given in TABLE 1.

DNA concentration in the hHTg group was up to 35% lower ($p < 0.05$) than in Wistar controls, and L-NAME increased it by 20% ($p < 0.05$). Treatment with captopril did not significantly change the DNA concentration (FIG. 2). There was no significant alteration of RNA concentration in the presented groups (FIG. 2).

Only minimal diffuse fibrosis was observed in hypertrophied LV of the hHTg group (TABLE 1). Administration of L-NAME induced extensive focal fibrosis of the LV, predominantly in the subendocardial area, and captopril added to L-NAME substantially reduced the fibrosis development (FIG. 3).

DISCUSSION

SBP increase by about 20 mmHg (measured by the tail-cuff method in our experiment) along with diastolic blood pressure enhancement by about 20–40 mmHg (measured invasively by other authors[4]) represent a hemodynamic overload that induced left ventricular hypertrophy (LVH) (weight increase by 25%). Surprisingly, LVH was associated only with minimal fibrosis. This result correlated with the decreased concentration of DNA in the LV, suggesting that LVH development took place predominantly on account of the enlargement of the cardiomyocytes, resulting in a relative decline of nuclear matter. Prominent LV fibrosis is regularly demonstrated in such models of hypertension, which are associated with activation of the circulating or local renin-angiotensin-aldosterone system (RAAS) since angiotensin II (Ang II) and aldosterone function as stimulators of fibrotic tissue growth.[5,6] How-

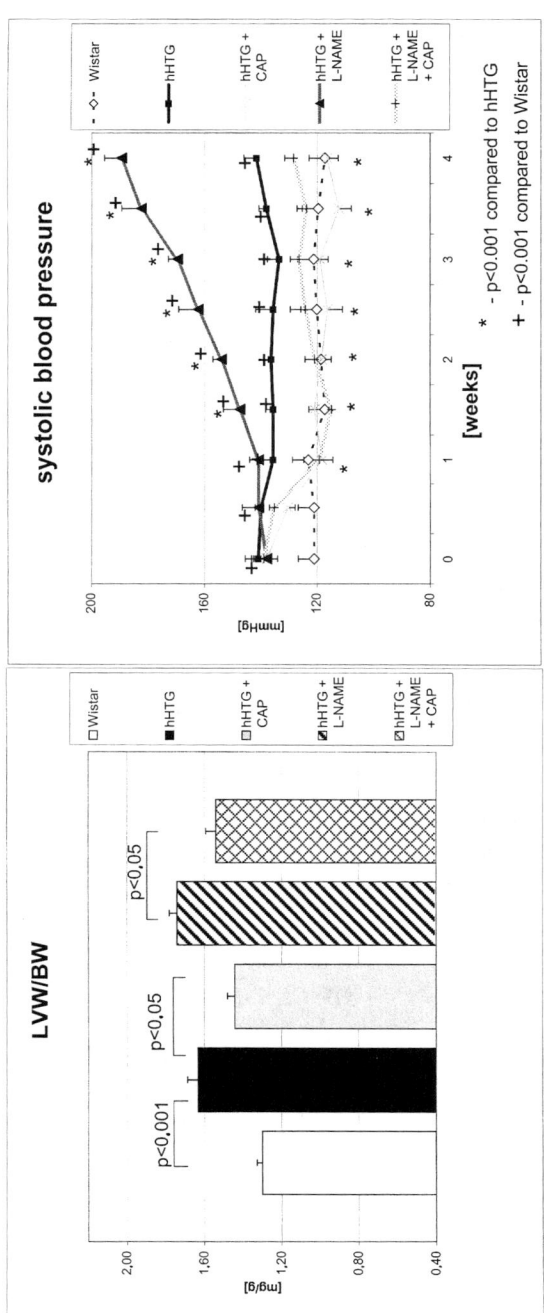

FIGURE 1. Effect of captopril and/or L-NAME treatment on left ventricle/body weight ratio (LVW/BW) and systolic blood pressure in hHTg rats.

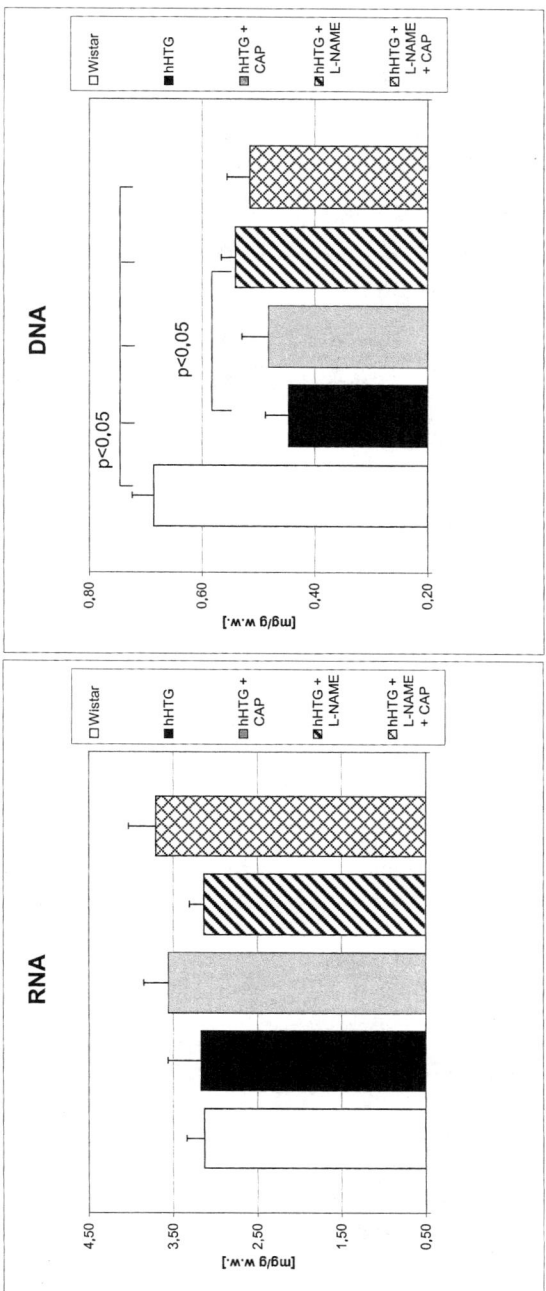

FIGURE 2. Effect of captopril and/or L-NAME treatment on RNA and DNA concentrations in the left ventricle of hHTg rats.

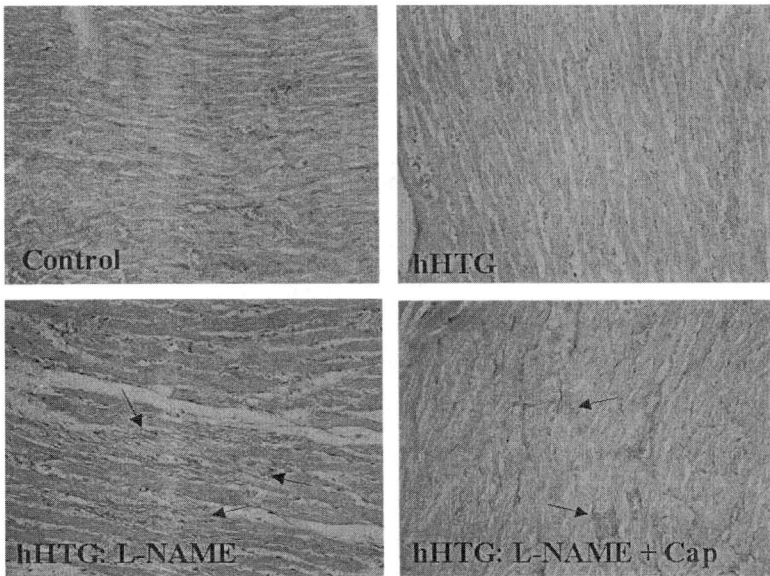

FIGURE 3. Fibrosis in the left ventricle myocardium of hHTg rats. Fibrosis indicated with arrows. Staining: van Gieson. Magnification: ×200. [Figure reduced to 66%.]

ever, in the hHTg model of hypertension, the neurohormonal activation has a specific character. Increased norepinephrine concentration reflecting sympathoadrenal system (SAS) activation is associated with normal levels of renin and aldosterone.[7] Thus, Ang II or aldosterone should not be considered to participate in LV remodeling in hHTg rats. Besides SAS activation, Na^+ retention followed by circulating volume enlargement also seems to support the hypertension[1] and LVH development. Relative hypervolemia provides a plausible explanation for the finding in this experiment where, besides hypertrophy of the LV, also prominent hypertrophy of the right chamber was observed.

Although the RAAS apparently is not activated in hHTg rats, the reduction of LVH by ACE inhibition was expected. It has been previously shown that ACE inhibition was able to reduce blood pressure and also the weight of the LV not only in states with hemodynamic overload, but also in control healthy individuals, in which no activation of the RAAS has been anticipated.[8] In this model, the reduction of activated SAS by the sympatholytic effect of ACE inhibition might have contributed to normalization of SBP and reduction of LVH.

The well-known vasodilatory effect of insulin in different vascular beds, especially in skeletal muscles and the coronary circulation,[1] is considered to be NO-dependent. As insulin can stimulate NO production from endothelium, the states with resistance to insulin can be linked to NO deficiency. Indeed, decreased acetylcholine-induced relaxation of the aorta was observed in hHTg rats in our laboratory.[9] NO deficiency may be considered as a disturbed counterregulatory system to activated SAS, contributing to hypertension development. Chronic treatment of hHTG rats with the

NO synthase inhibitor, L-NAME, additionally increased SBP by about 50 mmHg, similar to our previous observation in L-NAME-treated Wistar rats.[10,11] Thus, NO production seems to be only partly inhibited in hHTg rats and probably represents only one of the mechanisms leading to hypertension and LVH development. Interestingly, although the L-NAME-induced blood pressure increase did not stimulate additional hypertrophy of the LV in hHTg rats, it provoked fibrosis in the LV myocardium. Addition of captopril to L-NAME-treated hHTg rats substantially reduced fibrosis development and the weight of the LV. Similar profibrotic effects of L-NAME and protective effects of captopril as in hHTg rats were observed in Wistar rats,[10,11] suggesting that serious NO deficiency is associated with LV fibrosis by either NO deficiency itself or activation of the RAAS (or by both mechanisms).[12]

We conclude that hypertension in hHTg rats results in hypertrophic growth of the LV, which is associated with minimal fibrotic remodeling. This type of hypertrophy may be induced by SAS activation, circulating volume enlargement, and impairment of NO production rather than by activation of the RAAS.

ACKNOWLEDGMENTS

The support of VEGA Grant No. 1/7529/20 is highly appreciated. Captopril was a generous gift of EGIS Pharmaceuticals Limited (Budapest, Hungary).

REFERENCES

1. KLIMEŠ, I., J. ZICHA, J. KUNES et al. 1997. Hypertriglyceridemia, insulin resistance, and hypertension in rats: are they related? Endocr. Regul. **11:** 103–119.
2. CHOMCZYNSKI, P. & N. SACCHI. 1989. Single step method of RNA isolation by acid guanidinium thiocyanate–phenol–chloroform extraction. Anal. Biochem. **162:** 156–159.
3. SAMBROOK, J., E.F. FRITSCH & T. MANIATIS. 1989. Isolation of DNA from mammalian cells. In Molecular Cloning: A Laboratory Manual. Second edition, pp. 9.16–9.19. Cold Spring Harbor Laboratory Press. New York.
4. KLIMEŠ, I., E. ŠEBÖKOVÁ, D. GASPERIKOVA et al. 1996. Activity of the sympathoadrenal system of the hereditary hypertriglyceridemic and hypertensive rats during immobilization stress. In Molecular Genetic and Neurobiological Advances. Volume 1, pp. 343–353. Gordon & Breach. New York.
5. WEBER, K.T. 2000. Fibrosis and hypertensive heart disease. Curr. Opin. Cardiol. **15:** 264–267.
6. SIMKO, F. 1996. Left ventricular hypertrophy regression as a process with variable biological implications. Can. J. Cardiol. **12:** 507–513.
7. LICHARDUS, B., E. ŠEBÖKOVÁ, D. JEZOVA et al. 1993. Effect of low salt diet on blood pressure and vasoactive hormones in the hereditary hypertriglyceridemic rat. Ann. N.Y. Acad. Sci. **683:** 289–294.
8. SIMKO, F., O. PECHANOVA, I. BERNATOVA et al. 2000. Effect of angiotensin converting enzyme inhibitors on non-diseased myocardium of experimental animals: potential clinical implications. Med. Hypotheses **54:** 626–627.
9. TÖRÖK, J., P. BABÁL, J. MATUŠKOVÁ et al. 2002. Impaired endothelial function of the thoracic aorta in hereditary hypertriglyceridemic rats. This volume.
10. PECHANOVA, O., I. BERNATOVA, V. PELOUCH et al. 1977. Protein remodelling of the heart in NO-deficient hypertension: the effect of captopril. J. Mol. Cell. Cardiol. **29:** 3365–3374.

11. BERNATOVA, I., O. PECHANOVA & F. SIMKO. 1999. Effect of captopril in L-NAME-induced hypertension on the rat myocardium, aorta, brain, and kidney. Exp. Physiol. **84:** 1095–1105.
12. SIMKO, F. & J. SIMKO. 2000. The potential role of nitric oxide in the hypertrophic growth of the left ventricle. Physiol. Res. **49:** 37–46.

Energy Transfer in Acute Diabetic Rat Hearts

Adaptation to Increased Energy Demands due to Augmented Calcium Transients

A. ZIEGELHÖFFER,[a] T. RAVINGEROVÁ,[a] I. WACZULÍKOVÁ,[b] J. CÁRSKY,[c] J. NECKÁR,[d] B. ZIEGELHÖFFER-MIHALOVICOVÁ,[a,e] AND J. STYK[a]

[a]*Institute for Heart Research, Slovak Academy of Sciences, Bratislava, Slovak Republic*

[b]*Department of Biophysics and Chemical Physics, Faculty of Mathematics, Physics, and Informatics, Comenius University, Bratislava, Slovak Republic*

[c]*Department of Medical Chemistry, Biochemistry, and Clinical Biochemistry, Faculty of Medicine, Comenius University, Bratislava, Slovak Republic*

[d]*Institute of Physiology, Academy of Sciences of the Czech Republic and Center for Cardiovascular Research, Prague, Czech Republic*

ABSTRACT: *Objectives*—Hearts of rats with diabetes mellitus (DM) are characterized by energy demands exceeding their energy production, but they might also exhibit decreased vulnerability to ischemia and calcium overload. This indicates adaptation in cardiac energetics (CE), where energy transport is not rate-limiting. *Aim*—This study was designed to elucidate the functional significance of the DM-induced adaptation in CE by investigating the formation of mitochondrial contact sites (MiCS), facilitating the Ca-dependent/high-capacity energy transfer from mitochondria, in conjunction with testing the ischemic tolerance (IT) of hearts. *Methods*—After 1 week of streptozotocin-induced DM (45 mg/kg iv), the hearts of male diabetic and age-matched control rats (C) were isolated and Langendorff-perfused with either 1.6 or 2.2 mmol/L of $CaCl_2$. MiCS formation was assessed by cytochemical detection of mCPK octamers and was quantified stereologically as MiCS to mitochondrial surface ratio (S_S). IT was evaluated in anesthetized open-chest animals subjected to 30-min occlusion of the LAD coronary artery followed by 4-h reperfusion, by monitoring ischemic arrhythmias and by measuring the size of infarction (tetrazolium double staining). *Results*—In C hearts, increasing Ca^{2+} induced both positive inotropic response (*dP/dt* increase from 2270 ± 220 to 2955 ± 229, $p < 0.01$) and elevated MiCS formation (S_S increase from 0.070 ± 0.011 to 0.123 ± 0.012, $p < 0.01$). In DM hearts, basic MiCS formation was already comparable with that induced by elevated Ca^{2+} in C hearts and could not be further stimulated by Ca^{2+}. In C, ventricular tachycardia represented 55.4% of the total arrhythmias and occurred in 90% of the animals. In DM rats, the

Address for correspondence: A. Ziegelhöffer, Institute for Heart Research, Slovak Academy of Sciences, Dúbravská cesta 9, 84233 Bratislava, Slovak Republic. Voice: 4212 5477 4405; fax: 4212 5477 6637.
usrdzigy@savba.sk

[e]Present address: Institute of Physiology, University of Leipzig, Leipzig, Germany.

Ann. N.Y. Acad. Sci. 967: 463–468 (2002). © 2002 New York Academy of Sciences.

arrhythmia profile was similar to that in C, and the incidence of tachyarrhythmias and their severity were not enhanced (arrhythmia score: 3.18 ± 0.4 vs. 3.30 ± 0.3 in C). The infarct size normalized to the size of area at risk was smaller in the DM than in C hearts (52.3 ± 5.8% vs. 69.2 ± 2.2%, respectively; $p < 0.05$). *Conclusions*—Ca-signaling represents the link between energy delivery from mitochondria (via MiCS) and energy requirements of the heart. In DM hearts, energy transport via MiCS is elevated to the maximum value. This contributes to increased resistance of DM hearts to irreversible cell damage.

KEYWORDS: diabetic heart; adaptation of the myocardium; cardiac energetics; mitochondrial contact sites; calcium transient–mediated signaling; tolerance to ischemia; ischemic arrhythmias; size of infarction

INTRODUCTION

Numerous animal studies revealed that the metabolic disorganization resulting from diabetes causes myocardial remodeling and defective function of the cardiac subcellular membrane systems,[1] associated with abnormalities in the function of receptors, ion transport,[2,3] as well as Ca^{2+} handling.[1–3] In some cases, these alterations may lead to an increase in the level of free Ca^{2+} in the cytoplasm ($[Ca^{2+}]_i$),[1] but they always induce an amplification and prolongation of the Ca^{2+} transients.[4] In addition, hearts of rats with diabetes mellitus are also characterized by energy demands exceeding their energy production,[5–7] but they might also exhibit decreased vulnerability to ischemia,[6,8,9] calcium overload,[2,3,7,10] and reperfusion-induced ventricular arrhythmias.[11,12] These findings indicate that diabetes initiates an adaptation in the myocardial energy household. The latter is characterized by decreased, yet predominating, utilization of endogenous glycogen prior to exogenous glucose for energy production.[6] As concerns the mitochondria of diabetic hearts, they revealed decreased F1 ATPase activity and impairments in the function of the citric acid cycle and oxidative production of ATP,[13,14] but these alterations seemed to be, at least in part, compensated for by significantly increased formation of mitochondrial contact sites (MiCS) enabling high-capacity energy transport through the mitochondrial membrane.[4] The outcomes of all these alterations are proportionally lowered levels of adenosine triphosphate (ATP) and diphosphate (ADP), as well as creatine phosphate (CP).[6] Proportionality in the ATP, ADP, and CP decrease is manifested by ATP/ADP and ATP/CP ratios not differing considerably from those in the tissue of normal hearts.[5] This indicates that, in contrast to some expectations, at least the intracellular energy transfer may not be rate-limiting in the diabetic hearts.[4] The latter assumption was confirmed by our earlier results:[15] diabetic hearts, which are believed to have amplified Ca^{2+} transients, exhibited an MiCS formation identical with the maximal MiCS formation that could be achieved in healthy hearts by increased offers of calcium[4,15] and was coupled with a positive inotropic response.[15]

This study was designed to elucidate the functional significance of the diabetes-induced adaptation in cardiac energetics (CE) by investigating the formation of MiCS, which facilitate the Ca^{2+}-dependent, high-capacity energy transfer from the mitochondria, in conjunction with testing the ischemic tolerance (IT) of the diabetic heart.

METHODS

Isolated Heart Preparation

After 7 days of diabetes mellitus induced by streptozotocin (45 mg/kg iv), the hearts of male diabetic (DM) and age-matched control (C) Wistar rats (initial body weight: 230 ± 10 g) were isolated and Langendorff-perfused (perfusion pressure: 75 mmHg; 37°C; 15-min perfusion following 15-min stabilization perfusion without recirculation of perfusate) with Krebs-Henseleit solution containing either 1.6 or 2.2 mmol/L of $CaCl_2$ and gassed with 95% O_2 and 5% CO_2. Contractile function was assessed as dP/dt. Calcium paradox: 15-min stabilized perfusion followed by 3-min Ca^{2+} depletion and 12-min Ca^{2+} repletion. Inhibition of the Ca^{2+} signal for MiCS formation: heart arrest due to replacement of the 1.6 mmol/L of $CaCl_2$ by 5 mmol/L of $CdCl_2$. These groups of C and DM hearts were further processed for cytochemical determination of MiCS. For more details, see reference 7.

Cytochemical Detection of MiCS Formation

MiCS were assessed by cytochemical detection of the octameric form of the mitochondrial creatine phosphokinase.[4] The method is based on reduction of a thiocarbamyl nitro blue tetrazolium chloride salt in the presence of lactate and glucose-6-phosphate dehydrogenase. Thin sections of embedded tissue slices were examined by an electron microscope. MiCS formation was quantified stereologically as the ratio of MiCS surface to mitochondria surface (S_S). The testing grid was applied over the electron micrographs, and the ratio of the intersections of cycloids with MiCS and the intersections of cycloids with mitochondrial membranes was counted. For more details, see reference 16.

Assessment of Ischemic Tolerance (IT) of the Hearts

IT was evaluated in pentobarbital-anesthetized, artificially ventilated (Ugo Basile, Italy), open-chest DM as well as age-matched C animals. These rats were subjected to 30-min occlusion of the LAD coronary artery followed by 4-h reperfusion of the ischemic area. Evaluation involved the monitoring of ischemic arrhythmias (Hellige Servomed, Switzerland) and measuring the size of infarction by means of the tetrazolium staining technique. Arrhythmias were analyzed in accordance with the Lambeth Conventions. For more details, see references 9 and 12.

Assessment of Metabolic Status of the Animals

The metabolic status of the animals was checked by estimating the levels of glucose[17] (GLU), triacylglycerols[18] (TG), cholesterol[19] (CH), and glycohemoglobin[20] (GH) in the blood of DM and C animals prior to starting and at the end (on the eighth day) of the experiment. For more details, see references 3, 10, and 21.

Statistics

Data were expressed as means ± SEM. One-way analysis of variance (ANOVA) followed by Tukey's method were applied to test for any significant differences in normally distributed variables among the groups. Non-Gaussian distributed variables

TABLE 1. Metabolic variables in the blood of rats with acute (8 days) streptozotocin-diabetes

	GLU	TG	CH	GH
Controls	5.74 ± 0.15	1.17 ± 0.09	1.86 ± 0.11	4.36 ± 0.14
DM	17.20 ± 0.15*	4.62 ± 0.16*	2.83 ± 0.18*	5.95 ± 0.12*

NOTE: Abbreviations as in the METHODS. Data for GLU and TG are given in mmol/L, for CH in g/L, and for GH in % of the total hemoglobin content; means ± SEM; $n = 51$; *$p < 0.01$.

(incidences of ventricular tachycardias and ventricular fibrillations) were compared using Fisher's exact test. Differences were considered significant when $p < 0.05$.

RESULTS AND DISCUSSION

Metabolic Status

Metabolic status of healthy C animals was characterized by blood GLU, TG, and CH levels and the content of GH (TABLE 1), which all were fully within the range of normal metabolic values reported earlier.[2,3,21] In contrast, the DM group exhibited a significant ($p < 0.01$) elevation in the values of GLU, TG, CH, and GH, amounting to 199.7%, 394.9%, 52.2%, and 36.5%, respectively. Also, the latter findings were accompanied by enhanced Amadori product formation, manifested in a 31% ($p < 0.01$) increase in fructosamine content of heart sarcolemma (not shown), and confirmed the presence of DM. All these values are in agreement with biochemical findings reported in our earlier studies on the acute stage of the disease.[2,3,21]

Heart Inotropy and MiCS Formation

In healthy C hearts, elevation of Ca^{2+} concentration in the perfusate expectedly led to an enhanced calcium entry into the hearts and consequently augmented the intracellular Ca^{2+} transients that, in turn, propagated a signal for both an increase of heart contractility (dP/dt), coupled with increased utilization of ATP, and an increase in MiCS formation. The increase in MiCS formation is believed to augment energy delivery from the mitochondria to meet the increased energy demands of the heart.[4] This was manifested by a Ca^{2+}-induced positive inotropic response (dP/dt increase amounting to 30.2%, $p < 0.01$) and elevated MiCS formation (S_S increase by 75.7%, $p < 0.01$). On the other hand, replacement of Ca^{2+} by Cd^{2+} ions led to a weakening of the Ca^{2+} signal and caused a significant ($p < 0.01$) depression in S_S, proving its key role in MiCS formation (TABLE 2). In DM hearts, already the basic MiCS formation was identical with that induced by elevated Ca^{2+} in C hearts. Neither the elevated Ca^{2+} in perfusion medium nor the induction of the calcium paradox proved to be potent enough to stimulate considerably the S_S in DM hearts (TABLE 2). This indicated that the Ca^{2+} signal leading to augmented energy delivery through the mitochondrial membrane is already present in DM hearts and its effect is maximally developed.

TABLE 2. The effect of calcium on contractility and mitochondrial contact sites formation (dP/dt and S_S) in isolated perfused control (C) and diabetic (DM) rat hearts

	C 1.6 Ca^{2+}	C 2.2 Ca^{2+}	DM 1.6 Ca^{2+}	DM 2.2 Ca^{2+}	DM CaP	DM Cd
dP/dt	2270 ± 220*	2955 ± 229	2035 ± 191	2118 ± 208	–	–
S_S	0.070 ± 0.01*	0.123 ± 0.012	0.133 ± 0.011	0.134 ± 0.012	0.130 ± 0.013	0.060 ± 0.011*

NOTE: Abbreviations as in the METHODS. Further abbreviations—C 1.6 Ca^{2+}, C 2.2 Ca^{2+}, DM 1.6 Ca^{2+}, DM 2.2 Ca^{2+}: C and DM hearts perfused with either 1.6 or 2.2 mmol/L Ca^{2+}; DM CaP: DM hearts with calcium paradox; DM Cd: DM hearts with Ca^{2+} replaced by Cd^{2+}. Data represent means ± SEM, n = 10–12. Significances—For dP/dt: C 2.2 Ca^{2+} vs. C 1.6 Ca^{2+}, DM 1.6 Ca^{2+}, and DM 2.2 Ca^{2+}, *p < 0.05; all other possibilities, N.S.; for S_S: C 1.6 Ca^{2+} and DM Cd vs. C 2.2 Ca^{2+}, DM 1.6 Ca^{2+}, DM 2.2 Ca^{2+}, and DM CaP: *p < 0.05; all other possibilities, N.S.

TABLE 3. Acute coronary occlusion and reperfusion in hearts of rats with acute (8 days) streptozotocin-diabetes: the effect on arrhythmia score and infarct size

	Arrhythmia score	Infarct size
Controls	3.30 ± 0.30	69.20 ± 2.20
DM	3.18 ± 0.40	52.30 ± 5.80*

NOTE: Abbreviations as in the METHODS. Data represent means ± SEM; n = 12; *p < 0.05.

Ischemic Tolerance

In C hearts, ventricular tachycardia after LAD coronary artery ligation represented 55.4% of the total arrhythmias and occurred in 90% of the animals. In DM rats, the arrhythmia profile was similar to that in the C group, and the incidence of tachyarrhythmias and their severity were not enhanced (arrhythmia score: 3.18 ± 0.4 vs. 3.30 ± 0.3 in C; p > 0.05). The results in TABLE 3 indicate that, in comparison with C hearts, DM hearts are not more susceptible to ventricular arrhythmias induced by acute myocardial ischemia. On the other hand, the infarct size normalized to the size of area at risk was up to 16.9% smaller in the DM than in C hearts (p < 0.05).

Conclusions

(1) Ca-signaling represents the link between the energy delivery from mitochondria (via MiCS) and the energy requirements of the heart.
(2) In DM hearts that are permanently suffering from energy deficiency, the energy transport via MiCS seems to be involved in some adaptation changes and it is elevated to the maximum value. This contributes to the increased resistance of DM hearts to various forms of damage.
(3) In the acute phase of DM, rat hearts are more resistant to irreversible cell injury.

ACKNOWLEDGMENTS

This study was supported in part by VEGA Slovak Republic Grant Nos. 2/7157/21 and 2/6094/99.

REFERENCES

1. ZIEGELHÖFFER, A., T. RAVINGEROVÁ, J. STYK et al. 1998. Hearts with diabetic cardiomyopathy: adaptation to calcium overload. Exp. Clin. Cardiol. **3:** 158–161.
2. GOTZSCHE, O. 1991. Myocardial calcium uptake and catecholamine sensitivity in experimental diabetes. In The Diabetic Heart, pp. 199–208. Raven Press. New York.
3. ZIEGELHÖFFER, A., T. RAVINGEROVÁ, J. STYK et al. 1996. Diabetic cardiomyopathy in rats: biochemical mechanisms of elevated tolerance to calcium overload. Diabetes Res. Clin. Pract. **31:** S93–S103.
4. ZIEGELHÖFFER-MIHALOVICOVÁ, B., L. OKRUHLICOVÁ, N. TRIBULOVÁ et al. 1997. Mitochondrial contact sites detected by creatine phosphokinase activity in the hearts of normal and diabetic rats: is mitochondrial contact site formation a calcium dependent process? Gen. Physiol. Biophys. **16:** 329–338.
5. SEYMOUR, A-M. & M.J. BROSNAN. 1991. Nuclear magnetic resonance investigations of energy metabolism in diabetic cardiomyopathy. In The Diabetic Heart, pp. 371–382. Raven Press. New York.
6. STRÖDTER, D., P. WILLMANN, J. WILLMANN et al. 1991. Results of a balance in energy in the diabetic heart. In The Diabetic Heart, pp. 383–393. Raven Press. New York.
7. RAVINGEROVÁ, T., J. STYK, D. PANCZA et al. 1996. Diabetic cardiomyopathy in rats: alleviation of myocardial dysfunction caused by Ca^{2+} overload. Diabetes Res. Clin. Pract. **31:** S105–S112.
8. TANI, M. & J.R. NEELY. 1988. Hearts from diabetic rats are more resistant to in vitro ischemia: possible role of altered Ca^{2+} metabolism. Circ. Res. **62:** 931–940.
9. RAVINGEROVÁ, T., J. NECKÁR, F. KOLÁR et al. 2001. Ventricular arrhythmias following coronary artery occlusion in rats: is the diabetic heart less or more sensitive to ischemia? Basic Res. Cardiol. **96:** 160–168.
10. ZIEGELHÖFFER, A., T. RAVINGEROVÁ, J. STYK et al. 1997. Mechanisms that may be involved in calcium tolerance of the diabetic heart. Mol. Cell. Biochem. **176:** 191–198.
11. KUSAMA, Y., D.L. HEARSE & M. AVKIRAN. 1992. Diabetes and susceptibility to reperfusion-induced ventricular arrhythmias. J. Mol. Cell. Cardiol. **24:** 411–421.
12. RAVINGEROVÁ, T., R. ŠTETKA, D. PANCZA et al. 2000. Susceptibility to ischemia-induced arrhythmias and the effect of preconditioning in the diabetic rat heart. Physiol. Res. **49:** 607–616.
13. PIERCE, G.N. & N.S. DHALLA. 1984. Heart mitochondrial function in chronic experimental diabetes in rats. Can. J. Cardiol. **1:** 48–54.
14. KUO, H.T., K. MOORE, F. GIOMELLI & J. WIENER. 1983. Defective oxidative metabolism of heart mitochondria from genetically diabetic mice. Diabetes **32:** 181–187.
15. BAKKER, A., I. BERNAERT, M. DE BIE et al. 1994. The effect of calcium on mitochondrial contact sites: a study on isolated rat hearts. Biochim. Biophys. Acta **1224:** 583–588.
16. ZIEGELHÖFFER-MIHALOVICOVÁ, B., F. KOLÁR, W. JACOB et al. 1998. Modulation of mitochondrial contact sites formation in immature rat heart. Gen. Physiol. Biophys. **17:** 385–390.
17. TINDER, P. 1969. Determination of blood glucose using 4-amino phenazon as oxygen acceptor. J. Clin. Pathol. **22:** 246–253.
18. FOSSATI, P. & L. PRENCIPE. 1982. Serum triglycerides determined colorimetrically with an enzyme that produces hydrogen peroxide. Clin. Chem. **28:** 2077–2080.
19. WATSON, D. 1960. A simple method for determination of serum cholesterol. Clin. Chim. Acta **5:** 613–615.
20. BURRIN, J.M., R. WORTH, A.A. ASHWORTH et al. 1980. Automated colorimetric estimation of glycosylated hemoglobin. Clin. Chim. Acta **106:** 45–50.
21. ZIEGELHÖFFER, A., J. STYK, T. RAVINGEROVÁ et al. 1999. Prevention of processes coupled with free radical formation prevents also the development of calcium resistance in the diabetic heart. Life Sci. **65:** 1999–2001.

Impaired Endothelial Function of Thoracic Aorta in Hereditary Hypertriglyceridemic Rats

J. TÖRÖK,[a] P. BABÁL,[b] J. MATUŠKOVÁ,[c] I. LUPTÁK,[c] I. KLIMEŠ,[d] AND F. ŠIMKO[c]

[a]*Institute of Normal and Pathological Physiology, Slovak Academy of Sciences, Bratislava, Slovak Republic*

[b]*Department of Pathology, Comenius University School of Medicine, Bratislava, Slovak Republic*

[c]*Department of Pathophysiology, Comenius University School of Medicine, Bratislava, Slovak Republic*

[d]*Institute of Experimental Endocrinology, Slovak Academy of Sciences, Bratislava, Slovak Republic*

ABSTRACT: *Objective*—Hereditary hypertriglyceridemia (hHTG) in rats was found to be associated with metabolic abnormalities and elevation of blood pressure. There is controversy regarding the relation between hHTG and vascular function. The aim of this study was to determine the reactivity and accompanying structural changes in thoracic aorta from hereditary hypertriglyceridemic rats and hHTG rats that were given, for a long time, N^G-nitro-L-arginine methyl ester (L-NAME) with and without simultaneous captopril treatment. *Methods*—Isolated rings of thoracic aorta were mounted in organ chambers for isometric tension recording or for measurement of endothelium-dependent relaxation. Morphological changes of thoracic aorta (wall thickness, diameter) were measured using light microscopy. *Results*—Endothelium-dependent relaxation (EDR) to acetylcholine (ACh, 10^{-5} M) was significantly attenuated in the hHTG group compared to control Wistar rats (59.3 ± 8.5% vs. 95.8 ± 6.5%, $p < 0.001$), but normalized after pretreatment with captopril. EDR to ACh was further inhibited in hHTG rats treated with L-NAME (36.0 ± 2.3%, $p < 0.001$). Maximum residual relaxation was only partly restored with captopril treatment (72.4 ± 5.8%, $p < 0.001$). Hypertriglyceridemia did not significantly alter the sensitivity of the thoracic aorta to exogenous noradrenaline. The diameter/wall thickness (D/W) ratio in aortas of control Wistar rats averaged 16.25 ± 0.57. This ratio was significantly lower in hHTG rats (12.52 ± 0.38, $p < 0.01$) and was not altered after treatment with captopril. In the hHTG rats treated with L-NAME, the D/W ratio was further significantly decreased (8.25 ± 0.30, $p < 0.001$). Simultaneous captopril treatment attenuated the decrement of this ratio (9.80 ± 0.75, $p < 0.05$). *Conclusions*—Results showed that hHTG is accompanied by functional and morphological alterations in the rat thoracic aorta. These changes in hHTG and in hHTG rats treated with L-NAME could be, at least in part, protected by captopril treatment.

Address for correspondence: Jozef Török, M.D., Ph.D., Institute of Normal and Pathological Physiology, Slovak Academy of Sciences, Sienkiewiczova 1, 813 71 Bratislava, Slovak Republic. Voice: 4212 5292 6336; fax: 4212 5296 8516.

torok@unpf.savba.sk

KEYWORDS: hereditary hypertriglyceridemia (hHTG); rats; thoracic aorta; endothelial function

INTRODUCTION

Hereditary hypertriglyceridemia (hHTG) in rats was found to be associated with several metabolic abnormalities and elevation of blood pressure.[1] Hypertriglyceridemia, an important risk factor for atherosclerosis, interferes with endothelial cell functions.[2,3] However, there is no general agreement on the mechanisms that may be responsible for the inhibition of endothelium-dependent relaxation (vascular dilatation) and associated abnormalities in hHTG rats. Impaired endothelial function has been found within several hours after triglyceride load in both animals and humans.[3,4] On the other hand, Chowienczyk et al.[5] showed normal endothelium-dependent responsiveness to acetylcholine in patients with severe hypertriglyceridemia.

The aim of this study was to investigate the endothelial function and potential accompanying structural changes in thoracic aorta of hereditary hypertriglyceridemic rats and to evaluate the effect of the angiotensin converting enzyme (ACE) inhibitor, captopril. In addition, these parameters were studied also in hHTG rats treated with nitric oxide (NO) synthase inhibitor, N^G-nitro-L-arginine methyl ester (L-NAME), with and without simultaneous administration of captopril.

METHODS

Male 12-week-old inbred hypertriglyceridemic rats and age-matched control Wistar rats were used for experiments. Animals were housed in a controlled temperature (24°C) and lighting (12-h light/12-h dark) chamber with free access to tap water and pelleted food.

Rats were divided into five groups: control Wistar rats, hereditary hypertriglyceridemic (hHTG) rats, hHTG rats treated with captopril (100 mg/kg/day), hHTG rats treated with L-NAME (40 mg/kg/day), and hHTG rats simultaneously treated with L-NAME (40 mg/kg/day) and captopril (100 mg/kg/day). The substances were given in tap water for 4 weeks. Systolic blood pressure was measured weekly by the indirect tail-cuff method.

At the end of week 16, rats were killed by decapitation. The middle part of the thoracic aorta was rapidly removed, cleaned of excess of fat and connective tissue, and cut into rings. The rings were suspended in an organ bath containing modified Krebs solution and connected to a force-displacement transducer, Sanborn F10 (U.S.A.), to measure changes in isometric tension.

Morphological changes of the thoracic aorta (wall thickness, diameter) were measured using light microscopy as previously described.[6]

All values are means ± SEM. Statistical differences were analyzed with one-way ANOVA. Results are considered significant when $p < 0.05$.

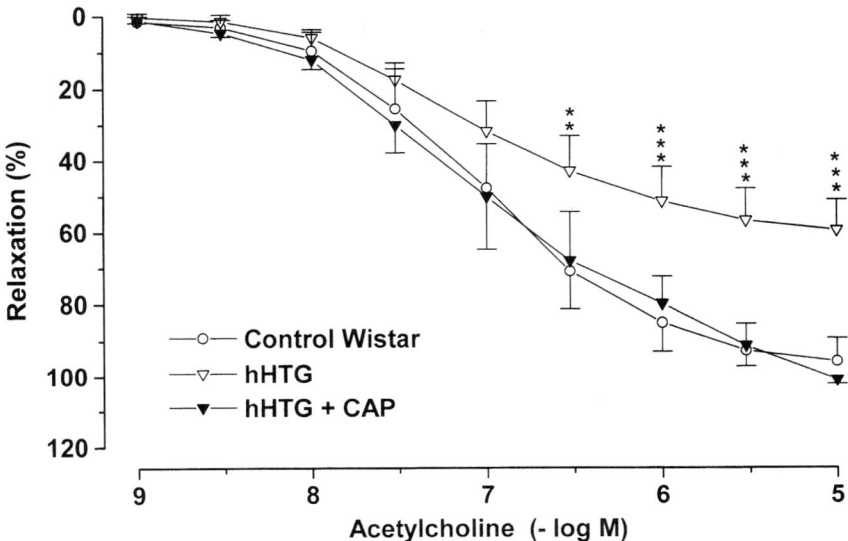

FIGURE 1. Endothelium-dependent relaxation of rat thoracic aorta induced by acetylcholine in control Wistar and hereditary hypertriglyceridemic (hHTG) rats with and without captopril treatment (hHTG + CAP). **$p < 0.01$, ***$p < 0.001$ with respect to the control and hHTG + CAP rats.

RESULTS

Systolic Blood Pressure

Systolic blood pressure in hHTG rats (142 ± 4 mmHg) was higher than in age-matched controls (117 ± 3 mmHg, $p < 0.05$). This was further increased in hHTG + L-NAME–treated rats to 190 ± 6 mmHg ($p < 0.001$). Captopril treatment in hHTG rats normalized systolic blood pressure and reduced it in hHTG + L-NAME–treated rats to 128 ± 7 mmHg ($p < 0.05$).

Vascular Reactivity

In the control Wistar rats, acetylcholine (ACh) at 10^{-5} M totally relaxed the aortic rings precontracted with phenylephrine (PE, 10^{-6} M). In hHTG rats, ACh-induced relaxation was significantly reduced ($59.3 \pm 8.5\%$ vs. $95.8 \pm 6.5\%$, $p < 0.001$). Captopril completely restored this response (FIG. 1). In the aortic rings of hHTG rats treated for 4 weeks with L-NAME, the maximum of the ACh-induced relaxation ($36.0 \pm 2.3\%$) was more reduced as compared to both the hHTG and control Wistar rats (FIG. 2). Simultaneous treatment with captopril significantly restored the ACh-induced response, but the maximum response remained reduced ($72.4 \pm 5.8\%$) compared with the aortic rings from control Wistar rats ($95.8 \pm 6.5\%$, $p < 0.001$).

Noradrenaline (NA) produced concentration-dependent contractions of the aorta that were comparable between the control Wistar, hHTG, and hHTG + L-NAME–

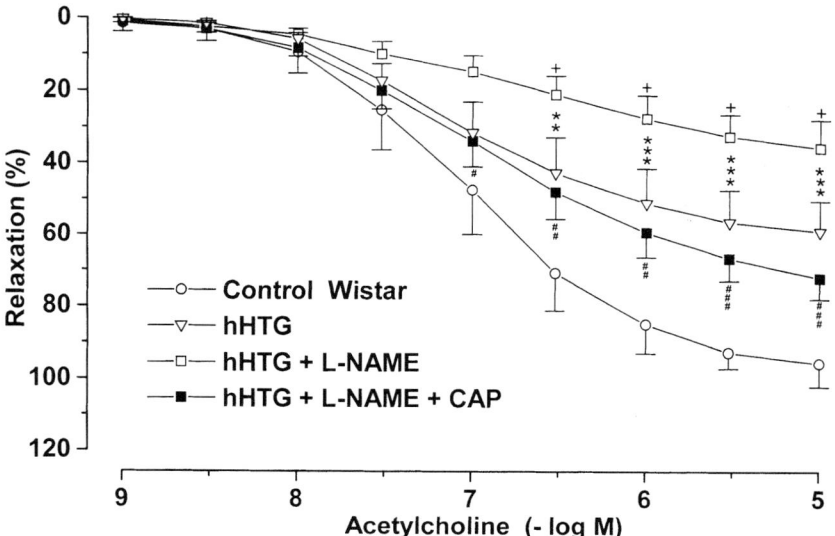

FIGURE 2. Acetylcholine-induced relaxation of thoracic aorta in control Wistar rats, in hereditary hypertriglyceridemic (hHTG) rats, and hHTG + L-NAME– and hHTG + L-NAME + captopril (CAP)–treated rats. $**p < 0.01$, $***p < 0.001$ with respect to the control rats. $^+p < 0.05$ with respect to the hHTG rats. $^\#p < 0.05$, $^{\#\#}p < 0.01$, $^{\#\#\#}p < 0.001$ with respect to the hHTG + L-NAME + CAP rats.

treated rats (data not shown). The concentration of NA producing 50% of the maximal effect (EC_{50}) was 7.82 ± 0.13 in control Wistar rats and was not significantly different in the aortas from other investigated groups. Captopril treatment in hHTG and hHTG + L-NAME–treated rats was without significant effect.

Morphometric Studies

The thickness of the aorta in control Wistar rats was 0.975 ± 0.014 mm, whereas in hHTG rats it increased to 1.120 ± 0.039 mm (FIG. 3, left) and in hHTG + L-NAME rats it reached a maximum value of 1.591 ± 0.060 mm (FIG. 4, left).

Changes in the structure of the aorta are better expressed by calculating the ratio of diameter to wall thickness (D/W). The D/W ratio in aortas from control rats averaged 16.25 ± 0.57. It was significantly lower in hHTG rats (12.52 ± 0.38, $p < 0.01$) and was not significantly altered after treatment with captopril (FIG. 3, right).

In the hHTG rats treated with L-NAME, the D/W ratio was further significantly decreased (8.25 ± 0.30, $p < 0.001$). Simultaneous captopril treatment attenuated the decrement of this ratio (9.80 ± 0.75, $p < 0.05$; FIG. 4, right).

DISCUSSION

In the present study, we demonstrated that hHTG in rats was associated with impaired endothelium-dependent relaxation induced by ACh in the thoracic aorta, which was accompanied by marked changes in vascular architecture.

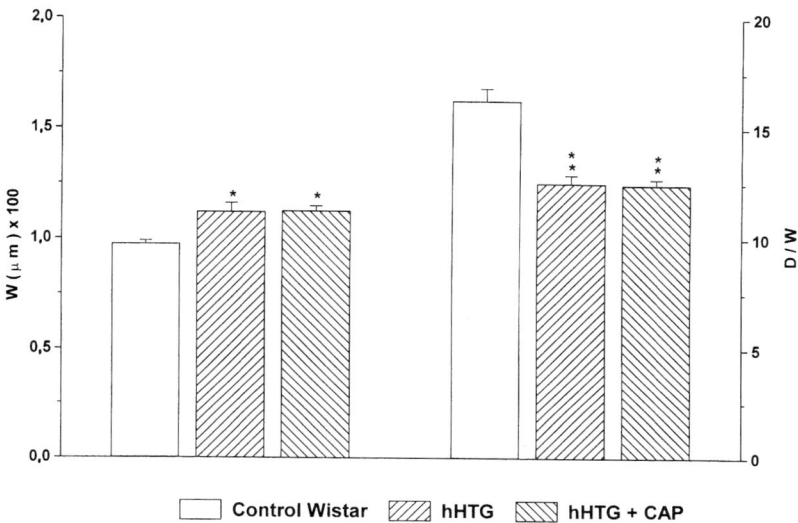

FIGURE 3. Changes in thoracic aorta wall thickness (W) and diameter/wall thickness ratio (D/W) in control Wistar and hHTG rats with and without captopril treatment (hHTG + CAP). $*p < 0.05$, $**p < 0.01$ with respect to the control rats.

The hHTG rats are characterized by hyperinsulinemia, insulin resistance, hypertriglyceridemia, and hypertension.[1] Each of these abnormalities has individually been shown to be associated with impaired endothelial function, as demonstrated by decreased endothelium-derived relaxation (vasodilation).[7-9]

Previous investigations have demonstrated that metabolic changes in hHTG rats were associated with modified vascular structure and function.[10,11] In this experiment, we have shown that hHTG rats have an impaired endothelium-dependent relaxation, mediated mainly by the NO pathway. ACE inhibition with captopril prevented impairment of aortic vascular relaxation, along with preserving normal systolic blood pressure. It suggests that NO production has improved by captopril treatment. The beneficial effect of captopril on functional changes in the thoracic aorta was probably nonspecific since, in a state of selective NO deficiency without hypertriglyceridemia, the ACE inhibitor effectively improved endothelial function as well.[12] On the other hand, this result is in contradiction with the findings of Bernátová et al.,[13] who showed that the effect of captopril on hypertension and on left ventricular hypertrophy reduction was not associated with improvement of L-NAME-reduced NO synthase activity in the aorta, myocardium, brain, or kidney.[13] At the moment, we are not able to explain this different response of the NO system to ACE inhibition in NO-deficient hypertensive Wistar rats and in hHTG rats.

Elevation of systolic blood pressure itself is probably not the cause of depressed ACh-induced relaxation of the aorta. Recently, we have compared endothelial function in different models of experimental hypertension[9] with the result that increased blood pressure and accompanying structural changes are not primarily responsible for impairment of endothelium-dependent relaxation in hypertension. Also, it seems

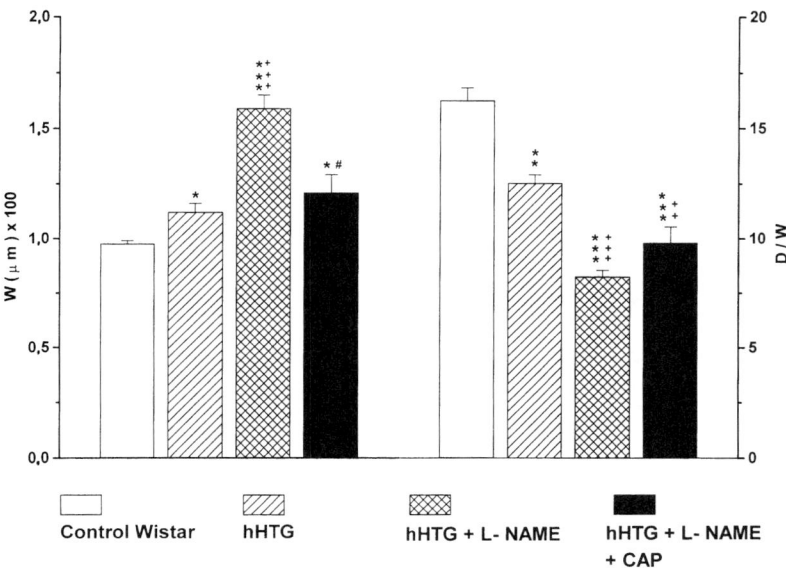

FIGURE 4. Changes in thoracic aorta wall thickness (W) and diameter/wall thickness ratio (D/W) in control Wistar and hHTG rats and in hHTG + L-NAME–treated rats with and without simultaneous administration of captopril (hHTG + L-NAME + CAP). $*p < 0.05$, $**p < 0.01$, $***p < 0.005$ with respect to the control rats. $^{++}p < 0.01$, $^{+++}p < 0.001$ with respect to the hHTG rats. $^{\#}p < 0.05$ with respect to the hHTG + L-NAME rats.

that HTG itself is not responsible for attenuation of endothelium-dependent relaxation since Chowienczyk et al.[5] showed unimpaired endothelium-dependent responsiveness in patients with severe HTG. On the other hand, a specific neurohumoral activation was described in hHTG rats when plasma catecholamine concentration was increased as compared to Wistar rats, and the levels of renin and aldosterone were normal.[14] Sympathoadrenal system activation, even without the activated renin-angiotensin-aldosterone system, may have participated in impairment of endothelial dysfunction in hHTG rats. This idea is supported by a finding in the models of passive cigarette smoking, where deterioration of aortic relaxation seems to be associated with increased plasma catecholamine level, yet with a finding of decreased activity of angiotensin converting enzyme.[15,16]

Ultrastructural changes in the hHTG aortic wall were described recently by Kristek et al.[11] These alterations were predominantly expressed as an accumulation of lipids and products of lipolysis in endothelial cells and in the subendothelial space and less as an accumulation of these substances in smooth muscle. In this paper, we demonstrated a significant thickening of the aortic wall in the hHTG rats. Captopril treatment did not influence these morphological changes. Arterial thickening could be induced by simultaneous action of several factors, including elevation of systolic blood pressure, NO-deficiency, and trophic effect of insulin. The considerable increase in systolic blood pressure in hHTG rats treated with L-NAME suggests that endothelial NO production may be only partly inhibited in hHTG rats and can be further depressed by artificial inhibition of NO synthase activity.

It is concluded that hHTG is accompanied by functional and morphological changes in the rat thoracic aorta. Systemic inhibition of NO synthase activity by L-NAME enhanced the observed changes in hHTG rats. Captopril administration had a beneficial effect on impaired endothelial function of the aorta in hHTG rats, while thickness of the aorta was only slightly reduced.

ACKNOWLEDGMENTS

This work was supported by VEGA Grant Nos. 1/7529/20 and 2/7240/21.

REFERENCES

1. KLIMEŠ, I., J. ZÍCHA, J. KUNEŠ & E. ŠEBÖKOVÁ. 1997. Hypertriglyceridemia, insulin resistance, and hypertension in rats: are they related? Endocr. Regul. **31:** 103–119.
2. DE MAN, F.H., A.W. WEVERLING-RIJNSBURGER, A. VAN DER LAARSE et al. 2000. No acute but chronic hypertriglyceridemia is associated with impaired endothelium-dependent vasodilation: reversal after lipid-lowering therapy by atorvastatin. Arterioscler. Thromb. Vasc. Biol. **20:** 744–750.
3. BAE, J.H., E. BASSENGE, K.B. KIM et al. 2001. Postprandial hypertriglyceridemia impairs endothelial function by enhanced oxidant stress. Atherosclerosis **155:** 517–523.
4. GUDMUNDSSON, G.S., C.A. SINKEY, C.A. CHENARD et al. 2000. Resistance vessel endothelial function in healthy humans during transient postprandial hypertriglyceridemia. Am. J. Cardiol. **85:** 381–385.
5. CHOWIENCZYK, P.J., G.F. WATS, A.S. WIERZBICKI et al. 1997. Preserved endothelial function in patients with severe hypertriglyceridemia and low functional lipoprotein lipase activity. J. Am. Coll. Cardiol. **29:** 964–968.
6. BABÁL, P., O. PECHÁNOVÁ, I. BERNÁTOVÁ & S. ŠTVRTINA. 1997. Chronic inhibition of NO synthesis produces myocardial fibrosis and arterial media hyperplasia. Histol. Histopathol. **12:** 623–629.
7. MILLER, A.W., M.E. HOENIG & M.R. UJHELYI. 1998. Mechanisms of impaired endothelial function associated with insulin resistance. J. Cardiovasc. Pharmacol. Ther. **3:** 125–134.
8. KUSTERER, K., T. POHL, H.P. FORTMEYER et al. 1999. Chronic selective hypertriglyceridemia impairs endothelium-dependent vasodilatation in rats. Cardiovasc. Res. **42:** 783–793.
9. TÖRÖK, J. & F. KRISTEK. 2001. Functional and morphological pattern of vascular responses in two models of experimental hypertension. Exp. Clin. Cardiol. **6:** 142–148.
10. EDELSTEINOVÁ, S., J. KYSELOVIC, I. KLIMEŠ et al. 1993. Effects of marine fish oil on blood pressure and vascular reactivity in the hereditary hypertriglyceridemic rat. Ann. N.Y. Acad. Sci. **683:** 353–356.
11. KRISTEK, F., E. EDELSTEINOVÁ, E. ŠEBÖKOVÁ et al. 1997. Structural changes in the aorta of the hereditary hypertriglyceridemic rat. Ann. N.Y. Acad. Sci. **827:** 514–520.
12. KÜNG, C.F., P. MOREAU, H. TAKASE & T.F. LÜSCHER. 1995. L-NAME hypertension alters endothelial and smooth muscle function in rat aorta. Hypertension **26:** 744–751.
13. BERNÁTOVÁ, I., O. PECHÁNOVÁ & F. ŠIMKO. 1999. Effect of captopril in L-NAME-induced hypertension on the rat myocardium, aorta, brain, and kidney. Exp. Physiol. **84:** 1095–1105.
14. LICHARDUS, B., E. ŠEBÖKOVÁ, D. JEŽOVÁ et al. 1993. Effect of a low salt diet on blood pressure and vasoactive hormones in the hereditary hypertriglyceridemic rat. Ann. N.Y. Acad. Sci. **683:** 289–294.
15. ŠIMKO, F., P. MARTINKA, J. BRASSANOVÁ et al. 2001. Passive cigarette smoking induced changes in reactivity of the aorta in rabbits: effect of captopril. Pharmazie **56:** 431–432.
16. TÖRÖK, J., A. GVOZDJÁKOVÁ et al. 2000. Passive smoking impairs endothelium-dependent relaxation of isolated rabbit arteries. Physiol. Res. **49:** 135–141.

A Novel Mammalian Homologue of a Bacterial Citrate-Metabolizing Enzyme

CHARLOTTE SÖDERBERG[a] AND PETER LIND[b]

[a]*Pharmacology, Biovitrum AB, SE-112 76 Stockholm, Sweden*
[b]*Target Validation, Biovitrum AB, SE-112 76 Stockholm, Sweden*

ABSTRACT: Mammals metabolize citrate to acetyl-CoA and oxaloacetate via the enzyme, ATP:citrate lyase. Bacteria lack this enzyme, but have the ability to cleave citrate in the form of citryl-CoA in an analogous manner using a structurally distinct enzyme. We have identified a novel mammalian gene that shows significant amino acid sequence homology to the bacterial CitE gene product that is responsible for cleavage of citryl-CoA. We propose that this gene encodes an enzyme that catalyzes cleavage of substrates related to CoA esters of citrate or an analogous intermediary metabolite. The product of this novel gene may represent a component of an unknown metabolic pathway in mammals.

KEYWORDS: citrate metabolism; citryl-CoA lyase; malyl-CoA lyase; CitE; mclA

INTRODUCTION

As a source of two carbon building blocks in fatty acid and cholesterol synthesis, mammals metabolize citrate into acetyl-CoA and oxaloacetate in an ATP-dependent reaction catalyzed by ATP:citrate lyase. In most mammals, this enzyme is mainly expressed in lipogenic tissues such as the liver and the adipose tissue.[1] Prokaryotes lack this system, but many species cleave citryl-CoA in an analogous manner during fermentation of citrate using a structurally different ATP-independent enzyme, citryl-CoA lyase, which is encoded by the *CitE* gene. This gene is part of an operon that encodes several proteins involved in transport of metabolism of citrate.[2] We hypothesized that the mammalian genome may encode homologues to metabolic enzymes previously known only in prokaryotes. In the search for candidates, we discovered a novel mammalian gene encoding a protein with similarity to bacterial *CitE* gene products.

METHODS

Transcript profiling was performed on white adipose tissue mRNA derived from normal male C57BL/6J mice (Bomholtsgaard, Denmark) using Mu11K and Mu19K oligonucleotide array GenChips™ (Affymetrix, Santa Clara, CA). 5'- and 3'-Rapid

---

Address for correspondence: Peter Lind, Biovitrum AB, Strandbergsgatan 49, SE-112 76 Stockholm, Sweden. Voice: +46-8-697 2912; fax: +46-8-618 5882.
peter.lind@biovitrum.com

Ann. N.Y. Acad. Sci. 967: 476–481 (2002). © 2002 New York Academy of Sciences.

Amplification of cDNA Ends (SMART™ RACE; Clontech, Palo Alto, CA) was used to derive partial cDNA clones from mouse and rat brain mRNA that were assembled into complete cDNAs. These sequences were used as queries in BLAST searches of expressed sequence tag (EST) data banks to search for human orthologues, and complete cDNA was obtained in a manner similar to the rodent genes. Tissue expression profiling of the mammalian genes was performed by mRNA blotting using multiple tissue Northern blots (Clontech, Palo Alto, CA).

RESULTS

Transcript profiling of mouse adipose tissue using Affymetrix oligonucleotide arrays was used to search for novel genes represented by ESTs.[3] We found expression of a gene (TIGR Mouse Gene Index TC20248) that showed significant homology with various prokaryotic *CitE* gene products when used as a query in BLAST searches on protein databases.[4] The full cDNA corresponding to the above ESTs was obtained by assembling cDNA fragments derived from mouse brain mRNA using 5′- and 3′-RACE.[5] Additional searches on public and proprietary EST databases revealed a number of sequences encoding the human and rat orthologues of the mouse gene and allowed molecular cloning of human and rat cDNAs using RACE methodology in an analogous manner. The human cDNA encodes a protein of 340 amino acids that shows strong similarity with the mouse and rat orthologues. BLAST searches using the human cDNA sequence against the human genomic sequence database revealed that the gene is located on chromosome 13 in the area of 13q33.1–13q33.2, and the coding region is divided into 8 exons spanning more than 100 kb (data not shown). Database searches and ClustalW[6] sequence alignments indicate that the human gene product shows 42% homology with the *C. elegans* hypothetical protein C01G10.7, but no obvious orthologues seem to be present in insects (*D. melanogaster*), plants (*A. thaliana*), or yeast (FIG. 1). In addition to its similarity to bacterial citryl-CoA lyase (CitE protein), the mammalian gene also shows significant homology with malyl-CoA lyase (mclA protein) from *Methylobacterium extorquens*, an enzyme that catalyzes cleavage of malyl-CoA to acetyl-CoA and glyoxylate in a chemical reaction similar to that of citryl-CoA lyase.[7] In contrast to the prokaryotic gene products, the mammalian and worm CitE-like proteins appear to possess a mitochondrial localization sequence on their N-terminus as revealed by analysis using PSORT.[8] mRNA blotting of various tissues revealed that the human gene encoding the CitE-like protein is expressed as a 1.5-kb transcript predominantly in liver and kidney, but also in heart and skeletal muscle, and to a lesser extent in brain, spleen, lung, and testis (FIG. 2). The mouse and rat gene products have a similar expression profile, with the exception that the liver mRNA in rats appears to be significantly less abundant than in humans and mice.

FIGURE 1. ClustalW alignment of the amino acid sequences of mammalian and worm (*C. elegans*; GenBank CAB02709) CitE-like proteins with bacterial (*E. coli*; GenBank BAA35252) and archeal (*Halobacterium* sp.; GenBank AAG19130) citryl-CoA lyases and methylobacterium (*M. extorquens*; GenBank AAB58884) malyl-CoA lyase. Amino acid residues conserved across all species are shown in bold with a gray background. The predicted mitochondrial targeting sequences in the mammalian and worm proteins are shown in italics.

Mouse CitE homol.	MALCVLRNTVRG--AAALPRLKASHVVSVYKPRYSSLSNHKYVPRRAVLY
Rat CitE homol.	MALCVLQNAVLG--AAALPRLKASLVASVCRPGYSSLSNHKYVPRRAVLY
Human CitE homol.	MALRLLRRAARGAAAAALLRLKASLAADIPRLGYSSSSHHKYIPRRAVLY
Worm CitE homol.	----------------MLPKTIIRHISQFAGVRDAA----KYVPRRALLY
Halobacterium CitE	----------------MPGRWTPPVKVAPETYKMGAAVVVHMTRRSVLF
E. coli CitE	------------------MISASLQQRKTR-------TRRSMLF
Methylobacterium mclA	------------------------MSFTLIQQAT-----PRLHRSELA
Mouse CitE homol.	VPGNDEKKIRKIPSLKVDCAVLDCEDGVAENKKNEARLRIAKTLEDFD--
Rat CitE homol.	VPGNDEKKIRKIPSLKVDCAVLDCEDGVAENKKNEARLRIAKTLEDFD--
Human CitE homol.	VPGNDEKKIKKIPSLNVDCAVLDCEDGVAANKKNEARLRIVKTLEDID--
Worm CitE homol.	VPASNQKMLDKVPMMQADSVVLELEDGVALTAKADARVRAAAALDKLPYH
Halobacterium CitE	SPGDQPSLMRKAPTTGADVVVFDLEDAVAPARKGDARRAVSDLLTAAD--
E. coli CitE	VPGANAAMVSNSFIYPADALMFDLEDSVALREKDTARRMVYHALQHPL--
Methylobacterium mclA	VPGSNPTFMEKSAASKADVIFLDLEDAVAPDDKEQARKNIIQALNDLD--
Mouse CitE homol.	-LGTTEKCVRINSVSSGLAEVDLETFLQARVLPSSLMLPKVEGPEEIRWF
Rat CitE homol.	-LGTTEKCVRINSVSSGLAEADLETFLQARVLPSSLMLPKVEGPEEIQWF
Human CitE homol.	-LGPTEKCVRVNSVSSGLAEEDLETLLQSRILPSSLMLPKVESPEEIQWF
Worm CitE homol.	TLACQELGLRVNSVSSGLLEDDIIAVSKAEKLPQAFMIPKVDCPEDLVTI
Halobacterium CitE	FDPDCEVCVRVNPVGSG-AGDDVDAVLSGGGF-DSVVVPKVTDADDVATV
E. coli CitE	-YRDIETIVRVNALDSEWGVNDLEAVVRGG--ADVVRLPKTDTAQDVLDI
Methylobacterium mclA	-WGNKTMMIRINGLDTHYMYRDVVDIVEACPRLDMILIPKVGVPADVYAI
Mouse CitE homol.	SDKFSLHLKG-RKLEQPMNLIPFVETAMGLLNFKAVCEETLKTGPQVGLC
Rat CitE homol.	SDKFSLHLKG-RKLEQPMNLIPFVETAMGLLNFKAVCEETLKIGPQVGLF
Human CitE homol.	ADKFSFHLKG-RKLEQPMNLIPFVETAMGLLNFKAVCEETLKVGPQVGLF
Worm CitE homol.	YNIFREHYGDERITNTNTRLVIWIESARALLDMPRIVSSTLNLHKQAGFF
Halobacterium CitE	DSLLAEHDAT-------SPVLALVESAAGVLNAHEIAAAG-P--------
E. coli CitE	EKEILRIEKACGREPGSTGLLAAIESPLGITRAVEIAHASER--------
Methylobacterium mclA	DVLTTQIEQAKKREKK-IGFEVLIETALGMANVEAIATSSKR--------
Mouse CitE homol.	-LDAVVFGGEDFRASIGATSNKDTQD---------------------I
Rat CitE homol.	-LDAVVFGGEDFRASIGATSNKDTQD---------------------I
Human CitE homol.	-LDAVVFGGEDFRASIGATSSKETLD---------------------I
Worm CitE homol.	KLDAVVFGSDDFCADIGATRSSHGTE---------------------T
Halobacterium CitE	-TDALVFGAEDLSADIGATRTDEGTE---------------------V
E. coli CitE	-LIGIALGAEDYVRNLRTERSPEGTE---------------------L
Methylobacterium mclA	-LEAMSFGVADYAASTRARSTVIGGVNADYSVLTDKDEAGNRQTHWQDPW
Mouse CitE homol.	LYARQKVVVTAKAFGLQAIDLVYIFRDEDGLLRQSREAAAMGFTGKQVI
Rat CitE homol.	LYARQKIVVTAKAFGLQAIDLVYIFRDEDGLLRQSREAAAMGFTGKQVI
Human CitE homol.	LYARQKVVIAKAFGLQAVDLVYIFRDGAGLLRQSREAAAMGFTGKQVI
Worm CitE homol.	LFARQKFVTCCKAFQLQAIDSVYIDIKDLDGLRRQSAEGWQWGFTGKQVI
Halobacterium CitE	LYARERVVLAASAAGVDAIDTVHTAIEDPSAVAADARFAADLGFDGKLAI
E. coli CitE	LFARCSILQAARSAGIQAFDTVYSDANNEAGFLQEAAHIKQLGFDGKSLI
Methylobacterium mclA	LFAQNRMLVACRAYGLRPIDGPFGDFSDPDGYTSAARRCAALGFEGKWAI
Mouse CitE homol.	HPNQIAVVQEQFTPTPEKIQWAEELIAAFKEHQQLGKGAFTFRGSMIDMP
Rat CitE homol.	HPNQIAVVQEQFTPTPEKIRWAEELIAAFKEHQQLGKGAFTFQGSMIDMP
Human CitE homol.	HPNQIAVVQEQFSPSPEKIKWAEELIAAFKEHQQLGKGAFTFQGSMIDMP
Worm CitE homol.	HPSQVSVVQEQFLPPKDRIEWAQELVHAYSEHEALGKGAFQFRGQMIDRP
Halobacterium CitE	HPAQVDPINDAYTPTEADTAWATRVLDAAADS---DAGVFRVDGEMIDAP
E. coli CitE	NPRQIDLLHNLYAPTQKEVDHARRVVEAAEAAAREGLGVVSLNGKMVDGP
Methylobacterium mclA	HPSQIDLANEVFTPSEAEVTKARRILEAMEEAAKAGRGAVSLDGRLIDIA
Mouse CitE homol.	LLKQAQNIVTLATS--IKEK
Rat CitE homol.	LLKQAQNIVMLATS--IKEK
Human CitE homol.	LLKQAQNTVTLATS--IKEK
Worm CitE homol.	LLLQALNIIQLVER--VQN-
Halobacterium CitE	LIAQAERIMRRARA--ADN-
E. coli CitE	VIDRARLVLSRAELSGIREE
Methylobacterium mclA	SIRMAEALIQKADA--MGGK

FIGURE 1. *See previous page for legend.*

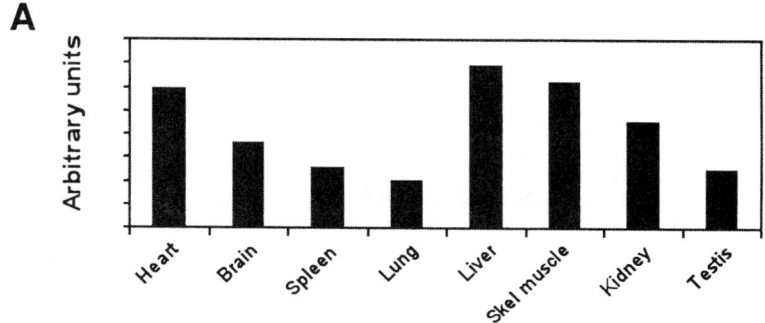
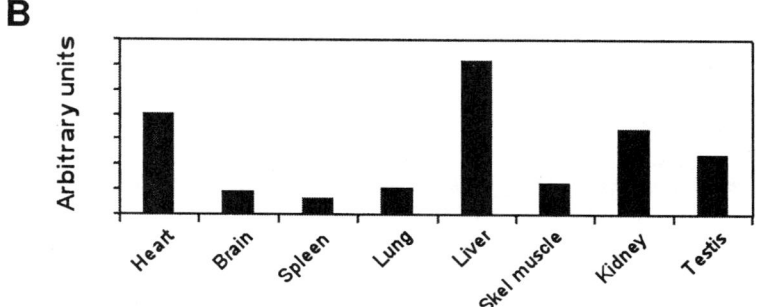
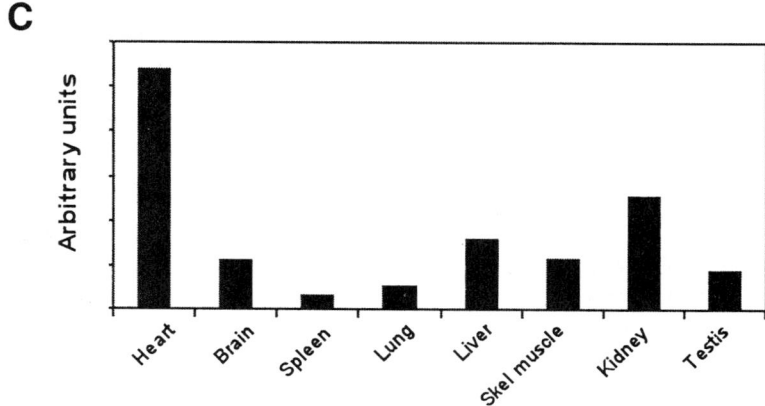

FIGURE 2. Multiple tissue mRNA blotting showing the expression pattern of the human (**A**), mouse (**B**), and rat (**C**) CitE-like protein transcripts.

DISCUSSION

The mammalian CitE-like protein represents a eukaryotic gene product, the function of which has been tentatively assigned based on homology with a bacterial protein with a known function. Another example of a mammalian gene that was originally defined by homology to a known bacterial counterpart is the methylmalonyl-CoA racemase gene, which was identified by searching for human homologues of proteins involved in bacterial propionyl-CoA metabolism.[9] The enterobacterial *CitCDEFG* gene cluster responsible for citrate metabolism encodes additional proteins apart from citryl-CoA lyase,[2] but these do not appear to have any homologous counterparts encoded in the mammalian genome (data not shown). The biochemical pathways in which the mammalian CitE-like protein might participate, or its physiological functions based on the structural homology to the prokaryotic citryl-CoA and malyl-CoA lyases, are not obvious. Based on the chemical similarity of the substrates and reaction products of the respective bacterial enzyme classes, one might hypothesize that the homologous mammalian protein is an enzyme that catalyzes a similar reaction.

From an evolutionary perspective, the acyl-CoA lyase class of enzymes represented by the bacterial CitE protein appears to be absent in plants and fungi, but is conserved in animals. There appear to exist a significant number of genes in bacteria and other lower organisms such as yeast, with various degrees of functional validation, that have corresponding unannotated homologues in the genome of higher eukaryotes.[10] Analysis of finished genomes of eukaryotic organisms, including the recently finished human genomic sequence, indicates that a significant number of genes appear to be laterally transferred between prokaryotes and eukaryotes.[11] It is a clear possibility that the mammalian CitE-like gene has been laterally transferred at an early stage of the metazoan evolution since it seems to occur in nematodes, although it appears to be absent in flies (*D. melanogaster*). In contrast to the bacterial CitE proteins, the nematode and mammalian CitE-like proteins appear to have a mitochondrial targeting sequence at the amino-terminus. This could be taken as an indication of functionality and as a feature that has been adopted during evolution to ensure proper function in a complex cellular environment. Considering that mammalian tissues predominantly expressing the CitE-like protein are known to be particularly active in metabolizing carbohydrate and lipid fuels,[12–14] this putative enzyme may possibly participate in pathways involving these substrates.

The mammalian homologue to bacterial citryl-CoA lyase could represent an enzymatic component of a novel mitochondrial pathway involved in metabolism of citrate or other carboxylic acid intermediates structurally reminiscent of citrate. Future work is necessary to validate its enzymatic activity and to determine its true function in animals, including humans.

[NOTE ADDED IN PROOF—After the submission of this manuscript, the human and mouse CitE-like proteins have been described: MORIKAWA, J., Y. NISHIMURA, A. UCHIDA *et al.* 2001. Molecular cloning of novel mouse and human putative citrate lyase beta-subunit. *Biochem. Biophys. Res. Commun.* **289**(5): 1282–1286.]

ACKNOWLEDGMENTS

We are grateful to Elisabeth Torstensson for excellent technical help throughout this work and to Cécile Martijn for performing the array hybridizations.

REFERENCES

1. ELSHOURBAGY, N.A., J.C. NEAR, P.J. KMETZ et al. 1990. Rat ATP citrate-lyase: molecular cloning and sequence analysis of a full-length cDNA and mRNA abundance as a function of diet, organ, and age. J. Biol. Chem. **265**(3): 1430–1435.
2. BOTT, M. 1997. Anaerobic citrate metabolism and its regulation in enterobacteria. Arch. Microbiol. **167**(2/3): 78–88.
3. LOCKHART, D.J., H. DONG, M.C. BYRNE et al. 1996. Expression monitoring by hybridization to high-density oligonucleotide arrays. Nat. Biotechnol. **14**(13): 1675–1680.
4. ALTSCHUL, S.F., T.L. MADDEN, A.A. SCHAFFER et al. 1997. Gapped BLAST and PSI-BLAST: a new generation of protein database search programs. Nucleic Acids Res. **25**(17): 3389–3402.
5. ROUX, K.H. 1995. Optimization and troubleshooting in PCR. PCR Methods Appl. **4**: 5185–5194.
6. THOMPSON, J.D., D.G. HIGGINS & T.J. GIBSON. 1994. CLUSTAL W: improving the sensitivity of progressive multiple sequence alignment through sequence weighting, position-specific gap penalties, and weight matrix choice. Nucleic Acids Res. **22**(22): 4673–4680.
7. ARPS, P.J., G.F. FULTON, E.C. MINNICH et al. 1993. Genetics of serine pathway enzymes in *Methylobacterium extorquens* AM1: phosphoenolpyruvate carboxylase and malyl coenzyme A lyase. J. Bacteriol. **175**(12): 3776–3783.
8. NAKAI, K. & P. HORTON. 1999. PSORT: a program for detecting sorting signals in proteins and predicting their subcellular localization. Trends Biochem. Sci. **24**(1): 34–36.
9. BOBIK, T.A. & M.E. RASCHE. 2001. Identification of the human methylmalonyl-CoA racemase gene based on the analysis of prokaryotic gene arrangements: implications for decoding the human genome. J. Biol. Chem. **276**(40): 37194–37198.
10. LANDER, E.S., L.M. LINTON, B. BIRREN et al. 2001. Initial sequencing and analysis of the human genome. Nature **409**: 860–921.
11. SALZBERG, S.L., O. WHITE, J. PETERSON & J.A. EISEN. 2001. Microbial genes in the human genome: lateral transfer or gene loss? Science **292**(5523): 1903–1906.
12. MEYER, C., M. STUMVOLL, V. NADKARNI et al. 1998. Abnormal renal and hepatic glucose metabolism in type 2 diabetes mellitus. J. Clin. Invest. **102**(3): 619–624.
13. MEYER, C., V. NADKARNI, M. STUMVOLL et al. 1997. Human kidney free fatty acid and glucose uptake: evidence for a renal glucose–fatty acid cycle. Am. J. Physiol. Endocrinol. Metab. **273**(3, part 1): E650–E654.
14. ZAMMIT, V.A. 1995. Insulin and the partitioning of hepatic fatty acid metabolism. Biochem. Soc. Trans. **23**(3): 506–511.

Effect of BRX-220 against Peripheral Neuropathy and Insulin Resistance in Diabetic Rat Models

MARIA KÜRTHY,[a] TAMÁS MOGYORÓSI,[a] KÁROLY NAGY,[a]
TIBOR KUKORELLI,[b] ANDREA JEDNÁKOVITS,[a] LÁSZLÓ TÁLOSI,[a]
AND KATALIN BÍRÓ[a]

[a]*Biorex Research and Development Company, Veszprém, Hungary*

[b]*Eötvös Lóránd University, Budapest, Hungary*

ABSTRACT: Bimoclomol (BML), a symptomatic antidiabetic agent, has been developed by Biorex R&D Co. to treat diabetic neuropathy and retinopathy. BRX-220, an orally active member of the BRX family, has been developed to treat diabetic complications and insulin resistance (IR) as a follow-up compound. The effect of BRX-220 on peripheral neuropathy was examined in rats with diabetes (type 1) induced by administration of a β-cell toxin, streptozotocin (STZ, 45 mg/kg iv). Nerve functions were evaluated by electrophysiological measurements of muscle motor and sensory nerve conduction velocities (MNCV and SNCV, respectively). MNCV and SNCV decreased in diabetic rats by 25% ($p < 0.001$). A 1-month preventive treatment with BRX-220 (2.5, 5, 10, and 20 mg/kg po) dose-dependently improved diabetes-related deficits in MNCV (51.3%, 71.3%, 86.1%, and 91.3%) and SNCV (48.9%, 68.5%, 86.1%, and 93.2%). Insulin sensitivity was measured using the insulin tolerance test (ITT), both in STZ diabetic and in Zucker diabetic fatty (ZDF) rats (model of type 2 diabetes). Severe IR was detected in STZ diabetic and ZDF rats. This resistance was significantly ($p < 0.05$) reduced by BRX-220 treatment.

KEYWORDS: BRX-220; diabetic neuropathy; STZ diabetes; ZDF rats; insulin action

INTRODUCTION

Insulin resistance (IR) has a main causal role in the development of several diseases,[1–19] for example, dyslipidemia, hypertension, atherosclerosis, and type 2 diabetes.[11] Type 2 diabetes mellitus is a common, underdiagnosed, and largely incurable condition associated with a number of serious complications, for example, neuropathy, retinopathy, and nephropathy. Diabetic neuropathy frequently appears among type 1 and type 2 diabetic patients (30–70% of diabetic patients according to the epidemiological studies). The most common form of diabetic neuropathy, the

---

Address for correspondence: Maria Kürthy, Biorex Research and Development Company, P. O. Box 348, Veszprém-Szabadságpuszta H-8201, Hungary. Voice: +36-88-545-232; fax: +36-88-545-201.

Maria.Kurthy@biorex.hu

distal symmetrical sensorimotor neuropathy, affects the sensory nerve fibers and can result in loss of normal sensations, such as pain and temperature. There is great interest in developing new agents capable of reducing both IR and complications of diabetes. BRX-220, a hydroxylamine derivative and a member of the BRX family, was developed by Biorex as a follow-up compound of bimoclomol (BML)[2,3] against diabetic complications and IR. BRX-220 proved to be effective in diabetic small-fiber sensory neuropathy and neuroregeneration,[4] as well as in high-fat diet–induced IR of rats[12] after long-term oral treatment. The present study examines the beneficial effects of BRX-220 on diabetic peripheral sensory and motor neuropathy and on IR in different animal models of diabetes in comparison with metformin (Met),[18] aminoguanidine (AG),[5-7] and pioglitazone (PG).[1,7]

MATERIALS AND METHODS

Peripheral Neuropathy Study

Induction of Diabetes and Treatment

Overnight fasted male (Charles River Wistar) rats (250–330 g) were treated with 45 mg/kg iv STZ, dissolved freshly in physiological saline solution. Animals were considered diabetic when their nonfasted blood glucose level was >15 mmol/L at 24 h after induction of diabetes. Treatment of diabetic rats with BRX-220 (doses: 2.5, 5, 10, and 20 mg/kg/day po) for 1 month started after verification of hyperglycemia.

Measurement of MNCV and SNCV

Nerve conduction velocities (NCVs) were measured using the electrophysiological method of Stanley,[15] modified by De Koning and Gispen.[8] Briefly, under urethane (400 mg/kg) and α-chloralose (80 mg/kg) ip anesthesia, the left sciatic and tibial nerves were electrically stimulated (square-wave impulses of 0.03 ms) at the sciatic notch or ankle, respectively. Electromyograms (EMGs) were recorded from the plantar muscles. Each EMG consists of two components: (1) the short-latency direct motor response (M) is due to stimulation of the A-α motor fibers; (2) the monosynaptically elicited long-latency sensory response (H, the Hoffmann reflex) is due to the activation of the proprioceptive afferent. M and H latencies were measured, and MNCV and SNCV were calculated from the following formulas: MNCV = distance between the sciatic- and tibial-stimulating points divided by differences of latencies for $M_{sciatic}$ and M_{tibial}; SNCV = distance between the sciatic- and tibial-stimulating points divided by differences of latencies for H_{tibial} and $H_{sciatic}$. The diabetes-related reductions in NCVs are presented as deficits (%). Improvement in NCVs caused by the drug treatments is expressed as a percentage of diabetic deficit.

Statistical Analyses

Results are expressed as means ± SE, with significance set at $p < 0.05$. Student's *t* test was used for comparisons.

Intravenous Insulin Tolerance Test (ITT) in STZ Diabetic Rats

Diabetes was induced as above. Rats were treated po by gavage with BRX-220 (5, 10, and 20 mg/kg/day) or Met (250 mg/kg/day) for 4 weeks. The treatment period started 4 weeks after induction of diabetes. ITT was used to test insulin sensitivity.[10] Short-acting insulin (Actrapid; 0.25 U/kg, iv) was administered to the overnight fasted rats. Blood samples were taken from the tail vein of the conscious rats before and 5, 10, 20, 30, 40, and 60 min after insulin administration. Serum was separated by centrifugation (2500 rpm, 4°C, 20 min), and glucose concentration was measured by the Vitros 250 automatic analyzator system. When plotting plasma glucose concentration regularly from 0 to 60 min after the iv insulin injection on a scale, a reasonable linear decline was observed. By means of the equation of these lines, plasma glucose half-life ($t_{1/2}$) and glucose disappearance rate [$K_g = (0.693 \times 100)/t_{1/2}$] were calculated and used to estimate insulin sensitivity. Serum triglyceride (TG) and cholesterol (Chol) levels were determined in the 0-min samples as well (Vitros 250).

Intravenous ITT in Zucker Diabetic Fatty (ZDF) Rats

Drug Treatment

PG at 6 mg/kg/day and BRX-220 at 20 mg/kg/day were administered po from the 7th until the 12th week of age. ITT was performed as described above. Blood samples were taken before and 6, 10, 14, 18, 22, and 30 min after insulin injection. Glucose concentration was measured as described above. Serum TG and Chol levels were determined in the 0-min samples.

Statistical Analyses

Results are expressed as means ± SEM; $t_{1/2}$ and K_g values were determined in all individual lines. ANOVA followed by the Kruskal-Wallis U test was used for comparison. Differences between groups were considered significant at $p < 0.05$.

RESULTS

Neuropathy Study

The motor and sensory response structures of EMGs were altered, and nerve conduction slowed in diabetic rats. MNCV decreased from 62.7 ± 0.4 to 46.9 ± 0.9 m/s and SNCV from 64.9 ± 0.7 to 48.4 ± 1 m/s (by 25%, $p < 0.001$). BRX-220 dose-dependently improved both the motor and sensory functions and in the highest dose achieved an almost complete normalization (FIG. 1). In another series of experiments, where similar NCV deficits were produced, the effects of BRX-220, BML, and AG were investigated. In a dose of 2.5 mg/kg, BRX-220 was as effective as 50 mg/kg AG; likewise, a dose of 5 mg/kg was as effective as 20 mg/kg BML. Moreover, 20 mg/kg BRX-220 almost completely prevented the development of diabetes-induced NCV deficits (FIG. 1).

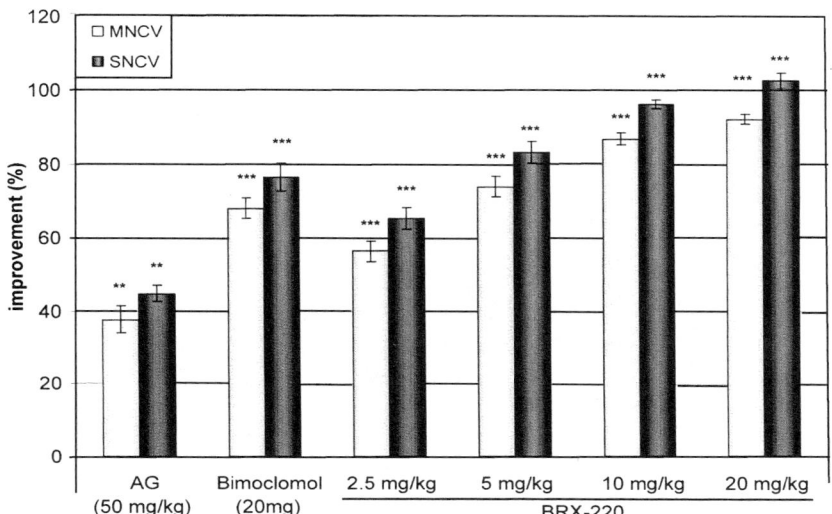

FIGURE 1. Improvement (in %) of motor nerve conduction velocity (MNCV) and sensory nerve conduction velocity (SNCV) deficit by aminoguanidine (AG), bimoclomol, and different doses of BRX-220 treatments in streptozotocin (STZ) diabetic rats. Each bar represents the mean ± SE of the measurements of 8–10 rats; $**p < 0.01$, $***p < 0.001$.

ITT and TG Measurements in STZ Diabetic Rats

In this experiment, Met served as a reference. STZ-induced diabetes caused an eightfold elevation in fasted serum glucose level (control: 6.35 ± 0.11; diabetic: 38.34 ± 1.46 mmol/L; $p < 0.001$). A moderate, but significant ($p < 0.01$) reduction in fasted serum glucose level was observed in rats treated with 20 mg/kg BRX-220 (28.61 ± 2.11 mmol/L; $p = 0.01$). Insulin sensitivity was measured by ITT. The time to evoke a 50% fall in serum glucose levels ($t_{1/2}$) was significantly longer in diabetic rats, while K_g was significantly decreased, indicating that STZ diabetic animals were not only insulin-deficient, but also insulin-resistant (FIGS. 2 and 3). In STZ diabetic rats, insulin produced a moderate, slowly developing serum glucose reduction. In BRX-220-treated animals, the insulin-induced glucose decay was proportional to the administered doses. However, insulin was able to normalize the serum glucose levels in Met-treated animals (FIGS. 2 and 3). A moderate, but statistically significant ($p < 0.01$) serum TG elevation occurred in fasted diabetic rats compared to controls (1.74 ± 0.26 vs. 0.93 ± 0.09 mmol/L; $p \leq 0.01$). Curative treatment with 20 mg/kg BRX-220 significantly decreased the serum TG level (0.94 ± 0.1 mmol/L; $p < 0.025$).

Results of ITT in ZDF Rats

The serum glucose level in ZDF rats was three times higher (14.94 ± 1.16 vs. 5.18 ± 0.1 mmol/L; $p < 0.01$) and the serum TG level was almost tenfold higher (5.24 ± 0.09 vs. 0.65 ± 0.05 mmol/L; $p < 0.01$) than those of the lean littermates. BRX-220 did not reduce the fasted serum glucose and TG levels in the fasted state, but insulin-induced glucose level decay was improved. PG-treated rats were almost

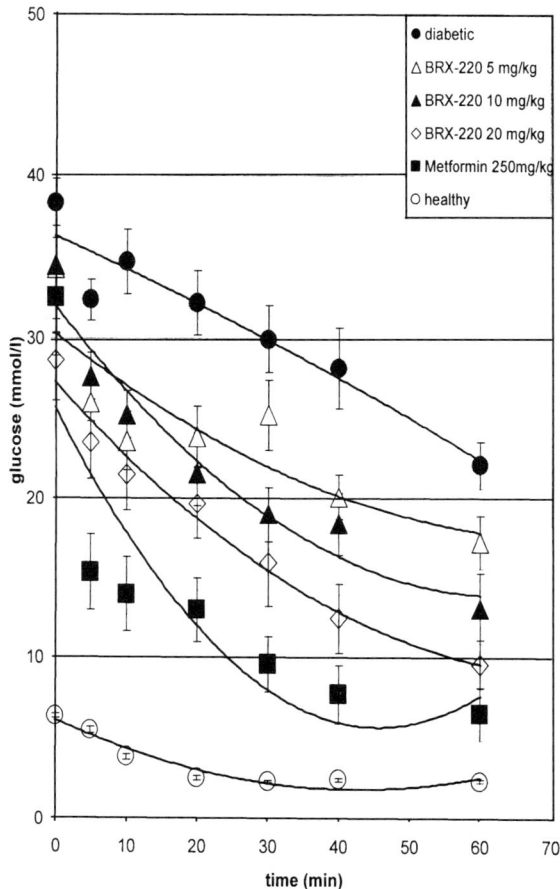

FIGURE 2. Serum glucose levels decay in response to bolus injection of insulin (0.25 U/kg iv) as a function of time in STZ diabetic rats. Each point represents the mean ± SE of the measurements of 8–10 rats.

normoglycemic (FIG. 4) and serum TG level was also reduced (3.76 ± 0.28; $p < 0.001$). The effect of PG was manifested mainly in the reduction of fasted glucose level. Insulin treatment produced only a temporal effect in these rats (FIG. 4). PG increased food intake and weight gain (FIG. 5) compared to BRX-220.

DISCUSSION

BRX-220 is an orally active new agent capable of improving diabetes-induced peripheral neuropathy of motor and sensory fibers in diabetic rats; these effects seem to be stronger than those of other agents (Bimoclomol and AG). In STZ diabetic

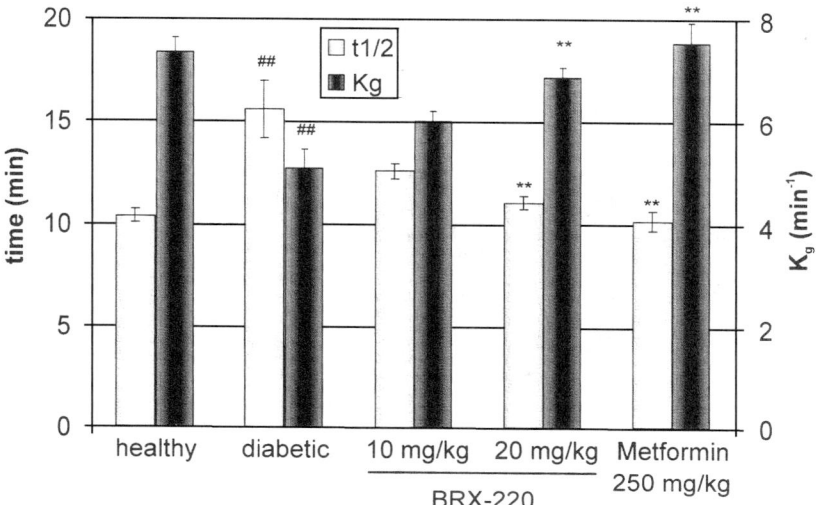

FIGURE 3. Plasma glucose half-life ($t_{1/2}$) and glucose disappearance rate (K_g) in STZ diabetic rats. Results are expressed as mean ± SE ($n = 6–8$ animals/group); ##$p < 0.01$ compared to healthy rats; **$p < 0.01$ compared to diabetic rats.

animals, insulin-stimulated whole-body glucose uptake decreased significantly as early as 1 day after STZ injection, according to the euglycemic, hyperinsulinemic clamp study of Youn et al.,[19] indicating the development of IR in STZ diabetic rats. In STZ diabetic animals, BRX-220 has a dose-dependent insulin-sensitizing capacity, which was comparable with that of metformin. This effect was accompanied by the reduction of serum TG level in the highest applied dose. Long-term oral administration of 20 mg/kg BRX-220 and/or 6 mg/kg pioglitazone led to an improvement of insulin sensitivity in high-fat diet–induced IR of Wistar rats and, at the same time, serum TG level and TG accumulation in liver and skeletal muscle were reduced. However, the improvement of IR was less pronounced in BRX-220-treated rats than in pioglitazone-treated ones in these experiments.[14]

Decreased expression of HSP-72 had been found in skeletal muscle from patients with type 2 diabetes, and the level of HSP-72 mRNA correlates positively with the rate of insulin-stimulated glucose uptake and inversely with the degree of glucose tolerance.[13] BML, a predecessor of this molecule, is a nontoxic heat shock protein (HSP) coinducer.[18] The HSP-coinductive properties of BRX-220 evoked by different stressors have been demonstrated on NIH/3T3 cell lines.[9] Several members of the BRX family induce a simultaneous membrane fluidization and membrane protection.[17] Taken together, BRX-220 improves lipid metabolism, preserves membrane integrity, and promotes HSP induction in stress conditions. Due to these effects, BRX-220 may provide a better nutritional environment to vasa nervorum and other tissues.

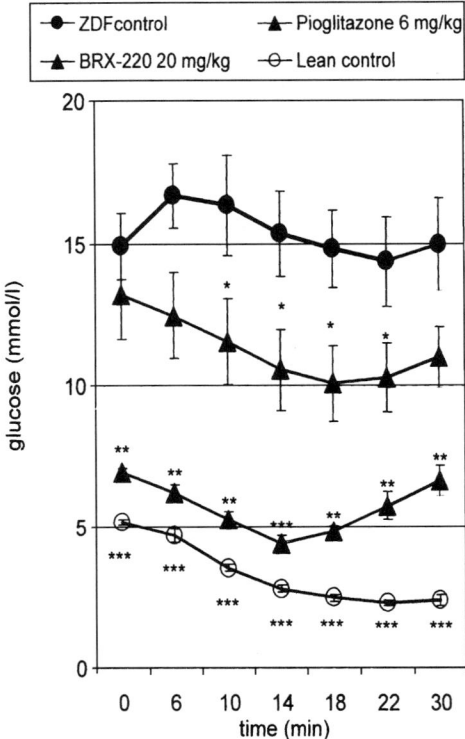

FIGURE 4. Serum glucose levels of Zucker diabetic fatty (ZDF) rats after bolus injection of 0.25 U/kg insulin as a function of time. Results are expressed as mean ± SE (n = 6–8 animals/group); *p < 0.05, **p < 0.01, ***p < 0.001 compared to ZDF controls.

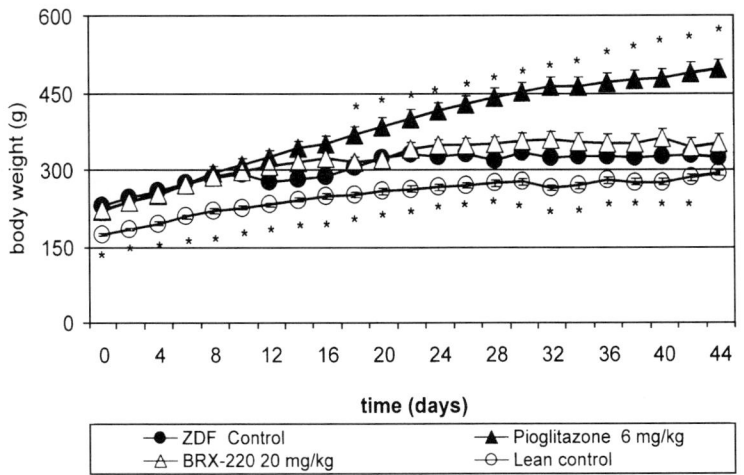

FIGURE 5. Body weights (g) of ZDF rats treated with BRX-220 or pioglitazone. Each point represents the mean ± SE of the measurements of 8–10 rats; *p < 0.05 compared to ZDF controls.

REFERENCES

1. ACTOS. 2002. ACTOS complete prescribing information (http://www.actos.com/pi.htm).
2. BÍRÓ, K., et al. 1997. Bimoclomol (BRLP-42) ameliorates peripheral neuropathy in streptozotocin diabetic rats. Brain Res. Bull. **44:** 259–263.
3. BÍRÓ, K., et al. 1998. Bimoclomol improves early electrophysiological signs of retinopathy in diabetic rats. Neuroreport **9:** 2029–2033.
4. BÍRÓ, K., et al. 2000. Presented at the XVth Conference of the Hungarian Diabetes Society (April 13–16, 2000), Tihany, Hungary.
5. BROWNLEE, M. 1992. Glycation end products and the pathogenesis of diabetic complications. Diabetes Care **15:** 1835–1843.
6. BROWNLEE, M. 2000. Negative consequences of glycation. Metabolism **49:** 9–13.
7. DAY, C. 1999. Thiazolidinediones: a new class of antidiabetic drugs. Diabet. Med. **16:** 179–192.
8. DE KONING, P. & W.H. GISPEN. 1987. Org. 2766 improves functional and electrophysiological aspects of regenerating sciatic nerve in the rats. Peptides **8:** 415–442.
9. HARGITAI, J., et al. 2001. BRX-345-PRE-EB-005 study report. Biorex Research and Development Company. Veszprém, Hungary.
10. HORGAARD, A. & T.E.H. THAYSSEN. 1929. Clinical investigation into the effect of the intravenous injection of insulin. Acta Med. Scand. **72:** 92–95.
11. KAHN, B.B. & J. FLIER. 2000. Obesity and insulin resistance. J. Clin. Invest. **106:** 473–481.
12. KLIMEŠ, I., et al. 2000. Presented at the XVth Conference of the Hungarian Diabetes Society (April 13–16, 2000), Tihany, Hungary.
13. KURUCZ, I., et al. 2002. Decreased expression of heat shock protein 70 in skeletal muscle of twins discordant for type 2 diabetes correlates with reduced insulin stimulated glucose disposal. Diabetes. In press.
14. ŠEBÖKOVÁ, E., et al. 2002. Comparison of the extrapancreatic action of BRX-220 and pioglitazone in the high-fat diet–induced insulin resistance. This volume.
15. STANLEY, E.F. 1981. Sensory and motor nerve conduction velocities and the latency of the H reflex during the growth of the rat. Exp. Neurol. **71:** 497–506.
16. VIGH, L. 1997. Bimoclomol: a non-toxic, hydroxylamine derivative with stress protein–inducing activity and cytoprotective effects. Nat. Med. **3:** 1150–1154.
17. VIGH, L. 2002. Personal communication.
18. WIERNSPERGER, N.F., et al. 1999. The antihyperglycaemic effect of metformin: therapeutic and cellular mechanisms. Drugs **58:** 31–39.
19. YOUN, J.H., et al. 1994. Time courses of changes in hepatic and skeletal muscle insulin action and GLUT4 protein in skeletal muscle after STZ injection. Diabetes **43:** 564–571.

Terguride Treatment Attenuated Prolactin Release and Enhanced Insulin Receptor Affinity and GLUT 4 Content in Obese Spontaneously Hypertensive Female, but Not Male Rats

S. ZORAD,[a] V. GOLDA,[b] M. FICKOVA,[a] L. MACHO,[a] L. PINTEROVA,[a] AND J. JURCOVICOVA[a]

[a]*Institute of Experimental Endocrinology, Slovak Academy of Sciences, Bratislava, Slovak Republic*

[b]*Institute of Experimental Neurosurgery, Hradec Kralove, Czech Republic*

ABSTRACT: Glucose tolerance, serum insulin, insulin receptors in epididymal fat tissue, and GLUT 4 content in muscle, as well as serum prolactin, were studied in obese and lean spontaneously hypertensive rats (SHRs) of both sexes. Obese animals displayed insulin resistance and decreased capacity of high-affinity binding sites of insulin receptors in fat tissue plasma membranes. GLUT 4 content in *musculus* quadriceps was diminished only in obese females. Terguride treatment lowered prolactin serum levels, which was concomitant with ameliorated insulin sensitivity in obese animals of both sexes. Similarly, only in obese females, terguride significantly increased the affinity of high-affinity insulin-binding sites and normalized GLUT 4 content. Our results document downregulation of insulin receptors and GLUT 4 in obesity and suggest a role for prolactin in obesity-induced insulin resistance, particularly in female rats.

KEYWORDS: spontaneously hypertensive rats; glucose tolerance; serum insulin; terguride

INTRODUCTION

The obese spontaneously hypertensive rat (SHR), a rat strain originally developed by Koletsky,[1] represents an animal model of obesity and hyperinsulinemia with a genetically hypertensive background. As we reported previously,[2,3] both obese and lean Koletsky SHRs display impaired glucose tolerance. The glucose intolerance is accompanied by decreased insulin binding and internalization in hepatocytes and erythrocytes when compared to normotensive Wistar rats.[4] On the other hand, hyperinsulinemia is present only in obese SHRs. In addition, the obese rats show markedly

Address for correspondence: Stefan Zorad, Ph.D., Metabolic Regulation Research Group, Institute of Experimental Endocrinology, Slovak Academy of Sciences, Vlarska 3, 833 06 Bratislava, Slovak Republic. Voice: 421-7-373-800; fax: 421-7-374-243.
ueenstef@savba.savba.sk.

elevated levels of plasma triglycerides.[3] Since adipose tissue is the main source of plasma free fatty acids for synthesis of triglycerides in the liver, it was of interest to study the state of insulin receptors in fat tissue. Insulin receptors could reflect the sensitivity of adipose tissue to insulin, which is an important negative regulator of free fatty acid production. In spite of recently reported data on reduced insulin receptor signaling in muscle and liver of obese SHRs,[5] the data describing insulin receptors in adipose tissue of these rats were lacking until now.

Since hyperprolactinemia in Koletsky SHRs has been observed,[6] the further aim of our study was to examine the effect of prolactin inhibition by a partial dopaminergic agonist, terguride, on serum insulin and prolactin levels, glucose tolerance, adipose tissue insulin receptors, and GLUT 4 content in muscle.

MATERIALS AND METHODS

Animals

Obese and lean SHRs of the NIH-derived substrain of Koletsky rats[7] were bred in the animal facility at the Institute of Experimental Neurosurgery, Hradec Kralove.[2] After weaning at the age of 30 days, the animals were kept in groups of four, having free access to a standard pelleted laboratory diet ST-1 (Velaz, Prague, Czech Republic) and tap water. The animals were used for terguride treatment at the age of 3 months. During the experiments, the animals were kept in groups of two.

Terguride Treatment

Terguride (*trans*-dihydrolisuride) maleate (Galena, Opava, Czech Republic) was applied ip in two daily doses (each 0.1 mg/kg) at 7:00 AM and 2:00 PM for 21 days for all determinations, with the exception of glucose tolerance. Glucose tolerance was determined in blood sampled from retrobulbar plexus after 11 days of terguride treatment.[3] All other determinations were performed after the termination of the experiment, that is, 21 days.

Serum Insulin and Prolactin

Three-month-old animals were sacrificed by decapitation after 14 h of starvation. Insulin was determined by a human radioimmunoassay kit (Cis Bio International, Solupharm) using rat insulin (Novo Nordisk) as standard. Prolactin was measured by radioimmunoassay using the double-antibody technique. ^{125}I-rPRL was purchased from NEN™ Life Science Product (CA). Reference preparation and specific antibodies were kindly provided by NHPP Ogden BioService (MD). The incubation was performed at room temperature for 24 h, followed by an overnight precipitation at 4°C. Results are expressed in terms of the rPRL-RP-3 standard. The mean error of the assay was 8.6%.

Glucose Tolerance

Blood was quickly sampled (within 30 s) to heparinized capillaries from the retrobulbar plexus under light ether anesthesia before glucose loading (basal glycemia), as well as 30, 60, 120, and 180 min after the loading. Glucose (3 g/kg BW, 30% solution) was applied intragastrically after 14 h of starvation. Our prior experiments showed the same blood glucose levels during glucose tolerance estimation when comparing blood sampling from the retrobulbar plexus under short-lasting anesthesia with sampling from tail or decapitation without anesthesia. Glycemia was estimated by a kit based on the oxochrome method (BIO-LA-TEST, Oxochrom, Lachema, Brno, Czech Republic). Glucose tolerance was expressed as the area under the glycemic curve (AUC).

Fat Tissue Plasma Membrane Preparation

Epididymal or perigonadal fat tissue and *musculus* quadriceps were dissected immediately after decapitation and weighed and stored in liquid nitrogen until use. Frozen tissue was homogenized in 10 volumes of 10 mM Tris-HCl, 250 mM sucrose, 1 mM phenylmethylsulfonyl fluoride (Sigma Chemicals), and 1 mM benzamidine (Sigma Chemicals) buffer, pH 7.4, using a Polytron homogenizer. All steps following homogenization were carried out at +4°C. The homogenate was centrifuged at 2000g for 15 min to sediment nuclei and cellular debris. The fat cake was removed from the top of the supernatant, and the supernatant was further centrifuged at 16,000g for 15 min. The resulting pellet designed as a crude plasma membrane preparation was resuspended in 50 mM Tris-HCl, pH 7.6, and used immediately for insulin binding and immunoblot.

Insulin Binding

The binding experiment was performed in Eppendorf tubes containing 50 µg of plasma membrane proteins, 100 µL of mono-^{125}I-(TyrA14)-insulin (0.2 nM final concentration), and 50 µL of native monocomponent porcine insulin (NOVO) in increasing concentrations (1 pM to 1 µM). This mixture was completed with 300 µL of 0.1 mM Tris-HCl (pH 7.6) binding buffer with 2 mM N-ethylmaleimide (Sigma), 1 mM CaCl$_2$, and 0.1% bovine serum albumin (proteinase-free) (Sigma). Total binding was determined in the absence of nonlabeled insulin. Incubation was carried out at 4°C for 21 h. The binding was stopped by centrifugation at 4°C, at 16,000g for 10 min, and immediate aspiration of supernatant. The bottom part of the Eppendorf tube containing the membrane pellet was cut off and counted in a gamma-counter. Obtained competition data were analyzed with the LIGAND program.[8] Mono-^{125}I-(TyrA14)-insulin was prepared using the lactoperoxidase method of iodination.[9,10]

Immunoblot of GLUT 4

Fresh muscle membrane preparation was solubilized with the Laemmli sample buffer and separated on 12% polyacrylamide gel. Proteins were then electrotransferred from the gel to the nitrocellulose membrane (Hybond, Amersham). The membranes were blocked with 5% nonfat dry milk in 50 mM Tris-HCl, 2 mM CaCl$_2$, 80 mM NaCl, pH 8, for 1 h at room temperature. After blocking, the membranes were incubated at 4°C with rabbit anti-GLUT 4 antibody (Charles River Pharm-Services, Wilmington, MA). Then, the membranes were washed 4× at 10 min with TEN buffer followed by 3× with a 10-min wash with PBS. At the end of the final wash, the membranes were incubated with secondary antibody linked to horseradish peroxidase (antirabbit IgG, Pierce) in PBS with 0.05% Tween 20 for 1 h at room temperature. Finally, the membranes were washed with PBS-Tween (5×, 10 min) and deionized water (3×, 5 min). The protein bands of GLUT 4 were visualized by exposing the membranes to an enhanced chemiluminescence reagent according to the manufacturer's instructions (Amersham). Autoradiography was carried out using Hyperfilm ECL (Amersham). The specific band intensities were quantitated by optic densitometry using a Kodak DS DC40 camera and ID Image Analysis Software (Eastman Kodak). The results are expressed in arbitrary units of signal intensity.

Statistical Analysis

Results are expressed as means ± SEM. For statistical analysis, the one-way ANOVA and the Bonferroni posttest were used.

RESULTS

Terguride treatment for 21 days significantly affected weight gain only in SHR females (control: 17.1 ± 3.1 g vs. control + terguride: 2.4 ± 0.5 g, $n = 6$, $p < 0.001$; obese: 91.3 ± 14.1 g vs. obese + terguride: 54.2 ± 8.3 g, $n = 6$, $p < 0.05$).

All groups of animals under study displayed comparable glycemia (TABLE 1.) Despite that, the obese SHRs had markedly elevated serum insulin levels (TABLE 1). Terguride treatment significantly diminished hyperinsulinemia only in obese females, although the same (but insignificant) tendency was seen also in males. The

TABLE 1. Serum glucose, insulin, prolactin, and glucose tolerance of experimental animals

	Glucose (mmol/L)	IRI (pmol/L)	AUC [(mmol/L) × h]	PRL (ng/mL)
Males				
Control (7)	5.45 ± 0.2	243 ± 26	29.3 ± 0.8	18.3 ± 2.8
Control + terguride (7)	4.90 ± 0.2	174 ± 19	26.5 ± 1.1	$1.6 \pm 0.2^{+++}$
Obese (8)	4.62 ± 0.3	$953 \pm 181^*$	$44.3 \pm 3.9^*$	13.3 ± 1.4
Obese + terguride (8)	4.07 ± 0.2	641 ± 87	$29.0 \pm 1.0^{++}$	$2.9 \pm 0.4^{++}$
Females				
Control (6)	4.84 ± 0.2	177 ± 6	29.2 ± 1.2	41.1 ± 12.4
Control + terguride (6)	5.28 ± 0.04	165 ± 9	26.7 ± 0.8	21.2 ± 5.3
Obese (6)	4.62 ± 0.4	$575 \pm 72^{**}$	34.8 ± 2.7	50.7 ± 16.4
Obese + terguride (6)	4.35 ± 0.4	$258 \pm 23^{++}$	$25.8 \pm 0.8^+$	$4.8 \pm 1.0^+$

NOTE: Control rats are lean SHRs of Koletsky type; obese animals are obese SHRs. The number of animals is indicated in the parentheses. Treatment of animals with terguride is described in detail in the text. IRI, serum immunoreactive insulin; AUC, glucose tolerance expressed as area under the glycemic curve; PRL, serum prolactin levels. Values are expressed as means ± SEM. Significant differences against the corresponding controls: $*p < 0.01$, $**p < 0.001$. Significant effect of terguride: $^+p < 0.05$, $^{++}p < 0.01$, $^{+++}p < 0.001$.

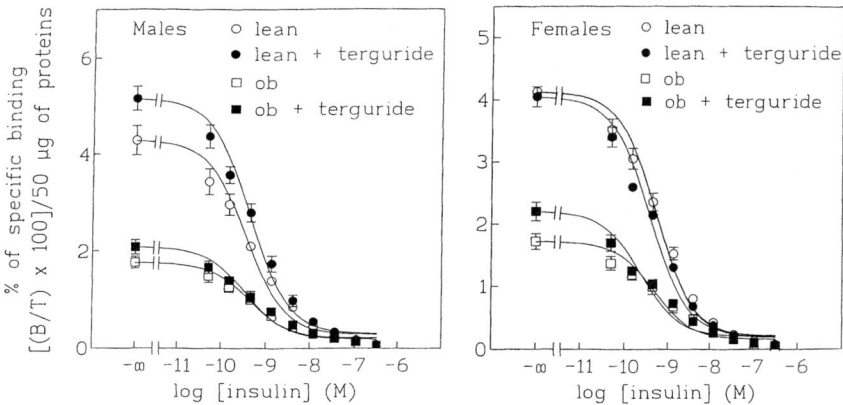

FIGURE 1. Averaged displacement curves for insulin binding to epididymal and perigonadal fat tissue plasma membranes. The points represent mean values ± SEM; ob, obese animals.

glucose tolerance expressed as the AUC parameter was significantly impaired in obese males in comparison to their nonobese controls. Tendency to a slight increase of AUC was observed also in obese females (TABLE 1). Terguride normalized the AUC in obese animals of both sexes (TABLE 1). Terguride was without any effect on glycemia, insulinemia, and AUC in lean animals.

Prolactin levels did not differ significantly between obese and lean SHRs of both sexes (TABLE 1). Considering sex differences, the obese females had significantly higher prolactin than obese males ($p < 0.05$). Terguride treatment significantly decreased serum prolactin in control and obese males and obese females, with the same tendency in nonobese females (TABLE 1).

Insulin binding to epididymal fat tissue plasma membranes was very distinct in lean and obese SHRs, as is documented by displacement curves for both sexes (FIG. 1). The total binding in obese animals decreased significantly by more than 50% (TABLE 2). Transformation of displacement data according to Scatchard[11] using the model of two independent binding sites with high and low affinity[8] revealed a profound decrease of capacity of high-affinity binding sites (R1) for insulin without changes of affinity in obese rats of both sexes (TABLE 2). GLUT 4 content in *musculus* quadriceps plasma membranes of male rats was the same in all experimental groups (TABLE 3). In females, however, a significant reduction of GLUT protein in obese animals was found (TABLE 3 and FIG. 2).

Terguride treatment resulted in a significant elevation of the affinity of high-affinity binding sites (K1) (TABLE 2). Terguride also tended to increase the affinity of insulin low-affinity binding sites (K2) in lean rats; however, those changes were statistically insignificant due to the large scatter of K2 data (TABLE 2). In addition, terguride normalized GLUT 4 content in obese females (TABLE 3 and FIG. 2).

TABLE 2. Parameters of insulin binding to adipose tissue plasma membranes

	SB (%)	R_1 (fmol/100 μg)	R_2 (fmol/100 μg)	K_1 (L × mol^{-1} × 10^8)	K_2 (L × mol^{-1} × 10^8)
Males					
Control (7)	4.3 ± 0.8	31.7 ± 5.8	470.7 ± 111.8	14.8 ± 4.3	0.053 ± 0.013
Control + terguride (7)	5.6 ± 0.6	23.1 ± 3.2	274.7 ± 37.3	21.3 ± 4.6	0.33 ± 0.16
Obese (8)	1.7 ± 0.3*	9.7 ± 1.3**	301.8 ± 64.2	14.7 ± 2.8	0.10 ± 0.02
Obese + terguride (8)	2.1 ± 0.4**	11.6 ± 2.9	250.4 ± 33.8	16.6 ± 2.4	0.12 ± 0.03
Females					
Control (6)	4.1 ± 0.1	30.9 ± 2.8	347.5 ± 52.8	11.6 ± 0.7	0.058 ± 0.015
Control + terguride (6)	4.1 ± 0.3	22.4 ± 2.0	218.8 ± 32.3	14.8 ± 0.9	0.14 ± 0.05
Obese (6)	1.8 ± 0.3***	16.5 ± 3.9*	304.1 ± 91.8	8.2 ± 1.0	0.11 ± 0.04
Obese + terguride (6)	2.2 ± 0.3**	11.9 ± 1.5	235.4 ± 18.0	14.9 ± 2.3$^+$	0.09 ± 0.03

NOTE: SB, insulin-specific binding; R_1 and R_2, capacity of high- and low-affinity insulin receptors, respectively; K_1 and K_2, affinity of high- and low-affinity receptors, respectively. The results are expressed as means ± SEM. Significant differences against the corresponding controls: *$p < 0.05$, **$p < 0.01$, ***$p < 0.001$. Significant effect of terguride: $^+p < 0.05$. For other information, see TABLE 1.

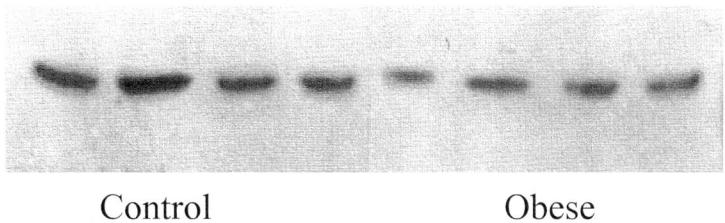

FIGURE 2. Representative immunoblot of GLUT 4 in female *musculus* quadriceps plasma membranes.

TABLE 3. GLUT 4 content in *musculus* quadriceps plasma membranes

	GLUT 4 content (rel. densitometric units)
Males	
Control (5)	31.6 ± 3.1
Control + terguride (4)	32.0 ± 1.1
Obese (4)	36.4 ± 3.9
Obese + terguride (4)	37.6 ± 4.3
Females	
Control (4)	57.8 ± 1.8
Control + terguride (4)	62.8 ± 4.7
Obese (4)	37.0 ± 2.5*
Obese + terguride (4)	56.7 ± 3.2+

NOTE: The data are expressed as means ± SEM. Significant differences against the corresponding controls: *$p < 0.05$. Significant effects of terguride: +$p < 0.05$. For other information, see TABLE 1.

DISCUSSION

Obese SHRs used in our experiments originated from an NIH-derived substrain of Koletsky rats.[7,12] However, taking into account the absence of significant differences in glycemia between obese and lean animals, our substrain resembles more the SHROB/Koletsky strain[5] than the NIH-derived Hansen corpulent SHR/N-cp one. Despite that, our data clearly indicate the presence of apparent signs of insulin resistance, such as significant hyperinsulinemia, higher values of AUC, and substantial downregulation of insulin receptors and GLUT 4 in our substrain of Koletsky rats. An inverse relationship between insulin concentration and the number of insulin receptors has been documented in several tissues in obesity, noninsulin-dependent diabetes, and other hyperinsulinemic states.[13] We found downregulation of insulin receptors and GLUT 4 in adipose tissue in rats with hyperinsulinemic and normoglycemic obesity induced by neonatal monosodium glutamate treatment.[14,15] Similarly, SHROB/Koletsky rats display about a 30% reduction in insulin receptor level in skeletal muscle and liver, presumably as a consequence of hyperinsulinemia.[5] Downregulated high-affinity insulin receptors in adipose tissue of obese SHRs may contribute to lower sensitivity of the antilipolytic action of insulin against stimulated lipolysis, leading to higher production of circulating free fatty acids in fat tissue and consequently increased triglyceride production in liver.

GLUT 4 content in *musculus* quadriceps is reduced significantly by 36% only in our obese females, in contrast to a moderate 23% decrease found by Friedman[5] in epitrochlearis muscle. Our obese females show the highest serum prolactin level significantly different from the corresponding male group. Therefore, we conclude that prolactin might contribute to downregulation of GLUT 4.

The presence of hyperprolactinemia in our obese and lean SHR/Koletsky rats in comparison to normotensive Wistar rats was described in our previous paper.[6] This fact is in agreement with the data of Cincotta[16,17] and implicates the possible role of prolactin in mechanisms of insulin resistance. Dopaminergic inhibition of prolactin release and prolactin gene expression[18] leads to a reduction of fat stores, decreases lipemia, and ameliorates other obesity-associated metabolic dysfunction in several animal species.[3,17] In turn, severe hyperprolactinemia in human patients is associated with decreased insulin binding and insulin resistance.[19] In our studies, prolactin inhibition by terguride improved insulin resistance and insulin binding and increased GLUT 4 content, mainly in obese females. Indeed, obese female SHRs display the highest prolactin levels and a very high degree of prolactin inhibition after terguride treatment. It seems that, particularly in obese females, prolactin may contribute to obesity-induced insulin resistance. This statement is strengthened by the fact that, in humans, the most profound effect of dopaminergic inhibition of prolactin on the degree of body fat reduction was documented in postmenopausal women.[20]

CONCLUSIONS

In summary, our results document insulin resistance and decreased insulin receptor capacity in adipose tissue of obese SHRs of both sexes. Lowered GLUT 4 content is present only in *musculus* quadriceps of obese females. Prolactin inhibition by terguride ameliorates insulin sensitivity in obese animals of both sexes, but it increases insulin receptor affinity and GLUT 4 content only in obese females. Whether the mechanism of terguride action is based solely on prolactin inhibition or involves also other consequences of dopamine receptor stimulation awaits further investigation. Nevertheless, terguride, aside from its classical indications (e.g., hyperprolactinemia and acromegaly) can be considered as a potential drug for alleviation of obesity-induced metabolic abnormalities.

ACKNOWLEDGMENTS

This work was supported by VEGA Grant Nos. 2/7213/20 and 2/7212/20 of the Slovak Republic. We wish to thank Carl T. Hansen, Animal Genetics Division, National Institutes of Health, Bethesda, Maryland, for providing the genetically hypertensive rats of the Koletsky type. We are also thankful to Susan Greenhut, NHPP, Ogden BioServices, for the supply of rPRL reagents.

REFERENCES

1. KOLETSKY, S. 1975. Pathologic findings and laboratory data in a new strain of obese hypertensive rats. Am. J. Pathol. **80:** 129–142.
2. GOLDA, V. & R. PETR. 1988. A genetically based animal model of Cushing's disease: glucose tolerance. Physiol. Bohemoslov. **36:** 366.
3. GOLDA, V. & L. CVAK. 1994. Terguride, but not bromocriptine alleviated glucose tolerance abnormalities and hyperlipidemia in obese and lean genetically hypertensive Koletsky rats. Physiol. Res. **43:** 299–305.
4. HILGERTOVA, J., L. KUMMEL, R. HOVORKA & V. GOLDA. 1989. Decreased insulin binding and internalization in hepatocytes and erythrocytes of SHR Koletsky rats depend on the presence of "f" gene and on sex: metabolic characteristics of hepatocytes from obese rats. *In* Insulin and the Cell Membrane, pp. 81–92. Harwood Academic Pub. London.
5. FRIEDMAN, J.E., T. ISHIZUKA, S. LIU et al. 1997. Reduced insulin receptor signalling in the obese spontaneously hypertensive Koletsky rat. Am. J. Physiol. (Endocrinol. Metab.) **273**(36)**:** E1014–E1023.
6. GOLDA, V. & J. JURCOVICOVA. 1998. Hyperprolactinemia in obese as well as in lean females of Koletsky rats: effect of long lasting terguride treatment. Acta Med. (Hradec Kralove) **41:** 159–162.
7. WEXLER, B.C., S.G. IAMS & J.P. MCMURTRY. 1980. Pathophysiological differences between obese and non-obese spontaneously hypertensive rats. Br. J. Exp. Pathol. **61:** 195–206.
8. MUNSON, P.J. & D. RODBARD. 1984. Computerized analysis of ligand binding data: basic principles and recent developments. *In* Computers in Endocrinology, pp. 117–145. Raven Press. New York.
9. JORGENSEN, K.H. & U.D. LARSEN. 1980. Homogenous mono-^{125}I-insulins: preparation and characterization of mono-^{125}I-(tyrA14) and mono-^{125}I-(tyrA19)-insulin. Diabetologia **19:** 546–554.
10. ZORAD, S., E. SVABOVA, I. KLIMEŠ & L. MACHO. 1985. Comparison of radiochemical purity and tissue binding of labelled insulin prepared by lactoperoxidase and chloramine T iodination. Endocrinol. Exp. **19:** 267–275.
11. SCATCHARD, G. 1949. The attraction of proteins for small molecules and ions. Ann. N.Y. Acad. Sci. **61:** 660–672.
12. HANSEN, C.T. 1983. Two new congenic rat strains for nutrition and obesity research. Fed. Proc. Fed. Am. Soc. Exp. Biol. **42:** 537.
13. GRUNBERG, G. 1988. Insulin receptors in cultured and circulated human monocytes. *In* Insulin Receptors. Part B: Clinical Assessment, Biological Responses, and Comparison to the IGF-I Receptor, pp. 39–54. Alan R. Liss. New York.

14. ZORAD, S., L. MACHO, D. JEZOVA & M. FICKOVA. 1997. Partial characterization of insulin resistance in adipose tissue of monosodium glutamate–induced obese rats. Ann. N.Y. Acad. Sci. **827:** 541–545.
15. MACHO, L., M. FICKOVA, D. JEZOVA & S. ZORAD. 2000. Late effects of postnatal administration of monosodium glutamate on insulin action in adult rats. Physiol. Res. **49**(suppl. 1)**:** S79–S85.
16. CINCOTTA, A.H. & A.H. MEIER. 1995. Bromocriptine inhibits *in vivo* free fatty acid oxidation and hepatic glucose output in seasonally obese hamster (*Mesocricetus auratus*). Metabolism **44:** 1349–1355.
17. CINCOTTA, A.H., E. TOZZO & P.W.D. SCISLOWSKI. 1997. Bromocriptine/SKF 38393 treatment ameliorates obesity and associates metabolic dysfunction in obese (ob/ob) mice. Life Sci. **61:** 951–956.
18. CHUANG, T.T., L. CACCAVELLI, C. KORDON & A. ENJALBERT. 1993. Protein kinase C regulation of prolactin gene expression in lactotroph cells: involvement in dopamine inhibition. Endocrinology **132:** 832–838.
19. SCHERNHANER, G., R. PRAGER, C. PUNZENGRUBER & A. LUGER. 1985. Severe hyperprolactinemia is associated with decreased insulin binding *in vitro* and insulin resistance *in vivo*. Diabetologia **28:** 138–142.
20. MEIER, A.H., A.H. CINCOTTA & W.C. LOVELL. 1992. Timed bromocriptine administration reduces body fat stores in obese subjects and hyperglycemia in type II diabetics. Experientia **48:** 248–253.

Endocrine Regulation of Subcutaneous Fat Metabolism during Cold Exposure in Humans

JURAJ KOSKA,[a] LUCIA KSINANTOVA,[a] ELENA ŠEBÖKOVÁ,[a] RICHARD KVETNANSKY,[a] IWAR KLIMEŠ,[a] GEORGE CHROUSOS,[b] AND KAREL PACAK[b]

[a]*Institute of Experimental Endocrinology, Slovak Academy of Sciences, Bratislava, Slovak Republic*

[b]*Pediatric and Reproductive Endocrinology Branch, National Institutes of Health, Bethesda, Maryland, USA*

ABSTRACT: Increased oxidation of carbohydrates and free fatty acids is a well-known phenomenon during cold stress. Nevertheless, sources of the fuels used have not been fully clarified as yet. Thus, the aim of our study was to evaluate the effect of acute cold exposure on lipid and carbohydrate metabolism in human subcutaneous adipose tissue and to identify the possible regulatory mechanisms involved. Ten volunteers were exposed for 30 min to an ambient temperature of 4°C. Interstitial metabolism was assessed with the aid of the microdialysis technique. Lipolysis intensity was evaluated from changes of glycerol concentration in plasma and in dialysate. Cold exposure induced a significant increase of glycerol concentration both in plasma (by $199 \pm 16\%$, $p < 0.01$) and in dialysate (by $308 \pm 58\%$, $p < 0.001$). No changes in glucose concentration were found whether in plasma or in the dialysate. Ethanol concentration in dialysate increased ($148 \pm 15\%$, $p < 0.01$), indicating a slower blood flow in the subcutaneous region. Plasma concentrations of various gluco- and/or lipid-regulatory hormones remained unaffected by the cold exposure, except for norepinephrine, which rose about threefold ($309 \pm 41\%$, $p < 0.001$). The data indicate an important role for subcutaneous adipose tissue in mobilization of free fatty acids during cold exposure. This process seems to be regulated by the sympathetic nervous system, whereas hormones involved in the regulation of lipid metabolism, such as epinephrine, insulin, cortisol, and growth hormone, may play a less significant role—at least under the conditions studied.

KEYWORDS: lipolysis; glycerol; glucose; catecholamines; cold; stress

INTRODUCTION

An important role for subcutaneous adipose tissue in fuel mobilization was reported during exposure to different stressors.[1,2] In accordance with the modern doctrine of stress, that is, each stressor has its own central neurochemical and periph-

---

Address for correspondence: Juraj Koska, M.D., Ph.D., Institute of Experimental Endocrinology, Slovak Academy of Sciences, Vlarska 5, 833 06 Bratislava, Slovak Republic. Voice: +421-2-5477-2800; fax: +421-2-5477-4247.

ueenjkos@savba.sk

eral neuroendocrine signature, exposure to cold stress induces a specific cascade of neurohumoral reactions to maintain the body temperature within the physiological range.[3,4] Studies in experimental animals exposed to cold stress have shown that increased rates of secretion and intravascular hydrolysis of triglycerides are mediated by the sympathetic nervous system via the beta-adrenoceptors.[5]

In spite of the fact that the potential for facultative thermogenesis in adult humans is rather limited, the principal thermoregulatory mechanisms are shivering thermogenesis and cutaneous vasoconstriction.[4] Shivering thermogenesis is accompanied by a marked increase of the metabolic rate. Thus, substrate mobilization has been suggested as being a limiting factor to maintain normothermia.[6] With respect to fuels used, glucose is derived from the plasma pool and from muscular glycogen, whereas free fatty acids might be derived from visceral fat as well as from subcutaneous adipose tissue. Enhanced triglyceride/free fatty acid cycling in humans during cold exposure involves both the nonoxidative and the oxidative pathways.[6]

Nevertheless, the evaluation of substrates in peripheral blood does not allow localizing their true source in organisms. Before entering the blood stream, the substances are released from cells into the interstitial space. Microdialysis allows sampling of small molecules directly from the interstitial space.[7,8] In our study, this technique has been applied for the first time to clarify the role of subcutaneous adipose tissue in the cold-induced fuel mobilization in humans. The results provide direct evidence that subcutaneous fat is an important site of triglyceride breakdown during cold stress, which is regulated mainly by the sympathetic nervous system.

SUBJECTS AND METHODS

Ten healthy nonobese subjects (5 men and 5 women in the follicular phase of the cycle; age, 25.6 ± 1.2 years; BMI, 21.2 ± 0.6 kg/m^2; mean $\pm$ SEM) volunteered for the study, which was approved by the local ethics committee. Starting with the evening before the trial, the subjects were asked to fast, to restrain from tobacco and alcohol, and to keep their physical activity at a minimum level. After arrival to the laboratory between 7:30 and 8:00 AM, an indwelling catheter was placed into each subject's antecubital vein. Thereafter, a polyamide microdialysis probe (20-kDa pore size, CMA 60, CMA Microdialysis) was inserted into abdominal subcutaneous tissue 5 cm paraumbilically. The probe was connected to a microinfusion pump (CMA 106, CMA Microdialysis) and continuously perfused (2 µL/min) with Ringer solution containing 50 mmol/L of ethanol for the blood flow measurement. Dialysate was collected into capped microvials (CMA Microdialysis). After placement of the probe, the subjects remained sitting in a room with temperature of 23°C for at least 30 min until control blood sampling and insertion of the first microvial. After the control sampling period (30 min), a second blood sample was taken. Thereafter, the subjects were transported in sitting position into the cold room where a constant temperature of 4°C was kept. They remained seated there for 30 min. Collection of dialysate started after a certain period of time required for washing out the dead space volume from the outlet tubing. Blood samples were taken after 15 and 30 min. Following cold exposure, subjects were transported back to the laboratory and had the third microvial inserted. Then, the subjects remained in a room at 23°C for 30 min. During this interval, the dialysate was collected and the last blood sample in

TABLE 1. Concentrations of growth hormone (GH), cortisol, insulin, and epinephrine (EPI) in plasma (means ± SEM) (at various temperatures and times)

	23°C		4°C			23°C, 30'
	−30'	0'	15'	30'		
GH (ng/mL)	1.8 ± 0.7	0.9 ± 0.3	0.5 ± 0.2	0.3 ± 0.2		0.8 ± 0.5
Cortisol (pmol/mL)	54 ± 6	39 ± 5	41 ± 4	41 ± 4		36 ± 4
Insulin (μU/mL)	6.1 ± 0.9	6.3 ± 1.3	5.8 ± 0.9	5.5 ± 0.7		4.5 ± 0.3
EPI (pg/mL)	38 ± 12	19 ± 4	20 ± 4	33 ± 5		18 ± 3

the 30th minute was taken. The aliquots of plasma and dialysate were kept frozen (at −70°C) until analysis. The samples were analyzed in parallel. Plasma catecholamines were determined radioenzymatically.[9] Commercially available (Immunotech, F) diagnostic kits were used to determine plasma cortisol (RIA) and growth hormone (IRMA) levels. An automatic analyzer with electrochemiluminescence detection was used for insulin determination (Elecsys, Roche, Ch). Glycerol concentrations in plasma and dialysate were measured colorimetrically (Randox, United Kingdom). Glucose levels were analyzed by electrochemical detection (Super-GL-Ambulance, Müller Gerätebau, Germany). The ethanol concentration in dialysate was determined enzymatically (Sigma Diagnostics). The data were statistically evaluated using the one-way ANOVA for repeated measurement with the consecutive post-hoc multiple comparison test. Data are expressed as mean ± SEM.

RESULTS

An acute exposure of volunteers to cold induced a significant increase of glycerol concentration both in plasma (by 199 ± 16%, $p < 0.001$, FIG. 1) and in the dialysate (by 308 ± 58%, $p < 0.001$, FIG. 1). Glucose levels were not changed in plasma or in the dialysate (FIG. 1). Ethanol concentration in dialysate increased significantly during the stay in the cold room (148 ± 15%, $p < 0.01$, FIG. 1), which indicates a decrease in the regional blood flow. Cold exposure led also to a threefold increase in plasma norepinephrine concentration ($p < 0.001$, FIG. 1). Plasma concentrations of the other lipid-regulatory hormones (e.g., epinephrine, cortisol, growth hormone, and insulin) remained unchanged during the cold exposure (TABLE 1).

DISCUSSION

Our data indicate that the subcutaneous adipose tissue plays an important role in mobilization of energetic substrates in response to cold exposure in humans. To the best of our knowledge, we were the first to use the microdialysis technique as reported in this study to investigate lipid and carbohydrate metabolism in subcutaneous

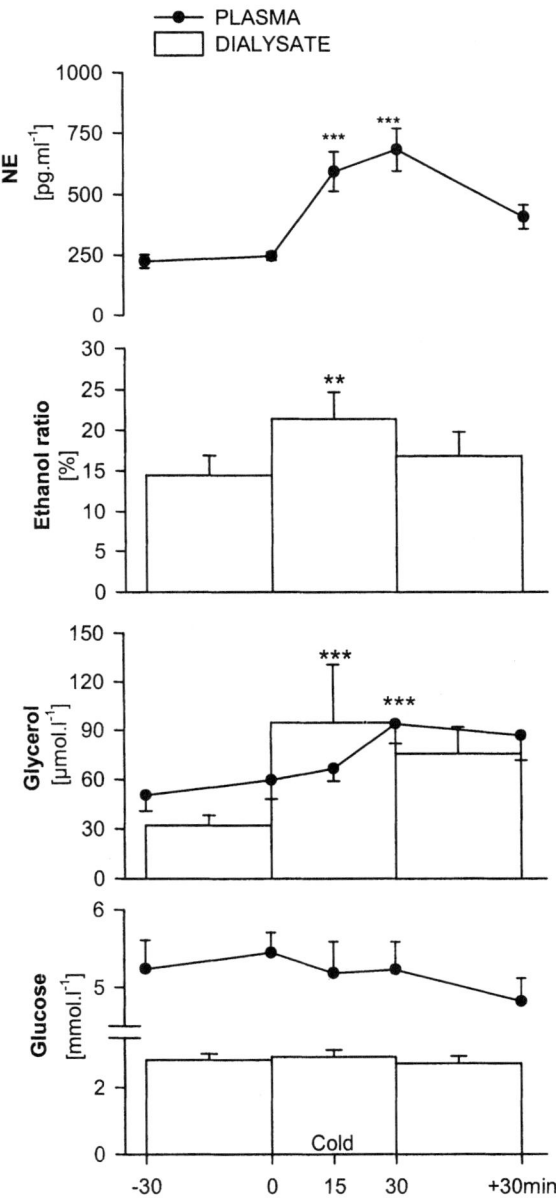

FIGURE 1. Concentration of glycerol, glucose, and norepinephrine (NE) in biological material (plasma: line plot; dialysate: bar plot) and ethanol outflow/inflow ratio (means ± SEM; statistical significance: ***$p < 0.001$, **$p < 0.01$).

adipose tissue of human volunteers exposed to low environmental temperature. The pronounced elevation of glycerol in both dialysate and plasma points to a significant activation of the adipose tissue lipolysis under these circumstances.

A similar activation of triglyceride metabolism in human subcutaneous adipose tissue has already been found in stress situations, such as physical exercise[1] or mental stress.[2] Both stressors are characterized by a pronounced activation of the sympathoadrenal system, the principal regulator of lipid metabolism in subcutaneous fat.[10,11] Nevertheless, these stress stimuli can induce the release of additional neurohumoral factors that are known to interfere at different steps with regulation of lipid and carbohydrate metabolism.[12] Under our given experimental setting, no changes in plasma concentration of these potentially prolipolytic (e.g., growth hormone) and/or antilipolytic (e.g., cortisol and insulin) hormones indicate rather their minor role in regulation of lipolysis.

The effect of the sympathoadrenal system on fat cells is mediated via five distinct subtypes of adrenoreceptors.[13] Meanwhile, the abundance of the beta-3-adrenoceptors in adult human adipose tissue is rather low; the thermogenic effect of catecholamines in human adipocytes is achieved indirectly via the beta-1- and beta-2-mediated stimulation of lipolysis.[10] It is to be stressed simultaneously that the antilipolytic action of catecholamines in fat cells also uses alpha-2-adrenoceptors.[10]

Elevated glycerol levels in subcutaneous adipose tissue of our healthy volunteers give supporting evidence for the prevailing role of a beta-adrenergic stimulation during an acute exposure to cold. The relative involvement of circulation borne from nerve endings' locally released norepinephrine is, however, unclear. Recently, it was reported that, for stimulation of lipolysis during physical exercise, the circulating catecholamines might be of greater importance than that locally released.[14]

Sympathetically mediated vasoconstriction is the principal defensive mechanism against hypothermia in humans.[4] Besides the direct stimulatory effect on lipolysis, catecholamines regulate metabolic processes in adipose tissue and also indirectly through changes in capillary permeability and in the local blood flow.[15] Thus, to measure regional blood flow, we have used the ethanol outflow/inflow technique, which provides qualitative information only on the processes studied.[16] In our study, local blood flow was significantly diminished during cold exposure. A diminished nutritional blood flow in adipose tissue was reported to increase dialysis recovery for both glucose[16] and glycerol.[17] In our study, the changes of glycerol and glucose levels in the dialysate correspond to those observed in plasma. When the substrate elevation in dialysate would have resulted from an increased recovery only (e.g., decreased uptake into system circulation), fewer significant changes of their plasma levels would have been observed. Nevertheless, an enhanced probe recovery due to a slower local blood flow might have been counteracted by the opposite effect of decreased local temperature as reported in the literature.[18]

Unchanged plasma epinephrine levels during hypothermia were observed also by others.[4] However, the same group recently reported a higher epinephrine release on the basis of analysis of arterial plasma in subjects exposed to body core cooling.[19] Thus, further studies are required in order to shed more light on regulation of lipolysis in subcutaneous adipose tissue. Details on the role of the arteriovenous difference of epinephrine and the relative role of circulatory versus locally released norepinephrine have yet to be clarified as well.

ACKNOWLEDGMENTS

This work was supported by VEGA Grant Nos. 2/7209/20 and 2/7210/20.

REFERENCES

1. ARNER, P., E. KRIEGHOLM, P. ENGFELDT et al. 1990. Adrenergic regulation of lipolysis in situ at rest and during exercise. J. Clin. Invest. **85:** 893–898.
2. HAGSTRÖM-TOFT, E., P. ARNER, H. WAHRENBERG et al. 1993. Adrenergic regulation of human adipose tissue metabolism in situ during mental stress. J. Clin. Endocrinol. Metab. **76:** 392–398.
3. PACAK, K., M. PALKOVITS, G. YADID et al. 1998. Heterogeneous neurochemical responses to different stressors: a test of Selye's doctrine of nonspecificity. Am. J. Physiol. **275**(4, part 2): 1247–1255.
4. FRANK, S.M., S.N. RAJA, C. BULCAO et al. 2000. Age-related thermoregulatory differences during core cooling in humans. Am. J. Physiol. Regul. Integr. Comp. Physiol. **279**(1): 349–354.
5. MANTHA, L. & Y. DESHAIES. 1998. Beta-adrenergic modulation of triglyceridemia under increased energy expenditure. Am. J. Physiol. **274**(6, part 2): 1769–1776.
6. VALLERAND, A.L., J. ZAMECNIK, P.J. JONES et al. 1999. Cold stress increases lipolysis, FFA Ra, and TG/FFA cycling in humans. Aviat. Space Environ. Med. **70**(1): 42–50.
7. UNGERSTEDT, U. & C. PYCOCK. 1974. Functional correlates of dopamine neurotransmission. Bull. Schweiz. Akad. Med. Wiss. **30**(1–3): 44–55.
8. LÖNNROTH, P., P.A. JANSSEN & U. SMITH. 1987. A microdialysis method allowing characterization of intercellular water space in humans. Am. J. Physiol. **253:** 228–231.
9. PEULER, J.D. & G.A. JOHNSON. 1977. Simultaneous single isotope radioenzymatic assay of plasma norepinephrine, epinephrine, and dopamine. Life Sci. **21**(5): 625–636.
10. ARNER, P. 1988. Control of lipolysis and its relevance to development of obesity in man. Diabetes Metab. Rev. **4:** 507–516.
11. DODT, C., P. LÖNNROTH, H.L. FEHM et al. 1999. Intraneural stimulation elicits an increase in subcutaneous interstitial glycerol levels in humans. J. Physiol. **521**(2): 545–552.
12. OTTOSSON, M., P. LÖNNROTH, P. BJÖRNTORP et al. 2000. Effects of cortisol and growth hormone on lipolysis in human adipose tissue. J. Clin. Endocrinol. Metab. **85**(2): 799–803.
13. LAFONTAN, M., P. BARBE, J. GALITZKY et al. 1997. Adrenergic regulation of adipocyte metabolism. Hum. Reprod. **12**(suppl. 1): 6–20.
14. STALLKNECHT, B., J. LORENTSEN, L.H. ENEVOLDSEN et al. 2001. Role of the sympathoadrenergic system in adipose tissue metabolism during exercise in humans. J. Physiol. **536**(part 1): 283–294.
15. SAMRA, J.S., E.J. SIMPSON, M.L. CLARK et al. 1996. Effects of epinephrine on infusion on adipose tissue: interactions between blood flow and lipid metabolism. Am. J. Physiol. **271:** E834–E839.
16. HICKNER, R.C., H. ROSDAHL, I. BORG et al. 1991. Ethanol may be used with the microdialysis technique to monitor blood flow changes in skeletal muscle: dialysate glucose concentration is blood flow dependent. Acta Physiol. Scand. **143:** 355–356.
17. ENOKSSON, S., J. NORDENSTRÖM, J. BOLINDER et al. 1995. Influence of local blood flow on glycerol levels in adipose tissue. Int. J. Obes. **19:** 350–354.
18. WAGES, S.A., W.H. CHURCH & J.B. JUSTICE, JR. 1986. Sampling considerations for on-line microbore liquid chromatography of brain dialysate. Anal. Chem. **58**(8): 1649–1656.
19. GOLDSTEIN, D.S. & S.M. FRANK. 2001. The wisdom of the body revisited: the adrenomedullary response to mild core hypothermia in humans. Endocr. Regul. **35**(1): 3–7.

Electrical Stimulation Improves Insulin Responses in a Human Skeletal Muscle Cell Model of Hyperglycemia

VIGDIS AAS,[a] SIRI TORBLÅ,[a] MERETHE H. ANDERSEN,[a] JØRGEN JENSEN,[b] AND ARILD CHR. RUSTAN[a]

[a]*Department of Pharmacology, School of Pharmacy, University of Oslo, Oslo, Norway*

[b]*Department of Physiology, National Institute of Occupational Health, Oslo, Norway*

ABSTRACT: Myoblasts from human skeletal muscle were isolated from needle biopsy samples of the *vastus lateralis* of young and healthy volunteers. Contaminating fibroblasts were removed, and myoblasts were fused into differentiated multinucleated myotubes. These myotubes manifested both basal and insulin-stimulated (1–100 nM) glucose transport and glycogen synthesis. Insulin increased 2-deoxyglucose uptake by 1.4-fold and glycogen synthesis by 2.1-fold. Measurements of impedance of cell-covered gold electrodes (ECIS system) showed increased micromotion of caffeine-stimulated cells, showing their ability to contract. Acute electrical stimulation of the myotubes increased 2-deoxyglucose uptake by about 30%. Treatment with high glucose concentrations (10–20 mM) for 2–8 days reduced both basal and insulin-stimulated glucose uptake. Maximal effect was seen after 2 days of treatment with 20 mM glucose. Baseline glucose uptake and glycogen synthesis were reduced by 35%, insulin-stimulated glucose uptake by 25%, and insulin-stimulated glycogen synthesis by 39%. Total cell content of glycogen was not changed by hyperglycemia. The insulin-stimulated glucose uptake in hyperglycemia-treated cells was improved by electrical stimulation of the cells. In conclusion, a model of hyperglycemia has been established, and electrical stimulation improved insulin responses.

KEYWORDS: electrical stimulation; hyperglycemia; 2-deoxyglucose uptake

INTRODUCTION

Insulin resistance is defined as a defect of insulin-stimulated glucose disposal in peripheral tissue. Reduced glucose uptake and glucose metabolism are observed in skeletal muscle where the major part of glucose is disposed during insulin stimulation. The reason for insulin resistance is not known, but several defects in insulin signaling have been reported.[1,2] Free fatty acids, fat feeding, and glucose infusion all cause insulin resistance.[3] Elevated glycogen concentration also decreases insulin-stimulated glycogen synthesis in skeletal muscles.[4]

Address for correspondence: Vigdis Aas, Department of Pharmacology, School of Pharmacy, University of Oslo, P. O. Box 1068, N-0316 Oslo, Norway. Voice: +47 22 85 65 61; fax: +47 22 85 44 02.

vigdisaa@farmasi.uio.no

TABLE 1. Clinical characteristics of biopsy donors

Clinical characteristics	Combined
Number (F/M)	13 (6/7)
Age (years)	26 (± 1)
Body mass index (kg/m^2)	23 (± 0.8)
Fasting plasma glucose (mmol/L)	4.8 (± 0.1)
Fasting plasma insulin (pmol/L)	92 (± 9.6)

NOTE: Data are means (± SEM).

Contractile activity stimulates glucose transport in skeletal muscles,[5] but more importantly contractile activity increases insulin sensitivity for both glucose uptake and glycogen synthesis.[6] The contribution of single components as free fatty acids of different kinds or other lipids,[7] hyperglycemia, or hyperinsulinemia[8] in insulin resistance is difficult to comprehend in the *in vivo* situation. Thus, we established a cell culture system of human skeletal muscle cells to study basic mechanisms involved in insulin resistance. Here, we show that muscle cells from healthy volunteers show reduced glucose uptake and glycogen synthesis after exposure to hyperglycemia and that electrical stimulation can improve insulin-stimulated glucose uptake in these cells and also can increase glucose uptake by itself.

MATERIALS AND METHODS

Human Muscle Cell Cultures

A cell bank of satellite cells was established from muscle biopsies of the *vastus lateralis* of 13 healthy volunteers (TABLE 1). The biopsies were obtained with informed consent and approval by the National Committee for Research Ethics, Oslo. Muscle cell cultures free of fibroblasts were established by the methods of Henry *et al.*[9] with minor modifications. Briefly, muscle tissue was dissected in Ham's F-10 media (Gibco BRL) at 4°C and dissociated by three successive treatments with 0.05% trypsin/EDTA (Gibco BRL), and satellite cells were resuspended in human skeletal growth media (SkGM Bullet kit without insulin, Clonetics) with 2% fetal calf serum (FCS, Gibco BRL). The cells were grown on extracellular matrix gel (ECM gel, Sigma-Aldrich)–coated culture wells (courtesy of Michael Gaster). After 2–3 weeks, at about 80% confluence, fusion of myoblasts into multinucleated myotubes was achieved by growth for 8 days in α-MEM (Gibco BRL) with 2% FCS. All cells used in the experiments were at passage 4 to 6.

Deoxyglucose Uptake

The cells were incubated for 60 min in serum-free α-MEM in a 95% O_2/5% CO_2 incubator at 37°C before addition of ^{3}H-deoxyglucose (0.1–1 µCi/mL, NEN) and insulin (1–100 nM) (Insulin Actrapid®, 100 IE/mL, Novo Nordisk). Deoxyglucose uptake was measured after incubation in medium containing 5.5 mM glucose for

60 min or in the presence of 10 μM unlabeled deoxyglucose for 15 min. After incubation, the cells were washed three times with ice-cold phosphate-buffered saline (PBS), lysed with 0.05 M NaOH, and measured for radioactivity. The protein content of each sample was measured according to Bradford,[10] and glucose uptake is presented as nmol glucose/mg cell protein/minute. Noncarrier-mediated uptake was determined in the presence of cytochalasin B (5 μM, Sigma-Aldrich) and subtracted from all presented values.

Glycogen Synthesis

The cells were incubated for 60 min in serum-free α-MEM in a 95% O_2/5% CO_2 incubator at 37°C before addition of ^{14}C-glucose (1–2 μCi/mL, NEN) and insulin (1–100 nM). After 60 min, the cells were washed three times with ice-cold PBS and lysed with 1 M KOH. Synthesized glycogen was measured as described by Franch et al.[11] and presented as nmol/mg cell protein/h.

Glycogen Content

The glycogen in muscle cells was hydrolyzed to glucose in a 0.3 M acetic acid buffer with amyloglucosidase (30 μg/mL, Boehringer), and the glucose units were measured fluorometrically by the method of Lowry and Passonneau.[12]

Electrical Stimulation of Muscle Cells

Multinucleated myotubes grown in 6-well plates were stimulated via platinum electrodes with 200-ms trains of 100 Hz delivered every 2 or 5 s.[13] Single bipolar square pulses with amplitudes of 10–30 V lasted 0.2 ms. The pulse trains were generated by a muscle stimulator built at the Institute of Occupational Health, Oslo. For measurements of electrically stimulated glucose uptake, the myotubes were stimulated for the first 5–15 min of the 60-min incubation period with ^{3}H-deoxyglucose.

Electric Cell-Substrate Impedance Sensing (ECIS)

Myoblasts were cultured on small gold electrodes evaporated on the bottom of standard tissue culture dishes. Myoblasts were differentiated into myotubes, as described above, and micromotion was measured as a change in impedance of the system.[14] Control cells were compared to cells stimulated with caffeine (10 mM, Sigma-Aldrich).

RESULTS

Cell Characteristics

Staining of fixed cells with a specific antibody against fibroblasts showed cell cultures practically free of fibroblasts. On day 8 of differentiation, when the cells were used in experiments, most myoblasts had fused into multinucleated myotubes, confirmed by fluorescent microscopy of Hoechst 33258–stained cells. These myotubes expressed the glucose transporter GLUT4 and responded well to insulin at the

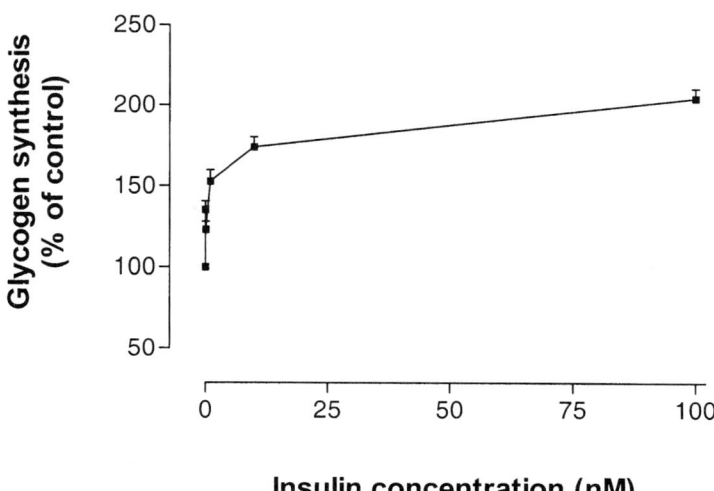

FIGURE 1. Glycogen synthesis in fused myotubes as a function of insulin concentration. Glycogen synthesis was measured for 60 min in the presence of 5.5 mM glucose. Data are presented as % of nonstimulated cells, and the average response of cells from 13 biopsy donors is illustrated (± SEM).

level of phosphorylation of insulin receptor substrate-1 (IRS-1), phosphorylation of protein kinase B (PKB), and IRS-1 association with the p85 subunit of phosphoinositide-3-kinase (PI3-K) (data not shown).

Insulin-stimulated glucose uptake and glycogen synthesis varied among the different biopsy donors. The average curves of the 13 donors showed maximal response at 100 nM insulin. Glucose uptake increased to 137% (± 13%) of control, and glycogen synthesis to 205% (± 43%) (FIG. 1). Basal deoxyglucose uptake was 3.4 (± 2.6) nmol/mg cell protein/min, and basal glycogen synthesis was 11.7 (± 4.9) nmol/mg cell protein/h.

Spontaneous contractions were not seen in these cells when observed by light microscopy. However, an increase in micromotion, assessed by cell-covered gold electrodes (ECIS), after addition of 10 mM caffeine, implied that our cells had an ability to contract (FIG. 2). In addition, deoxyglucose uptake was significantly increased by 29% after electrical stimulation of the myotubes (FIG. 3). In contrast to what has been observed *in vivo*, no additional effect of insulin and electrical stimulation was observed. Alone, insulin increased glucose uptake by 26%. We were unable to demonstrate a reduction in total cell glycogen content by electrical stimulation. After electrical stimulation for 15 min, the average cell glycogen content was 243 (± 3) nmol/mg cell protein ($n = 24$), compared to 244 (± 4) nmol/mg cell protein in controls ($n = 18$).

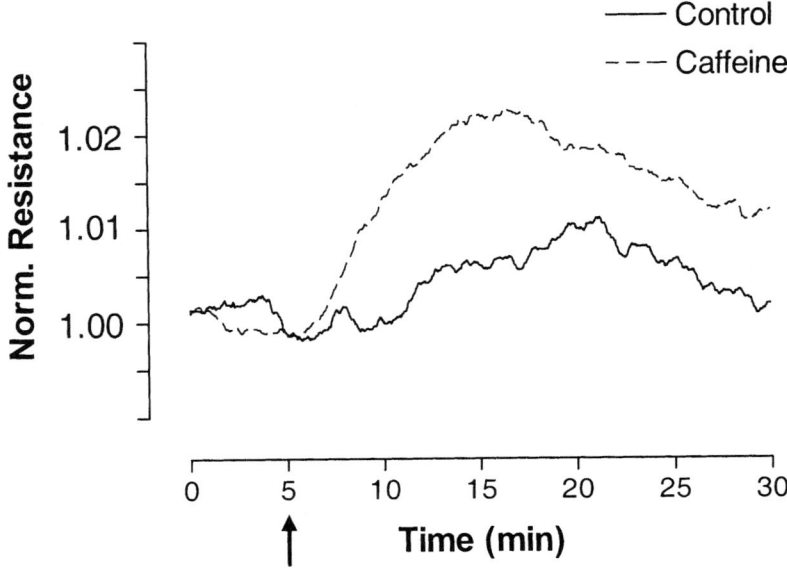

FIGURE 2. Caffeine (10 mM) stimulates micromotion of fused myotubes. Human skeletal muscle cells were grown on small gold electrodes carrying weak ac currents, and impedance was tracked as a function of time. The fluctuations observed are direct measures of cell motion. An average response from one well of cells is shown. This response is representative of three separate experiments performed on one randomly chosen biopsy donor.

Hyperglycemia

Time-course experiments showed that glucose uptake as well as glycogen synthesis were reduced already after 2 days of incubation in the presence of 10–20 mM glucose (data not shown). Incubation for 4 days with 20 mM glucose (the last 4 days of the 8-day differentiation period) was chosen as the standard hyperglycemic condition. Then, both baseline glucose uptake and glycogen synthesis were reduced by 35%. Insulin-stimulated (100 nM) glucose uptake was reduced by 25% and glycogen synthesis by 39%. Actually, the insulin responses, given as the percentage increase above baseline, were unchanged (FIG. 4). Total cell glycogen content was not affected by 4 days of incubation with 20 mM glucose (294 ± 22 vs. 291 ± 4 nmol/mg cell protein [$n = 3$]) in hyperglycemic and control cells, respectively.

Electrical Stimulation Improves Glucose Uptake

In control cells, both insulin and electrical stimulation increased glucose uptake, but the effects were not additive (FIG. 3). However, after preincubation with 20 mM glucose, the effect of insulin (100 nM) on glucose uptake in combination with electrical stimulation for 15 min was significantly higher than the effect of insulin alone (FIG. 5). Insulin increased glucose uptake by 44% and, in combination with 15-min electrical stimulation, glucose uptake was increased by 66%. No additional effect of 5-min electrical stimulation was observed.

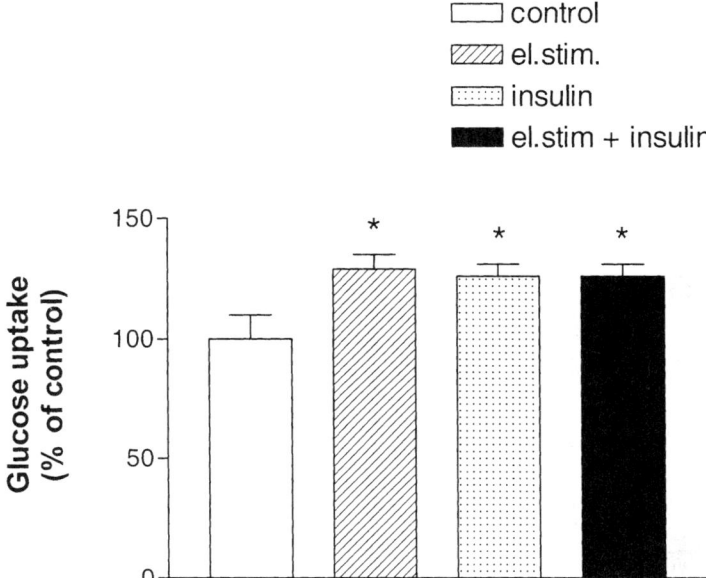

FIGURE 3. The myotubes were incubated with ^{3}H-deoxyglucose (in 5.5 mM glucose medium) for 60 min in the absence (control) or presence of 100 nM insulin (insulin). Electrical stimulation (30 V every 2 s) was performed during the first 5 min of the 60-min glucose uptake period. Average responses (± SEM) are shown ($n = 6$–15). Baseline glucose uptake was 1.1 nmol/mg cell protein/min (± 0.1) in these experiments. *Significantly increased from control ($p < 0.05$, Student's t test).

DISCUSSION

To our knowledge, electrically stimulated glucose uptake has not previously been shown in a human skeletal muscle cell (HSMC) culture system like ours. In agreement with experiments performed *in vivo* or on isolated skeletal muscles *in vitro*,[5,6] we observed that electrical stimulation of glucose uptake could occur independently of insulin. In HSMC, we did not find an additive effect of insulin and electrical stimulation, which is often reported in skeletal muscle,[15] but not always in all muscle fiber types.[5] Synchronous contraction could not be detected by light microscopy, so it is possible that electrical stimulation induced only local changes in micromotion and that the contractile machinery is not fully developed in these cells. Differentiation markers, such as sarcomeric actin, creatine kinase, and cross-striation, were not examined. An ability to change micromotion was, however, confirmed by the ECIS experiments. Rat myotubes are reported to contract spontaneously, while human cells possibly need coculturing with nerve cells or addition of nerve growth factor (NGF) to improve myogenesis.[16,17]

Hyperglycemia (10–20 mM glucose) decreased baseline as well as insulin-stimulated glucose uptake and glycogen synthesis in these cells. Actually, the reduction in baseline uptake could account for the impaired insulin responses. Thus, the

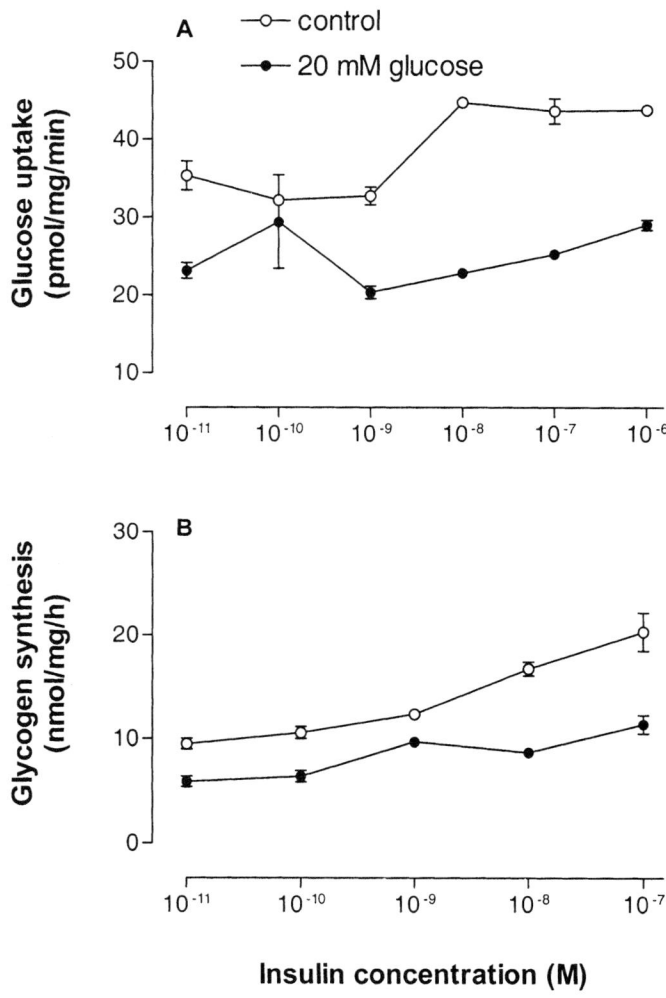

FIGURE 4. Hyperglycemia (20 mM glucose for 4 days) reduces basal as well as insulin-stimulated glucose uptake and glycogen synthesis. After a 1-h incubation in serum-free media, insulin-stimulated glucose uptake (**A**) and glycogen synthesis (**B**) were measured, as described (glucose uptake was measured in the presence of 10 µM deoxyglucose). Representative examples of three individual dose-response experiments are shown.

hyperglycemic cells are not actually insulin-resistant, but they display reduced glucose disposal. The cause of reduced glucose disposal in these cells is presently not known. Possibly, it may be a result of glucose toxicity, although there were no indications of toxic effects on the cells. The protein content of cells treated with high glucose–containing media and the morphology of those cells were similar to controls. In contrast to our findings, Ciaraldi *et al.* observed practically no effect of hyperglycemia alone on glucose uptake[18] and glycogen synthase activity.[8] In rat

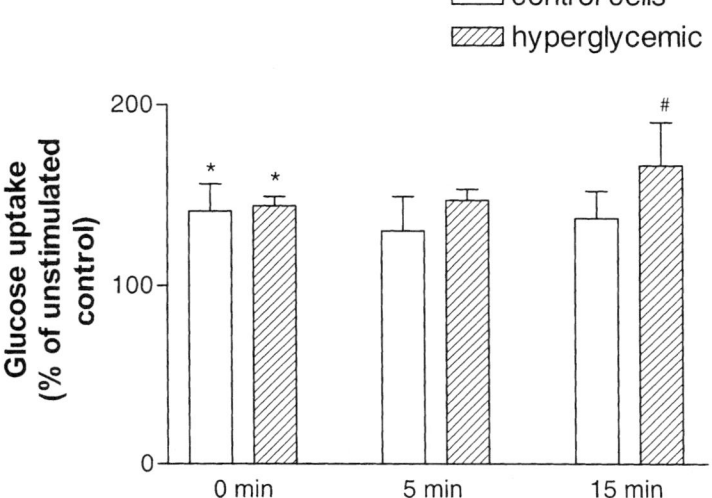

FIGURE 5. Electrical stimulation increases insulin-stimulated glucose uptake in myotubes exposed to hyperglycemia. Control or 20 mM glucose–treated cells were stimulated with 100 nM insulin alone or in combination with electrical stimulation (30 V every 2 s) for the first 5 or 15 min of the 60-min glucose uptake period. Baseline glucose uptake was 2.8 (± 0.2) nmol/mg cell protein/min in control cells and 1.8 (± 0.1) nmol/mg cell protein/min in cells treated with 20 mM glucose. The insulin responses are presented as % of unstimulated cells. Average responses (± SEM) are shown ($n = 6$). *Significantly increased from unstimulated cells ($p < 0.05$). #Significantly increased from stimulation with insulin alone ($p < 0.05$, Student's t test).

skeletal muscle, on the other hand, hyperglycemia reduced the basal glucose transport rate, while it increased insulin-stimulated glucose uptake.[19] The mechanism of autoregulated glucose transport is not known, but is probably mediated by changes in the number of glucose transporters in the plasma membrane.[20,21] Another possibility is that muscle cells exposed to glucose oversupply develop reduced fatty acid oxidation and hence accumulate malonyl-CoA,[22] long-chain acyl-CoA (LCA-CoA), and fatty acids as triglycerides.[23] Accumulation of LCA-CoA and diacylglycerol (DAG) has been suggested to play a role in the generation of insulin resistance in muscle.[24,25] So far, we have not studied fatty acid metabolism in these hyperglycemic cells, but the total cell glycogen content was not changed after 4 days of incubation with 20 mM glucose. Further experiments are necessary to explain the mechanisms by which glucose oversupply reduces glucose disposal.

The HSMC culture technique is well established[9,26] and used by several laboratories. Our method of satellite cell isolation, proliferation, and differentiation into multinucleated myotubes is slightly modified from Henry *et al.*[9] Insulin increased glucose uptake by 1.4-fold in our model, which is comparable to what others have reported;[26,27] likewise, there was a 2.1-fold increase in glycogen synthesis in accordance with previous observations.[9] The differences in the amount of glucose taken up by the cells can probably be explained by methodological differences. For instance, baseline glucose uptake was about 100-fold higher after 1-h uptake in the presence of 5.5 mM glucose (FIGS. 1 and 5) than after 15 min in the presence of 10 μM deoxyglucose (FIG. 4).

Interestingly, electrical stimulation improved insulin responses in the hyperglycemic cells. Exercise improves regulation of blood glucose homeostasis in type II diabetics, and at least part of the reason is that exercise increases insulin-stimulated glucose uptake in skeletal muscles.[28] Training increases insulin-stimulated translocation and glucose uptake in skeletal muscles from the obese Zucker rat.[29] To our knowledge, this is the first report showing the beneficial effect of contractile activity on insulin responses in muscle cell cultures. However, this finding will also require further examination. Anyway, we think that we have established a cell model that is convenient for such studies.

ACKNOWLEDGMENTS

We are grateful to Cecilie Røe for obtaining the muscle biopsies and to Mari-Ann Baltzersen for skilled technical assistance. The work was supported by grants from the Research Council of Norway.

REFERENCES

1. PESSIN, J.E. & A.R. SALTIEL. 2000. J. Clin. Invest. **106:** 165–169.
2. SALTIEL, A.R. 2001. Cell **104:** 517–529.
3. SHULMAN, G.I. 2000. J. Clin. Invest. **106:** 171–176.
4. YEAMAN, S.J., J.L. ARMSTRONG, S.M. BONAVAUD et al. 2001. Biochem. Soc. Trans. **29:** 537–541.
5. ASLESEN, R. & J. JENSEN. 1998. Am. J. Physiol. **38:** E448–E456.
6. HAYASHI, T., J.F.P. WOJTASZEWSKI & L.J. GOODYEAR. 1997. Am. J. Physiol. **36:** E1039–E1051.
7. SCHMITZ-PEIFFER, C. 2000. Cell. Signalling **12:** 583–594.
8. HENRY, R.R., T.P. CIARALDI, S. MUDALIAR et al. 1996. Diabetes **45:** 400–407.
9. HENRY, R.R., L. ABRAMS, S. NIKOULINA & T.P. CIARALDI. 1995. Diabetes **44:** 936–946.
10. BRADFORD, M.M. 1976. Anal. Biochem. **72:** 248–254.
11. FRANCH, J., R. ASLESEN & J. JENSEN. 1999. Biochem. J. **344:** 231–235.
12. LOWRY, O.H. & J.V. PASSONNEAU. 1972. A Flexible System of Enzymatic Analysis. Academic Press. New York.
13. WEHRLE, U., S. DUSTERHOFT & D. PETTE. 1994. Differentiation **58:** 37–46.
14. GIAEVER, I. & C.R. KEESE. 1991. Proc. Natl. Acad. Sci. U.S.A. **88:** 7896–7900.
15. NESHER, R., I.E. KARL & D.M. KIPNIS. 1985. Am. J. Physiol. **249:** C226–C232.
16. DELAPORTE, C., B. DAUTREAUX & M. FARDEAU. 1986. Biol. Cell **57:** 17–22.
17. DELAPORTE, C. & B. DAUTREAUX. 1987. Adv. Exp. Med. Biol. **209:** 15–17.
18. CIARALDI, T.P., L. ABRAMS, S. NIKOULINA et al. 1995. J. Clin. Invest. **96:** 2820–2827.
19. SASSON, S. & E. CERASI. 1986. J. Biol. Chem. **261:** 16827–16833.
20. SASSON, S., Y. ASHHAB, D. MELLOUL & E. CERASI. 1993. Adv. Exp. Med. Biol. **334:** 113–127.
21. KLIP, A., T. TSAKIRIDIS, A. MARETTE & P.A. ORTIZ. 1994. FASEB J. **8:** 43–53.
22. LAYBUTT, D.R., C. SCHMITZ-PEIFFER, A.K. SAHA et al. 1999. Am. J. Physiol. **277:** E1070–E1076.
23. PAN, D.A., S. LILLIOJA, A.D. KRIKETOS et al. 1997. Diabetes **46:** 983–988.
24. THOMPSON, A.L., M.Y.C. LIM-FRASER, E.W. KRAEGEN & G.J. COONEY. 2000. Am. J. Physiol. **279:** E577–E584.
25. MONTELL, E., M. TURINI, M. MAROTTA et al. 2001. Am. J. Physiol. **280:** E229–E237.

26. SARABIA, V., L. LAM, E. BURDETT et al. 1992. J. Clin. Invest. **90:** 1386–1395.
27. SARABIA, V., T. RAMLAL & A. KLIP. 1990. Biochem. Cell Biol. **68:** 536–542.
28. WOJTASZEWSKI, J.F.P., B.F. HANSEN, J. GADE et al. 2000. Diabetes **49:** 325–331.
29. ETGEN, G.J., C.M. WILSON, J. JENSEN et al. 1996. Am. J. Physiol. **34:** E294–E301.

Differences in Oral Temperature and Body Shape in Two Populations with Different Propensities for Obesity

B. VOZAROVA, C. WEYER, C. BOGARDUS, E. RAVUSSIN, AND P. A. TATARANNI

Clinical Diabetes and Nutrition Section, National Institute of Diabetes and Digestive and Kidney Diseases, National Institutes of Health, Phoenix, Arizona, USA

ABSTRACT: *Objective*—Body temperature is a function of heat production and heat dissipation. Substantial interindividual variability has been reported in healthy humans. We hypothesized that Pima Indians, a population with a high prevalence of abdominal obesity, may have a lower surface area relative to volume, that is, lower radiating area, and therefore a higher body temperature compared to Caucasians. *Methods*—Body composition, including volume (hydrodensitometry), and oral temperature were assessed in 69 nondiabetic Caucasian [age, 30 ± 7 years; body fat, 21 ± 8% (mean ± SD)] and 115 Pima Indian males [age, 27 ± 6 years; body fat, 28 ± 6%]. Surface area was estimated from height, weight, and waist circumference (Bouchard's equation). In 47 Pima Indians, measures of insulin sensitivity (M, hyperinsulinemic euglycemic clamp) were available. *Results*—Compared to Caucasians, Pima Indians had a higher oral temperature [36.4 ± 0.3°C vs. 36.3 ± 0.3°C (mean ± SD), $p < 0.04$] and lower surface area relative to volume (2.19 ± 0.05 vs. 2.23 ± 0.26 m^2, $p < 0.0001$). Surface area relative to volume was negatively correlated with oral temperature ($r = -0.14$, $p < 0.05$), but in a multiple linear regression model it did not entirely explain the ethnic difference in oral temperature. Oral temperature was inversely correlated with M ($r = -0.28$, $p < 0.05$). *Conclusions*—Pima Indians have higher oral temperature and lower surface area relative to volume than Caucasians. The ethnic difference in temperature does not seem to be entirely explained by differences in body composition and body shape. Interestingly, higher oral temperature was associated with insulin resistance, a risk factor for type 2 diabetes.

KEYWORDS: thermogenesis; ethnicity; surface area; insulin sensitivity

INTRODUCTION

Body temperature in homeotherms is regulated centrally and is a function of heat production and heat dissipation. Heat production is mostly related to body size and specific thermodynamic properties of metabolically active tissues. Heat dissipation,

Address for correspondence: Barbora Vozarova, M.D., Clinical Diabetes and Nutrition Section, National Institutes of Health, 4212 North 16th Street, Room 541-A, Phoenix, AZ 85016. Voice: 602-200-5328; fax: 602-200-5335.
bvozarov@mail.nih.gov

consisting of dry heat dissipation by convection and radiation and evaporative heat dissipation from evaporation of water from lungs and skin, is dependent to some extent on the body surface area.

In the present study, we examined the relationship between body temperature, body size, and body shape in thermoneutral conditions in a large group of Pima Indians and Caucasians. We hypothesized that Pima Indians, who have high prevalence of abdominal obesity[1] and thus have a different body shape when compared to Caucasians, may also have a lower surface area relative to volume, that is, a lower radiating area, and therefore a higher body temperature.

SUBJECTS

Since 1985, volunteers have been admitted to the Metabolic Ward of the Clinical Diabetes and Nutrition Section of the National Institutes of Health in Phoenix, Arizona, for an ongoing study of the pathogenesis of obesity. For the present analysis, 184 male subjects (69 Caucasians and 115 Pima Indians), who had measurements of oral temperature and total body volume by hydrodensitometry, were included. Females were excluded from the analyses because of missing information on the phase of the menstrual cycle. All subjects were nondiabetic (WHO, 1985) and, except for obesity, were healthy according to a physical examination and routine laboratory tests. The study protocol was approved by the Institutional Review Board of the National Institute of Diabetes and Digestive and Kidney Diseases and by the Tribal Council of the Gila River Indian Community, and all subjects provided written informed consent prior to participation.

METHODS

Upon admission, all subjects were fed a weight-maintaining diet (50%, 30%, and 20% of daily calories provided as carbohydrate, fat, and protein, respectively) and abstained from strenuous exercise. Total body volume and body composition were assessed by hydrodensitometry with simultaneous determination of residual lung volume by helium dilution. The waist and thigh circumferences were measured at the umbilicus and at the gluteal fold in the supine and standing positions, respectively, and the waist-to-thigh ratio was calculated as an index of body fat distribution. Surface area was calculated from Bouchard's prediction equation,[2] which uses waist circumference in addition to height and weight, and thus takes into account both body size and body shape. Oral temperature was measured using an electronic thermometer (Diatek, San Diego, CA) three times before and three times after the stay in the respiratory chamber. On both occasions, temperature was measured in the supine position and in the condition of thermoneutrality (24°C) immediately upon awakening.[3] In the subgroup of Pima Indians [$n = 47$; age, 27 ± 6 years; body fat, $28 \pm 6\%$; oral temperature, $36.4 \pm 0.2°C$], insulin action was assessed at physiologic (M-low) and supraphysiologic insulin concentrations (M-high) during a two-step hyperinsulinemic euglycemic glucose clamp as previously described.[4]

Statistical analyses were performed using the procedures of the SAS Institute (Cary, NC). Results are given as mean ± SD. Relationships of oral temperature to

TABLE 1. Physical and metabolic characteristics of the study population (mean ± SD)

	All (n = 184)		Caucasians (n = 69)	Pima Indians (n = 115)	p value
	Mean ± SD	Range			
Age (years)	29 ± 7	(18–48)	31 ± 7	28 ± 7	<0.01
Height (cm)	174 ± 7	(154–189)	177 ± 7	173 ± 7	<0.001
Body weight (kg)	88.4 ± 19.1	(54.5–137.5)	81.3 ± 16.9	92.8 ± 19.1	<0.0001
Body fat (%)	25 ± 8	(7–42)	21 ± 8	28 ± 7	<0.0001
Waist-to-thigh ratio	1.60 ± 0.13	(1.33–2.13)	1.55 ± 0.13	1.63 ± 0.13	<0.0001
Body volume (L)	88 ± 23	(50–157)	80 ± 21	93 ± 23	<0.001
Body surface area (m^2)	2.21 ± 0.27	(1.65–2.91)	2.13 ± 0.26	2.26 ± 0.27	<0.01
Adjusted body surface area (m^2)	2.21 ± 0.27	(1.65–2.91)	2.23 ± 0.26	2.19 ± 0.27	<0.0001
Oral temperature (°C)	36.3 ± 0.3	(35.6–37.0)	36.3 ± 0.3	36.4 ± 0.3	<0.05
M-low (mg/kg EMBS/min) (n = 47)	2.8 ± 1.2	(1.7–7.9)	—	2.8 ± 1.2	—
M-high (mg/kg EMBS/min) (n = 47)	9.2 ± 2.0	(5.2–13.3)	—	9.2 ± 2.0	—
Fasting insulin concentration (µU/mL) (n = 47)	40 ± 19	(16–92)	—	40 ± 19	—

anthropometric and metabolic variables were assessed by simple correlation analysis. Multiple linear regression models were used to calculate residuals (measured-predicted values) and to examine the relationship between variables after adjustment for covariates.

RESULTS

The anthropometric and metabolic characteristics of the study population are summarized in TABLE 1. Although body surface area in absolute terms was also higher in Pima Indians, surface area relative to volume was significantly lower.

Oral temperature was associated with fat-free mass ($r = -0.17$, $p < 0.03$) and surface area relative to volume (adjusted body surface area, $r = -0.14$, $p < 0.05$). Oral temperature was significantly higher in Pima Indians than Caucasians (36.4 ± 0.3°C vs. 36.3 ± 0.3°C, $p = 0.04$). The ethnic differences in oral temperature persisted after adjustment for fat-free mass ($p = 0.03$) or surface area ($p = 0.03$) in multivariate analysis (FIG. 1a). When both surface area relative to volume and race were in the model, neither one of the two variables was a significant determinant of oral temperature (both $p > 0.1$). However, comparison of the β-estimates of the former model and the model with race only indicates that approximately 40% of the ethnic effect on oral temperature was explained by body shape. Oral temperature was inversely correlated with M-low (FIG. 1b) and M-high ($r = -0.35$, $p = 0.02$) and only weakly with fasting insulin concentrations ($r = 0.27$, $p = 0.06$).

DISCUSSION

In the present study, we hypothesized that the Pima Indians of Arizona, a population with one of the highest reported prevalence rates of obesity in the world[5] and with a predominantly abdominal body fat distribution,[1] may have a lower surface area relative to volume, that is, a lower radiating area, and therefore higher body temperature compared to Caucasians. Although Pima Indians have higher oral temperature and lower surface area relative to their volume compared to Caucasians, we found that the ethnic differences in oral temperature were only in part explained by differences in body shape.

We have previously shown that Pima Indians have a lower body core temperature during sleep and higher temperature (although not significantly) during daytime compared to Caucasians.[6] The small, but significant ethnic difference in morning oral temperature described in the present study represents 7% of the physiologic range of body temperature (35.5–37°C). The pathophysiological consequences of this ethnic difference cannot be elucidated in this study. However, it has been described in the literature that increase in body temperature during sepsis is associated with decrease in insulin sensitivity.[7] Our data in 47 Pima Indians, who were characterized for degree of insulin sensitivity, show that a 0.1°C change in oral temperature is associated with a 36% decrease of insulin sensitivity. Having established that Pima Indians have higher oral temperature and a lower surface area to volume ratio, the question was if the difference in body shape explains the difference in oral

a)

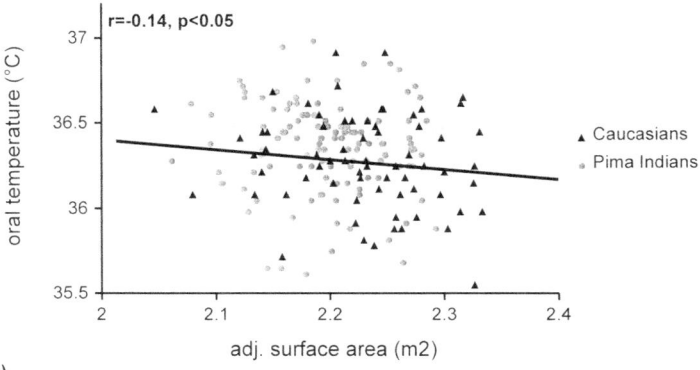

b)

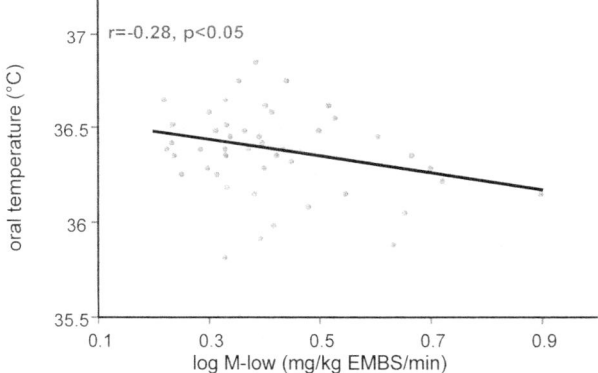

FIGURE 1. (a) Relationship between oral temperature and surface area (adjusted for volume) in nondiabetic, healthy Caucasian (black triangles) and Pima Indian males (gray circles). **(b)** Relationship between oral temperature and insulin sensitivity (M-low) in 47 Pima Indian males.

temperature. Our data seem to suggest that, in resting conditions, ethnic differences in oral temperature are not entirely explained by differences in body shape.

Alternative explanations for the higher oral temperature in Pima Indians may involve a different setting of the hypothalamic thermoregulatory center or thermogenic processes associated with increased adiposity and/or insulin resistance. As to the latter aspect, a relationship between body fat, insulin resistance, and cytokines has been reported previously, suggesting an association between obesity and/or insulin resistance and the proinflammatory state.[8] In the present study, we showed an association between a high oral temperature and insulin resistance in Pima Indians, both expressions of a subclinical activation of the immune system.

In summary, the present study demonstrates that Pima Indians have higher oral temperature and lower surface area relative to volume compared to Caucasians. However, at least under resting conditions, this ethnic difference in oral temperature is not entirely explained by differences in body shape. Interestingly, oral temperature was associated with insulin resistance, a risk factor for type 2 diabetes.

REFERENCES

1. SAAD, M.F., W.C. KNOWLER, D.J. PETTITT et al. 1990. Insulin and hypertension: relationship to obesity and glucose intolerance in Pima Indians. Diabetes **39:** 1430–1435.
2. BOUCHARD, C. 1897. Consideration sur l'etat statique du corps. La Semaine Medicale **17:** 89–91.
3. RISING, R., A. KEYS, E. RAVUSSIN & C. BOGARDUS. 1992. Concomitant interindividual variation in body temperature and metabolic rate. Am. J. Physiol. **263:** E730–E734.
4. LILLIOJA, S., D.M. MOTT, B.V. HOWARD et al. 1988. Impaired glucose tolerance as a disorder of insulin action: longitudinal and cross-sectional studies in Pima Indians. N. Engl. J. Med. **318:** 1217–1225.
5. KNOWLER, W.C., D.J. PETTITT, M.F. SAAD & P.H. BENNETT. 1990. Diabetes mellitus in the Pima Indians: incidence, risk factors, and pathogenesis. Diabetes Metab. Rev. **6:** 1–27.
6. RISING, R., A.M. FONTVIEILLE, D.E. LARSON et al. 1995. Racial difference in body core temperature between Pima Indian and Caucasian men. Int. J. Obes. Relat. Metab. Disord. **19:** 1–5.
7. RAYMOND, R.M., G.A. ROSENFELD & T.E.J. EMERSON. 1982. Direct effects of insulin and endotoxin on glucose uptake by skeletal muscle during high cardiac index sepsis in the dog. Surg. Gynecol. Obstet. **155:** 881–887.
8. PICKUP, J.C. & M.A. CROOK. 1998. Is type II diabetes mellitus a disease of the innate immune system? Diabetologia **41:** 1241–1248.

Effects of Selected Anthropometric Parameters on Plasma Lipoproteins, Fatty Acid Composition, and Lipoperoxidation

ALEŠ ŽÁK, EVA TVRZICKÁ, MAREK VECKA, SEVERYN ROMANIV, MIROSLAV ZEMAN, AND MARTA KONÁRKOVÁ

Fourth Department of Medicine, First Faculty of Medicine, Charles University, Prague, Czech Republic

ABSTRACT: Four anthropometric parameters (body mass index, waist-to-hip circumference ratio, waist circumference, and ratio of subscapularis to triceps skinfold thickness), cutoff values being set according to the WHO guidelines, were compared as discriminative parameters for anthropometric, hemodynamic, and metabolic data relevant as risk factors of coronary heart disease. Waist circumference reflected differences between below and above cutoff values throughout all parameters, with the highest "effect size" values being for fat mass; blood pressure; glucose homeostasis parameters; plasma and LDL cholesterol; plasma, LDL, and VLDL apo B; and LDL oxidability (cumulative effect size value: 2777). Body mass index reflected most significantly plasma, LDL, and VLDL triglyceride; HDL and VLDL cholesterol; and content of linoleic acid in LDL phosphatidylcholine (cumulative effect size value: 2016). Waist-to-hip ratio dominated in effect size value only in VLDL oxidability (cumulative effect size value: 1497). Subscapularis to triceps skinfold thickness ratio had the lowest discriminating ability in all data.

KEYWORDS: anthropometric parameters; lipoproteins; lipoperoxidation; fatty acids

It is generally accepted that not only overweight and obesity per se, expressed as body mass index (BMI), but mainly upper body obesity are risk factors (RF) of coronary heart disease (CHD) in men and women.[1,2] Upper body obesity is closely associated with hemodynamic, hemostatic, and metabolic disturbances, which lead to an increased risk of CHD.[3] It has been shown that intra-abdominal adipose tissue (AT) is highly lipolytic;[4] increased nonesterified fatty acid (NEFA) concentrations in the portal vein cause a number of metabolic consequences in the liver, skeletal muscle, and pancreatic β-cells.[3,4]

Address for correspondence: Aleš Žák, Fourth Department of Medicine, First Faculty of Medicine, Charles University, Prague, Czech Republic. Voice: +420 2 2496 2506; fax: +420 2 2492 3524.
azak@vfn.cz

Oxidatively modified LDL play an important role in atherogenesis and connected metabolic disorders;[5,6] however, their connection with obesity was not intensively studied. Dyslipidemia that comprises hypertriglyceridemia (HTG), low HDL-C, and increased levels of small-dense LDL has been proposed as a potential link between central adiposity, insulin resistance, and other risk factors of CHD.[3,7]

New imaging techniques (CT, MRI, DEXA) have been used to measure the proportion of trunk fat and to localize upper body fat to intra-abdominal and subcutaneous depots.[4] Despite this fact, simple anthropometric methods are still used as clinical tools to assess the accumulation of AT in the upper part of the body.

The aim of this study was to evaluate relationships between BMI, waist-to-hip circumference ratio (WHCR), waist circumference, ratio of subscapularis (SSC) to triceps (TRC) skinfold thicknesses (SFT), and composition of plasma lipoproteins (LP) and lipoperoxidation markers.

MATERIALS AND METHODS

The studied group consisted of 127 (67 males/60 females; age, 40.3 ± 5.7 years, mean ± SEM; range, 28–67 years) consecutive outpatients of our lipid department. Subjects with atherosclerotic vascular disease, DM type 1 and 2, severe obesity (BMI > 35.0 kg/m^2), or hormonal, renal, and liver disorders were not included in the study. In all subjects, anthropometric measurements and measurements of body fat content were carried out according to WHO guidelines[8] and Durnin and Womersley.[9] We measured the waist, WHCR, and SSC/TRC ratio. Subjects were further divided into subgroups with respect to cutoff values: BMI = 25.0 kg/m^2, WHCR = 1.0 (males) or 0.85 (females), waist circumference = 94 cm (males) or 80 cm (females), and SSC/TRC = 1.33.[8] All persons were advised to follow the AHA Step-1 diet[10] at least 6 weeks before blood collection.

Blood samples were obtained after overnight fast. Concentrations of plasma total cholesterol (TC), triglycerides (TG), NEFA, glucose, and uric acid were measured using enzymatic-colorimetric methods, HDL-C was determined in supernatant after precipitation of LP-B with PTA/Mg^{2+} (Boehringer, Mannheim, Germany; NEFA, Randox Laboratories, United Kingdom). Plasma immunoreactive insulin (IRI) was assessed by RIA (Insulin, IRMA, Imunotech, Prague, Czech Republic). Concentrations of apolipoproteins (apo) were measured by Laurell rocket electroimmunoassay [apo B, apo A-I, apo A-II (Behring Werke AG, Marburg, Germany)]. Lipoproteins of very low density (VLDL) and low density (LDL) were separated by the two-step sequential ultracentrifugation,[11] and their lipids and apo B were measured directly. Composition of fatty acids (FA) in LDL phosphatidylcholine (PC) was determined by capillary GC as described previously.[12] Parameters of LDL oxidation were measured by the method of Cu^{2+}-mediated kinetics of conjugated dienes.[13]

Statistical analyses were performed with Mann-Whitney U test and chi-square test using commercial software (BMDP Statistical Software). Differences of means between below and above cutoff values for individual subgroups were tested using a parameter, "effect size" (ES), which was calculated by

$$|x_1 - x_2| \cdot (SD_1 \cdot SD_2)^{-1/2} \cdot 100$$

where x_1 and x_2 are the means of below and above cutoff values for individual subgroups, and SD_1 and SD_2 are the respective standard deviations.

RESULTS AND DISCUSSION

TABLE 1 summarizes anthropometric and hemodynamic data, parameters of glucose homeostasis, concentrations of plasma lipids and LP, parameters of lipoperoxidation, and linoleic acid (LA) content in groups reflecting cutoff values of individual indices. Data reflecting cutoff values of the SSC/TRC ratio are not shown. Compared subgroups did not differ significantly in age and M/F ratio.

Waist had, among all three indices, the highest power to discriminate subjects' below and above cutoff values in fat mass, systolic and diastolic BP, plasma TC and apo B, LDL-C and LDL apo B, VLDL apo B, and LDL lag time. Concomitantly, waist revealed the highest power to discriminate between above and below cutoff values in IRI, glucose, and NEFA. On the other hand, BMI revealed the highest power to discriminate between above and below cutoff values of TG in plasma, VLDL, and LDL, as well as HDL-C and content of LA in LDL PC. Finally, WHCR was the best discriminator only for VLDL lag time. Cumulative ES values (sum of all ES) for waist, BMI, and WHCR were 2777, 2016, and 1497, respectively.

The fourth mentioned index (SSC/TRC) had the lowest discriminating power. Subjects with above and below cutoff value of 1.33 differed significantly in body weight (+11%, $p < 0.01$), BMI (+13%, $p < 0.001$), waist (+13%, $p < 0.001$), WHCR (+7%, $p < 0.001$), fat mass (+29%, $p < 0.001$), and SBP (+7%, $p < 0.05$). In biochemical parameters, differences were found only for plasma TG (+132%, $p < 0.05$), LDL-C (+10%, $p < 0.05$), VLDL TG (+39%, $p < 0.05$), and VLDL apo B (+66%, $p < 0.05$). In lipoperoxidation parameters, the subjects' above cutoff value revealed only the increased basal absorbance in LDL (+12%, $p < 0.05$).

Levels of uric acid were significantly differentiated by all indices ($p < 0.01$), and no differences were observed in the levels of apo A-I and apo A-II.

Our results are consistent with those of Pouilot *et al.*, who described WHCR as a poorer discriminator of intra-abdominal fat accumulation than waist, which fails in its prediction, especially in women.[1] These results are in concordance with metabolic studies describing the highest TG turnover in visceral fat in comparison with gluteofemoral and abdominal subcutaneous fat.[4] Increased waist circumference combined with fasting HTG is a risk factor for CHD in men and implicates the presence of low HDL-C, high apo B, and small-dense LDL.[14]

The phenomenon of increased oxidative stress was described in hereditary HTG rats (which represent an experimental model of insulin resistance syndrome[15]), in diabetes mellitus and hypertension,[6] and in the presence of the small-dense LDL phenotype.[16]

Decreased content of LA in LDL PC may be connected with its increased peroxidative degradation.[17] Obese children with the metabolic syndrome revealed alteration of LA metabolism due to stimulated activity of Δ6 desaturase and decreased activity of Δ5 desaturase.[18]

In conclusion, waist circumference reflected most significantly the observed pathological changes concerning anthropometric (fat mass), hemodynamic (systolic and diastolic BP), and metabolic (BG, IRI, NEFA) parameters, and plasma lipid and

TABLE 1. Clinical and biochemical data with respect to individual indexes of fat accumulation

Parameter	BMI Below (n = 55)	BMI Above (n = 72)	BMI ES[c]	WHCR Below (n = 77)	WHCR Above (n = 49)	WHCR ES	Waist Below (n = 57)	Waist Above (n = 70)	Waist ES
Body weight (kg)	66.1 ± 1.2[a]	86.7 ± 1.4***	<u>204</u>	75.1 ± 1.5	81.4 ± 2.5*	42	68.8 ± 1.4	85.3 ± 1.8***	133
BMI (kg/m^2)	23.5 ± 0.5	28.3 ± 0.7***	104	24.7 ± 0.4	27.9 ± 0.6***	85	22.9 ± 0.3	28.7 ± 0.4***	<u>214</u>
Fat mass (kg)	13.9 ± 0.6	25.2 ± 1.2***	171	16.8 ± 0.8	24.6 ± 1.2***	104	13.2 ± 0.5	25.8 ± 0.8***	<u>254</u>
Waist (cm)	79.2 ± 1.3	102.2 ± 1.4***	219	84.9 ± 1.3	99.9 ± 2.3***	113	78.2 ± 1.1	101.8 ± 1.3***	<u>252</u>
WHCR (ratio)	0.87 ± 0.01	0.96 ± 0.01***	116	0.87 ± 0.01	1.01 ± 0.01***	<u>182</u>	0.86 ± 0.01	0.97 ± 0.01***	140
SBP (mmHg)	125.7 ± 2.2	138.3 ± 2.1***	75	125.8 ± 1.8	141.8 ± 2.9***	91	122.3 ± 1.9	139.3 ± 2.2***	<u>106</u>
DBP (mmHg)	77.6 ± 1.1	83.9 ± 0.8***	86	77.8 ± 1.20	86.6 ± 1.6***	82	76.1 ± 1.2	85.0 ± 1.4***	<u>88</u>
TC (mmol/L)	5.53 ± 0.20	6.40 ± 0.18**	59	5.55 ± 0.19	6.34 ± 0.20**	53	5.02 ± 0.18	6.55 ± 0.18***	<u>108</u>
TG (mmol/L)	1.67 ± 0.26	3.86 ± 0.50***	<u>78</u>	2.01 ± 0.27	3.81 ± 0.52***	62	1.59 ± 0.29	3.46 ± 0.36***	74
HDL-C (mmol/L)	1.38 ± 0.03	1.14 ± 0.04***	<u>89</u>	1.31 ± 0.03	1.17 ± 0.05*	47	1.37 ± 0.03	1.16 ± 0.04***	77
Apo B (mg/dL)	81.9 ± 3.8	100.5 ± 4.1***	60	83.1 ± 4.3	96.8 ± 4.2**	41	72.9 ± 3.2	103.7 ± 4.7***	<u>101</u>
NEFA (mmol/L)	0.710 ± 0.070	1.280 ± 0.092***	91	0.818 ± 0.084	1.197 ± 0.103***	53	0.587 ± 0.058	1.318 ± 0.098***	<u>124</u>
Glucose (mmol/L)	4.88 ± 0.21	5.93 ± 0.24***	60	5.03 ± 0.18	5.95 ± 0.38*	46	4.6 ± 0.1	6.2 ± 0.3***	<u>118</u>
IRI (μU/mL)	6.44 ± 0.64	10.44 ± 1.02**	<u>64</u>	7.06 ± 0.68	10.88 ± 1.35**	52	5.59 ± 0.57	11.10 ± 1.05***	<u>91</u>
LDL-C (mmol/L)	3.10 ± 0.15	3.42 ± 0.13	29	3.08 ± 0.14	3.43 ± 0.19	28	2.78 ± 0.14	3.71 ± 0.15***	<u>82</u>
LDL-TG (mmol/L)	0.49 ± 0.03	0.72 ± 0.07***	<u>64</u>	0.53 ± 0.04	0.68 ± 0.05***	44	0.47 ± 0.04	0.69 ± 0.05***	63
LDL-apo B (mg/dL)	69.1 ± 4.7	83.2 ± 5.6*	35	70.1 ± 4.8	80.7 ± 6.5	25	58.8 ± 3.2	87.5 ± 6.1***	<u>83</u>
VLDL-C (mmol/L)	0.52 ± 0.08	1.13 ± 0.09***	<u>92</u>	0.64 ± 0.08	1.14 ± 0.12***	66	0.51 ± 0.09	1.10 ± 0.09***	84
VLDL-TG (mmol/L)	0.92 ± 0.17	2.18 ± 0.23***	<u>82</u>	1.15 ± 0.16	2.25 ± 0.31***	64	0.89 ± 0.18	2.06 ± 0.21***	77
VLDL-apo B (mg/dL)	7.4 ± 1.2	15.8 ± 1.7**	76	8.4 ± 1.2	16.5 ± 2.5***	61	6.7 ± 1.2	15.4 ± 1.7***	<u>78</u>
LDL-BA (arb. unit)[b]	0.292 ± 0.009	0.294 ± 0.009	3	0.300 ± 0.007	0.292 ± 0.014	10	0.297 ± 0.009	0.293 ± 0.010	5
LDL lag time (min)	98.9 ± 4.3	94.4 ± 3.4	15	99.7 ± 3.9	88.6 ± 4.2	36	105.9 ± 4.7	87.7 ± 3.4**	<u>277</u>
VLDL-BA (arb. unit)	0.481 ± 0.032	0.428 ± 0.026	24	0.475 ± 0.031	0.428 ± 0.036	18	0.489 ± 0.038	0.434 ± 0.030	20
VLDL lag time (min)	302.1 ± 14.1	258.0 ± 14.5	40	320.6 ± 14.5	243.9 ± 15.6*	<u>66</u>	322.2 ± 15.7	248.8 ± 14.7*	62
LDL-PC LA (M%)[d]	24.58 ± 0.36	21.87 ± 0.63***	<u>80</u>	23.64 ± 0.49	22.62 ± 0.52	26	24.41 ± 0.39	21.90 ± 0.61***	66

[a]Mean ± SEM; statistical analysis (Mann–Whitney U test): *$p < 0.05$, **$p < 0.01$, ***$p < 0.001$. [b]Arbitrary unit (A_{234nm}). [c]Effect size: for calculation, see text; bold underlined numbers indicate the highest value. [d]Content of linoleic acid in LDL phosphatidylcholine (molar %).

LP composition (with the exception of plasma, LDL, and VLDL TG; HDL-C; VLDL-C; and LDL PC LA content). Waist circumference was the best discriminator of LDL resistance to oxidation as well. The results indicated that persons with intra-abdominal fat accumulation revealed not only disturbed glucose homeostasis, but also unfavorable composition of LDL and its increased oxidability.

ACKNOWLEDGMENTS

This study was supported by Research Project No. J 13/98 1111 0000 2-1 from the Ministry of Education, Czech Republic.

REFERENCES

1. POUILOT, M.C., J.P. DESPRÉS, S. LEMIEUX et al. 1994. Waist circumference and abdominal sagittal diameter: best simple anthropometric indexes of abdominal visceral adipose tissue accumulation and related cardiovascular risk in men and women. Am. J. Cardiol. **73:** 460–468.
2. RIMM, E.B., M.J. STAMFER, E. GIOVANNUCCI et al. 1995. Body size and fat distribution as predictors of coronary heart disease among middle-aged and older US men. Am. J. Epidemiol. **141:** 1117–1127.
3. HOPKINS, P.N., S.C. HUNT, L.L. WU et al. 1996. Hypertension, dyslipidemia, and insulin resistance: links in a chain or spokes on the wheel? Curr. Opin. Lipidol. **7:** 241–253.
4. CAREY, D.G.P. 1998. Abdominal obesity. Curr. Opin. Lipidol. **9:** 35–40.
5. STEINBERG, D. & A. LEWIS. 1997. Oxidative modification of LDL and atherogenesis. Circulation **18:** 1062–1071.
6. KOJDA, G. & D. HARRISON. 1999. Interaction between NO and reactive oxygen species: pathophysiological importance in atherosclerosis, hypertension, diabetes mellitus, and heart failure. Cardiovasc. Res. **43:** 562–571.
7. SEIDELL, J.C. & G. BOUCHARD. 1997. Visceral fat in relation to health: is it major culprit or simply an innocent bystander? Int. J. Obes. **21:** 626–631.
8. WHO. 1987. Measuring obesity: classification and description of anthropometric data. *In* Report on WHO Consultation on the Epidemiology of Obesity, Warsaw 1987. Regional Office of WHO. Copenhagen.
9. DURNIN, J.V.G.A. & J. WOMERSLEY. 1974. Body fat assessed from the total body density and its estimation from skinfold thickness: measurements on 481 men and women aged from 16 to 71 years. Br. J. Nutr. **32:** 77–97.
10. THE EXPERT PANEL. 1988. Report of the National Cholesterol Education Program Expert Panel on detection, evaluation, and treatment of high cholesterol in adults. Arch. Intern. Med. **148:** 36–69.
11. SCHUMACKER, V.N. & D.L. PUPPIONE. 1986. Sequential flotation ultracentrifuge. Methods Enzymol. **128:** 155–170.
12. TVRZICKÁ, E., E. CVRCKOVÁ, B. MÁCA & M. JIRÁSKOVÁ. 1994. Changes in the liver, kidney, and heart fatty acid composition following ibuprofen administration in mice. J. Chromatogr. **656:** 51–57.
13. PUHL, H., G. WAEG & H. ESTERBAUER. 1994. Methods to determine oxidation of low density lipoproteins. Methods Enzymol. **233:** 425–441.
14. LEMIEUX, I., A. PASCOT, C. COUILLARD et al. 2000. Hypertriglyceridemic waist: a marker of the atherogenic metabolic triad (hyperinsulinemia; hyperapolipoprotein B; small, dense LDL) in men? Circulation **102:** 179–184.
15. KAZDOVÁ, L., A. ŽÁK & A. VRÁNA. 1997. Increased lipoprotein oxidability and aortic lipid peroxidation in an experimental model of insulin resistance syndrome. Ann. N.Y. Acad. Sci. **827:** 521–525.

16. CHAIT, A., R.L. BARZG & D.L. TRIBBLE. 1993. Susceptibility of small, dense, low-density lipoproteins to oxidative modification in subjects with the atherogenetic lipoprotein phenotype, pattern B. Am. J. Med. **94:** 350–356.
17. ŽÁK, A., M. ZEMAN, S. ŠTÍPEK & E. TVRZICKÁ. 1998. Insulin resistance syndrome: fatty acid composition and lipoperoxidation of LDL. *In* Abstract Book of the 13th International Symposium on Drugs Affecting Lipid Metabolism, Florence (May 30 to June 3), p. 120.
18. DECSI, T., G. CSABI, K. TÖRÖK *et al.* 2000. Polyunsaturated fatty acids in plasma lipids of obese children with and without metabolic cardiovascular syndrome. Lipids **35:** 1179–1184.

Intima-Media Thickness and Atherosclerotic Plaques in Familial Defective Apolipoprotein B-100 and Familial Hypercholesterolemia

M. KAISER, T. TEMELKOVA-KURKTSCHIEV, AND M. HANEFELD

Center for Clinical Studies, Technical University Dresden, D-01307 Dresden, Germany

ABSTRACT: Familial defective apolipoprotein B-100 (FDB) is a genetic disorder characterized by a decreased binding of low-density lipoprotein (LDL) particles to the LDL receptor due to defective apo B-100. Impaired LDL clearance could also be due to defects of the LDL receptor (familial hypercholesterolemia, FH). FDB was suggested to be clinically indistinguishable from classical FH. The measurement of the intima-media thickness (IMT) is an accepted method for the direct evaluation of early atherosclerosis. Thus, the aim of this study was to examine the IMT in patients with FDB in comparison to FH. Our data indicate that IMT in FDB does not differ from IMT in FH.

KEYWORDS: intima-media thickness (IMT); familial defective apolipoprotein B-100 (FDB); familial hypercholesterolemia (FH); common carotid artery

INTRODUCTION

Familial defective apolipoprotein B-100 (FDB) is a dominantly inherited lipid disorder characterized by a decreased binding of low-density lipoprotein (LDL) particles to the LDL receptor (LDLR) due to defective apo B-100. References 1 and 2 described an adenine for guanine mutation in exon 26 of the apo-B gene that causes a glutamine for arginine substitution at amino acid position 3500 in apo B-100. It was shown by a fibroblast LDLR binding assay that isolated LDL with defective binding only possessed about 10% of normal affinity for the LDLR.[3] FDB is one of the most common single gene mutations responsible for primary hypercholesterolemia and coronary artery disease (CAD), with a prevalence in the United States, Canada, and Europe of about 1/700 to 1/500.[4,5]

Familial hypercholesterolemia (FH), an autosomal dominantly inherited disease, can be due to different genetic defects in the LDLR that also lead to impaired LDL clearance. More than 600 mutations of the LDLR gene have been identified.[6] Both FDB and FH[1,4,5] are associated with significantly elevated plasma total and LDL cholesterol levels, elevated extravascular cholesterol deposits, and an elevated risk

---

Address for correspondence: M. Kaiser, Center for Clinical Studies, Technical University Dresden, Fiedlerstrasse 34, D-01307 Dresden, Germany. Voice: +49-351-4400580; fax: +49-351-4400581.
KaiserMarion77@aol.com

TABLE 1. Baseline and lipid characteristics

	FDB ($n = 24$) (mean ± SD)	FH ($n = 67$) (mean ± SD)	p
Age (years)	51.8 ± 16.2	53.0 ± 13.6	n.s.
Sex (M/F)	9/15	29/38	n.s.
BMI (kg/m^2)	27.2 ± 4.8	25.6 ± 3.2	n.s.
WHR	0.86 ± 0.07	0.87 ± 0.07	n.s.
Smoking ($n/\%$)	2 (9%)	18 (27%)	n.s.
Blood pressure (mmHg) systolic	129 ± 12	130 ± 13	n.s.
Blood pressure (mmHg) diastolic	78 ± 11	82 ± 7	n.s.
Hypertension ($n/\%$)	8 (33%)	29 (43%)	n.s.
Average therapy period (years)	12.4 ± 7.6	13.5 ± 9.9	n.s.
Total cholesterol (mmol/L) [TC]	7.86 ± 1.17	8.64 ± 2.24	n.s.
LDL cholesterol (mmol/L) [LDL-C]	5.76 ± 1.19	5.89 ± 1.83	n.s.
Triglycerides (mmol/L) [TG]	1.64 ± 0.76	2.1 ± 1.17	<0.05
HDL cholesterol (mmol/L) [HDL-C]	1.28 ± 0.36	1.29 ± 0.32	n.s.
TC/HDL-C ratio	5.92 ± 1.68	6.27 ± 2.16	n.s.
TC/LDL-C ratio	4.38 ± 1.40	4.48 ± 1.93	n.s.
Apo B (g/L)	1.48 ± 0.36	1.65 ± 0.57	n.s.
Lp(a) (mg/L) (log)	518 ± 636	543 ± 724	n.s.
Fibrinogen (g/L)	3.8 ± 1.0	3.6 ± 1.0	n.s.

of premature CAD.[7] FDB is suggested to be clinically indistinguishable from FH;[8,9] however, there are controversial reports. The phenotypic expression of many mutations may vary more than previously thought. Studies demonstrated that the combination of tendon xanthoma, arcus lipoids, and premature coronary and carotid atherosclerosis is justified as well for FH as it is in FDB.[4] By way of contrast, reference 8 showed that hypercholesterolemia and premature atherosclerosis were more common in LDLR patients (FH) than in FDB. All in all, the data pool presents only a few possibilities for comparing FDB and FH. Besides, the two genetic lipid disorders have not been compared so far with respect to early atherosclerosis. Therefore, we examined the intima-media thickness (IMT) of the common carotid and superficial femoral artery and the atherosclerotic plaques by ultrasound in subjects with FDB in comparison to patients with FH.

MATERIALS AND METHODS

A total of 24 FDB subjects and 67 FH patients were identified among outpatients with hypercholesterolemia from the Institute for Clinical Metabolic Research (Technical University Dresden). No homozygous FDB or FH subjects were included. FDB was detected by amplification of exon 26 of the apo B gene using polymerase chain reaction, followed by digestion with Msp I and DNA fragment separation by electrophoresis. Criteria used to identify individuals with heterozygous FH included genetic defects (mutations in the LDLR), family (or personal) history of early CAD, and/or biochemical parameters (i.e., plasma LDL cholesterol). None of the examined patients had diabetes mellitus, renal or hepatic damage, or other diseases that could be the cause of secondary hyperlipoproteinemia. Eighty-seven (96%) of the patients were under treatment with lipid-lowering drugs. One patient with FDB and 3 with FH received no medication. The baseline characteristics of the examined subjects are shown in TABLE 1.

All participants answered a questionnaire on medical history, drug use, and lifestyle. With respect to smoking behavior, subjects were classified as "ever" smokers and nonsmokers. Measurement of systolic and diastolic blood pressure was conducted after a rest of at least 10 min in a sitting position. Subjects with an SBP > 160 mmHg or a DBP > 95 mmHg or who were taking antihypertensive medication were considered hypertensive. In addition, body mass index (BMI) and waist-to-hip ratio (WHR) were measured.

LABORATORY METHODS

Total cholesterol and plasma triglycerides were measured enzymatically using the PAP method. LDL cholesterol and HDL cholesterol were also determined enzymatically by commercially available test kits. Apo B was examined by a nephelometric immunoassay on a Behring nephelometric analyzer. The Lp(a) level was assessed by an immunolatex-enhanced immunoassay. Fibrinogen was examined by the Clauss method.

ULTRASONOGRAPHY

IMT of the common carotid artery (CCA) and the superficial femoral artery (SFA) was determined by B-mode ultrasonography.[10] Acuson 128XP with a linear 10/5 MHz transducer was used. B-mode scanning of the right and left CCA was conducted longitudinally, 0.5–1 cm proximal of the bulb. The mean of these four values was used to define the CCA-IMT mean. The carotid bifurcation was examined transversally. The maximal IMT of the CCA was determined by the investigator, independent of the localization. A double measurement of both SFAs (longitudinally) was performed, and the mean of these values presented the SFA-IMT mean. Plaque occurrence (carotid and femoral) was defined as a distinct area with an IMT that was more than 50% thicker than neighboring sites.

RESULTS

No difference was found between FDB and FH with respect to age, sex, BMI, WHR, systolic and diastolic blood pressure, hypertension, average therapy period,

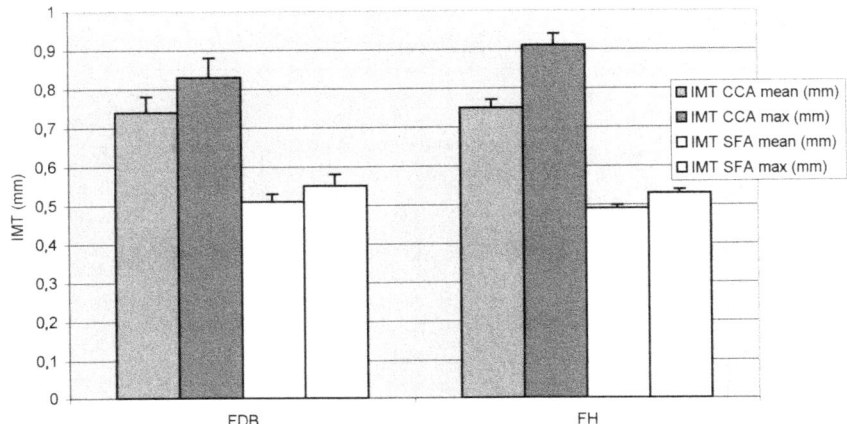

FIGURE 1a. Intima-media thickness in CCA and SFA in FDB and FH patients (mean and SEM).

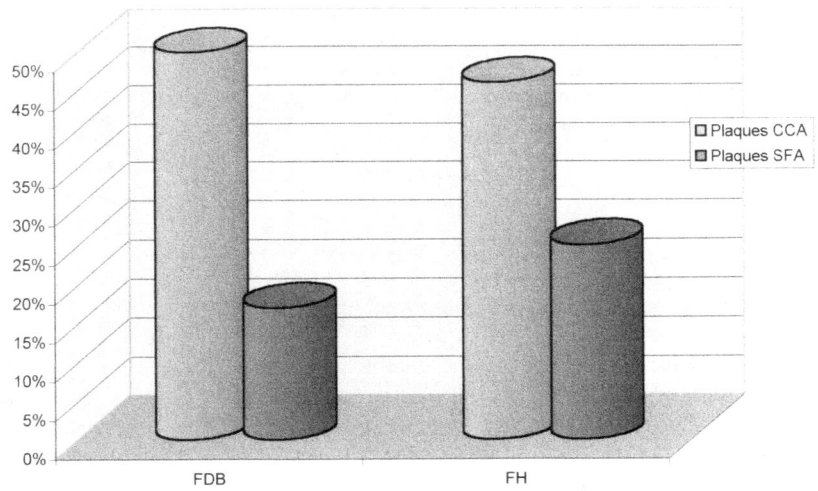

FIGURE 1b. Plaque occurrence in CCA and SFA in FDB and FH patients.

smoking habits, fibrinogen, and CRP. Similarly, the two groups did not differ with respect to lipoprotein parameters, except for plasma triglyceride level, which was significantly higher in the FH subjects (TABLE 1).

No significant difference was observed between the FDB group and the FH group for IMT and plaque occurrence of the CCA and SFA (FIGS. 1a and 1b). All IMT measurements of CCA and SFA correlate significantly with the plaque occurrence in these vessels ($p < 0.01$). The plaques in the CCA correlate significantly with

TABLE 2. Correlation between IMT and plaque occurrence

	CCA mean	CCA max	SFA mean	SFA max	Plaque-CCA
FDB					
CCA max	0.815 (<0.001)				
SFA mean	n.s.	n.s.			
SFA max	n.s.	n.s.	0.862 (<0.001)		
Plaque-CCA	0.446 (<0.05)	0.577 (<0.01)	0.479 (<0.05)	n.s.	
Plaque-SFA	n.s.	n.s.	0.538 (<0.01)	n.s.	0.513 (<0.01)
FH					
CCA max	0.838 (<0.001)				
SFA mean	n.s.	n.s.			
SFA max	n.s.	n.s.	0.929 (<0.001)		
Plaque-CCA	0.248 (<0.05)	n.s.	n.s.	n.s.	
Plaque-SFA	n.s.	n.s.	n.s.	n.s.	0.270 (<0.05)

plaque occurrence in the SFA. The mean IMT of the CCA and of the SFA are also significantly correlated ($p < 0.01$) (TABLE 2).

In both groups, the IMT and plaque occurrence correlate with age. The average therapy period also shows a significant relationship to IMT-CCA (mean, max) in FDB and FH.

DISCUSSION

FDB is a common, genetically determined lipid disorder, associated with hypercholesterolemia and accelerated atherosclerosis. Measurement of the arterial IMT by B-mode of ultrasound is a highly reproducible and suitable method for monitoring the early stages of atherosclerosis.[10–12] In the present study, we analyze for the first time the IMT and plaque occurrence of the CCA and SFA in FDB versus FH patients. No significant difference was found between the two groups, which were matched for age, sex, BMI, and known atherosclerosis risk factors. The results were confirmed after adjustment for triglyceride level, which was the only significantly different parameter between the FDB and FH subjects. Carotid IMT was somewhat higher in the FH than in the FDB group, but this difference did not reach statistical significance. Since the majority of the patients were under lipid-lowering agents (96% in each of the groups), the IMT measured is certainly reduced in comparison to untreated subjects. Under treatment, FH subjects had a slightly higher cholesterol level than FDB patients, but this difference was also not statistically significant. Our finding that triglyceride level was increased in FH, but not in FDB, is consistent with other reports.[9] This could be due to the fact that in FDB there is a normal clearance of the LDL precursors since the VLDL and IDL particles bind to the LDL receptor normally, with apo E as a ligand, whereas the receptor defect in FH affects the binding of the LDL precursors.

Several studies indicate that hypercholesterolemia is associated with IM thickening, which may be reduced by cholesterol lowering.[13] It is known that the mean carotid IMT is significantly greater in patients with FH than in healthy controls.[14] The same results are shown for the common femoral artery.[15] Additionally, the study group found a clear relationship between IMT in the carotid artery and the prevalence of plaque occurrence in the carotid and femoral arteries.[16] So far, there have been no data on IMT in FDB subjects.

The association of the CCA-IMT mean with plaque occurrence is observed in the examined hypercholesterolemic patients and represents a predictor of carotid plaque occurrence. No significant association between SFA-IMT and plaques in the same arterial system was found. It seems likely that some risk factors could be specifically associated with increased IMT alone or with plaques alone. A recent report of the EVA study[17] showed that a parental history of premature death from coronary heart disease was strongly associated with the presence of plaques, but not with increased IMT.

In conclusion, our data indicate that subjects with FDB do not differ significantly from those with FH with respect to early atherosclerosis when they receive cholesterol-lowering therapy.

REFERENCES

1. INNERARITY, T.L., *et al.* 1987. Familial defective apolipoprotein B-100: low density lipoproteins with abnormal receptor binding. Proc. Natl. Acad. Sci. U.S.A. **84:** 6919–6923.
2. SORIA, L.F., *et al.* 1989. Association between a specific apolipoprotein B mutation and familial defective apolipoprotein B-100. Proc. Natl. Acad. Sci. U.S.A. **86:** 587–591.
3. ARNOLD, K.S., *et al.* 1994. Isolation of allele-specific, receptor-binding defective low density lipoproteins from familial defective apolipoprotein B-100 subjects. J. Lipid Res. **35:** 1469–1476.
4. SCHUSTER, H., *et al.* 1990. Familial defective apolipoprotein B-100: comparison with familial hypercholesterolemia in 18 cases detected in Munich. Arteriosclerosis **10:** 577–581.
5. RAUH, G., *et al.* 1992. Familial defective apolipoprotein B100: clinical characteristics of 54 cases. Atherosclerosis **92:** 233–241.
6. HEALTH, K.E., *et al.* 2001. Low-density lipoprotein receptor gene (LDLR) world-wide website in familial hypercholesterolemia. Atherosclerosis **154:** 243–246.
7. INNERARITY, T.L., *et al.* 1990. Familial defective apolipoprotein B-100: a mutation of apolipoprotein B that causes hypercholesterolemia. J. Lipid Res. **31:** 1337–1349.
8. BRUGGER, D., H. SCHUSTER & N. ZOLLNER. 1996. Familial hypercholesterolemia and familial defective apolipoprotein B-100: comparison of the phenotypic expression in 116 cases. Eur. J. Med. Res. **1:** 383–386.
9. MISEREZ, A.R. & U. KELLER. 1995. Differences in the phenotypic characteristics of subjects with familial defective apolipoprotein B-100 and familial hypercholesterolemia. Arterioscler. Thromb. Vasc. Biol. **15:** 1719–1729.
10. PIGNOLI, P., E. TREMOLI & A. POLI. 1986. Intimal plus medial thickness of the arterial wall: a direct measurement with ultrasound imaging. Circulation **74:** 1399–1406.
11. ZUREIK, M., *et al.* 2000. Common carotid intima-media thickness predicts occurrence of carotid atherosclerotic plaques. Arterioscler. Thromb. Vasc. Biol. **20:** 1622–1629.
12. HANDA, N., M. MATSUMOTO & H. MAEDA. 1990. Ultrasonic evaluation of early carotid atherosclerosis. Stroke **21:** 1567–1572.
13. SMILDE, T.J., *et al.* 2000. The effect of cholesterol lowering on carotid and femoral artery wall stiffness and thickness in patients with familial hypercholesterolaemia. Eur. J. Clin. Invest. **30:** 473–480.

14. WENDELHAG, I., et al. 1992. Arterial wall thickness in familial hypercholesterolemia. Arterioscler. Thromb. **12:** 70–77.
15. WENDELHAG, I., et al. 1993. Atherosclerotic changes in the femoral and carotid arteries in familial hypercholesterolemia. Arterioscler. Thromb. **13:** 1404–1411.
16. WENDELHAG, I., et al. 1996. On quantifying plaque size and intima-media thickness in carotid and femoral arteries. Arterioscler. Thromb. Vasc. Biol. **16:** 843–850.
17. ZUREIK, M., et al. 1999. Differential association of common carotid intima-media thickness and carotid atherosclerotic plaques with parental history of premature death from coronary heart disease: the EVA study. Arterioscler. Thromb. Vasc. Biol. **19:** 366–371.

Acute Elevation of NEFA Causes Hyperinsulinemia without Effect on Insulin Secretion Rate in Healthy Human Subjects

BEATE BALENT, GAYOTRI GOSWAMI, GEORGE GOODLOE,
EDUARD ROGATSKY, OLIMPIA RAUTA, ROBERT NEZAMI, LISA MINTS,
RUTH HOGUE ANGELETTI, AND DANIEL T. STEIN

Albert Einstein College of Medicine, Yeshiva University, Bronx, New York, USA

ABSTRACT: Increased circulating levels of nonesterified free fatty acids (NEFA) have been observed in such hyperinsulinemic states as obesity, impaired glucose tolerance, diabetes, and dyslipidemia where they have been causally linked to the development of insulin resistance and hyperinsulinemia. The concentration of NEFA in plasma is believed to have direct modifying effects on insulin secretion and clearance. It remains controversial whether acute increases in NEFA potentiate insulin secretion in human subjects. We studied the effect of an acute elevation of NEFA during lipid-heparin infusion compared to a glycerol-only control on glucose-stimulated insulin secretion and clearance during a 120-min hyperglycemic (10 mM) clamp in 7 healthy normoglucose-tolerant volunteers. The metabolic clearance rate of C-peptide (MCR_{CP}) was measured in each subject during the study by simultaneous infusion of C-peptide. Insulin secretion rate (ISR) was calculated from deconvolution of C-peptide data after correction for the rate of C-peptide infusion. Clearance rate of insulin (MCR_{INS}) was calculated based upon endogenous ISR. Plasma glucose (mg/dL): basal (90–115 min) 90.2 ± 2.8 vs. 90.2 ± 2.3; clamp (150–240 min) 180.5 ± 2.8 vs. 180.9 ± 1.3. Plasma insulin (pmol/L): prebasal (fasting) 29.6 ± 10.0 vs. 29.8 ± 10.6; basal (90–115 min) 30.1 ± 9.2 vs. 34.5 ± 12.1; second phase clamp (210–240 min) 127.6 ± 18.2 vs. 182.5 ± 17.3*. Plasma NEFA (mM): prebasal 0.47 ± 0.08 vs. 0.52 ± 0.09; basal 0.35 ± 0.05 vs. 0.98 ± 0.02*; clamp (122–240 min) 0.06 ± 0.02 vs. 0.77 ± 0.06*. ISR (pmol/min): prebasal 72.7 ± 7.5 vs. 72.0 ± 7.9; second phase clamp (210–240 min) 268.5 ± 27.2 vs. 200.2 ± 23.7. MCR_{INS} (mL/min): prebasal 3393 ± 488 vs. 3370 ± 511; clamp 2284 ± 505 vs. 1214 ± 153* (*$p < 0.05$ glycerol vs. intralipid/heparin). This study demonstrates that acute NEFA elevation causes hyperinsulinemia due to a significant decrease in systemic insulin clearance without increasing rates of insulin secretion.

KEYWORDS: NEFA; free fatty acids; insulin secretion; C-peptide

Increased circulating levels of nonesterified free fatty acids (NEFA) have been observed in such hyperinsulinemic states as obesity, impaired glucose tolerance, diabetes, and dyslipidemia where they have been causally linked to the development

---

Address for correspondence: Daniel T. Stein, Department of Endocrinology, Albert Einstein College of Medicine, Yeshiva University, 1300 Morris Park Avenue, Bronx, NY 10461. Voice: 718-430-2446; fax: 718-430-8998.

dstein@aecom.yu.edu

of insulin resistance and hyperinsulinemia.[1–4] The concentration of NEFA in plasma is believed to have direct modifying effects on insulin secretion and clearance.[5–9] Studies with isolated beta cells or beta cell lines from rodents as well as humans, and in perfused pancreas models both *ex vivo* as well as *in vivo*, have clearly defined a stimulatory role for fatty acids on nutrient-stimulated insulin secretion. Nevertheless, despite this substantial body of data, the issue of whether *acute* increases in NEFA potentiate insulin secretion in human subjects continues to remain controversial. Studies in human subjects have variably found that acute elevations in plasma NEFA increase or have no effect on glucose-stimulated insulin secretion.[7–14] Differences in outcome may have been due to different experimental paradigms or methods for estimating insulin secretory rates, although similar methods have resulted in contradictory outcomes. Since insulin and C-peptide are secreted in a 1:1 molar ratio and the latter is not extracted by liver, the gold standard method of measuring insulin secretory rates has been to use deconvolution of C-peptide pharmacokinetics after measurement of C-peptide clearance rates by C-peptide bolus. No data are currently available as to whether insulin secretory rates differ when C-peptide clearance is measured during the study in the setting of normal or elevated free fatty acids. The aim of our study, therefore, was to determine the effect of acute elevations in plasma NEFA with lipid/heparin vs. glycerol/saline control on glucose-stimulated insulin secretion during a 10 mM (180 mg/dL) hyperglycemic clamp. Due to the importance of C-peptide clearance kinetics for each individual, the C-peptide clearance rate was specifically measured in each subject during the course of each study.

STUDY DESIGN AND METHODS

Subjects

Seven healthy research volunteers (3 males/4 females; age, 36.2 ± 3.7 years; BMI, 22.9 ± 0.8 kg/m^2; BSA, 1.76 ± 0.07) underwent two separate studies—intralipid/heparin versus glycerol/saline control—in random order with at least a 7-day interstudy interval. Each study subject had normal glucose tolerance as determined by a 75-g oral glucose tolerance test. All women were premenopausal. None of the study subjects were on any medication. After being informed of the goals and risks of the study, written informed consent was obtained from all participants prior to participation. The study was approved by the Committee on Clinical Investigations (IRB) of the Albert Einstein College of Medicine.

Hyperglycemic Clamp Studies

Subjects were asked to follow a high-carbohydrate diet containing at least 50 g carbohydrate for 3 days prior to study. No other restrictions were placed on dietary fat intake. All studies were performed in the morning after a 14-h overnight fast. Two basal blood samples were taken before any infusions were started. Each study consisted of two blocks of 120 min each. Beginning at time −10 min, a 20% intralipid (Baxter, Deerfield, IL) emulsion/heparin (50 mL/m^2/h, 1000 U bolus followed by 1000 U/m^2/h) or a glycerol/saline (2 g/m^2/h) infusion was started. At time 0, boluses of porcine insulin tracer (3.2 mU [19.2 pmol]/m^2) and C-peptide (60 pmol/kg), both over 2 min, were given simultaneously, followed by constant infusion rates of 0.16 mU/m^2/min (0.96 pmol) for insulin and 3.2 pmol/kg/min for C-peptide. The infused C-peptide was produced in-house by solid phase synthesis and purified by reverse phase HPLC. It was tested for pyrogenicity and sterility prior to use. Full details will be reported separately. At time 120 min, the hyperglycemic clamp started with a glucose bolus (6.5 g/m^2) and another porcine insulin tracer bolus (8 mU/m^2)/[48 pmol], followed by a variable infusion of 20% dextrose to clamp plasma glucose at 10 mM (180 mg/dL) and a constant infusion rate of 0.56 mU/m^2/min [3.36 pmol] for the porcine insulin tracer. C-peptide infusion rate remained the same during the study.

Plasma samples were taken for NEFA, TG, glycerol, C-peptide, and insulin at time −20, −10, −5, 90, 105, 110, 115, 122, 125, 130, 150, 180, 210, 220, 230, and 240 min and immediately chilled on ice. A variable 20% glucose infusion was administered to clamp the glucose at 180 mg/dL after beginning the hyperglycemic clamp (time, 120 to 240 min) with plasma glucose monitored by a glucose analyzer every 5 min.[8]

Laboratory Methods

All samples were kept on ice until centrifuged (within 30 min). Plasma samples were stored frozen at −20°C for later assays. The blood samples for NEFA and glycerol were collected into EDTA tubes containing the lipase inhibitors, tetrahydrolipstatin and paraoxon, to prevent ongoing *in vitro* lipolysis.[11,15] Plasma total insulin was measured by ELISA kit according to the manufacturer's instructions (Alpco, Inc., Windham, NH). The insulin antibody is specific for intact insulin and is known to cross-react between human and porcine insulin. C-peptide concentrations were measured by in-house radioimmunoassay using specific antibody from Linco, Inc. (St. Charles, MO), and WHO reference standards (NIBSC, Herts, United Kingdom). Increases in plasma NEFA and TG do not interfere with quantitative measurement of either C-peptide or insulin in these assays (data not shown). NEFA, glycerol, and TG were measured by an enzymatic colorimetric method as previously described.[8]

Calculations

The metabolic clearance rate of C-peptide (MCR$_{CP}$) was measured in each study subject during the basal period of each study by simultaneous infusion of C-peptide. The C-peptide infusion rate divided by the increment in C-peptide concentration was used to calculate the clearance rate of C-peptide:[16]

MCR$_{CP}$ = C-peptide infusion rate/Δ[C-peptide]$_{\text{[basal mean (90–120 min) – prebasal mean]}}$.

Insulin secretion rate (ISR) was calculated from deconvolution of C-peptide data after correction for the increase in C-peptide concentration due to exogenous infusion:[16,17]

ISR = MCR$_{CP}$ × [CP$_{net}$]

where the [CP$_{net}$] was calculated as the total concentration of C-peptide at time t minus the increment due to constant C-peptide infusion,

[CP]$_t$ − [ΔCP]$_{\text{[basal mean (90–120 min) – prebasal mean]}}$.

TABLE 1. Glucose, insulin, NEFA, and TG during glycerol/saline control study and intralipid/heparin study

	Glycerol/saline control	Intralipid/heparin
Plasma glucose (mg/dL)		
basal (90–115 min)	90.2 ± 2.8	90.2 ± 2.3
clamp (150–240 min)	180.5 ± 2.8	180.9 ± 1.3
Plasma insulin (pmol/L)		
prebasal (fasting, before study started)	29.6 ± 10.0	29.8 ± 10.6
basal (90–115 min)	28.1 ± 9.2	34.5 ± 12.1
first phase clamp (122–130 min)	157 ± 22	154 ± 22
second phase clamp (210–240 min)	131 ± 18	165 ± 17*
Plasma NEFA (mM)		
prebasal	0.47 ± 0.08	0.52 ± 0.09
basal	0.35 ± 0.05	0.98 ± 0.02*
clamp (122–240 min)	0.06 ± 0.02	0.77 ± 0.06*
Plasma TG (mg/dL)		
prebasal	68.8 ± 5.1	56.8 ± 8.3
basal	72.9 ± 6.2	172.6 ± 27.5*
clamp (122–240 min)	67.2 ± 4.4	229.4 ± 32.8*

NOTE: *$p < 0.05$.

Prebasal insulin secretion rates from both studies were calculated using the MCR_{CP} measured during the glycerol-only study to avoid potential bias from the lipid infusion on C-peptide clearance. Individual C-peptide clearance rates were applied to the hyperglycemic period for the individual studies to calculate ISRs. Clearance rate of insulin was calculated based upon endogenous ISR without correction for porcine insulin tracer since the tracer infusion rate was less than 5% of the ISR:

$$\text{insulin clearance} = \text{ISR}/[\text{insulin}].$$

No estimate of insulin clearance was attempted during the basal period due to the potential confounding effect of simultaneous lipid and C-peptide infusion in the setting of low endogenous C-peptide concentrations.

Statistical Analysis

Results are expressed as means ± SE. All results were compared by a two-tailed t test, with $p < 0.05$ being significant.

RESULTS

Profiles of plasma glucose, insulin, TG, and NEFA concentrations achieved during glycerol/saline or intralipid/heparin studies, both in basal state and during hyperglycemic clamp, are shown in TABLE 1. The average levels of glycemia during the basal period as well as during the hyperglycemic clamp were well matched (90.2 ± 2.8 vs. 90.2 ± 2.3, and 180.5 ± 2.8 vs. 180.9 ± 1.3, for the control and lipid studies, respectively). The amount of glucose infusion required to maintain 10 mM

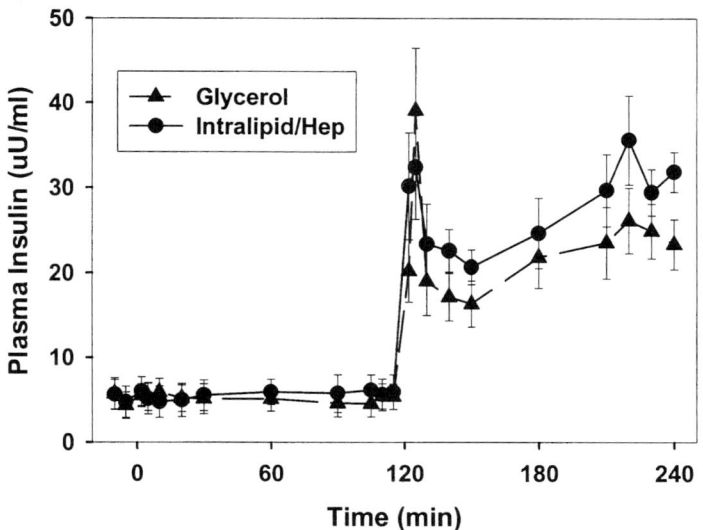

FIGURE 1. Plasma insulin response to elevated NEFA during hyperglycemic clamp.

plasma glucose was also similar: 258.3 ± 36.6 mg/m^2/min for control vs. 229.7 ± 55.4 mg/m^2/min for the lipid study. The intralipid/heparin infusion raised plasma NEFA approximately 3-fold over control levels during the basal period. During the hyperglycemic clamp, NEFA levels declined as expected to 0.06 mM due to the known antilipolytic effects of insulin, while they were maintained at 0.77 mM during lipid/heparin infusion. TG levels remained approximately at the same level throughout the control study (~70 mg/dL), but were elevated 2- to 3-fold during intralipid/heparin infusion.

As shown in FIGURE 1 (and TABLE 1), the fasting plasma insulin levels were identical (29.6 ± 10.0 vs. 29.8 ± 10.6 pmol/L), but tended to be higher during the steady-state phase of the baseline period of the intralipid study; however, this difference was not statistically significant (30.0 ± 9.2 vs. 35.0 ± 12.1 pmol/L). Plasma insulin levels during the first phase (122–130 min) of the hyperglycemic clamp were similarly elevated in the two groups (157 ± 22 pmol/L in the control study vs. 154 ± 22 pmol/L in the intralipid study). During the second phase of the glucose clamp, plasma insulin increased significantly more in the lipid-infused group and was 31% higher than during the control study over the last 30 min of the hyperglycemic clamp (135 ± 18 vs. 176 ± 17 pmol/L; $p < 0.05$).

Plasma C-peptide concentrations are presented in FIGURE 2. Fasting (prebasal) C-peptide concentrations were similar (441 ± 17 vs. 410 ± 25 pmol/L), as were the increments in C-peptide concentrations during the basal periods (data not shown). Net C-peptide concentrations were also similar during hyperglycemic clamp, during both first and second phases, with a trend to being lower during the latter second phase (FIG. 2).

TABLE 2. Insulin secretion rate and metabolic clearance rates of insulin and C-peptide levels during glycerol/saline control study and intralipid/heparin study

	Glycerol/saline control	Intralipid/heparin
ISR (pmol/min)		
prebasal (fasting, before study started)	72.7 ± 7.5	72.0 ± 7.9
first phase clamp (122–130 min)	201.7 ± 21.4	192.2 ± 24.9
second phase clamp (210–240 min)	268.5 ± 27.2	200.2 ± 23.7*
MCR_{INS} *(mL/min)*		
prebasal	3393 ± 488	3370 ± 511
clamp	2284 ± 505	1214 ± 153*
MCR_{CP} *(mL/min)*		
basal (0–120 min)	154.8 ± 13.0	152.2 ± 13.0

NOTE: $*p < 0.05$.

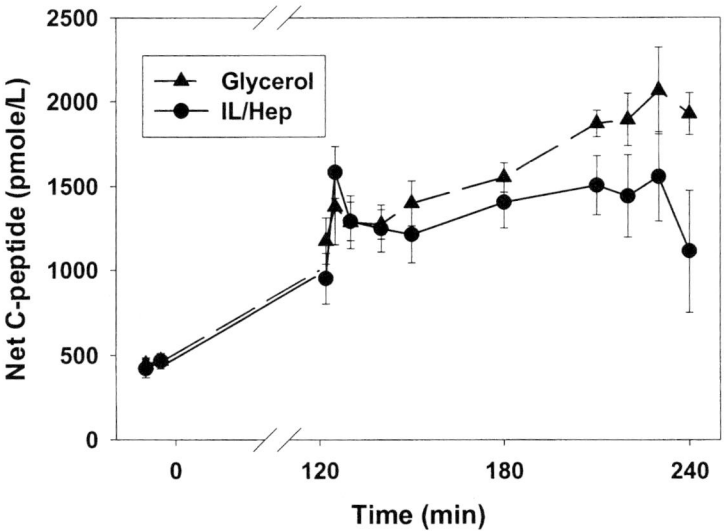

FIGURE 2. Net plasma C-peptide response over prebasal values to elevated NEFA during hyperglycemic clamp. Values from 122–240 min represent the increment due to stimulated output.

TABLE 2 shows data on rates of endogenous insulin secretion, metabolic clearance of insulin, and metabolic clearance of C-peptide levels during the glycerol/saline control and intralipid/heparin studies. The ISR was calculated from deconvolution of net C-peptide concentrations using the measured C-peptide clearance rates. Fasting (prebasal) ISR was comparable during the control and lipid infusion studies (72.7 ± 7.5 vs. 72.0 ± 7.9 pmol/min). When the plasma glucose was clamped at 10 mmol/L, the ISR rose to 268.5 ± 27.2 pmol/min during the control day, but only to 200.2 ± 23.7 pmol/min during the intralipid study.

Insulin clearance was similar in the fasting (prebasal) period during the two study days (3393 ± 488 vs. 3370 ± 511 mL/min). In contrast, insulin clearance during hyperglycemic clamp was 47% lower during the intralipid study compared to control (2284 ± 505 vs. 1214 ± 153 mL/min, $p < 0.05$). Metabolic clearance rate of C-peptide was similar for both studies (155 ± 13 vs. 152 ± 13 mL/min).

DISCUSSION

This is the first study to examine the acute effects of free fatty acids on insulin secretion and insulin clearance where C-peptide clearance rates have been measured during the study day in question, that is, under identical conditions of elevated or suppressed plasma fatty acids. To our surprise, and in strong contrast to several reports in animals as well as humans showing that elevated NEFA augment glucose-stimulated insulin secretion, the present study clearly demonstrates a strong effect to decrease endogenous insulin clearance that totally accounted for the degree of hyperinsulinemia since insulin secretion was not only not increased, but there was a trend toward decreased second phase output. These results are consistent with those reported by Hennes et al.[10] They found that acute increases in NEFA for 3–6 h during an 11 mmol/L hyperglycemic clamp in lean normal female volunteers resulted in a 50% increase in plasma insulin concentrations, determined to be entirely due to reduction in insulin clearance. This group measured C-peptide clearance by bolus injection in each subject on a separate occasion prior to the secretory studies. Interestingly, this group also examined the effect of elevated NEFA with more modest hyperglycemia (7 mmol/L), where they observed a 34% increase in insulin concentration that was due in equal measures to modest increases in insulin secretion and decreased insulin clearance.

Frias et al. performed a 12 mmol/L hyperglycemic clamp in nonoverweight women after 5.5–7 h of lipid/heparin infusion to raise NEFA and found no differences in either stimulated insulin or C-peptide concentrations. Interestingly, basal insulin and C-peptide concentrations were higher during the intralipid/heparin study, although this was not significant. Even though this group did not measure C-peptide clearance rates, they concluded that elevated NEFA for 4-7 h had no net effect on glucose-stimulated insulin secretion or clearance.

In contrast to the present study and that of Hennes et al. described above, several other studies have observed hyperinsulinemia consistent with increased insulin secretion in the setting of hyperlipacidemia.[7,8,11,14] Paolisso et al. observed a significant relationship between the increase in plasma NEFA concentration during 6 h of intralipid/heparin and the acute insulin response to glucose during an intravenous glucose tolerance test.[14] This study, however, did not measure C-peptide; thus, the changes observed might have been influenced by alterations in insulin clearance. Several other studies have examined the effects of elevated NEFA during hyperglycemia on insulin and C-peptide and found changes consistent with lipid-mediated increases in insulin secretion.[7–9,11] Boden and colleagues observed a strong, but nonsignificant trend toward hyperinsulinemia during 48 h of lipid/heparin infusion in conjunction with clamping at approximately 9 mmol/L glycemia. Interestingly, they found no changes in insulin clearance even after 24 h, but only a strong trend toward increased insulin secretion. These investigators specifically measured C-peptide clearance by bolus injection on a separate occasion. Dobbins *et al.* observed a significant increase in C-peptide output during IVGTT when NEFA were increased with lipid/heparin infusion for 4 h compared to normal basal NEFA or when NEFA were decreased with the antipolytic agent, nicotinic acid. Similarly, Carpentier *et al.*[11] found that acute exposure for 3–5 h to increased NEFA with lipid/heparin increased calculated ISRs by 17% during a 10 mmol/L hyperglycemic clamp without any change in insulin clearance. Subsequent elevations in plasma glucose to 20 mmol/L resulted in a 50% increase in calculated ISRs, but only a 7% increase in insulin concentrations. These paradoxical results of calculated insulin secretion increasing out of proportion to circulating insulin concentrations are consistent with either lipid-induced increases in insulin clearance rates or potentially a decrease in C-peptide clearance rates. The former possibility seems unlikely as NEFA within the physiological range have been shown to decrease insulin clearance in a concentration-dependent manner in primary rat hepatocytes[18] or when perfused into the portal vein of rat livers.[19] Similar results have demonstrated that elevated fatty acids decrease both peripheral as well as first-pass hepatic insulin extraction *in vivo* in dogs.[20] It has not previously been addressed whether elevated increases in NEFA alter C-peptide concentrations.

Our results suggest that, if there is an effect of acute changes in NEFA on C-peptide clearance, it is a small one. The increment in C-peptide concentration was increased by <5% during the intralipid/heparin study, resulting in the calculated decrease in C-peptide clearance of similar degree. Using the present methodology, it is not possible to differentiate a <5% decrease in C-peptide clearance from a <5% increase in total C-peptide appearance rate due to lipid-induced increases in insulin secretion during the basal period. Nevertheless, the error introduced is small; thus, this would not change the main conclusion of the present study, namely, that acute changes in plasma NEFA cause hyperinsulinemia without changing ISRs.

An obvious question that arises in the context of the discrepant results of acute NEFA effects on insulin secretion among these well-controlled studies is whether they are the result of additional experimental factors that have not been previously identified. An important technical variable may be matrix effects of increased plasma lipid on insulin and C-peptide immunoassays.[21] Hydrophobic plasma lipids, similar to surfactants, can weaken antigen-antibody affinity interactions and could potentially result in higher or lower peptide values depending upon the specifics of the assay. While we found no lipid matrix effects in the ELISA and RIA assays used in this study, this has not to our knowledge been reported for other insulin and C-peptide assays. Another potential variable is gender. Studies of lean men have found increased NEFA-mediated insulin secretion,[7,11] while those restricted to nonoverweight women[10,13] have suggested no change in insulin secretion and, in at least one study, a decrease in insulin clearance.[10] Precedent for a gender effect on NEFA-mediated insulin sensitivity has recently been demonstrated, with healthy women being relatively nonsusceptible to lipid-induced insulin resistance.[22] Whether this and other experimental variables will provide an improved framework for understanding lipid-insulin pharmacokinetic interactions remains to be established.

In summary, we demonstrate for the first time that, when C-peptide pharmacokinetics are measured during hyperlipacidemia, acute (4 h) elevations in plasma NEFA cause marked defects in insulin clearance with no increase in insulin secretion. These results challenge the recently popular conception that NEFA acutely stimulate insulin secretion in healthy human subjects. In conclusion, this study demonstrates that acute NEFA elevation causes hyperinsulinemia without increasing rates of insulin secretion in healthy normal-weight subjects.

REFERENCES

1. BODEN, G. 1996. Fatty acids and insulin resistance (review). Diabetes Care **19:** 394–395.
2. CHEN, Y.D., *et al.* 1987. Resistance to insulin suppression of plasma free fatty acid concentrations and insulin stimulation of glucose uptake in noninsulin-dependent diabetes mellitus. J. Clin. Endocrinol. Metab. **64:** 17–21.
3. BODEN, G., *et al.* 1994. Mechanism of fatty acid induced inhibition of glucose uptake. J. Clin. Invest. **93:** 2438–2446.
4. RODEN, M., *et al.* 1996. Mechanism of free fatty acid–induced insulin resistance in humans. J. Clin. Invest. **97:** 2859–2864.
5. STEIN, D.T., *et al.* 1996. Essentiality of circulating fatty acids for glucose-stimulated insulin secretion in the fasted rat. J. Clin. Invest. **97:** 2728–2735.
6. DOBBINS, R.L., *et al.* 1998. A fatty acid–dependent step is critically important for both glucose- and non-glucose-stimulated insulin secretion. J. Clin. Invest. **101:** 2370–2376.
7. BODEN, G., *et al.* 1995. Effects of a 48-h fat infusion on insulin secretion and glucose utilization. Diabetes **44:** 1239–1242.
8. DOBBINS, R.L., *et al.* 1998. Circulating fatty acids are essential for efficient glucose-stimulated insulin secretion after prolonged fasting in humans. Diabetes **47:** 1613–1618.
9. BODEN, G., *et al.* 1998. Acute lowering of plasma fatty acids lowers basal insulin secretion in diabetic and non-diabetic subjects. Diabetes **47:** 1609–1612.
10. HENNES, M.M., *et al.* 1997. Effects of free fatty acids and glucose on splanchnic insulin dynamics. Diabetes **46:** 57–62.
11. CARPENTIER, A., *et al.* 1999. Acute enhancement of insulin secretion by NEFA in humans is lost with prolonged FFA elevation. Am. J. Physiol. **276:** E1055–E1066.
12. AMERY, C.M., *et al.* 2000. Elevation of plasma fatty acids by ten-hour intralipid infusion has no effect on basal or glucose-stimulated insulin secretion in normal man. Metabolism **49:** 450–454.
13. FRIAS, J.P., *et al.* 2000. Lack of effect of a physiological elevation of plasma non-esterified fatty acid levels on insulin secretion. Diabetes Metab. **26:** 133–139.
14. PAOLISSO, G., *et al.* 1995. Opposite effects of short- and long-term fatty acid infusion on insulin secretion in healthy subjects. Diabetologia **38:** 1295–1299.
15. ZAMBON, A., *et al.* 1993. Analysis of techniques to obtain plasma for measurement of levels of free fatty acids. J. Lipid Res. **34:** 1021–1022.
16. POLONSKY, K., *et al.* 1986. Use of biosynthetic human C-peptide in the measurement of insulin secretion rates in normal volunteers and type 1 diabetic patients. J. Clin. Invest. **77:** 98–105.
17. KAHN, S.E., *et al.* 1997. Assessment of beta cell function in humans: approach and interpretation. *In* Clinical Research in Diabetes and Obesity. Volume 1, pp. 3–22. Humana Press. Totowa, New Jersey.
18. WIESENTHAL, S.R., *et al.* 1999. Free fatty acids impair hepatic insulin extraction *in vivo*. Diabetes **48:** 766–774.
19. SVEDBERG, J., *et al.* 1990. Free fatty acid inhibition of insulin binding, degradation, and action in isolated rat hepatocytes. Diabetes **39:** 570–574.
20. SVEDBERG, J., *et al.* 1991. Fatty acids in the portal vein of the rat regulate hepatic insulin clearance. J. Clin. Invest. **88:** 2054–2058.
21. CHARD, T. 1987. An introduction to radioimmunoassay and related techniques. *In* Lab Techniques in Biochemistry and Molecular Biology. Volume 6, part 2. Elsevier. Amsterdam/New York.
22. FRIAS, J.P., *et al.* 2001. Decreased susceptibility to fatty acid–induced peripheral tissue insulin resistance in women. Diabetes **50:** 1344–1350.

Trans Fatty Acids in Subcutaneous Fat of Pregnant Women and in Human Milk in the Czech Republic

PAVEL DLOUHÝ,[a] EVA TVRZICKÁ,[a] BARBORA STANKOVÁ,[a] MARTA BUCHTÍKOVÁ,[a] RAJMUND POKORNÝ,[a] OLGA WIEREROVÁ,[b] DIANA BÍLKOVÁ,[c] JOLANA RAMBOUSKOVÁ,[a] AND MICHAL ANDEL[a]

[a]Charles University, Prague, Czech Republic

[b]Královské Vinohrady University Hospital, Prague, Czech Republic

[c]University of Economics, Prague, Czech Republic

ABSTRACT: Using capillary gas chromatography, we determined total content of *trans* fatty acids (TFA) and C18:1 *trans* fatty acids in human milk and subcutaneous fat in 35 healthy Prague women. The average content of TFA in human milk fat was 4.22% (SD = 1.87%) of all fatty acids, and the value of *trans* C18:1 isomers was 3.63% (SD = 1.81%). The average concentration of total *trans* fatty acids in subcutaneous fat was 4.41% (SD = 0.79%) and the average content of C18:1 *trans* isomers was 2.81% (SD = 0.61%).

KEYWORDS: *trans* fatty acids; human milk; subcutaneous fat; pregnancy; nutrition

INTRODUCTION

Trans fatty acids are unsaturated fatty acids, which have one or more double bonds in *trans* configuration. Their dietary sources are hardened fats and fats of ruminants. It is suspected that *trans* fatty acids could be one of the coronary heart disease risk factors[1,2] and that they have adverse effect on the metabolism of essential fatty acids and fetus development.[3]

OBJECTIVES

The aim of this study was to identify the total content of *trans* fatty acid isomers and C18:1 *trans* isomers in human milk and subcutaneous fat samples from women immediately after delivery, as the indicator of dietary exposure.

Address for correspondence: Pavel Dlouhý, Third Faculty of Medicine, Charles University, Ruská 87, 100 00 Prague 10, Czech Republic. Fax: 420-2-67102620.

Pavel.Dlouhy@lf3.cuni.cz

TABLE 1. Content of total *trans* fatty acids and C18:1 *trans* fatty acids in subcutaneous fat and fat of human milk

Sample	n	Σ total *trans* fatty acids				Σ *trans* C18:1			
		x	SD	Min	Max	x	SD	Min	Max
Subcutaneous fat	35	4.41%	0.79%	2.30%	5.96%	2.81%	0.61%	1.30%	4.24%
Fat of human milk	35	4.22%	1.87%	1.84%	9.78%	3.63%	1.81%	1.39%	9.19%

METHODS

We collected samples of subcutaneous fat and human milk from 35 healthy Prague women. All women were informed about the aim of the study and signed the informed consent. A sample of subcutaneous fat (~0.5 g) from perineum was taken after episiotomy. Samples of human milk (late colostrum or transitional milk) were obtained during the first week after delivery. After preseparation by thin-layer chromatography, fatty acids were esterified by methanol and quantified by capillary gas chromatography. The analyses themselves were carried out on a capillary gas chromatograph, Chrompack CP 9001 (Chrompack, Middelburg, the Netherlands), equipped with a split/splitless injector and a flame-ionizing detector. Columns: CP-Sil 88 (Chrompack), 100 m long, internal diameter of 0.25 mm, and stationary phase thickness of 0.25 µm; CP-WAX 52 CB (Chrompack), 25 m long, i.d. of 0.25 mm, and df of 0.25 µm.

Thirty-five fatty acids in total were identified in the individual samples, including unsaturated fatty acids containing one or several double bonds in a *trans* configuration.

The obtained results were compared with published results from other countries.

RESULTS

The average concentration of total *trans* fatty acids (TFA) in subcutaneous fat in women ($n = 35$) was 4.41% (SD = 0.79%) and the average content of C18:1 *trans* was 2.81% (SD = 0.61%).

Total *trans* fatty acids and C18:1 *trans* fatty acids in subcutaneous fat were not related to birth weight or gestational age.

The average total *trans* fatty acid content in human milk fat was 4.22% (SD = 1.87%) and the value of *trans* C18:1 isomers was 3.63% (SD = 1.81%).

The results are shown in TABLE 1.

DISCUSSION

For comparison, the *trans* fatty acid content in maternal milk was found to be 1.2% in Spain, 4.4% in Germany, and 7.19% in Canada.[4-6]

In the United States, the content of TFA in adipose tissue was approximately 4% (t C18:1 + ct, ts, and tt C18:2 from partially hydrogenated vegetable oils: 3.5%).[2] In

TABLE 2. *Trans* fatty acids in subcutaneous fat and human milk (results of previous studies from other countries)

Country	Sample	Total TFA	*Trans* C18:1	Reference
The Netherlands	Adipose tissue	4.9%	3.1%	Katan et al.[7]
The Netherlands	Adipose tissue	4.9%	3.1%	Van Staveren et al.[8]
United States	Adipose tissue	4.1%	2.7%	Hudgins et al.[9]
United States	Adipose tissue	4.3%	2.9%	London et al.[10]
European countries	Adipose tissue		1.6%	Aro et al.[11]
Spain	Human milk	1.2%	0.95%	Boatella et al.[4]
Germany	Human milk	4.4%		Koletzko et al.[5]
Canada	Human milk	7.2%		Chen et al.[6]

the Netherlands, Katan et al. and Van Staveren et al. found 4.9% of TFA average content in subcutaneous fat (3.1% C18:1 *trans*).[7,8] In the United States, Hudgins et al. found 4.1% of total TFA in human adipose tissue, out of which 2.7% was C18:1 *trans*.[9] Similarly, London et al. found 4.3% of total TFA, out of which 2.9% was C18:1 *trans*.[10] Aro et al. presents lower values in nine countries (1.6% TFA) than those found in the Czech Republic.[11] The results of previous studies from other countries are shown in TABLE 2.

Content of TFA in subcutaneous fat and human milk has been related to the types of fats present in the diet of pregnant women or nursing mothers. In the Czech Republic, most spread margarines now have only a trace or very low TFA content. The average TFA content in half-hard margarines is, however, still high. Out of nine products analyzed from various producers, six had more than 20% of TFA in fat and only two had their content of TFA under 2%. As for shortenings, the situation is even worse. Only one out of four products analyzed had its TFA content in fat lower than 1%. The remaining three products had 27.0%, 27.9% and even 38% of TFA in fat.[12] TFA content in other food products (waffles, biscuits, instant soups, etc.) might also be a problem as hardened fat is used in their production. The same applies to fast food that is becoming more and more popular in the Czech Republic.

Trans fatty acid content in subcutaneous fat reflects long-term dietary intake of TFA.[2,7,13] Levels of *trans* fatty acids in human milk fat are rather variable. They vary with fat composition of the previous day's diet, with liberation of TFA from adipose tissue, and with the amount of fat produced by the mammary gland.[2,14–17]

CONCLUSIONS

The human milk and subcutaneous fat samples contained *trans* fatty acids. The content of these isomers in milk from Czech women is higher than the concentration found in human milk in Spain. The concentration of *trans* fatty acids in human milk in Germany and the Czech Republic is similar, with higher levels being identified in Canadian human milk.

Our absolute values of *trans* fatty acid content in subcutaneous fat are similar to the figures reported elsewhere (in the Netherlands and United States).

ACKNOWLEDGMENTS

This study was supported by the research program of the Third Faculty of Medicine, Charles University: "Initial phases of diabetes mellitus, metabolic, endocrine, nutritional, and toxic damage of organism" (MSMT 1112200001).

REFERENCES

1. BRITISH NUTRITION FOUNDATION. 1995. Trans Fatty Acids—The Report of the British Nutrition Foundation Task Force. London.
2. KRIS-ETHERTON, P.M., D.B. ALLISON, M.A. DENKE et al. 1995. Trans fatty acids and coronary heart disease risk: report of the expert panel on trans fatty acids and coronary heart disease. Am. J. Clin. Nutr. 62: 655S–708S.
3. KOLETZKO, B. 1992. Trans fatty acids may impair biosynthesis of long chain polyunsaturates and growth in man. Acta Paediatr. 81: 302–306.
4. BOATELLA, J., M. RAFECAS, R. CODONY et al. 1993. Trans fatty acids content of human milk in Spain. J. Pediatr. Gastroenterol. Nutr. 16: 432–434.
5. KOLETZKO, B., M. MROTZEK, B. ENG et al. 1988. Fatty acid composition of mature human milk in Germany. Am. J. Clin. Nutr. 47: 954–959.
6. CHEN, Z.Y., G. PELLETIER, R. HOLLYWOOD et al. 1995. Trans fatty acids isomers in Canadian human milk. Lipids 30: 15–21.
7. KATAN, M.B., W.A. VAN STAVEREN & P. DEURENBERG. 1986. Linoleic and trans unsaturated fatty acids content of adipose tissue biopsies as objective indicators of the dietary habits of individuals. Prog. Lipid Res. 25: 193–195.
8. VAN STAVEREN, W.A., P. DEURENBERG, M.B. KATAN et al. 1986. Validity of the fatty acid composition of subcutaneous fat tissue microbiopsies as an estimate of the long term average fatty acid composition of the diet of separate individuals. Am. J. Epidemiol. 123: 455–463.
9. HUDGINS, L.C., J. HIRSCH & E.A. EMKEN. 1991. Correlation of isomeric fatty acids in human adipose tissue with clinical risk factors for cardiovascular disease. Am. J. Clin. Nutr. 53: 474–482.
10. LONDON, S.J., F.M. SACKS, J. CAESAR et al. 1991. Fatty acid composition of subcutaneous adipose tissue and diet in postmenopausal US women. Am. J. Clin. Nutr. 54: 340–345.
11. ARO, A., A.F. KARDINALL, I. SIMINEN et al. 1995. Adipose tissue isomeric trans fatty acids and risk of myocardial infarction in nine countries: the EURAMIC study. Lancet 345: 273–278.
12. DLOUHÝ, P., E. TVRZICKÁ, M. ANDEL et al. 1999. Obsah trans forem mastných kyselin v jedlých tucích na českém trhu. Hygiena 44: 110–116.
13. EMKEN, E.A. 1984. Nutrition and biochemistry of trans and positional fatty acids isomers in hydrogenated oils. Annu. Rev. Nutr. 4: 339–376.
14. SCHMIDT, M.C., J.D. WEETE, S.A. FAIRCLOTH et al. 1984. The effect of hydrogenated fat in the diet of nursing mothers on lipid composition and prostaglandin content of human milk. Am. J. Clin. Nutr. 39: 778–786.
15. CHAPPELL, J.E., M.T. CHLANDININ & C. KEARNEY-VOLPE. 1985. Trans fatty acids in human milk lipids: influence of maternal diet and weight loss. Am. J. Clin. Nutr. 42: 49–56.
16. GARZA, C., R. SCHARD, J. HOPKINSON et al. 1986. Feeding the premature infant: methods to assess lactation performance. In Human Lactation 2: Maternal and Environmental Factors, pp. 253–262. Plenum. New York.
17. HACHEY, D.L., M.R. THOMAS, E.A. EMKEN et al. 1987. Human lactation: maternal transfer of dietary triglycerides labeled with stable isotopes. J. Lipid Res. 28: 1185–1192.

Detection of a Promoter Polymorphism in the Gene of Intestinal Fatty Acid Binding Protein (I-FABP)

K. GESCHONKE,[a] M. KLEMPT,[a] N. LYNCH,[b] S. SCHREIBER,[b] S. FENSELAU,[a] AND J. SCHREZENMEIR[a]

[a]*Institute of Physiology and Biochemistry of Nutrition, Federal Research Center, 24103 Kiel, Germany*

[b]*Medizinische Klinik, University of Kiel, 24105 Kiel, Germany*

ABSTRACT: Postprandial fat absorption is supposed to be a major factor in the development of the metabolic syndrome. In recent years, the assimilation of plasma triglycerides has been the focus of several groups, revealing a number of specific fat or fatty acid transporters. The intestinal fatty acid binding protein, I-FABP-2, participates in the absorption of nutritional fats. The influence of a coding polymorphism has been investigated intensively. However, it remains still unclear whether this polymorphism has a major impact on postprandial TG levels in humans. We found a polymorphism in the promoter of FABP-2, which might involve the retinoid receptor in the transcriptional activity. In functional analysis, we have been able to demonstrate that the various promoter alleles develop different activities in the human intestinal epithelial cells and that the postprandial appearance of plasma TGs in healthy subjects also depends on their genotype. Since the distribution of the identified promoter polymorphism does not differ in subjects suffering from type 2 diabetes, the overall influence on the development of the metabolic syndrome seems to be minor.

KEYWORDS: I-FABP; metabolic syndrome; postprandial triglycerides

INTRODUCTION

Altered postprandial triglyceride levels are known to play a role in the development of the metabolic syndrome.[1–5]

Triglyceride uptake is not passive, but controlled by a variety of different fatty acid binding, transporting, and transferring proteins. The intestinal fatty acid binding protein (I-FABP or FABP-2) belongs to a family of 15-kDa proteins that are able to bind lipid ligands. The expression of I-FABP is limited to the enterocytes of jejunum and ileum and shows a strong affinity to long-chain fatty acids.[6] I-FABP is therefore thought to be involved in the transport of dietary fatty acids in the small intestine.

Address for correspondence: M. Klempt, Institute of Physiology and Biochemistry of Nutrition, Federal Research Center, Hermann-Weigmann-Strasse 1, 24103 Kiel, Germany. Fax: 0431-6092472.

klempt@bafm.de

An alanine to threonine substitution at codon 54 of this protein has been first identified in Pima Indians, a population with an extraordinary high rate of type 2 diabetes.[7] However, there are some contradictory reports regarding the involvement of this polymorphism in the development of obesity, type 2 diabetes, and altered postprandial triglyceride levels.[8–17]

Due to technical reasons, the amount of I-FABP in humans within the intestinal mucosa was to date not a topic of investigation. We thus used cell-culture systems to investigate the role of polymorphism within the promoter region of the I-FABP gene.

METHODS

DNA Analyzing

DNA of 55 unaffected and 45 diabetic (NIDDM) subjects was extracted from whole blood samples using standard procedures. The I-FABP promoter region was amplified (primers: 5′-GGT CTA CCT TCC AAG TGC TGT CA-3′ and 5′-CAA GCT TGG CCT GTT CCT G-3′), subcloned into pGEMTeasy (Clonetech), and sequenced.

Promoter Gene Construction

Both promoter alleles were subcloned into the CAT-reporter gene plasmid pCAT3 (Promega) using PCR-directed restriction (primer: 5′-GCC TCG AGC AAG CTT GGC CTG TTC C-3′ and 5′-ATG GAT CCT ATT ATA TTT CTT AT-3′). This promoter construct has 870 bp upstream of the transcription initiation site. Subsequent restriction with BglII and religation revealed a promoter construct that contained 175 bp upstream of the first transcription initiation site.

Cell Culture Experiments

INT 407 cells (human embryonic intestinal, ATTC CCL 006) were grown in MEM supplemented with 15% fetal calf serum. For differentiation, the cells were stimulated with bezafibrate (0.357 µM) at about 80% of confluence. After 8 days of incubation, cells were transfected with the reporter gene plasmids using the Gene Porter system (Peqlab). After an additional 24 h of incubation, cells were harvested and the protein extract was assayed for CAT activity.[18]

Oral Fat Loading Test

Eighty-four male nondiabetic subjects (age: 25.0 ± 0.3 years; BMI: 22.7 ± 0.2 kg/m^2) had to drink a standardized manufactured test meal[14] after 12 h of fasting. The meal had a volume of 500 mL with 1012 kcal (4221 kJ). It contained 30 g protein, 75 g carbohydrates (93% sucrose, 7% lactose), 10 g ethanol, 58 g fat (65% saturated, 35% unsaturated fatty acids), and 600 mg cholesterol. It was ingested within 15 min. Over the next 6 h, subjects were not allowed to eat, but could drink water ad libitum.

Blood samples were taken during this test at times of 0, 0.5, 1, 2, 3, 4, 5, and 6 h after the meal. Triglyceride, glucose, and insulin blood concentrations were measured at all these points using standard methods.

TABLE 1. DNA of 55 unaffected and 45 diabetic subjects analyzed for the I-FABP promoter allele

Phenotype	Genotype			Allele frequency	
	A/A	B/B	A/B	A	B
Controls	31%	20%	49%	55%	45%
Diabetics	40%	20%	40%	60%	40%

```
           -130         -120         -110
A     AGAACGAGAA   TTAAGAATTA   ATAAGAATAA
B     AGAACGAGAA   TTAAGAATTA   AT   TAATAA

      -100         -90          -80          -70
A  GAATTAATTA  ATTGCTTGAC   ATAGAGTAGT   TAGGTGATTT
B  GAATTAATTA  ATTGCTTGAC   AT      AGT  TAGGTGATTT

      -60         -50          -40          -30
A  CCTGAACTTT  AAGCTTCCAC   ATCACAGTAT   GAAGTTGGTT
B  CCTGAACTTT  AAGCTTCCAC   ATCACAGTAT   GAAGTTGGTT

      -20         -10          1            10           20
A  C AAGATAAG   AAATATAATA   AATTCTCGCC   CAAGGACAGA
B  CTAAGATAAG   AAATATAATA   AATTCTCGCC   CAAGGACAGA

       30           40
A  CCTGAATCTC   TCTAGCTGCC
B  CCTGAATCTC   TCTAGCTGCC
```

FIGURE 1. Sequence of the I-FABP promoter. This figure shows a polymorphism of the promoter region that consists of three linked alterations at position -108, -78, and -19 of the transcription initiation site.

RESULTS

A yet undescribed allele of the FABP-2 promoter is shown in FIGURE 1. It consists of three polymorphisms located within 110 bp 5' of the transcription initiation site (FIG. 1). Allele frequency of the 200 sequenced alleles was 0.57 for allele A and 0.43 for allele B, which were not different between diabetic and unaffected subjects (TABLE 1).

Both promoter alleles were cloned in a reporter gene vector at a length of 870 bp. Transient reporter gene analysis in INT 407 cells revealed that promoter B is 1.3 times more potent ($p < 0.05$) compared to the promoter A (FIG. 2). However, the shorter fragments (175 bp), when responsible for reporter gene expression, showed no differences in promoter activity, although they include the entire polymorphism (FIG. 2).

An influence of the promoter polymorphism on the kinetics of the postprandial increase of triglycerides was detected. In subjects with A/A genotype, the triglyceride maxima were attained at about 2.70 ± 0.26 h (mean $\pm$ SEM), while the peaks were found 3.93 ± 0.4 h after the meal in the subjects with the B/B genotype ($p < 0.05$;

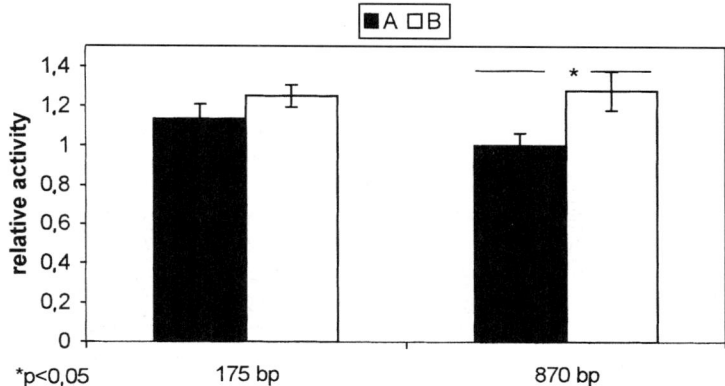

FIGURE 2. I-FABP promoter activity in INT 407 cells. INT 407 cells were transfected with reporter gene plasmids (pCAT3) containing I-FABP promoter fragments (870 bp and 175 bp upstream of the transcription initiation site). Afterwards, CAT activity was analyzed. The figure shows the activity in relation to the activity of the 870-bp allele A activity.

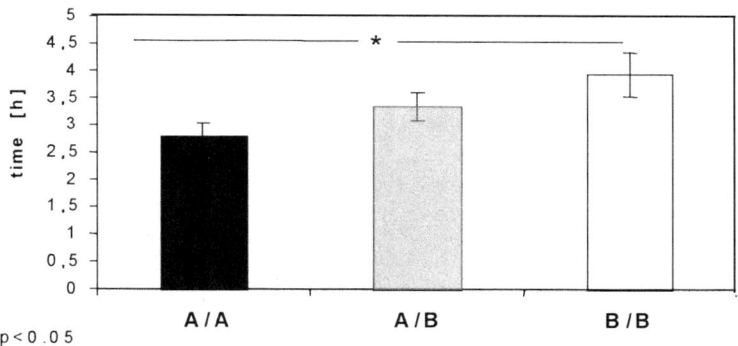

FIGURE 3. Time until the maximal plasma triglyceride concentration is reached compared for different genotypes of the I-FABP promoter. An oral fat load test was applied to 84 unaffected subjects after 12 h of fasting. At 0, 0.5, 1, 2, 3, 4, 5, and 6 h after the meal, blood was analyzed for triglyceride concentration.

FIG. 3). The maximal triglyceride levels tended to be higher in subjects with A/A than with B/B genotype. Other parameters like plasma glucose and insulin concentrations did not differ significantly between genotypes.

DISCUSSION

The *in vitro* experiments reveal a higher activity of the promoter allele B when a long promoter fragment was used. Enterocytes of subjects with this promoter will most probably express more I-FABP protein, resulting in a higher intracellular binding capacity for fatty acids. Consecutively, the triglyceride remodeling will be slower in subjects with the B/B genotype compared to the A/A genotype, resulting in a delayed exocytosis of triglycerides. This will then result in a slower postprandial triglyceride increase in plasma, which could be observed in our study (FIG. 3). Therefore, I-FABP may be responsible for retaining the fatty acids in the intestine epithelium. This theory is sustained by the observation that I-FABP-deficient mice showed an increase in the transfer of dietary fat.[16]

It has been proposed that the increase of triglycerides after a meal contributes to the metabolic syndrome.[1-5,17] We therefore sequenced 55 unaffected and 45 diabetic subjects, but found no significant difference in the allele frequency, although the allele A was slightly more abundant in diabetic compared to unaffected subjects. To verify the assumption further, investigations are currently performed.

It remains unclear if this detected polymorphism within the I-FABP promoter is responsible for the different activity of the two alleles. The shorter fragment does not show a significant difference in activity, although it contains all three parts of the alteration. A possible explanation for this might be an altered binding of transcription factors within the site of variation. Another possibility is that there are more alterations belonging to the polymorphism that have not been found yet.

In conclusion, we suggest that I-FABP does play a major role in dietary fatty acid uptake. We speculate that this protein is not responsible for transporting the fatty acids from the intestine epithelium to the blood. I-FABP is more likely a fatty acid storage protein in the intestine epithelial cell. Therefore, a larger amount of I-FABP protein might protect from fast high postprandial triglyceride levels that are supposed to contribute to the metabolic syndrome.

REFERENCES

1. GRUNDY, S.M. 1999. Hypertriglyceridemia, insulin resistance, and the metabolic syndrome. Am. J. Cardiol. **83:** 25F–29F.
2. KARAMANOS, B.G., A.C. THANOPOULOU & D.P. ROUSS-PENESI. 2001. Maximal postprandial triglyceride increase reflects post-prandial hypertriglyceridaemia associated with the insulin resistance syndrome. Diabet. Med. **18:** 32–39.
3. BYRE, C.D., N.J. WAREHAM, D. PHILLIPS et al. 1997. Is an exaggerated postprandial triglyceride response associated with the component features of the insulin resistance syndrome? Diabet. Med. **14:** 942–950.
4. SCHREZENMEIR, J., S. FENSELAU, I. KEPPLER et al. 1997. Postprandial triglyceride high response and the metabolic syndrome. Ann. N.Y. Acad. Sci. **827:** 353–368.
5. SCHREZENMEIR, J., I. KEPLER, S. FENSELAU et al. 1993. The phenomenom of a high triglyceride response to an oral lipid load in healthy subjects and its link to the metabolic syndrome. Ann. N.Y. Acad. Sci. **683:** 302–314.
6. SWEETSER, D.A., E.H. BIRKENMEIER, I.J. KLISAK et al. 1987. The human and rodent intestinal fatty acid binding protein genes. J. Biol. Chem. **262:** 16060–16071.
7. BAIER, L.J., J.C. SACCHETTINI, W.C. KNOWLER et al. 1995. An amino acid substitution in the human intestinal fatty acid–binding protein is associated with increased fatty acid binding, increased fat oxidation, and insulin resistance. J. Clin. Invest. **95:** 1281–1287.
8. YAMADA, K., X. YUAN, S. ISHIYAMA et al. 1997. Association between Ala54Thr substitution of the fatty-acid-binding protein gene 2 with insulin resistance and intraabdominal fat in Japanese men. Diabetologia **40:** 706–710.
9. MITCHELL, B.D., C.M. KAMMERER, P. O'CONNELL et al. 1995. Evidence for linkage of post-challenge insulin levels with intestinal fatty acid–binding protein (FABP2) in Mexican Americans. Diabetes **44:** 1046–1053.

10. HAYAKAWA, T., Y. NAGAI, E. NOHARA et al. 1999. Variation of the fatty acid–binding protein 2 gene is not associated with obesity and insulin resistance in Japanese subjects. Metabolism **48:** 655–657.
11. ITO, K., K. NAKATANI, M. FUJII et al. 1999. Codon 54 polymorphism of the fatty acid–binding protein gene and insulin resistance in the Japanese population. Diabet. Med. **16:** 119–124.
12. RISSANEN, J., J. PIHLAJAMAKI, S. HEIKKINEN et al. 1997. The Ala54Thr polymorphism in the fatty acid–binding protein 2 does not influence insulin sensitivity in Finnish nondiabetic and NIDDM subjects. Diabetes **46:** 711–712.
13. PRATLEY, R.E., L. BAIER, D.A. PAN et al. 2000. Effects of an Ala54Thr polymorphism in the intestinal fatty acid–binding protein on responses to dietary fat in humans. J. Lipid Res. **41:** 2002–2008.
14. FENSELAU, S. & J. SCHREZENMEIR. 1998. Effects of diet on postprandial lipaemia: a suggestion for methodological standardization. Nutr. Metab. Cardiovasc. Dis. **8:** 68–69.
15. FENSELAU, S., J. SCHREZENMEIR & L. BAIER. 2001. Impact of a I-FABP gene polymorphism on insulin resistance in young men with postprandial hypertriglyceridemia. This conference.
16. VASSILEVA, G., L. HUWYLER, K. POIRIER et al. 2000. The intestinal fatty acid binding protein is not essential for dietary fat absorption in mice. FASEB J. **14:** 2040–2046.
17. SCHREZENMEIR, J. 1996. Hyperinsulinemia, hyperproinsulinemia, and insulin resistance in the metabolic syndrome. Experientia **52:** 426–432.
18. SAMBROOK, J., E.F. FRITSCH & T. MANIATIS. 1989. Molecular Cloning: A Laboratory Manual. Cold Spring Harbor Laboratory Press. Cold Spring Harbor, New York.

Glucose Transporter 4 Gene

Association Studies Pertaining to Alleles of Two Polymorphisms in Extremely Obese Children and Adolescents and in Normal and Underweight Controls

SUSANN FRIEDEL,[a] BENJAMIN ANTWERPEN,[a] ANNE HOCH,[a] CONSTANZE VOGEL,[a] WOLFGANG GRASSL,[a] FRANK GELLER,[b] JOHANNES HEBEBRAND,[a] AND ANKE HINNEY[a]

[a]*Clinical Research Group, Child and Adolescent Psychiatry, Philipps-University of Marburg, 35039 Marburg, Germany*

[b]*Institute of Medical Biometry and Epidemiology, Philipps-University of Marburg, 35039 Marburg, Germany*

ABSTRACT: The human insulin-responsive glucose transporter 4 gene (*GLUT4*) has been related to non-insulin-dependent diabetes mellitus (NIDDM) in several studies. Obesity is commonly found in patients with NIDDM. Hence, genes involved in NIDDM might also be relevant for obesity. We have analyzed 212 extremely obese children and adolescents, 82 normal-weight students, and 94 underweight students for two single nucleotide polymorphisms (SNPs: promoter −30G/A; exon 4a: silent 2061T/C) in the vicinity of the *GLUT4* by polymerase chain reaction with subsequent restriction fragment length polymorphism analyses (PCR-RFLP) or single-strand conformation polymorphism analyses (SSCP). Allele and genotype distributions were similar in all study groups (all p values > 0.05). Hence, we did not detect association of any of the analyzed SNP alleles in the *GLUT4* to different weight extremes, so these seem not to be involved in weight regulation in our study groups.

KEYWORDS: *GLUT4*; weight regulation

INTRODUCTION

Obesity is a multifactorial disease that is influenced by both environmental and genetic factors. Patients with non-insulin-dependent diabetes mellitus (NIDDM) are often obese, so genes involved in NIDDM might also be relevant for obesity. The human insulin-responsive glucose transporter 4 gene (*GLUT4*) has been analyzed in several studies pertaining to NIDDM.[1] Additionally, two recent genome-wide scans for phenotypes related to diet and the metabolic syndrome identified a region on chromosome 17p13 that harbors the *GLUT4*.[2,3]

Address for correspondence: Dr. A. Hinney, Clinical Research Group, Child and Adolescent Psychiatry, Philipps-University of Marburg, Schützenstrasse 49, 35039 Marburg, Germany. Voice: +49-(0)6421/28-65361; fax: +49-(0)6421/28-63056.
Anke.Hinney@med.uni-marburg.de

Additional evidence implicates GLUT4 in body weight regulation: GLUT4 knockout mice have greatly reduced fat depots.[4] Female mice overexpressing GLUT4 (transgenic) have increased adipose cell size and tissue weight.[5] GLUT4 promoter activity is increased in obese compared to lean rats.[6]

MATERIALS AND METHODS

Study Subjects

Briefly, blood samples were collected[7] from 212 extremely obese German children and adolescents (mean BMI: 32.8 ± 6.41 kg/m^2; mean age: 14 ± 2.48 years), 82 normal-weight students (mean BMI: 21.84 ± 1.05 kg/m^2; mean age: 24.8 ± 2.62 years), and 94 underweight students (mean BMI: 18.51 ± 1.15 kg/m^2; mean age: 25.35 ± 3.76 years). Sixty-seven percent of the obese children and adolescents had an age- and gender-specific BMI ≥ 99th percentile, as previously determined in a representative German population sample.[8] The BMI of the underweight students was below the 15th percentile and between the 40th and 60th percentile for the normal-weight students.

Written informed consent was given by all participants and, in the case of minors, their parents. This study was approved by the Ethics Committee of the University of Marburg.

Molecular Analyses

We investigated two single nucleotide polymorphisms (SNPs) in the vicinity of GLUT4: one SNP in the promoter region (−30G/A) and another one in exon 4a (silent Asp-130).[9–11]

For the promoter (−30G/A) SNP, we performed standard polymerase chain reaction (PCR) and subsequent restriction fragment length analysis (RFLP) with *Bam HI*. PCR primers were as follows: 5′-GGGCTTCTCGCGTCTTTT-3′ (forward) and 5′-TGAAAGAACCGATCCTGGAG-3′ (reverse). The amplicon (189 bp) with the A-allele was digested by *Bam HI* (124 bp/65 bp). PCR-RFLP products were run on ethidium bromide–stained 2.5% agarose gels. Positive controls were run on each gel.

To detect the alleles of the SNP in exon 4a, standard nonisotopic single-strand conformation polymorphism analysis (SSCP) (15% acrylamide gel run at 600 V for 2.5 h at ambient temperature and subsequent silver staining) was performed. Primers were 5′-AAAGAGGAAGGGAGCCACTG-3′ (forward) and 5′-GTGCCCGTGAGTACCTGAGT-3′ (reverse). The amplified segment was 203 bp in length.

Statistics

Pearson's χ^2 test (asymptotic, two-sided) and Cochran-Armitage's trend test (exact, two-sided) were used to investigate for association.

TABLE 1. Genotype and allele distributions of a single nucleotide polymorphism (–30G/A) in the promoter region of *GLUT4*

Study group	Genotypes			Alleles	
	GG (%)	GA (%)	AA (%)	G (%)	A (%)
Extremely obese children and adolescents ($n = 212$)	40 (18.9)	92 (43.4)	80 (37.7)	172 (40.6)	252 (59.4)
Underweight students ($n = 94$)	13 (13.8)	39 (41.5)	42 (44.7)	65 (34.6)	123 (65.4)
Normal-weight students ($n = 82$)	15 (18.3)	32 (39.0)	35 (42.7)	62 (37.8)	102 (62.2)

NOTE: Genotype frequencies are not different from Hardy-Weinberg equilibrium.

TABLE 2. Genotype and allele distributions of a single nucleotide polymorphism (silent Asp-130) in exon 4a of *GLUT4*

Study group	Genotypes			Alleles	
	TT (%)	TC (%)	CC (%)	WT	Mut
Extremely obese children and adolescents ($n = 212$)	24 (11.3)	99 (46.7)	89 (42.0)	147 (34.7)	277 (65.3)
Underweight students ($n = 94$)	7 (7.5)	49 (52.1)	38 (40.4)	63 (33.5)	125 (66.5)
Normal-weight students ($n = 82$)	11 (13.4)	35 (42.7)	36 (43.9)	57 (34.8)	107 (65.2)

NOTE: Genotype frequencies are not different from Hardy-Weinberg equilibrium.

RESULTS

The genotype and allele distributions of both SNPs are given in TABLES 1 and 2. The genotype frequencies were not different from Hardy-Weinberg equilibrium. No significant differences in genotype and allele distributions were found between extremely obese children and adolescents, underweight students, and normal-weight students. All nominal p values were >0.2.

CONCLUSIONS

This study analyzed a possible association of body weight with two SNPs in the insulin-stimulated *GLUT4* in a German Caucasian population. We investigated a total of 388 probands: 212 extremely obese children and adolescents, 94 underweight students, and 82 normal-weight students.

We did not detect association of any of the analyzed SNP alleles in the vicinity of *GLUT4* to different weight categories. Hence, the analyzed polymorphisms are not involved in weight regulation in our study groups. Our results do not exclude the occurrence of relevant mutations in the *GLUT4*.

ACKNOWLEDGMENTS

We thank the probands and their families for their participation. This work was supported by the Institut Danone für Ernährung and the BMBF (01GS0118 and 01KW0006).

REFERENCES

1. JUN, H., et al. 1999. Pathogenesis of non-insulin-dependent (type II) diabetes mellitus (NIDDM)—genetic predisposition and metabolic abnormalities. Adv. Drug Delivery Rev. **35:** 157–177.
2. KISSEBAH, A.H., et al. 2000. Quantitative trait loci on chromosomes 3 and 17 influence phenotypes of the metabolic syndrome. Proc. Natl. Acad. Sci. U.S.A. **97:** 14478–14483.
3. LEE, J.H. 1999. Genome scan for human obesity and linkage to markers in 20q13. Am. J. Hum. Genet. **64:** 196–209.
4. KATZ, E.B., et al. 1995. Cardiac and adipose tissue abnormalities, but not diabetes in mice deficient in GLUT4. Nature **377:** 151–155.
5. KATZ, E.B., et al. 1996. The metabolic consequences of altered glucose transporter expression in transgenic mice. J. Mol. Med. **74:** 639–652.
6. HAINAULT, I., et al. 1995. Fatty genotype-induced increase in GLUT4 promoter activity in transfected adipocytes: delineation of two fa-responsive regions and glucose effect. BBRC **209:** 1053–1061.
7. HINNEY, A., et al. 1999. Several mutations in the melanocortin-4 receptor gene including a nonsense and a frameshift mutation associated with dominantly inherited obesity in humans. J. Clin. Endocrinol. Metab. **84:** 1483–1486.
8. HEBEBRAND, J., et al. 1996. Use of percentiles for the body mass index in anorexia nervosa: diagnostic, epidemiological, and therapeutic considerations. Int. J. Eating Disord. **19:** 359–369.
9. BJORBAEK, C., et al. 1994. Genetic variants in promoters and coding regions of the muscle glycogen synthase and the insulin-responsive GLUT4 genes in NIDDM. Diabetes **43:** 976–983.
10. BUSE, J.B., K. YASUDA, T.P. LAY et al. 1992. Human GLUT4/muscle-fat glucose-transporter gene: characterization and genetic variation. Diabetes **41:** 1436–1445.
11. CHOI, W.H., S. O'RAHILLY, J.B. BUSE et al. 1991. Molecular scanning of insulin-responsive glucose transporter (GLUT4) gene in NIDDM subjects. Diabetes **40:** 1712–1718.

Association of Insulin Gene VNTR Polymorphism with Polycystic Ovary Syndrome

MARKÉTA VANKOVÁ,[a] JANA VRBÍKOVÁ,[a] MARTIN HILL,[a] ONDREJ CINEK,[b] AND BELA BENDLOVÁ[a]

[a]*Institute of Endocrinology, Prague, Czech Republic*

[b]*Second Department of Pediatrics, Second Medical Faculty, Charles University, Prague, Czech Republic*

ABSTRACT: Variability in the number of tandem repeats of the insulin gene (INS VNTR) is known to influence several phenotypes, including polycystic ovary syndrome (PCOS), diabetes mellitus type 1, diabetes mellitus type 2, and birth weight. The presence of the class III allele of INS VNTR has been reported to be protective in diabetes mellitus type 1, but in contrary it increases the disease risk of PCOS and diabetes mellitus type 2. PCOS is a very common endocrinopathy in women of reproductive age. The etiology of PCOS is uncertain, but family history of this syndrome suggests a major genetic cause. The aim of this pilot study was to investigate the possible association of INS VNTR polymorphism with PCOS in Czech women. In PCOS, significantly higher WHR, BMI, G_0, G_{180}, I_{30}, Cp_0, Cp_{30}, Cp_{60}, AUC-I, AUC-Cp, and insulinogenic index and significantly lower AUC-G/AUC-I were found. No significant differences in INS VNTR genotype, phenotype, or allele frequencies were found between PCOS and controls. In spite of several differences in anthropometric and biochemical parameters (abdominal fat localization, increased β-cell function, and lower insulin sensitivity in PCOS women), no effect of INS VNTR polymorphism was found on insulin secretion, insulin action, or any other screened parameter.

KEYWORDS: polycystic ovary syndrome; INS VNTR; genetic association

INTRODUCTION

Polycystic ovary syndrome (PCOS) is one of the most common endocrine disorders in women of reproductive age that results in reduced fertility. PCOS is a heterogeneous disorder that is characterized by hyperandrogenism and is also very often associated with chronic anovulation, hirsutism, obesity, insulin resistance, and disturbances in insulin secretion.[1,2]

Heterogeneity of clinical and biochemical features suggests that PCOS represents numerous disorders rather than one. Although the clinical and biochemical features

Address for correspondence: Mgr. Markéta Vanková, Institute of Endocrinology, Národní 8, 116 94 Prague 1, Czech Republic. Voice: +4202 24905301; fax: +4202 24905325.
marketa.vankova@email.cz

in women with PCOS are variable, there are certain abnormalities that are consistently found in PCOS. The most typical biochemical abnormality is hypersecretion of androgens. Recent studies show that hyperandrogenism in PCOS is primarily the result of ovarian androgen hypersecretion rather than androgen hypersecretion as a result of abnormal gonadotropins.[3]

Many cases of this syndrome are characterized by significant metabolic abnormalities, such as insulin resistance and hyperinsulinemia; thus, it is also associated with an increased risk of diabetes mellitus type 2[4] and cardiovascular disease[5] in later life. Despite the clinical and biochemical determination of PCOS status, its etiology remains uncertain, but characteristic family clustering of cases suggests that there is a major genetic cause. Recent studies show that much of the clinical and biochemical variability within PCOS can be explained by the interaction of environmental (notably nutritional) factors with a small number of major causative genes, which include those involved in androgen production and the secretion and/or action of insulin.[6]

The aim of our pilot study was to elucidate the relationship between the human INS VNTR regulatory polymorphism (INS VNTR = polymorphism in variable number of tandem repeats of the insulin gene) and insulin secretion and/or insulin action in PCOS.

VNTRs include high polymorphic loci in the human genome. The INS VNTR has been intensively analyzed for association with diabetes mellitus type 1,[7] diabetes mellitus type 2,[8] birth size,[9] and PCOS.[10] These associations result from the influence of the VNTR on transcriptional regulation of the insulin gene. The INS VNTR lies 5′ to the insulin gene on chromosome 11p15.5. The polymorphism arises from tandem repetition of 14- to 15-bp oligonucleotides. Allelic variations in the size have a bimodal distribution in Caucasians. The shorter class I alleles (28–44 repeats) are associated with diabetes mellitus type 1, and the longer class III alleles (138–159 repeats) are associated with PCOS, diabetes mellitus type 2, and size at birth. The results of the published studies are inconsistent and often controversial. Class II alelles of intermediate size are rare in Caucasian populations. The INS VNTR polymorphism is difficult to determine directly due to extremely different allele lengths. Therefore, the INS VNTR is often tested by using −23 Hph I as a surrogate marker. The Hph I polymorphism is a single nucleotide (T/A) polymorphism and is at the fifth nucleotide of the polypyrimidine tail of the splice site between intron 1 and exon 2 of the insulin gene. In Europeans, the Hph I "T" polymorphism is in complete linkage disequilibrium with the class I allele of the neighboring INS VNTR, as is the Hph I "A" polymorphism with the class III allele.[11] (See FIG. 1.)

STUDY SUBJECTS

We examined 38 anovulatory women with PCOS matching NIH criteria. None of the women had taken oral contraceptives during the preceding 3 months. The control group consisted of 22 healthy women without family history of PCOS and diabetes mellitus type 2. Ten of them were using hormonal contraception at the time of the examination. As there was no difference in parameters of insulin resistance and β-cell function between women using or not using hormonal contraception, we analyzed them together.

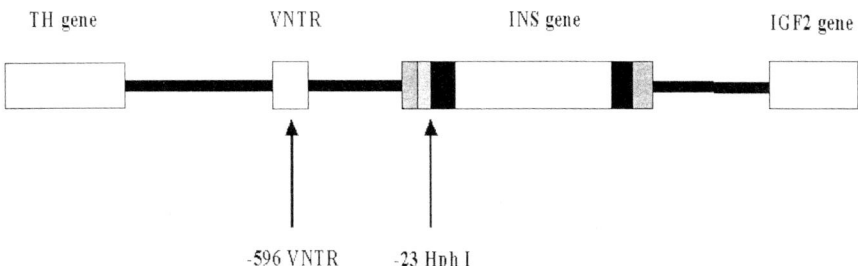

FIGURE 1. The insulin gene region on chromosome 11p15.5. Polymorphisms are designated by their position with respect to the first base of the initiating AGT (+1).[20]

TABLE 1. The used indices for evaluation of insulin resistance and β-cell function derived from fasting and oGTT measurements

Index	Formula	Reference
Beta cell function		
HOMA-F (mIU/mmol)	$20 \times I_0/(G_0 - 3.5)$	Matthews[12]
Insulinogenic index $\Delta I/\Delta G_{30}$ (mIU/mmol)	$(I_{30} - I_0)/(G_{30} - G_0)$	Seltzer[13]
Insulin resistance		
HOMA-R (mIU · mmol/L^2)	$I_0 \times G_0/22.5$	Matthews[12]
FIRI (mIU · mmol/L^2)	$I_0 \times G_0/25$	Duncan[14]
Suma I (mIU/L)	$I_0 + I_{60} + I_{120}$	Rosolová[15]
Insulin sensitivity		
FGIR (mg/10^{-4} IU)	G_0/I_0	Legro[16]
Matsuda index (1/mmol · 1/mIU · L^2)	$10^4/\sqrt{[(\text{mean I} \times \text{mean G}) \times G_0 \times I_0]}$	Matsuda[17]
Cederholm index (mg · L^2/mmol/mIU/min)	$M/(\text{mean G} \times \log \text{mean I}) =$ $[75{,}000/120 + (G_0 - G_{120}) \times$ $1.15 \times 180 \times 0.19 \times (m/120)]/$ $[\text{mean G} \times \log \text{mean I}]$	Cederholm[18]
AUC-G/AUC-I	AUC-G/AUC-I	Drivsholm[19]

METHODS

A three-hour oGTT (oral glucose tolerance test) with 75 g of glucose was carried out. Estimations of glucose, insulin, C-peptide, and proinsulin (marked as G_x, I_x, Cp_x, and PI_x) and anthropometric parameters were determined. Glucose was measured using the glucose-oxidase method (Beckman Glucose Analyzer, Fullerton, CA). Serum C-peptide and insulin levels were determined by immunoradiometric assays (Immunotech, Marseille, France). Proinsulin was measured by ELISA (IBL, Hamburg, Germany). (See TABLE 1.)

To estimate β-cell function and insulin resistance, we have used various indices derived from fasting values of insulin and glucose or from oGTT measurements. All values are given as mean ± SEM. Differences between controls and PCOS were tested using the nonparametric Mann-Whitney test (two-tailed). Significance was defined at $p < 0.05$ (resp. 0.01 and 0.001). For statistical analyses, the NCSS 2000 program was used (Statistical Solutions, Saugus, MA).

Probands were typed as having INS VNTR class I and/or class III alleles using Hph I digestion of PCR products. Conditions of PCR reaction: 50 ng genomic DNA in 25 µL PCR volume containing 3 mM $MgCl_2$ (PE), 200 µM dNTPs (Takara), 0.24 µM primers, 0.03 U AmpliTaq Gold (PE), 10× PCR buffer (PE), and 11.75 µL ddH_2O. Primer sequences: forward AGC AGG TCT GTT CCA AGG; reverse CTT GGG TGT GTA GAA GAA GC. Thermocycling was at 94°C 1 min 1×, 95°C 30 s 1×, 95°C 20 s, touch down 65°C to 59°C 30 s, 72°C 1 min 10×, 95°C 20 s, 59°C 30 s, 72°C 1 min 27×, and 72°C 10 min. Then, 5 µL of PCR product was digested with 4 U Hph I restriction endonuclease (Fermentas) in a 20-µL reaction at 37°C overnight, followed by resolution of fragments on a 2.5% MetaPhor agarose gel (Sigma) in 0.5× TBE electrophoresis buffer and ethidium bromide staining.

The positivity for a class I allele was characterized by presence of 122-, 185-, and 53-bp-long specific fragments (presence of restriction site), while the positivity for a class III allele was characterized by presence of 122- and 238-bp-long specific fragments (absence of restriction site). Fisher's exact test was used to reveal differences in allele and genotype frequencies between PCOS and controls. For evaluation of the differences in anthropometric and biochemical parameters between class III allele carriers and noncarriers, the nonparametric Mann-Whitney test was used. Risk of PCOS status for class III allele carriers was expressed as odds ratios, and their confidence intervals were calculated according to Woolf's formula, with comparisons being made by testing the 2×2 table using a two-tailed Fisher's exact test. Significance was defined at $p < 0.05$.

RESULTS

The biochemical and anthropometric data of control and PCOS women are given in TABLE 2. The body mass index (BMI) of the PCOS women differed significantly from the controls ($p < 0.01$). The PCOS women had a much higher waist/hip ratio (WHR) ($p < 0.0001$). In PCOS, significantly higher fasting G ($p < 0.05$) and fasting Cp ($p < 0.001$) were found. During the oGTT, G_{180} ($p < 0.05$), I_{30} ($p < 0.01$), Cp_{30} ($p < 0.01$), Cp_{60} ($p < 0.05$), AUC-I ($p < 0.05$), and AUC-Cp ($p < 0.05$) values were higher in PCOS women than in controls (AUC = area under the curve). The index of insulin sensitivity derived from oGTT (AUC-G/AUC-I, $p < 0.05$) was lower in PCOS women than in controls. The index of β-cell function derived from oGTT ($\Delta I/\Delta G_{30}$) ($p < 0.01$) was significantly higher in PCOS women than in controls.

No significant differences in INS VNTR genotype, phenotype, or allele frequencies were found between PCOS women and controls. No association of INS VNTR polymorphism with screened parameters (within PCOS, controls, and combined groups) was found. (See TABLE 3.)

TABLE 2. Comparison of anthropometry and oGTT between PCOS and controls

	Controls ($n = 22$)	PCOS ($n = 38$)	p
Age	28.3 ± 1.5	25.8 ± 1.1	NS
Weight	65.8 ± 2.8	73.0 ± 3.0	NS
BMI	22.6 ± 0.7	26.1 ± 1.0	<0.01
WHR	79.1 ± 1.3	94.3 ± 1.3	<0.0001
G_0	4.6 ± 0.1	4.9 ± 0.1	<0.01
G_{30}	7.3 ± 0.3	7.3 ± 0.2	NS
G_{60}	6.3 ± 0.4	6.7 ± 0.3	NS
G_{120}	4.7 ± 0.2	5.1 ± 0.2	NS
G_{180}	3.9 ± 0.1	4.3 ± 0.1	<0.05
AUC-G	965.7 ± 2.3	1030.0 ± 27.2	NS
I_0	9.2 ± 1.4	13.4 ± 2.2	NS
I_{30}	47.6 ± 4.3	82.8 ± 8.2	<0.01
I_{60}	57.13 ± 4.6	93.1 ± 12.5	NS
I_{120}	26.3 ± 3.6	53.9 ± 10.8	NS
I_{180}	8.7 ± 1.3	17.7 ± 3.2	NS
AUC-I	5972.8 ± 429.6	10,682 ± 1411.7	<0.05
$\Delta I / \Delta G_{30}$	15.8 ± 1.9	31.8 ± 3.5	<0.01
HOMA-R	2.0 ± 0.3	2.9 ± 0.5	NS
HOMA-F	181.1 ± 26.2	218.2 ± 42.3	NS
Matsuda	133.5 ± 13.4	115.5 ± 12.7	NS
FGIR	11.6 ± 1.2	14.3 ± 2.3	NS
FIRI	1.8 ± 0.3	2.6 ± 0.4	NS
AUC-G/AUC-I	0.18 ± 0.01	0.16 ± 0.02	<0.05
Cederholm	75.5 ± 4.5	67.9 ± 3.9	NS
Suma I	92.6 ± 6.9	160.4 ± 22.7	NS
Cp_0	0.59 ± 0.06	0.92 ± 0.06	<0.001
Cp_{30}	2.28 ± 0.14	3.0 ± 0.17	<0.01
Cp_{60}	2.92 ± 0.16	3.5 ± 0.18	<0.05
Cp_{120}	2.21 ± 0.14	2.6 ± 0.2	NS
Cp_{180}	1.04 ± 0.1	1.41 ± 0.13	NS
AUC-Cp	372.7 ± 17.5	460.7 ± 24.6	<0.05
PI_0	2.8 ± 0.3	2.43 ± 0.35	NS
PI_{30}	6.7 ± 1.1	6.4 ± 0.7	NS
PI_{180}	7.5 ± 1.0	7.14 ± 1.0	NS

NOTE: G_x [mmol/L], I_x [mIU/L], Cp_x [mIU/L], PI_x [pmol/L]; NS, not significant; mean ± SEM.

TABLE 3. Genotype, phenotype, and allele frequencies of the INS VNTR polymorphism in PCOS and controls

	PCOS ($n = 38$)		Controls ($n = 22$)		OR	95% CI	p
Genotype frequencies							
I/I	16	42%	12	55%	0.61	0.21–1.75	0.43
I/III	19	50%	9	41%	1.44	0.50–4.18	0.59
III/III	3	8%	1	5%	1.80	0.18–18.44	1.00
Phenotype frequencies							
I+	35	92%	21	95%	0.56	0.05–5.69	1.00
III+	22	58%	10	45%	1.65	0.57–4.75	0.43
Allelic frequencies							
I	51	67%	33	75%	0.68	0.30–1.56	0.41
III	25	33%	11	25%	1.47	0.64–3.38	0.41

NOTE: OR, odds ratio; p values were obtained by testing the 2×2 table using a two-tailed Fisher's exact test; CI, confidence intervals were calculated according to Woolf's formula.

CONCLUSIONS

The association of the INS VNTR with PCOS was tested in 38 affected women and 22 healthy controls. On the basis of anthropometric and biochemical parameters, PCOS women differed from the healthy controls. PCOS women had a significantly higher BMI and WHR. WHR values suggest the risk of abdominal fat localizations in PCOS women. The biochemical differences between PCOS women and controls include higher values of insulin and C-peptide during oGTT, significantly higher insulinogenic index, and significantly lower AUC-G/AUC-I, indicating increased β-cell function and lower insulin sensitivity in PCOS. Hyperinsulinemia and insulin resistance are the common features of the metabolic syndrome in women with PCOS. In recent data, whereas insulin resistance was reversible by weight reduction, an abnormal first-phase insulin secretion persisted.[21] This suggests that there is an abnormal pancreatic β-cell function. Thus, the insulin gene is a candidate in PCOS etiology.

Strong evidence associating the INS VNTR with PCOS has been published. Waterworth *et al.*[22] not only found that INS VNTR may be a PCOS susceptibility gene, but also that it predisposes to hyperinsulinemia. Waterworth and colleagues also found preferential transmission of the class III allele of the INS VNTR from heterozygous fathers, but not from mothers. Franks *et al.*[23] described strong evidence for both linkage and association of the INS VNTR with PCOS. In contrast, Urbanek *et al.*[24] found no evidence for any association and they observed no parentally preferential transmissions. We also found no association between the INS VNTR regulatory polymorphism and PCOS status in Czech women. The negative results of our pilot study reflect that the INS VNTR polymorphism is not a likely candidate for PCOS etiology in Czech women.

ACKNOWLEDGMENTS

The work was supported by the following grants: IGA MH CR NB/6696-3, NB/5395-5, and COST OC B17.10 MSMT CR.

REFERENCES

1. LEGRO, R.S., A.R. KUNSELMAN, W.C. DODSON et al. 1999. Prevalence and predictors of risk for type 2 diabetes mellitus and impaired glucose tolerance in polycystic ovary syndrome: a prospective, controlled study in 254 affected women. J. Clin. Endocrinol. Metab. **84:** 165–169.
2. DUNAIF, A. 1997. Insulin resistance and polycystic ovary syndrome: mechanism and implications for pathogenesis. Endocr. Rev. **18:** 774–800.
3. GILLING-SMITH, C., E.H. STORY, V. ROGERS et al. 1997. Evidence for a primary abnormality of thecal cell steroidogenesis in the polycystic ovary syndrome. Clin. Endocrinol. **47:** 93–99.
4. HOLTE, J. 1996. Disturbances in insulin secretion and sensitivity in women with polycystic ovary syndrome. Clin. Endocrinol. Metab. **10:** 221–247.
5. DAHLGREN, E., P.O. JANSSON, S. JOHANSSON et al. 1992. Polycystic ovary syndrome and risk for myocardial infarction. Acta Obstet. Gynecol. Scand. **71:** 599–604.
6. FRANKS, S., N. GHARANI, D. WATERWORTH et al. 1997. The genetic basis of polycystic ovary syndrome. Hum. Reprod. **12:** 2641–2648.
7. OWERBACH, D. & K.H. GABBAY. 1993. Localization of a type I diabetes susceptibility locus to the variable tandem repeat region flanking the insulin gene. Diabetes **42:** 1708–1714.
8. HUXTABLE, S.J., P.J. SAKER, L. HADDAD et al. 2000. Analysis of parent-offspring trios provides evidence for linkage and association between the insulin gene and type 2 diabetes mediated exclusively through paternally transmitted class III variable number tandem repeat alleles. Diabetes **49:** 126–130.
9. DUNGER, D.B., K.K.L. ONG, S.J. HUXTABLE et al. 1998. Association of the INS VNTR with size at birth. Nat. Genet. **19:** 98–100.
10. FRANKS, S., N. GHARANI, D. WATERWORTH et al. 1998. Current developments in the molecular genetics of the polycystic ovary syndrome. TEM **9:** 51–54.
11. LUCASSEN, A.M. 1993. Susceptibility to insulin dependent diabetes mellitus maps to a 4.1 kb segment of DNA spanning the insulin gene and associated VNTR. Nat. Genet. **14:** 305–310.
12. MATTHEWS, D.R., J.P. HOSKER, A.S. RUDENSKI et al. 1985. Homeostasis model assessment: insulin resistance and beta-cell function from fasting plasma glucose and insulin concentrations in man. Diabetologia **28:** 412–419.
13. SELTZER, H.S., E.W. ALLEN, A.L. HERRO et al. 1967. Insulin secretion in response to glycemic stimulus: relation of delayed initial release to carbohydrate intolerance in mild diabetes mellitus. J. Clin. Invest. **46:** 323–335.
14. DUNCAN, M.H., B.M. SINGH, P.H. WISE et al. 1995. A simple measure of insulin resistance. Lancet **346:** 120–121.
15. ROSOLOVÁ, H., O. MAYER, J. SIMON et al. 1998. Detection of risk of insulin resistance in the population. Cas. Lek. Cesk. **137:** 80–83.
16. LEGRO, R.S., D. FINEGOOD & A. DUNAIF. 1998. A fasting glucose to insulin ratio is a useful measure of insulin sensitivity in women with polycystic ovary syndrome. J. Clin. Endocrinol. Metab. **83:** 2694–2698.
17. MATSUDA, M. & R.A. DEFRONZO. 1999. Insulin sensitivity indices obtained from oral glucose tolerance testing: comparison with the euglycemic insulin clamp. Diabetes Care **22:** 1462–1470.
18. CEDERHOLM, J. & L. WIBELL. 1990. Insulin release and peripheral sensitivity at the oral glucose tolerance test. Diabetes Res. Clin. Pract. **10:** 167–175.

19. DRIVSHOLM, T., T. HANSEN, S.A. URHAMMER et al. 1999. Assessment of insulin sensitivity and beta cell function in an oral glucose tolerance test. Diabetologia **42**(suppl. 1): A185.
20. BELL, G.I., L.J. SELBY & J.W. RUTTER. 1982. The highly polymorphic region near the human insulin gene is composed of simple tandemly repeating sequences. Nature **295:** 31–35.
21. HOLTE, J., T. BERGH, C. BERNE et al. 1995. Restored insulin sensitivity, but persistently increased early insulin secretion after weight loss in obese women with polycystic ovary syndrome. J. Clin. Endocrinol. Metab. **80:** 2586–2593.
22. WATERWORTH, D.M., T.S. BENNETT, N. GHARANI et al. 1997. Linkage and association of insulin gene VNTR regulatory polymorphism with polycystic ovary syndrome. Lancet **349:** 986–989.
23. FRANKS, S., N. GHARANI & M. MCCARTHY. 1999. Genetic abnormalities in polycystic ovary syndrome. Ann. Endocrinol. **60:** 131–133.
24. URBANEK, M., R.S. LEGRO, D.A. DRISCOLL et al. 1999. Thirty-seven candidate genes for polycystic ovary syndrome: strongest evidence for linkage is with follistatin. Proc. Natl. Acad. Sci. U.S.A. **96:** 8573–8579.

Association between a Variant at the GABA$_A$α6 Receptor Subunit Gene, Abdominal Obesity, and Cortisol Secretion

ROLAND ROSMOND,[a,b] CLAUDE BOUCHARD,[b] AND PER BJÖRNTORP[c]

[a]*Department of Clinical Chemistry, Sahlgrenska University Hospital, S-413 45 Göteborg, Sweden*

[b]*Pennington Biomedical Research Center, Baton Rouge, Louisiana 70808, USA*

[c]*Department of Heart and Lung Diseases, Sahlgrenska University Hospital, S-413 45 Göteborg, Sweden*

> ABSTRACT: In the present study, we examined the potential impact of a T-to-C substitution at nucleotide 1519 of the GABA$_A$α6 receptor subunit gene (*GABRA6*) on obesity and obesity-related phenotypes as well as circulating hormones, including salivary cortisol, in 284 unrelated Swedish men born in 1944. The subjects were genotyped by using PCR amplification followed by digestion with the restriction enzyme *Alw*NI. The frequency of allele T was 0.54 and that of allele C was 0.46. Carriers for the T allele ($n = 211$) had higher waist-to-hip ratio ($p = 0.094$) and abdominal sagittal diameter ($p = 0.084$) compared to homozygotes for the C allele ($n = 56$). The homozygotes for the T allele had, in comparison to heterozygotes, significantly ($p = 0.004$–0.024) higher mean cortisol levels at 11:45 AM; at 30, 45, and 60 min after a standardized lunch; and finally at 5:00 PM. In addition, T/T subjects had significantly ($p = 0.031$) higher diurnal cortisol secretion compared to T/C subjects. Leptin, insulin, and glucose were not different across the genotype groups. In conclusion, these findings suggest a role of the point substitution (T-to-C) at nucleotide 1519 of *GABRA6* in the predisposition to hypercortisolism and perhaps abdominal obesity. The pathophysiology may involve various environmental factors, particularly stress, that destabilize the GABA-hypothalamic-pituitary-adrenal systems in those with genetic vulnerability.
>
> KEYWORDS: cortisol; GABA; gene; obesity; polymorphism

INTRODUCTION

The hypothalamic-pituitary-adrenal (HPA) axis plays a central role in homeostatic processes.[1] However, if the HPA axis becomes poorly regulated with disruption of central regulatory systems and increased cortisol secretion, the development of visceral obesity, insulin resistance, dyslipidemia, and hypertension may occur.[2]

Address for correspondence: Roland Rosmond, M.D., Ph.D., Department of Clinical Chemistry, Sahlgrenska University Hospital, S-413 45 Göteborg, Sweden. Voice: (+46) 31 342 62 72; fax: (+46) 31 82 84 58.
rolandrosmond@hotmail.com

The hypothalamus has extensive and complex neural connections and receives afferent regulatory signals from different parts of the brain. The inhibitory effects are exerted mainly by γ-aminobutyric acid (GABA).[3] GABA is the major inhibitory neurotransmitter in the vertebrate brain and acts by binding to the GABA$_A$ receptors, which are hetero-oligomeric chloride channels that are modulated among others by cortisol and benzodiazepines.[4] To date, at least 16 human GABA$_A$ receptor cDNAs have been cloned, including the GABA$_A$α6 receptor subunit.[4] Recently, a novel single nucleotide polymorphism (SNP) in the 3′ noncoding region of the GABA$_A$α6 receptor subunit gene (*GABRA6*) has been described.[5] DNA sequencing has shown that it is a T→C substitution resulting in the loss of an *Alw*NI restriction site at nucleotide 1519.[5] The *GABRA6* is located on chromosome 5q31.1-q35.[6]

With this background, we addressed the hypothesis that the point substitution (T→C) of the *GABRA6* may influence obesity and obesity-related phenotypes as well as circulating hormones, including salivary cortisol.

SUBJECTS AND METHODS

For the present study, the subjects ($n = 284$) were randomly selected from a larger geographically defined total population cohort of men born in Göteborg, Sweden, in 1944. The design has been described elsewhere.[7] The subjects gave written informed consent before participating in the study, which was approved by the local ethics committee.

Body mass index (BMI, kg/m^2), waist-to-hip ratio (WHR), and abdominal sagittal diameter were measured as described before.[7] Leptin, insulin, and glucose were measured in an overnight fasting state as described.[7] Salivary cortisol was measured repeatedly over a random working day as described previously.[7]

Genotyping was performed on genomic leukocyte DNA (150 ng in a final volume of 10 μL) and carried out by PCR technique (annealing temperature, 53°C) using primers described previously.[5] The PCR products were digested overnight at 37°C with 5 U of *Alw*NI.

All statistical analyses were performed using SPSS for Windows, release 10.0 (SPSS Inc., Chicago). The *p* values are two-sided throughout, and $p < 0.05$ was considered significant. The results are presented as mean and standard deviation (SD). Data comparisons were carried out with the General Linear Model, with genotype as independent factors and BMI and WHR as covariates. All *p* values were adjusted for multiple tests by using the Spjotvoll-Stoline posthoc correction.

RESULTS

The frequency of allele T was 0.54 and that of allele C was 0.46. The genotype frequencies were 28.8%, 50.2%, and 21.0% for T/T, T/C, and C/C, respectively. The observed genotype frequencies were in a Hardy-Weinberg equilibrium.

TABLE 1 presents the differences in obesity-related phenotypes among the *GABRA6* genotypes. Since these variables are highly influenced by obesity and body fat distribution, the results were adjusted for BMI and WHR whenever appropriate. Carriers for the T allele had higher WHR ($p = 0.094$) and abdominal sagittal diameter

TABLE 1. GABA$_A$α6 receptor subunit genotypes and variables of the study

	Genotypes			
	T/T (n = 77)	T/C (n = 134)	C/C (n = 56)	p
BMI (kg/m^2)	26.0 (3.6)	26.3 (4.0)	26.1 (4.2)	>0.20
WHR	0.94 (0.08)^a	0.94 (0.06)	0.92 (0.08)^b	0.094
Abdominal sagittal diameter (cm)	22.5 (3.7)	23.0 (3.5)^a	22.2 (4.0)^b	0.084
Leptin (μg/L)	6.1 (4.4)	6.3 (4.1)	6.0 (4.6)	>0.20
Fasting insulin (mU/L)	11.7 (8.6)	13.4 (12.9)	12.2 (8.4)	>0.20
Fasting glucose (mmol/L)	4.4 (0.7)	4.6 (1.0)	4.7 (1.2)	0.197
Cortisol level (nmol/L) in the morning	15.1 (8.4)	14.2 (6.1)	15.9 (8.4)	>0.20
Cortisol level (nmol/L) at 11:45 AM	9.0 (8.8)^a	6.3 (2.3)^b	6.7 (2.8)^b	0.004
Cortisol level (nmol/L) at 30 min after lunch	10.5 (13.4)^a	7.2 (3.4)^b	7.8 (3.8)	0.023
Cortisol level (nmol/L) at 45 min after lunch	9.0 (8.7)^a	6.7 (2.7)^b	7.3 (3.4)	0.023
Cortisol level (nmol/L) at 60 min after lunch	8.5 (8.9)^a	6.0 (2.0)^b	6.7 (3.0)	0.010
Cortisol level (nmol/L) at 5:00 PM	5.3 (3.1)^a	4.3 (1.9)^b	5.2 (2.4)	0.024
Cortisol level (nmol/L) before bedtime	4.1 (6.6)	3.0 (2.7)	3.2 (3.3)	>0.20
Diurnal cortisol secretion (nmol/L)	8.4 (5.9)^a	6.8 (1.8)^b	7.6 (2.5)	0.031

NOTE: Values (mean ± SD) with different superscript letters (a, b) are significant at the 0.05 level; p values are adjusted for BMI and WHR.

($p = 0.084$) compared to homozygotes for the C allele. The homozygotes for the T allele had, in comparison to heterozygotes, significantly ($p = 0.004$–0.024) higher mean cortisol levels at 11:45 AM; at 30, 45, and 60 min after a standardized lunch; and finally at 5:00 PM. In addition, T/T subjects had significantly ($p = 0.031$) higher diurnal cortisol secretion compared to T/C subjects. Leptin, insulin, and glucose were not different across the genotype groups (all adjusted for BMI and WHR).

DISCUSSION

In developed societies, metabolic and circulatory disorders such as abdominal obesity, type 2 diabetes, dyslipidemia, and hypertension rarely occur in isolation, but are typically part of a complex phenotype of metabolic abnormalities usually referred to as the Metabolic Syndrome.[8] The Metabolic Syndrome is usually defined as a set of risk factors that cluster in some people.[9,10] These risk factors include elevated insulin levels (insulin resistance), abdominal obesity, high levels of both LDL cholesterol and triglycerides, and hypertension. In addition to insulin resistance, abdominal obesity in adults is a key contributor to the development of the Metabolic Syndrome. There are currently several important theories that attempt to explain the causes of abdominal obesity. Studies of twins and families suggest that about 40–80% of the variation in human obesity can be ascribed to genetic factors.[11] Most variation in human obesity-related phenotypes is due to a limited number of common genetic variations that interact with the environment to produce the final phenotype.[11] The number of genes, markers, and chromosomal regions that have been associated or linked with human obesity-related phenotypes is now well above 200 and continues to increase.[12]

In the present study, we found a new polymorphism potentially associated with abdominal obesity. Carriers for the T allele tended to have both elevated WHR and abdominal sagittal diameter (TABLE 1). In animal models, the obese Zucker rat possesses altered brain GABAergic mechanisms that contribute to their overeating.[13,14] In subjects with Prader-Willi syndrome, the mean level of plasma GABA was found to be 2 to 3 times higher than in retarded nonobese controls.[15] Although a role of GABA in human obesity has not, to our knowledge, been previously described, these studies suggest that GABAergic tonic inhibition may be potentially involved.

In addition to GABA, complex interactions among various neurotransmitters and neuromodulators involved in the regulation of energy balance, such as leptin, neuropeptide Y, and several hormones, including cortisol, may directly or indirectly participate in the GABAergic regulation. Interestingly, carriers for the T allele had also elevated diurnal cortisol secretion (TABLE 1). These results indicate that the SNP at nucleotide 1591 of *GABRA6* might be involved in the dysfunction of the hypothalamic cortisol homeostasis-controlling mechanisms within the central nervous system. It is well documented that cortisol contributes to the regulation of adipose tissue differentiation, function, and distribution, and in excess causes abdominal obesity.[2] Inadequate cortisol secretion is associated with abnormalities in glucose, insulin, and lipid metabolism, bringing the importance of the HPA axis in the development of abdominal obesity to the forefront.[2]

In summary, we found that a T/C polymorphism at nucleotide 1519 of the *GABRA6* is associated with elevated salivary cortisol levels and potentially abdominal obesity. As this polymorphism is not located in a coding region of the $GABA_A\alpha6$ receptor subunit gene, its functional role, if any, is uncertain. Further studies are required to replicate these preliminary findings and to define the biological effects of this polymorphism.

ACKNOWLEDGMENTS

This study was supported by the Swedish Medical Research Council (Grant No. K97-19X-00251-35A) and the Pennington Biomedical Research Center. R. Rosmond would also like to acknowledge the Henning and Johan Throne-Holst Foundation for the support of a postdoctoral fellowship at the Pennington Biomedical Research Center. C. Bouchard is partially supported by the George A. Bray Chair in Nutrition.

REFERENCES

1. CHROUSOS, G.P. 1998. Stressors, stress, and neuroendocrine integration of the adaptive response: the 1997 Hans Selye Memorial Lecture. Ann. N.Y. Acad. Sci. **851:** 311–335.
2. BJÖRNTORP, P. & R. ROSMOND. 2000. Obesity and cortisol. Nutrition **16:** 924–936.
3. CALOGERO, A.E., W.T. GALLUCCI, G.P. CHROUSOS *et al.* 1988. Interaction between GABAergic neurotransmission and rat hypothalamic corticotropin-releasing hormone secretion *in vitro*. Brain Res. **463:** 28–36.
4. CHEBIB, M. & G.A. JOHNSTON. 1999. The "ABC" of GABA receptors: a brief review. Clin. Exp. Pharmacol. Physiol. **26:** 937–940.
5. LOH, E.W., I. SMITH, R. MURRAY *et al.* 1999. Association between variants at the $GABA_A\beta2$, $GABA_A\alpha6$, and $GABA_A\gamma2$ gene cluster and alcohol dependence in a Scottish population. Mol. Psychiatry **4:** 539–544.
6. HICKS, A.A., M.E. BAILEY, B.P. RILEY *et al.* 1994. Further evidence for clustering of human GABAA receptor subunit genes: localization of the alpha6-subunit gene (GABRA6) to distal chromosome 5q by linkage analysis. Genomics **20:** 285–288.
7. ROSMOND, R., M.F. DALLMAN & P. BJÖRNTORP. 1998. Stress-related cortisol secretion in men: relationships with abdominal obesity and endocrine, metabolic, and hemodynamic abnormalities. J. Clin. Endocrinol. Metab. **83:** 1853–1859.
8. BJÖRNTORP, P. & R. ROSMOND. 1999. Hypothalamic origin of the metabolic syndrome X. Ann. N.Y. Acad. Sci. **892:** 297–307.
9. BJÖRNTORP, P. 1988. The associations between obesity, adipose tissue distribution, and disease. Acta Med. Scand. **723:** 121–134.
10. REAVEN, G.M. 1988. Role of insulin resistance in human disease. Diabetes **37:** 1595–1607.
11. BOUCHARD, C. 1995. Genetics of obesity: an update on molecular markers. Int. J. Obes. Relat. Metab. Disord. **19:** S10–S13.
12. PERUSSE, L., Y.C. CHAGNON, S.J. WEISNAGEL *et al.* 2001. The human obesity gene map: the 2000 update. Obes. Res. **9:** 135–169.
13. COSCINA, D.V., T.W. CASTONGUAY & J.S. STERN. 1992. Effects of increasing brain GABA on the meal patterns of genetically obese vs. lean Zucker rats. Int. J. Obes. Relat. Metab. Disord. **16:** 425–433.
14. OROSCO, M., C. JACQUOT, J. WEPIERRE *et al.* 1981. GABA levels and synthesis index in different brain areas of two obese rat models. Gen. Pharmacol. **12:** 369–371.
15. EBERT, M.H., D.E. SCHMIDT, T. THOMPSON *et al.* 1997. Elevated plasma gamma-aminobutyric acid (GABA) levels in individuals with either Prader-Willi syndrome or Angelman syndrome. J. Neuropsychiatry Clin. Neurosci. **9:** 75–80.

Increased Abdominal Obesity in Subjects with a Mutation in the 5-HT$_{2A}$ Receptor Gene Promoter

ROLAND ROSMOND,[a,b] CLAUDE BOUCHARD,[b] AND PER BJÖRNTORP[c]

[a]Department of Clinical Chemistry, Sahlgrenska University Hospital, S-413 45 Göteborg, Sweden

[b]Pennington Biomedical Research Center, Baton Rouge, Louisiana 70808, USA

[c]Department of Heart and Lung Diseases, Sahlgrenska University Hospital, S-413 45 Göteborg, Sweden

ABSTRACT: In the present study, we examined the potential impact of the 5-HT$_{2A}$ −1438G/A promoter polymorphism on obesity and estimates of insulin, glucose, and lipid metabolism as well as circulating hormones, including salivary cortisol, in 284 unrelated Swedish men born in 1944. The subjects were genotyped by using PCR amplification of the promoter region of the gene for 5-HT$_{2A}$ followed by digestion with the restriction enzyme MspI. The frequencies were 0.39 for allele −1438A and 0.61 for allele −1438G. Homozygotes for the −1438G allele had, in comparison with −1438A/A subjects, higher body mass index (BMI), waist-to-hip ratio (WHR), and abdominal sagittal diameter. Moreover, cortisol escape from 0.25 mg dexamethasone suppression was found in subjects with the −1438A/G genotype. Serum leptin, fasting insulin and glucose, as well as serum lipids were not different across the −1438G/A genotype groups. From these results, we suggest the possibility that an abnormal production rate of the 5-HT$_{2A}$ gene product might lead to the development of abdominal obesity. The pathophysiology could involve stress factors that destabilize the serotonin-hypothalamic-pituitary-adrenal systems in those with genetic vulnerability in the serotonin receptor gene.

KEYWORDS: abdominal obesity; dexamethasone; genes; polymorphism; salivary cortisol; serotonin receptor

INTRODUCTION

Serotonin (5-hydroxytryptamine; 5-HT) is a neurotransmitter that contributes to the regulation of many physiologic processes, such as sleep, appetite, and hormone secretion. Over the years, a multitude of 5-HT receptors have been identified, including the 5-HT$_{2A}$ receptor.[1]

Address for correspondence: Roland Rosmond, M.D., Ph.D., Department of Clinical Chemistry, Sahlgrenska University Hospital, S-413 45 Göteborg, Sweden. Voice: (+46) 31 342 62 72; fax: (+46) 31 82 84 58.
 rolandrosmond@hotmail.com

A considerable amount of evidence has emerged regarding the pathogenic role of cortisol in abdominal obesity.[2] That 5-HT is involved in regulating cortisol secretion has long been recognized, and recent evidence suggests that cortisol secretion is regulated by central 5-HT$_{2A/2C}$ receptors.[3]

Lately, an *Msp*I restriction fragment length polymorphism in the promoter region of the 5-HT$_{2A}$ gene (−1438G/A) has been described.[4] The human 5-HT$_{2A}$ gene is located on 13q14-q21, consists of 3 exons separated by 2 introns, and spans over 20 kb. More recent data suggest that the *Msp*I polymorphism of the 5-HT$_{2A}$ gene may influence food and alcohol intake in obese subjects.[5] In addition, some, but not all studies have indicated a role for the −1438G/A in the pathogenesis of anorexia nervosa.[4]

In this study, we addressed the hypothesis that the −1438G/A promoter variant of the 5-HT$_{2A}$ gene is involved in the pathogenesis of abdominal obesity and related perturbations in insulin, glucose, and lipid metabolism as well as circulating hormones, including salivary cortisol.

SUBJECTS AND METHODS

For the present study, the subjects ($n = 284$) were randomly selected from a larger geographically defined total population cohort of men born in Göteborg, Sweden, in 1944. The design has been described elsewhere.[6] The subjects gave written informed consent before participating in the study, which was approved by the local ethics committee.

Body mass index (BMI, kg/m^2), waist-to-hip ratio (WHR), and abdominal sagittal diameter were measured as described before.[7] Venous blood was obtained after overnight fasting. Commercial RIA kits were used for the determination of serum testosterone, insulin-like growth factor I, insulin, and leptin. Glucose and serum lipids were determined enzymatically as detailed previously.[7] Diurnal cortisol secretion was measured by a series of saliva sampling over an ordinary working day, in which cortisol levels were measured. Additionally, a dexamethasone suppression test was done at home, using 0.25 mg dexamethasone. The details of the procedures have been described previously.[7]

Genotyping was performed on genomic leukocyte DNA (100 ng in a final volume of 10 µL) and carried out by PCR technique (annealing temperature, 55°C) using primers described previously.[4] The PCR products were digested overnight at 37°C with 5 U of *Msp*I.

Statistical differences between genotypes were tested using a one-way ANCOVA model including genotype as independent factors and BMI and WHR as covariates. To determine which mean values differed ($p < 0.05$), we used the Duncan multiple-range test (posthoc). All statistical analyses were performed using the SAS System for Windows, release 8.0 (SAS Institute Inc., Cary, NC).

RESULTS

The frequency of allele −1438A was 0.39 and that for allele −1438G was 0.61. The observed genotype frequencies were 35.6%, 51.6%, and 12.9% for −1438A/A,

TABLE 1. Differences in anthropometric, endocrine, and metabolic measurements by 5-HT$_{2A}$ −1438G/A genotypes

	Genotypes			
	−1438A/A (n = 94)	−1438A/G (n = 136)	−1438G/G (n = 34)	p
Body mass index (kg/m^2)	25.1 (3.5)^a	26.7 (4.0)^a	26.9 (4.3)^b	0.017
Waist-to-hip ratio	0.92 (0.06)^a	0.94 (0.07)^b	0.96 (0.06)^b	0.015
Abdominal sagittal diameter (cm)	21.9 (3.4)^a	22.9 (3.8)^b	23.6 (3.2)^b	0.039
DEX suppression test (nmol/L)	16.6 (18.3)^a	5.0 (5.7)^b	7.9 (2.3)	0.021
Testosterone (nmol/L)	20.2 (5.0)	19.6 (5.9)	19.0 (5.1)	>0.20
Insulin-like growth factor I (µg/L)	208.4 (56.4)	204.0 (71.0)	203.3 (62.1)	>0.20
Leptin (µg/L)	5.4 (4.1)	6.4 (4.3)	7.3 (4.4)	>0.20
Fasting insulin (mU/L)	10.4 (6.1)	13.5 (13.0)	13.8 (10.1)	>0.20
Fasting glucose (mmol/L)	4.4 (0.7)	4.6 (1.1)	4.4 (0.6)	>0.20
Triglycerides (mmol/L)	1.8 (0.9)	1.8 (1.2)	1.8 (0.7)	>0.20
Total cholesterol (mmol/L)	6.1 (1.0)	6.2 (1.1)	6.2 (1.2)	>0.20
HDL cholesterol (mmol/L)	1.2 (0.3)	1.3 (0.4)	1.2 (0.3)	0.113
LDL cholesterol (mmol/L)	4.0 (1.0)	4.2 (1.0)	4.2 (1.1)	>0.20

NOTE: Values (mean ± SD) with different superscript letters (a, b) are significant at the 0.05 level; p values are adjusted for BMI and WHR. Abbreviations: DEX, dexamethasone; HDL, high-density lipoprotein; LDL, low-density lipoprotein.

−1438A/G, and −1438G/G, respectively. Genotype frequencies were in a Hardy-Weinberg equilibrium.

TABLE 1 presents the differences in anthropometric, endocrine, and metabolic measurements among the 5-HT$_{2A}$ −1438G/A genotypes. Homozygotes for the −1438G allele had, in comparison with −1438A allele homozygotes, higher BMI, WHR, and abdominal sagittal diameter. Subjects with the −1438A/G genotype showed significantly less suppression of cortisol with 0.25 mg dexamethasone compared to −1438A/A homozygotes (p = 0.021). Serum leptin, fasting insulin and glucose, as well as serum lipids were not different across the −1438G/A genotype groups.

DISCUSSION

The examined men were selected from an ongoing cohort study and 80% volunteered to participate in the first part of the study. The second part, which was laboratory-based, attracted fewer participants (63%), but there was no difference between nonresponders and responders concerning the prevalence of hypertension, diabetes mellitus, myocardial infarction, stroke, and angina pectoris, as well as education level, housing conditions, smoking, and alcohol habits.[7] We thus believe that the participating men are representative of men at this age in the city of Göteborg, Sweden.

The assessment of the hypothalamic-pituitary-adrenal (HPA) axis function in the present study included measurements of salivary cortisol levels under basal conditions and after dexamethasone suppression. The assessment of cortisol in saliva is specific for the detection of unbound, free cortisol, and the concentration of cortisol in saliva is independent of the saliva flow.[8] The test is sufficiently sensitive to measure cortisol levels in normal subjects and to distinguish normal secretory pattern from hypo- and hypercortisolism.[9] Moreover, cortisol in saliva reflects accurately the free fraction of cortisol in plasma,[8] which has also been confirmed in our laboratory ($r > 0.90$, unpublished). From a practical point of view, the assessment of cortisol from saliva samples represents less of a burden for the subjects.

The major findings of the present study are that homozygotes for the −1438G allele had increased body mass and abdominal distribution of body fat (WHR and abdominal sagittal diameter) along with less suppression of cortisol with 0.25 mg dexamethasone compared to other $MspI$ 5-HT$_{2A}$ genotypes. These results might be important in understanding the pathophysiology of abdominal obesity. Several features of abdominal obesity suggest that a relative deficiency of serotoninergic effects is involved. These include traits of depression and anxiety,[10] carbohydrate craving,[11] alcohol, and smoking.[10]

During the course of antidepressant pharmacotherapy of depression, the relative resistance to serum cortisol suppression by exogenous glucocorticoids seems to retreat.[12,13] Furthermore, long-term treatment with antidepressant drugs in rats results in decreased basal as well as stress-related plasma levels of corticotropin and cortisol.[12,13] Besides normalization of the HPA axis activity, serotoninergic drugs increase energy expenditure in humans.[12,13] To a considerable extent, serotonin agonists, such as fluoxetine, produce a resumption of the normal eating patterns by diminishing meal size and disrupting night-eating behavior.[12,13] Recent data indicate that serotoninergic agents decrease abdominal fat mass and improve glucose as well as insulin metabolism.[12,13]

In summary, the results of our study suggest that it would be constructive to put more emphasis on the serotonergic system in treatment. Moreover, there is evidence that some selective serotonin reuptake inhibitors are effective as antiobesity drugs. Since abdominal obesity is an unsolved therapeutic problem associated with a high mortality and morbidity, new approaches are clearly warranted. This is particularly important for the individuals who are at risk as a result of a genetic predisposition. The example of the 5-HT$_{2A}$ gene promoter polymorphism provided here appears to represent one way through which such a predisposition may occur.

ACKNOWLEDGMENTS

This study was supported by the Swedish Medical Research Council (Grant No. K97-19X-00251-35A) and the Pennington Biomedical Research Center. R. Rosmond would also like to acknowledge the Henning and Johan Throne-Holst Foundation for the support of a postdoctoral fellowship at the Pennington Biomedical Research Center. C. Bouchard is partially supported by the George A. Bray Chair in Nutrition.

REFERENCES

1. MENESES, A. 1998. Physiological, pathophysiological, and therapeutic roles of 5-HT systems in learning and memory. Rev. Neurosci. **9:** 275–289.
2. BJÖRNTORP, P. & R. ROSMOND. 2000. Obesity and cortisol. Nutrition **16:** 924–936.
3. RITTENHOUSE, P.A., E.A. BAKKUM, A.D. LEVY et al. 1994. Evidence that ACTH secretion is regulated by serotonin$_{2A/2C}$ (5-HT$_{2A/2C}$) receptors. J. Pharmacol. Exp. Ther. **271:** 1647–1655.
4. COLLIER, D.A., M.J. ARRANZ, T. LI et al. 1997. Association between 5-HT$_{2A}$ gene promoter polymorphism and anorexia nervosa. Lancet **350:** 412.
5. AUBERT, R., D. BETOULLE, B. HERBETH et al. 2000. 5-HT$_{2A}$ receptor gene polymorphism is associated with food and alcohol intake in obese people. Int. J. Obes. Relat. Metab. Disord. **24:** 920–924.
6. ROSMOND, R., M.F. DALLMAN & P. BJÖRNTORP. 1998. Stress-related cortisol secretion in men: relationships with abdominal obesity and endocrine, metabolic, and hemodynamic abnormalities. J. Clin. Endocrinol. Metab. **83:** 1853–1859.
7. ROSMOND, R. & P. BJÖRNTORP. 1998. Endocrine and metabolic aberrations in men with abdominal obesity in relation to anxio-depressive infirmity. Metabolism **47:** 1187–1193.
8. KIRSCHBAUM, C. & D.H. HELLHAMMER. 1994. Salivary cortisol in psychoneuroendocrine research: recent developments and applications. Psychoneuroendocrinology **19:** 313–333.
9. CASTRO, M., P.C. ELIAS, A.R. QUIDUTE et al. 1999. Out-patient screening for Cushing's syndrome: the sensitivity of the combination of circadian rhythm and overnight dexamethasone suppression salivary cortisol tests. J. Clin. Endocrinol. Metab. **84:** 878–882.
10. ROSMOND, R. 2001. Visceral obesity and the metabolic syndrome. In The International Textbook of Obesity, pp. 339–350. Wiley. New York/Chichester.
11. STRÖMBOM, U., M. KROTKIEWSKI, K. BLENNOW et al. 1996. The concentrations of monoamine metabolites and neuropeptides in the cerebrospinal fluid of obese women with different body fat distribution. Int. J. Obes. Relat. Metab. Disord. **20:** 361–368.
12. WURTMAN, R.J. & J.J. WURTMAN. 1995. Brain serotonin, carbohydrate-craving, obesity, and depression. Obes. Res. **3:** 477S–480S.
13. ROSMOND, R. & P. BJÖRNTORP. 2000. The role of antidepressants in the treatment of abdominal obesity. Med. Hypotheses **54:** 990–994.

Lipids and Insulin Resistance: What We've Learned at the Fourth International Smolenice Symposium

E. RAVUSSIN,[a] I. KLIMEŠ,[b] E. ŠEBÖKOVÁ,[b] AND B. V. HOWARD[c]

[a]*Pennington Biomedical Research Center, Department of Health and Performance Enhancement, Baton Rouge, Louisiana 70808-4124, USA*

[b]*Diabetes and Nutrition Research Laboratory, Institute of Experimental Endocrinology, Slovak Academy of Sciences, SK-833 06 Bratislava, Slovak Republic*

[c]*MedStar Research Institute, Washington, District of Columbia 20010, USA*

ABSTRACT: A summary of the Fourth International Smolenice Symposium on Lipids and Insulin Resistance focusing on "The Role of Fatty Acid Metabolism and Fuel Partitioning" is provided. Highlights and issues of the conference are mentioned, as well as strategies for the future.

KEYWORDS: lipids; insulin resistance; obesity; diabetes; fat

The Fourth International Smolenice Symposium on Lipids and Insulin Resistance focused on "The Role of Fatty Acid Metabolism and Fuel Partitioning" in the "Metabolic Syndrome". This symposium was a logical continuation of the three previous ones, which primarily dealt with issues such as (1) the role of lipids in intracellular signaling pathways of insulin action in 1988, (2) the role of dietary lipids on insulin action and the different facets of the insulin resistance syndrome in 1992, and (3) molecular mechanisms of insulin resistance and its clinical consequences in 1996.

In face of the emerging new information on the role of lipids in insulin resistance, the 2001 theme for this symposium was quite timely. The event focused not only on the role of circulating lipids, but also on the role of lipids within the cells of insulin-sensitive tissues. It is now clear that the adipocyte reemerged as a pivotal causal factor in the increasing prevalence of type 2 diabetes around the globe. In 1997, the World Health Organization (WHO) published an alarming report entitled "Obesity: Preventing and Managing the Global Epidemic". This represented the first step towards the recognition that obesity and related diseases will become major burdens

---

Address for correspondence: E. Ravussin, Pennington Biomedical Research Center, Department of Health and Performance Enhancement, 6400 Perkins Road, Baton Rouge, LA 70808-4124. Voice: 225-763-3186; fax: 225-763-3030.
ravusse@pbrc.edu

for health care systems in the twenty-first century. Globally, more than a half billion people are obese or overweight and this increased prevalence is worse in developing countries. Throughout humankind's history, most populations have evolved in restrictive environments in which large amounts of physical effort were required to obtain food in limited supply. It is, therefore, no surprise that the human genome harbors more genes that predispose to positive energy balance for protection against famine than genes that protect against the effects of affluence. Because of the known association between obesity and type 2 diabetes, it is not so surprising that the new data from the WHO predict a 46% increase in the worldwide population of diabetics in this decade (from 151 million in 2000 to 221 million in 2010). Can our societies mentally and economically cope with this frightening increase? Is our scientific knowledge sufficient to design pharmacological strategies to treat, or better prevent, type 2 diabetes? When will novel public health measures be in place to reverse the "obesogenic" environment that caused such a rapid explosion of the prevalence of chronic diseases linked to affluence? Against this background, the symposium provided an ideal forum for scientists from more than 15 countries to share their newest data. Such exchange was facilitated by assembling more than 130 participants in a beautiful part of the Slovak Republic where we all experienced "la vie de chateau".

The program was organized around five major themes:

(1) the genetics of complex diseases such as obesity and type 2 diabetes;
(2) the regulation of energy balance and fuel partitioning;
(3) the role of the adipocyte in insulin resistance and the importance of transcription factors for both adipogenesis and insulin action;
(4) the effect of fat stored in the wrong place, that is, skeletal muscle and liver;
(5) present intervention strategies to decrease the galloping epidemic of obesity and diabetes.

A quite coherent story emerged from the different plenary sessions and poster presentations. Even though genetic and physiological studies have not yet delineated the causes of obesity and/or diabetes, nevertheless the adipocyte is now recognized as a major culprit in a lot of the obesity-associated chronic diseases. When adipocytes become full because of excess fat intake, reduced fat oxidation, or impaired adipogenesis, the overflow of fat increasingly accumulates in tissues not meant to accommodate excessive storage, such as skeletal muscle, liver, and pancreas. At the meeting, a new understanding emerged concerning the molecular mechanisms by which ectopic fat deposition seems to impact insulin resistance. This information will certainly provide new pharmacological means to increase fat oxidation in insulin-dependent tissue and therefore decrease insulin resistance. Review of the recent intervention strategies to prevent or delay the development of type 2 diabetes in people at risk provided hope that changing lifestyle (dietary modification and increased physical activity) can be effective in preventing diabetes. The question is now how to successfully implement such strategies for the general public living into an environment so conducive to overconsumption of unhealthy food and sedentary lifestyles. Only new public health policies in concert with a change in the mind-set of food manufacturers could help to curb down the alarming increasing prevalence of diseases of affluence.

GENETICS OF COMPLEX DISEASE

A few presentations, including a "state-of-the-art lecture", dealt with some of the results from genome-wide scans conducted to identify novel susceptibility genes for obesity and/or type 2 diabetes. However, despite new information, there is, thus far, almost no success story in the identification of new genes for complex diseases, with the exception of the identification of calpain 10 for diabetes in Mexican-Americans. The various presentations stressed the importance of potential false-positive or false-negative findings from genome scan studies and the need for replication of the results in the same and/or different populations. More importantly, speakers emphasized the fact that, from the time of discovery and confirmation of new genetic loci harboring susceptibility genes, there is an enormous amount of work to positionally clone the genes through strategies of linkage disequilibrium using numerous single nucleotide polymorphisms. Many groups are presently working intensively on cloning such new genes using not only positional cloning, but in combination with gene expression strategies. The polygenic nature of complex diseases will render the hunt for new genes extremely difficult. Progress in this search will be much faster if more research funds were directed toward the analysis of QTLs (quantitative trait loci) in mice and rats.

REGULATION OF ENERGY BALANCE AND FUEL PARTITIONING

Many speakers reminded us that obesity is always the result of a positive energy balance, a condition in which food intake is exceeding energy expenditure. However, growing evidence indicates that some animal models of obesity and humans alike are predisposed to weight gain because of an impaired fat oxidation and energy expenditure. It is indeed not surprising that species evolving in an environment rather restrictive for energy may have developed survival mechanisms in which energy expenditure and fat oxidation would be as thrifty as possible. We were all reminded of the importance of the work of late Dennis McGarry who discovered and championed the pivotal role of malonyl-CoA as a fuel-sensing mechanism that can tilt the fat balance towards lipogenesis or towards fat oxidation. The important roles of acetyl-CoA carboxylases and carnitine palmitoyl transferase I for tissues' fat balance were reemphasized and proposed as potential targets for pharmacological therapy. In this group of presentations, results from classical physiological studies were provided emphasizing that 40% of insulin resistance may be due to impaired insulin transport in the insulin-sensitive cells and 60% to intracellular defects. Emphasis was also given to the importance of visceral fat. The basis to study a novel axis, that is, the visceral hepatic axis (VHA), was provided. Compared to previous symposia, much less importance was given to circulating free fatty acids and the Randle cycle. In contrast, much more focus was given to the fat and fat metabolites accumulating in the cells. One surprising outcome from the Smolenice presentations was the finding that food intake, but not energy metabolism, is usually precisely measured in rodents. The reverse, however, is true in humans. It is now important that investigators involved in studies of body weight regulation begin to "close the loop" by investigating both arms of the macronutrients' balances. Some novel imaging techniques may prove to become very important in tracing the fate of energy substrate from ingestion and absorption in the digestive system to oxidation and storage in tissues.

ADIPOCYTE AND INSULIN RESISTANCE

In the past few decades, the predominant paradigms used to explain the link between increasing adiposity and insulin resistance have been the Randle cycle and the visceral adiposity hypothesis. However, over the past 2 or 3 years, a new paradigm has emerged that may explain the established link between adiposity and diseases. A failure to develop adequate adipose tissue mass may result in severe insulin resistance. This is apparent in human lipodystrophies or in "engineered fatless" animals, both characterized by ectopic fat deposition. Fat that cannot be stored into the adipocyte will find new sites of storage in liver, skeletal muscle, and pancreatic cells. Since large fat cell size seems to be a predictor of the development of diabetes, it was hypothesized that a potential failure of the adipose tissue to expand may be an important factor in the pathophysiology of type 2 diabetes. Moreover, research over the past 3–5 years has provided new evidence of the role of the adipose tissue as an "endocrine organ" since adipocytes secrete a variety of proteins such as leptin, interleukins, TNF-α, adiponectin, and resistin, all known to modulate insulin action and therefore insulin resistance. Many of the presentations discussed our present understanding of the role of the adipocyte and its secreted proteins in the development of insulin resistance in both animals and humans.

IMPACT OF FAT DEPOSITION IN SKELETAL MUSCLE

Since most of the ingested glucose is disposed in liver and skeletal muscle, increased emphasis has been given to these tissues over the past decades. There is now strong physiological and molecular evidence that fat itself, as well as its metabolites, are major determinants of insulin resistance in the skeletal muscle tissue. Many recent studies using imaging techniques, such as CT scan and spectral magnetic resonance or *in situ* biochemistry, have consistently provided evidence that an increase in intracellular fat is the best determinant of insulin resistance in this tissue. Physiological studies of energy balance across the leg have confirmed the crucial role of fat and its metabolism as determinants of insulin resistance. However, more importantly, data obtained in skeletal muscle cell cultures are now providing more evidence that not only environmental factors can impact skeletal muscle insulin resistance, but also genetic factors. Cells cultured for more than 6 weeks in a "neutral" environment seem to retain at least part of their initial phenotype of insulin action measured *in vivo* in the donor of the cells. In the field of skeletal muscle insulin resistance, new preliminary strategies and results using microarrays for gene expression were provided. Although consistent results among laboratories are being obtained, the full realization of this technology will only come with improved bioinformatic tools and an increasing understanding of the human genome. Gene expression studies are likely to provide valuable information on the determinants of health and disease in a not too distant future.

INTERVENTION STRATEGIES

The first three Smolenice symposiums on lipids and insulin resistance provided much data on the potential importance of dietary intervention strategies (especially dietary fat) to prevent or alleviate diseases of the Metabolic Syndrome. Even if data are not clear-cut, it is generally well accepted that a decrease in dietary saturated fat and an increase in polyunsaturated fat will improve insulin resistance in animals and humans alike. The few recent randomized clinical trials in which lifestyle was changed to decrease food intake and/or increase physical activity have now provided convincing evidence that such changes can be beneficial at improving insulin sensitivity and delaying the development of type 2 diabetes. These costly trials have provided sufficient information to convince our politicians to implement novel public health policies designed at helping people to comply to the necessary changes in lifestyle if the epidemic of diseases of affluence are to be reversed. In the United States, the recent surgeon general's call to action to prevent an increase in overweight and obesity is certainly a step in the right direction. However, a concerted effort involving politicians and industrials alike will be necessary to make a dent in this huge public health problem.

CONCLUSIONS

One more time, the now classic International Smolenice Insulin Symposium was a unique opportunity for scientists and students to discuss some of the important issues about the relationship between lipids and insulin resistance and, therefore, the potential causes of the recent epidemic of the diseases of affluence. More importantly, the idyllic site of the Smolenice Castle offered many occasions to all the participants not only to discuss scientific matters, but also to socialize. Many new collaborative scientific endeavors have been established and some young scientists from central Europe have now found new homes to conduct their postdoctoral training. This is truly the spirit of the symposium, which will continue to prevail in future meetings.

Index of Contributors

Aas, V., 506–515
Andel, M., 544–547
Andersen, M.H., 506–515
Andersson, A., 183–195
Angeletti, R.H., 535–543
Antwerpen, B., 554–557
Augert, G., 274–282, 403–413
Auwerx, J., 28–33

Babal, P., 454–462, 469–475
Baier, L., 1–6, 258–264
Balent, B., 535–543
Bardová, K., 88–101
Bendlová, B., 265–273, 558–565
Bernatova, I., 454–462
Bílková, D., 544–547
Bíró, K., 482–489
Björntorp, P., 566–570, 571–575
Boberg, M., 183–195
Bocher, V., 7–18
Bogardus, C., 1–6, 258–264, 516–521
Bouchard, C., 566–570, 571–575
Brauner, P., 88–101
Buchtíková, M., 544–547

Cameron-Smith, D., 274–282
Carr, R.D., 414–423
Cársky, J., 463–468
Carter, L., 66–70
Cawthorne, M.A., 112–119
Cha, B-S., 66–70
Chehab, S., 274–282
Chen, K., 389–397
Chrousos, G., 500–505
Ciaraldi, T.P., 66–70
Cinek, O., 558–565
Clapham, J.C., 112–119
Clarke, S.D., 283–298
Collier, G.R., 274–282, 403–413
Cooksey, R.C., 102–111
Cooney, G.J., 196–207
Cray, K., 258–264

Dagher, Z., 43–51

De Natale, C., 329–335
de Silva, A., 403–413
Del Parigi, A., 389–397
Demcáková, E., 71–79, 424–430, 446–453
Demmelmair, H., 299–310
Divišová, J., 440–445
Dlouhý, P., 544–547
Dobešová, Z., 352–362
Dobrzyn, A., 236–248
Donnelly, R., 176–182
Dransfeld, O., 208–216
Drevon, C.A., 71–79

Eckel, J., 208–216

Facchini, F.S., 342–351
Fenselau, S., 548–553
Fickova, M., 490–499
Flachs, P., 88–101
Friedel, S., 554–557
Fruchart, J-C., 7–18
Furler, S.M., 158–175, 196–207

Gašperíková, D., 71–79, 283–298, 446–453
Gautier, J-F., 389–397
Geller, F., 554–557
Geschonke, K., 548–553
Giacobino, J-P., 398–402
Golda, V., 490–499
Goodloe, G., 535–543
Górski, J., 236–248
Goswami, G., 535–543
Grassl, W., 554–557
Gray, S., 176–182
Gustafsson, I-B., 183–195

Hainer, V., 265–273, 311–323
Hanefeld, M., 528–534
Hanson, R., 1–6
Hargreaves, M., 274–282
Hebebrand, J., 554–557

Heird, W.C., 283–298
Henry, R.R., 66–70
Hill, M., 265–273, 558–565
Hinney, A., 554–557
Hoch, A., 554–557
Howard, B.V., 324–328, 576–580
Hubová, M., 440–445
Hulin, I., 454–462

Ido, Y., 43–51
Idris, I., 176–182

Jednákovits, A., 482–489
Jensen, J., 506–515
Jéquier, E., 379–388
Jeukendrup, A.E., 217–235
Jørgensen, C., 431–439
Jurcovicova, J., 490–499

Kaiser, M., 528–534
Kantham, L., 403–413
Kazdová, L., 440–445
Kelley, D.E., 135–145
Kim, H-J., 34–42
Klempt, M., 548–553
Klimeš, I., xiii–xv, 71–79, 424–430, 446–453, 469–475, 500–505, 576–580
Knudsen, J., 431–439
Koletzko, B., 299–310
Konárková, M., 336–341, 522–527
Kopecký, J., 88–101
Koranyi, L., 424–430
Koska, J., 500–505
Kotzka, J., 19–27
Koutnikova, H., 28–33
Kovacs, P., 258–264
Kozak, L.P., 80–87
Kraegen, E.W., 196–207
Kratchmarova, I., 431–439
Kristiansen, K., 431–439
Krogsdam, A-M., 431–439
Ksinantova, L., 500–505
Kukorelli, T., 482–489
Kunešová, M., 265–273, 311–323
Kürthy, M., 424–430, 482–489
Kvetnansky, R., 500–505

Lapillonne, A., 283–298
Larque, E., 299–310
Larsen, M.O., 414–423
Lilli, S., 329–335
Lin, Y-S., 43–51
Lind, P., 476–481
Luo, Z., 43–51
Luptak, I., 454–462, 469–475
Lynch, N., 548–553

Macho, L., 490–499
Man, W.C., 34–42
Mandrup, S., 431–439
Matušková, J., 454–462, 469–475
McClain, D.A., 102–111
Menzel, S., 249–257
Meschišvili, E., 440–445
Mints, L., 535–543
Miriam Hubová, M., 440–445
Miyazaki, M., 34–42
Mogyorosi, T., 424–430, 482–489
Mudaliar, S.R., 66–70
Muller, Y.L., 258–264
Müller-Wieland, D., 19–27

Nagy, K., 424–430, 482–489
Neckár, J., 463–468
Nelson, C., 283–298
Nezami, R., 535–543
Ntambi, J.M., 34–42

Oakes, N.D., 158–175

Pacak, K., 500–505
Parízková, J., 311–323
Park, K-S., 66–70
Pechanova, O., 454–462
Pelikánová, T., 352–362
Permana, P., 1–6
Phinney, S., 311–323
Pineda-Torra, I., 7–18
Pinterova, L., 490–499
Pokorný, R., 544–547
Pražák, T., 88–101
Prochazka, M., 1–6

Rakatzi, I., 208–216

INDEX OF CONTRIBUTORS

Rambousková, J., 544–547
Rauta, O., 535–543
Ravingerová, T., 463–468
Ravussin, E., 363–378, 389–397, 516–521, 576–580
Reiman, E., 389–397
Reseland, J.E., 71–79
Rivellese, A.A., 329–335
Rogatsky, E., 535–543
Rolin, B., 414–423
Romaniv, S., 336–341, 522–527
Rosmond, R., 566–570, 571–575
Rossmeisl, M., 80–87
Ruderman, N.B., 43–51
Rustan, A.C., 506–515

Saha, A., 43–51
Salbe, A.D., 389–397
Sanigorski, A., 403–413
Sasson, S., 208–216
Saylor, K.L., 342–351
Schemidt, A., 258–264
Schmitz-Peiffer, C., 146–157
Schreiber, S., 548–553
Schrezenmeir, J., 548–553
Šeböková, E., xiii–xv, 71–79, 424–430, 446–453, 500–505, 576–580
Segal, D., 274–282, 403–413
Shen, G-Q., 258–264
Simko, F., 454–462, 469–475
Smith, S.R., 363–378
Söderberg, C., 476–481
Šponarová, J., 88–101
Šrámková, D., 265–273
Staels, B., 7–18
Stanková, B., 544–547
Stein, D.T., 535–543
Štich, V., 311–323
Stunkard, A., 311–323
Styk, J., 463–468
Suchánková, G., 352–362
Sutherland, J., 258–264
Svendsen, O., 414–423

Tálosi, L., 482–489
Tataranni, P.A., 389–397, 516–521
Temelkova-Kurktschiev, T., 528–534
Tengblad, S., 183–195

Tenne-Brown, J., 403–413
Thompson, A.L., 196–207
Thuillez, P., 258–264
Tomás, E., 43–51
Torblå, S., 506–515
Török, J., 469–475
Traurig, M., 258–264
Tvrzická, E., 311–323, 336–341, 522–527, 544–547

Ukropec, J., 71–79, 424–430, 446–453

Vanková, M., 558–565
Vcelák, J., 265–273
Vecka, M., 336–341, 522–527
Vessby, B., 183–195
Vlasáková, Z., 352–362
Vogel, C., 554–557
Vokurková, M., 352–362
Vozarova, B., 516–521
Vrbíková, J., 558–565

Waczulíková, I., 463–468
Walder, K., 274–282, 403–413
Wallberg-Henriksson, H., 120–134
Wang, S., 112–119
Weyer, C., 516–521
Wiedrich, C., 258–264
Wiererová, O., 544–547
Wilken, M., 414–423
Willson, T.M., 431–439
Wolford, J., 1–6

Ye, J., 196–207

Žák, A., 311–323, 336–341, 522–527
Zammit, V.A., 52–65
Zeman, M., 336–341, 522–527
Zendzian-Piotrowska, M., 236–248
Zicha, J., 352–362
Ziegelhöffer, A., 463–468
Ziegelhöffer-Mihalovicová, B., 463–468
Zierath, J.R., 120–134
Zorad, S., 490–499

OHIO UNIVERSITY LIBRARY

Please return this book as soon as you have finished with it. In order to avoid a fine it must be returned by the latest date stamped below. All books are subject to recall after two weeks or immediately if needed for reserve.

NOV 0 7 2004

CF